P9-CEY-842

PRINCIPLES OF ENGINEERING ECONOMY

PRINCIPLES OF ENGINEERING ECONOMY

SEVENTH EDITION

EUGENE L. GRANT
Stanford University

W. GRANT IRESON
Stanford University

RICHARD S. LEAVENWORTH
University of Florida

John Wiley & Sons
NEW YORK • CHICHESTER • BRISBANE • TORONTO • SINGAPORE

Library of Congress Cataloging in Publication Data:

Grant, Eugene Lodewick, 1897–
Principles of engineering economy.

Bibliography: p.
Includes index.
1. Engineering economy. I. Ireson, William
Grant, 1914– II. Leavenworth, Richard S.
III. Title.

TA177.4.G7 1982 658.1'5 81-10399
ISBN 0-471-06436-X AACR2

Printed in the United States of America

10 9 8 7 6 5 4 3 2 1

PREFACE

We are indeed grateful for the widespread acceptance of this book over a period of more than 50 years. It has been used in many hundreds of colleges and universities and scores of training programs in business and industry. The tendency of students to regard it as part of their basic equipment, for retention and later reference in professional or business life, has been a source of particular satisfaction.

Through successive editions we have emphasized that this is a book about a particular type of decision making. It explains the principles and techniques needed for making decisions about the acquisition and retirement of capital goods by industry and government. Normally, such decisions should be made on grounds of long-run economy. Because engineers make many such decisions and make recommendations for many others, the body of principles and techniques relating to them has been called *engineering economy.*

The same concepts and methods that are helpful in guiding decisions about investments in capital goods are useful in certain kinds of decisions between alternative types of financing (for example, ownership versus leasing) and in many personal decisions. Applications to these other areas of decision making are also discussed.

As in the past, our book may be used both as a text and as a reference. Experience has shown that its material is appropriate not only for engineering students but also for many students whose major interests are in economics, accounting, finance, or management. At the same time, it will serve as a working manual for engineers, management personnel, government officials, and others whose duties require them to make decisions about investments in capital goods.

The underlying philosophy regarding comparisons of alternatives is the same as in previous editions, and throughout, continued emphasis is placed on the following two important points:

It is prospective differences between alternatives that are relevant in their comparison.

The fundamental question regarding a proposed investment in capital goods is whether the investment is likely to be recovered plus a return commensurate with the risk and with the return obtainable from other opportunities for the use of limited resources. The purpose of calculations that involve the time value of money should be to answer this question.

Just as in previous editions, the changes from the preceding edition have been made in part to improve the presentation of basic principles and in part to try to keep the treatment of various topics up to date. Some of the major changes are as follows:

1. The explanation of the relationship between economy studies and income taxation has been made more general. Now our intent is to present the subject in a way that will be useful to readers wherever income taxes are levied. Nevertheless, we continue to discuss relevant tax laws and regulations of the United States. In this edition we have added a new appendix that aims to give an up-to-date coverage of certain pertinent federal tax legislation.
2. The discussion of prospective price inflation has been expanded, particularly with reference to the interaction between inflation and economy studies.
3. Chapter order has been rearranged so that our introductory treatment of income taxes now precedes *all* of the chapters that explain techniques for economy studies.
4. The chapter on benefit–cost analysis has been expanded to introduce the subject of studies based on the concept of cost effectiveness.
5. The chapter on public regulation has been expanded to introduce the subject of government regulation of all types of private business. Special consideration still is given to the influence of regulatory policies on economy studies for privately owned public utility companies in the United States.
6. In the chapter on economy studies for replacements, our treatment of some of the older mathematical models has been shortened. This has made room for more detailed comments on certain popular types of errors in reasoning about replacement economy.
7. Our discussion has been expanded on the fallacy of using two interest rates during the same period of time in analysis of a cash flow time series.
8. In the 1980s in the United States and elsewhere, many college students who study engineering economy can look forward to paying substantial income taxes in the fairly near future. With this in mind we have expanded our coverage of the impact of income taxes on decision making by individuals.
9. There are now 500 problems, an increase of 18 per cent from the preceding edition. More than half are entirely new; many of the others have been substantially modified. Just as in the previous editions, answers are given to a number of representative problems, with the thought that this may be helpful to those persons who use the book for home study.

Our arrangement of chapters continues to be influenced by the fact that some introductory courses are too short to permit a full coverage of the subject. Most of the material in Chapters 1 through 17 is fundamental and should be included in any presentation of basic principles. The subject matter of Chapters 18 through 21 is appropriate for an elementary course if time permits but might also be deferred until an advanced course. Appendixes A through C cover topics that might be omitted in elementary college courses but should be included in advanced courses and in any presentation to persons in industry.

The authors wish to thank many of the users of the *Sixth Edition* for helpful suggestions for changes and improvements. In this connection we want to make special mention of help from Professors James Burns, Timothy Lowe, Ralph Swalm, and Robert C. Waters, and from Messrs. Lawrence F. Bell, John F. Roberts, and William M. Vatavuk.

In preparing this *Seventh Edition*, we have been mindful of the responsibility imposed by the success of its predecessors and have made every effort to provide the reader with a body of knowledge that can be carried well into the future.

Eugene L. Grant
W. Grant Ireson
Richard S. Leavenworth

CONTENTS

PRINCIPLES OF ENGINEERING ECONOMY

PART I

SOME BASIC CONCEPTS IN ENGINEERING ECONOMY

1

DEFINING ALTERNATIVES AND PREDICTING THEIR CONSEQUENCES

*As the correct solution of any problem depends primarily on a true understanding of what the problem really is, and wherein lies its difficulty, we may profitably pause upon the threshold of our subject to consider first, in a more general way, its real nature; the causes which impede sound practice; the conditions on which success or failure depends; the directions in which error is most to be feared.—A. M. Wellington**

The practice of engineering involves many choices among alternative designs, procedures, plans, and methods. Since the available alternative courses of action involve different amounts of investment and different prospective receipts and disbursements, the question "Will it pay?" is nearly always present. This question may be broken down into subsidiary questions.

For example, there are the often-quoted three questions that were asked by General John J. Carty when he was chief engineer of the New York Telephone Company in the early years of the present century. He applied these questions to the many engineering proposals that came before him for review:

1. Why do this at all?
2. Why do it now?
3. Why do it this way?

Why do this at all? Shall a proposed new activity be undertaken? Shall an existing activity be expanded, contracted, or abandoned? Shall existing standards or operating procedures be modified?

Why do it now? Shall we build now with excess capacity in advance of demand, or with only sufficient capacity to satisfy the demand immediately in prospect? Are the costs of capital and other business conditions favorable to a present development?

*A. M. Wellington, *The Economic Theory of Railway Location*, 2d ed. (New York: John Wiley & Sons, Inc., 1887), p. 1.

Why do it this way? This choice among alternative ways of doing the same thing is common to all types of engineering activity.

This book deals with certain principles and techniques that are useful in securing rational answers to questions of this type. The central problem discussed in the book is how we may judge whether any proposed course of action will prove to be economical in the long run, as compared to other possible alternatives. Such judgment should not be based on an unsupported "hunch"; it calls for an economy study. An economy study may be defined as a comparison between alternatives in which the differences between the alternatives are expressed so far as practicable in money terms. Where technical considerations are somehow involved, such a comparison may be called an engineering economy study. In most cases, the engineering economy studies discussed in this book deal with the evaluation of proposed investments.

Management's Responsibility for Decisions on Plant Investment

The earliest book on engineering economy was Wellington's *The Economic Theory of Railway Location*. Wellington wrote in a missionary spirit in a day when investments in railway plant in the United States were greater than the aggregate of all other investments in industrial assets. Railway location obviously is a field in which many alternatives are likely to be available. Nevertheless, Wellington observed what seemed to him to be an almost complete disregard by many locating engineers of the influence of their decisions on the prospective costs and revenues of the railways. In his first edition (1877) he said of railway location, "And yet there is no field of professional labor in which a limited amount of modest incompetency at $150 per month can set so many picks and shovels and locomotives at work to no purpose whatever."

Although salary rates and many other things have changed since Wellington's time, the type of problem that he recognized is an ever-present one in an industrialized civilization. If, in a business enterprise or in government, many important decisions that in the aggregate can have a major influence on the success (and sometimes on the survival) of the enterprise are badly made by persons of "modest incompetence," these bad decisions are not primarily the fault of those persons; they are the fault of management.

A Conceptual Framework for the Presentation of Engineering Economy

The first two chapters of this book introduce a number of concepts that the authors believe are important in decision making—particularly so with reference to decisions about proposed investments in physical assets. The reader will doubtless observe that these concepts are not mutually exclusive; some of them overlap a bit. Throughout the remainder of the book, the application of these concepts is discussed and illustrated in various ways, often with reference to specific examples that involve numerical solutions in the comparison of alternatives.

In these two initial chapters, each concept is first stated in italics and then expanded by means of a short discussion. Some of the discussions contain descriptions of cases chosen to illustrate specific points. These early examples are intentionally brief; the desired points in this initial presentation can be made without giving all the details needed for a formal analysis and a numerical solution.

Recognizing and Defining Alternatives

1. *Decisions are among alternatives; it is desirable that alternatives be clearly defined and that the merits of all appropriate alternatives be evaluated.*

There is no need for a *decision* unless there are two or more courses of action possible. However, many decisions are, in effect, made by default; although many alternatives exist, the decision maker fails to recognize them and considers only one possible course of action.

In many other instances, formal consideration is given to several alternatives. Nevertheless, an unwise decision is finally made (or recommended) because of an analyst's failure to examine an alternative that is superior to any of the ones selected. It is obvious that a poor alternative will appear to be attractive if it is compared with alternatives that are even worse.

Frequently, one alternative is to do nothing, i.e., maintain the existing conditions. This alternative is sometimes overlooked or ignored.

A Case of a Failure to Recognize the Existence of Any Alternatives. An industrial concern owned a "total energy system" which had been operated for many years to furnish steam for heating, for operation of steam-driven pumps and steam-driven air compressors, and for generation of the electricity needed in the plant for lighting and for the operation of small motors. An increase in the concern's volume of business finally increased the demands for compressed air and for electric energy above the capacity of the existing plant.

Without any engineering survey of the situation, the general manager of the plant contracted to purchase from the local electric light and power company the excess of his needs for electric energy above the amount that could be generated in his existing plant. To meet the increased needs for compressed air, he bought a large electrically driven air compressor.

His decision proved to be a costly one. The air compressor purchased was a large single unit; it turned out that it was operated most of the time at a very small fraction of its capacity and at a correspondingly very low efficiency. The amount of energy purchased was too low to bring the unit rate into the lower blocks of the power company's rate schedule. No reduction was possible in labor cost for operation of boilers, prime movers, compressors, and pumps, or in the fuel and maintenance costs of the old and inefficient prime movers.

Finally, after a long period of uneconomical operation, an engineering economy study was made to discover and evaluate the possible alternatives. This study made it clear that there were a number of plans that would have been more economical than the one that had been adopted. Steam capacity might have been increased by the addition of new boilers and the old inefficient prime movers might have been replaced by modern efficient ones of greater capacity. Or the old prime movers might have been shut down with all electric energy requirements purchased from the power company, permitting all the steam generated to be used for operation of steam-driven compressors and pumps and for heating. Or the compressors and pumps might have been electrified and all electric energy requirements then purchased, with boilers operated for steam heat only in cold weather.

Some Cases of Failure to Consider Appropriate Alternatives. In a certain study of alternate highway locations, Proposal A required a major improvement of an existing interstate highway. Proposal B called for an entirely new location that would relegate the existing road chiefly to the service of local traffic. A prospective favorable consequence of the new location was to make possible the development of new economic activity in a certain area not now served by an adequate highway. This consequence, included in the economic analysis as a "benefit" for B but not for A, was a major factor in the analyst's recommendation favorable to proposal B. The analyst failed to recognize that the same benefit could be obtained by making a relatively small additional investment to add to Proposal A a low-cost secondary road that would serve the new area.

In another case, an irrigation district was having great difficulty with the maintenance of a number of flumes in its main canal. The district's consulting engineer estimated a cost of $1,200,000 for a proposed plan of flume replacement. When the district's commissioners tried to sell the district's bonds for this amount, the bond house that they approached sent its engineer to investigate. This engineer suggested that the investment might be reduced and a more permanent ditch obtained by substituting earth fills for many of the low flumes that needed replacement. This plan was later carried out at a cost of about $400,000.

Improved Analytical Procedures as a Possible Alternative to Investments. Sometimes when an unsatisfactory condition is under review and an investment in fixed assets is proposed to correct this condition, no thought is given to possible methods of improving the condition without a substantial investment.

For example, new machinery may be proposed to reduce high labor costs on a certain operation. Work simplification methods based on motion study may provide an alternative way to reduce these costs. As another example, new machinery may be proposed to reduce the percentage of spoilage of a manufactured product that must meet close tolerances. Possibly the same result might be obtained through the use of the techniques of statistical quality control.

A number of organizations have reported that the analysis of procedural problems preparatory to the purchase or lease of a large, high-speed computer has resulted in the improvement of existing procedures to the point that the computer could not be justified. The introduction of a computer always requires the careful analysis of the problems to be solved on it in order to translate each problem into language the computer can understand. Such analysis frequently reveals flaws in the current procedures that could have been eliminated without waiting until the lease or purchase of a computer was proposed.

In the public works field, also, proposed investments may have alternatives that are not obvious at first glance. For instance, the cost of flood damage may be reduced by investment in flood protection reservoirs, levees, and channel improvement. This cost may also be reduced by a system of flood zoning that prevents certain types of land use where there is a likelihood of flooding. Moreover, the cost of flood damages often may be reduced by an improved system of flood forecasting accompanied by an effective system of transmitting the forecasts to people in the area subject to flood.

Imperfect Alternatives Are Sometimes the Most Economical. The satisfaction of the engineer's sense of perfection is not a necessary prerequisite for the most economical alternative. Sometimes it happens that a careful study will show that an alternative that at first was summarily rejected affords the most economical solution of a given problem.

An illustration is the case of a geographically diversified group of public utility companies that needed to buy a great many poles. Poles came in a number of classes, AA, A, B, C, D, E, F, and G, depending upon the top diameter and the butt diameter. The past practice of these companies in pole selection had been based on their experience of what had proved to be satisfactory rather than on any considerations of theoretical design, and usually had involved purchasing no poles below class B.

It was then decided to analyze pole requirements on the basis of such factors as the expected storm loads in different areas and the importance of each pole line to the entire system. This analysis showed that many of the lighter grades of poles that had not previously been purchased were satisfactory for certain conditions. Savings were effected because cheaper poles were used in many cases. Additional savings resulted because the distribution of pole requirements among all of the classes made it possible to use a "wood's run" of poles, so that lumber companies were able to set a lower price on poles A and B than was possible when they had trouble in selling their lower classes.

The Common Condition of the Existence of Major Alternatives and Subsidiary Alternatives. Earlier in this chapter, we cited a "hunch" decision by the manager of a factory that had a total energy system. When it was finally recognized that this was an uneconomical decision, an engineer was called in to survey the situation and make recommendations.

The engineer examined the factory's past requirements for the relevant services—heat, pumping, compressed air, and electricity. After consideration of the trend of growth of production and of some changes in production methods that seemed to be in prospect, a forecast was made of the needs for these services for several years to come. With this forecast as a basis, preliminary designs were then made for meeting the expected needs by each of the possible alternatives recognized. For each alternative, approximate estimates were made of the immediate investment required and of the annual expenditures necessary in the future. With these estimates, the engineer was able to make a preliminary comparison of the long-run economy of the several alternative plans (including as one plan a continuance of the existing scheme), and to select those that seemed to justify detailed study.

Of all the alternatives given preliminary study, two appeared to be much more economical than any of the others. One was a plan to modernize the power plant by the purchase of one or more steam turbines; the other was a plan for the electrification of the compressors and pumps and the purchase of all power. Each of these plans was given detailed study with complete designs, and with careful estimates of investment costs and operation and maintenance costs.

As is characteristic of all economy studies to determine general policy, each of these designs involved numerous subsidiary alternatives, and each selection between subsidiary alternatives required a subsidiary economy study. For example, in the first alternative what type and size of turbines should be selected? How many should there be? What boiler pressures should be used? Should the power generation be combined with other steam requirements by selecting turbines that exhaust at pressures that permit the use of their exhaust steam for other purposes? In the second alternative, several different possible rates were offered by the power company. "Primary" power could be taken at 23,000 volts, requiring the customer to install transformers for stepping down the voltage and, of course, to take the transformer losses involved; a variation of this was an off-peak rate that restricted the power that could be used in certain specific hours of certain months. Two other different rates were available under which the power company supplied electricity at the voltages at which it was ultimately to be used. The electrification of compressors and pumps involved several possible alternative designs.

What Alternatives Are "Appropriate" for Evaluation? Costs for design, estimating, and analysis are involved whenever an additional alternative is to be reviewed. Clearly a balance is required between the possible advantages to be gained from evaluating more alternatives in any given case and the expenses (including sometimes the adverse consequences of time delays) incident to the evaluation. It is reasonable to try to avoid the costs of examining alternatives that have no merit.

The statement (on page 5) that the merits of all appropriate alternatives should be evaluated obviously leaves room for interpretation of the word *appropriate*. In most instances, as in the case cited of the engineer's study of the total energy system, preliminary economic evaluation will serve to eliminate many alternatives that are physically possible. Detailed study can be limited to those alternatives that have appeared to be best in the preliminary comparison.

The Need to Consider Consequences

2. *Decisions should be based on the expected consequences of the various alternatives. All such consequences will occur in the future.*

If there were no basis for estimating the results of choosing one proposed alternative rather than another, it would not matter which alternative was chosen. But in most proposals that involve investments in physical assets by business enterprises or by governmental bodies, it is possible to make estimates of differences in consequences. Clearly, rational decisions among alternatives should depend on prospective consequences to the extent that consequences can be anticipated. Because the consequences of a decision are necessarily *after* the moment of decision, this estimation always applies to the future.

As will be brought out by a variety of examples throughout this book, some types of consequences of investment decisions can be forecast with a fair amount of justifiable confidence, particularly if adequate effort is devoted to the job of estimation. In contrast, other consequences are inherently difficult to forecast. For example, other things being equal, consequences that are expected in the fairly distant future are harder to forecast than those in the near future. Moreover, some types of prospective consequences are fairly easy to quantify in a way that makes them commensurable with one another. But there may also be prospective consequences that are difficult or impossible to quantify in this way.

The Critical Issue of Consequences to Whom

3. *Before establishing procedures for project formulation and project evaluation, it is essential to decide whose viewpoint is to be adopted.*

The majority of the examples and problems in this book deal with economy studies for competitive business enterprise. In these studies we shall assume that the analysis is to be made primarily from the point of view of the owners of the enterprise.

The matter is a bit more complicated in economy studies for privately owned public utility companies that operate under the rules of rate regulation that have developed in the United States. We explain in Chapter 20 that the common practice of trying to minimize "revenue requirements" when choosing among design alternatives is, in effect, taking the viewpoint of the customers of a utility company. We also note certain circumstances in which such a customers' viewpoint may be insufficient or inappropriate.

The choice of viewpoint is much more complicated in economy studies made for governmental bodies. In many types of economy study for public works, it clearly is undesirable to restrict the viewpoint to the financial position of the particular governmental unit that is involved. We shall see first in Chapter 9 and later in Chapter 19 that, in principle, many such economy studies ought to be made considering prospective consequences "to whomsoever they may accrue." We shall also see that, because of their diffused nature, consequences are often much harder to estimate and evaluate in government projects than in private enterprise.

Commensurability

4. *In comparing alternatives, it is desirable to make consequences commensurable with one another insofar as practicable. That is, consequences should be expressed in numbers and the same units should apply to all the numbers. In economic decisions, money units are the only units that meet the foregoing specification.*

Money units at different times are not commensurable without calculations that somehow reflect the time value of money; in some cases they are not commensurable without adjustment for prospective changes in purchasing power of the monetary unit.

Words Versus Monetary Figures as a Basis for Economic Decisions. Often a good way to start an economy study is to use words to itemize the expected differences among the alternatives being compared. But it is easy to reach unsound conclusions if differences expressed in words are not later converted into units that make the differences comparable.

To illustrate this point, consider a proposal that incandescent lamps be replaced by fluorescent fixtures in a certain industrial plant. The engineer who made this proposal listed the advantages and disadvantages as follows:

Advantages:

1. More light for the same amount of energy
2. Smaller number of fixtures
3. Less frequent lamp replacement
4. Lower maintenance costs
5. Better light
6. Less heat to be dissipated
7. Improvement of lighting without having to install new, larger conductors
8. Improved working conditions
9. Less eye fatigue for employees
10. Better-quality product
11. Better employee morale

Disadvantages:

1. Higher investment in fixtures
2. Higher unit lamp cost
3. Labor cost of installation
4. Interruption of work during installation
5. "Flicker" may occur and be very annoying

A common characteristic of such a tabulation is that the listed differences are not mutually exclusive; the same thing may be listed more than once with different words. Moreover, because the stated differences are not commensurable with one another, trivial differences between the alternatives tend to be given the same weight as important differences. A hazard of this verbal technique of comparing alternatives is that the decision will be unduly influenced by sheer weight of words.

Cash Flow and the Time Value of Money. Two steps generally are required before differences stated in words can be made commensurable with one another. First, the differences must be expressed in their appropriate physical units. Then the physical units must be converted to money units. For example, wire and conduit of various lengths and diameters, lamps and lamp fixtures of various types and sizes, hours of installation and maintenance labor, and kilowatt hours of electric energy are not commensurable with one another until they are converted to a common unit, namely, money.

But it is not sufficient to estimate the *amounts* of money receipts and disbursements influenced by a decision; it is also necessary to estimate the *times* of the cash flows. Because money has a time value, a monetary unit (dollar, peso, pound, yen, etc.) at one date is not directly comparable with the same monetary unit at another date. The relevance of the timing of cash flow is discussed at the start of Chapter 2 and developed throughout the remainder of this book.

Prospective Price Level Changes. It is not always enough to make the diverse consequences of a decision commensurable by converting them into money and then making a further conversion to reflect the time value of money. Where a change in general price levels is anticipated, a further analysis may be needed.

In Chapter 14 we introduce the subject of the relationship of economic decision making to the expectation of price-level changes. The point is made there that this topic is fairly simple when it is expected that the various prices relevant to a particular decision will rise (or fall) at approximately the same rate. Complications are introduced when it is expected that there will be a substantial difference in the rates at which various relevant prices are expected to change.

The Issue of the Validity of Market Prices as a Basis for Decisions Among Investment Alternatives. In the long run, a competitive enterprise cannot survive without money receipts that more than offset its money disbursements. The case for evaluating proposals in terms of their influence on prospective cash flow to and from the enterprise is therefore quite clear and straightforward in competitive business. Market prices, in the sense used here, are simply the prices that the enterprise expects to pay for its various inputs and expects to receive for its various outputs.

It is explained in Chapters 9 and 19 that the matter is not so simple in economy studies for governments. Part of the difficulty is that no market may exist to establish prices for certain types of desired outputs from government projects. (Representative examples of such desired outputs are reduction of air pollution, reduction of deaths from aircraft and motor vehicle accidents, savings of time to operators of noncommercial motor vehicles.) Also, in certain cases where market valuations are available, some persons may deem such valuations to be unsuitable for assigning monetary figures to project outputs. At this point in our exposition, we shall merely note that we favor the use of market prices, where available, in the initial formulation and evaluation of the economic aspects of proposed government projects. However, final decisions may reasonably give weight to relevant matters that were omitted by such use of market prices. Some aspects of the pricing of the inputs and outputs from proposed public works projects are discussed in Chapter 19.

Irrelevance of Matters Common to All Alternatives

5. *Only the differences among alternatives are relevant in their comparison.*

This generally useful concept needs to be recognized in a number of ways that will be developed throughout this book. For example, the past is common to all alternatives for the future; there can be no consequences of a decision before the moment of decision. Also, it is easy to draw incorrect conclusions by basing estimates on average costs or allocated costs rather than on cost differences. Thus, cost consequences of an alternative should not be compared solely by using figures originating in a cost accounting system; allocated costs that were properly included in the accounts often should be omitted in an engineering economy study. Although this concept is illustrated by many examples throughout this entire book, it is given particular attention in Chapters 12, 16, and 17.

Separation of Decisions

6. *Insofar as practicable, separable decisions should be made separately.*

In most instances (although not all), decisions on the financing of physical plant are independent of decisions on the specific assets to be selected. The combining of a particular plant investment with a particular scheme of financing in a single analysis may lead to unsound decisions. This point is developed particularly in Chapter 18.

Many proposed engineering projects have a number of different possible levels of investment that are physically possible. Each separable increment of investment that might either be included or left out ought to be evaluated on its own merits. Each increment of investment should produce favorable results sufficient to pay its own way. Otherwise, the fact that certain proposed separable increments are economically unproductive may be concealed by evaluation of a project only as a whole. This point is developed in various places throughout this book and is particularly emphasized in Chapter 13.

2

THE NEED FOR CRITERIA AND ANALYTICAL PROCEDURES IN MAKING DECISIONS ABOUT PROPOSED INVESTMENTS

I often say that when you can measure what you are speaking about, and express it in numbers, you know something about it; but when you cannot measure it, when you cannot express it in numbers, your knowledge is of a meagre and unsatisfactory kind; it may be the beginning of knowledge, but you have scarcely, in your thoughts, advanced to the stage of Science, whatever the matter may be.—Lord Kelvin

Although investments in physical assets were involved in the actual cases that were described briefly in Chapter 1, the *concepts* presented were widely applicable to many types of decision making. In the present chapter, we look more specifically at concepts relevant to the investment-type decisions that so often are needed when engineering alternatives are present.

Need for Decision Criteria

7. *It is desirable to have a criterion for decision making, or possibly several criteria.*

Clearly, the criteria should be applied to the differences in consequences that are anticipated from the choice among the different alternatives.

With reference to proposals for alternative investments in physical assets, it has been pointed out that the consequences of a choice should be expressed as far as practicable in terms of cash flows (or other monetary figures) at stated points in time. It is suggested in this chapter that there should always be a primary criterion applied to such monetary figures. In dealing with certain types of proposals, it may also be desirable to have one or more supplementary or secondary criteria applied to the monetary figures.

It was pointed out in Chapter 1 that not all prospective consequences of decisions about investments in physical assets are reducible to monetary terms. Weight often needs to be given to such irreducible data. It follows that there may also be secondary criteria for decision making that are related to differences in estimated consequences that have not been expressed as monetary figures.

Choice of a Primary Criterion

8. *The primary criterion to be applied in a choice among alternative proposed investments in physical assets should be selected with the objective of making the best use of limited resources.*

Whether one thinks of an individual, a family, a business enterprise, or a governmental unit such as a city, state, or nation, it is generally true that at any given time there is a limitation on the available resources that can be devoted to investment in physical assets. The resources that are limited may be of many types, such as land, labor, or materials. But because the market gives us money valuations on most resources, it usually is reasonable to express the overall limitation in terms of money.

In evaluating proposed investments, the question should be asked whether the investment will be productive enough, all things considered. "Productive enough" can be interpreted as yielding a sufficient rate of return as compared with one or more stated alternatives. Throughout this book, we shall assume that a decision can and has been made on the minimum rate of return that is attractive in any given setting and that this is the basis for the primary criterion to be used for investment decisions. Chapter 10 introduces some of the "all-things-considered" issues that may arise in selecting a minimum attractive rate of return. These issues are examined further in the remaining chapters of this book, particularly Chapters 14, 18, 19, 20, and 21.

We shall see in Chapters 3 through 9 that it is necessary to use the mathematics of compound interest to apply a primary criterion based on a stipulated minimum attractive rate of return. Four different ways to implement this criterion are explained in Chapters 6, 7, 8, and 9. It is pointed out in these chapters that with the same input data, all four methods lead to the same conclusions as to whether or not the primary criterion has been met.

Wherever income taxes are levied, such taxes tend to reduce the rate of return from investments in physical assets. However, income tax laws and regulations often favor certain types of investment as compared to other types; for various reasons related to the technicalities of income taxation, the after-tax rate of return is not a fixed percentage of the before-tax rate. It follows that, in principle, the income tax consequences of proposed investments ought to be estimated, and that a stipulated minimum attractive rate of return in private industry should be an after-tax rate rather than a before-tax rate.

The use of the after-tax minimum attractive rate of return is illustrated in examples starting with Chapter 6 using certain simplified assumptions about the relationship between before-tax and after-tax cash flow. A critical look at the complexities of the ways in which income taxes influence the relative merits of proposed investments is deferred until Chapter 12 and subsequent chapters.

Secondary Criteria Applied to Consequences Expressed in Monetary Terms

9. *Even the most careful estimates of the monetary consequences of choosing*

different alternatives almost certainly will turn out to be incorrect. It often is helpful to a decision maker to make use of secondary criteria that reflect in some way the lack of certainty associated with all estimates of the future.

Such secondary criteria are illustrated in a number of examples and problems in Part III of this book.

"Irreducible Data of the Problem of Investment"

10. *Decisions among investment alternatives should give weight to any expected differences in consequences that have not been reduced to money terms as well as to the consequences that have been expressed in terms of money.*

In the second edition of his *Engineering Economics*, published in 1923, Professor J. C. L. Fish coined the phrase "irreducible data of the problem of investment" to apply to prospective differences between alternatives that are not reduced to estimated receipts and disbursements for purposes of analysis. Some other words or phrases that have been applied to such nonmonetized differences are "judgment factors," "imponderables," and "intangibles."

There is no word or short phrase that by itself conveys the precise idea of something that is relevant in a particular decision but has not been expressed in terms of money for one reason or another. Often it is not so much that an analyst believes that the difference in question will not eventually influence receipts and disbursements as that there is no satisfactory basis for estimating how much the influence on cash flow will be and when this influence will occur.

In this book we have elected to use *irreducibles* or *irreducible data* in the special technical sense of relevant differences in the expected consequences of a decision that have not been reduced to money terms. There are two reasons for this election. One reason is historical, namely, that this usage dates back to 1923. The other reason is that the word or phrase is not generally used in everyday speech and therefore does not have a different popular meaning that might be a cause of its misinterpretation.

Irreducibles may play a particularly large part in personal investment decisions. For example, consider a choice between home ownership and renting. It is obvious that many aspects of the matter should be estimated in terms of money as a guide to intelligent decision making. Nevertheless, matters of personal taste that cannot be expressed in money terms should reasonably be given weight in such a decision. Thus, the pride of home ownership may be of great importance to one family and a matter of complete indifference to another.

For reasons discussed in Chapter 19, irreducibles may also be given considerable weight in many governmental decisions.

The Need for a "System Viewpoint"

11. *Often there are side-effects that tend to be disregarded when individual decisions are made. To consider such side-effects adequately, it may be necessary to examine the interrelationships among a number of decisions before any of the individual decisions can be made.*

The basic question here is whether too narrow a view is being taken of the alternatives that are being compared. If the side-effects of a particular decision are sufficiently trivial, presumably a study of them would not change the decision. However, a study of the interrelationships among a group of decisions may be needed to provide a basis for judgment on whether the side-effects are trivial or important. Comments on this topic are made particularly in Chapters 13 and 21.

An Example of the Application of Certain Concepts

Near the start of Chapter 1 and again in the middle of that chapter, there was reference to the case of a company that owned a total energy system that had become inadequate. This story is now continued to its conclusion.

The reader may recall that when management finally recognized that a number of alternatives needed to be evaluated, a preliminary economy study narrowed the comparison to two alternatives, which were then studied in detail. One of these alternatives called for continuing the concept of a total energy system by the purchase and installation of a modern power plant. This plant would have greater thermal efficiency and greater capacity than the old inadequate plant. The other alternative involved the purchase of electric energy, the electrification of certain equipment, and the continued generation of steam only during the colder months of the year for purposes of space heating.

When the cash flow estimates for these two alternatives were assembled, it was evident that the total disbursements over a 20-year analysis period would be somewhat lower with the new power plant. However, the initial disbursement was of course much greater with the power plant. The question at issue (in relation to our primary criterion for investment decisions) was whether the annual operating savings over a 20-year period would be sufficient to recover the extra investment required for the new power plant and its related facilities plus a rate of return that was high enough to be attractive, all things considered.

In this company, experience had shown that there always seemed to be plenty of proposals for internal investment that promised an after-tax rate of return of 10% or more. Such projects absorbed all the funds that could be made available for internal investment. The consequence of accepting a proposal that had a prospective after-tax yield of less than 10% was the turning down of some other proposal that had a prospective yield of more than 10%. In effect, 10% after taxes was the minimum attractive rate of return in this company even though this particular phrase was not used.

An analysis was made of the difference in year-by-year estimated cash flow between the alternative with the new power plant and the alternative calling for the purchase of electricity. The prospective after-tax rate of return on the extra investment in the power plant alternative was about 6%. Because 6% was less than the stipulated 10%, the application of our primary criterion for decision making favored the alternative involving purchased electricity.

Secondary criteria based on certain modifications of the original cash flow estimates also favored the purchase of electricity. If one or more business recessions should occur during the analysis period, expenses for purchased electricity could be decreased more readily than expenses for the operation of a power plant; in the past, with power generated in the factory's own plant, it had been necessary to keep the power plant operating with a full labor force even when the factory was operating at only a small fraction of its full capacity. Moreover, if there should be an unanticipated rapid growth in the plant's demand for total energy, the need could be met more rapidly by purchasing more electricity than by expanding the power plant.

On the other hand, there was an irreducible factor adverse to the alternative involving purchased electricity. The change in policy would create certain personnel problems. An engine-room force would no longer be needed and a boiler-room force would not be required during the summer months. However, a survey of personnel requirements elsewhere in the plant indicated that these employees could be retrained and transferred to other jobs.

In all decision making it finally becomes necessary for some person (or possibly a group of persons) to make a choice. Other things being equal, a choice between investment alternatives should be based on the chosen primary criterion. In the frequent case where other things are not entirely equal, weight should be given to secondary criteria including irreducibles. In this example, the analysis clearly favored the alternative involving the purchase of electricity, and this was the one actually chosen.

Some general points may be illustrated by this example. One point is that this was an *engineering* economy study because so many technical matters were involved that only an engineer could be expected to recognize the appropriate alternatives and make a competent analysis of them. (In this case the engineer who made the study *recommended* the decision and someone else made the decision to follow the recommendation.) Another point is that the setting of this decision was unique; no other plant would have exactly the same existing physical facilities, present and prospective needs for various forms of energy, personnel, financial circumstances, and so on. (Many economic comparisons that superficially seem to be similar actually turn out to be quite different because of differences in the surrounding circumstances.)

Still another point that we shall emphasize throughout this book is that only the *differences* between the two best alternatives are relevant in their comparison and each increment of investment must meet the primary criterion. Either of these two alternatives as well as several others that were discarded would have shown an after-tax rate of return of more than 10% as compared to the uneconomic alternative of *continuing the present condition*. It would not have been sufficient for an analyst to show that the proposed new power plant would yield say, an 18% rate of return as compared to the present way of doing things. The proposed new power plant was not a sound investment unless it could sustain its challenge against the best of the other possible alternatives as well as a continuation of the present condition.

The events described in the foregoing story took place in the United States prior to the 1970s. Therefore the analysis was not influenced by certain matters that we discuss in Chapter 14, including the prospect of rapid inflation and of adverse differential changes in prices associated with energy.

Summary of Part I

The eleven concepts given in these chapters are here repeated for emphasis:

1. Decisions are among alternatives; it is desirable that alternatives be clearly defined and that the merits of all appropriate alternatives be evaluated.
2. Decisions should be based on the expected consequences of the various alternatives. All such consequences will occur in the future.
3. Before establishing procedures for project formulation and project evaluation, it is essential to decide whose viewpoint is to be adopted.
4. In comparing alternatives, it is desirable to make consequences commensurable with one another insofar as practicable. That is, consequences should be expressed in numbers and the same units should apply to all the numbers. In economic decisions, money units are the only units that meet the foregoing specification.

   Money units at different times are not commensurable without calculations that somehow reflect the time value of money; in some cases they are not commensurable without adjustment for prospective changes in purchasing power of the monetary unit.
5. Only the differences among alternatives are relevant in their comparison.
6. Insofar as practicable, separable decisions should be made separately.
7. It is desirable to have a criterion for decision making, or possibly several criteria.
8. The primary criterion to be applied in a choice among alternative proposed investments in physical assets should be selected with the objective of making the best use of limited resources.
9. Even the most careful estimates of the monetary consequences of choosing different alternatives almost certainly will turn out to be incorrect. It often is helpful to a decision maker to make use of secondary criteria that reflect in some way the lack of certainty associated with all estimates of the future.
10. Decisions among investment alternatives should give weight to any expected differences in consequences that have not been reduced to money terms as well as to the consequences that have been expressed in terms of money.
11. Often there are side-effects that tend to be disregarded when individual decisions are made. To consider such side-effects adequately, it may be necessary to examine the interrelationships among a number of decisions before any of the individual decisions can be made.

PROBLEMS

2–1. You are a college student in your final year. You have been forced by a housing shortage to take a room eight miles from your campus. You can ride a public bus to and from your campus; this travels at 40-minute intervals from 6 A.M. to 8 P.M. An alternative is to buy an automobile for your transportation. If you buy a car you expect to dispose of it at the end of the school year. What prospective cash disbursements and receipts seem to you to be relevant in relation to your decision of whether or not to buy a car? What irreducibles do you think are important?

2–2. In the preceding problem assume that you have just purchased the car. Now there has been an adverse change in your financial circumstances that makes you consider whether or not you should dispose of the car at once. What differences can you see between your analysis now and your original analysis before you purchased the car?

2–3. You have traveled from your home to the city of Q on a 30-day-limit round-trip bus ticket for which you paid $83. The one-way bus fare between the two places is $50; if you do not use the return half of your ticket, the bus company will redeem it at the difference between the round-trip price and the one-way fare.

(a) Assume that it becomes necessary for you to stay in Q until after the return date limit. What is the lowest price at which you could afford to sell the return half of your ticket to a friend rather than turn it in for credit to the bus company?

(b) If you plan to return home by bus before the return date, what is the lowest price at which you could afford to sell the return half of your ticket?

2–4. A ranch owner who is five miles from the nearest public power line wishes to consider providing electricity for the ranch. What alternatives do you see that might be possible? Discuss the problem of choice among these alternatives with reference to the concepts stated in Chapters 1 and 2.

2–5. The owner of a party fishing boat has lost a number of days of operation because its engine has broken down several times, requiring a day of overhaul each time. The owner is afraid that too frequent breakdowns will cause a reputation for unreliable operation. What alternatives do you see that might be available to improve this condition? How would you expect future receipts and disbursements to be influenced by the choice among these alternatives? What irreducibles might reasonably be given weight in this decision?

2–6. In a growing city decisions must be made from time to time as to whether various physical means of traffic control are to be used at specific street intersections to control motor vehicle traffic and pedestrian traffic. The various possibilities at an intersection may include:

1. No control devices
2. Arterial stop sign on one street but not on the other
3. Arterial stop signs on both streets
4. Traffic lights with duration of signals in each direction controlled by preset timing devices
5. Traffic lights in which the frequency and duration of the traffic signals are determined by traffic-activated devices on one of the intersecting streets
6. Traffic lights in which the frequency and duration of the traffic signals are determined by traffic-activated devices on both of the intersecting streets

Decisions by city officials among such alternatives often are made largely by intuition, although weight may be given to such matters as traffic counts of motor vehicles, the current state of the city budget, the occurrence of recent accidents at a particular intersection, and pressures from residents living near the various intersections. It has been suggested that decision making might be improved by the practice of using formal economy studies requiring an evaluation (in money terms as far as practicable) of the various benefits and costs associated with each alternative.

Consider some street intersection in your vicinity that is now in condition (2) above. If possible, select an intersection for which some residents have proposed a change to (3), (4), (5), or (6).

Discuss the problems of making the decision between continuing condition (2) and changing to one of the other conditions. Tie your discussion to the eleven concepts stated in Chapters 1 and 2. Which of these concepts, if any, seem to you to be of particular importance in a decision of this type?

2–7. What are some alternative methods that a college might consider for the provision of janitorial service for the college buildings? Discuss the problem of choice among these alternatives with reference to the concepts stated in Chapters 1 and 2.

2–8. What are some alternatives that a married student might consider for the provision of family housing while the student is in college? Discuss the problem of choice among these alternatives with reference to the concepts stated in Chapters 1 and 2.

PART II

JUDGING THE ATTRACTIVENESS OF PROPOSED INVESTMENTS

3

EQUIVALENCE

[The] growth of money in time must be taken into account in all combinations and comparisons of payments.—*J. C. L. Fish**

Most problems in economy involve determining what is economical in the long run, that is, over a considerable period of time. In such problems it is necessary to recognize the time value of money; because of the existence of interest, a dollar now is worth more than the prospect of a dollar next year or at some later date.

Definition of Interest

Interest may be defined as money paid for the use of borrowed money. Or, broadly speaking, interest may be thought of as the return obtainable by the productive investment of capital. While the point of view required in dealing with problems in engineering economy is the one implied in the latter broader definition, it is, nevertheless, desirable to start the discussion of interest by considering situations in which money is actually borrowed. The broader viewpoint is developed in Chapter 6 and thereafter.

Interest Rate

The rate of interest is the ratio between the interest chargeable or payable at the end of a period of time, usually a year or less, and the money owed at the beginning of that period. Thus, if $9 of interest is payable annually on a debt of $100, the interest rate is $9/$100 = 0.09 per annum. This is customarily described as an interest rate of 9%, the "per annum" being understood unless some other period of time is stated.

Even though interest is frequently payable more often than once a year, the interest rate per annum is usually what is meant when an interest rate is stated. Thus, rates of 0.0075 payable monthly, 0.0225 payable quarterly, or 0.045 payable semiannually are all described as 9%. The difference between the payment of interest annually and its payment more frequently is dis-

*J. C. L. Fish, *Engineering Economics*, 2d ed. (New York: McGraw-Hill Book Co., Inc., 1923), p. 20.

cussed briefly in the next chapter under the heading "Nominal and Effective Interest Rates."

Plans for Repayment of Borrowed Money

Consider the plans shown in Table 3–1 by which a loan of $10,000 might be repaid in 10 years with interest at 9% payable annually. In the tables showing these plans, the date of the loan is designated as 0 years and time is measured in years from that date. The $10,000 is called the *principal* of the loan.

TABLE 3–1
Four plans for repayment of $10,000 in 10 years with interest at 9%

	End of year	*Interest due (9% of money owed at start of year)*	*Total money owed before year-end payment*	*Year-end payment*	*Money owed after year-end payment*
	0				$10,000
	1	$ 900	$10,900	$ 900	10,000
	2	900	10,900	900	10,000
	3	900	10,900	900	10,000
	4	900	10,900	900	10,000
Plan I	5	900	10,900	900	10,000
	6	900	10,900	900	10,000
	7	900	10,900	900	10,000
	8	900	10,900	900	10,000
	9	900	10,900	900	10,000
	10	900	10,900	10,900	0
	0				$10,000
	1	$ 900	$10,900	$ 1,900	9,000
	2	810	9,810	1,810	8,000
	3	720	8,720	1,720	7,000
	4	630	7,630	1,630	6,000
Plan II	5	540	6,540	1,540	5,000
	6	450	5,450	1,450	4,000
	7	360	4,360	1,360	3,000
	8	270	3,270	1,270	2,000
	9	180	2,180	1,180	1,000
	10	90	1,090	1,090	0
	0				$10,000.00
	1	$ 900.00	$10,900.00	$ 1,558.20	9,341.80
	2	840.76	10,182.56	1,558.20	8,624.36
	3	776.19	9,400.55	1,558.20	7,842.35
	4	705.81	8,548.16	1,558.20	6,989.96
Plan III	5	629.10	7,619.06	1,558.20	6,060.86
	6	545.46	6,606.32	1,558.20	5,048.12
	7	454.33	5,502.45	1,558.20	3,944.25
	8	354.98	4,299.23	1,558.20	2,741.03
	9	246.69	2,987.72	1,558.20	1,429.52
	10	128.66	1,558.18	1,558.18	0.00

TABLE 3–1 *(continued)*

	0				$10,000.00
	1	$ 900.00	$10,900.00	$ 0.00	10,900.00
	2	981.00	11,881.00	0.00	11,881.00
	3	1,069.29	12,950.29	0.00	12,950.29
	4	1,165.53	14,115.82	0.00	14,115.82
Plan IV	5	1,270.42	15,386.24	0.00	15,386.24
	6	1,384.76	16,771.00	0.00	16,771.00
	7	1,509.39	18,280.39	0.00	18,280.39
	8	1,645.24	19,925.63	0.00	19,925.63
	9	1,793.31	21,718.94	0.00	21,718.94
	10	1,954.70	23,673.64	23,673.64	0.00

Characteristics of Repayment Plans

The student of engineering economy should carefully examine these four plans because they are representative of various schemes in common use for the repayment of money borrowed for a term of years. Plan I involves no partial payment of principal; only interest is paid each year, and the principal is paid in a lump sum at the end of the period. A cash flow diagram to represent the borrower's receipts and payments for Plan I is shown in Figure 3–1a*.

Plans II and III involve systematic reduction of the principal of the debt by uniform repayment of principal with diminishing interest in Plan II and by a scheme that makes the sum of the interest payments and the principal payment uniform in Plan III. The cash flow diagram for Plan II is shown in Figure 3–1b. The cash flow diagram for Plan III is shown in Figure 3–1c.

Plan IV, on the other hand, involves no payment of either principal or interest until the single payment of both at the end of the tenth year. Its diagram is shown in Figure 3–1d.

The advantages and disadvantages of the various plans from the standpoint of different classes of borrowers will be discussed in a later chapter. The point for immediate consideration is their relation to compound interest and to the time value of money.

Compound Interest

In Plan IV, interest is *compounded*; that is, the interest each year is based on the total amount owed at the end of the previous year, a total amount that included the original principal plus the accumulated interest that had not been paid when due. The formulas and tables explained in the next chapter are all based upon the compounding of interest.

*The cash flow diagram is drawn from the borrower's viewpoint. Upward vectors represent receipts and are considered to be positive. Downward vectors represent payments (disbursements) and are considered to be negative. Horizontal distance represents the relative time periods. See Chapter 5 and Figure 5–1.

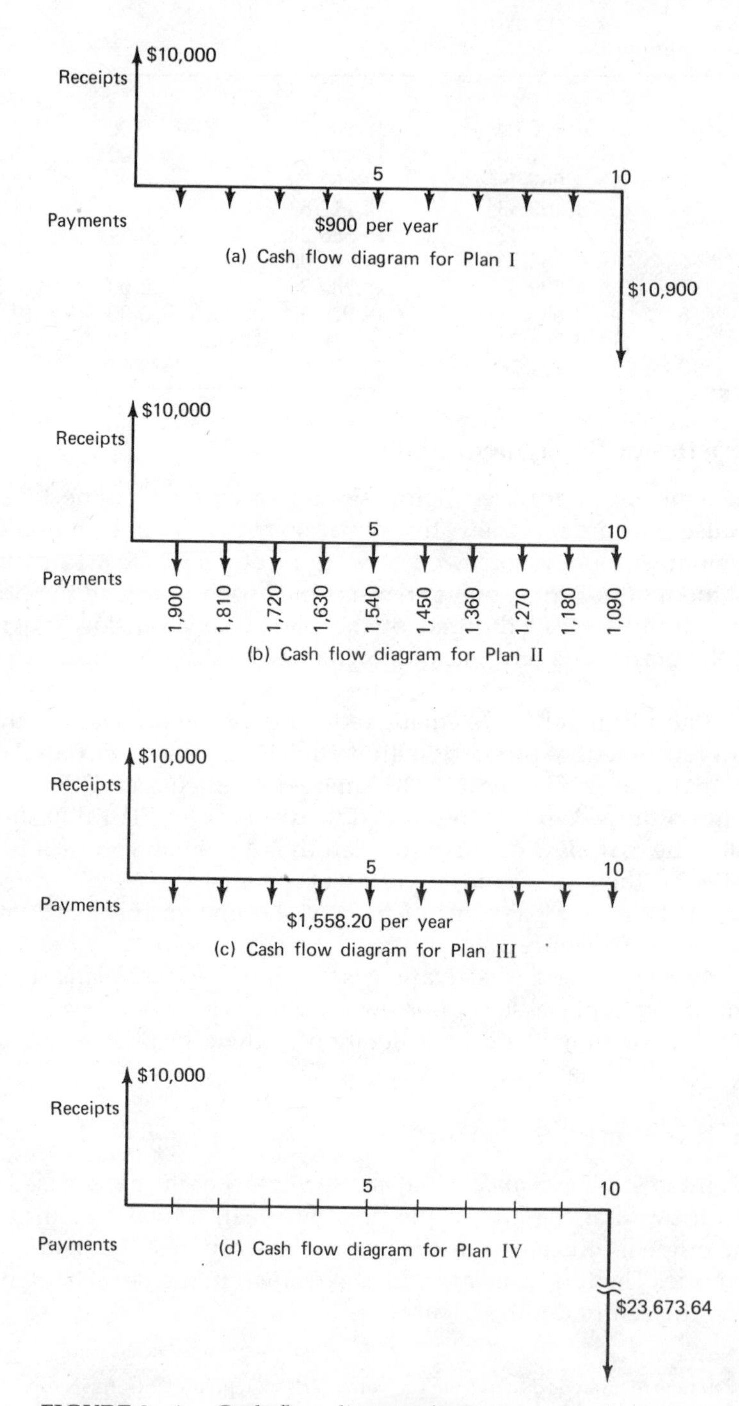

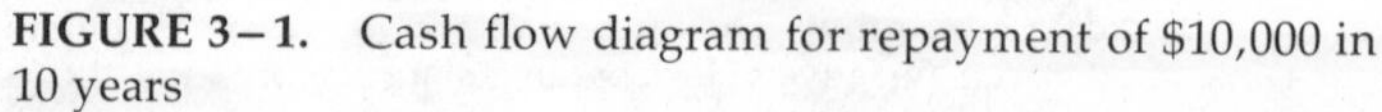

FIGURE 3–1. Cash flow diagram for repayment of $10,000 in 10 years

The distinction is made, both in the literature of the mathematics of investment and in the law, between *compound interest* and *simple interest*. If in Plan IV a lump-sum payment of principal and interest had been called for at simple rather than compound interest, the only interest payable would have been that charged on the original principal of $10,000. The total payment required under simple interest would thus have been $19,000 instead of the $23,674 that is required under compound interest.

Where money is borrowed for a period of years, the usual business practice is for interest to be due—and nearly always actually paid—annually or more often. This practice, in effect, involves compound interest, whether considered from the viewpoint of the lender or of the borrower. Where interest is paid each year, the lender receives a payment that can be used immediately. Also, the borrower, in paying interest annually, is foregoing the opportunity to use the money paid to the lender.

Thus, compound interest is the general practice of the business world, and Plans I, II, and III involve what is, in effect, compound interest, because interest is paid annually in each plan. It will be shown in the following chapters how compound interest formulas may be used in dealing with all these various types of repayment situations. Problems in engineering economy generally require consideration of compound interest; simple interest (chiefly of importance in connection with loans for periods of a year or less) will not be discussed further in these pages. Wherever the term "interest" is used, compound interest (i.e., interest due annually or more often) is implied.

Equivalence

The concept that payments that differ in total magnitude but that are made at different dates may be equivalent to one another is an important one in engineering economy. In developing this concept, we may place side by side for comparison, as in Table 3–2, the $10,000 borrowed and the four series of money payments that we have seen would repay it with interest at 9%. (Payments are shown to four significant figures only.)

If interest is at 9%, these five sets of payments are equivalent to one another. They are equivalent from the standpoint of a prospective lender (investor) with $10,000, because with that sum the lender can get any one of the four future series of payments (or more precisely, someone's promise to make the payments) in exchange for the present $10,000. Similarly, from the standpoint of the prospective borrower who needs $10,000 (perhaps to invest productively in a business enterprise), the four future series are equivalent to each other and to $10,000 now, because, by agreeing to pay any one of these future series, the needed present sum may be secured.

Obviously, we might think of any number of series of payments that would serve to just repay $10,000 with interest at 9%. All of these would be equivalent to $10,000 now, and to each other.

TABLE 3–2
Five equivalent series of payments

Year	*Investment*	*I*	*II*	*III*	*IV*
0	$10,000				
1		$ 900	$1,900	$1,558	
2		900	1,810	1,558	
3		900	1,720	1,558	
4		900	1,630	1,558	
5		900	1,540	1,558	
6		900	1,450	1,558	
7		900	1,360	1,558	
8		900	1,270	1,558	
9		900	1,180	1,558	
10		10,900	1,090	1,558	$23,670

The meaning of equivalence may be explained by using an analogy from algebra: If a number of things are equal to one thing, then they are equal to each other. Given an interest rate, we may say that any payment or series of payments that will repay a present sum of money with interest at that rate is equivalent to that present sum. Therefore, *all future payments or series of payments that would repay the present sum with interest at the stated rate are equivalent to each other*.

Present* Worth

From the foregoing we found that a loan of $10,000 can be repaid with interest in four different ways, involving different amounts of money and at different times over a period of 10 years. To the lender (investor) the loan (investment) is the amount necessary to secure the promise of the future payment or series of payments, with interest at the given rate. The investment necessary to secure the promise of the future payment or payments is the present worth of the future payments. To the borrower, the present worth may be thought of as the present sum that may be secured in exchange for the promise to make specified future payments or series of payments. To both the lender and the borrower the repayment series of payments is equivalent to the present worth. Also, since the four different series of payments are equivalent to the same present worth, the series of payments are equivalent to each other.

Significance of Equivalence in Engineering Economy Studies

The five columns in Table 3–2 show equivalent series of payments; however, the total payments called for are quite different, totaling $10,000, $19,000, $14,950, $15,580, and $23,670, respectively. The longer the repayment period, the greater this apparent difference. Thus, if the repayment period were 20 years, the corresponding total payments for similar equivalent series would be $10,000, $28,000, $19,450, $21,910, and $56,044.

*Of course the "present" may be moved in the imagination of the analyst to any convenient date. Thus, we may speak in 1980 of the present worth in 1985 of a payment to be made in 1995.

Engineering economy studies usually involve making a choice of several alternative plans for accomplishing some objective of providing a given service. If a given service could be provided by five alternative methods requiring payments as shown in the five columns of Table 3–2, all of the alternative plans would be equally economical with interest at 9%; that is, they could each be financed by a present sum of $10,000. This fact would not be evident from a comparison of the total payments called for in the various plans; it would be clear only if the different money series were converted either to equivalent single payments (e.g., present worth) or to equivalent uniform series.

Engineering economy studies usually require some conversion as a basis for intelligent decision. A comparison of total payments involved in alternative plans, without the use of interest factors to convert the two series to make them comparable, is nearly always misleading.

Equivalence Depends on the Interest Rate

It was shown that the five payment series in the columns of Table 3–2 were equivalent because each series would repay an original loan of $10,000 with *interest at* 9%. These series payments would not be equivalent if the stated interest rate had been anything other than 9%. If the rate is less than 9%, the series of payments in the four plans will repay a greater amount than the present worth of $10,000; if the rate is greater than 9%, they will all repay a lesser amount. Table 3–3 shows the *present worths* of the four series of payments at *various interest rates*.

A present sum is always equivalent at some interest rate to a larger future sum of payments. In many engineering economy problems the answer desired is the interest rate that will make two series equivalent to each other; this is often described as the *rate of return* obtainable on a proposed extra investment. This viewpoint is developed in Chapter 8.

TABLE 3–3
Present worths at various interest rates of payment series shown in Table 3–2

Interest rate	*Plan I*	*Plan II*	*Plan III*	*Plan IV*
0%	$19,000	$14,950	$15,580	$23,670
3%	15,118	12,939	13,290	17,616
6%	12,208	11,320	11,468	13,219
9%	10,000	10,000	10,000	10,000
12%	8,305	8,912	8,804	7,623
15%	6,989	8,008	7,821	5,852

Summary

The main points of this chapter may be stated as follows:

One definition of interest is money paid for the use of borrowed money. The rate of interest may be defined as the ratio between the interest chargeable or payable at the end of a stipulated period of time and the money owed at the beginning of that period. The general practice of the business world is for interest to be chargeable or payable annually or more often.

Any future payment or series of payments that will exactly repay a present sum with interest at a given rate is *equivalent* to that present sum; all such future payments or series of payments that will repay the given present sum are equivalent to one another. The present sum is the *present worth* of any future payment or series of payments that will exactly repay it with interest at a given rate.

Equivalence calculations are necessary for a meaningful comparison of different money time series; they are thus usually required in engineering economy studies. Economy studies, however, generally imply a broader definition of interest as the return obtainable by the productive investment of capital.

PROBLEMS

3–1. Prepare a table similar to Plans II and III of Table 3–1 showing the annual repayment (principal and interest) of a loan of $10,000 in 5 years with interest at 9%. The uniform annual payment for Plan III is $2,570.90. Compare the total repayments required in these cases with those required for Plans II and III in Table 3–1.

3–2. Prepare a table similar to Plan II of Table 3–1 showing the annual repayment schedule (principal and interest) of a loan of $4,000 in 5 years with interest at 11%.

3–3. Prepare a table similar to Plan III of Table 3–1 showing the annual repayment schedule (principal and interest) of a loan of $4,000 in 5 years with interest at 11%. The uniform annual payment is $1,082.28.

3–4. Prepare a table similar to Table 3–1 showing four plans for the repayment of $2,000 in 4 years with interest at 12%. The uniform annual payment in the plan corresponding to Plan III is $658.46.

3–5. Prepare tables similar to Plan II of Table 3–1 showing the annual repayment schedule (principal and interest) of a loan of $5,000 in 5 years with interest at 9% and 15%. Compare the total repayments required under each schedule.

3–6. Prepare tables similar to Plan II of Table 3–1 showing the annual repayment schedule (principal and interest) of a loan of $5,000 in 4 years and in 10 years with interest at 12%. Compare the total repayments required under each schedule.

3–7. Prepare a table similar to Table 3–1 showing four plans for the repayment of $2,000 in 4 years with interest at 15%. The uniform annual payment in the plan corresponding to Plan III is $700.54.

3–8. Prepare a table similar to Plans II and III of Table 3–1 showing the annual repayment schedule (principal and interest) of a loan of $10,000 in 10 years with interest at 6%. The uniform annual payment for the plan corresponding to Plan III is $1,358.68. Compare the total repayments required in these cases with those required for Plans II and III in Table 3–1.

4

INTEREST FORMULAS

Some engineers may be discouraged when they find that financial analysis generally involves compound interest concepts and terms. To most, compound interest would seem to be something for the banker rather than the engineer. The mastery of the use of compound interest in cost studies is not in itself really difficult. Perhaps more difficult is the process of becoming convinced that it is needed at all.—American Telephone and Telegraph Company, Engineering Department*

Symbols

These symbols are used in the following explanation of interest formulas:

- i represents an interest rate per interest period.
- n represents a number of interest periods.
- P represents a present sum of money.
- F represents a sum of money at the end of n periods from the present date that is equivalent to P with interest i.
- A represents the end-of-period payment or receipt in a uniform series continuing for the coming n periods, the entire series equivalent to P at interest rate i.

Although a one-year interest period is used in most of the illustrations in this book, the formulas presented apply to interest periods of any length.

The various symbols are chosen so that each is an initial letter of a key word associated with the most common meaning of the symbol. Thus, i applies to *interest*, n applies to *number* of periods, P applies to *present* worth, F applies to *future* worth, and A applies to *annual* payment or *annuity*.

Formulas

The fundamental interest formulas that express the relationship between P, F, and A in terms of i and n are as follows.

**Engineering Economy*, 2d ed., p. 74, copyright American Telephone and Telegraph Company, 1963. While this text was written basically for internal Bell System use, it was made available outside the System through the Graybar Electric Company. The current third edition (1977) is available through McGraw-Hill Book Company.

Given P, to find F. $$F = P\,(1 + i)^n$$

Given F, to find P. $$P = F\left[\frac{1}{(1 + i)^n}\right] \qquad (2)$$

Given F, to find A. $$A = F\left[\frac{i}{(1 + i)^n - 1}\right] \qquad (3)$$

Given P, to find A. $$A = P\left[\frac{i(1 + i)^n}{(1 + i)^n - 1}\right] \qquad (4)$$

or $$A = P\left[\frac{i}{(1 + i)^n - 1} + i\right] \qquad (4)$$

Given A, to find F. $$F = A\left[\frac{(1 + i)^n - 1}{i}\right] \qquad (5)$$

Given A, to find P. $$P = A\left[\frac{(1 + i)^n - 1}{i(1 + i)^n}\right] \qquad (6)$$

or $$P = A\left[\frac{1}{\dfrac{i}{(1 + i)^n - 1} + i}\right] \qquad (6)$$

The following explanation of the formulas assumes the interest period as one year; the explanation can be made general by substituting "period" for "year."

Development of Formulas for Single Payments

If P is invested at interest rate i, the interest for the first year is iP and the total amount at the end of the first year is $P + iP = P(1 + i)$.

The second year the interest on this is $iP(1 + i)$, and the amount at the end of this year is $P(1 + i) + iP(1 + i) = P(1 + i)^2$. Similarly, at the end of the third year the amount is $P(1 + i)^3$; at the end of n years it is $P(1 + i)^n$.

This is the formula for the compound amount, F, obtainable in n years from a principal, P,

$$F = P(1 + i)^n \qquad (1)$$

If we express P in terms of F, i, and n,

$$P = F\left[\frac{1}{(1 + i)^n}\right] \qquad (2)$$

P may then be thought of as the principal that will give a required amount F in n years; in other words, P is the present worth of a payment of F, n years hence.

The expression $(1 + i)^n$ is called the *single payment compound amount factor* (F/P). Its reciprocal $1/(1 + i)^n$ is called the *single payment present worth factor* (P/F). A cash flow diagram illustrating the relationship between P and F is shown in Figure 4–1(*a*).

Development of Formulas for Uniform Annual Series of End-of-Year* Payments

If A is invested at the end of each year for n years, the total amount at the end of n years will obviously be the sum of the compound amounts of the individual investments. The money invested at the end of the first year will earn interest for $(n - 1)$ years; its amount will thus be $A(1 + i)^{n-1}$. The second year's payment will amount to $A(1 + i)^{n-2}$; the third year's to $A(1 + i)^{n-3}$; and so on until the last payment, made at the end of n years, which has earned no interest. The total amount F is $A[1 + (1 + i) + (1 + i)^2 + (1 + i)^3 \ldots + (1 + i)^{n-1}]$. The relationship between A and F is illustrated in Figure 4–1(*b*).

This expression for F in terms of A may be simplified to its customary form by the following algebraic manipulations:

$$F = A\,[1 + (1 + i) + (1 + i)^2 \ldots + (1 + i)^{n-2} + (1 + i)^{n-1}]$$

$$= A \sum_{t=1}^{n} (1 + i)^{t-1}$$

Multiplying both sides of the equation by $(1 + i)$,

$$(1 + i)F = A\,[(1 + i) + (1 + i)^2 + (1 + i)^3 \ldots$$

$$+ (1 + i)^{n-1} + (1 + i)^n] = A \sum_{t=1}^{n} (1 + i)^t$$

Subtracting the original equation from this second equation

$$iF = A[(1 + i)^n - 1]$$

*Throughout the main body of this book, the so-called end-of-period convention has been used. This means that a future value, F, and all uniform series amounts, A, occur at the end of the period, or periods, in question. It is very easy to convert these values to beginning-of-period amounts, as will be shown in Chapter 5.

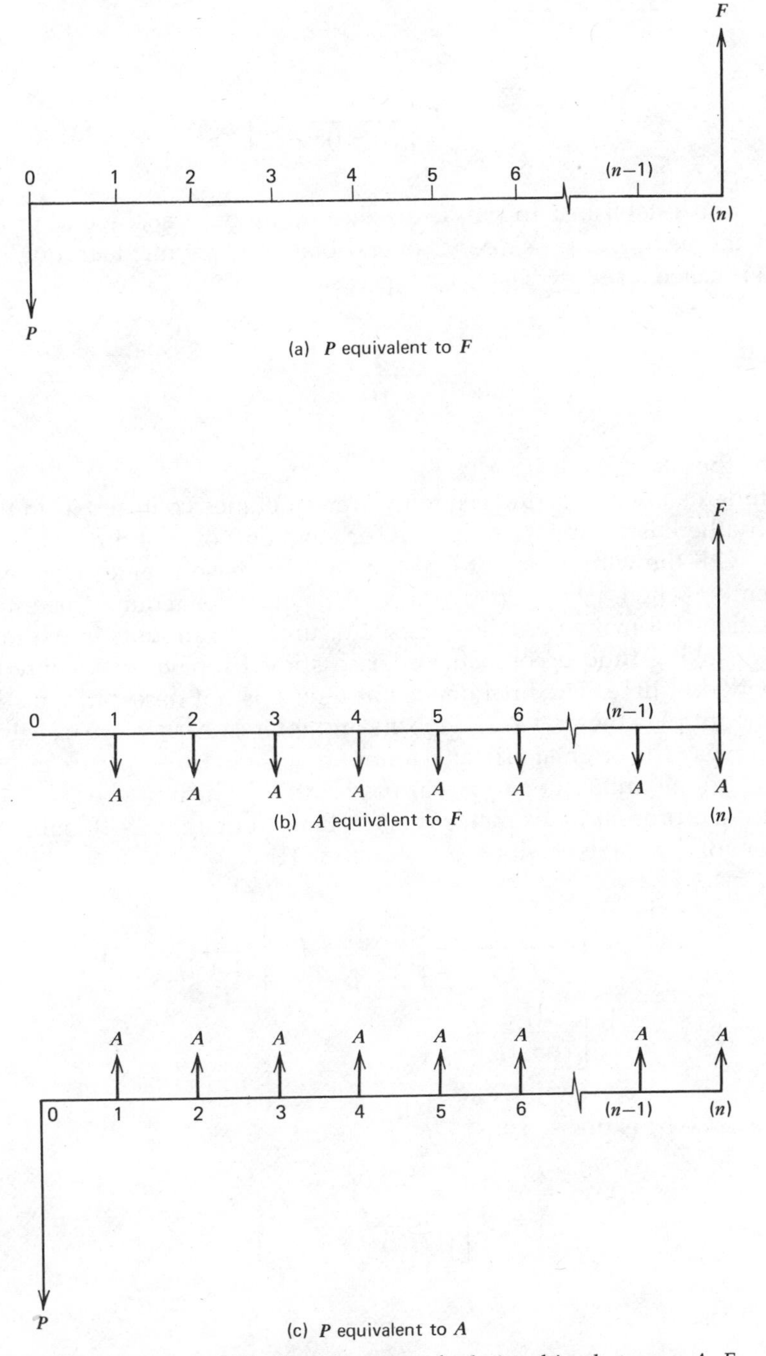

FIGURE 4–1. Cash flow diagram of relationships between A, F, and P

Then

$$A = F\left[\frac{i}{(1+i)^n - 1}\right] \tag{3}$$

A fund established to produce a desired amount at the end of a given period of time by means of a series of end-of-period payments throughout the period is called a *sinking fund*. The expression

$$\frac{i}{(1+i)^n - 1}$$

is called the *sinking fund factor* (A/F).

Students of engineering economy are sometimes confused as to the reason why there is an apparent deposit (or payment) of A at time n coinciding exactly with the withdrawal of F at time n. The reason comes from finance. Frequently, when a large loan principal payment is required at time n, a corporation will make periodic deposits of uniform amounts into a interest-earning sinking-fund account in order to spread repayment evenly over a long period of time. The final deposit at time n is not necessarily made, but that amount plus the previous deposits and interest earned are exactly sufficient to repay the original principal amount.

To find the uniform end-of-year payment, A, which can be secured for n years from a present investment, P (as in Plan III of Table 3–1), substitute in equation (3) the value given for F in equation (1):

$$A = F\left[\frac{i}{(1+i)^n - 1}\right] = P(1+i)^n\left[\frac{i}{(1+i)^n - 1}\right]$$

$$= P\left[\frac{i(1+i)^n}{(1+i)^n - 1}\right] \tag{4}$$

This may also be expressed as

$$A = P\left[\frac{i}{(1+i)^n - 1} + i\right] \tag{4}$$

This expression

$$\frac{i(1+i)^n}{(1+i)^n - 1}$$

is called the *capital recovery factor* (*A*/*P*). As shown by its identity with

$$\left[\frac{i}{(1+i)^n - 1} + i\right]$$

it is always equal to the sinking fund factor plus the interest rate. When multiplied by a present debt (which, from the point of view of the lender, is a present investment), it gives the uniform end-of-year payment necessary to repay the debt (the lender's investment) in n years with interest rate i. This factor, or an approximation to it, is used in the solution of many problems in engineering economy.

Formulas (3) and (4) may be reversed to show F and P in terms of A as follows:

$$F = A\left[\frac{(1+i)^n - 1}{i}\right] \tag{5}$$

$$P = A\left[\frac{(1+i)^n - 1}{i(1+i)^n}\right] \tag{6}$$

The expression

$$\left[\frac{(1+i)^n - 1}{i}\right]$$

is called the *uniform series compound amount factor* (*F*/*A*). This is usually abbreviated to *series compound amount factor*.

The expression

$$\left[\frac{(1+i)^n - 1}{i(1+i)^n}\right]$$

is called the *uniform series present worth factor* (*P*/*A*). Similarly, this is usually abbreviated to *series present worth factor*.

Functional Symbols

Throughout the later chapters of this book, many equations will be written using symbols that show the conversion to be made rather than the corresponding mathematical expressions involving i and n. These functional symbols are as follows:

$(F/P,i\%,n)$ is the single payment compound amount factor

$$(1 + i)^n$$

$(P/F,i\%,n)$ is the single payment present worth factor

$$\frac{1}{(1 + i)^n}$$

$(A/F,i\%,n)$ is the sinking fund factor

$$\frac{i}{(1 + i)^n - 1}$$

$(A/P,i\%,n)$ is the capital recovery factor

$$\frac{i(1 + i)^n}{(1 + i)^n - 1}$$

$(F/A,i\%,n)$ is the uniform series compound amount factor

$$\frac{(1 + i)^n - 1}{i}$$

$(P/A,i\%,n)$ is the uniform series present worth factor

$$\frac{(1 + i)^n - 1}{i(1 + i)^n}$$

Interest Tables

The solution of problems in equivalence is greatly facilitated by the use of interest tables. Tables D–1 through D–29 in Appendix D give values of the single payment compound amount factor, single payment present worth factor, the sinking fund deposit factor, the capital recovery factor, the uniform series compound amount factor, and the uniform series present worth factor for each value of n from 1 to 35, and for values of n that are multiples of 5 from 40 to 100. Each interest rate has a separate table; the interest rates given are 1%, 1½%, 2%, 2½%, 3%, 3½%, 4%, 4½%, 5%, 5½%, 6%, 7%, 8%, 9%, 10%, 11%, 12%, 13%, 14%, 15%, 16%, 18%, 20%, 25%, 30%, 35%, 40%, 45%, and 50%.

If the payment given (P, F, or A) is unity (in any units desired, dollars, pounds, pesos, francs, etc.), the factor from the interest tables gives directly the payment to be found. Thus, the respective columns in each table might have been headed:

Compound amount of 1
Present worth of 1
Uniform series that amount to 1
Uniform series that 1 will purchase
Compound amount of 1 per period
Present worth of 1 per period

Interest tables in books on the mathematics of investment commonly use these headings or some variations of them.

Relationship Between Interest Factors

In using the interest tables, it is desirable that the student be familiar with the simple relationships that exist between the different factors for a given value of i and n. The illustrations given all apply to $i = 0.06$ and $n = 5$. The factors have been rounded off to agree with the number of places in Table D–11, Appendix D.

Single payment compound amount factor and single payment present worth factor are reciprocals. Thus

$$\frac{1}{(F/P,6\%,5)} = (P/F,6\%,5) = \frac{1}{1.3382} = 0.7473$$

Sinking fund factor and uniform series compound amount factor are reciprocals. Thus

$$\frac{1}{(A/F,6\%,5)} = (F/A,6\%,5) = \frac{1}{0.17740} = 5.637$$

Capital recovery factor and series present worth factor are reciprocals. Thus

$$\frac{1}{(A/P,6\%,5)} = (P/A,6\%,5) = \frac{1}{0.23740} = 4.212$$

Series compound amount factor equals 1.000 plus sum of first $(n - 1)$ terms in the column of single payment compound amount factors. Thus

$$(F/A,6\%,5) = 1.000 + (F/P,6\%,1) + (F/P,6\%,2) + (F/P,6\%,3) + (F/P,6\%,4)$$

$$= 1.0000 + 1.0600 + 1.1236 + 1.1910 + 1.2625 = 5.6371$$

Series present worth factor equals sum of first n terms of single payment present worth factors. Thus

$$(P/A,6\%,5) = (P/F,6\%,1) + (P/F,6\%,2) + (P/F,6\%,3) + (P/F,6\%,4) + (P/F,6\%,5)$$

$$4.212 = 0.9434 + 0.8900 + 0.8396 + 0.7921 + 0.7473$$

Capital recovery factor equals sinking fund factor plus interest rate. Thus

$$(A/P,6\%,5) = (A/F,6\%,5) + i$$

$$0.23740 = 0.17740 + 0.06$$

Nominal and Effective Interest Rates

Many loan transactions stipulate that interest is computed and charged more often than once a year. For example, interest on deposits in savings banks may be computed and added to the deposit balance four times a year; this is referred to as interest "compounded quarterly." Interest on corporate bond issues usually is payable every 6 months. Building and loan associations, automobile finance companies, and other organizations making personal loans often require that interest be computed monthly.

Consider a loan transaction in which interest is charged at 1% per month. Usually such a transaction is described as having an interest rate of 12% per annum. More precisely, this rate should be described as a *nominal* 12% per annum compounded monthly.

It is desirable to recognize that there is a real difference between 1% per month compounded monthly and 12% per annum compounded annually. Assume that \$1,000 is borrowed with interest at 1% per month. Using Table D–1, the amount owed at the end of 12 months may be calculated as follows:

$$F = \$1,000(1.01)^{12}$$

$$= \$1,000(F/P,1\%,12)$$

$$= \$1,000(1.1268) = \$1,126.8$$

If the same \$1,000 had been borrowed at 12% per annum compounded annually, the amount owed at the end of the year would have been only

\$1,120, \$6.80 less than \$1,126.8. The monthly compounding at 1% has the same effect on the year-end compound amount as the charging of a rate of 12.68% compounded annually. In the language of financial mathematics, the *effective* interest rate is 12.68%.

The phrases *nominal interest rate* and *effective interest rate* may be defined more precisely as follows:

Let interest be compounded m times a year at an interest rate $\frac{r}{m}$ per compounding period.

$$\text{The nominal interest rate per annum} = m\left(\frac{r}{m}\right) = r.$$

$$\text{The effective interest rate per annum} = \left(1 + \frac{r}{m}\right)^m - 1.$$

Nominal rates of interest for different numbers of annual compoundings are not comparable with one another until they have been converted into the corresponding effective rate. The more frequent the number of compoundings at a given nominal rate, the greater the difference between the effective and nominal rates. (For example, a nominal rate of 12% compounded semiannually yields an effective rate of 12.36% in contrast to the 12.68% for a nominal 12% compounded monthly.) The higher the nominal rate for a given m, the greater both the absolute and the relative difference between effective and nominal rates. (For example, a nominal rate of 24% compounded monthly yields an effective rate of 26.82%, 2.82% more than the nominal rate. In contrast, the effective rate is only 0.68% above the nominal rate with interest at a nominal 12% compounded monthly.)

In engineering economy studies it usually is preferable to deal with effective interest rates rather than nominal rates.

Continuous Compounding of Interest

The mathematical symbol e (the base of natural or "Napierian" logarithms) may be defined as the limit approached by the quantity $\left(1 + \frac{1}{k}\right)^k$ as k increases indefinitely. It is shown in textbooks on calculus that $e = 2.71828+$. The common logarithm of e is 0.43429.

If the sum P is invested for n years with a nominal interest rate r and with m compounding periods a year, the compound amount F may be expressed as follows:

$$F = P\left(1 + \frac{r}{m}\right)^{mn}$$

If we designate $\frac{m}{r}$ by the symbol k, $m = rk$, and

$$F = P\left(1 + \frac{1}{k}\right)^{rkn} = P\left[\left(1 + \frac{1}{k}\right)^{k}\right]^{rn}$$

As the number of compounding periods per year, m, increases without limit, so also must k. It follows that the bracketed quantity in the foregoing formula approaches the limit e. Therefore, the limiting value of F is Pe^{rn}.

With continuous compounding, the single payment compound amount factor is e^{rn}. The single payment present worth factor is, of course, e^{-rn}.

Although continuous compounding formulas assume that interest is computed and added to principal at every moment throughout the year, the results obtained using continuous compounding are very close to the results obtained using monthly compounding with a nominal rate r.

Application of Continuous Compounding in Engineering Economy

Although continuous compounding is rarely used in actual loan transactions,* the topic is of importance in connection with certain problems of decision making. Two types of application are important.

In some economy studies it may be desired to recognize that certain receipts or disbursements will be spread throughout a year rather than concentrated at a particular date. Continuous compounding is well adapted to the assumption of a continuous flow of funds at a uniform rate throughout a stated period of time. Formulas and tables that may be used for this purpose are explained in Appendix A. The interest factors for the uniform series case are given in Appendix D, Tables D–30 and D–31. These tables use effective rather than nominal rates.

In the development of certain mathematical models intended as aids to decision making, the mathematical treatment is facilitated by the use of continuous compounding rather than periodic compounding. In these cases, the user should be careful to distinguish between the use of nominal and effective interest rates. See Appendix A for a more complete explanation of effects of using nominal and effective interest rates.

Interest Tables for Uniform Gradient

Engineering economy problems frequently involve disbursements or receipts that increase or decrease each year by varying amounts. For example, the maintenance expense for a piece of mechanical equipment may tend to increase somewhat each year. If the increase or decrease is the same every year,

*Many banks and savings and loan institutions use daily compounding, which is very close to continuous compounding, for the computation of interest on savings accounts and short-term loans.

the yearly increase or decrease is known as a *uniform arithmetic gradient*. Even when it is reasonable to believe that the annual expenses or receipts will increase or decrease somewhat irregularly, a uniform gradient may be the best and most convenient way to estimate the changing condition.

Since the amount of money is different each year, the uniform series interest factors previously discussed cannot be used and each year's disbursement or receipt must be handled by means of the single payment factors. This time-consuming hand computation can be avoided by deriving simple formulas for the equivalent uniform annual value of a gradient and for the present worth of a gradient for n years.

Figure 4–2 gives the cash flow diagram for a uniform gradient. Using end-of-year payments, the payment the second year is greater than the first year by G, the third is G greater than the second year, and so on. Thus, the payments by years are as follows:

End of year	Payment
1	0
2	G
3	$2G$
4	$3G$
....	....
$(n-1)$	$(n-2)G$
n	$(n-1)G$

These payments may be thought of as a set of payments that will accumulate to an amount F at the end of the nth year, and that amount can be converted to a uniform series of payments by multiplying F by the sinking fund factor. For convenience, it can be assumed that a series of annual payments of G is started at the end of the second year, another series of G is started at the end of the third year, and so on. Each of these series terminates at the same time, the end of the nth year. The series compound amount factor can be applied to each series of G per year to determine its compound amount on this terminal date:

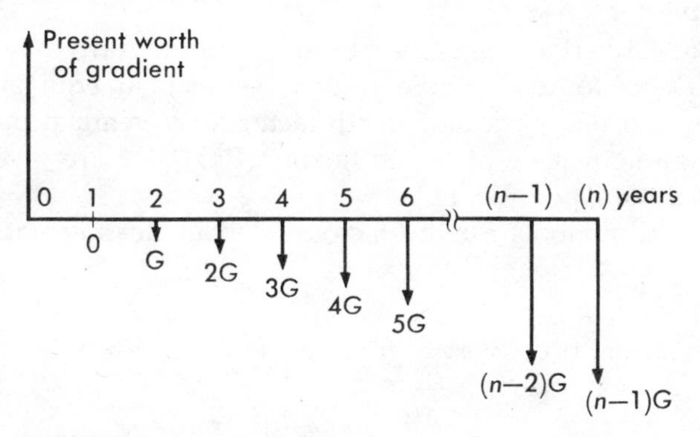

FIGURE 4–2. Cash flow diagram for a gradient

f compound amounts

$$F = G\left[\frac{(1+i)^{n-1}-1}{i} + \frac{(1+i)^{n-2}-1}{i} \ldots + \frac{(1+i)^2-1}{i} + \frac{(1+i)-1}{i}\right]$$

$$= \frac{G}{i}[(1+i)^{n-1} + (1+i)^{n-2} \ldots + (1+i)^2 + (1+i) - (n-1)]$$

$$= \frac{G}{i}[(1+i)^{n-1} + (1+i)^{n-2} \ldots + (1+i)^2 + (1+i) + 1] - \frac{nG}{i}$$

The expression in brackets is the compound amount of a sinking fund of 1 for n years. Hence

$$\text{sum of compound amounts} = \frac{G}{i}\left[\frac{(1+i)^n - 1}{i}\right] - \frac{nG}{i}$$

The equivalent uniform annual figure for n years may be found by multiplying this sum of the compound amounts by the sinking fund factor for n years. Hence

$$A = \frac{G}{i}\left[\frac{(1+i)^n - 1}{i}\right]\left[\frac{i}{(1+i)^n - 1}\right] - \frac{nG}{i}\left[\frac{i}{(1+i)^n - 1}\right]$$

$$= \frac{G}{i} - \frac{nG}{i}\left[\frac{i}{(1+i)^n - 1}\right]$$

This is a general expression applicable to any value of n. Values for the equivalent uniform annual cash flow of a uniform gradient (A/G) of unity for the interest rates previously listed and for n from 2 to 100 years are given in Tables D–1 through D–29, Appendix D. The use of the gradient factors will be illustrated in Chapter 5.

The factor to convert a gradient series to a present worth may be obtained by multiplying the factor to convert a gradient series to an equivalent uniform annual series by the series present worth factor for n years at interest i. The values for the gradient present worth factors (P/G) also are given in Tables D–1 through D–29 of Appendix D.

The functional symbols for the factors used in dealing with arithmetic gradients are:

$(A/G, i\%, n)$ = factor to convert a gradient series to an equivalent uniform annual series.

$(P/G, i\%, n)$ = factor to convert a gradient series to a present worth.

The relationship between these two factors can be shown as:

$$(P/G, i\%, n) = (A/G, i\%, n)(P/A, i\%, n)$$

The Geometric Gradient Series

It is sometimes desired to estimate certain cash flows which are expected to grow geometrically through time at some constant rate of g percent per year. Given a known present cost, for example, of A_0, the next-year cost would be A_1, found by multiplying A_0 by $(1 + g)$; the second year cost, A_2, would be $A_0(1 + g)^2$; and so on. The reader will note that the compound growth, or escalation, factor is identical to the single payment compound amount factor replacing i with g in the formula. Thus, any future annual cash flow, A_t, may be found from:

$$A_t = A_0(F/P, g\%, t)$$

While escalation factors mathematically may be combined with interest factors to produce a single factor to handle both cash flow escalation and interest equivalence conversions in a single step, the authors recommend against such practice, for reasons discussed in Chapter 14.

Finding Unknown Interest Rates

Frequently, the sum to be invested (or loaned) is known, and the prospective future series of money receipts (or plan of repayment) is known, and it is desired to find the interest rate that will be earned on the investment.

When a single payment and a single receipt are involved, when n is known and i is wanted, the problem is quite simple. Formula (1) becomes

$$i = \sqrt[n]{\frac{F}{P}} - 1 \qquad (7)$$

This may be solved by logarithms or by the use of one of the special pocket calculators available.

When a single payment and a uniform series are involved, the problem becomes more complicated to solve directly. When nonuniform payments or receipts are involved, the only reasonable method of hand solution is by interpolation. The use of the interest tables in Appendix D makes the solution of unknown interest problems relatively simple by the interpolation method.

Approximate methods using interpolation are recommended for engineering economy studies and are illustrated and explained in Chapters 5 and 8.

Interest Formulas and Tables in Relation to Engineering Economy

"Will it pay?" (which is the central question of engineering economy) usually means "Will an investment pay?" An investment will not pay unless it can ultimately be repaid with interest. Thus, interest enters into most problems in engineering economy.

But the engineer's decision as to whether a proposed investment will pay must be based on estimates of future events rather than on certain knowledge. Even the most careful estimates are almost certain to vary from the actual amounts as they occur; also, engineers' decisions regarding economy must be based on preliminary estimates made in advance of design that necessarily have a considerable danger of large errors. The subjects of sensitivity analysis and probabilities and their usefulness in analyses for economy are treated in Chapters 14 and 15.

For this reason great precision is not usually required in interest calculations made for economy studies. For instance, where cost estimates are subject to errors of 5% or 10%, there is no justification for carrying out interest calculations to seven significant figures, as is sometimes done, especially when computers are used to solve for the unknown interest rate. Interest tables giving three significant figures are adequate for purposes of most economy studies. Similarly, the difference between paying interest once a year and paying it more often (i.e., the difference between compounding annually and compounding semiannually, quarterly, or monthly—a difference of considerable importance to the financier) is usually neglected in economy studies; the interest rate used in economy studies will frequently be considerably higher than the cost of borrowed money for a number of reasons that are explained in Chapter 10.

PROBLEMS

4–1. What effective interest rate per annum corresponds to a nominal rate of 12% compounded semiannually? Compounded quarterly? Compounded monthly?

4–2. The interest rate charged by various banks that operate credit card systems is 1½% per month. What nominal and effective rates per annum correspond to this?

4–3. What effective rate per annum corresponds to a nominal rate of 15% compounded monthly? Use logarithms.

4–4. What effective interest rate per annum corresponds to a nominal rate of 15% compounded daily? Compounded continuously? Use logarithms.

4–5. Develop a formula for the *beginning-of-period* payment, B, into a sinking fund to amount to F at the end of n periods with interest at i percent per period.

4–6. The factor developed in Problem 4–5, $(B/F, i\%, n)$, may be found by the use of the tables in Appendix D by taking the product of two of those factors. Calculate the value of the factor when $n = 10$ and $i = 15\%$ using the formula derived in Problem 4–5. Using functional symbols show the factors to be used, their values from Table D–20, and compare their product with the previous result obtained for Problem 4–5.

4–7. How would you determine the factor to convert a uniform gradient series to an equivalent uniform annual series (A/G) if you had only a table of sinking fund factors? Compare your results using the formula for (A/G) and the value of the sinking fund factor from Table D–16 when $i = 11\%$ and $n = 10$.

4–8. How would you determine a desired *uniform series compound amount factor* if you had only:

(a) a table of *single payment compound amount factors*?

(b) a table of *sinking fund factors*?

Illustrate your solution using the 16% table, Table D–21, to obtain the factor when $n = 10$.

4–9. How would you determine a desired *capital recovery factor* if you had only:

(a) a table of *single payment present worth factors*?

(b) a table of *uniform series present worth factors*?

Illustrate your solution using the 9% table, Table D–14, to obtain the factor for $n = 6$.

4–10. A grandfather deposited \$1,000 in a savings account for a baby on the day of her birth to help support her college education. Over the years the interest rate paid by the bank as well as the term of compounding have changed considerably, but when the young lady collected the accumulated fund on her 18th birthday, she received \$2,689.37. Use logarithms to determine the effective annual interest rate she received on her savings account over the years.

4–11. If you wished to have an effective annual interest rate of 12%, but use continuous compounding of annual periodic cash flows, what would the nominal interest rate need to be? Use logarithms.

4–12. If you wished to have an effective annual interest rate of 12%, but use daily compounding of annual cash flows, what would the nominal annual interest rate need to be? Use logarithms and assume 365 days per year. How does your solution to this problem compare to that of Problem 4–11?

4–13. Use the formula for the *capital recovery factor* to find values for $n = 5$,

10, and 20 when $i = 22\%$. Use Tables D–23 and D–24 to make a linear interpolation between 20% and 25% for the values of this factor. Compare your results.

4–14. Use the formula for the *uniform series present worth factor* to find values for $n = 5$, 10, and 20 when $i = 22\%$. Use Tables D–23 and D–24 to make a linear interpolation between 20% and 25% for the values of this factor. How do these results compare?

4–15. Comment on the difference in results obtained in Problems 4–13 and 4–14. Does either factor appear to be superior to the other in making linear interpolations of this type?

4–16. Use Tables D–3 and D–15 to make a linear interpolation between 2% and 10% for values of the *uniform series present worth factor* when $n = 5$, 10, and 20 with an interest rate of 6%. Compare your results with the values given in the 6% table, Table D–11.

5

SOLVING INTEREST PROBLEMS

*The mere accumulation of savings . . . does not necessarily lead to technological progress. There must also be investment of savings in capital facilities, either directly by purchase of capital facilities, or through purchase of the securities of business representing capital facilities. But the investment of savings under the private enterprise system will occur in adequate volume only when there is adequate prospect of profit therefrom.—Machinery and Allied Products Institute**

This chapter illustrates the use of interest tables and formulas in the solution of practical problems related to economy. Each illustrative example is numbered for convenient reference.

Many of the illustrative examples are stated in several different ways. It is desirable that the student of engineering economy recognize the variety of questions that may be answered by the same equivalence calculation.

In solving any interest problem, it is necessary to note which of the various elements of such problems (i, n, P, F, and A) are known and which are wanted. This is the first step in any solution and follows immediately after the problem statement in each illustrative example.

Unless otherwise stated, it is to be assumed in the following examples that interest is payable (or compounded) annually.

Examples Illustrating the Use of the Interest Factors Relative to Time

The concept of equivalence was introduced in Chapter 3, and Table 3–1 gave four plans for the repayment of a loan with interest. All these plans were equivalent to each other. Examples 5–1 through 5–7 illustrate the use of the interest factors to compute equivalence in terms of both single payments and series of payments over different periods of time. An initial amount of $1,000 is assumed on January 1, 1981 and is converted to equivalent amounts at different times, and finally, the original $1,000 is obtained by converting a series to a single amount as of January 1, 1981. These examples show that "now" or zero time can be assumed at any date, and an equivalent amount or

**Capital Goods and the American Enterprise System*, Machinery and Allied Products Institute (1939) p. 42.

a series of amounts can be obtained for dates either preceding or following the assumed zero time. The cash flow diagrams for Examples 5–1 through 5–7 are shown in Figure 5–1.

EXAMPLE 5–1

If $1,000 is invested at 6% compounded interest on January 1, 1981, how much will be accumulated by January 1, 1991? (Figure 5–1a.)

Solution:

$i = 0.06;\ n = 10;\ P = \$1{,}000;\ F = ?$

$$F = P(F/P,6\%,10)$$
$$= \$1{,}000(1.7908) = \$1{,}791$$

EXAMPLE 5–2

How much would you have to invest at 6% interest on January 1, 1985, in order to accumulate $1,791 on January 1, 1991? (Figure 5–1b.)

Solution:

$i = 0.06;\ n = 6;\ F = \$1{,}791;\ P = ?$

In this case zero time is assumed to be January 1, 1985.

$$P = F(P/F,6\%,6) = \$1{,}791(0.7050)$$
$$= \$1{,}263$$

EXAMPLE 5–3

What is the present worth on January 1, 1978, of $1,263 on January 1, 1985, if interest is at 6%? (Figure 5–1c.)

Solution:

$i = 0.06;\ n = 7;\ F = \$1{,}263;\ P = ?$

$$P = F(P/F,6\%,7) = \$1{,}263(0.6651)$$
$$= \$840$$

EXAMPLE 5–4

If $840 is invested at 6% on January 1, 1978, what equal year-end withdrawals can be made each year for 10 years, leaving nothing in the fund after the tenth withdrawal? (Figure 5–1d.)

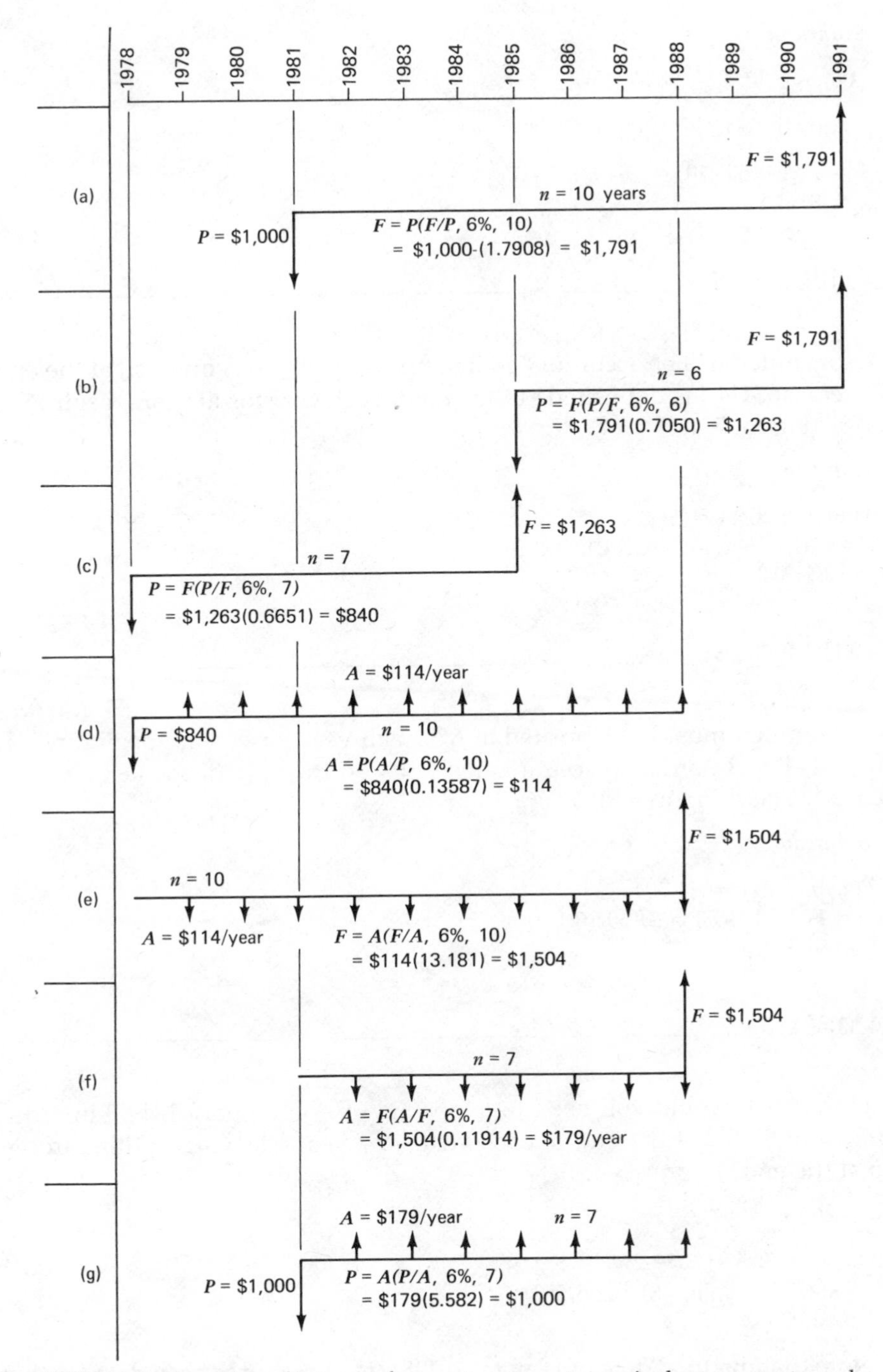

FIGURE 5–1. Use of interest factors to compute equivalent amounts and equivalent series

Solution:

$i = 0.06; n = 10; P = \$840; A = ?$

Now zero time is January 1, 1978.

$$A = P(A/P,6\%,10) = \$840(0.13587) = \$114.1$$

EXAMPLE 5–5

How much will be accumulated in a fund, earning 6% interest, at the end of 10 years if $114.1 is deposited at the end of each year for 10 years, beginning in 1978? (Figure 5–1e.)

Solution:

$i = 0.06; n = 10; A = \$114.1; F = ?$

$$F = A(F/A,6\%,10) = \$114.1(13.181) = \$1,504$$

EXAMPLE 5–6

How much must be deposited at 6% each year for 7 years beginning on January 1, 1982 in order to accumulate $1,504 on the date of the last deposit, January 1, 1988? (Figure 5–1f.)

Solution:

$i = 0.06; n = 7; F = \$1,504; A = ?$

$$A = F(A/F,6\%,7) = \$1,504(0.11914) = \$179.2$$

EXAMPLE 5–7

How much would you need to deposit at 6% on January 1, 1981 in order to draw out $179.2 at the end of each year for 7 years, leaving nothing in the fund at the end? (Figure 5–1g.)

Solution:

$i = 0.06; n = 7; A = \$179.2; P = ?$

$$P = A(P/A,6\%,7) = \$179.2(5.582) = \$1,000$$

Note that the final date of January 1, 1981, is the same as zero date for this series of seven examples. The final figure of $1,000 is the same as the initial figure. All multiplications have been carried to only four significant figures because some of the interest factors used in the series of examples have only this number of significant figures.

Comments on Examples 5–1 Through 5–7

These examples illustrate several important points. First, in each simple problem four of the five elements i, n, P, F, and A are present and three of the four elements must be known. In these examples i, n, and one other element were always known. In solving the problem, the known elements are first identified and the unknown or desired element is specified. Then the unknown element is equated to the proper interest factor times the known monetary amount. (If i or n is unknown, then the value of the interest factor must be determined by solving the equation, and the unknown is found by referring to the interest tables and by interpolating, if necessary, between two factors found there. This will be illustrated later in this chapter.)

Next, these examples demonstrate that, with a given interest rate, equivalent single amounts or series of amounts can be found at many different relative times. Thus, all of the following are equivalent to $1,000 now, assuming that "now" is January 1, 1981.

$1,791 10 years hence

$1,263 4 years hence

$840.0 3 years previously

$114.1 a year (year-end payments) for 10 years beginning 3 years ago

$1,504 7 years hence

$179.2 a year for the next 7 years

Since these amounts or series of amounts are all equivalent to $1,000 now, they must be equivalent to each other. The difference in the timing of the payments or receipts is the significant element in these computations, and it is very important to determine precisely the n for each problem. Dates have been used for these examples, but the n was determined in the solution of each example. Normally it is more convenient to select the starting point in time, calling it date zero or year zero, and to determine the number of years from that selected time. Thus, in Example 5–2, the desired quantity was that necessary investment in 1985 to obtain $1,791 in 1991, and 1985 became a convenient starting point for the example. Year zero is 1985 and 1991 is 6.

Another point that is emphasized in Examples 5–4 through 5–7 is that conventional interest tables and formulas are based upon uniform payments made at the *end* of each period (not at the beginning of the period). The only time this convention may become confusing occurs when one is converting from the compound amount, F, to uniform periodic payments or vice versa. The confusion arises from the fact that the date of the compound amount is the same date as the last of the uniform series of payments or receipts. Practice in using the interest tables will soon eliminate this source of possible confusion. (See p. 36.)

Examples Involving Conversion of Single Payments at One Date to Equivalent Single Payments at Another Date

Examples of this type involve i, n, P, and F, and any one of these four elements may be unknown. Examples 5–1, 5–2, and 5–3 were of this type, with either P or F unknown. The following examples illustrate other aspects of such problems.

EXAMPLE 5–8

If $2,000 is invested now, $1,500 2 years hence, and $1,000 4 years hence, all at 8%, what will the total amount be 10 years hence?

or

What is the compound amount of $2,000 for 10 years plus $1,500 for 8 years plus $1,000 for 6 years with interest at 8%?

or

What must be the prospective saving 10 years hence in order to *justify* spending $2,000 now, $1,500 in 2 years, and $1,000 in 4 years, if money is worth 8%?

Solution:

It is evident from the cash flow diagram (Figure 5–2a) that this involves a non-uniform, non-periodic series of investments and requires three separate calculations: in the first the "present" is now; in the second it is 2 years hence; in the third it is 4 years hence:

$$i = 0.08; \quad \left\{ \begin{array}{l} n_1 = 10;\ P_1 = \$2{,}000 \\ n_2 = 8;\ P_2 = \$1{,}500 \\ n_3 = 6;\ P_3 = \$1{,}000 \end{array} \right\}; \quad \sum F = F_1 + F_2 + F_3 = ?$$

$$\begin{aligned} F_1 &= P_1(F/P,8\%,10) = \$2{,}000(2.1589) = \$4{,}318 \\ F_2 &= P_2(F/P,8\%,\ 8) = \$1{,}500(1.8509) = \$2{,}776 \\ F_3 &= P_3(F/P,8\%,\ 6) = \$1{,}000(1.5869) = \underline{\$1{,}587} \\ \sum F &= \qquad\qquad\qquad\qquad\qquad\qquad\ \ \$8{,}681 \end{aligned}$$

Another approach is to consider "now" to be at the time of the first deposit and to compute F by the following method:

$$P_1' = P_1(P/F,\ 8\%,\ 0) = \$2{,}000(1) \qquad = \$2{,}000$$

$$P_2' = P_2(P/F, 8\%, 2) = \$1{,}500(0.8573) = \$1{,}286$$
$$P_3' = P_3(P/F, 8\%, 4) = \$1{,}000(0.7350) = \$\ \ 735$$
$$\sum P' = \$4{,}021$$
$$F = \sum P'(F/P, 8\%, 10) = \$4{,}021(2.1589) = \$8{,}681$$

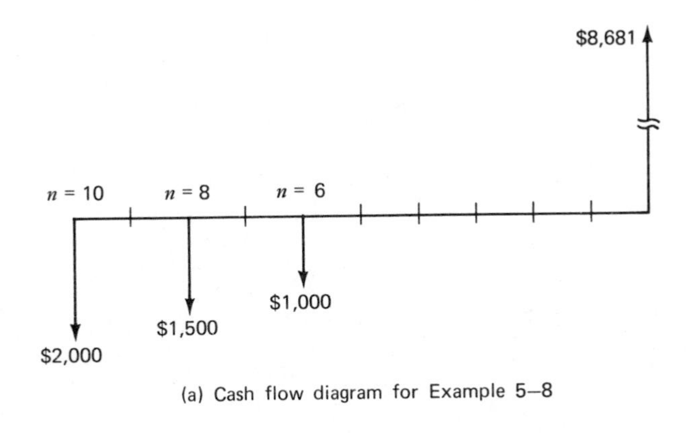

(a) Cash flow diagram for Example 5–8

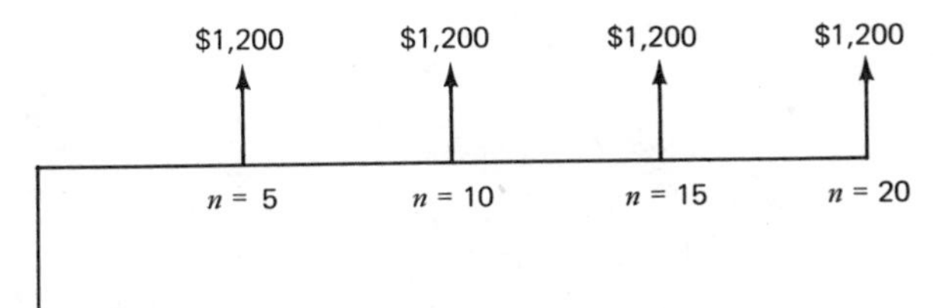

(b) Cash flow diagram for Example 5–11

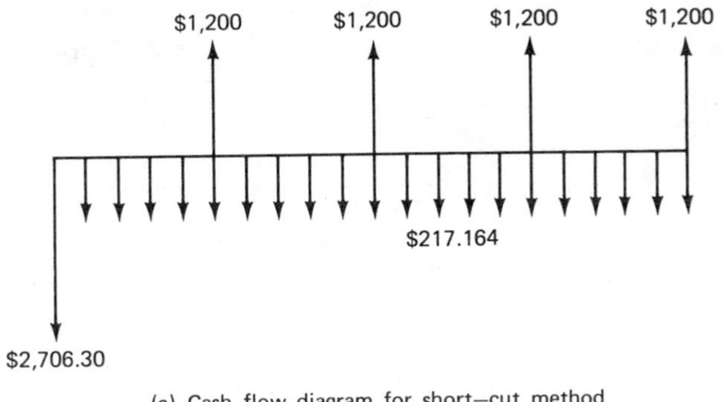

(c) Cash flow diagram for short–cut method in Example 5–11

FIGURE 5–2. Cash flow diagrams for Examples 5–8 and 5–11

EXAMPLE 5–9

What is the compound amount of \$3,500 for 18 years with interest at 4.25%?

Solution:

$i = 0.0425$; $n = 18$; $P = \$3{,}500$; $F = ?$

$$F = P(F/P, 4.25\%, 18) = \$3{,}500(1.0425)^{18}$$

As the tables do not give compound amount factors for 4.25%, the problem must be solved by logarithms unless an appropriate calculator is available:

Log 1.0425 = 0.018076

18 Log 1.0425	= 0.32537
Log 3,500	= 3.54407
Log F	= 3.86944
F	= \$7,404

An approximate solution may be obtained by linear interpolation between the compound amount factors for 4% and 4½% as follows:

$$(F/P, 4\%, 18) = (1.04)^{18} = 2.0258;$$
$$(F/P, 4\tfrac{1}{2}\%, 18) = (1.045)^{18} = 2.2085$$

Approximate value of

$$(1.0425)^{18} = 2.0258 + \frac{25}{50}(2.2085 - 2.0258) = 2.1172$$

Approximate $F = \$3{,}500(2.1172) = \$7{,}410$

This particular approximation involves an error of less than one-tenth of 1%. The percent of error involved in such interpolations increases with an increase in n and is greater with higher interest rates. This is an approximation because it is obvious that graphs of $(1 + i)^n$ for different i's and n's will not be linear. Assuming a straight line between two points is only an approximation of the actual curve.

EXAMPLE 5–10

What is the compound amount of \$1,000 for 64 years with interest at 6%?

Solution:

$i = 0.06$; $n = 64$; $P = \$1{,}000$; $F = ?$

By noting that the compound amount factor for 64 years, which is not given in our tables, is the product of the respective factors for 60 years and 4 years, this can be solved without recourse to logarithms:

$$F = P(F/P, 6\%, 60)(F/P, 6\%, 4)$$
$$= \$1{,}000\ (32.9877)(1.2625) = \$41{,}647$$

EXAMPLE 5–11

How much invested now at 5% would be just sufficient to provide $1,200 5 years hence, $1,200 10 years hence, $1,200 15 years hence, and $1,200 20 years hence?

or

What payment now is acceptable in place of prospective payments of $1,200 at the end of 5, 10, 15, and 20 years, if interest is at 5%?

or

How much is it justifiable to spend now in order to save prospective expenditures of $1,200 at the end of 5, 10, 15, and 20 years if money is worth 5%?

Solution:

This cash flow series is shown in Figure 5–2b.

$$i = 0.05; \quad \left\{ \begin{array}{l} n_1 = \ \ 5;\ F_1 = \$1{,}200 \\ n_2 = 10;\ F_2 = \ \ 1{,}200 \\ n_3 = 15;\ F_3 = \ \ 1{,}200 \\ n_4 = 20;\ F_4 = \ \ 1{,}200 \end{array} \right\}; \quad \sum P = P_1 + P_2 + P_3 + P_4 = ?$$

$$\begin{aligned} P_1 &= F_1(P/F,5\%,5) \ = \$1{,}200(0.7835) = \$\ \ 940.20 \\ P_2 &= F_2(P/F,5\%,10) = \ \ 1{,}200(0.6139) = \ \ \ 736.70 \\ P_3 &= F_3(P/F,5\%,15) = \ \ 1{,}200(0.4810) = \ \ \ 577.20 \\ P_4 &= F_4(P/F,5\%,20) = \ \ 1{,}200(0.3769) = \ \ \ \underline{452.30} \\ \sum P &= \qquad\qquad\qquad\qquad\qquad\qquad \$2{,}706.40 \end{aligned}$$

Since the F's are all equal and the intervals between F's are constant, 5 years, a short-cut method can be used. Each F can be converted to an equivalent of 5 year-end amounts, and the present worth of 20 such amounts found:

$$\begin{aligned} P &= F(A/F,5\%,5)(P/A,5\%,20) \\ &= \$1{,}200(0.18097)(12.462) = \$2{,}706.30 \end{aligned}$$

Figure 5–2c explains this short cut by showing that each 5 year period is just like all the others. The difference in the two results is due to rounding off of the interest factors.

EXAMPLE 5–12

In how many years will an investment of $1,000 now, increase to $2,000 with interest at 7%?

or

Within how many years must a prospective expenditure of $2,000 be required in order to justify spending $1,000 now to prevent it, if money is worth 7%?

Solution:

$i = 0.07; P = \$1{,}000; F = \$2{,}000; n = ?$

$$(F/P,7\%,n) = \frac{F}{P} = \frac{\$2{,}000}{\$1{,}000} = 2.000$$

The value of n can be determined by examining the compound amount factors, single payment, in the 7% table, and interpolating between the next higher and next lower values:

$n = 10, (F/P,7\%,10) = 1.9672$
$n = 11, (F/P,7\%,11) = 2.1049$

Therefore

$n =$ approximately 10.2 years*

EXAMPLE 5–13

A savings certificate costing $80 now will pay $100 in 5 years. What is the interest rate?

or

At what interest rate will $80 accumulate to $100 in 5 years?

or

Spending $80 now to avoid spending $100 5 years hence is, in effect, securing what interest rate?

Solution:

$n = 5; P = \$80; F = \$100; i = ?$

$$(F/P,i\%,5) = \frac{F}{P} = \frac{\$100}{\$80} = 1.2500$$

By interpolation between the next higher and the next lower single payment compound amount factors for 5 years in the interest tables, the approximate interest rate can be determined:

$i = 0.045, (F/P,4.5\%,5) = 1.2462$
$i = 0.050, (F/P,5\%,5) \;= 1.2763$

*This problem is related to the "rule of 70" often quoted in the literature of finance. This rule-of-thumb states that you may find the period of years in which a sum of money will double by dividing 70 by the interest rate in percent. Thus, at 7% interest, a sum of money will double in 10 years. It is a reasonably accurate approximation for interest rates below 12 percent.

$$i = 0.045 + 0.005 \left[\frac{1.2500 - 1.2462}{1.2763 - 1.2462}\right]$$

$$= 0.0456 \text{ or } 4.56\%$$

Examples Involving Conversions to or from Uniform Series of Payments

The general technique for solving a problem involving a single payment and a uniform series is similar to the technique used in the previous examples. The four elements of such problems are either i, n, A, and P or i, n, A, and F. Note whether it is P or F that enters into the given problem; note which three elements are known and their values. If the unknown, the value of which is desired, is A, P, or F, and the given values of i and n are values for which factors are available in the tables, use the interest tables. Otherwise, solve by interpolation. If i or n is the unknown, an exact solution is seldom possible using tables, and interpolation is employed to obtain an approximate solution. For engineering economy purposes, the errors introduced by straight-line interpolation are usually acceptable, and the solutions obtained thereby are adequate for the decision-making function.

EXAMPLE 5–14

How much must be invested at the end of each year for 30 years in a sinking fund which is to amount to $200,000 at the end of 30 years, if interest is at 6%?

or

What uniform year-end annual expenditure for 30 years is justifiable in order to avoid having to spend $200,000 30 years hence, if money is worth 6%?

Solution:

$i = 0.06$; $n = 30$; $F = \$200{,}000$; $A = ?$

$$A = F(A/F,6\%,30)$$
$$= \$200{,}000(0.01265) = \$2{,}530$$

EXAMPLE 5–15

How much would be accumulated in the sinking fund of Example 5–14 at the end of 18 years?

or

If $2,530 is invested at the end of each year for 18 years with interest at 6%, how much will have accumulated at the end of that time?

or

What must be the prospective saving 18 years hence in order to justify spending $2,530 a year for 18 years, if money is worth 6%?

Solution:

$$i = 0.06;\ n = 18;\ A = \$2{,}530;\ F = ?$$
$$F = A(F/A,6\%,18)$$
$$= \$2{,}530(30.906) = \$78{,}192$$

EXAMPLE 5–16

A present investment of $50,000 is expected to yield receipts of $7,000 a year for 15 years. What is the approximate rate of return that will be obtained on this investment?

Solution:

$$n = 15;\ P = \$50{,}000;\ A = \$7{,}000;\ i = ?$$
$$A = P(A/P,i\%,15)$$

$$(A/P,i\%,15) = \frac{A}{P} = \frac{\$7{,}000}{\$50{,}000} = 0.1400$$

This can be solved approximately by interpolation in the interest tables for capital recovery factors:

For $i = 11\%, (A/P,11\%,15) = 0.13907$
$i = 12\%, (A/P,12\%,15) = 0.14682$

By interpolation,

$$i = 0.11 + \left[\frac{0.1400 - 0.13907}{0.14682 - 0.13907}\right]0.01$$
$$= 0.111 \text{ or } 11.1\%$$

EXAMPLE 5–17

How much can one afford to spend each year for 15 years to avoid spending $1,000 at zero date, $1,500 after 5 years, and $2,000 after 10 years, if money is worth 8%?

or

What is the equivalent uniform annual cost for 15 years of disbursements of $1,000 at once, $1,500 5 years hence, and $2,000 10 years hence if interest is at 8%?

Solution:

The first step in solving problems of this type is to convert all disbursements either to present worth at zero date or compound amount at the final date. Then the present worth or compound amount can be converted into an equivalent uniform annual series A.

$$i = 8\%; \quad \left\{\begin{array}{l} n_1 = 0;\ P_1 = \$1{,}000; \\ n_2 = 5;\ P_2 = \$1{,}500; \\ n_3 = 10;\ P_3 = \$2{,}000; \\ n_4 = 15; \end{array}\right\} A = ?$$

Method 1:

$$\begin{aligned} A &= [\$1{,}000 + \$1{,}500(P/F,8\%,5) + \$2{,}000(P/F,8\%,10)](A/P,8\%,15) \\ &= [\$1{,}000 + \$1{,}500(0.6806) + \$2{,}000(0.4632)](0.11683) = \$344.33 \end{aligned}$$

Method 2:

$$\begin{aligned} A &= [\$1{,}000(F/P,8\%,15) + \$1{,}500(F/P,8\%,10) + \$2{,}000(F/P,8\%,5)](A/F,8\%,15) \\ &= [\$1{,}000(3.1722) + \$1{,}500(2.1589) + \$2{,}000(1.4693)](0.03683) \\ &= \$344.33 \end{aligned}$$

Illustration of the Use of Gradient Factors

Just as in other compound interest problems, a first step in solving problems involving arithmetic gradients is to identify what is known and what is wanted. Usually i, n, and G are known. The unknown element in the problem may be either A or P depending on whether it is desired to find an equivalent uniform annual series or the present worth of the gradient series. The appropriate factor from Tables D–1 through D–29 for the given i and n is multiplied by the gradient G.

EXAMPLE 5–18

A piece of construction equipment will cost $6,000 new and will have an expected life of 6 years, with no salvage value at the end of its life. The disbursements for taxes, insurance, maintenance, fuels, and lubricants are estimated to be $1,500 for the first year, $1,700 the second, $1,900 the third, and to continue to increase by $200 each year thereafter.

What is the equivalent uniform annual cost of this piece of equipment if the rate of interest is 12%?

Solution:

This example is representative of many engineering economy problems because it involves several different patterns of disbursement. All disbursements should be converted to their respective equivalent uniform annual amounts and added together.

$$i = 0.12;\ n = 6;\ \begin{cases} P = \$6{,}000;\ A_1 = ? \\ \qquad\qquad A_2 = \$1{,}500 \\ G = \ \$200;\ A_3 = ? \end{cases}$$

$$\begin{aligned} \textstyle\sum A &= A_1 + A_2 + A_3 \\ A_1 &= \$6{,}000(A/P,12\%,6) = \$6{,}000(0.24323) = \$1{,}459 \\ A_2 &= \$1{,}500 \end{aligned}$$

The problem stated that the annual disbursements would be $1,500 plus a gradient of $200 per year. The $1,500 portion requires no conversion because it is already a uniform annual figure. In this case the gradient is in the same "direction" as the $1,500 portion; therefore it is an increasing gradient. The next example illustrates a decreasing gradient, that is, the gradient is in opposite "direction" to the initial annual disbursement.

$$\begin{aligned} A_3 &= G(A/G,12\%,6) = \$200(2.172) = \$434 \\ \textstyle\sum A &= \$1{,}459 + \$1{,}500 + \$434 = \$3{,}393 \end{aligned}$$

EXAMPLE 5–19

A machine tool company offers to lease you a machine on the following terms: Pay initially $2,000 and a rental fee at the end of each year. The first year rental fee is $2,400, the second year is $2,100, and each successive year's fee is $300 less the previous year's. At the end of six years the machine will be returned to the company.

What is the equivalent annual cost of the lease if interest, i, is 8%?

Find the present worth of the total lease expenditures for 6 years.

$$i = 0.08;\ n = 6\ \begin{cases} P_1 = \$2{,}000;\ A_1 = ? \\ \qquad\qquad A_2 = \$2{,}400 \\ G = -\$300;\ A_3 = -\ ? \end{cases}$$

Solution:

$$\textstyle\sum A = A_1 + A_2 + A_3$$

$$\begin{aligned} A_1 &= \$2{,}000(A/P,8\%,6) = \$2{,}000(0.21632) = &\$433 \\ A_2 &= &2{,}400 \\ A_3 &= -\$300(A/G,8\%,6) = -\$300(2.276) = &-683 \\ \textstyle\sum A &= &\$2{,}150 \end{aligned}$$

To find the present worth of the payments, the $\sum A$ could be multiplied by the factor $(P/A,8\%,6)$, or the amount can be found by direct calculation:

$$\begin{aligned} \text{PW} &= \$2{,}000 + \$2{,}400(P/A,8\%,6) - \$300(P/G,8\%,6) \\ &= \$2{,}000 + \$2{,}400(4.623) - \$300(10.523) \\ &= \$2{,}000 + \$11{,}095 - \$3{,}157 = \$9{,}938 \end{aligned}$$

EXAMPLE 5–20

What is the present worth at 12% of the disbursements described in Example 5–18?

Solution:

The initial cost of the equipment is already at zero date. However, the annual outlays must be reduced to their present worths.

$$i = 0.12;\ n = 6;\ \begin{cases} P_1 = \$6{,}000 \\ A_2 = \$1{,}500;\ P_2 = ? \\ G = \$200;\ P_3 = ? \end{cases}$$

$$\sum P = P_1 + P_2 + P_3$$
$$P_2 = A_2(P/A,12\%,6) = \$1{,}500(4.111) = \$6{,}166$$
$$P_3 = G(P/G,12\%,6) = \$200(8.930) = \$1{,}786$$
$$\sum P = \$6{,}000 + \$6{,}166 + \$1{,}786 = \$13{,}952$$

EXAMPLE 5–21

A bank offers the following personal loan plan called "The Seven Percent Plan."

The bank adds 7% to the amount borrowed; the borrower pays back one-twelfth of this total at the end of each month for a year. On a loan of \$1,000, the monthly payment is \$1,070/12 = \$89.17.

What is the true interest rate per month? What are the nominal and effective rates per annum?

Solution:

$n = 12;\ P = \$1{,}000;\ A = \$89.17;\ i = ?$

$A = P(A/P,i\%,12)$

Solve by interpolation.

$$(A/P,i\%,12) = \frac{A}{P} = \frac{\$89.17}{\$1{,}000} = 0.08917$$

Try $i = 0.01$

$(A/P,1\%,12) = 0.08885$

Try $i = 0.015$

$(A/P,1.5\%,12) = 0.09168$

Interpolating,

$$i = 0.01 + \left[\frac{0.08917 - 0.08885}{0.09168 - 0.08885}\right](0.0050)$$

$$= 0.01 + \left[\frac{0.00032}{0.00283}\right](0.0050)$$

$$= 0.01 + 0.00057 = 0.01057 \text{ or } 1.057\%$$

The *nominal interest rate per annum* corresponding to this monthly rate is:

$12(0.01057) = 0.1268$ or 12.68%

The *effective interest rate per annum* corresponding to this monthly rate is:

$(1.01057)^{12} - 1 = 0.1345$ or 13.45%

Deferred Annuities

Frequently it is necessary to deal with a series of uniform annual payments or receipts that begin sometime in the future and continue for some number of years. Such a deferred series of uniform annual payments or receipts is known as a *deferred annuity*. It may be desired to convert a deferred annuity to a present worth, to an equivalent annual series over some different period of years, or to a compound amount at some future time. Two methods of solving such a problem are illustrated in Example 5–22.

EXAMPLE 5–22

How much could you afford to spend each year for the next 6 years to avoid spending $500 a year for 10 years, beginning 5 years hence, if money is worth 8%?

or

What is the equivalent annual cost over a period of the next 6 years of spending $500 a year for 10 years beginning 5 years hence, if interest is at 8% per annum?

Solution:

$i = 0.08;\ n_1 = 6;\ A_1 = ?$
$n_2 = 10;\ A_2 = \$500$
$n_3 = 5;$

Method 1: Subtract uniform series present worth factor for 4 years from uniform series present worth factor for 14 years and multiply by the capital recovery factor for 6 years.

$$A_1 = \$500[(-P/A,8\%,14) - (P/A,8\%,4](A/P,8\%,6)$$
$$= \$500(8.244 - 3.312)(0.21632)$$
$$= \$500(4.932)(0.21632) = \$533.40$$

Method 2: Multiply the uniform series present worth factor for 10 years by the single payment present worth for 4 years and multiply that by the capital recovery factor for 6 years.

$$\begin{aligned} A_1 &= \$500(P/A,8\%,10)(P/F,8\%,4)(A/P,8\%,6) \\ &= 500(6.710)(0.7350)(0.21632) \\ &= \$533.40 \end{aligned}$$

EXAMPLE 5–23

What amount must be deposited at the *beginning* of each year for the next 20 years in a savings fund that earns 6% interest in order to accumulate $20,000 at the end of 20 years?

$i = 0.06$; $n_1 = 20$; $F = \$20{,}000$
$n_2 = 1$; $A = ?$

Solution:

$$\begin{aligned} A &= F(A/F,6\%,20)(P/F,6\%,1) \\ &= \$20{,}000(0.02718)(0.9434) = \$512.80 \end{aligned}$$

Note: An end-of-period series of uniform amounts can be converted to a beginning-of-period series by multiplying the end-of-period amount by the single payment present worth factor for one period at the same interest rate.

Summary

The general technique for solving a problem in compound interest equivalence is to determine first which elements of the problem are known and which element is unknown. If the problem is one to which available interest tables apply, it will usually be most convenient to use them in its solution. Otherwise, it is necessary to substitute the known elements of the problem in the appropriate interest formula, and to solve for the unknown one.

PROBLEMS

5–1. It is desired to make an initial lump sum investment that will provide for a withdrawal of $500 at the end of year 1, $600 at the end of year 2, and amounts increasing $100 per year to a final $2,400 at the end of year 20. How great an initial investment will be required if it earns 5% compounded annually? (*Ans*. $16,080)

5–2. Solve Problem 5–1 assuming that the $2,400 withdrawal will be at the end of year 1 and that withdrawals will decrease by $100 a year to a final $500 at the end of year 20. (*Ans*. $20,060)

5–3. $100,000 is borrowed at a nominal 7% compounded semiannually, to be repaid by a uniform series of payments, partly principal and partly interest, to be made at the end of each 6-month period for 30 years. How much of the principal of the loan will have been repaid at the end of 10 years, just after the 20th payment has been made? (*Ans*. $14,388)

5–4. How much must be deposited on January 1, 1984, and every 6 months thereafter until July 1, 1992, in order to withdraw $1,000 every 6 months for 5 years starting January 1, 1993? Interest is at a nominal 11% compounded semiannually. (*Ans*. $255.70)

5–5. At 10% interest, what uniform annual payment for 10 years is equivalent to the following irregular series of disbursements: $10,000 at zero date, $5,000 at date 5, $1,000 at the end of year 1, $1,500 at the end of year 2, and year-end payments increasing by $500 a year to $5,500 at the end of year 10?

5–6. A loan of $600 is to be repaid in 15 equal end-of-month payments computed as follows:

Principal of loan	$600
Interest for 15 months at "1.5% per month"	135
Loan fee of 5%	30
	$765

Monthly payment = $765 ÷ 15 = $51

What nominal and effective interest rates per annum are actually paid? (*Ans*. 38.4%; 46.0%)

5–7. Using interest tables and interpolation as far as possible, determine the approximate rates of interest indicated by the following valuations of prospective series of future cash receipts:

(a) $8,000 now for $1,300 at the end of each year for 10 years (*Ans*. 10%)

(b) $6,000 now for $680 at the end of each year for 20 years (*Ans*. 9.5%)

(c) $6,000 now for $400 at the end of each year for 15 years (*Ans*. 0%)

(d) $10,000 now for $300 at the end of the first year, $350 at the end of the second year, and receipts increasing by $50 at the end of each year to a final receipt of $1,250 at the end of the 20th year (*Ans*. 3.7%)

(e) $5,000 now for $225 at the end of each year forever (*Ans*. 4.5%)

5–8. A company can either buy certain land for outdoor storage of equipment or lease it on a 15-year lease. The purchase price is $80,000. The annual rental is $5,000 payable at the *start* of each year. In either case, the company must pay property taxes, assessments, and upkeep. It is estimated that the land will be needed for only 15 years and will be salable for $100,000 at the end of the 15-year period. What rate of return before income taxes will the company receive by buying the land instead of leasing it? (*Ans*. 7.7%)

5–9. The landowner in Problem 5–8 also offers a 15-year lease for a prepaid rental of $55,000. If the company has decided to lease rather than to buy, what interest rate makes the prepaid rental equivalent to the annual rental? (*Ans*. 4.8%) How should this rate be interpreted as a basis for the company's choice between prepaid and annual rental?

5–10. A person engaged in making small loans offers to lend $200 with the borrower required to pay $14.44 at the end of each week for 16 weeks to extinguish the debt. By appropriate use of your interest tables, find the approximate interest rate per week. What is the nominal interest rate per annum? What is effective interest rate per annum? (*Ans.* 1.75%; 90.9%; 146.3%)

5–11. The purchase of certain unimproved city lots is under consideration. The price is $20,000. The owner of this property will pay annual property taxes of $400 the first year; it is estimated that these taxes will increase by $40 each year thereafter. It is believed that if this property is purchased, it will be necessary for the investor to wait for 10 years before he can sell it at a favorable price. What must the selling price be in 10 years for the investment to yield 15% before income taxes? (*Ans.* $91,781)

5–12. What uniform annual payment for 30 years is equivalent to spending $10,000 immediately, $10,000 at the end of 10 years, $10,000 at the end of 20 years, and $2,000 a year for 30 years? Assume an interest rate of 9%. (*Ans.* $3,558.22)

5–13. If $5,000 is deposited in a savings account earning 6% interest per year compounded quarterly, how much will be in the account at the end of 20 years? How much more is this than the amount that would be in the account if interest were compounded annually?

5–14. Assume in Problem 5–13 that $5,000 is withdrawn from the account at the end of 10 years. How much will be in the account at the end of 20 years with interest at 6% per year compounded quarterly?

5–15. Using interest tables and interpolation as far as possible, determine the approximate interest rates indicated by the following valuations for prospective series of future payments:

(a) $5,000 now for $600 a year at the end of each of the next 20 years.

(b) $3,000 now for $500 a year at the end of each of the next 8 years.

(c) $50,000 now for $8,500 a year at the end of each of the next 15 years plus $15,000 at the end of year 15.

5–16. A co-worker offers to loan you $40 if you will repay $50 at the end of one week. What interest rate is this person asking? What is the nominal interest rate per year? the effective interest rate per year?

5–17. Solve the following assuming interest at 11% compounded annually:

(a) A payment now of how much is acceptable in place of a payment of $4,000 at the end of 12 years?

(b) A loan of $6,000 now will require equal annual payments of how much at the end of each of the next 20 years?

(c) The present worth of $50,000 40 years hence is how much?

(d) A present investment of $10,000 will secure a perpetual income of how much a year?

5–18. You are offered a mortgage loan of $50,000 at 9% per year to be repaid in equal annual installments. You may repay the loan in either 20 or 30 years. What would be the annual payment in each case? What is the difference in total amount paid between these two terms of payment?

5–19. In Problem 5–18, what principal balance will remain after the 15th annual payment has been made in each case?

5–20. In 1626, the Indians traded Manhattan Island for $24 worth of trade goods. Had they been able to deposit $24 into a savings account paying 6% interest per year, how much would they have today (1985)?

5–21. Maintenance expenditures for a structure with a 25-year life are expected to come as periodic outlays of $1,000 at the end of the 5th year, $3,000 at the end of the 10th year, and $5,000 at the end of the 15th and 20th years. With interest at 13%, what is the equivalent annual cost for a 25-year period?

5–22. A savings account is to be set up into which $2,000 will be deposited immediately, $200 three months from now, $220 six months from now, and with quarterly deposits increasing by $20 per quarter for the next 3 years. If interest is 6% per year compounded quarterly, how much will be in the account at the end of three years? What will be the amount of the last quarterly deposit?

5–23. An investor paid $750 for 10 shares of stock 8 years ago. Dividends of $5 per share were paid the first year increasing by $2 per share per year until the latest dividend of $19 per share just received. The stock can now be sold for $1,500. If sold, what rate of return will have been earned on the investment?

5–24. A purchaser of furniture on a time payment plan agrees to pay $84 at the end of each month for 24 months. Later it is discovered that the same furniture could have been purchased for $1,500 cash. What interest rate per month is the purchaser paying? What are the nominal and effective interest rates per year?

5–25. As a special privilege to professional people and business executives, a loan corporation offers to loan $6,000 with repayment to be made monthly in the amount of $304.07 for 24 months. The total repayment is $7,297.68, which includes $1,207.04 in finance charges plus $90.49 for a required life insurance policy covering the amount of the loan. What nominal and effective annual interest rates are being charged for this executive credit?

5–26. An annuity fund is to be created to provide $2,500 at the end of each quarter year from ages 65 through 90 beginning at the end of the first quarter after age 65. Quarterly deposits are to be made to an account paying 8% per year compounded quarterly. These deposits will be stopped at age 65.

(a) How much should each quarterly deposit be if the person is now 25 years old?

(b) How much should the deposits be if the person is now 40 years old?

5–27. In Problem 5–26, the quarterly withdrawals were to stop at age 90. Recalculate the required deposits assuming the withdrawals are to continue in perpetuity. Compare your results with those obtained in Problem 5–26.

5–28. A person receives a bonus each year of at least $1,000. It is planned to deposit $1,000 each year into a savings account that earns 6% per year compounded quarterly. How much will be in the account at the end of 20 years? The first deposit will be made immediately.

6

EQUIVALENT UNIFORM ANNUAL CASH FLOW

Every engineering structure, with few exceptions, is first suggested by economic requirements; and the design of every part, excepting few, and of the whole is finally judged from the economic standpoint.

It is therefore apparent that the so-called principles of design are subordinate to the principles which underlie economic judgment.—*J. C. L. Fish**

Proposed investments in industrial assets are unattractive unless it seems likely they will be recovered with interest; the rate of interest should be at least the minimum rate of return that is attractive in the particular circumstances. This rate is designated throughout the remainder of this book as i^* (this might be pronounced "eye-star").

The introduction of the time value of money into economy studies reflects this requirement that capital be recovered with a return. This chapter and the three following chapters explain four possible ways to compare proposed alternatives that involve different series of prospective receipts and disbursements. These ways are:

1. Equivalent uniform annual cash flow, with a stipulated minimum attractive rate of return i^* used as an interest rate.
2. Present worth, with a stipulated minimum attractive rate of return i^* used as an interest rate.
3. Prospective rate of return, with the calculated rate of return compared with the stipulated minimum attractive rate i^*.
4. Benefit-cost ratio (applicable chiefly to governmental projects), with a stipulated minimum attractive rate of return i^* used as an interest rate.

In our initial explanation of these four methods in Chapters 6 through 9, we shall consider alternatives in pairs and shall limit our examples and problems to cases where there are no more than three proposals. In Chapter 13, we

*J. C. L. Fish, *Engineering Economics*, 1st ed. (New York: McGraw-Hill Book Co., Inc., 1915), first two paragraphs of preface, p. v.

shall examine certain special aspects of analysis where many alternatives are being compared. The discussion in Chapter 13 will be related to all four of the methods of analysis introduced in Chapters 6 through 9.

As we develop the subject, it will become evident to the reader that, correctly applied with the same minimum attractive rate of return, the four methods will lead to the same decision among alternative designs in the common type of case where it is physically possible to choose only one of the alternatives. However, we shall also see that each method has certain advantages and disadvantages as a guide to judgment. Because it often is necessary for a decision maker to give weight to matters that have not been expressed in money terms, it is not necessarily a matter of indifference which method is to be used.

It also will become evident that a critical matter is the choice of the minimum attractive rate of return. Proposals that look good at values of i^* of, say, 4% or 5%, will be decidedly unattractive at rates of, say, 12% or 15%. In Chapters 6 through 9, the value of i^* will be stipulated in all examples and problems without discussion of why a particular i^* was chosen. Chapter 10 examines a number of aspects of the selection of i^*. It stresses the point, already mentioned in Chapter 2, that an important element in selecting a minimum attractive rate of return is to make the best possible use of the limited resources that can be devoted to capital investment.

Cash Flow

This chapter introduces the equivalent uniform annual cash flow method as applied to relatively simple circumstances. The method is illustrated by a series of simple examples. Applications of equivalent uniform annual cash flow to more complex situations are developed in later chapters.

The data for all of the examples and problems in Chapters 6, 7, and 8 are given in terms of prospective cash flow (i.e., receipts and disbursements) associated with the stated alternatives. In the examples and problems in the present chapter the differences between the alternatives are almost entirely in disbursements; the only differences in receipts entering into the comparisons are in receipts from salvage values.

Many of the problems in economy that confront the engineer are of this type; the prospective receipts from the sale of a product or service are unaffected by the engineer's choice among the various alternatives available. Economy studies in such cases must start with estimates of the amounts and dates of the disbursements for each alternative. It also is necessary either to estimate the full period of service from each alternative or to concentrate attention in the economy study on some shorter period that might be described as the study period or analysis period.

Once such estimates have been made, a mere inspection of the figures may settle the question of relative economy; one alternative may involve less disbursements both initially and subsequently. But in the many cases where this is not true, the common situation is for one alternative to involve a higher first cost that leads to some future advantages, such as lower annual disbursements or longer life or higher salvage value. The question at issue in such cases is whether these future advantages are sufficient to justify the greater initial investment.

"Annual Cost" Means Equivalent Uniform Annual Net Disbursements

To compare nonuniform series of money disbursements where money has a time value, it is necessary somehow to make them comparable. One way to do this is by reducing each to an equivalent uniform annual series of payments. In general, the phrase *annual cost* when used in connection with economy studies is simply a short way of saying *equivalent uniform annual net disbursements*. As a practical matter, however, it sometimes is expedient to use various approximations to the desired equivalent uniform annual figure. These approximations may also be described as *annual cost*. A brief description of some common types of approximation and an evaluation of their merits is given in Chapter 11.*

Income Tax Considerations in Investment Evaluation

In private enterprise in most industrialized countries and in many developing countries, prospective income taxes may be influenced by decisions regarding investments in physical assets. If decisions are to be made from the viewpoint of the owners of an enterprise, the expected income tax effects of each decision need to be recognized.

Most of the examples and problems in this book are based on analysis of differences in estimated cash flows associated with alternatives that are being compared. Prospective differences in disbursements for income taxes constitute one of the elements of cash flow. In principle, therefore, differences in cash flow for income taxes ought to be included in any analysis based on cash flow.

In general, there is no entirely satisfactory answer to the question: "At what point and in what way should income tax matters be introduced in the study of engineering economy?" The difficulty is that, although under certain circumstances the estimation of the income tax effects of decisions is fairly simple, under other circumstances this estimation is a complex matter requiring an understanding of accounting and a familiarity with income tax laws and regulations. Doubtless, the best that can be done is to give the student of engineering economy a basis for identifying the numerous routine types of economy studies in which it is a fairly simple matter for an analyst to consider

*It is important to recognize that there are several reasons why the comparable equivalent uniform annual net disbursements computed by the method of Chapter 6 cannot be compared with the "costs" that will be reported by the cost accountants.

income taxes in a way that is good enough for practical purposes; also the student should have a basis for recognizing the cases where an analyst may need advice from a tax specialist.

The more complex aspects of estimating the income tax aspects of decisions are presented in Chapter 12. However, starting with the examples and problems in the present chapter, we shall include in many cash flow series, the estimated difference between alternatives in cash flow for income taxes. Usually this income tax difference will be a separately identified item, although occasionally it may be combined with other cash flow differences. Where the income tax difference is separately identified for cases in competitive industry, it is computed under fairly simple assumptions that are explained in Chapter 11.*

There are two main reasons for using the *after-tax* minimum attractive rate of return. One reason is that its use in examples and problems is correct in principle; prospective differences between alternatives in the more distant future are discounted too greatly when the higher *before-tax* values of i^* are used. The other reason is that the important subject of the impact of income tax considerations on business decisions receives its appropriate emphasis when specific figures are given for differences in income tax disbursements.

As pointed out in Chapter 20, economy studies for regulated public utilities in the United States generally are different in principle from economy studies for competitive industry in relation to the way prospective income tax differences ought to be estimated. Under certain assumptions explained in Chapter 20, it is appropriate to estimate the portion of public utility income taxes affected by a choice among proposed alternatives as a percentage of the first costs of the respective proposed assets.

Income taxes as such are not a factor in the usual economy studies for governments. However, it is pointed out in Chapter 19 that income taxes (and other taxes) forgone ought to be considered as one element in any governmental decision on the question of whether certain activities should be carried out by government or by private industry.

EXAMPLE 6–1

A Proposed Investment to Reduce Labor Costs

Statement of Alternatives

At present a certain materials-handling operation in the warehouse of a manufacturing company is being done by hand labor. Annual disbursements for

*See the paragraph "Economy-Study Estimates of Cash Flow for Income Taxes Based on Certain Simplified Assumptions" in Chapter 11. They include the assumptions that: the lives and salvage values of depreciable assets are the same for economy studies as for tax purposes; straight-line item depreciation accounting will be used; and the incremental tax rate on income is 50%.

this labor and for certain closely related expenses (such "labor extras" as social security taxes, industrial accident insurance, paid vacations, and various employees' fringe benefits) are $9,200. The proposal to continue materials handling by the present method is called Plan A.

An alternative proposal, Plan B, is to build certain equipment that will reduce this labor cost. The first cost of this equipment will be $15,000. It is estimated that the equipment will reduce annual disbursements for labor and labor extras to $3,300. Annual payments for power, maintenance, and property taxes and insurance are estimated to be $400, $1,100, and $300 respectively. Extra annual disbursements for income taxes over those required with Plan A are estimated to be $1,300.

It is expected that the need for this particular operation will continue for 10 years and that because the equipment in Plan B is specially designed for the particular purpose, it will have no salvage value at the end of that time. It is assumed that the various annual disbursements will be uniform throughout the 10 years. The minimum attractive rate of return after income taxes is 9%.

Annual Cost Comparison

Everything but the initial $15,000 is already assumed to be a uniform annual disbursement. The only compound interest calculation needed is a conversion of this $15,000 to its equivalent uniform annual cost of capital recovery. The equivalent uniform annual disbursements may then be tabulated for each plan and their totals may be compared.

Plan A		Plan B	
Labor and labor extras	$9,200	CR = $15,000(*A*/*P*,9%,10)	
		= $15,000(0.15582)	= $2,337
		Labor and labor extras	3,300
		Power	400
		Maintenance	1,100
		Property taxes, insurance	300
		Extra income taxes	1,300
Comparative equivalent uniform annual disbursements	$9,200	Comparative equivalent uniform annual disbursements	$8,737

Plan B is therefore more economical than Plan A.

Simplicity of a Uniform Annual Series of Disbursements

Even without inflation, experience indicates that it is almost inevitable that certain disbursements will vary from year to year.* Maintenance costs fluctuate and tend to increase with age; wage rates change; property tax rates and assessed valuations change; and so forth. Nevertheless, it often happens that

*The impact of inflation and its proper treatment in economy studies is discussed in Chapter 14. The introduction of this complicated subject at this point would only serve to detract from a solid understanding of the basic techniques and principles.

there is no rational basis for making different estimates for each year. Even where there is some basis for making separate year-by-year estimates, the prospective differences in year-by-year totals may be so small that it is good enough for practical purposes merely to estimate average annual disbursements and to treat the average figures as if they were uniform.

Whatever may be the reason for estimates of uniform annual disbursements, it is evident that such estimates simplify the comparison of equivalent uniform annual costs; only the capital costs require conversion by appropriate compound interest factors.

Tabulation of Cash Flow

A useful tool in many economy studies is a year-by-year tabulation of estimated disbursements and receipts associated with each of two alternatives, followed by a tabulation of the differences in cash flow between the alternatives. Table 6–1 shows such a tabulation for Example 6–1.

Many such tabulations appear throughout this book. In all of them a net disbursement is preceded by a minus sign and a net receipt is preceded by a plus sign. The final column of Table 6–1 recognizes that a reduction of disbursements is, in effect, an increase in receipts. That is, spending $2,800 less has the same effect on the company's cash as receiving $2,800 more.

The figures for totals at the bottom of cash flow tables provide a check on the arithmetic in the table. Although this check seems unnecessary in Table 6–1, it will prove useful in more complicated types of circumstances such as are discussed later in this book. It should be noted by the reader that the totals from any such cash flow tables disregard the time value of money; therefore, these totals do not by themselves provide a satisfactory basis for choosing between the alternatives being compared. Such totals may be thought of as giving the present worth of the cash flow using an interest rate of 0%.

TABLE 6–1
Tabulation of comparative cash flow, Example 6–1

Year	*Plan A*	*Plan B*	*B – A*
0		–$15,000	–$15,000
1	–$9,200	–6,400	+2,800
2	–9,200	–6,400	+2,800
3	–9,200	–6,400	+2,800
4	–9,200	–6,400	+2,800
5	–9,200	–6,400	+2,800
6	–9,200	–6,400	+2,800
7	–9,200	–6,400	+2,800
8	–9,200	–6,400	+2,800
9	–9,200	–6,400	+2,800
10	–9,200	–6,400	+2,800
Totals	–$92,000	–$79,000	+$13,000

In a tabulation such as Table 6–1 that is intended to show how receipts and disbursements will be influenced by a particular decision, the source of an estimated receipt or disbursement is immaterial. For example, a dollar spent for one purpose in a particular year has the same effect on cash flow as a dollar paid out for some other purpose; in a cash flow analysis, it is a matter of indifference whether the dollar is spent for labor, for power, for property taxes, for income taxes, or for anything else. Although this point seems obvious in Example 6–1, it is not always so evident; we shall have need to mention it again when we take a more critical look at the income tax effects of decisions in Chapter 12 and when we discuss incremental costs and sunk costs in Chapter 16.

Moreover, in a tabulation such as Table 6–1 showing expected cash flow for two alternatives, it is the right-hand column—the column of differences in cash flow—that is significant. This column of differences serves to clarify the question at issue in the choice between the alternatives. In Example 6–1, the question—quite obviously—is whether it will pay to spend $15,000 at once in order to save $2,800 a year (after income taxes) for the next 10 years.

The difference shown in the third column, B–A, illustrates principle number 5 of Chapter 1, in that only differences between the alternatives are relevant. Items, such as supervision, control, floor space, and so on, which are not affected by the choice were not included.

Interpretation of Annual Cost Comparison in Example 6–1

Our annual cost comparison has answered the foregoing question in the affirmative. The smaller equivalent annual cost computed for Plan B means "Yes, it will pay to spend the proposed $15,000."

This answer should properly be viewed as a qualified "Yes," subject to the appropriateness of the 9% i^* used in the equivalence conversion and to the weight, if any, to be given to irreducible data. If a 9% return after income taxes is high enough to be attractive, all things considered, and if other matters not reflected in the cash flow estimates do not favor Plan A, it is clear that Plan B is better.

It should also be recognized that the answer "Yes, it will pay" refers only to the relative merits of Plans A and B. Conceivably there may be some other possible plan that, if considered, would prove to be superior to both A and B.

The Borrowed Money Point of View

Assume that all the $15,000 to be invested in the new equipment is to be borrowed at 9% interest. Assume also that this borrowing will be repaid by uniform annual end-of-year payments over the 10-year life of the equipment in the manner illustrated in Plan III, Table 3–1. These annual payments would then be $2,337, the computed capital recovery cost for Plan B in our solution of Example 6–1.

Although this point of view is helpful in understanding the calculation of equivalent uniform annual costs, it is only rarely that borrowing takes place in this manner. Moreover, for reasons that are explained in Chapter 10, the appropriate minimum attractive rate of return for use in an economy study is nearly always higher than the bare cost of borrowed money. In addition, as explained in Chapter 18, borrowed money and equity funds are treated differently for income tax purposes. But regardless of whether the first cost is to be borrowed 100%, or to be financed 100% out of the funds of the prospective owner, or to be financed by some combination of borrowed funds and of owner's funds, and regardless of the plans for the repayment of any borrowed funds, the calculated equivalent uniform annual costs provide an entirely valid method of comparing the long-run economy of the two alternatives, once a particular interest rate is accepted as a standard.

Nevertheless, questions of financing, separate from those of long-run economy, arise whenever proposed assets are to be financed by borrowing that must be paid back rapidly. The decision between the alternatives is based on the long-run economy. The decision regarding the method of financing the proposed plan is a completely separate decision, illustrating principle number 6 in Chapter 1. Here a separate question is always whether the repayment obligation can be met. This topic is discussed in Chapter 18.

The End-of-Year Convention in Economy Studies

In Plan B the $2,337 capital recovery cost was an end-of-year series for 10 years. If the date of the $15,000 investment is designated as zero (0) date, the ten $2,337 figures apply to dates 1 through 10, respectively.

When we add this $2,337 year-end figure to the $6,400 that we expect will actually be paid out each year, we are, in effect, assuming that the $6,400 is also a year-end figure. As a matter of fact the disbursements included in the $6,400 are expected to occur throughout each year. Some of the disbursements (such as wages of labor) will doubtless take place at a fairly uniform rate during the year; others (such as property taxes and income taxes) will occur at regular intervals; still others (such as maintenance) will occur irregularly.

It is convenient in economy studies to treat receipts and disbursements that occur throughout a year as if they took place at year end. This end-of-year convention is used in nearly all of the examples and problems in this book; unless otherwise stated, it is implied in cash flow tables and in calculations of annual costs, present worths, and rates of return. The convention greatly simplifies the required compound interest conversions. In most cases the assumption is good enough for practical purposes in the sense that it will not lead to errors in decisions between alternatives.

If continuous compounding of interest is assumed, a different convention may be adopted—namely, that all receipts and disbursements occur uniformly throughout each year. This latter convention is discussed and illustrated in Appendix A. Its chief advantages occur where the study period is short and the prospective rate of return is high.

EXAMPLE 6–2

A Proposed Investment that has a Salvage Value

Statement of Alternatives

Plan C, an alternative to Plan B in Example 6–1, calls for the purchase of certain general-purpose materials-handling equipment. The first cost of this equipment will be $25,000, a considerable increase over the $15,000 first cost in Plan B. However, it is estimated that this general-purpose equipment will have a $5,000 net salvage value at the end of the 10-year period of service. (The net salvage value may be defined as the gross receipts from the sale of the equipment minus any disbursements required by its removal and sale.) This equipment, which has more automatic features than the equipment in Plan B, is expected to reduce annual disbursements for labor and labor extras to $1,450. Estimated annual disbursements for power, maintenance, and property taxes and insurance are $600, $1,500, and $500, respectively.

The extra annual disbursement for income taxes in Plan C as compared to Plan A is estimated as $1,575. (The reader will recall that Plan A was used as the base of comparison in computing the extra income tax payments to be made under Plan B.) It is now desired to compare Plan C with Plan B using an i^* of 9%.

Annual Cost of Plan C

Using Plan A as a base, the annual cost of Plan B has already been calculated as $8,737. With the same base, equivalent annual net disbursements for Plan C are as follows:

Plan C

CR = $25,000 $(A/P,9\%10)$ − $5,000$(A/F,9\%,10)$	
= $25,000(0.15582) − $5,000(0.06582)	
= $3,895 − $329	= $3,566
Labor and labor extras	1,450
Power	600
Maintenance	1,500
Property taxes, insurance	500
Extra income taxes	1,575
Comparative equivalent uniform annual net disbursements	$9,191

It is evident that although Plan C is slightly more economical than Plan A, it is less economical than Plan B. The savings of $775 in annual disbursements promised by Plan C are not enough to offset its higher capital recovery cost.

The Influence of Salvage Value on the Annual Cost of Capital Recovery

Let P = first cost of a machine or structure, n = the life, study period, or analysis period in years, S = prospective net salvage value or net terminal value at the end of n years, and $i^\star$ = minimum attractive rate of return. The equivalent uniform annual cost of capital recovery may then be expressed as follows:

$$\text{CR} = P(A/P,i^\star\%,n) - S(A/F,i^\star\%,n)$$

Following the cash flow methodology introduced in Chapter 5, this method converts the first cost to a uniform annual series over the life by multiplying it by the capital recovery factor. From this product there is subtracted a uniform annual figure obtained by multiplying the salvage value by the sinking fund factor.

Three other correct methods of computing capital recovery cost where salvage values exist are described and illustrated in the following paragraphs.

The first, and perhaps most used, of these formulas calculates the annual cost of capital recovery as the sum of two figures. One figure is the product of first cost minus estimated salvage, $(P - S)$, and the capital recovery factor. The other is the product of salvage, (S), and interest rate $i^\star$. This may be expressed as

$$\text{CR} = (P - S)(A/P,i^\star\%,n) + Si^\star$$

This formula, considered algebraically, is also applicable with a zero salvage value or a negative salvage value.

The borrowed money viewpoint may be helpful in examining the rational basis of this formula. Let us apply this viewpoint to the $25,000 initial disbursement, the prospective $5,000 receipt from salvage value at the end of 10 years, and the 9% $i^\star$ in Plan C. Assume that the $25,000 is borrowed at 9% interest and that it is anticipated that $5,000 of the debt will be repaid from the proceeds of the salvage value. In effect, the $25,000 debt may be divided into two parts. One part of the debt, $20,000 (i.e., $P - S$ or $25,000 – $5,000), must be repaid by uniform annual payments of $3,116 for 10 years (obtained as the product of $20,000 and the 10-year capital recovery factor, 0.15582). On the other part of the debt, $5,000, it is necessary merely to pay interest of $450 each year because the investment itself will generate the $5,000 salvage necessary to repay the principal. The total annual payment on the debt is therefore $3,116 + $450 = $3,566.

Another method converts salvage value to its present worth at zero date by multiplying salvage by the single payment present worth factor. The difference between the first cost and the present worth of the salvage is then multiplied by the capital recovery factor. This may be expressed as

$$\text{CR} = [P - S(P/F,i^\star\%,n)](A/P,i^\star\%,n)$$

As applied to the data of Plan C, the present worth of the salvage is $5,000(0.4224) = $2,112.

$$CR = (\$25{,}000 - \$2{,}112)(0.15582) = \$3{,}566$$

In a third method the capital recovery cost is calculated as the sum of the product of first cost and interest rate, and the product of the difference between first cost and salvage value and the sinking fund factor. This may be expressed as

$$CR = Pi^{\star} + (P - S)(A/F, i^{\star}\%, n)$$

As applied to the data of Plan C

$$\begin{aligned} CR &= \$25{,}000(0.09) + (\$25{,}000 + \$5{,}000)(0.06582) \\ &= \$2{,}250 + \$1{,}316 = \$3{,}566 \end{aligned}$$

Cash Flow for Example 6–2

Table 6–2 compares cash flow in Plans B and C. The tabulation makes it evident that the question at issue is whether it is desirable to pay out $10,000 at once in order to receive $5,000 at the end of 10 years and to avoid disbursements of $775 a year throughout the 10-year period. Although the extra $10,000 investment will ultimately be recovered with $2,750 to spare by the combination of reduced annual disbursements and receipts from salvage, our annual cost analysis has told us that this recovery will not be rapid enough to yield the stipulated 9%.

EXAMPLE 6–3

Comparing Alternatives That Have Different Lives

Statement of Alternatives

In the design of certain industrial facilities, two alternative structures are under consideration. We shall call them Plans D and E. The receipts from the sale of goods and services will not be affected by the choice between the two plans. Estimates for the plans are:

	Plan D	Plan E
First cost (P)	$50,000	$120,000
Life (n)	20 years	40 years
Salvage value (S)	$10,000	$20,000
Annual O & M disbursements	$9,000	$6,000

The estimated annual O & M disbursements include operation, maintenance, property taxes, and insurance. Extra annual disbursements for income taxes with Plan E are estimated as $1,250.

It is desired to compare these alternatives using a minimum attractive rate of return of 11% after income taxes.

TABLE 6–2
Tabulation of comparative cash flow, Example 6–2

Year	*Plan B*	*Plan C*	*C–B*
0	−$15,000	−$25,000	−$10,000
1	−6,400	−5,625	+775
2	−6,400	−5,625	+775
3	−6,400	−5,625	+775
4	−6,400	−5,625	+775
5	−6,400	−5,625	+775
6	−6,400	−5,625	+775
7	−6,400	−5,625	+775
8	−6,400	−5,625	+775
9	−6,400	−5,625	+775
10	{ −6,400	{ −5,625 +5,000	{ +775 +5,000
Totals	−$79,000	−$76,250	+$2,750

Annual Cost Comparison
The comparative equivalent uniform annual disbursements are as follows:

Plan D

CR = $50,000 $(A/P,11\%,20)$ − $10,000$(A/F,11\%,20)$	
= $50,000 (0.12558) − $10,000(0.01558)	
= $6,279 − $156	= $ 6,123
Annual O & M disbursements	9,000
Comparative equivalent uniform annual disbursements	$15,123

Plan E

CR = $120,000$(A/P,11\%,40)$ − $20,000$(A/F,11\%,40)$	
= $120,000(0.11172) − $20,000(0.00172)	
= $13,406 − $34	= $13,372
Annual O & M disbursements	6,000
Extra annual income taxes	1,250
Comparative equivalent uniform annual disbursements	$20,622

Plan D has the lower annual cost.

Some Considerations in Annual Cost Comparisons When Two Alternatives Have Different Lives

When annual cost comparisons such as the preceding one are made, an objection along the following lines is sometimes raised: "Plan E has an important advantage over Plan D in that it has a much longer prospective life. How is this advantage reflected in your annual cost comparison? Doesn't your $15,123 a year give you service for only 20 years, whereas your $20,622 a year gives you service for 40 years?"

Such an objector doubtless missed a fundamental point in the mathematics of compound interest. Nevertheless, an interesting and important topic has been introduced. We shall discuss various facets of this topic in different places throughout this book.

A direct answer to our objector may be made by pointing out that the 20-year estimated life in Plan D was reflected by the use of the 20-year capital recovery factor, 0.12558, in obtaining the $6,123 CR cost, and the 40-year estimated life in Plan E was reflected by the use of the 40-year factor, 0.11172, in obtaining the CR cost of $13,372. The answer may be amplified by pointing out that the estimate of a 40-year life in Plan E implied that a service of at least this long would be required. Although the $15,123 a year for Plan D is for 20 years only, the service must be continued after the 20-year structure is retired. Presumably, although not necessarily, the annual costs of continuing the service will be of the same order of magnitude. If it is assumed that as good an estimate as any is that the replacement structure will have the same first cost, life, salvage value, and annual disbursements as the initial structure, the $15,123 annual cost in Plan D will be repeated during the second 20 years. This point is illustrated in Figure 6–1.

A somewhat more sophisticated view of the matter recognizes that the present decision between a long-lived and a short-lived alternative is simply a decision as to what to do *now*. In aiming to have the decision now turn out to be the best decision in the long run, it is appropriate to consider what may happen after the end of the life of the shorter-lived alternative. In Example 6–3, a forecast that the replacement structure in the second 20 years will have

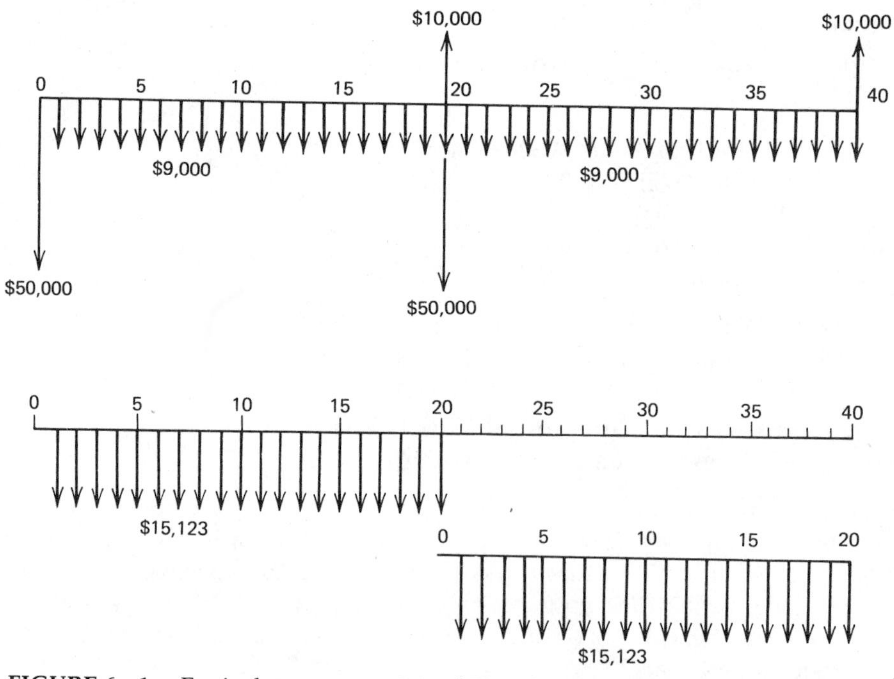

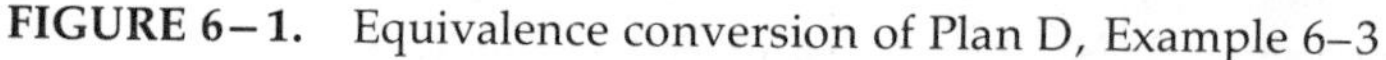

FIGURE 6–1. Equivalence conversion of Plan D, Example 6–3

much higher annual costs than the initial structure is favorable to Plan E. Similarly, the prospect that a replacement structure with much lower annual costs will be available should be given weight in the present choice as a factor favoring Plan D. In general, prospects for price increases and for extra costs incident to replacement are favorable to the selection of longer-lived alternatives; prospects for technological improvements, changes in service requirements, and price reductions are favorable to the selection of shorter-lived alternatives. The extent to which such prospects may be evaluated numerically in the cost comparison and the extent to which they must be considered only as irreducible data will naturally depend on circumstances. This topic is explored further in Chapter 14 and thereafter. In this chapter and in the three that follow, wherever some specific assumption is necessary for the cost comparison, it will be *assumed that replacement assets will repeat the costs that have been forecast for the initial asset.*

Cash Flow Tabulations When Alternatives Have Different Lives

If cash flow for Plans D an E is to be compared for a 40-year period, some assumptions must be made regarding disbursements in the final 20 years of Plan D. In Table 6–3 it is assumed that the final 20 years will repeat the costs of the first 20. The table has been shortened by using a single line for each series of years in which cash flow is uniform.

The final column of the table shows that the initial extra outlay of $70,000 will ultimately be recovered plus an additional $50,000. Nevertheless, our annual cost comparison has indicated that this recovery of capital is too slow to yield 11% after income taxes and that, by the standard we have set, this extra $70,000 outlay for Plan E is undesirable.

Because 20 years is evenly divisible into 40 years, it was a simple matter to tabulate prospective cash flow for a period that would give the same number of years of service for the two plans. All that was required was our assumption regarding the disbursements during the second 20 years. In most comparisons of alternatives with different lives, the matter is more complicated

TABLE 6–3
Tabulation of comparative cash flow, Example 6–3

Years	*Plan D*	*Plan E*	*E–D*
0	−$50,000	−$120,000	−$70,000
1-19	−9,000 per year	−7,250 per year	+1,750 per year
20	{ −9,000 +10,000 −50,000	{ −7,250	{ +1,750 +40,000
21-39	−9,000 per year	−7,250 per year	+1,750 per year
40	{ −9,000 +10,000	{ −7,250 +20,000	{ +1,750 +10,000
Totals	−$440,000	−$390,000	+$50,000

than in Example 6–3; the total years tabulated must be the least common multiple of the estimated lives of the two alternatives. For example, if one alternative had a 10-year life and the other a 25-year life, it would be necessary to consider a 50-year period, with 5 life cycles for one alternative and 2 for the other. If lives were, say, 13 and 20 years, a 260-year period would have to be tabulated before reaching the point where the alternatives gave equal years of service. To avoid such complications it is possible to make specific estimates of what cash flows will occur if the project is terminated at some specific future time. This approach is explained fully in Chapter 20.

We shall see that a disparity in lives of alternatives creates the same difficulty in present worth comparisons and in calculation of rates of return that we now observe in the tabulations of comparative cash flow. Various methods of dealing with this difficulty are developed in subsequent chapters. Because no one method is completely satisfactory, the method selected is properly influenced by the circumstances of the economy study.

EXAMPLE 6–4

Comparing Alternatives That Have Perpetual Lives

Statement of Alternatives

In the design of an aqueduct that is assumed to have a perpetual period of service, two alternative locations are proposed for a certain section.

Location J involves a tunnel and flume. The tunnel is estimated to have a first cost of $200,000 and is assumed to be permanent. Its annual upkeep costs are estimated as $500. The flume will cost $90,000, has an estimated life of 20 years, and is expected to have annual maintenance costs of $2,000.

Location K involves a steel pipeline and several miles of concrete-lined earth canal. The pipeline has an estimated first cost of $70,000, an estimated life of 50 years, and an estimated annual maintenance cost of $700. The earth canal will cost $80,000 and is assumed to be permanent. During the first 5 years it is estimated that maintenance on the earth canal will be $5,000 a year; thereafter, it is estimated as $1,000 a year. The concrete lining will cost $40,000; it has an estimated life of 25 years with annual maintenance cost of $300.

All salvage values are assumed to be negligible. The stipulated i^* is 8%. Because this is a government project, no income tax differences are involved.

Annual Cost Comparison

Annual costs may be compared as follows:

Location J

Tunnel
  Interest = $200,000(0.08) = $16,000
  Maintenance 500
Flume
  CR = $90,000(*A*/*P*,8%,20) = $90,000(0.10185) = 9,167

Maintenance	2,000
Total equivalent uniform annual disbursements	$27,667

Location K

Pipeline	
CR = $70,000(*A*/*P*,8%,50) = $70,000(0.08174)	= $ 5,722
Maintenance	700
Earth canal	
Interest on first cost = $80,000(0.08)	= 6,400
Interest on PW of extra early maintenance = $4,000(*P*/*A*,8%,5)(0.08) = $4,000(3.993)(0.08)	= 1,278
Maintenance	1,000
Concrete lining	
CR = $40,000(*A*/*P*,8%,25) = $40,000(0.09368)	= 3,747
Maintenance	300
Total equivalent uniform annual disbursements	$19,147

Selection of Location K will result in a saving of $8,520 in annual costs.

Two new points arise in this solution. One deals with the annual cost associated with a perpetual life for a structure. Here interest (or return) on investment takes the place of capital recovery; as n approaches infinity, the capital recovery factor approaches the interest rate. If one adopts the borrowed money point of view, it is as if the $200,000 for the tunnel, for example, were borrowed under terms that permitted a perpetual debt with $16,000 interest paid every year.*

The extra maintenance of $4,000 a year for the first 5 years for the earth canal is a nonrecurring expenditure somewhat comparable to first cost. To translate this into an equivalent perpetual annual cost, it first must be converted into its present worth of $15,972 on zero date. Then, like the $80,000 investment in the canal itself, it is converted into an equivalent perpetual series by multiplying by the interest rate.

The Small Difference in Annual Cost Between Very Long Life and Perpetual Life

Forever is a long time! The estimator whose economy studies imply permanence of certain constructions does not really have the illusion that these projects will last forever; they are merely expected to last a very long time, possibly 100 years or more.

*Because of the common practice of bond refunding in the public utility industry, many business debts are, in effect, perpetual. This matter is discussed in Chapter 18. But if this were an aqueduct for a municipal water supply or for an irrigation district, there would nearly always be a public requirement that the debt be paid off within a specified number of years, possibly 40 or less. Such a requirement would not affect the validity of the annual cost calculations based on perpetual life. However, as brought out in Chapter 18, it would create an additional question of the actual total disbursements for debt service and other purposes with each of the proposed alternative locations.

In an economy study the difference between 100 years and forever is very small indeed. For example, at 8% interest the capital recovery factor for 100 years is 0.08004. That is, an increase of the interest rate from 8% to 8.004% would have the same effect on annual cost as reducing the estimated life of a structure from forever to 100 years. Even with the interest at 3%, the difference between 100 years and forever has the same influence on annual cost as a difference of 1/6 of 1% in the interest rate.

In many long-lived projects, economy studies are made and costs are computed as if the expected life were 50 years. This is common both in public works projects, such as federal river basin projects in the United States (see Chapter 19) and in private projects.

EXAMPLE 6–5

Comparing Alternatives in Which Annual Disbursements Have a Uniform Gradient

Statement of Alternatives

Many tractors of a particular type but of different ages are being used by a company engaged in large-scale farming operations. Although there have been no formal rules on replacement policy, the usual practice has been to replace tractors when they were about 10 years old. The first cost of a tractor is $18,000. Records have been kept of maintenance costs under conditions of fairly uniform use from year to year. There is clearly a marked upward tendency in maintenance costs as tractors get older, even though the maintenance costs at a given age differ from tractor to tractor. An analysis indicates that, on the average, maintenance costs will be $1,600 the first year, $2,000 the second, $2,400 the third, and will increase by $400 a year for each year of age.

It is desired to compare the equivalent annual costs for an average tractor assuming a 7-year life with such costs for an average tractor assuming a 10-year life. Estimated salvage value will be $6,000 for a 7-year-old tractor and $4,200 for a 10-year-old one. It is believed that costs other than maintenance and capital recovery costs can be disregarded in this comparison as these costs will be practically the same whether tractors are retired after 7 years or after 10 years. The stipulated i^* is 8%. Long-run differences in income taxes are assumed to be so small that they can be neglected in the analysis.

Annual Cost Comparison

The year-by-year maintenance costs may be treated as if they are made up of two parts. The first part is a uniform series of $1,600 a year. The second part is $0 the first year, $400 the second, $800 the third, and so on, increasing $400 each year. The second part constitutes a gradient series, which may be converted to an equivalent uniform annual series by the use of the appropriate gradient factor.

Tractor retired after 7 years

CR = \$18,000(A/P,8%,7) − \$6,000(A/F,8%,7)
= \$18,000(0.19207) − \$6,000(0.11207) = \$3,457 − \$672 = \$2,785
Equivalent uniform annual maintenance cost
= \$1,600 + \$400(A/G,8%,7)
= \$1,600 + \$400(2.694) = \$1,600 + \$1,078 = 2,678
Total of equivalent annual net disbursements compared \$5,463

Tractor retired after 10 years

CR = \$18,000(A/P,8%,10) − \$4,200(A/F,8%,10)
= \$18,000(0.14903) − \$4,200(0.06903) = \$2,683 − \$290 = \$2,393
Equivalent uniform annual maintenance cost
= \$1,600 + \$400(A/G,8%,10)
= \$1,600 + \$400(3.871) = \$1,600 + \$1,548 = 3,148
Total of equivalent annual net disbursements compared \$5,541

The equivalent annual costs over the life of the 7-year tractor will be somewhat lower than those over the life of the 10-year one.

Comments on Example 6–5

Where it seems reasonable to estimate that disbursements (or receipts) will increase or decrease by a uniform amount each year, the gradient factor is helpful in computing an equivalent uniform annual figure. It often happens that the estimate of a uniform gradient is as good an estimate as it is practicable to make in cases where upward or downward trends are expected.

Where disbursements are expected to vary irregularly from year to year and the figures for each year are predicted for use in the economy study, the gradient approach is not suitable; it is necessary first to convert all disbursements to present worth before converting to equivalent uniform annual cost. This type of calculation is illustrated in the next chapter.

Although Example 6–5 introduces the subject of the time at which it is economical to replace an asset that is physically capable of being continued in service, this example omits consideration of a number of matters that are important in replacement economy. A more adequate consideration of this interesting and important subject is deferred until Chapter 17 and thereafter.

The Usefulness of Calculations of Equivalent Uniform Annual Cash Flow is not Limited to Comparisons of Alternatives Involving Different Series of Disbursements

Although the examples given in this chapter have all dealt with comparisons of alternatives of the "Why this way?" type, a number of other uses of equivalent uniform annual figures are illustrated in later chapters. Where proposals involve differences in estimated annual receipts, comparisons may be made of equivalent uniform annual net positive cash flow (rather than of equivalent annual net negative cash flow as in the examples in this chapter). Where for some reason, such as public regulation of certain prices or agreement between the parties concerned, the pricing of a product or service is to be based on an equivalent uniform figure, the purpose of such a calculation may be to find a justified selling price. In economy studies for public works, as introduced in Chapter 9 and discussed at greater length in Chapter 19, equivalent uniform annual costs may be computed for comparison with equivalent uniform annual benefits.

Omission of Discussion of Irreducibles in This Chapter

Economy studies are generally undertaken to arrive at a decision on action or at a recommendation for action. Insofar as possible, it is helpful to reduce alternative courses of action to money terms in order to have a common unit to measure the differences between alternatives. Nevertheless, as pointed out in Part I, it often happens that important matters for consideration simply cannot be reduced to terms of money in any satisfactory way. Such irreducibles should be considered along with the money figures in arriving at any decision among alternatives. In those borderline cases where the money comparisons are close, the irreducibles are likely to control the decision.

In this chapter and the three that follow, we deal chiefly with the calculations necessary to reflect the time value of money in comparing alternatives that involve different estimated cash flows. It is desirable that persons responsible for decisions on matters of economy understand the principles involved in such interest conversions. These are definite principles that exist regardless of the irreducibles entering into any particular situation. Experience shows that they frequently are not clearly understood by engineers and other persons who need to understand them. In order to permit concentration of attention on those principles, the subject matter has been deliberately de-emotionalized in these chapters by references to Plans A and B, Locations J and K, and so forth, and by the omission of irreducibles as far as possible. A more complete discussion of economy studies in a realistic setting involving irreducibles is deferred until Part III of this book.

Summary

One way of reflecting in economy studies the desirability of recovering invested capital with a return is to compare alternatives on the basis of equivalent uniform annual net disbursements, using as an interest rate the minimum attractive rate of return. Because many of the estimated disbursements are the same year after year, such annual cost comparisons are likely to be convenient. The basic data for such comparisons consist of estimated cash flows associated with the alternatives being compared. Conversions into equivalent uniform annual figures require the use of appropriate factors obtained from compound interest tables or formulas.

PROBLEMS

6–1. Compare the equivalent annual costs of perpetual service for the following two plans for a government project using an i^* of 9%:

Plan I involves an initial investment of $150,000. Of this, $75,000 is for land (assumed to be permanent) and $75,000 is for a structure that will require renewal, without salvage value, at an estimated cost of $75,000 every 30 years. Annual disbursements will be $10,000 for the first 10 years and $7,000 thereafter.

Plan II involves an initial investment of $250,000. Of this, $100,000 is for land and $150,000 is for a structure that will require renewal, with a $30,000 salvage value, every 50 years. Assume that the net outlay for each renewal is $120,000. Annual disbursements will be $4,000. (*Ans*. Plan I, $22,784; Plan II, $26,648)

6–2. A university has pumped its water supply from wells located on the campus. The falling water table has caused pumping costs to increase greatly, the quantity of available water to decrease, and the quality of the water to deteriorate. A public water company has now built a large main carrying water of a satisfactory quality to a point within 3 miles of the university's present pumping station. The decision has been made to build a pipeline connecting to the water company's main and to purchase water. Two alternative types of pipe are considered to supply the needs for a 60-year period, with estimates as follows:

	Type A	Type B
Initial investment in pipe	$120,000	$80,000
Estimated life of pipe	60 years	30 years
Initial investment in pumping equipment	$15,000	$20,000
Estimated life of pumping equipment	20 years	20 years
Annual energy cost for pumping in first year	$3,000	$4,000
Yearly increase in energy cost for pumping (each year for 60 years)	$60	$80

Using a 60-year study period, compare the equivalent uniform annual costs that will be influenced by the choice of the type of pipe. Use an i^* of 6%. (Because the university is a nonprofit organization, no income taxes are involved and 6% is the average rate of return on its endowment funds.) Assume zero net terminal salvage values for pipe and pumping equipment and assume that renewal costs during the 60-year period will be the same as the initial investment. (*Ans*. Type A, $12,621; Type B, $12,739)

6–3. A school district considers two alternative plan for an athletic stadium. An engineer makes the following cost estimates for each:

Concrete Bleachers. First cost, $350,000. Life, 90 years. Annual upkeep cost, $2,500.
Wooden Bleachers on Earth Fill. First cost of entire project, $200,000. Painting cost every 3 years, $10,000. New seats every 15 years, $40,000. New bleachers every 30 years, $100,000. Earth fill, which accounts for the $50,000 balance of first cost, will last for the entire 90-year period.

Compare equivalent uniform annual costs for a 90-year period using an i^* of 7%. (*Ans*. Concrete $26,556; Wood, $19,769)

6–4. Two types of heat exchanger are to be compared for service in a chemical plant. Type Y has a first cost of $8,400, an estimated life of 6 years with zero salvage value, and annual operating costs of $1,700. Type Z has a first cost of $10,800, an estimated life of 9 years with zero salvage value, and annual operating costs of $1,500. Estimated extra annual income taxes with Type Z are $200.

Compare equivalent uniform annual costs using an after-tax i^* of 16%. (*Ans*. Type Y, $3,980; Type Z, $4,044)

6–5. Annual disbursements (other than driver's wages) for operation and maintenance of certain trucks under particular operating conditions tend to increase by $400 a year for the first 5 years of operation; first-year disbursements are $2,400. The first cost of a truck is $8,400. The estimated salvage value after 4 years is $2,400; after 5 years it is $1,500. In the long run, it is estimated income taxes will be about the same with a 4-year of 5-year life. Using an i^* of 10%, compare the equivalent uniform annual costs of a truck held for 4 years with one held for 5 years. (*Ans*. 4 years, $5,085; 5 years, $5,094)

6–6. Machine J has a first cost of $50,000, an estimated service period of 12 years, and an estimated salvage value of $14,000 at the end of the 12 years. Estimated annual disbursements for operation and maintenance are $6,000 for the first year, $6,300 the second year, and will increase $300 each year thereafter. An alternate is Machine K, which has a first cost of $30,000 and an estimated zero salvage value at the end of the 12-year service period. Estimated annual disbursements for operation and maintenance are $8,000 for the first year, $8,500 the second year, and will increase $500 each year thereafter. Estimated extra income taxes with Machine J are $750 the first year, $850 the second, and will increase $100 each year thereafter. Using an after-tax i^* of 12%, compare the equivalent uniform annual costs of a 12-year service from Machines J and K. (*Ans*. Machine J, $15,918; Machine K, $14,938)

6–7. A manufacturer proposes to build a new warehouse. A reinforced concrete building will cost $116,000, whereas the same amount of space can be secured in a frame and galvanized metal building for $60,000. The life of the concrete building is estimated as 50 years; average annual maintenance cost is estimated as $1,000. The life of the frame building is estimated to be 25 years; average annual maintenance cost is estimated as $1,800. Fire insurance will be carried on the building and its contents in either case; the annual rate will be \$1.50 per \$1,000 of insurance for the concrete building and \$4.00 per \$1,000 of insurance for the frame building. Assume that the average amount of insurance will be on contents of \$400,000 plus 75% of first cost of the building. Average annual property taxes are estimated at 1.5% of the first cost. The deductible expenses for income taxes for the frame building will exceed those for the concrete building, resulting in the payment of extra income taxes of \$545 per year if the concrete building is chosen. Find the comparative equivalent uniform annual costs for the two types of warehouse using an after-tax i^* of 9%. Neglect any possible salvage values at the ends of the lives of the buildings. (*Ans*. Concrete, \$14,598; Frame, \$10,589)

6–8. Make a comparison of annual costs before income taxes for the two proposed warehouses in Problem 6–7, using a before-tax i^* of 18%. (*Ans*. Concrete, \$24,356; Frame, \$15,455)

6–9. A Taxi Cab Company has been using two different automobile models, H and K, for its fleet. It has kept good records on fuel mileage, maintenance, repairs, and so on, but has never performed an economy study on the different models. The Company now needs to buy 10 new cabs and has employed you to make an engineering economy study. The drivers prefer Model H, but the manager feels Model K is more economical. You find that the primary differences are in fuel mileage, maintenance, and lost time due to repairs. Such costs as license fees, registration, drivers' wages, and insurance are the same for either model.

Model H has a first cost of \$9,000. Average operation and maintenance costs have been \$2,200 the first year, \$2,500 the second, and \$2,800 the third. This model is usually retired at a salvage value of \$2,500 after 3 years' service because it requires a substantial overhaul after that time.

Model K has a first cost of \$11,000 and a salvage value of \$1,500 after 4 years' service. Annual operating and maintenance costs have been averaging \$1,700 the first year increasing by \$300 each year during the second through fourth years.

Find the equivalent uniform annual disbursements for each model using an i^* before income taxes of 20%. Which model is preferable?

6–10. Compare the equivalent uniform annual costs of service for two emergency power plants. Unit A has a first cost of \$40,000, an expected life of 8 years, and no net salvage value. Annual disbursements for operation and maintenance are expected to be \$1,800.

Unit B has a first cost of \$50,000, and expected 10 year life, and an estimated net salvage value of \$5,000. Annual operation and maintenance is expected to be \$1,200. If Unit B is chosen, an additional annual income tax payment of \$550 will be required. Use an i^* of 14% after income taxes.

6–11. Compare the equivalent uniform annual costs of the two units in Problem 6–10 on a before-income-tax basis using an i^* of 25%.

6–12. Compare the equivalent uniform annual costs of two materials handling systems, E and F. Use an i^* of 15% after income taxes.

	Unit E	Unit F
First cost	$22,000	$36,000
Life in years	6	10
Salvage value	$1,000	$3,000
Annual O & M costs	$8,400	$4,900
Extra annual income tax		$1,850

6–13. Solve Problem 6–12 using an i^* of 30% before income taxes.

6–14. A certain automatic testing operation can be performed by either Unit X or Unit Y. Unit X has a first cost of $16,000 and a $2,000 salvage value. Annual disbursements are expected to be $9,000 in year 1, $9,030 in year 2, and to increase by $30 each year thereafter. Unit Y has a first cost of $24,000 and a salvage value of $3,000. Annual disbursements are expected to be a uniform $7,000 each year. If Unit Y is chosen, additional annual income taxes will be $650 in year 1, $665 in year 2, and will increase by $15 each year thereafter. Each unit has an estimated life of 10 years. Find the comparative equivalent uniform annual costs for each unit using an i^* of 20% after income taxes.

6–15. Pages 79 and 80 give four formulas for calculating capital recovery (CR) cost. Using the mathematical formulas for the factors, show that each is equal to $P(A/P,i\%,n) - S(A/F,i\%,n)$, the first formula given.

6–16. A company has a choice between the two compressors, A and B, for installation in its plant. Compare the equivalent uniform annual costs for 12 years service using an i^* of 13% after income taxes.

	Type A	Type B
First cost	$6,000	$7,800
Salvage value after 12 years	$600	$600
Annual operation and repair costs	$1,800	$1,500
Extra annual income taxes		$75

6–17. Two types of curing furnace are being considered for a certain operation. Type P has a first cost of $50,000 and no net salvage value at the end of its 12-year expected life. Operating costs, except labor, are expected to be $1,600 the first year and $1,750 the second, and to increase by $150 each year thereafter. Maintenance costs are expected to be $800 each year. The furnace will have to be relined at an additional cost of $6,000 every 4 years. Furnace type Q has a first cost of $65,000 and a net salvage value of $1,000 at the end of its 20-year expected life. Operating costs, except labor, are expected to be $800 the first year and $850 the second, and to increase by $50 each year thereafter. Maintenance costs are expected to be $400 each year. This furnace will have to be relined only once at an additional cost of $6,000 at the end of 10 years. Labor costs are expected to be the same in either case.

Compare the equivalent uniform annual costs of the two furnaces using an i^* of 20% before income taxes.

6–18. The company in Problem 6–17 suspects that the curing operation will be needed for only the next 10 to 12 years. After that time, a new method of manufacture and new materials will eliminate that step. Compare the equivalent uniform annual costs of the two furnaces assuming a 12-year study period. The salvage value of type Q will remain $1,000.

6–19. A city is considering the development of a recreational complex. Two plans have been proposed, one of which is described as a "low-maintenance" facility because of the use of heavier, more durable and weather-resistant materials. Costs for each plan are estimated as follows:

	Plan I	**Plan II**
Land development (permanent)	$65,000	$65,000
Buildings	$125,000	$175,000
Recreational facilities	$200,000	$250,000
Annual maintenance	$12,000	$4,000

Under Plan I, buildings will have to be refurbished at a cost of $50,000 every 15 years and facilities will have to be refurbished at a cost of $100,000 every 10 years. Under Plan II, buildings will be refurbished every 25 years and facilities every 20 years at a cost equal to the original cost of construction.

Compare the equivalent uniform annual cost of perpetual service assuming that the periodic refurbishing make the complex "as good as new." The city can borrow adequate funds for either plan at 9% interest.

6–20. The economy study in Problem 6–19 assumes that the refurbishing of buildings and facilities restores them to new condition. Assume that the facilities will have to be replaced every 40 years and that buildings will have to be replaced every 75 years. These replacements will be at the same cost as that when originally constructed. Naturally, the refurbishments at the dates of replacement will not take place. Compare the equivalent uniform annual cost of perpetual service for the two plans.

7

PRESENT WORTH

*It would be difficult to exaggerate the economic and social importance of engineering economy. The innumerable decisions that are made each day in this field in private industry determine whether proposals for investment in new plant and equipment are accepted or rejected. These decisions have far-reaching effects on our national standard of living.—Paul T. Norton**

Two uses of present worth calculations in engineering economy are explained and illustrated in this chapter, namely:

1. Comparison of alternative series of estimated money receipts and disbursements.
2. Placing a valuation on prospective net money receipts.

A third important use of present worth (PW), discussed in Chapter 8, is for trial-and-error calculations to determine unknown rates of interest or return.

Because calculation of present worth is often called *discounting*, writers on economics often refer to an interest rate used in present worth calculations as a *discount rate*.

Use of Present Worth to Compare Plans A, B, and C of Examples 6–1 and 6–2

It will be recalled that Plans A, B, and C involved, respectively, annual disbursements of $9,200, $6,400, and $5,625 for a 10-year period. Plan A had no first cost; Plan B had $15,000 first cost and zero salvage value; Plan C had $25,000 first cost and $5,000 salvage value. The stipulated i^* was 9%. The data for the following calculations consist of these estimates of cash flow for the three plans; no use is made of the annual costs that were calculated in Chapter 6.

*Paul T. Norton, Sec. 3, *Handbook of Industrial Engineering and Management*, 2d ed., Engineering Economy, W. G. Ireson, and E. L. Grant, eds. (Englewood Cliffs, N.J.: Prentice-Hall, Inc., copyright 1971); p. 124.

Plan A

PW of annual disbursements	
= $9,200(*P/A*,9%,10)	
= $9,200(6.418)	= $59,050

Plan B

PW of annual disbursements	
= $6,400(*P/A*,9%,10)	
= $6,400(6.418)	= $41,080
First cost	15,000
PW of all disbursements for 10 years	$56,080

Plan C

PW of annual disbursements	
= $5,625(*P/A*,9%,10)	
= $5,625(6.418)	= $36,100
First cost	25,000
PW of all moneys paid out for 10 years	$61,100
Less	
PW of salvage value	
= $5,000(*P/F*,9%,10)	
= $5,000(0.4224)	= 2,110
PW of net disbursements for 10 years	$58,990

Present worths are calculated as of the zero date of the series of payments being compared. Because first costs are already at zero date, no interest factors need to be applied to first cost. Where an estimated salvage value occurs, as in Plan C, the present worths of the salvage value must be subtracted to obtain the present worth of the net disbursements.

Simplicity of Conversion from Present Worth to Annual Cost and Vice Versa

In Chapter 6 the equivalent uniform annual costs for these three plans were calculated directly from the estimated cash flows. An alternate way to find annual costs would be to calculate them from the present worths. Each present worth can be converted into an equivalent uniform annual series by multiplying it by 0.15582, the capital recovery factor (*A/P*, 9%,10).

Annual cost, Plan A	= $59,050(0.15582) = $9,201
Annual cost, Plan B	= $56,080(0.15582) = $8,738
Annual cost, Plan C	= $58,990(0.15582) = $9,192

These figures, of course, are substantially identical with the annual cost figures obtained in Chapter 6. (There is a $1 difference in each plan due to loss of significant figures by rounding.) In a similar manner an alternate way to have computed the present worths for the three plans would have been to multiply each annual cost calculated in Chapter 6 by 6.418, the series present worth factor (*P*/*A*,9%, 10).

The general statement may be made that annual cost can be calculated from present worth by multiplying present worth by the appropriate capital recovery factor; present worth can be calculated from annual cost by multiplying annual cost by the appropriate series present worth factor (or dividing by the capital recovery factor). The alternative that is favored in an annual cost comparison is also favored in a present worth comparison, and by the same proportion. For example, the annual cost of Plan B is 5.0% below the annual cost of Plan C; the present worth of B is also 5.0% below that of C. Of course this convertibility between annual cost and present worth depends on the use of the same interest rate in both calculations and on the use of identical cash flow estimates for the same period of years.

Where alternatives involve irregular series of payments, the first step in computing annual costs should be to find the present worths.

EXAMPLE 7–1

Comparison of Alternatives Involving Irregular Series of Disbursements

Statement of Alternatives

Engineers for a public utility company have proposed two alternate plans to provide a certain service for the next 15 years. Each plan includes sufficient facilities to take care of the expected growth of the demand for the particular utility service during this period.

Plan F calls for a three-stage program of investment in facilities; $60,000 will be invested at once, $50,000 more after 5 years, and $40,000 more after 10 years. Plan G, a two-stage program, calls for a $90,000 immediate investment followed by a $30,000 investment at the end of 8 years. In both plans estimated annual income taxes are 3% of the investment that has been made up to date and estimated annual property taxes are 2% of the investment to date. Annual maintenance costs in Plan F are estimated as $1,500 for the first 5 years, $2,500 for the second 5 years, and $3,500 for the final 5 years. For Plan G, annual maintenance costs are estimated as $2,000 for the first 8 years and $3,000 for the final 7 years. Salvage value at the end of 15 years is estimated to be $45,000 for Plan F and $35,000 for Plan G.

Table 7–1 presents the foregoing estimates as a tabulation of cash flows. For example, the −$4,500 cash flow for Plan F in years 1 through 5 is made up of $1,500 estimated annual maintenance cost plus the estimated tax percentages, 3% income tax plus 2% property tax, times the investment up to that point, $60,000.

For purposes of the economy study, a minimum attractive rate of return of 7% is to be used.

TABLE 7-1
Tabulation of cash flow, Example 7-1

Year	*Plan F*	*Plan G*	*G–F*
0	−$60,000	−$90,000	−$30,000
1	−4,500	−6,500	−2,000
2	−4,500	−6,500	−2,000
3	−4,500	−6,500	−2,000
4	−4,500	−6,500	−2,000
5	−4,500	−6,500	−2,000
	−50,000		+50,000
6	−8,000	−6,500	+1,500
7	−8,000	−6,500	+1,500
8	−8,000	−6,500	+1,500
		−30,000	−30,000
9	−8,000	−9,000	−1,000
10	−8,000	−9,000	−1,000
	−40,000		+40,000
11	−11,000	−9,000	+2,000
12	−11,000	−9,000	+2,000
13	−11,000	−9,000	+2,000
14	−11,000	−9,000	+2,000
15	−11,000	−9,000	+2,000
	+45,000	+35,000	−10,000
Totals	−$222,500	−$200,000	+$22,500

Comparison of Present Worths

The present worths of the respective net disbursements for 15 years may be computed as follows:

Plan F

Initial investment	$ 60,000
PW of investment made after 5 years	
= $50,000(*P/F*,7%,5)	
= $50,000(0.7130)	= 35,650
PW of investment made after 10 years	
= $40,000(*P/F*,7%,10)	
= $40,000(0.5083)	= 20,330
PW of annual disbursements, years 1 to 5	
= $4,500(*P/A*,7%,5)	
= $4,500(4.100)	= 18,450
PW of annual disbursements, years 6 to 10	
= $8,000(*P/A*,7%,10 minus *P/A*,7%,5)	
= $8,000(7.024 − 4.100)	
= $8,000(2.924)	= 23,390
PW of annual disbursements, years 11 to 15	
= $11,000(*P/A*,7%, 15 minus *P/A*,7%,10)	

= \$11,000(9.108 − 7.024)	
= \$11,000(2.084)	= 22,920
PW of all moneys paid out for 15 years	\$180,740
Less	
PW of salvage value	
= \$45,000(*P*/*F*,7%,15)	
= \$45,000(0.3624)	= 16,310
PW of net disbursements for 15 years	\$164,430

Plan G

Initial investment	\$ 90,000
PW of investment made after 8 years	
= \$30,000(*P*/*F*,7%,8)	
= \$30,000(0.5820)	= 17,460
PW of annual disbursements, years 1 to 8	
= \$6,500(*P*/*A*,7%,8)	
= \$6,500(5.971)	= 38,810
PW of annual disbursements, years 9 to 15	
= \$9,000(*P*/*A*,7%,15 minus *P*/*A*,7%,8)	
= \$9,000(9.108 − 5.971)	
= \$9,000(3.137)	= 28,230
PW of all moneys paid out for 15 years	\$174,500
Less	
PW of salvage value	
= \$35,000(*P*/*F*,7%,15)	
= \$35,000(0.3624)	= 12,680
PW of net disbursements for 15 years	\$161,820

Comparison of Annual Costs

To compute equivalent uniform annual costs for the 15-year period, the respective present worths must be multiplied by the capital recovery factor (*A*/*P*,9%,10).

Plan F

Annual cost = \$164,430(0.10979) = \$18,050

Plan G

Annual cost = \$161,820(0.10979) = \$17,770

Which Plan Should Be Selected?

The present worth (and annual cost) of Plan G is only slightly less than that of Plan F. Here is a case where a small change in the basic estimates for the two plans (either in amount or timing of cash flows or in the assumed minimum attractive rate of return) could have altered the present worths enough to shift the balance in favor of Plan F. For example, at the time this study was made, an i^* of 7% was considered appropriate by the utility management. More recent money market conditions would suggest the use of a higher rate, which would favor the deferred investment.

Generally speaking, when a comparison is as close as this one, the decision should be made on the basis of any differences between the plans that have not been reduced to money estimates—on the so-called irreducible data entering into the problem of choice. Further comment is made in Chapter 8 and Appendix B regarding certain other aspects of this example.

Present Worth Comparisons When Alternatives Have Different Lives

There is no point in converting two or more alternative cash flow series into present worth and then comparing the present worths unless the cash flow series relate to a provision of a needed service for the same number of years.

Some of the difficulties arising in comparing alternatives with different lives were discussed in Chapter 6. In our discussion there, it was pointed out that a convenient simple assumption is that replacement assets will repeat the costs that have been forecast for the initial asset. In the present chapter we shall continue to make this assumption; a more critical look at the matter will be deferred until Chapter 14 and thereafter. Sometimes it is desirable to choose an arbitrary analysis period that is shorter than the expected service period, with valuations assigned to all assets at the end of the analysis period; this technique is illustrated in Example 9–1 and discussed further in Chapter 14.

If the assumption is made that costs will be repeated for replacement assets, a study period can be selected that is the least common multiple of the lives of the various assets involved. Or in some cases a present worth study may be for a perpetual period. Usually present worths for an assumed perpetual period of service are referred to as *capitalized costs*.

Illustration of a Comparison of Present Worths for a Least Common Multiple of the Lives of the Alternatives

It will be recalled that in Example 6–3, Plan D had $50,000 first cost, 20-year life, $10,000 salvage value, and $9,000 annual disbursements; Plan E had $120,000 first cost, 40-year life, $20,000 salvage value, and $7,250 annual disbursements. Table 6–3 and Figure 6–1 show the estimated cash flows for a 40-year period. The interest rate was 11%. A comparison of the present worths of a 40-year service, computed directly from the cash flow data, is as follows:

Plan D

First cost		$50,000
PW of net disbursements for renewal in 20 years		
= ($50,000 − $10,000)($P/F$,11%,20)		
= $40,000 (0.1240)	=	4,960

PW of annual disbursements	
= $9,000($P/A$,11%,40) = $9,000(8.951)	= 80,560
Total PW of disbursements	$135,520
Less	
PW of receipt from final salvage value after 40 years	
= $10,000($P/F$,11%,40) = $10,000(0.0154)	= $ 150
PW of net disbursements for 40 years	$135,370

Plan E

First cost	$120,000
PW of annual disbursements	
= $7,250($P/A$,11%,40) = $7,250(8.951)	= 64,890
Total PW of disbursements	$184,890
Less	
PW of receipt from salvage value after 40 years	
= $20,000($P/F$,11%,40) = $20,000(0.0154)	= 310
PW of net disbursements for 40 years	$184,580

It will be noted that in Plan D, the $10,000 receipt from salvage of the first asset was subtracted from the $50,000 investment in the renewal asset in order to find the net disbursement required after 20 years; the $40,000 figure thus obtained was then multiplied by the 20-year present worth factor (P/F). However, it was necessary to make a separate calculation of the present worth of the receipt from the final salvage value after 40 years and to subtract this present worth from the present worth of all disbursements.

In Chapter 14, we shall discuss the subject of the sensitivity of the conclusions of an economy study to moderate changes in the estimates. It is evident from the present worth comparison of Plans D and E that the estimates of salvage values after 40 years have very little influence on the choice between the plans; a dollar 40 years hence is equivalent to only 1.54 cents today when interest is at 11%. In general, it may be stated that comparisons of economy are not sensitive to changes in estimated distant salvage values unless the interest rate used is very low.

Moreover, in comparing Plans D and E, the choice is relatively insensitive to the estimated cost of the renewal asset in Plan D. For example, if the estimated net disbursement for the renewal asset 20 years hence should be doubled, the total present worth for Plan D would be increased by only $4,960; Plan D would still be considerably more economical than Plan E.

The present worths for Plans D and E may be converted into annual costs by multiplying them by the capital recovery factor (A/P,11%,40).

Annual cost, Plan D = $135,370(0.11172) = $15,124
Annual cost, Plan E = $184,580(0.11172) = $20,621

These annual cost figures check within one dollar with the ones that were computed directly from the cash flow series in Example 6–3.

Capitalized Cost

The calculation of the present worth of perpetual service may be illustrated by comparing Plans D and E, as follows:

Plan D

First cost	$ 50,000
PW of infinite series of renewals	
= ($50,000 − $10,000)(*A/F*,11%,20) ÷ 0.11	
= $40,000(0.01558) ÷ 0.11	= 5,665
PW of perpetual annual disbursements	
= $9,000 ÷ 0.11	= 81,818
Total capitalized cost	$137,483

Plan E

First cost	$120,000
PW of infinite series of renewals	
= ($120,000 − $20,000)(*A/F*,11%,40) ÷ 0.11	
= $100,000(0.00172) ÷ 0.11	= 1,564
PW of perpetual annual disbursements	
= $7,250 ÷ 0.11	65,909
Total capitalized cost	$187,473

As illustrated in Figure 7–1, the calculation of the present worth of an infinite series of renewals starts with the conversion of the periodic renewal cost into an equivalent perpetual uniform annual series. Thus, $40,000 at the end of any 20-year period is multiplied by the 20-year sinking fund factor (*A/F*) to convert it to $623, a uniform annual figure throughout the 20-year period. Therefore, $40,000 at the end of every 20th year is equivalent to $623 a year forever. The present worth of this infinite series in Plan D is then $623 ÷ 0.11 = $5,665.

It will be noted that the figures for the present worth of perpetual service in Plans D and E are only slightly greater than the previous figures for the present worths of 40 years' service. The $49,990 advantage for Plan D in capitalized cost is not much more than the $49,210 advantage in present worth of 40 years' service. Considered from the viewpoint of present worth at 11%, the difference between 40 years and forever is small.

Capitalized costs are simply annual costs divided by the interest rate. The annual cost in Plan D was $15,123; the capitalized cost is $15,123 ÷ 0.11 = $137,482.

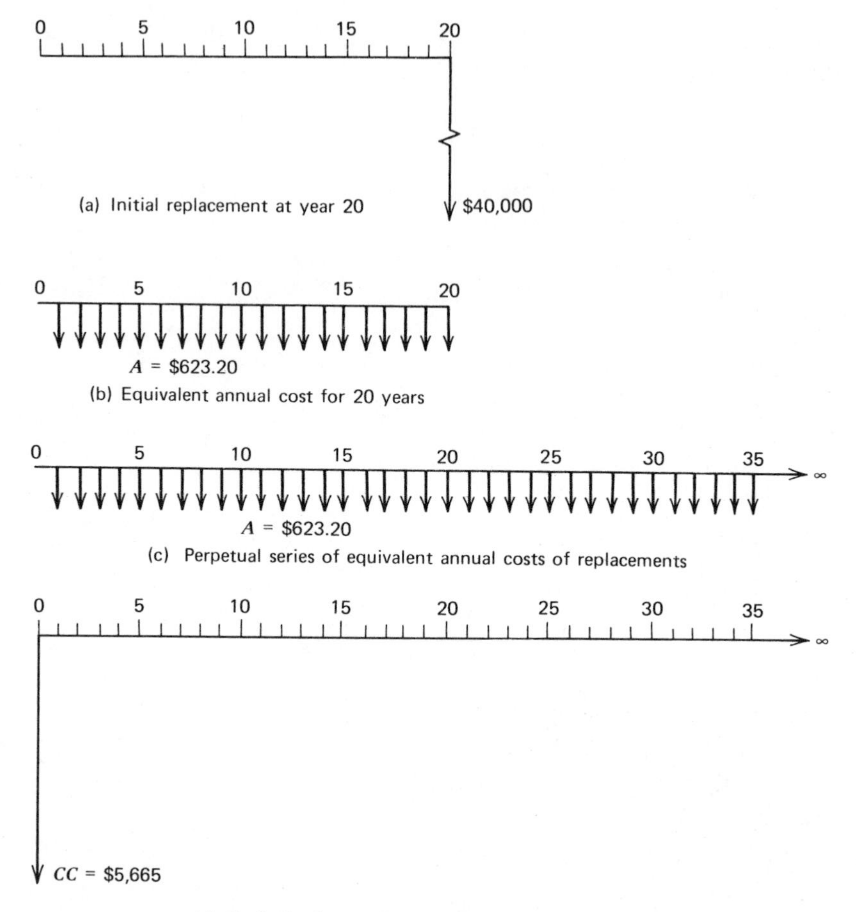

FIGURE 7–1. Equivalence conversion of perpetual series of replacements, Plan D

Effects of Anticipated Growth of Demand on the Choice Between an Immediate and a Deferred Investment

An engineer is often concerned with trying to get something done that must be completed tomorrow at two o'clock or that has some similar urgency concerned with it. In such circumstances it is sometimes better to make a second-rate decision immediately than to make a first-rate one at some later time. Nevertheless, decisions made only with a view to meeting immediate emergencies are likely to prove costly in the long run.

Piecemeal construction, in which each piece of apparatus is installed to meet a present emergency without regard to its adequacy under prospective future conditions, will ordinarily be less economical than development of a long-range planned expansion. Where planning is neglected, mistakes are likely to be made that will cost a great deal more to correct than they would have cost to avoid. A forecast of growth is an essential part of engineering designs.

In a program of planned development the installation of extra capacity as part of the original construction is likely to require less money outlay than will be required to add this capacity when needed in the future. This saving in money outlay may be due to the inherently lower costs per unit of capacity that are often associated with larger units. It may be due to the fact that future changes involve expenses that are avoidable when excess capacity is provided initially. As an example, each addition to the capacity of any underground conduit in a city's streets requires a ditch that must be refilled and repaved.

This necessity of providing excess capacity against an expected growth of demand exists whenever that growth is reasonably certain to occur and where its rate may be predicted with some degree of confidence in the forecast. This is usually the situation in the engineering of most public utility equipment. Example 7–2 deals with the calculations that are appropriate once forecasts of growth have been made.

EXAMPLE 7–2

Comparing an Immediate with a Deferred Investment

A Preliminary Solution Considering Investments Only

Cost comparisons to determine whether a proposed investment in capacity that is in excess of present needs is economically justifiable are usually made on a present worth basis. The degree of complexity of the calculations will depend on the assumptions made by the estimator. An example of an economy study in which these assumptions are very simple is as follows:

In the design of an aqueduct for municipal water supply, a tunnel is necessary. It is estimated that a tunnel built to half the ultimate capacity of the aqueduct will be adequate for 20 years. However, because of certain fixed elements in the cost of tunnel construction, it is estimated that a full-capacity tunnel can be built now for $300,000 as compared with $200,000 for a half-capacity tunnel. The problem is whether to build the full-capacity tunnel now, or to build a half-capacity tunnel now—supplementing it by a parallel half-capacity tunnel when needed.

At first glance it appears as if the disbursements to be compared are as follows:

Full capacity now	**Half capacity now**
$300,000 now	$200,000 now
	$200,000 20 years hence

Since the $200,000 present investment necessary in either case can be canceled out as irrelevant, this appears to be the question of whether it is better to spend $100,000 now or $200,000 in 20 years. If interest is taken at 8%, the present worth of $200,000 20 years hence is $200,000(0.2145) = $42,900. This indicates an advantage of $57,100 in present worth for the half-capacity plan.

This solution to the problem implies "all other things being equal." Two matters that might not be equal are the expected service lives with the two plans, and their respective operation and maintenance costs.

A Solution Considering Capitalized Operation and Maintenance Costs
In these circumstances it might be reasonable to assume the expected service lives as perpetual for both plans; if this assumption is made, the two plans do not differ in expected service life. However, there does appear to be a prospective difference in operation and maintenance costs. (In studies of this character there is sometimes a tendency to neglect the possibility of such differences; this tendency should be resisted by an examination of the circumstances to see whether differences are likely to occur.)

Let us assume that in this situation the estimator notes that the two half-capacity tunnels will involve a larger area of tunnel lining with correspondingly greater periodic costs of lining repairs. Lining repair cost for the full-capacity tunnel is estimated at $10,000 every 10 years; for each half-capacity tunnel it is estimated as $8,000 every 10 years. The estimator also notes that friction losses will be somewhat greater in the half-capacity tunnel; it is estimated that this will increase pumping costs in the aqueduct line by $1,000 a year so long as a single tunnel is in use, and by $2,000 a year after the second tunnel has come into use. It now appears as if the disbursements to be compared for perpetual service with the two plans are as follows:

Full capacity now	**Half capacity now**
$300,000 now	$200,000 now
10,000 10 years hence and every 10th year thereafter	8,000 10 years hence and every 10th year thereafter
	1,000 a year forever
	200,000 in 20 years
	8,000 30 years hence and every 10th year thereafter
	1,000 a year, starting 20 years hence

With interest at 8%, a capitalized cost comparison of these alternatives is as follows:

Full capacity now

Investment	$300,000
Lining repairs $10,000 $\left(\dfrac{0.06903}{0.08}\right)$ =	8,629
Total capitalized cost	$308,629

Half capacity now

First tunnel:	
Investment	$200,000
Lining repairs \$8,000 $\left(\frac{0.06903}{0.08}\right)$	6,903
Extra pumping costs $\left(\frac{\$1,000}{0.08}\right)$	12,500
Second tunnel:	
Investment \$200,000(0.2145)	42,900
Lining repairs \$6,903(0.2145)	1,481
Extra pumping costs \$12,500(0.2145)	2,681
Total capitalized costs	$266,465

The recognition of the higher operation and maintenance costs associated with the half-capacity plan therefore does not shift the advantage to the full-capacity plan. The rate of return on the extra $100,000 investment in the full-capacity plan is less than the 8% interest rate assumed in this calculation; that is, a somewhat lower interest rate would result in the two plans having the same capitalized cost.

Annual Cost Versus Present Worth for Comparing Alternative Series of Disbursements

Historically, present worth methods have been advocated by a number of writers on engineering economy.* Capitalized costs were widely used for many years, particularly by civil engineers. The widespread use of capitalized costs probably had its origin in Wellington's classic work *The Economic Theory of Railway Location* (1887). This—in a day in which most engineers worked for railways during at least part of their careers—influenced the thinking of the entire engineering profession. Wellington—considering that many elements of the railway had perpetual life—would divide an expected saving by the interest rate to determine the justifiable increase in first cost to bring about that estimated saving.

For most economy studies comparing mutually exclusive design alternatives, the authors of this book prefer comparisons of equivalent uniform annual costs to comparisons of present worths. The most important advantage is that, generally speaking, people seem to understand annual costs better than they understand present worths. A relatively minor advantage is that annual costs are usually somewhat easier to compute except in circumstances such as Examples 7–1 and 7–2, where irregular series of disbursements are involved.

*Goldman's *Financial Engineering* (New York: John Wiley & Sons, Inc., 1921) developed the subject of engineering economy through capitalized cost comparisons; the term coined for capitalized cost in this book was "vestance." Johannesson in his *Highway Economics* (New York: McGraw-Hill Book Co., Inc., 1931) based nearly all his comparisons on capitalized cost, assuming perpetual life for most highway improvements.

Nevertheless, when present investments are large in proportion to other disbursements, it seems natural to make comparisons on a present worth basis. This often is the case in a comparison of immediate and deferred investments such as was illustrated in Example 7–2.

It already has been emphasized that, given the same interest rate and the same estimated series of disbursements, comparisons by annual cost lead to the same conclusions as comparisons by present worth. Nevertheless, certain serious errors in the basic data of economy studies seem to have been more common when analysts have used present worth methods.

Some Common Errors When Present Worth Comparisons Are Used

Three errors that the authors have often observed in present worth studies are as follows:

1. Perpetual lives are assumed, particularly in capitalized cost comparisons, even though the observed facts indicate that the lives of the proposed machines and structures are likely to be fairly short.

2. The dates when accounting charges are to be made are substituted for the dates of expected disbursements before the present worths are computed. As an example, assume a proposed structure with a first cost of $100,000 and an estimated life of 50 years with zero salvage value. If straight-line depreciation accounting (explained in Chapter 11) is used, the annual depreciation charge in the accounts will be $2,000 (i.e., $100,000 ÷ 50) for 50 years. An uncritical analyst might add this $2,000 a year to the expected disbursements during each year and then convert his annual total to present worth at zero date at, say, 5% interest. As the present worth of $2,000 a year for 50 years at 5% is $36,500, he has, in effect, converted $100,000 at zero date to $36,500 at zero date. This peculiar result comes from the use of two interest rates in a series of conversions, one forward through 50 years at 0%, the other backward through the same years at 5%. By such confused conversions using two interest rates, it is possible to appear to prove that three is equal to one—or to any other figure!

3. Interest rates used in present worth comparisons are frequently too low.

The pointing out of these common errors is an indictment of analysts using present worth methods rather than of the methods themselves. But the evidence obtained from examination of many economy studies is that many of the cost comparisons that have been made by present worth methods have, in fact, led to incorrect conclusions for the reasons that have been outlined.

What is a "Conservative" Interest Rate to Use in a Present Worth Comparison?

The purpose of the interest calculations in economy studies is to determine whether proposed investments are justified—whether it seems likely that they will be recovered with at least a stipulated minimum attractive rate of return. For reasons that are explained in Chapter 10, this rate often should be considerably higher than the bare cost of borrowed money.

If the question were asked, "How large an endowment fund is required to endow a scholarship of $1,000 a year?" it is obvious that the answer should be obtained by capitalizing $1,000 a year at an interest rate that can be obtained on securities involving a minimum risk of loss. This will be a relatively low interest rate.

The present worth of the disbursements required for a given service may be described as the sum of money necessary to endow that service. This description of present worth sometimes leads analysts to the selection of an interest rate appropriate to an endowment fund; such a rate is viewed as a "conservative" one. Such analysts fail to recognize that the economy-study calculations are made to guide a decision between alternatives and that no actual endowment is contemplated.

The use in economy studies of a low interest rate appropriate to endowment funds has the effect of making alternatives requiring higher investments appear to be desirable even though they show the prospect of yielding a relatively small return. The interest rate used for present worth conversions in economy studies usually should be considerably higher than an endowment-type rate; it should be the rate of return required to justify an added investment considering all of the circumstances of the case.

An Estimate of the Value of an Income-Producing Property to a Present or Prospective Owner Implies a Present Worth Calculation

Many different kinds of property are acquired for the purpose of securing prospective future money receipts in excess of the future disbursements, if any, associated with the ownership of the property.

To determine the maximum amount that it is reasonable to pay for any such income-producing property, the following are required:

1. An estimate of the amounts and dates of prospective money receipts resulting from ownership of the property
2. An estimate of the amounts and dates of prospective money disbursements resulting from ownership of the property
3. A decision on the minimum attractive rate of return required to justify this investment in the light of its risk and in the light of returns available from other prospective investments
4. A calculation to determine the net present worth, that is, the excess of the present worth of the prospective receipts over the present worth of the prospective disbursements

If the purchase price of the income-producing property is greater than this computed net present worth, the estimates indicate that the property will fall short of earning its minimum attractive rate of return. Hence if reliance is to be placed on these estimates, the purchase of the property is not attractive.

The same set of estimates and calculations is appropriate in judging the value of a property to its present owner. Generally speaking, a property is worth to its owner at least the net amount for which he can sell it (i.e., the amount to be received from the sale minus the disbursements incident to the sale). In other words, market value in the sense of the net price for which a property could actually be sold tends to place a *lower* limit on the value of a property to its owner. If the sole service of a property to its owner is to produce future net money receipts, an *upper* limit on the value of a property to its owner is generally placed by the excess of present worth of prospective receipts over prospective disbursements. If this present worth is less than the net price for which the property can be sold, it will ordinarily pay to sell the property.

Other factors entering into the estimate of the value of a property to its owner are discussed in later chapters. One important factor is the cost of replacement with an equally desirable substitute.

The remainder of this chapter gives examples of the calculation of net present worth for three different types of income-producing property, as follows:

1. A series of uniform annual payments for a limited number of years
2. A perpetual series of uniform annual payments
3. A government bond

The general subject of the present worth aspects of valuation is explored further by a number of problems and examples throughout the remainder of this book.

Valuation of Uniform Annual Series

It was shown in Plan III of Table 3–1 that the present worth of $1,558.20 a year for 10 years was $10,000 with interest at 9%. The influence of interest rate on value may be observed if we consider the question of how much an investor would be willing to pay for the prospect of these payments if a 10% return on investment were required. Here $P = \$1,558.20(P/A,10\%,10) = \$1,558.20(6.144) = \$9,573.58$. If 8% were considered to be a satisfactory return, $\$1,558.20(P/A,8\%,10) = \$1,558.20(6.710) = \$10,455.52$ would be paid.

Valuation of Perpetual Annual Series

As has already been pointed out in Chapter 5, the present worth of a perpetual uniform annual series is the annual payment divided by the interest rate; this stretches the definition of present worth to mean the investment that will provide—as interest on it—a desired annual payment forever.

Whenever an investment is made where the termination of the income series at a definite time is not contemplated, valuation calculations are likely to be made on this basis. Thus, a stock paying an annual dividend of $6 would be valued at $120 on a 5% basis. The implications of such valuations should be recognized by those who make them; it is unlikely that the corporation will continue its existence forever—or that dividends will always be paid at exactly the present rate—or that a given investor's ownership of the stock will continue indefinitely. Consideration should be given to the question of whether or not the assumption of perpetuity is a fairly close approximation to what is anticipated, before using calculation methods that assume perpetuity.

This comment is also pertinent with respect to comparisons of economy on the basis of capitalized cost, such as are described in this chapter. This limitation particularly applies to the determination of the investment that may be justified by a prospective annual saving.

EXAMPLE 7–3

Valuation of a Bond

Expected Cash Flow

Most corporation bonds and many bonds issued by governments are promises to pay interest, usually semiannually, at a given rate, and to pay the principal of the bond at a definite future date. Consider a 7%, $10,000 government bond due after 20 years. This calls for a payment of $350 every 6 months during the 20-year period and a payment of $10,000 at the end of the 20 years. Assume that this bond is to be valued so that the purchaser will realize a nominal 9% return compounded semiannually.

Calculation of Present Worth

Since 9% is a nominal rate and the interest period is half a year, the present worth would be calculated using an i of 4.5%, i.e., half of 9%. Similarly, there will be 2 (20), or 40, interest periods. Therefore $n = 40$.

PW of 40 interest payments
= $350($P/A$,4.5%,40)
= $350(18,402) = $6,441

PW of principal payment
= $10,000($P/F$,4.5%,40) = $10,000(0.1719) = 1,719

Value of bond to yield a nominal 9%, compounded semiannually $8,160

This 9% is, of course, the *before-tax* yield to a buyer at this price; we have not considered the effect of the bond ownership on the payment of income taxes.

The valuation of this bond at a number of interest rates could be used to plot a graph of valuation, or present worth, as a function of the nominal interest rate. Such a plot is shown in Figure 7–2. Note that at a nominal interest rate of 7%, the valuation is exactly $10,000. At interest rates below 7%, the valuation is greater than $10,000.

There are published bond value tables that give the relation between price and yield to the investor for bonds with various coupon rates and years to maturity. In the common case where such tables are available, calculations of the type illustrated in this example are unnecessary.

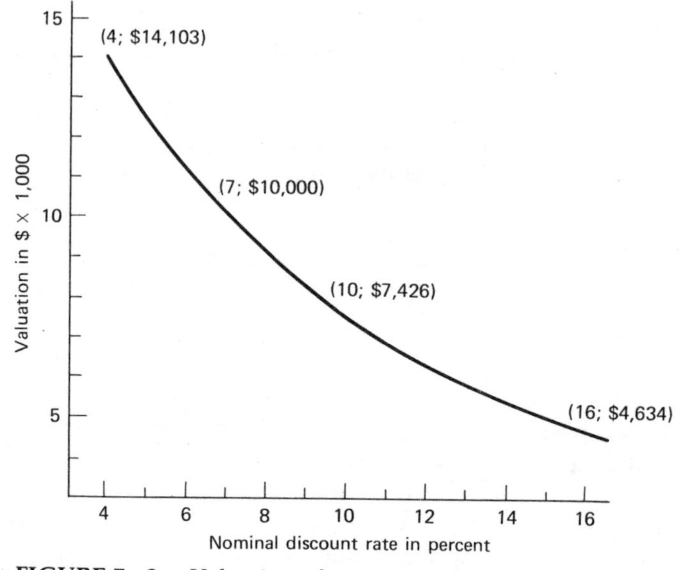

FIGURE 7–2. Valuation of a certain 7% bond as a function of discount rate

Compound Amount Comparisons in Economy Studies

Money time series may be compared by converting them to equivalent single payments at some specified date; the present usually is the most convenient date. With respect to economy studies regarding proposed investments, the "present" is usually the beginning of the period of time under consideration. An alternative possible date is the end of this period; this requires compound amount conversions rather than present worth conversions.

Although such compound amount comparisons are occasionally used in economy studies, the use of compound amount would seem to have no merit not possessed by present worth. On the other hand, small changes in the interest rate have a deceptively large effect on the differences in compound amount. Hence, the chances of misinterpretation of a compound amount study would seem much greater than the chances of misinterpretation of a study based on present worth.

Summary

Valuation of prospective future series of net money receipts is a problem in present worth. In economy studies the comparison of estimated disbursements for alternative plans may be done by present worth conversions. If such comparisons are for a limited number of years of service, the present worth of the cost of the same number of years of service should be calculated for each alternative. When such comparisons are of the present worth of the cost of perpetual service, they are termed capitalized cost comparisons.

The interest rate used in conversions for economy studies should be the rate of return required to justify an investment; this applies to present worth conversions as well as to annual cost conversions. For various reasons, present worth comparisons in economy studies seem more difficult to interpret than annual cost comparisons. Except in certain special situations to which present worth comparisons seem particularly adapted, it is recommended that annual cost comparisons be preferred.

PROBLEMS

7–1. Compare Plans C and D for a proposed public works project on the basis of the capitalized cost of perpetual service using an i^* of 7%. Plan C calls for an initial investment of $500,000, with disbursements of $20,000 a year for the first 20 years and $30,000 a year thereafter. It also calls for the expenditure of $200,000 at a date 20 years from the date of the initial investment and every 20th year thereafter. Plan D calls for an initial investment of $800,000 followed by a single investment of $30,000 30 years later. It also involves annual expenditures of $10,000. (*Ans.* Plan C, $892,317; Plan D, $946,799)

7–2. A construction company must set up a temporary office building at a construction site. Two alternate schemes are proposed for heating this building. "Bottled gas" can be used for floor-type furnaces, or electric radiant panels can be installed in the walls and ceiling. It is estimated that the building will be used for 5 years before being dismantled.

The gas installation will require an investment of $6,000. It is believed its net realizable value will be zero at the end of the 5 years. The estimated annual fuel and maintenance cost is $1,100.

The electric radiant panel installation will require an investment of $8,000; it has an estimated salvage value of $1,000. Estimated annual energy and maintenance cost is $700. Choice of the electric installation will cause an estimated extra payment for income taxes of $100 a year.

Compare the present worths of the costs of these two alternatives using an after-tax i^* of 10%. (*Ans.* Gas, $10,170; Electric, $10,412)

7–3. Two plans are under consideration to provide certain facilities for a publicly owned public utility. Each plan is designed to provide enough capacity during the next 18 years to take care of the expected growth of load during that period. Regardless of the plan chosen now, it is forecast that the facilities will be retired at the end of 18 years and replaced by a new plant of a different type.

Plan I requires an initial investment of $50,000. This will be followed by an investment of $25,000 at the end of 9 years. During the first 9 years, annual disbursements will be $11,000; during the final 9 years, they will be $18,000. There will be a $10,000 salvage value at the end of the 18th year.

Plan II requires an initial investment of $30,000. This will be followed by an investment of $30,000 at the end of 6 years and an investment of $20,000 at the end of 12 years. During the first 6 years annual disbursements will be $8,000; during the second 6 years they will be $16,000; during the final 6 years they will be $25,000. There will be no salvage value at the end of the 18th year.

Using an i^* of 9%, compare the present worths of the net disbursements for the two plans. (*Ans.* Plan I, $175,033; Plan II, $173,562)

7–4. Interest on a 5%, $10,000 bond, due in 20 years, is payable semiannually with the first payment 6 months from now. What should be the price of this bond to have a before-tax yield of a nominal 7% compounded semiannually? (*Ans.* $7,865)

7–5. An investor is considering the purchase of a rental property. The excess of receipts over disbursements is estimated as $3,540 a year for 15 years. It is estimated that the property can be sold for $25,000 at the end of the 15 years. At what price for this property would an investor just recover his investment with an 18% rate of return before incomes taxes? (*Ans.* $20,113)

7–6. The owner of a patent has made a contract with a corporation that is given the exclusive right to use the patent. The corporation has agreed to pay him $1,000 a year at the end of each of the first 4 years during the period of developing a market for the invention, $5,000 at the end of each year for the next 8 years, and $2,000 at the end of each year for the final 5 years of the 17-year life of the patent. If the corporation wished to buy the patent outright and there are 15 years remaining in the contract, how much is the maximum offer the corporation could afford to make to the owner at this time if it requires a 16% rate of return on such investments before income taxes? (*Ans.* $19,229)

7–7. Two schemes of partial federal subsidy of public works projects have been used in the United States. In certain types of projects, outright grants have been made to local governmental units for a portion of the construction cost of approved projects, with the remainder of the cost to be repaid by the local units with interest over a period of years. For example, a grant of 30% might be made, with the remaining 70% to be paid at interest over a 20-year period. Another plan of subsidy has been to require the repayment of all the construction cost without any interest. This latter plan is suggested for a proposed college dormitory project. The government is asked to put up $2,000,000 for this project, with the college paying back this amount at $50,000 a year for 40 years.

Assume that, all things considered, an appropriate interest rate to charge on such a "loan" is 7%. On this assumption, the plan to make the $2,000,000 repayment in 40 years without interest really amounts to a subsidy of how many dollars? (*Ans.* $1,333,400)

7–8. The XYZ Tile Co. secures its tile clay from property owned by John Doe, adjacent to the tile plant. Some years ago the company made a royalty contract with Doe on which it pays royalties of $1.00 per ton for all clay removed from his property. This contract has 5 years to run. It is estimated that Doe's holdings will supply the company's needs of 20,000 tons per year for the next 15 years before the clay is exhausted. The company owns a large deposit of clay at some distance from the plant; in relation to the company's needs, the deposit may be viewed as practically inexhaustible. Costs of removing the clay would be substantially the same as from Doe's holdings; however, the cost of transporting the clay to the plant would be greatly increased. Doe is aware of this fact; it is believed that a new royalty contract (5 years hence) for the final 10 years would need to provide a royalty rate of $2 per ton. At this royalty rate, it will continue to be advantageous to use Doe's clay rather than the company's more distant holdings.

The president of the XYZ Tile Co. has just learned that Doe would consider an outright sale of his land to the company. By purchasing this land, the company would no longer have to pay royalty for the clay removed. It is believed that at the end of 15 years, when the clay is exhausted, the land can be sold for $30,000.

At what price for this property would the XYZ Co. have an investment that would yield 14% before income taxes as compared to the alternative of continuing to pay royalties? (*Ans*. $181,220)

7–9. John Doe, whose current and probable future contracts for the sale of the clay on his land are described in Problem 7–8, is in relatively poor health. He feels that he cannot manage and check the tonnage removed himself and that he will need to employ a part-time engineer to survey the clay removal at a cost of about $2,000 a year. In trying to decide how much he should ask for the land, should he decide to sell it, he considers 11% before income taxes as his i^*. What should his asking price be? Do you think that Doe and the XYZ Company will be able to reach an agreement? Why or why not? (*Ans*. $205,610)

7–10. Compare the capitalized costs of perpetual service of the two plans in Problem 6–1 (page 89). Use the given i^* of 9%. (*Ans*. Plan I, $253,149; Plan II, $296,084)

7–11. Compare the present worths of 60 years of service of the two alternatives in Problem 6–2 (page 89). Use the given i^* of 6%. (*Ans*. Type A, $203,961; Type B, $205,875)

7–12. Compare the present worth of 12 years' service for the two models of Taxi Cab in Problem 6–9 (page 91). Use the given i^* of 20% before income taxes.

7–13. Compare the present worth of 40 years' service for the two power plants in Problem 6–10 (page 91). Use the given i^* of 14% after income taxes.

7–14. Compare the present worth of 10 years' service for the two testing units in Problem 6–14 (page 92). Use the given i^* of 20% after income taxes.

7–15. Compare the capitalized cost of perpetual service for the two recreational development plans in Problem 6–19 (page 93). Use the given i^* of 9%.

7–16. A donor wishes to endow a scholarship at a certain university. The endowment may be made by a lump-sum deposit to a special foundation set up by the university for such purposes. The foundation director believes that its funds will earn at least 8% interest per year, tax free, for the indefinite future.

The scholarship is to provide $5,000 the first year increasing by $200 each year thereafter to a maximum of $10,000 per year. On the assumption that the scholarship will start at the end of the first year and continue forever, what endowment must the donor make now?

7–17. In Problem 7–16, assume that the donor wishes the scholarship to continue as described for 30 years, at which time the principal amount of the endowment is to be used up. How much must the endowment be in this case?

7–18. The donor in Problem 7–16 has exactly $80,000 to give. If the scholarship is to operate exactly as described, for how many years can the $10,000 scholarship level be supported before the fund is exhausted? What will be the total length of the scholarship period?

7–19. Two possible types of road surface are being considered with cost estimates per mile as follows:

	Type X	**Type Y**
First cost	$36,000	$44,000
Resurfacing period	10 years	15 years
Resurfacing cost	$18,000	$24,000
Average annual upkeep cost	$ 3,400	$ 1,800

The periodic resurfacings will involve replacement only of the wearing surface and not the base or subsurface. Compare these alternatives on the basis of the present worth of the cost of 30 years' service. Assume a zero terminal salvage value for both types of surface and use an i^* of 9%.

7–20. Two alternative water supply systems are being considered for a small community. Compare the present worths of the costs of 20 years' service using an i^* of 11%.

System A requires an initial investment of $58,000 with the replacement of certain elements at the end of 10 years at an expected cost of $16,000. Annual operation and maintenance costs will be $12,000 the first year and are expected to increase by $500 each year thereafter.

System B requires an initial investment of $72,000 and is expected to last the full 20 years without major replacements. Annual disbursements are expected to be $10,000 the first year and to increase by $350 each year thereafter. There will be no net residual value for either system at the end of the 20-year period.

7–21. An investor is considering the purchase of a six-unit apartment house offered for sale for $90,000. Because of its proximity to a local university, the apartment is expected to remain fully rented. Annual receipts are estimated to be $10,800 and disbursements for maintenance are estimated to be $2,500. The investor estimates that the property can be sold for $65,000 at the end of 15 years. Use an i^* before income taxes of 18% to determine if the investor should offer $90,000 for this property.

7–22. An investor wishes to sell $100,000 in corporate bonds paying 7% compounded semiannually in order to take advantage of another investment opportunity. Similar bonds are currently yielding a nominal 9% per year. A 6-month interest payment has just been received and the bonds have 8½ years to go to maturity. What price should they bring on the market?

7–23. The lining of a chemical tank in a certain manufacturing operation is replaced every 2 years at a cost of $4,500. A new material and process is now available which will extend the service life of the lining to 4 years. The cost of this process will be $10,000. Relining takes nearly 3 weeks to complete at an estimated net loss due to down time of $3,000. It is expected that the entire tank will have to be replaced in 15 years. Use an i^* of 15% before income taxes to compare the present worth of costs for each process. Assume that the next relining must be done one year from now.

7–24. In Problem 7–23, it is suggested that the tank will probably have to be replaced in 12 years rather than in 15 years as stated. How does this somewhat shorter life affect the solution found in Problem 7–23?

7–25. The following alternatives are being considered for a government service. Compare the present worths of 30 years' service using an i^* of 9%.

	Structure C	**Structure D**
First cost	$26,000	$50,000
Estimated life	15 years	30 years
Estimated salvage value	$4,000	$5,000
Annual disbursements	$4,800	$2,600

7–26. Estimates for alternative plans in the design of a certain materials-handling facility are:

	Plan X	**Plan Y**
First cost	$120,000	$185,000
Estimated life	20 years	30 years
Salvage value	none	$15,000
Annual disbursements	$19,000	$14,000

Estimated extra annual income taxes if Plan Y is adopted are $2,667. Use an i^* after income taxes of 10% to compare the present worths of the net disbursements for 60 years with the two plans. Assume that replacement costs, estimated lives, and annual disbursements will remain constant during the study period.

7–27. An investment opportunity offers these expected cash flows: The initial investment is $140,000 and is expected to have a net salvage value of $20,000 at the end of 15 years. Annual receipts are expected to be $37,000 the first year and $37,500 the second year, and to increase by $500 each year until the 10th year. In years 11 through 15, receipts are expected to decline by $300 each year. Disbursements are expected to be $13,500 the first year and to increase by $200 each year thereafter. Use an i^* of 15% before income taxes to calculate the net present worth of this investment opportunity. Is this opportunity acceptable based on the stated criterion?

7–28. A $10,000 bond with interest payable semiannually at a nominal rate of 7% per year is due to mature in 15 years. The first interest payment will be made 6 months from now. What will the price of this bond be to have a before-tax yield of a nominal 6%, compounded semiannually? a nominal 8%? a nominal 9%? a nominal 10%? Plot the market price as a function of the nominal interest rate.

7–29. Acquisition of a certain small business will require an investment of $60,000. Annual receipts are estimated to be $46,000 and annual disbursements to be $38,000. A potential investor anticipates that, if acquired, the business will be sold in 15 years at a net price of $40,000. The investor is unsure what value of i^* to use, as previous investments of this type have yielded anywhere from 6% to 15% before income taxes. Plot the net present worth of the cash flows in the stated range of interest rates as an assist to the investor in making this decision.

8

CALCULATING AN UNKNOWN INTEREST RATE

Discounted-cash-flow analysis makes three contributions to top management thinking:

1. *An explicit recognition that time has economic values to the corporation; hence that near money is more valuable than distant money*
2. *A recognition that cash flows are what matter; hence capitalization accounting and the resulting book depreciation are irrelevant for capital decisions except as they affect taxes*
3. *A recognition that income taxes have such an important effect upon cash flow that their amount and timing must be explicitly figured into project worth.—Joel Dean**

Our examples in Chapters 6 and 7 started with the assumption of an interest rate or minimum attractive rate of return. Calculations were then made to provide a basis for judgment as to whether proposed investments would meet this standard.

Often it is a good idea to compute the prospective rate of return on an investment rather than merely to find out whether the investment meets a given standard of attractiveness. Usually this calculation is carried out to best advantage by a trial-and-error method unless an appropriate computer program or special calculator is available. Two or more interest rates are assumed, present worths or equivalent uniform annual cash flows are calculated, and the rate of return is found by interpolation.

Moreover, in a number of instances where money is borrowed, the circumstances are such that the cost of borrowed money (expressed as an interest rate) cannot be found without similar trial-and-error calculations. Often, it turns out that the true cost of borrowed money is considerably higher than it is believed to be by the prospective borrower.

Although economy studies necessarily deal with prospective investments, an example of the finding of the rate of return on a terminated invest-

*Joel Dean, Sec. 2, Managerial Economics, in *Handbook of Industrial Engineering and Management*, 2d ed., W. G. Ireson and E. L. Grant, eds. (Englewood Cliffs, N.J.: Prentice-Hall, Inc., copyright 1971), p. 110.

ment may throw light on certain aspects of the subject that need to be understood. Our first example is of this type. It is followed by a number of examples dealing with prospective investments and prospective borrowings. Where disbursements for income taxes can be introduced in a relatively simple way (as in Examples 8–3, 8–5, 8–6, and 8–7), examples and problems involve calculation of rates of return *after* income taxes. However, in cases where the income tax aspects are fairly complicated (Examples 8–1, 8–2, and 8–4), the analysis is for rate of return *before* income taxes. The reasons for the complications in such cases are discussed in Chapter 12. All examples and problems involving the cost of borrowed money expressed as an interest rate deal with this cost *before* income taxes.

EXAMPLE 8–1

Rate of Return on a Past Investment

Facts of the Case

This example relates to a completed 7-year period. Our assumed zero date is January of the first year when an investor purchased a commercial rental property for $109,000. In December of the seventh year, the property was sold for a gross sales price of $220,000, from which a 5% broker's commission had to be paid, leaving a net receipt of $209,000. The second column of Table 8–1 shows the receipts that occurred during the period of ownership. The third column shows all disbursements (other than income taxes), including maintenance costs, property taxes, and insurance. The fourth column combines these figures to give the year-by-year net cash flow before income taxes. Two lines are devoted to the final year (year 7), one showing the receipts and disbursements in connection with rentals and the other showing the receipts and disbursements in connection with the sale of the property.

TABLE 8–1
Cash flow from a terminated investment in rental property

Year	*Receipts*	*Disbursements*	*Net cash flow*
0		−$109,000	−$109,000
1	+$15,000	−5,000	+10,000
2	+18,000	−5,500	+12,500
3	+18,000	−5,700	+12,300
4	+18,000	−4,500	+13,500
5	+18,000	−3,600	+14,400
6	+18,000	−4,300	+13,700
7	+17,000	−4,100	−12,900
7	+220,000	−11,000	+209,000
Totals	+$342,000	−$152,700	+$189,300

Calculation of Rate of Return

The rate of return is the interest rate at which the present worth of the net cash flow is zero. Here this is the interest rate at which the present worth of the net receipts that occurred in years 1 to 7 is just equal to the $109,000 disbursement that was made at zero date.

In the present worth calculations in Table 8–2, the end-of-year convention has been used. (It will be recalled that this convention has been used throughout Chapters 6 and 7.) The cash flow figures for each year have been multiplied by the respective present worth factors for interest rates of 18% and 20%. The sum of the present worths is +$3,924 at 18% and −$6,051 at 20%. The following linear interpolation between these values indicates that this investment yielded a return of a little less than 19% before income taxes.

$$\text{Rate of return} = 18\% + \frac{\$3{,}924}{\$9{,}975}(20\% - 18\%) = 18.8\%$$

TABLE 8–2
Present worth calculations for trial-and-error determination of rate of return on a terminated investment in rental property

Year	Net cash flow	(P/F,18%,n)	PW at 18%	(P/F,20%,n)	PW at 20%
0	−$109,000		−$109,000		−$109,000
1	+10,000	0.8475	+8,475	0.8333	+8,333
2	+12,500	0.7182	+8,978	0.6944	+8,680
3	+12,300	0.6086	+7,486	0.5787	+7,118
4	+13,500	0.5158	+6,963	0.4823	+6,511
5	+14,400	0.4371	+6,294	0.4019	+5,787
6	+13,700	0.3704	+5,074	0.3349	+4,588
7	+12,900	0.3139	+4,049	0.2791	+3,600
7	+209,000	0.3139	+65,605	0.2791	+58,332
Totals	+189,300		+3,924		−6,051

EXAMPLE 8–2

Rate of Return on a Prospective Investment in a Rental Machine

Estimates Relative to a Proposed Investment

Mary Smith has received an inheritance of $25,000. A friend engaged in the business of rental of construction machinery suggests that Smith invest this $25,000 in a tractor of a certain type. The friend will serve as Smith's agent in the rental of this asset on a commission basis and will remit to Smith the net receipts from each year's rentals at the end of the year. He estimates that for rental purposes the machine will have a useful life of 8 years, with a 10% final salvage value. His estimates of year-by-year receipts from rentals and of disbursements for all purposes are as shown in Table 8–3. Disbursements are chiefly for repairs and maintenance but also include storage, rental commissions, property taxes, and insurance.

Calculation of Rate of Return

Table 8–4 indicates that if the friend's estimates of cash flow turn out to be correct, Smith's return before income taxes will be 15.4%, obtained by linear interpolation between 15% and 16%. In Table 8–4 and in subsequent tables of

TABLE 8–3
Estimated cash flow from purchase and ownership of a rental machine

Year	*Receipts*	*Disbursements*	*Net cash flow*
0		−$25,000	−25,000
1	+$8,000	−1,100	+6,900
2	+7,800	−1,300	+6,500
3	+7,500	−1,500	+6,000
4	+7,100	−1,700	+5,400
5	+6,700	−1,900	+4,800
6	+6,300	−2,200	+4,100
7	+6,000	−2,400	+3,600
8	+5,600	−2,600	+3,000
8	+2,500*		+2,500
Totals	+$57,500	−$39,700	+$17,800

*Receipt from salvage value.

TABLE 8–4
Present worth calculations to determine prospective rate of return, Example 8–2

Year	*Estimated cash flow*	*Present worth* at 14%	at 15%	at 16%
0	−$25,000	−$25,000	−$25,000	−$25,000
1	+6,900	+6,053	+6,000	+5,948
2	+6,500	+5,002	+4,915	+4,831
3	+6,000	+4,050	+3,945	+3,844
4	+5,400	+3,197	+3,088	+2,982
5	+4,800	+2,493	+2,389	+2,285
6	+4,100	+1,868	+1,772	+1,683
7	+3,600	+1,439	+1,353	+1,274
8	+5,500	+1,928	+1,798	+1,678
Totals		+$1,030	+$260	−$475

this type, the present worth factors are omitted. In Table 8–2 these factors were shown for each year and for each interest rate.

In deciding whether or not to undertake this particular investment, Smith should compare the prospective return of 15.4% before income taxes with the prospective return before income taxes obtainable from alternative investments that she believes are of comparable risk. She must also decide whether she is willing to undertake the risks associated with this type of investment.

EXAMPLE 8–3

Rate of Return When the Investment Period Extends Over Several Years

Estimates Relative to a Proposed New Product Investment

A chemical company is considering a proposal to buy land and build a plant to manufacture a new product. The required investment of $300,000 in land must be made 2 years before the start of operation of the plant. Of the $1,500,000 estimated investment in plant and equipment, $800,000 will have been spent one year before the plant starts to operate, and the remaining $700,000 in the year just preceding the start of operations. For purpose of analysis, the date of starting operations will be adopted as zero date on the time scale.

It is estimated that the life of the plant will be 15 years from the date of the start of operation. (It is believed the life will be terminated by product obsolescence.) Throughout the period of operation, it is estimated that $200,000 will be invested in working capital, chiefly in the inventories of raw materials, work in process, and finished product, and in the excess of accounts receivable over accounts payable. (See Chapter 16 for a discussion of certain problems that arise in estimating such working capital requirements.) For purposes of the economy study, this investment will be assumed to be a negative cash flow of $200,000 at zero date, finally recoverable by a positive $200,000 cash flow at date 15. It will also be assumed that the land will be sold for its $300,000 original cost at date 15. The plant and equipment is assumed to have zero net salvage value at date 15.

Receipts from the sale of the chemical product and disbursements in connection with its production and sale are estimated for each of the 15 years. It is expected that there will be an initial 3-year period involving start-up costs and development of markets before the full earning power of the project is developed, and that there will be a period of declining earning power in the final 4 years. The specific estimates of net positive before-tax cash flow from operations year by year are: 1, $100,000; 2, $300,000; 3, $400,000; 4 through 11, $500,000; 12 and 13, $400,000; 14, $300,000; and 15, $200,000. These positive cash flows will be diminished by yearly disbursements for income taxes estimated to be 50% of the amount that each year's figure exceeds $100,000.

Calculation of Rate of Return

Table 8–5 tabulates the before-tax and after-tax cash flows from year minus 2 to year 15. The equivalent sum of money at zero date is computed at 9% and 10% interest. (Although this might loosely be described as a present worth calculation at zero date, it will be noted that compound amount calculations are required for the conversions from years minus 1 and minus 2.) Interpolation between the sums of the equivalent amounts at zero date indicates a prospective after-tax return of slightly over 9.1%.

This 9.1% interpolated figure is independent of the choice of reference date. That is, if present worths had been calculated at year minus 2, or if compound amounts had been calculated at year 15, the interpolated rate of return would still have turned out to be 9.1%. However, in systematic evaluation of a number of projects of this type, it seems reasonable that zero date should always be chosen as the date of start of operations. Chapter 21 illustrates industrial capital budgeting forms that are based on this assumption.

The $300,000 land investment at date minus 2 and the $200,000 working capital investment at date 0 have no effect on prospective taxable income or on income taxes because it is assumed that they will be recovered at date 15

TABLE 8–5
Present worth calculations to determine prospective rate of return, Example 8–3
(All figures in thousands of dollars)

Year	Estimated before-tax cash flow	Estimated after-tax cash flow (and PW at 0%)	Equivalent amount at zero date	
			at 9%	at 10%
−2	−$300	−$300	−$356	−$363
−1	−800	−800	−872	−880
0	−900	−900	−900	−900
1	+100	+100	+92	+91
2	+300	+200	+168	+165
3	+400	+250	+193	+188
4	+500	+300	+213	+205
5	+500	+300	+195	+186
6	+500	+300	+179	+169
7	+500	+300	+164	+154
8	+500	+300	+151	+140
9	+500	+300	+138	+127
10	+500	+300	+127	+116
11	+500	+300	+116	+105
12	+400	+250	+89	+80
13	+400	+250	+82	+72
14	+300	+200	+60	+53
15	+700	+650	+178	+156
Totals	+$4,600	+$2,300	+$19	−$136

without either increase or decrease. The tabulated cash flow at date 15, both before and after taxes, includes the recovery of this total of $500,000.

If we had specified a minimum attractive after-tax rate of return of, say, 7%, this new-product proposal would clearly be acceptable. If we had specified, say, 15%, it would clearly be unacceptable. However, a common state of affairs in industry is for the available capital funds and other resources to be insufficient to permit accepting all the major investment proposals that seem likely to have fairly high rates of return. For this reason, decision making about such proposals is rarely quite as simple as reaching either a "Yes" or "No" decision by comparing a figure such as our 9.1% with a stipulated i^*. Chapters 10, 14, 15, 18, and 21 include comments on various aspects of the problem of choice among major proposed investments in relation to the common need for capital rationing.

Some Comments on Examples 8–1, 8–2, and 8–3

Example 8–1 is adapted from an actual case with receipts and disbursements rounded off to multiples of $100. Examples 8–2 and 8–3 make use of assumed data with figures that were chosen to be useful in illustrating several different facets of engineering economy.

It will be noted that the method used for finding rate of return is the same in all examples, even though one deals with a past investment and the others with proposed investments.

One aspect of these examples is that a single project constitutes a separate activity for which all receipts and disbursements are assumed to be identifiable. It is rarely true of the capital goods of modern industry that specific receipts can be identified with individual machines or structures. For example, the receipts from the sale of a manufactured product cannot be identified with, say, the factory building, or with a specific machine used in a sequence of operations, or with a specific item of materials-handling equipment. In this particular respect these three examples are not typical of the usual rate-of-return analysis for purposes of decision making in industry. Subsequent examples throughout this book will continue to emphasize the point brought out in earlier chapters that it is the prospective differences in cash flow between alternatives that need to be forecast and analyzed as a basis for choosing between any given alternatives.

Different Names Applied to the Computation of Rate of Return by the Methods Illustrated in This Chapter

Rate of return calculations of the type illustrated in Examples 8–1, 8–2, and 8–3 are as old as writings on the mathematics of finance. Since the early 1950s, however, the use of such calculations by industrial companies in the United States has greatly increased. Various names have been applied to this method of calculation. One name is the *discounted cash flow method;* another is the *Investor's Method*. Rate of return calculated in this way has been called the

Profitability Index (sometimes abbreviated to PI), *interest rate of return, solving rate of return,* and *internal rate of return*. More recently two terms, *measure of worth* and *figure of merit*, have come into popular usage in financial journal articles.*

Of the eight phrases mentioned, "discounted cash flow" seems to be the one most widely used in industry. It describes the data required and the method of calculation illustrated in Tables 8–2, 8–4, and 8–5. *Cash flow* refers to the fact that the required data must be given as the amounts and dates of receipts and disbursements. *Discounted* applies to the calculation of present worth.

But the phrase "discounted cash flow" is properly applied to *any* calculation to find the present worth of cash flow whether or not the calculation is to be used in computing a rate of return. Thus, the phrase is applicable generally to present worth comparisons for economy studies; it also applies to annual cost comparisons for which a finding of present worth has been an intermediate computational step. Moreover, the calculation of rates of return by correct methods often may be done without employing present worths; frequently the use of equivalent uniform annual figures is equally satisfactory.

Therefore it is misleading to assume either that a discounted cash flow calculation yields a rate of return or that a correct calculation of rate of return necessarily requires an analyst to discount cash flow. In general, the phrase "rate of return" used in this book means a rate found by applying appropriate compound interest analysis to past or prospective cash flow.

A number of methods in common use in industry purport to be rates of return, but all of these methods are approximations and they usually give figures that differ widely from one another as well as from the correct rate of return. These so-called rate of return methods are described in Chapter 11.

The calculations illustrated in Tables 8–2, 8–4, and 8–5 required a fair amount of computation since each year's cash flow had to be multiplied by the single payment present worth factor for two or more interest rates. Very frequently, the cash flows are uniform over the project lives or involve arithmetic gradients. In these cases, the computation can be simplified by the use of the uniform series factors and gradient factors. Since future cash flows are usually estimates, it may be just as accurate to assume uniform or gradient flows as to try to estimate each year's flows separately. Many relatively simple digital computer programs are available to solve for the rate of return on a proposed investment.

The examples in the remainder of this chapter illustrate methods of computing unknown interest rates where the facts are simpler than in Examples 8–1, 8–2, and 8–3.

*The phrase "discounted cash flow" is associated with the writings of Joel Dean. The phrase "Investor's Method" (usually capitalized) is associated with the writings of Horace G. Hill, Jr., and John C. Gregory. The phrase "Profitability Index" (also usually capitalized) is associated with the writings of Ray I. Reul. The phrase "interest rate of return" is associated with the writings of J. B. Weaver and R. J. Reilly.

EXAMPLE 8–4

Determining the Prospective Rate of Return on a Bond Investment

Facts of the Case
A $10,000 7% government bond due in 20 years was described in Example 7–3. This bond can be purchased for $8,000. On the assumption that a buyer at this price will hold the bond to maturity, it is desired to find his before-tax rate of return. In accordance with the usual practice in stating bond yields, this rate is to be expressed as a nominal interest rate assuming semiannual compounding.

Calculation of Bond Yield
In Example 7–3, it was found that with a semiannual i of 4.5%, the present worth of the payments was $8,160. A similar calculation using a semiannual i of 5% gives a present worth of $7,426. Linear interpolation between these figures gives an i of 4.6%. The nominal rate per annum compounded semiannually is therefore 2(4.6%) = 9.2%.

As explained in Example 7–3, the foregoing type of approximate calculation is unnecessary in the common case where published tables of bond yields are available.

EXAMPLE 8–5

Determining the Prospective Rate of Return from a Uniform Annual Series of Net Receipts

Data from Example 6–1
The final column of Table 6–1 (page 75) showed that the differences in cash flow between Plans B and A consisted of a negative cash flow of $15,000 at zero date and positive after-tax cash flow of $2,800 a year for 10 years.

Calculation of Rate of Return
The present worth of the net cash flow may be computed for various interest rates using the series present worth factors for 10 years.

The objective is to find the value of i such that the following equation is satisfied:

$$\text{PW} = 0 = -\$15{,}000 + \$2{,}800(P/A,i\%,10)$$

As explained earlier, this can be accomplished by trial and error, substituting different i's until two closest interest rates for which interest factor tables are available give a positive and a negative result. Then the i is estimated by linear interpolation between the two rates. In this example, 13% and 14% "bracket" the desired rate. Interpolation indicates a rate of 13.3%.

In the special case where the prospective positive cash flow from an investment, P, constitutes a uniform annual series, A, an alternate solution is to compute P/A or A/P and to interpolate between the appropriate factors in the interest tables. In this instance, $(P/A,i\%,10) = \$15{,}000/\$2{,}800 = 5.357$. Interpolation between 5.426 and 5.216, the respective series present worth factors for 13% and 14%, also indicates an after-tax rate of return of about 13.3%.

In Example 6–1, the i^* was stipulated to be 9%. It should be understood that the $15,000 investment in Plan B is optional; whether or not it is attractive depends on whether the prospective rate of return is at least i^*. Since this is approximately 13.3%, which is more than 9%, the proposed investment is acceptable by the firm's criterion. This example illustrates the principle that each increment (optional) of investment must meet the primary criterion established for the analysis.

EXAMPLE 8–6

Prospective Rate of Return in Example 6–2

Data from Example 6–2

Plan B in Example 6–2 had a first cost of $15,000, annual disbursements of $6,400, a 10-year life, and zero salvage value. Plan C had a first cost of $25,000, annual disbursements of $5,625, a 10-year life, and a $5,000 salvage value. (The foregoing disbursements included extra income taxes as compared to Plan A.) The difference in cash flow between C and B (shown in the final column of Table 6–2, page 81) was −$10,000 initially, +$775 a year for 10 years, and +$5,000 at the end of 10 years.

Calculation of Rate of Return

A calculation of present worth of the net difference in cash flow using 3% and 3.5% interest indicates that the extra $10,000 investment in Plan C will be recovered with an after-tax rate of return of about 3.5%.

$$\text{PW at } 3\% = -\$10{,}000 + \$775(P/A,3\%,10) + \$5{,}000(P/F,3\%,10)$$
$$= -\$10{,}000 + \$775(8.530) + \$5{,}000(0.7441) = +\$331$$
$$\text{PW at } 3.5\% = -\$10{,}000 + \$775(8.317) + \$5{,}000(0.7089) = -\$10$$

In Example 8–5 and in the foregoing calculation, we dealt only with the differences in cash flow between the two alternatives that were being compared. Of course the same conclusion is reached when present worths of *all* disbursements are computed for both alternatives. The following tabulation shows that the two alternatives have the same present worth at about 3.5%.

	Plan B	Plan C	Difference (B − C)
PW at 3%	$69,592	$69,261	+$331
PW at 3.5%	$68,229	$68,238	−$9

The rate of return on extra investment must also be the interest rate at which the alternatives have the same equivalent uniform annual cost. The following tabulation shows that this occurs at about 3.5%.

	Plan B	Plan C	Difference (B − C)
Annual cost at 3%	$8,158	$8,120	+$38
Annual cost at 3.5%	$8,203	$8,205	−$2

It will be noted that although the elements of cash flow common to Plans B and C have been included in the foregoing present worth and annual cost figures, it is only the differences in cash flow that have any influence on the conclusion regarding rate of return. The common elements of cash flow will contribute equally to the present worths (or to the annual costs) of both alternatives regardless of the interest rate assumed.

A Possible Misinterpretation of Rate of Return When Three or More Alternatives Are Being Compared

It will be recalled that in Examples 6–1 and 6–2, Plan A was to continue the present method of carrying out a certain materials-handling operation. Plan B required a $15,000 investment in equipment intended to reduce labor costs; Plan C required a $25,000 investment in such equipment. The minimum attractive rate of return after income taxes was 9%. Plan B was favored by the annual cost comparison in Chapter 6 and by the present worth comparison in Chapter 7.

Our calculations of rate of return should also lead us to the conclusion that Plan B is the best of the three plans. The $15,000 investment in Plan B promises a 13.3% rate of return after income taxes as compared to the present method, Plan A. Because 13.3% is higher than our 9% standard, B is superior to A. On the other hand, the extra $10,000 investment required for Plan C will yield only 3.5% after income taxes as compared to B. Since 3.5% is less than our 9% standard, the $10,000 investment is not justified and Plan C should be rejected. The $10,000 presumably can be invested elsewhere for at least 9%.

Once a particular i^* is selected for the comparison of alternatives, a correct analysis of relevant rates of return will invariably lead to the same conclusion that will be obtained from a correct annual cost comparison or a correct present worth comparison.

Nevertheless, incorrect conclusions are sometimes reached by computing rates of return from inappropriate pairs of alternatives. For example, if Plan C is compared with Plan A, the prospective rate of return on the $25,000 investment is 9.05% after income taxes. Someone favoring Plan C might argue that because it promises 9.05% as compared to the present method of doing things and because 9.05% exceeds the stipulated i^* of 9%, the full $25,000 investment is justified.

If Plan B were not available, it is true that the 9.05% rate of return would indicate the justification of Plan C. But because of the availability of B, C is unattractive. The $10,000 increment of investment in C over B will not pay its way, yielding only 3.5% after income taxes. In general, each separable increment of proposed investment ought to be considered separately in relation to its justification. Unsound proposals often appear to be justified because they are improperly combined with sound ones from which they should be separated.

The prospective return on Plan C as compared to Plan A has no useful meaning as a guide to decision making in this case. The viewpoint presented here is expanded in Chapter 13, which deals with comparisons of multiple alternatives.

Demonstrating the Validity of a Rate of Return Computed by Compound Interest Methods

In our introductory discussion of compound interest in Chapter 3, Table 3–1 illustrated four cases where an investment of $10,000 was recovered in 10 years with a return of 9%. The four cases involved four quite different series of year-end cash receipts by the investor of the original $10,000. The figures in Table 3–1 showed the year-by-year unrecovered balances in each case and demonstrated that the four cases were alike in providing complete capital recovery of the original $10,000 with interest at 9%. Whenever correct compound interest methods are used to compute an unknown interest rate, a similar tabulation may be used to show the validity of the computed rate.

Where someone questions the meaning of a computed rate of return, a tabulation such as Table 3–1 will sometimes help to clarify matters. Where persons are suspicious of compound interest methods and prefer some other "approximate" method of computing rate of return (such as one of those discussed later in Chapter 11), such a tabulation may be useful in demonstrating the correctness of the compound interest method and the incorrectness of the other methods.

Rate of Return Calculations Assume the Termination of the Consequences of an Investment Decision

Example 8–1 viewed an investment in retrospect: that is, it considered a past investment in income-producing property from the date of acquisition of the property until the date of its final disposal. With full information about the money receipts and disbursements that were associated with this investment, it was possible to compute the rate of return obtained. Although Examples 8–2 to 8–6 dealt with prospective investments rather than with past ones, the viewpoint was really the same as in Example 8–1; the transactions were viewed from the date of prospective acquisition to the prospective date of termination of ownership.

Although an economy study regarding the desirability of a proposed investment may properly take the point of view of a terminated transaction, this viewpoint is hardly possible with regard to a past investment not yet terminated. For this reason, all judgments regarding profits or losses in business enterprises not yet terminated should really be thought of as preliminary estimates that in the long run may turn out to be either too favorable or not favorable enough. For example, the relatively high return of 18.8% in Example 8–1 was caused by the sale of the rental property for approximately twice its cost; if there had been accounts for this enterprise, a conventional analysis of the accounts at any time before the property was sold would have indicated a return of much less than 18.8%.

Conclusions regarding the profitability of investments not yet terminated are usually drawn from the accounts of business enterprises. Some aspects of the difference in viewpoint between an economy study to determine whether or not to make a proposed investment and the accounting procedures relating to the same investment once it is actually made are explored in Chapter 11.

Chapters 6 and 7 explained that when alternatives deal with assets having different estimated lives, a convenient simple assumption is that replacement assets will repeat the cycle of disbursements and receipts that have been forecast for the initial asset. Example 8–7 illustrates the use of this assumption in computing rate of return.

EXAMPLE 8–7

Rate of Return Calculations When Alternatives Have Different Lives

Data from Example 6–3

It will be recalled that Plan D in Example 6–3 had a first cost of $50,000, a life of 20 years, a $10,000 terminal salvage value, and annual O & M disbursements of $9,000. Plan E had a first cost of $120,000, a life of 40 years, a $20,000 terminal salvage value, annual O & M disbursements of $6,000, and extra annual income tax disbursements of $1,250. These plans were compared in Chapters 6 and 7 by annual costs, present worths, and capitalized costs using an after-tax i^* of 11%.

Calculation of Rate of Return

To compute a prospective rate of return on the $70,000 extra investment in Plan E, it is necessary that the two plans apply to service for the same number of years. Assume that the first 20 years' estimated disbursements for Plan D will be repeated in the second 20 years as shown in the cash flow tabulation, Table 6–3 (page 83).

One way to find the rate of return on extra investment is to compute the present worths of the net disbursements for 40 years assuming different interest rates. Interpolation between the following differences in present worth shows that the two plans have the same present worth at about 2.7%.

	Plan D	Plan E	Difference (D − E)
PW at 2.5%	$296,620	$294,550	+ $2,070
PW at 3%	$277,120	$281,450	− $4,330

The same 2.7% rate of return is found by interpolating between the differences in equivalent uniform annual net cash flow using different interest rates.

	Plan D	Plan E	Difference (D − E)
Equivalent annual cost at 2.5%	$11,816	$11,734	+ $82
Equivalent annual cost at 3%	$11,989	$12,126	− $137

Still another method, not illustrated here, would be to compute the capitalized costs of perpetual service using interest rates of 2.5% and 3% and to interpolate between the differences.

Although any of the foregoing methods of solution will give the correct answer, the point that 2.7% is really the prospective rate of return on extra investment may be somewhat clearer if the problem is approached using only the differences in cash flow between the two plans. These differences were tabulated in the final column of Table 6–3 (page 83). Plan E requires an extra disbursement of $70,000 at zero date, offset by reduced after-tax disbursements of $1,750 a year for 40 years and $40,000 at the end of the 20th year, and by an increased receipt of $10,000 from the larger salvage value at the end of the 40th year. Present worths of the differences in cash flow are as follows:

$$\begin{aligned}
\text{PW at 2.5\%} &= -\$70{,}000 + \$1{,}750(P/A,2.5\%,40) + \$40{,}000(P/F,2.5\%,20) \\
&\quad + \$10{,}000(P/F,2.5\%,40) \\
&= -\$70{,}000 + \$1{,}750(25.103) + \$40{,}000(0.6103) \\
&\quad + \$10{,}000(0.3724) \\
&= +\$2{,}070 \\
\text{PW at 3\%} &= -\$70{,}000 + \$1{,}750(23.115) + \$40{,}000(0.5537) \\
&\quad + \$10{,}000(0.3066) \\
&= -\$4{,}330
\end{aligned}$$

Again interpolation gives us the 2.7% figure.

Guessing the Rate of Return Before a Trial-and-Error Calculation

In computing unknown rates of return by compound interest methods, it usually is necessary to compute present worths (or equivalent uniform annual costs) at two or more interest rates. The time needed for calculation will be minimized if the first interest rate tried is fairly close to the correct rate. Frequently a simple inspection of the cash flow series will tell whether to start by guessing a fairly low rate or a fairly high one.

Where the cash flow series is irregular, it sometimes saves time to make a preliminary calculation before deciding on the first guessed rate. In such a calculation the cash flow series being analyzed may be changed in a way that makes it possible to find the interest rate quickly with the help of interest tables. The following paragraphs illustrate how such preliminary calculations might have been made for the cash flow series of Examples 8–1 and 8–2.

In Example 8–1 most of the positive net cash flow is concentrated in the 7th or final year. A first guess might assume the entire amount, +$298,300, in the seventh year. In effect, the cash flow series is approximated so that a P of

\$109,000 leads to an F of \$298,300 at the end of 7 years. The corresponding $(P/F,i^*,7)$ = \$109,000/\$298,300 = 0.365. Examination of the P/F factors for 7 years reveals that the approximated factor lies between the factors for 15% and 16%. Since the approximation concentrated all of the net positive cash flows at the end of the life of the investment, it is obvious that the real rate of return will be higher than 15%. This is true because one-third of the net positive cash flows actually occurred in the first 6 years. Good tactics in this case would be to make the first trial at 18%; the second trial could then be at either 16% or 20%, depending on the result of the first trial.

In Example 8–2 the net positive cash flow is not concentrated at the end of the period; it is well distributed throughout the 8 years. Substitute a cash flow series with the same net positive cash flow spread uniformly over the 8 years. In such a series, A is \$18,050 ÷ 8 = \$2,256. The corresponding series present worth factor is then $P \div A$ = \$12,000 ÷ \$2,256 = 5.32. For $n = 8$, P/A at 10% is 5.33 and P/A at 12% is 4.97. With uniform cash flow the return would have been about 10%. Because there is more positive cash flow in the early years than in the later years, the return must be more than 10%. Good tactics in this case would be to make the first trial at 12%; the second trial could then be at either 11% or 13%, depending on the result of the first trial.

Minor Errors Introduced by Linear Interpolation in Computing Rates of Return

In the examples in this chapter, the rates of return have been computed by the use of linear interpolation between two interest rates for which compound interest tables are provided in Appendix D. Since all of the factors involve the basic formula $(1 + i)^n$, it is obvious that the use of linear interpolation introduces some minor errors. It should be equally obvious that the closer together the two interest rates are, the smaller will be the error introduced by linear interpolation. Recall Example 8–4, in which the investment of \$8,000 in a 7% government bond would yield \$350 interest every 6 months for 20 years and the recovery of \$10,000 at the end of 20 years. Using the present worth method, it was determined that the rate of return would be between 4.5% and 5.0%. By linear interpolation between the values obtained at those rates, the solving rate was 4.638%. A complete solution by means of a computer showed the solving rate to be 4.603%.

The difference in this case was due to a combination of the use of linear interpolation *and* the fact that the interest tables are computed to three or four decimal places while the computer used eight decimal places. Figure 8–1 shows the present worth of future income by 0.1% intervals. It can easily be seen that the error would have been greater if the interpolation had been between 4% and 5% and even larger if it had been between 3% and 6%.

In this book the practice is to use linear interpolation and to state the interpolated rates of return to the nearest tenth of a percent. Advantages of linear interpolation are that the calculation can be made quickly by slide rule

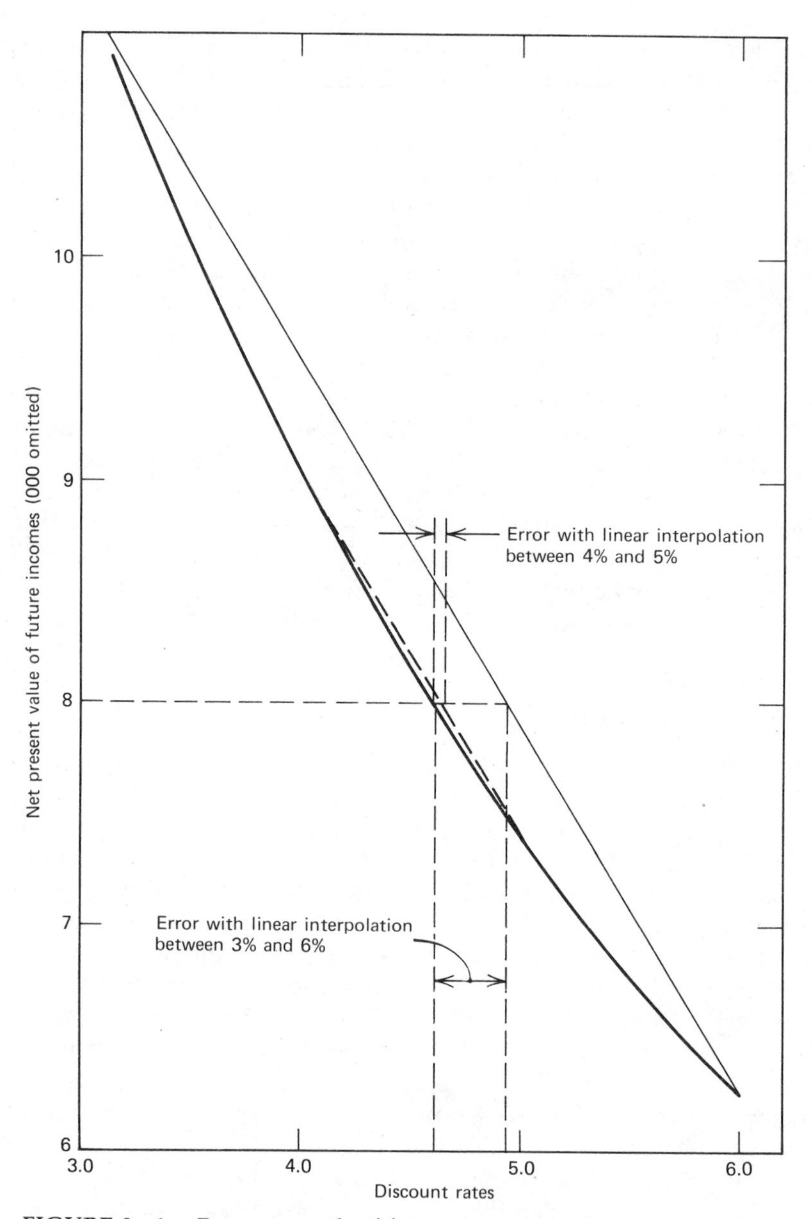

FIGURE 8–1. Present worth of future income at discount rates from 3% to 6% for Example 8–4

or by handheld calculators and that present worths need be calculated at only two interest rates. Analysts should recognize that linear interpolation sometimes causes an error in the rate of return of one or two tenths of a percent.

In general, the calculations of rates of return in economy studies are made to influence decisions among alternatives. The errors introduced by a linear interpolation in any compound interest method are usually so small as to have no appreciable influence on the decision making.

Use of Graphical Solutions

The examples in this chapter have been solved by the trial-and-error method, with linear interpolation between two interest values. Such a solution can be compared with the specified i^* and the proposed project can be declared acceptable or not acceptable. Frequently there are compelling reasons why the decision maker should know how a change in the i^* would affect the comparative advantages and disadvantages of the competing proposals. A graph of the net present worth of each cash flow stream for interest rates from 0% to some reasonable upper value can be very helpful.

Figure 8–2 shows the graphs of three cash flow streams plotted against varying interest rates from 0% to 50%. The cash flow streams are shown in Table 8–6 and are the results of projecting the future consequences of using each of three different production systems to manufacture a specified number of units of a product each year for 6 years. The before-tax i^* for this project was specified as 25%. The graphs clearly show that the solving rates will be

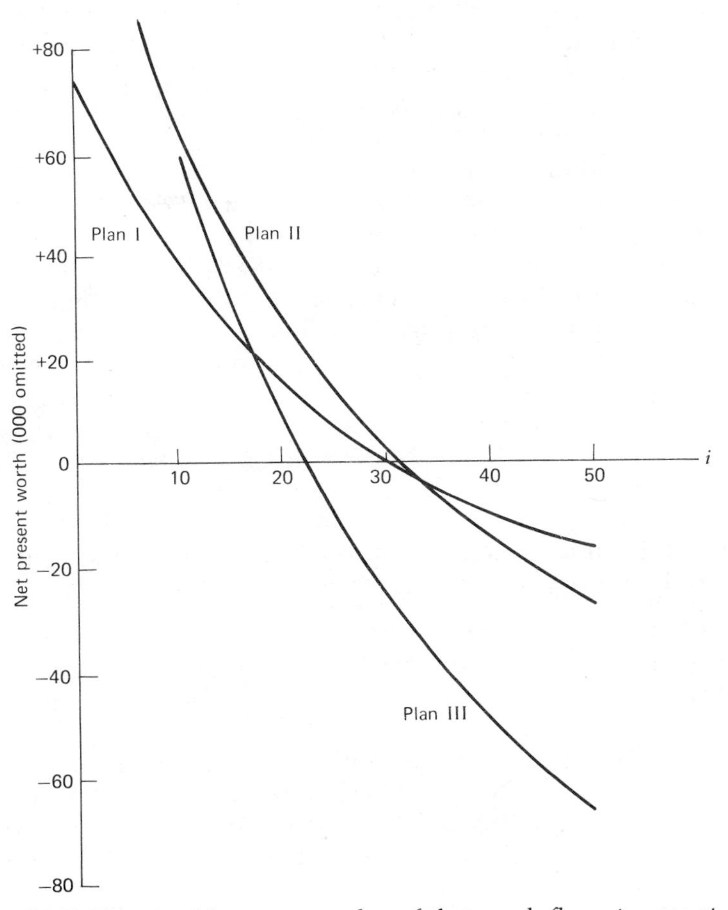

FIGURE 8–2. Net present value of three cash flow streams at interest rates from zero to 50%

TABLE 8–6
Prospective cash flows for three plans for the production of a specified quantity of a product each year for 6 years
(All figures in thousands of dollars)

Year	*Plan I*	*Plan II*	*Plan III*
0	−$50	−$90	−$150
1	+7	+16.5	+21
2	+24	+49	+64
3	+29	+52	+64
4	+23	+41	+54
5	+19	+34	+51
6	+11	+7	+27
6 S =	+10	+10	+10
Totals	+$73	+$119.5	+$141
Solving rate, i =	30.3%	31.2%	22.0%
Incremental rate		32.4%	8.1%

about 30% for Plan I, 31% for Plan II, and 22% for Plan III. If those prospective rates of return had been obtained by solving the present worth equations the decision maker would know that Plan II is better than Plan I and Plan III. An incremental analysis of the extra investment in Plan II over Plan I and Plan III over Plan II will verify that conclusion. More information may be very valuable to the decision maker.

Since the i^* is usually set in anticipation of other investment opportunities (as will be explained in Chapter 10), it may turn out to be overly optimistic or overly pessimistic. The graphs of the present worths of the three cash flow streams show the decision maker what the net present worths will be at different interest rates and clearly demonstrate that Plan II is preferable to Plan III at all interest rates above about 8%. At no interest rate below 32% is Plan I preferable to Plan II. Obviously if i^* is 32% or greater, none of the plans would be acceptable. The graphs give an excellent visual sensitivity analysis of the effects of varying the interest rates. (See Chapter 14 for other sensitivity analysis techniques.)

Use of Interest Tables Based on Continuous Compounding in Computing Unknown Rates of Return

The end-of-year convention in economy studies was explained in Chapter 6. This convention has been employed in the calculations of rates of return in the present chapter and is used in such calculations throughout the main body of this book.

Table D–30 and D–31 provide continuous compound interest factors. Their use in connection with rate of return calculations is explained in Appendix A. This appendix also discusses the circumstances under which the uniform-flow convention may be preferable to the end-of-year convention.

Rate of Return from a Combination of Two Separable Proposed Investments That Have Different Prospective Rates of Return

A certain mining property is for sale for $1,500,000. The engineer for the prospective purchaser estimates the remaining life of the mine as 8 years. For each of these years it is estimated that the excess of receipts over disbursements will be $391,000. The prospective rate of return on the investment is desired.

In mining enterprises there is a traditional method of determining rate of return known as Hoskold's method.* In this method it is assumed that uniform annual deposits will be made into a conservatively invested sinking fund that will earn interest at a relatively low rate. The annual deposits are to be just sufficient to replace the original investment at the end of the life of the property. The rate of return is computed by dividing the investment into the annual amount remaining after setting aside the sinking fund deposit.

To illustrate Hoskold's method, assume 4% interest on the sinking fund. The annual deposit in the fund to recover $1,500,000 at the end of 8 years is $1,500,000 (*A*/*F*,4%,8) = $1,500,000(0.10853) = $162,800. The annual cash remaining for the owners of the property after they have made the sinking fund deposit will be $391,000 − $162,800 = $228,200. As $228,200 ÷ $1,500,000 = 0.152, this project is viewed as one promising a 15.2% return.

If the same proposal is analyzed by correct compound interest methods, the computed rate of return is 20%; $391,000 (*P*/*A*,20%,8) = $391,000(3.937) = $1,500,000. The difference between the viewpoints underlying these 15.2% and 20.0% figures deserves some comment here, particularly because the viewpoint leading to the 15.2% figure is by no means restricted to the mineral industries.

If the purchaser of the mining property actually makes the two investments contemplated in the 15.2% calculation and if receipts and disbursements turn out as forecast, it is true that his combined rate of return will be 15.2%. From the combination of the two investments he will have a cash flow of −$1,500,000 at zero date, +$228,200 a year for 8 years, and +$1,500,000 at the end of the 8th year.

The important point to recognize here is that this 15.2% return is the result of *two* separate investments, one yielding 20% and the other yielding only 4%. Presumably the decision to make the investment with the 20% yield does not require that there also be a decision to make the 4% investment. If not, the 4% investment has no relevance in making the decision about the proposed investment with the 20% yield. It is the 20% figure, not the 15.2% one, that should be used as the index of attractiveness of the proposed investment in the mining property.

*This method is similar in some respects to the "Explicit Reinvestment Rate of Return Method" suggested by some writers and discussed in Appendix C of this book.

This Hoskold-type viewpoint on computing rate of return is rarely, if ever, appropriate as a basis for decision making on proposed investments. It seems particularly indefensible in the common case where it is used when no actual sinking fund is contemplated. Further comment on this topic is made in Appendix C.

Determining the Before-Tax Cost of Long-Term Debt

In finding the true interest rate paid by a corporation that borrows money by the sale of bonds, the calculations are similar to those indicated for computing rate of return on a bond investment. However, it is necessary to recognize that borrowing causes the corporation to make certain disbursements that are not receipts to the bond investor.

For instance, a $50,000,000 bond issue of 9.2%, 20-year bonds that was sold by investment bankers to the ultimate investor at 97½ (i.e., $9,750 per $10,000 bond) might have been sold by the issuing corporation to an investment banking syndicate at 94 9/10. Thus, the corporation would receive $47,450,000 for its promise to pay $4,600,000 a year for 20 years and $50,000,000 at the end of that time. These payments would repay the amount received with interest at about 9.8%.

If the initial expenses to the corporation in connection with the bond issue were $850,000 (for such items as engraving bonds, preparing a registration statement for the Securities and Exchange Commission, accounting and legal expenses in connection with the issue), and if the annual disbursements involved in fees for registrar and trustee, costs of making interest payments, and the like were $90,000, the true cost of this borrowed money would be even greater. The corporation is really receiving a net sum of $46,600,000 at zero date in exchange for an obligation to pay $4,690,000 a year for 20 years and $50,000,000 at the end of that time. Present worth calculations and interpolation indicate that the true cost of this borrowed money (before income taxes) is about 10.2%.

Thus a given loan may appear to have different interest rates depending on the point of view. In the foregoing example, the coupon rate on the bonds was 9.2%; the yield to the bond investor was about 9.5%; the bonds were sold by the corporation to the investment bankers at a price giving a yield of about 9.8%; considering the cost incidental to the borrowing of money, the actual cost of money to the corporation was about 10.2%.

The foregoing analysis disregards income tax considerations. Because interest payments and other expenses incident to borrowing are normally viewed as deductions from taxable income, borrowers who pay income taxes may find that their after-tax costs of borrowed money are much less than the before-tax costs that are obtained from the foregoing type of calculation. Certain aspects of the relationship between income taxes and borrowing are discussed in Chapter 18.

Interest Rates Are Not Always What They Seem

There are many cases where a superficial look at the facts may lead to an underestimate of the interest rate being paid by a borrower. One such case has just been described. Another was described in Example 5–21, where a "Seven Percent Plan" turned out to involve an interest rate of nearly 14%.

Another exists whenever, in a purchase of a property "on terms," there is a difference between the selling price to a cash buyer and one to a buyer who agrees to pay most of the purchase price in periodic installments with interest.

For instance, a residential property is for sale for $48,000 under the following arrangements; $8,000 cash and the balance of $40,000 to be repaid with interest at 8% in uniform installments for 15 years. Investigation discloses that a buyer on these terms must also pay $1,200 immediately for various expenses incidental to securing the loan. By inquiry, a prospective buyer discovers that the same property could be purchased for $45,800 cash.

For the sake of simplicity in our calculations, let us assume uniform *annual* payments rather than the monthly payments that would customarily be required. The uniform payment, A, to repay a P of $40,000, with $i = 0.08$ and $n = 15$, is $40,000(0.11683) = $4,673.

It is evident that a buyer for cash will pay $45,800 at once and thus conclude the transaction. A buyer on borrowed money will pay out $9,200 at once and $4,673 a year for 15 years. This $4,673 a year is clearly an alternative to a $36,600 immediate cash payment; if the buyer had the $36,600 he could substitute it for the promise to pay $4,673 a year for 15 years. To find the real cost of borrowed money to him, it is necessary to find the interest rate at which his annual payments of $4,673 for 15 years would repay $36,600. As A/P = $4,673/$36,600 = 0.12768, interpolation between the capital recovery factors for 9% and 10% shows that this interest rate is approximately 9.5%.

The difference between the apparent 8% interest and the actual 9.5% interest paid by the buyer on credit was concealed in the difference between the cash price and the credit price, and in the initial charges incident to the loan.

Certain Cases in Which Two or More Solutions Are Possible in Computing an Unknown Interest Rate

Some proposals involving estimated prospective cash flows combine one or more periods of time that are, in effect, investment periods, with one or more periods of time that are, in effect, borrowing or financing periods. Proposals of this type occur from time to time in the production operations of the petroleum industry and occasionally in other industries. They may be identified by the fact that the series of estimated cash flows has two or more reversals of sign.

This chapter has illustrated conventional types of trial-and-error calculations to find an unknown interest rate. Such calculations are appropriate for the common types of proposals that are solely investment or solely borrowing/financing. In an investment-type proposal, an initial negative cash flow is succeeded by one or more prospective positive cash flows. In a borrowing/financing-type proposal, an initial positive cash flow is succeeded by one or more prospective negative cash flows. For mixed proposals, where the cash flow series has two or more reversals of sign, such conventional calculations may give two or more values for the "solving" interest rate or, sometimes, no values at all. In such mixed proposals, misleading conclusions often will be reached from *any* conventional type of compound interest analysis, whether the analysis is based on net present worth, equivalent uniform annual cash flow, or the calculation of an unknown interest rate or rate of return. This subject is explored in Appendix B.

In that appendix it is explained that the key to an analysis of a mixed proposal that is primarily an investment proposal is the use of an auxiliary interest rate during the borrowing or financing period. For a proposal that is primarily a borrowing or financing proposal, the key is using an auxiliary interest rate during the investment period. It also is explained that in many cases, the conclusions of an analysis are relatively insensitive to large changes in the value of the assumed auxiliary interest rate; in such cases it usually is good enough for practical purposes to make conventional compound interest analyses even though the prospective cash flow series being analyzed has two or more reversals of sign.

Another Meaning for "Rate of Return on Investment"

Throughout this book, the phrase "rate of return on investment" is used in the meaning illustrated in this chapter, as the rate of "interest" at which an investment is repaid by an increase in net cash receipts. However, it should be pointed out that another meaning is sometimes given to this phrase. In this other use, "rate of return" is taken to mean the excess of the return over the current interest rate on borrowed capital, or the excess of the return over the going interest rate on conservative investments. A return described in this book as 9% would be described as a 3% return in a case where the going rate of interest was assumed to be 6%.

This other meaning for "rate of return" corresponds somewhat to the economic theorist's concept of "profit," just as the meaning adopted in this book corresponds more closely to the accountant's concept of profit. Either meaning is a possible one, but it is obvious that both cannot be used without confusion. The meaning used here has been chosen because it seems better adapted to practical business situations. Thus, wherever "rate of return" is used throughout these pages, it means a figure to be compared with the interest obtainable on investments elsewhere, rather than a figure in excess of such interest.

Summary

A comparison between alternatives involving money payments and receipts of different amounts at different dates may be expressed by an interest rate, the rate that makes the two alternatives equivalent. When one alternative involves a higher present investment and higher future net receipts (possibly as a result of lower future disbursements), this interest rate may be called the prospective rate of return on the extra investment. Its calculation provides one of the several methods of determining in an economy study whether a proposed investment will be recovered with a return commensurate with the risk—in other words, one of the methods of considering the time value of money in economy studies.

The actual rate of return realized by an investor from an investment cannot be determined until his association with the investment has terminated; it may differ substantially from the apparent rate of return at some intermediate period. The viewpoint of an engineering economy study for a proposed investment, involving as it does estimates for the full expected economic life of a machine or structure, implies calculations of rate of return of the same type as would be required to judge the actual return realized from terminated investments.

PROBLEMS

General Notes Regarding Problems for Chapter 8

Unless otherwise stated, the end-of-year convention is to be assumed for receipts and disbursements occurring during the year. Where alternative assets have different lives, it is to be assumed that replacement assets will have the same first costs, lives, salvage values, and annual disbursements and receipts as the assets they replace. Answers to Problems 8–1 through 8–11 are approximate, determined by interpolation to the nearest tenth of a percent.

8–1. Five years ago a man purchased a small office building for a net cost of \$100,000. The rent receipts exceeded his disbursements, including his income tax payments on the taxable income, by the following amounts:

Year 1	+\$ 9,500	Year 3	+\$10,500	Year 5	+\$11,500
Year 2	+\$10,000	Year 4	+\$11,000		

He has just sold the building for \$120,000, but must pay \$12,500 capital gains tax, leaving him a net recovery of \$107,500. What after-income-tax rate of return did he earn on his original investment? (*Ans.* 11.59%)

8–2. A certain municipal (tax exempt) bond which pays a nominal 6%, with semiannual interest payments, will pay its owner $10,000 at the end of 8 years. The bond can be purchased at a current market value of $9,200.

(a) What is the prospective nominal rate of return that will be earned on this investment?

(b) What will the effective annual rate be? (*Ans.* (a) 7.3%; (b) 7.5%)

8–3. A newly designed electronic component to be used in an assembly can be produced by either of two methods, A or B. Different investments and annual operating expenses are involved. The following table gives the comparative data:

Method	**A**	**B**
First cost	$40,000	$55,000
Economic life	5 years	5 years
Salvage value	0	$5,000
Operating disbursements:		
Year 1	$12,000	$ 9,000
Year 2	14,000	10,000
Year 3	16,000	11,000
Year 4	16,000	11,000
Year 5	16,000	11,000

The company's before-income-tax i^* is 18%. Should the company invest in A or B, and what is the prospective before-tax rate of return on the extra investment in B over A? (*Ans.* Invest in B; i = 19.0%)

8–4. A corporation receives a net $9,300,000 as a result of an issue of $10,000,000 of 10%, 20-year bonds on which interest is payable semiannually. There will be estimated semiannual expenditures of $40,000 for trustee's and registrar's fees and clerical and other expenses in connection with interest payments. What nominal interest rate compounded semiannually expresses the true before-tax cost of this borrowed money to the corporation? (*Ans.* 11.7%)

8–5. It is proposed to purchase a machine to be used for rental purposes. The first cost is $40,000. For the first year of ownership, $8,400 is estimated as the excess of receipts over disbursements for everything except income taxes. Considering declining rental receipts with age and increased upkeep costs, it is believed that this figure will decline by $500 each year and will be $7,900 in the second year, $7,400 in the third, and so on. It is estimated that the machine will be retired after 15 years with a $4,000 salvage value. Estimated disbursements for income taxes are $3,000 the first year and $2,750 the second, and will decrease by $250 each year thereafter. What is the prospective after-tax rate of return? (*Ans.* 5.4%)

8–6. Refer to Problem 6–1 (page 89). What is the prospective rate of return on the extra investment in Plan II over Plan I? If the i^* is 9%, should the government invest in Plan II? (*Ans.* 5.2%; no)

8–7. Refer to Problem 6–2 (page 89). What is the prospective rate of return on the extra investment in Plan A over Plan B? Which plan should the University select if its i^* is 6%? (*Ans.* 6.2%; Plan A)

8–8. Refer to Problem 6–6 (page 90). What is the prospective rate of return on the extra investment in Machine J over Machine K? If the company's $i^* = 12\%$, which machine should be selected? (*Ans.* 6.9%; select Machine K)

8–9. Refer to Problem 7–6 (page 112). The owner of the patent in that problem needs immediate cash in order to buy a home and offers to sell the patent to the corporation for $18,000. The patent has a remaining life of 15 years. If the corporation buys the patent now it will avoid the remaining 15 payments under the existing contract. What is the corporation's prospective rate of return on that investment? If the corporation's i^* before income tax is 16%, is this an attractive investment? (*Ans.* 17.3%; yes)

8–10. In problem 7–2 (page 111), what is the prospective after-tax rate of return on the extra investment required for the electric radiant panels? (*Ans.* 6.2%)

8–11. Two alternate designs are to be evaluated for a certain new project that has been proposed. Design Y involves a present investment of $100,000. Estimated annual receipts for 20 years are $45,000; estimated annual disbursements for everything except income taxes are $20,000. Design Z involves a present investment of $140,000, estimated annual receipts for 20 years of $64,500, and annual disbursements for everything except income taxes of $36,000. It is expected that there will be no value remaining in the project after 20 years regardless of the choice between the two designs. Estimated annual income taxes will be $10,000 with design Y and $10,750 with Z. Compute the prospective after-tax rate of return on the project with design Y and with design Z. Compute the prospective after-tax rate of return on the extra investment required for design Z. If the after-tax i^* is 10%, would you recommend Y, Z, or neither? (*Ans.* Y = 13.9%; Z = 11.2%; Z − Y = 3.3%; select Y)

8–12. An unimproved commercial tract of land is being considered for purchase as a speculation. The land will cost $100,000 now and the owner will have to pay property taxes of about $1,200 a year. The investor knows that commercially zoned land has increased in price at a rate of about 20% per year for the last 5 years. Thus, if it is retained for 5 years it should bring about $250,000. Sales commissions and title insurance will reduce the selling price by about $16,000. There will be no income from the land during the period of ownership. If these estimates are correct, what before-tax rate of return will the investor earn?

8–13. A company is considering the introduction of a new product which is believed to have a 10-year life. An initial investment of $120,000 will be required with an estimated salvage value of $40,000 at the end of 10 years. It is estimated that the gross receipts will be about $30,000 a year, but disbursements including extra income taxes will be about $15,000 a year. What is the prospective rate of return on this proposal? If the company's i^* is 12% after income taxes, should this project be undertaken?

8–14. In Problem 8–13, assume that the annual receipts will start at $24,000 the first year and increase by $1,500 a year, while the disbursements will start at $12,000 the first year and increase by $500 each year. Compute the prospective rate of return on the investment with these changed conditions. Is this an attractive investment?

8–15. Due to recent increases in interest rates, a bond that originally sold for $1,000 (face value) can be purchased for $850. The stated interest rate is 7.5%, with semiannual payments of interest. The bond will mature in 6 years. What is the effective annual rate of return for an investor who buys the bond and holds it until maturity?

8–16. The Joneses plan to build a new home in about 4 years and wish to consider the purchase of a city lot now to avoid the inflation of land prices. A lot in an area which they consider desirable at this time can be purchased for $25,000. The only cash flows during the period will be annual property taxes of about $300 (after income tax deductions). No income will be obtained while it is owned. The realtor assures them that in 4 years the lot will cost $38,000. The Joneses have $25,000 currently invested in tax-exempt municipal bonds that pay 8.5% per year. They believe that they can sell the bonds for at least $25,000 at any time. If they decide to sell the lot in 4 years rather than build on it they will have to pay long-term capital gains tax of about 20% of the difference between the purchase price and the selling price. Determine the prospective rate of return on the investment in the lot if the Joneses build a home on it and if they sell the lot 4 years hence. Analyze the proposal for the family, including consideration of irreducibles.

8–17. Refer to Problem 7–8 (page 112). Doe spends about $2,000 a year for a part-time engineer to check the tonnage of clay removed and he also considers that the land will only be worth $25,000 after the removal of the clay 15 years hence. The XYZ Company has offered him $185,000 for the land, and he believes that he can invest the money at 12% before income taxes. What rate of return will he earn on his "investment" if he does not sell the land to the XYZ Company? Should he sell now or hold it? Discuss the irreducibles involved in Doe's decision.

8–18. A commercial building containing stores and offices is for sale at $200,000. A prospective investor estimates that annual receipts from rentals will be about $32,000 and that annual disbursements other than income taxes will be $12,400. He also estimates that the building will be salable for a net of $180,000 at the end of 20 years. What is the investor's before-tax rate of return if the property is purchased and the estimates hold true for 20 years?

8–19. If a person had purchased 100 shares of Foy Manufacturing stock in

1972 the price would have been $5,300. The stock split 2 for 1 in 1975, and could have been sold in 1980 for $9,600. During the time of ownership the dividends were $318 per year for the first 3 years and $480 a year after that. What rate of return would the investor have obtained on the investment?

8–20. Jane Doe wants to purchase a car from a local dealer but does not have enough ready cash to pay the full price. The dealer offers her the following terms: Price of car = $6,500; down payment = $500; monthly payments for 36 months = $216.91. In addition she will have to take collision, fire, and theft insurance from the dealer at a cost of $400 a year, paid at the beginning of each year.

Jane thinks this is rather high, so she goes to the bank and asks for a car loan of $6,000. The bank's terms require a monthly payment of $207.99 and require that she carry collision, fire, and theft insurance with an acceptable company. She finds that she can obtain the same insurance coverage from a mutual company for about $325 a year (paid at the beginning of each year).

Considering the additional cost of insurance, what rate of interest (effective annual rate) will Jane be paying to finance her car purchase through the dealer? What interest will she be paying the bank to finance the purchase?

8–21. Two pumps are being considered for a certain service of 25 years. The Toltec pump has a first cost of $5,000 and an estimated annual cost of $3,200 for electric energy used for pumping purposes. The Mandan pump has a first cost of $3,800. Because it is less efficient its annual costs for pumping energy will be $140 higher than those of the Toltec pump. Estimated annual income taxes will be $42 higher with the Toltec pump. It is anticipated that there will be no other cost differences and no salvage values. What is the prospective after-tax rate of return on the extra investment required to purchase the Toltec pump?

8–22. A corporation sells an issue of $20,000,000 of 10%, 20-year bonds to an investment banking concern for $19,200,000. The corporation's initial disbursements for fees of lawyers, accountants, trustee, and other outlays in connection with the bond issue are $400,000. Each year the disbursements for fees to the trustee and registrar of the bonds and the clerical and other expenses in connection with interest payments are $250,000. What is the before-tax cost of this borrowed money expressed as an interest rate? To simplify calculations, assume that interest is payable annually.

8–23. Two mutually exclusive proposals are Projects X and Y. Project X requires a present investment of $250,000. Estimated annual receipts for 25 years are $88,000; estimated annual disbursements other than income taxes are $32,000; estimated annual income taxes are $24,000. Project Y requires a present investment of $325,000. Estimated annual receipts for 25 years are $100,000; estimated annual disbursements other than income taxes are $40,000; estimated annual income taxes are $23,500. Each project is estimated to have a $50,000 salvage value at the end of 25 years. Assuming an after-tax

i^* of 11%, make the necessary calculations to determine whether to recommend Project X, Project Y, or neither. Make a specific recommendation and explain why you made it.

8–24. John Brown bought 10 shares of K & L Co. stock for $10,000. He held this stock for 10 years. For the first 5 years he received annual dividends of $500. For the next 5 years he received annual dividends of $400. At the end of the 10th year, he sold his stock for $12,000. What before-tax rate of return did he make on this investment?

8–25. Refer to Problem 6–14 (page 92). What is the prospective rate of return on the extra investment in Unit Y over Unit X? If the i^* is 20% after income taxes, which unit should be purchased?

8–26. A well-known monthly magazine sells for $2 a copy on the news stands, but a person can subscribe to it at any one of the following rates:

1-year subscription	$21
2-year subscription	$38
3-year subscription	$52

Payments are made at the beginning of the year. Mary Smith is sure she will continue to read the magazine for at least 3 years. What effective annual rate of return will she obtain by subscribing for 1 year instead of buying single copies; 2 years; and 3 years? Assume that she must pay for a subscription 1 month before receiving the first issue.

8–27. A city hospital has an emergency power supply driven by a gasoline engine. The engine is 5 years old and requires a lot of maintenance to keep it in reliable condition. The engineer has made a study of replacement engines and has concluded that the best choice is between a K-400 gasoline engine or an E-450 diesel engine. The data for the decision are as follows:

	K-400	**E-450**
First cost	$4,000	$8,000
Estimated economic life	4 years	8 years
Estimated salvage value	$1,000	$2,000
Annual fuel and maintenance	$1,850	$1,400

Since this is a municipally owned facility, no income taxes are involved. The hospital's management uses 10% as its minimum attractive rate of return on investment decisions. What is the prospective rate of return on the extra investment in the E-450 over the K-400? Which engine should be selected?

8–28. A company that sells home appliances also provides repair and maintenance service for its customers. In order to provide fast service, each service man is supplied with a radio-equipped van. The company is expanding its service facilities and needs to add four new vans to its fleet. The question is, "Shall we buy or lease the vans?"

The company has selected a specific model and obtained the following information: To purchase the van will require a $7,000 investment, with an estimated economic life of 3 years. At that time, the estimated salvage value is $1,000. A leasing company will rent the van to the company for $200 a month (paid at the beginning of each month).

In both cases the company will pay for insurance, maintenance, fuel, repairs, and so on, so that there are no differences in those disbursements between owning and leasing the vans.

(a) What is the prospective effective annual rate of return before income taxes that will be obtained by purchasing the vans over leasing them?

(b) If the company buys the vans it will pay more income taxes by $324 a year than if it leased the vans. (Assume that income taxes are paid at the end of the year.) What is the after-tax effective annual rate of return on the investment in the vans over leasing?

(c) The leasing company can buy the vans for only $6,400 and estimates the salvage value after 3 years at $1,400. What before-tax rate of return will it earn on its investment?

8–29. An electronics manufacturer has been purchasing a certain component from an outside supplier. It is believed that the total cost of this component will be $10,000 next year and that the price will increase by $1,000 a year thereafter. A production engineer has proposed that the company buy some specialized equipment for $25,000, which would enable the company to manufacture the components in-house. Labor, materials, energy, property taxes, and insurance will be about $4,000 the first year and increase by about $500 a year. It is believed that the company will need these components for at least 5 years and that the equipment could be sold for $5,000 at the end of the 5 years. If the company makes its own components, it will incur an additional income tax of $1,000 next year and the extra income tax will increase by $250 a year thereafter.

Determine the prospective rate of return on the investment both before and after income taxes. The company's after-tax i^* is 9%. Should the investment be made?

9

BENEFIT–COST RATIO

It is hereby recognized that destructive floods upon the rivers of the United States, upsetting orderly processes and causing loss of life and property, including the erosion of lands, and impairing and obstructing navigation, highways, railroads, and other channels of commerce between the States, constitute a menace to national welfare; that it is the sense of Congress that flood control on navigable waters or their tributaries is a proper activity of the Federal Government in cooperation with States, their political subdivisions, and localities thereof; that investigations and improvements of rivers and other waterways, including watersheds thereof, for flood-control purposes are in the interest of the general welfare; that the Federal Government should improve or participate in the improvement of navigable waters or their tributaries, including watersheds thereof, for flood-control purposes if the benefits to whomsoever they may accrue are in excess of the estimated costs, and if the lives and social security of people are otherwise adversely affected.—Flood Control Act of June 22, 1936*

We started this book with General Carty's questions "Why at all?" "Why now?" "Why this way?" These questions are just as relevant in government as in private enterprise. The basic procedures in decision making about proposed governmental outlays for fixed assets ought to be the same as in proposed outlays in private business, namely:

1. Define alternatives clearly and try to determine the differences in consequences of various alternatives.
2. Insofar as practicable, make these differences commensurable by expressing them in terms of money.
3. Apply some criterion to the monetary figures to provide a basis for judgment whether proposed investments are justified. The time value of money should be recognized in establishing this criterion.
4. Choose among alternatives applying the foregoing criterion but also giving consideration to the differences among alternatives that were not reduced to money terms.

**United States Code*, 1940 ed. (Washington, D.C.: U.S. Government Printing Office), p. 2964.

Economic Evaluation of Proposed Public Works in Terms of "Benefits" and "Costs"

In the famous passage quoted at the start of this chapter, the United States Flood Control Act of 1936 stipulated that "benefits to whomsoever they may accrue" should exceed "estimated costs." Although this phrasing of a primary criterion for project evaluation applied officially only to flood control projects in the United States that were to be financed entirely or partially by the federal government, the phrasing gradually was adopted in the economic evaluation of various other types of federal public works projects. Then, as time went on, it was also applied to the evaluation of many proposed local public works projects within the United States. During the 1950s and 1960s, the same type of formulation of project evaluation standards was introduced in many other countries throughout the world.

Some Difficulties in Identifying and Measuring Relevant Consequences in Economy Studies for Proposed Government Projects

Conceptually, matters are more complicated in evaluating proposed public works projects than in evaluating similar projects in private enterprise. Also, the application of concepts is more difficult where governmental activity is involved. Certain differences between economy studies for private enterprise and economy studies for governments may be illustrated by contrasting the evaluation of a proposed relocation of a portion of a privately owned railway with the evaluation of a proposed relocation of a section of highway.

In both cases an economic evaluation calls for consideration of the required investment and of the influence of the project on roadway maintenance costs and on the costs associated with the movement of traffic. But there is an important point of difference. The railway company makes the investment and expects itself to recover the investment plus an adequate return through savings in costs of moving traffic and, in some cases, also through increased revenues and/or reduced maintenance costs. In contrast, although governmental agencies make the investment and pay highway maintenance costs, the savings in costs of moving traffic are made by the general public, not by any government agency. Therefore, the answer to the question "Whose viewpoint should be adopted?" is different for two projects that physically are similar. Clearly, the difficulty in estimating and valuing consequences is much less for the railway project than for the highway project.

Moreover, in looking at the railway decision from the viewpoint of the company's owners, an economy study can be based on cash flow, that is, on receipts and disbursements by the railway company. In contrast, when one looks at the highway decision from the viewpoint of consequences "to whomsoever they may accrue," it is not practicable to look at *all* cash flows. If one considers the entire population, every positive cash flow (receipt) by someone is a negative cash flow (disbursement) by someone else. Therefore, the algebraic sum of the cash flows would be zero and there would be no basis for an economic evaluation.

For example, if a project is expected to reduce highway accidents, there will be reduced *disbursements* by highway users for automobile repairs, medical and hospital services, and legal services. But the reduction in these outlays by the highway users will be accompanied by an equal reduction in *receipts* by the automobile repair shops, physicians, hospitals, and lawyers.

The foregoing illustration brings out the point that it becomes necessary in the economic analysis of a public works project to make a judgment as to which consequences are to be counted and which are to be disregarded. In the case of a proposed reduction in highway accidents, this judgment is not a difficult one. From the public viewpoint, highway accidents are deemed to be undesirable; one of the purposes of a proposed highway improvement may be to reduce them. When we take a more critical look at the subject in Chapter 19, we shall see that it is difficult to get a fully satisfactory monetary evaluation of all their adverse consequences. Nevertheless, a way to start such an evaluation is to estimate the reduction in monetary outlays by accident victims associated with a prospective reduction in the number and severity of highway accidents. Such a monetary saving is reasonably counted as a project *benefit*. The prospective reduction in receipts by repair shops, physicians, hospitals, and lawyers is disregarded in a project analysis; such consequences should not be included as "negative benefits" or "dis-benefits" or "costs" of a proposed highway project.

Purpose of This Chapter

Although most persons would doubtless agree that the foregoing reasoning is acceptable in the evaluation of highway accident reduction, decisions separating consequences considered to be relevant in an economy study from consequences considered to be irrelevant are not always so clear and noncontroversial. Chapter 19 deals with economy studies for governmental activities. We shall put off until that chapter our discussion of a number of conceptual issues and practical difficulties involved in identifying relevant benefits and costs and placing money values on them. In examples and problems before Chapter 19, we shall assume that suitable year-by-year money figures have been obtained to measure the relevant benefits and costs.

The main objective of the present chapter is to show the relationship between the various benefit–cost techniques and the methods of analysis that have already been explained. We shall see that with the same input data (including of course a stipulated minimum attractive rate of return), the *decision* reached by comparing benefits with costs is the same decision that will be reached by the methods explained in Chapters 6, 7, and 8. However, we shall also see that the same proposed project may have several different values of the benefit–cost *ratio* depending on whether certain adverse items are subtracted from benefits or added to costs.

Examples 9–1 and 9–2 involve three alternatives, one of which is a continuation of a present condition. Example 9–3 introduces *cost-effectiveness* analysis as an aid to decision making in cases wherein the benefits may not be reducible to money terms. Examples and problems involving more than three alternatives are introduced in Chapter 13.

Importance of the Interest Rate Selected in Economy Studies for Governments

Various names have been applied to the value of i^* used in the economic analysis of proposed public works. Whether this is called an imputed interest rate, a discount rate, a vestcharge, or given some other name, it is, in effect, the chosen minimum attractive rate of return. Moreover, even though the primary criterion for decision making may be phrased in terms of benefits and costs, it is the chosen value of i^* that really sets the standard for the investment decision. A discussion of the issues involved in selecting i^* in governmental economy studies is deferred until Chapter 19. Examples 9–1 and 9–2 assume values of 7% and 6%, respectively.

EXAMPLE 9–1

Application of Several Methods of Analysis to a Simple Highway Economy Study

Facts and Estimates

A certain stretch of rural highway is in such bad condition that its resurfacing or relocation is required. The present location is designated as H; two possible new locations that will shorten the distance between the terminal points are designated as J and K. Location K is somewhat shorter than J but involves a considerably higher investment for grading and structures. An economy study comparing the proposals for the three locations is to use a study period of 20 years and an i^* of 7%.

The estimated initial investment to be made by government highway agencies would be $110,000 at location H, $700,000 at J, and $1,300,000 at K. If location H is abandoned now, it will have no net salvage value; also it is assumed to have no residual value at the end of 20 years if this location is kept in service. However, because the estimated useful lives of the works that would be constructed at J or K are longer than the 20-year analysis period, residual values are estimated at the end of 20 years. These are $300,000 for J and $550,000 for K. Estimated annual maintenance costs, also to be paid by the government, are $35,000 for location H, $21,000 for J, and $17,000 for K.

It is forecast that the traffic on this section of highway will increase by a uniform amount each year until year 10 and will then continue at a constant level until year 20. It is not expected that the volume of traffic will be influenced by this decision on the location of the highway. For location H, annual road user costs deemed to be relevant in the economic analysis are estimated as $210,000 in year 1 and $220,000 in year 2, increasing by $10,000 each year until they reach $300,000 in the 10th year; thereafter they will continue at $300,000. For location J, which is shorter, the corresponding estimates are $157,500 for year 1, an increase of $7,500 a year until year 10, and a constant annual figure of $225,000 thereafter. For the even shorter location K, the corresponding figures are $136,500, $6,500, and $195,000. There are no differences in nonuser consequences that need to be considered.

Comparison of Present Worths of Relevant Costs
A tabulation of the present worths of the foregoing costs using an i^* of 7% is given in Table 9–1. This tabulation does not separate the capital and maintenance costs paid by the government from the road user costs paid by the general public. It is evident that location J has the lowest total present worth.

Comparison of Equivalent Uniform Annual Costs
Table 9–2 shows a comparison of the equivalent uniform annual figures using an i^* of 7%. Of course the three totals are in the same proportions to one another as the present worths totals in Table 9–1; in fact, each total annual cost figure is the product of total present worth and 0.09439, the capital recovery factor for 7% and 20 years.

Calculation of Prospective Rates of Return on Extra Investments
Trial and error calculations of the type illustrated in Chapter 8 will show that, as compared to location H, the extra investment in location J will yield a prospective rate of return of approximately 12.8%. That is, the prospective savings in maintenance costs and road user costs plus the greater residual value are just sufficient to recover the extra $590,000 initial outlay with an interest rate of 12.8%. Because 12.8% exceeds the stipulated i^* of 7%, location J is economically superior to location H.

Similar calculations comparing K with J will show that the extra investment in K over J has a prospective rate of return of approximately 3%. Because 3% is less than the stipulated i^* of 7%, the extra investment of $600,000 needed for K is not economically justified. Of course, the rate-of-return

TABLE 9–1
Tabulation of present worths, Example 9–1

	Location H	*Location J*	*Location K*
Investment less PW of residual value	$ 110,000	$ 622,000	$1,158,000
PW of maintenance costs	371,000	223,000	180,000
PW of road user costs	2,823,000	2,117,000	1,835,000
Total	$3,304,000	$2,962,000	$3,173,000

TABLE 9–2
Tabulation of equivalent uniform annual costs, Example 9–1

	Location H	*Location J*	*Location K*
Capital recovery cost	$ 10,400	$ 58,800	$109,300
Annual maintenance cost	35,000	21,000	17,000
Equivalent uniform annual road user cost	266,500	199,900	173,200
Total	$311,900	$279,700	$299,500

analysis reaches the same conclusion as the present worth analysis and the annual cost analysis; it favors J as compared to H or K.

Calculation of the Excess of Benefits over Costs
In the solutions given in this example, it will be assumed that a prospective saving in road user costs (a favorable consequence to road users) is classified as a "benefit" and that the initial investment and the annual outlay for maintenance (both paid by the government) are classified as "costs." An analysis comparing benefits with costs can be made using either present worths or equivalent uniform annual figures.

Using present worths and designating benefits by **B** and costs by **C**:
To compare location J with location H,

$$\begin{aligned} \mathbf{B} &= \$2{,}823{,}000 - \$2{,}117{,}000 &&= \$706{,}000 \\ \mathbf{C} &= (\$622{,}000 + \$223{,}000) - (\$110{,}000 + \$371{,}000) &&= \underline{364{,}000} \\ \mathbf{B} - \mathbf{C} & &&= +\$342{,}000 \end{aligned}$$

To compare location K with location J,

$$\begin{aligned} \mathbf{B} &= \$2{,}117{,}000 - \$1{,}835{,}000 &&= \$282{,}000 \\ \mathbf{C} &= (\$1{,}158{,}000 + \$180{,}000) - (\$622{,}000 + \$223{,}000) &&= \underline{493{,}000} \\ \mathbf{B} - \mathbf{C} & &&= -\$211{,}000 \end{aligned}$$

The foregoing differences of course are the same differences in total present worths that could have been calculated from Table 9–1. Naturally, the conclusion is favorable to location J just as in Table 9–1.

Using equivalent annual figures:
To compare J with H,

$$\begin{aligned} \mathbf{B} &= \$266{,}500 - \$199{,}900 &&= \$66{,}600 \\ \mathbf{C} &= (\$58{,}800 + \$21{,}000) - (\$10{,}400 + \$35{,}000) &&= \underline{34{,}400} \\ \mathbf{B} - \mathbf{C} & &&= +\$32{,}200 \end{aligned}$$

To compare K with J,

$$\begin{aligned}
\mathbf{B} &= \$199{,}900 - \$173{,}200 &&= \quad \$26{,}700 \\
\mathbf{C} &= (\$109{,}300 + \$17{,}000) - (\$58{,}800 + \$21{,}000) &&= \quad \underline{46{,}500} \\
\mathbf{B} - \mathbf{C} & &&= -\$19{,}800
\end{aligned}$$

Here we have the same differences that could have been computed from the three totals in Table 9–2.

Calculation of Benefit–Cost Ratios

The reader will recall that the 1936 flood control act quoted at the start of this chapter merely stated that benefits should exceed costs. Nevertheless, analysts who compare benefits with costs nearly always compute a ratio of benefits to costs. The stipulation $\mathbf{B} - \mathbf{C} > 0$ can also be expressed as $\mathbf{B/C} > 1$. The **B/C** ratios in this example will of course be the same whether they are computed from present worths or from equivalent uniform annual figures. To compare J with H,

$$\mathbf{B/C} = \frac{\$706{,}000}{\$364{,}000} \text{ or } \frac{\$66{,}600}{\$34{,}400} = 1.94$$

To compare K with J,

$$\mathbf{B/C} = \frac{\$282{,}000}{\$493{,}000} \text{ or } \frac{\$26{,}700}{\$46{,}500} = 0.57$$

Like all the other types of analysis that have been illustrated, the analysis using the **B/C** ratios favors location J. However, later in this chapter we shall see that the same input data can give different values of the **B/C** ratio depending on whether certain items are viewed as affecting the numerator or the denominator of the **B/C** fraction.

EXAMPLE 9–2

Benefit–Cost Analysis of Flood Control Alternatives

Facts and Estimates

Just before Willow Creek has its outlet into a salt water bay, it goes through an urban area. Because there have been occasional floods that have caused damage to property in this area, a flood control project has been proposed. Estimates have been made for two alternative designs, one involving channel improvement (CI) and the other involving a dam and reservoir (D & R). Economic analysis is to be based on an estimated 50-year project life assuming zero terminal salvage value and using an i^* of 6%.

The "expected value" of the annual cost due to flood damages is $480,000 with a continuation of the present condition of no flood control (NFC). The alternative CI will reduce this figure to $105,000; the alternative D & R will reduce it to $55,000. (Obviously, some years will have no flood damage; other years may have considerable damage. It is not possible to predict the specific dates of the years that will have the severe floods; the best that can be done is to predict the long-run relative frequencies of flood magnitudes and their related damages. The "expected value" is obtained from such estimates by the use of the mathematics of probability; methods of computation are illustrated in Chapter 15. In that chapter we shall see why it is reasonable in this type of economic analysis to treat the respective figures of $480,000, $105,000, and $55,000 as if they were uniform annual figures.)

The CI alternative has an estimated first cost of $2,900,000 and estimated annual maintenance costs of $35,000. Both of these require disbursements by the government.

The D & R alternative has an estimated first cost of $5,300,000, and estimated annual operation and maintenance costs of $40,000, both requiring disbursements by the government. This alternative also has two types of adverse consequences related to the conservation of natural resources. These are to be treated in the economic analysis as disbenefits (sometimes called "negative benefits" or "malefits"). The dam will cause a damage to anadromous fisheries; this is priced at $28,000 a year. The reservoir will cause a loss of land for agricultural purposes including grazing and crop raising; this is priced at $10,000 a year.

Benefit–Cost Analysis

Example 9–1 made the point that it was possible to use equivalent uniform annual figures in a benefit–cost analysis. Because most of the estimates in the present example consist of annual figures, our analysis will be based on annual benefits and annual costs.

In comparing the alternative of channel improvement (CI) with a continuation of the present condition of no flood control (NFC), the annual benefits are due to the reduction in the expected value of flood damages. The annual costs are the annual capital recovery costs and maintenance costs of CI.

$$\begin{aligned} &\mathbf{B}\ (\text{CI} - \text{NFC}) = \$480{,}000 - \$105{,}000 &&= \quad \$375{,}000 \\ &\mathbf{C}\ (\text{CI} - \text{NFC}) = \$2{,}900{,}000(A/P,6\%,50) + \$35{,}000 &&= \quad \underline{219{,}000} \\ &\mathbf{B} - \mathbf{C} &&= +\$156{,}000 \end{aligned}$$

$$\mathbf{B/C} = \frac{\$375{,}000}{\$219{,}000} = 1.71$$

In comparing the dam and reservoir (D & R) with channel improvement, the extra annual benefit is the further reduction in the expected value of flood damages minus the disbenefits associated with the loss of fisheries and agricultural resources. The extra annual cost is the extra capital recovery cost due to the greater investment plus the extra annual operation and maintenance cost.

$$\mathbf{B}\,(\text{D \& R} - \text{CI}) = (\$105{,}000 - \$55{,}000) - (\$28{,}000 + \$10{,}000) = \$\ 12{,}000$$

$$\mathbf{C}\,(\text{D \& R} - \text{CI}) = [\$5{,}300{,}000(A/P,6\%,50) + \$40{,}000] - [\$2{,}900{,}000(A/P,6\%,50) + \$35{,}000] = 157{,}000$$

$$\mathbf{B} - \mathbf{C} = -\$145{,}000$$

$$\mathbf{B/C} = \frac{\$12{,}000}{\$157{,}000} = 0.08$$

With the given input data associated with a criterion stated either as $\mathbf{B} - \mathbf{C} > 0$ or $\mathbf{B/C} > 1$, it is evident that channel improvement is economically justified whereas the proposed dam and reservoir are not economically justified.

The Fallacy of Merely Comparing All Proposals for Change with a Continuation of a Present Condition

In both of the foregoing examples, one alternative was to continue a present condition (location H in 9–1 and no flood control in 9–2).

In Example 9–1, location J was economically superior to both locations H and K. Similarly in Example 9–2, channel improvement was economically superior both to no flood control and to the proposed dam and reservoir. Nevertheless, the higher investments required for location K and for the dam and reservoir would have appeared to be justified if comparison had been made only with a continuation of the present condition. If K should be compared to H, the **B/C** ratio would be 1.15. If D & R should be compared with NFC, the **B/C** ratio would be 1.03.

It should be clear to the reader that where there are more than two alternatives, it is *never* sufficient to compare a proposed alternative only with the least attractive of the remaining alternatives. This is elaborated in Chapter 13 and discussed further in some of the later chapters.

The Arbitrary Aspects of the Classification of Certain Items in a Benefit–Cost Analysis

The present-worth and annual-cost solutions in Example 9–1 treated all input data as affecting costs; road user costs were included as one of the components of all the relevant costs. Subsequently, in the solutions using benefits and costs, reductions in road user costs were classified as *benefits*. The capital recovery cost of the highway investment and the highway maintenance cost were classified as *costs*. In the D & R alternative in Example 9–2, the adverse consequences associated with the loss of certain natural resources were classified as *disbenefits*.

Our practice in the benefit–cost analyses in these two examples was consistent with a rule that relevant consequences to the general public should be classified as benefits or disbenefits and that consequences involving disbursements by governmental units should be classified as costs. However, such a rule of classification is entirely arbitrary; different rules are used by different analysts. For example, annual maintenance costs are sometimes deducted from benefits rather than added to costs. Certain adverse consequences to the general public (such as the loss of natural resources in Example 9–2) are sometimes added to costs rather than treated as disbenefits.

The Influence on the Benefit–Cost Ratio of the Decision on Whether Certain Items Are Classified as Costs or as Disbenefits

It is a deficiency of the benefit–cost *ratio* as a scheme of project evaluation that legislators, administrative officials, and concerned members of the general public often have the view that the higher the ratio the better the project and vice versa. Actually, although (**B** – **C**) is unaffected by the decision as to whether an item is classified as a cost or as a disbenefit, the ratio **B/C** can be considerably influenced by this arbitrary decision.

For example, consider a project that has $300,000 of benefits $100,000 of costs, and a $90,000 adverse item that some analysts would classify as a cost and others would classify as a disbenefit. If this $90,000 item should be classified as a cost:

$$\mathbf{B/C} = \frac{\$300{,}000}{\$100{,}000 + \$90{,}000} = \frac{\$300{,}000}{\$190{,}000} = 1.58$$

However, if it should be classified as a disbenefit:

$$\mathbf{B/C} = \frac{\$300{,}000 - \$90{,}000}{\$100{,}000} = \frac{\$210{,}000}{\$100{,}000} = 2.10$$

With either classification, (**B** – **C**) = $110,000. The real merits of the project are unrelated to the classification of the $90,000 adverse item. Nevertheless, to an uncritical observer the apparent merits of the project may be greatly influenced by this classification.

Cost-Effectiveness Analysis

In many cases involving the investment of public funds, and in certain cases involving private industry investment, the primary benefits or consequences are not reducible to money terms. Examples include the evaluation of weapons systems, certain aspects of social health and welfare programs, and pollution control measures. In those cases wherein some nonmonetary measure (or measures) of effectiveness can be established, *cost-effectiveness analysis* provides management with a quantitative guide to decision making.

Costs, in this case, may include research, development, acquisition (construction or manufacture), operation, maintenance, and salvage costs for an expected number of required units, including replacements, over the expected operational life of the system. Usually these costs will be expressed as a present worth or an equivalent uniform annual cost for a specified number of years. The present worth is frequently termed the *life-cycle cost* of the system.

EXAMPLE 9–3

Cost-Effectiveness Analysis of Air Pollution Alternatives

A coastal city must install air pollution control equipment at one of its coal-fired electric generating stations to reduce the amount of sulphur dioxide (SO_2) emission. The system must be capable of reducing the SO_2 level at a pollution control monitoring station one-quarter mile downwind from the plant from a maximum recorded level of 23 parts per million (ppm) to a maximum of 4 ppm in any 8-hour period as measured on the monitor's collector.

Three bids have been received from companies with equally good reputations for the design and installation of reliable stack scrubber systems. Each bid was to be made on a 20-year life-cycle cost basis using an interest rate of 12%. That is, each company was asked to assume a service period of 20 years and to present its bid in terms of the present worth of all costs for that period. The three bids are as follows:

Company	Initial cost	Life-cycle cost
A	$118,000	$268,000
B	96,000	294,000
C	136,000	312,000

Companies A and B guarantee that their systems will meet or be less than the specified maximum of 4 ppm of SO_2 in any 8-hour period. Company C guarantees that its system will meet or better a 2 ppm specification.

Although the initial cost of the system proposed by Company A is higher than that for the system proposed by Company B, the life-cycle cost is lower. Both systems provide the same level of protection in terms of the specified measure of effectiveness. Therefore, Company A's bid is preferred to Company B's.

The choice between A's bid and C's is not as clear because there is no market price for the greater effectiveness guaranteed by Company C. As is the case when applying benefit–cost analysis, we must consider the incremental improvement in effectiveness compared to the increased life-cycle cost. In this case, the improvement is a guaranteed reduction of 2 ppm of SO_2 at a life-cycle cost of $44,000. If this is an acceptable cost, or if it is anticipated that the standard may be reduced, Company C's bid would be preferred.

Discussion of Example 9–3

The decision process in cost-effectiveness analysis is, in some respects, quite similar to that used in benefit–cost analysis. In other respects it is quite different. Usually, only the consequences measurable in money terms are costs. The beneficial consequences usually cannot be reduced to monetary values.

When a target value of the effectiveness measure is given and each alternative meets the target, the problem reduces to that of choosing the minimum cost alternative. This type of choice has been illustrated in many of the examples discussed so far. The choice between Company A's and Company B's bids further illustrated this principle. In cases where the life-cycle costs are the same but the effectiveness measures are different, the alternative with the greatest effectiveness should be selected.

This *rule of dominance* may reduce the number of alternatives from which a choice must be made, but it will not necessarily lead to a final selection. In our example, the comparison of Company C's bid with that of Company A offered improved effectiveness but at an additional cost. Unlike the benefit–cost analyses of the previous examples, no clear choice is indicated by this incremental comparison.

Considering the 50% reduction in *inefficiency* accomplished by choosing C's bid over A's as the appropriate measure of effectiveness to be compared with the life-cycle cost of $44,000 can only confuse the decision process because the 50% figure is an elastic measuring stick. It would apply whether the reduction were from 4 ppm to 2 ppm, or 2 ppm to 1 ppm, or 60 ppm to 30 ppm. Someone must decide whether the reduction of 2 ppm is worth $44,000.

Some General Comments on the Subject Matter of This Chapter

In public works, just as in private enterprise, valid economy studies can be made without using the word "benefits." In fact, we illustrated such studies in Chapters 6 and 7 (Examples 6–4 and 7–2). Moreover, even in Example 9–1 we noted that present worth, annual cost, and rate of return analyses gave the same conclusion yielded by the benefit–cost analysis. It is not necessary or desirable to phrase decision criteria in terms of benefits and costs in the economic evaluation of all proposed governmental investments.

The authors of this book have observed that the attempt sometimes made to formulate the economic analysis of *all* government projects in terms of so-called benefits and costs has occasionally been an obstacle to sound thinking. In cases where all the relevant differences expressed in monetary terms can be described as estimated cash flows by a governmental body, as was the case in Examples 6–4 and 7–2, there is no good reason for a benefit–cost type of formulation. The special useful concept emphasized by a formulation using the word *benefits* is that it is desirable to examine prospective consequences "to whomsoever they may accrue."

In some cases where certain favorable or unfavorable consequences cannot be reduced to money terms, other objective measures of effectiveness may exist to aid in decision making. Where either the costs or the effectiveness measures are the same, the decision process is relatively simple. However, when incremental effectiveness may be improved at an incremental cost, the decision may have to be based on a *subjective* evaluation of that incremental difference.

It was pointed out near the start of this chapter that we are putting off until Chapter 19 our discussion of certain troublesome and controversial matters that often arise in the evaluation of public-sector proposals. Among these matters are the choice of a minimum attractive rate of return, the selection of a method for placing a money valuation on consequences to the general public, and the decision as to *which* prospective consequences to the public are deemed to be relevant in any given economic evaluation. In that chapter, also, we shall take a brief look at certain institutional factors that sometimes create obstacles to sound decision making in the public sector of the economy.

PROBLEMS

9–1. Example 9–1 followed the practice, which has been fairly common in highway economy studies, of classifying highway maintenance expenditures as "costs" in computing **B/C** ratios. The respective **B/C** ratios for locations J over H, K over J, and K over H, were 1.94, 0.57, and 1.15.

It has been suggested that outlays for maintenance ought to be classed as disbenefits. Compute the three **B/C** ratios making this change in classification. (*Ans*. 1.67; 0.61; 1.13)

9–2. In Example 9–2, the $38,000 a year figure for the prospective loss of natural resources that would be caused by the D & R project was classed as a disbenefit. The respective **B/C** ratios for D & R as compared to CI and NFC were 0.08 and 1.03. What would these **B/C** ratios have been if the loss of natural resources had been classified as a cost rather than as a disbenefit? (*Ans*. 0.26; 1.03)

9–3. Make the necessary calculations to check the present worth figures given in Table 9–1. (Do not use the annual cost figures from Table 9–2.)

9–4. Make the necessary calculations to check the annual cost figures given in Table 9–2. (Do not use the present worth figures from Table 9–1.)

9–5. In Example 9–2, what is the prospective rate of return on the extra investment in CI as compared to NFC? What is the prospective rate of return on the extra investment in D & R as compared to CI? (*Ans*. 11.7%; negative)

9–6. A relocation of a stretch of rural highway is to be made. Alternate new route locations are designated as M and N. The initial investment by government highway agencies will be $3,000,000 for M and $5,000,000 for N. Annual highway maintenance costs will be $120,000 for M and $90,000 for the shorter location N. Relevant annual road user costs are estimated as $880,000 for M and $660,000 for N. Compute the **B/C** ratio or ratios that you believe to be relevant for an economy study comparing the two locations. Use an i^* of 8%, a 20-year study period, and assume residual values equal to 60% of first cost. (*Ans*. 1.49)

9–7. For a certain proposed government project, annual capital costs to the government are $300,000 and annual operation and maintenance costs to the government are $700,000. Annual favorable consequences to the general public of $1,100,000 are partially offset by certain annual adverse consequences of $300,000 to a portion of the general public. What is the **B/C** ratio if all consequences to any of the general public are counted in the numerator of the ratio and all consequences to the government are counted in the denominator? What will it be if the classification of the $300,000 adverse consequences to the general public is changed from a disbenefit to a cost? If the $700,000 operation and maintenance cost to the government is changed from a cost to a disbenefit? If both changes are made? (*Ans*. 0.80; 0.85; 0.33; 0.67)

Regardless of the classification of these two items, the benefits are $200,000 less than the costs. However, this is a project that will be of great advantage to certain persons in the general public who are promoting it enthusiastically. These persons have argued that irreducibles favorable to the project are so important that the project should be undertaken even though the estimated benefits are somewhat less than the estimated costs. Which classification of these two items will make the project appear to be the most favorable by giving it a **B/C** ratio nearest to unity? Which classification will be the least favorable to the project?

9–8. Black Creek flows through a suburban area near a medium-sized city into a large river. During recent years, annual floods have damaged homes in its vicinity at an increasing rate. Average annual flood damages have been estimated to be $300,000 at present and are expected to increase by $20,000 each year.

A proposal has been made to do some channel improvement (CI) and build a storage dam and small reservoir (D & R) to contain excessive runoff. Initial cost of CI is $400,000 and annual maintenance is expected to be $30,000. The D & R will be built only if the CI is made. Initial cost of D & R is $1,600,000; annual operation and maintenance is estimated to be $45,000.

If the CI is made, average annual flood damages are expected to be

$180,000 at present and to increase by $12,000 each year. Certain losses to the ecology are believed to be offset by increased recreational benefits due to improved navigability. If the D & R is built, average annual flood damages are expected to be reduced to $30,000 at present and to increase by $5,000 each year. The D & R will remove certain land from agriculture and possible future development. This disbenefit, valued at $20,000 per year, is partially offset by improved recreational fishing valued at $8,000 per year.

Using an i^* of 9%, find the appropriate **B/C** ratios for CI and D & R. Consider all consequences to the government as costs and all consequences to the general public as benefits (or disbenefits). Use a 30-year study period and assume no residual value at the end of that time.

9–9. Solve Problem 9–8 for the present worth of costs for each alternative. Consider no flood control (NFC) as the base alternative. Which alternative has the lowest present worth?

9–10. Calculate the appropriate **B/C** ratios in Problem 9–8 assuming an i^* of 5% rather than 9%. Does this change in i^* affect the conclusions drawn in Problem 9–8?

9–11. In Problem 9–8, it is suggested that recreational benefits of the dam and reservoir (D & R) can be substantially enhanced by performing certain clearing operations and by the provision of an access road and boat launching ramp. This modification will cost $100,000 initially and increase annual maintenance costs by $5,000. Timber salvaged from the reservoir basin can be sold for $15,000. The additional recreational benefit is valued at $18,000 per year. How does this modification affect the conclusions reached in Problem 9–8?

9–12. Solve Problem 9–6 assuming that the stated road user costs are for the first year of service and that these costs will increase by $30,000 each year at location M and $25,000 each year at location N throughout the 20-year analysis period.

9–13. A community has been directed to reduce by 75% the amount of solid and chemical waste effluent discharged from its sewage treatment plant into a small river. Three companies have bid on systems to modify the plant. Each system is expected to last for the remaining life of the plant, 15 years. Initial costs and annual increases in plant operation and maintenance costs for each of the three bids are:

	Bid A	Bid B	Bid C
Initial cost	$125,000	$160,000	$190,000
Annual O & M	15,000	12,000	9,200

If Bid A is accepted, certain elements of the system will have to be replaced at 5-year intervals at a cost of $25,000. The contractor offering Bid C guarantees that the proposed system will reduce waste discharged by 90%.

Contractors offering Bids A and B guarantee only that their systems will meet or exceed specifications.

Assuming an i^* of 9%, which bid should be accepted? Is this the system with the lowest initial cost? What is the present worth of the costs of the greater efficiency guaranteed by Bid C? What factors might lead the community to select Bid C?

9–14. A government agency offers a loan subsidy program to communities to encourage local pollution control measures. Assume in Problem 9–13 that the community can borrow necessary construction funds for its project at 4%. How does this fact affect the decision as to which bid should be accepted?

9–15. Solve Problem 9–6 assuming that the stated road user costs are for the first year of service, and that these costs will increase by $15,000 each year at M and $11,500 each year at N throughout the 20-year analysis period.

10

CHOICE OF A MINIMUM ATTRACTIVE RATE OF RETURN

"Well, in our country," said Alice, still panting a little, "You'd generally get to somewhere else—if you ran very fast for a long time as we've been doing."

*"A slow sort of country!" said the Queen. "Now, here, you see, it takes all the running you can do, to keep in the same place. If you want to get somewhere else, you must run twice as fast as that!"—Lewis Carroll**

In many examples and problems up to this point, we have stipulated a figure for i^*, the minimum attractive rate of return, without any discussion of why the particular value of i^* was selected. In this chapter we examine some of the troublesome issues that arise in selecting a minimum attractive rate of return.

The Minimum Attractive Rate of Return Is a Matter for Policy Determination

The minimum attractive rate of return to be used in judging the attractiveness of proposed investments is normally a policy matter to be determined by the top management of the organization, and the bases for setting it are quite varied. Since the i^* will be used by engineers, designers, and managers at all levels of the organization, the i^* must be set only after careful consideration of all the factors, because many decisions that will affect the long-term welfare of the organization will be based on the comparison of the prospective rate of return on proposed investments with the i^*. The determination of an i^* to be used in decision-making process at all levels helps to assure that the decisions are all based on the same primary criterion and to assure the best use of available funds.

Factors Considered in Setting i^*

The factors usually considered in the determination of the i^* to be used during any period of time include:

*Lewis Carroll, *Through the Looking Glass*, chap. 2.

1. Availability of funds for investment and their sources—equity or borrowing.
2. Competing investment opportunities.
3. Differences in the risk involved in the different competing investment opportunities.
4. Differences in the time required for recovery of the investment with the desired rate of return—short-lived versus long-lived investments.
5. The "going price of money" as represented by the interest rates paid or charged on such investments as FDIC insured savings accounts, the "prime rate" used by large banks, and the government short- and long-term notes and bonds.

Various aspects of these factors will be discussed in this chapter, but it should be recognized that there is no way for one person to tell another person or an organization what its minimum attractive rate of return *should be*. The importance placed on the different factors varies widely among individuals and changes over time. For example, some persons are "risk averse" and place more importance on avoiding risk than others who are frequently termed "risk seekers." The former will normally seek investments involving very little probability of failure even if that decision means that the prospective rate of return is quite low. The risk seeker will look for investment with very high prospective rates of return, even knowing that the probability of failure is significant; for example, oil exploration.

Some investors rely heavily on the theoretical economist's view that investment should continue, even if the funds must be obtained by borrowing, as long as the marginal rate of return promises to be equal to or greater than the marginal cost of borrowed capital. Other investors, and they tend to be a majority, are not willing to borrow capital for investment unless their prospective rate of return on the investment is substantially higher than the cost of the borrowed capital. This is viewed not only as a safety factor, but also as the incentive to assume the fixed obligation to repay the borrowed capital with a specified interest, even if the investment does not turn out as predicted. How large the difference between the cost of borrowing and the prospective return must be to justify that action is very dependent on the individual making the decision.

Some investors assume that there is no limitation on the amount of capital that can be obtained by borrowing and try to base their decision rules, and their i^*s, on some differential between the borrowing rate and the prospective rate of return. On the other hand, there is usually a practical limit to the amount that any organization can borrow because as the debt-to-equity ratio increases for the organization, the lenders will require higher and higher interest rates for the borrowed funds. At the same time, it becomes increasingly difficult for the organization to raise more equity capital by selling new shares of stock. To do so requires the promise of much higher earnings (dividends and growth in stock value.) This condition, if for no other reason, makes the capital rationing approach to the determination of the i^* a practical necessity.

These introductory remarks help to show that there is no "one best way" to arrive at the minimum attractive rate of return to use in any organization. Nor is it possible to state "typical i^*s for typical industries" with any degree of confidence. An examination of the *Fortune* magazine's annual report of the "Fortune 500" will indicate very quickly that the actual after-tax rates of return on the total invested capital in the largest corporations in the United States vary widely from losses in some years to very large returns in others. These so-called internal rates of return are not good guides to the i^*s that should be used by those corporations. Economic conditions, constraints on business, tax laws, investment opportunities, availability of money, prime rates, and the individual's estimates of the future economic climate are subject to considerable fluctuation over time. The best that can be done is to try to clarify the significance of the various factors that affect the decision factors entering into the determination of the i^*s and to indicate how these factors can be used in the decision process.

The following examples help to emphasize the importance of these factors in establishing the decision criteria.

EXAMPLE 10–1

Criteria for Investment Evaluation in a Certain Closely Held Corporation

Facts of the Case

Three members of a family held slightly more than half of the stock of a successful small company which we shall call the ABC Manufacturing Company. This company had no long-term debt. These controlling stockholders were active officers of the corporation. As the company made capital goods, its profits fluctuated considerably; although its overall profit record had been excellent, there had been occasional loss years.

The proposal was made to expand by manufacturing certain new products. An analysis of the proposal indicated that there was a good prospective rate of return. However, all the moneys for plant investment that became available each year from retained earnings and depreciation charges were being absorbed by replacements and plant modernization in connection with the present product line. A substantial investment in plant and equipment was needed to undertake the new product line. Therefore, it was necessary to raise new capital if this proposal for business expansion were to be accepted.

Investigation disclosed that this capital could be raised either by the sale of new stock to certain persons interested in the company or by a 10-year loan from an individual investor. The proposal to make the new products was finally rejected by the three controlling stockholders on the grounds that neither type of financing was acceptable to them.

The objection to the sale of stock was that the three stockholders would no longer have a majority stock interest that made it certain they could control their company's affairs. The 10-year loan also involved certain restrictions on their control through stipulations (such as one limiting dividend payments while the loan was outstanding). However, their chief objection to the loan was that the required annual payments of principal and interest made them much more vulnerable to any business recession that might cause one or more loss years.

In effect, this decision, based on considerations related to financing, caused the rejection of an investment proposal (plant expansion) yielding a high prospective rate of return even though other investment proposals (replacements and modernization) were being accepted yielding lower rates of return.

EXAMPLE 10–2

Criteria For Investment Decisions in an Underfinanced Manufacturing Business

Facts of the Case

Two partners purchased a small manufacturing enterprise. They used their entire personal savings for a payment to the former proprietor of half the purchase price. The remainder of the purchase price was to be paid from a stipulated percentage of the profits.

During its initial years, the partnership was always short of cash. The partners saw many chances to reduce production costs by moderate outlays for new equipment or for changes in existing equipment. Because of the cash shortage and because it was impracticable to bring new money into the business, every proposal had to be judged primarily with relation to its effect on the short-term cash position of the business. During the first year it was not possible to adopt any proposal that—in terms of cash flow—would not "pay for itself" in three months.

EXAMPLE 10–3

Changing Minimum Attractive Rates of Return as Opportunities Change

Facts of the Case

After the end of World War II, a large multiplant, international company recognized the opportunity to add another major product line to its very large line of products. Machine tools, general equipment, skilled personnel, and building materials were in very short supply, and the company did not see how it could build and equip a new plant in time to take advantage of the pent-up demand for the new product. Therefore, it started looking for an existing plant which could be purchased and quickly converted to the new product line.

Several existing plants were available, and after careful analysis of each relative to equipment, location relative to market areas, and personnel available, the company selected a rather old plant for purchase. Most of the equipment in the plant was in good condition and well maintained but was rather old and somewhat obsolete, requiring a disproportionate amount of skilled labor. The plant had been used for war supplies production and had enjoyed the privilege of deferring employees from the draft. Therefore, it had a large number of highly skilled workers who wished to remain in the same area.

Each of the company's factories was operated more or less autonomously, and after an initial period of investment, the factory was expected to be self-sustaining in every respect and to earn a rate of return on the total current investment equal to or greater than that for the company as a whole. When this plant was acquired, the Board of Directors made $400,000 available to the plant manager for capital investments in the first year of operation. Any improvements requiring capital investments after the first year would have to come from internally generated funds. Very shortly after opening the plant, the manager asked his department managers and engineers to propose investments which would increase the output and reduce operating costs. Within a matter of weeks he had received proposals for over $1,500,000, and most of the proposals were very attractive. Rather than review every one of so many proposals, he designed a special evaluation form requiring the estimation of the prospective rate of return to be earned on each proposal, and stated that "no proposal will be considered unless it has the prospect of earning a return of 45% after income taxes during its conservatively estimated economic life." He was successful in obtaining enough proposals which met the criterion to use all the funds at his disposal.

In successive years the internally generated funds for capital investment tended to remain about the same as the initial allocation, but the opportunities for very attractive investments decreased as the major problem areas were corrected. Within 4 years the manager reduced the minimum attractive rate of return after income taxes to 12% at a time when the overall internal earning rate on all investments for the parent company was a little less than 10%.

The Concept of Capital Rationing in Relation to the Minimum Attractive Rate of Return

Resources available for new investment in capital assets during any given period of time usually are limited even though the constraints may not be as severe as those in Examples 10–1 and 10–2. Management usually makes an annual or periodic estimate of the funds that will be available for capital investments, and it desires to invest those funds in such a way as to maximize the prospective rate of return on them, observing certain internal or policy restraints. Also, technological progress and an expanding economy make it common for the total of the proposals for investment in new assets to be considerably greater than the total of available funds. For example, assume that the manufacturing company of Example 10–1 had $90,000 of funds for its capital budget for a given year and that there were investment proposals for plant modernization and expansion totaling $207,000 as listed in Table 10–1.

In this simple example it is evident that the $90,000 of available funds would have been exhausted by Projects U, Y, Z, and S. It follows that the minimum rate of return that was attractive was 15%, the prospective rate on Project S. If an investment had been made in any project yielding less than 15% (X, T, V, W, or some other project not tabulated), the effect would have been to eliminate the possibility of investing in some project that was expected to yield 15% or more.

Example 10–3 is an illustration of a fairly typical situation in which the overall business plan of a large organization allocates a certain amount of capital for investment to the different manufacturing plants or divisions. Top level corporate management generally decides the amount of investment capital to be allocated to each operating division. In contrast, a plant manager often can decide among the proposals competing for the investment capital made available to that particular plant. In this case the availability of extremely attractive opportunities was due more or less to the obsolescent conditions of the plant and its equipment, and not to any special skill or knowledge of the manager. Knowledge and skill in management were demonstrated by setting a very high i^* in order to be sure that the available money was allocated to the most productive proposals. These investments no doubt helped to make more capital available in future years by showing very high

TABLE 10–1
Proposals for capital expenditures in ABC Manufacturing Company for a certain year
(Available funds limited to $90,000)

Project	*Investment required*	*Prospective rate of return after taxes*	*Cumulative total of investments*
U	$12,000	40%	$ 12,000
Y	45,000	20%	57,000
Z	8,000	18%	65,000
S	25,000	15%	90,000
X	22,000	12%	112,000
T	30,000	10%	142,000
V	55,000	9%	197,000
W	10,000	8%	207,000

profitability indexes. The gradual reduction of the i^* to 12% over four years represents the common situation wherein the unusually good investment opportunities are chosen first, and gradually the i^* is reduced to conform to the opportunities available as well as to other criteria.

Since this was a new plant the manager might have decided to hold some of the available capital in reserve, in highly liquid form earning a low rate of return, rather than commit it all to less attractive proposals (less than 45% in the first year.) Such decisions are often made when the management believes that better opportunities will become available in a short time and wants to have the resources available to take advantage of the situation.

Validity of the Capital Rationing Concept

Matters are rarely as simple as implied by the foregoing discussion. For example, the alternative of securing new outside capital often is available. Moreover, it may be impracticable to array all the proposals for capital expenditures during the coming year and to be sure that no other good proposals—now unforeseen—will develop during the year. Even without outside financing, the total funds that can be made available for capital expenditures may not be fixed absolutely but may be related to the attractiveness of the proposed projects. In choosing among available projects, it often is desirable to apply supplementary criteria in addition to prospective rate of return. The differences in the duration of the consequences of the various proposals may be deemed to be an important consideration. There may be no group of projects that will exactly absorb the available funds. Various aspects of these topics are discussed throughout the remainder of this book and the whole subject is considered more critically in Chapter 21.

Nevertheless, the principle is entirely sound that the minimum attractive rate of return ought to be chosen with the objective of making the best possible use of a limited resource. This resource is, of course, the money that can be made available for investment in capital assets and closely related items. If the consequence of making an investment yielding 10% is to forgo some other investment that would yield 20%, it is not sensible to make the 10% investment. The high figures for i^* that so often are used in competitive industry are based in part on this principle.

Moreover, as emphasized throughout this book, it is prospective *differences* among alternatives that are relevant in their comparison. The prospective rate of return from a proposed investment should be based on the difference between making the investment and not making it.

Explanation of Interest in Economic Theory

We can answer the practical question, "Why consider the time value of money in decision making?" by saying that interest is a business fact. It is also desirable to consider the answer to the more fundamental question, "Why does interest exist as a business fact?"

In answering this question, economists explain interest, as they explain any other sort of price, by examining the supply and demand situations for investment funds. On the supply side they point out that interest is necessary as an incentive to saving; on the demand side they point out that interest is possible because capital is productive.

Explanation of the Possibility of and the Necessity for Interest

There probably is not a person over three or four years of age who does not know that the use of a tool will enable him or her to do a specific job more quickly, or with less energy, or of better quality, all with very desirable and valuable results. The same persons will also know that the tools are acquired at the sacrifice of money for their purchase. The concept of investing in capital or durable goods in order to improve productivity is a well-known and well-understood fact. Economists explain that one must forgo immediate consumption in order to save for investment, and the ordinary person commonly "invests" by depositing some unused income in savings accounts in banks. The banks pay interest for the use of this money, which is in turn loaned to business, individuals, or governments for investment in durable goods. The interest the bank pays the depositor is more or less the result of the supply of money and the demand for money to be invested in durable goods, property, or business assets. The bank provides a convenient vehicle by which the savings of many persons can be consolidated and used in large amounts by other investors.

The person or corporation wishing to have money for investment in capital goods goes to the lending institutions to borrow money. Again, the interest the borrower must pay fluctuates with the supply and demand situation. The rate is always higher than the rate the lending institution pays to its depositors, because it is rendering a beneficial service which involves operating expenses and it is also entitled to some reward for its services.

Thus, interest is possible because investment capital is productive; it can provide a business enterprise with better tools, better buildings, rental property, machinery and equipment for lease, reduced labor requirements, better products, and so on. Moreover, there needs to be an incentive for individuals to forgo immediate consumption so that investment opportunities will be attractive to them. The amount of the incentive will naturally affect the amount available for investment.

When engineering economists review alternative ways of accomplishing some task or goal, they usually are concerned with the required amount of capital investment in relation to the benefits to be derived from the investment. Therefore they must always consider the earning power of money within their own organizations as well as in the money markets. They must justify their proposals on the basis that each proposed investment has the prospect of being recovered with a rate of return that is sufficient in light of the risk involved and the going cost of money. Thus, when i^* is being determined for an organization for a specific period of time, the current and foreseeable conditions of the money market should be taken into consideration. An obvious *lower limit* for i^* in a private organization is the rate of return available from relatively risk-free opportunities such as investments in government bonds, money market certificates, treasury notes, and guaranteed bank deposits. In most cases the appropriate i^* will be considerably above this lower limit because of the competition among good investment opportunities within the organization.

This statement of the reason for recognizing interest in decision making seems to imply a slight difference in the explanation between the situation in which money is actually borrowed and the situation in which it is available without borrowing. Since businessmen sometimes reason differently about these two situations, it is worthwhile to examine them separately at this point.

Distinction Between Equity Funds and Borrowed Funds

Assume that you buy an $80,000 home by paying $20,000 in cash and securing a long-term loan for the remaining $60,000. Your *equity* in the $80,000 property is then $20,000. Your home ownership has been financed 25% by equity funds and 75% by borrowed funds.

In business the equity funds are the funds provided by the owners of the enterprise. In corporate business the owners are the stockholders of the corporation. A corporation may also finance in part by long-term borrowings, frequently through the sale of bonds. The bondholders are *creditors* of the corporation; their legal relationship to the corporation differs greatly from that of the stockholders. Where funds are borrowed, there is generally an agreement to pay interest and principal at stipulated dates. No such obligation exists in connection with equity funds.

Some of the aspects of doing business on borrowed money are discussed in Chapter 18. At the present point in our discussion, it needs to be brought out that the less stable the earning power of a prospective borrower, the less desirable it is to borrow. Many business enterprises in competitive industry are financed largely or entirely from equity funds. Authorities on finance recognize that it is appropriate for many regulated public utilities to secure from one-third to one-half of their capital from long-term borrowing. In certain types of government projects, the entire first cost of a project is financed by borrowing.

Considering the Time Value of Money for Proposals to Be Financed by Equity Funds

Engineering structures and machines may be built or acquired by individuals, partnerships, private corporations, and governmental bodies. Unless they are financed by borrowing, they must necessarily be financed out of money belonging to the owners of the enterprise.

These equity funds may come from various sources. In private corporations, for instance, they may come from the sale of stock, or from profits that are "plowed back" into the business rather than paid out as dividends to the stockholders, or from the recovery of capital previously invested in other machines and structures. In governmental bodies, equity funds may come from direct assessments or taxation (e.g., the gasoline tax used to finance highway improvement).

Where capital assets are financed entirely by equity funds, it is not necessary to pay out interest to any creditor. Here interest is a cost in the economists' sense of *opportunity cost*. When funds are invested in any particular capital goods, the opportunity is forgone to obtain a return from the investment of the funds elsewhere. Interest is a cost in the sense of an opportunity forgone.

In deciding whether to invest equity funds in specific capital assets, an important question is how good an opportunity will be forgone. In other words, if the investment in the specific capital assets is not made, what return is likely to be obtainable from the same funds invested elsewhere? In principle the interest rate (minimum attractive rate of return) used in an economy study ought to be equal to or greater than the rate of return obtainable from the opportunity forgone, as nearly as can be determined.

What Investment Opportunity is Being Forgone?

The opportunity forgone may be either within the business enterprise or outside of it. In Table 10–1, the minimum attractive rate of return was 15% because any investment yielding less than 15% would cause the ABC Manufacturing Company to forgo the opportunity to earn 15% in Project S.

The appropriate figure for minimum attractive rate of return is generally higher when the opportunity forgone is within the enterprise. Two circumstances, both illustrated in Table 10–1, are generally present when a within-the-enterprise figure is controlling. One circumstance is the presence of many good opportunities for investment within the enterprise. The other circumstance is the limitation of available funds. High minimum attractive rates of return are common in competitive industry because both of these circumstances occur so frequently.

If new equity capital cannot be secured for a business enterprise and no new money is to be borrowed, the available funds for investment in new fixed assets are usually limited to current earnings retained in the business (if any) and to capital recovered from previous investments in fixed assets. (This latter source of funds is discussed in Chapter 18.) But even where new equity funds *can* be secured, the management of a business enterprise may deem it unwise to obtain them. In many small and moderate-sized enterprises that are owned by a few individuals, the securing of new equity capital may involve a sacrifice of control by the present owners. This condition was illustrated in Example 10–1. In large corporations in competitive industry where no question of control is involved, boards of directors often find other reasons that influence them against the raising of new equity capital.

In determining a minimum attractive rate of return in a given business enterprise or other sphere of activity, it always is appropriate to consider possible opportunities for return that may be forgone outside of the enterprise as well as within it. For example, corporate stockholders have opportunities for personal investments outside of their corporation; the board of directors should not withhold part of the current earnings from the stockholders unless the prospective return from the reinvestment of these earnings within the enterprise is as great as the return the stockholders could obtain from personal investments elsewhere. A similar line of reasoning may be applied to government projects financed by current taxation; the collection of these taxes requires the taxpayers to forgo an opportunity to earn a return from personal investment of the moneys collected. Determination of appropriate personal investment opportunities is not a simple problem for either boards of directors or government administrators. (See Problem 10–7.)

The Element of Risk in Relation to the Minimum Attractive Rate of Return

In some types of loan transactions the risk of loss is recognized to be greater than in other types. (A good measure of risk of loss would be obtained by finding the actual losses sustained by lenders on different types of loans over a long period of years.) The risk of loss influences the interest rate. Generally speaking, the poorer the credit rating of a borrower, the greater the interest rate he or she will have to pay.

In a similar way the standard of attractiveness applied to proposals for capital expenditures in industry may be related to estimated risk of loss. For example, there are four major divisions in the petroleum industry: production, refining, transportation, and marketing. There are obvious differences in risk associated with investment proposals in the different divisions. A large integrated oil company once recognized these differences by requiring a minimum attractive rate of return of 18% after income taxes for certain types of proposals in the production division, 14% for proposals in the refining division, and 10% for proposals in the transportation and marketing divisions.

All future estimates of the consequences of a decision are just that—estimates. Consequently, some decision makers assume a long-lived proposal involves more risk than a short-lived project simply because it is more difficult to estimate the consequences accurately. They might, therefore, accept a 5-year project with a prospective rate of return of 12% over a 15-year project with a prospective rate of return of 16%.

Often the element of risk is recognized at the level of decision making by top management without the use of any such formal rules. For example, in Table 10–1 the management of the ABC Manufacturing Company might deem

Project X to be considerably less risky than Project S. Project X might therefore be preferred even though its rate of return is only 12% compared to the 15% estimated for Project S.

Analysts disagree on the question of whether it is better to recognize this element of risk of loss in setting a minimum attractive rate of return or to introduce the matter into economy studies in some other way. This topic is discussed further in Chapters 14, 15, and 21.

The "With or Without" Viewpoint with Reference to Prospective Rates of Return on Investments in Competitive Industry

The president of a large manufacturing company in a highly competitive industry was discussing various matters with the engineer responsible for review and analysis of investment proposals. The president made comments along the following line:

> In our company, we approve many proposals for investments aimed at cost reduction. Generally speaking, the proposals approved show prospective rates of return of 16% or more after income taxes. When we post-audit the results of these investments, we conclude that the cost reductions realized have been, on the average, somewhat greater than we forecast in computing the prospective rates of return. Nevertheless, our average overall rate of return on investment is only about 8% after taxes and does not seem to be improved by these numerous cost reduction investments that individually seem to be so successful.

The president was, in fact, describing a condition that reasonably may be expected to be the normal state of affairs in competitive industry. The difference between *making* the investments in cost reduction equipment and *not making* these investments was measured by the 16% rate of return. However, under the stress of competition, the favorable consequences of these good investments were shared among the owners of the enterprise, its employees, and its customers. Wage and salary rates to the company's employees were increased from year to year; prices of the industry's product was reduced (if measured in monetary units of constant purchasing power) and the quality of the product was improved.

In spite of the fact that the owners of the enterprise did not keep all the return yielded by the cost-reduction equipment, the 16% rate was a valid measure of the productivity of this equipment from their point of view. If their company had *not* installed the modern equipment but its competitors had done so, competition would still have made it necessary to reduce prices and improve product quality; it would have been necessary to raise wage and salary rates because such rates are responsive to industry-wide conditions, not merely to conditions in one particular company. Under conditions of competition the overall rate of return of 8% could not have been maintained unless there had been cost reductions that, considered on a "with or without" basis, yielded rates of return of much more than 8%.

Relationship of the Minimum Attractive Rate of Return to the Cost of Borrowed Money

Consider an economy study to judge the justification of a project to be financed entirely by borrowing.

Persons who have not given the matter much critical thought often assume that the interest rate to be used in such a study ought to be the bare cost of borrowed money. Although this view is particularly common in governmental projects and in personal economy studies, the same view is sometimes advanced in connection with projects in competitive industry. The following paragraphs relate solely to competitive industry.

The reasons why the minimum attractive rate of return to be used in such an economy study should be greater than the cost of borrowed money may be summarized as follows:

1. Decisions made for business enterprises engaged in competitive industry are presumably made from the viewpoint of the owners of the enterprise. If the prospective return to be obtained from investing borrowed funds in capital assets is just equal to the cost of the borrowed money, the owners will gain no advantage from the borrowing. The debt and interest must be repaid regardless of the success or failure of the proposed investment. The prospective rate of return on the proposed investment must be greater than the cost of the funds in order to provide an incentive to assume the risk.

2. Even though it may *seem* as if certain types of assets can be financed entirely by borrowing (e.g., certain machinery purchased on the installment plan), the amount of possible borrowing by any business enterprise depends on the amount of equity capital in the enterprise. Generally speaking, the cost of new capital to an enterprise ought to be viewed as a weighted average of the cost of borrowed capital and equity capital. This weighted average will nearly always be considerably higher than the cost of borrowed money.

3. If there is a limit on the total funds available for investment in capital assets from all sources including borrowing, and if there are many proposals for investments in assets that seem likely to yield high returns, the type of reasoning illustrated in Table 10–1 is applicable. If the $90,000 of available funds in Table 10–1 had come entirely from borrowing at, say 7% after income taxes, rather than from equity sources, the minimum attractive rate of return for the ABC Manufacturing Company would still have been 15% after income taxes. The controlling element in determining the minimum attractive rate would still have been the fact that the selection of any project yielding less than 15% would cause the elimination of some project that would yield 15% or more; the net 7% cost of the borrowed money would not have been relevant.

Shall Minimum Attractive Rate of Return in Competitive Industry Be Before or After Income Taxes?

Decisions in business enterprises engaged in competitive industry are presumably made from the viewpoint of the owners of the enterprise. Obviously it is to the owners' advantage to obtain the best possible rate of return *after* income taxes rather than *before* income taxes.

Where analysts responsible for economy studies in competitive industry make studies before income taxes, the implication is that the same choices among alternatives will be made by studies made before taxes or after taxes.

If it were invariably true that an array of projects in order of rate of return would be the same before and after income taxes, the conclusions of economy studies would not depend on whether the studies were made before or after taxes. Under such circumstances the greater simplicity of making studies before taxes would be a valid basis for always using before-tax studies and merely increasing the minimum attractive rate of return (or interest rate used) enough to recognize the effect of income taxes.

However, it frequently happens that the best projects after income taxes are not the same as the best ones before income taxes. Usually this circumstance arises because of differences in rate of write-off for tax purposes applicable to different investments or because of different tax rates applicable to different investments. A number of such cases are described in Chapter 12 and thereafter. For this reason, it is desirable that most economy studies in competitive industry be made after income taxes.

Economy Studies When Inflation Is Expected

Certain difficulties in choosing among proposed investments are caused when the general price level is expected to change, either upward or downward. During the 1970s, inflation, an upward movement of prices, was common throughout the world. We shall put off until Chapter 14 our discussion of the relationship between economy studies and the expectation of inflation.

Here we merely point out that when price levels rise and the interest and principal payments on loans are paid in currency units of decreasing purchasing power, lenders may find that their rates of return are very low indeed if an analysis is made in units of constant purchasing power. (This point will be illustrated in Examples 14–1 and 14–2.) Therefore, interest rates generally rise in periods of inflation. In effect, part of the stated interest rate on a loan is compensation to the lender for the reduced purchasing power of future receipts of principal and interest. The high interest rates associated with prospective inflation may or may not be relevant in choosing a minimum attractive rate of return for an economy study; this depends on matters that are discussed in Chapter 14.

Need for a Uniform Criterion of Attractiveness of Proposed Investments at All Levels of Decision Making Throughout an Enterprise

A common condition in industry is described by Robert F. Barrell as follows:*

> Most companies have some definite sum of money available for investment purposes. Since the amount of available capital is limited, decisions must be made

*In an unpublished paper, "Analog Computers for Calculating the Rate of Return on Added Investment," submitted to fulfill the requirements of the management training program of the University of Buffalo. Mr. Barrell is an engineer for a large manufacturing company.

> with respect to alternative uses such that the company will maximize its earnings on this added investment. Furthermore, a company's budget director and executives must set up some base for evaluating the returns on alternative uses for capital before such decisions can be made. Generally, a company will establish some minimum acceptable rate of return, below which it feels the return is insufficient to justify the risk assumed by the company on that particular venture.
>
> It has been observed that under present practices in industry, decision making pertaining to investments is based upon time-rate-use of money only in the higher echelons of top management. For example, decisions pertaining to amounts set aside for expansion programs, research and development, are usually made on this basis. However, the decisions which are made by the lower echelons of management and by engineering, scientific and manufacturing personnel, pertaining to the actual expenditure of these funds, are not based on time-rate-use of money, but rather on hunch decisions and value judgments. Although these decisions may be fortified by rough computations based upon reasonable assumptions, they often yield misleading results. Furthermore, due to the lack of uniformity in such approximations from one person to the next, these decisions cannot possibly reflect top management policy for the optimum use of invested capital.
>
> It would appear that this is a serious defect in our industrial planning structure. Decisions are based on sound calculations up to the point where the detailed expenditures are made, and at this point the basis for decision making suddenly changes. What would it mean to a company if each of the purchases of machines, tools, molds, dies, or plant modifications were all judged on a minimum acceptable rate of return for the additional investment set by top management for that particular plant or division?

Our Table 10–1 represented an analysis of a group of separate projects such as might be prepared by a budget director for submission to top management. As pointed out in the foregoing quotation, it is desirable that the economic decisions within each project be made on substantially the same basis as the top management decisions among the projects. Otherwise, desirable elements that would yield high returns may be eliminated from some projects and other projects may be overdesigned in the sense that they include unjustifiable increments of investment.

Representative Values for Minimum Attractive Rate of Return

Whether one wishes to discuss what in fact is being done or what ought to be done in selecting a value of i^*, it is helpful to give separate consideration to the following four types of activity:

1. Business enterprises in competitive industry where, for one reason or another, the capital budget is limited to funds generated within the enterprise.
2. Business enterprises in competitive industry that regularly or occasionally acquire new funds for plant investment either by acquiring new

equity capital (in the case of a corporation, selling new stock) or by making new long-term borrowings, or both.

3. Regulated public utility companies (selling electricity, gas, water, telephone and telegraph service, transportation service), particularly those operating under the rules of rate regulation that have developed in the United States.
4. Governmental activities.

The distinctions among these four groups are not always clear, but for the purpose of the following discussion we shall assume that they are four distinct categories.

1. In principle, opportunity costs within the enterprise normally should determine the choice of i^* when the capital budget is limited to internally generated funds. Clearly, no generalizations can be made about where the cut-off point is likely to come under these circumstances; it is a matter of the estimated productivity of the investment proposals and the capital budget available. As in Example 10–3, the i^* might be quite high at certain times and low at others.

If internal investment opportunities are not particularly attractive, opportunity costs outside the enterprise may establish the appropriate value of i^*.

2. This is the type of case about which many writers on economics and finance argue that the i^* should be the weighted average "cost of capital" to the enterprise, considering both long-term borrowings and equity capital. For example, if the cut-off point based on internally generated funds is 15%, as in Table 10–1, and the overall cost of capital is, say, 9%, there should be enough new financing to permit the approval of all projects with prospective rates of return greater than 9%. One difficulty, discussed in Chapter 21, is that there is seldom any clearly defined and agreed-upon figure for the "cost" of equity capital to be used in the computation of the weighted average cost of capital. Another difficulty in trying to determine common practices in this type of industry is that industry uses many different definitions of rate of return.

It seems to the authors that no one really knows the relative frequencies of values of i^* selected by enterprises in the United States that make annual cost studies, present worth studies, or rate of return studies by the compound interest methods. However, the authors have discussed this question with a number of analysts in competitive enterprises that do use these methods. The impression obtained from these conversations is that the use of an after-tax i^* of 12% or higher is more common than values lower than 12%. Furthermore, these rates have tended to rise during the late 1970s.

3. Chapter 20 deals with some of the problems of the regulated public utilities, including the matter of an appropriate i^*. The "fair return" permitted by the regulatory agencies tends to be an absolute lower limit for i^*, and that has tended to range between 6.5% and 15% depending on various circumstances. On the other hand, the regulated utilities must maintain a constant supply of new capital in order to keep up with the service demands, and this

new capital must be a combination of borrowed and new equity capital. In order to attract the new equity capital there must be the prospect of higher returns to the stockholder than the cost of borrowed money (interest on its bonds). Therefore, the $i^\star$ usually is several points higher than the "fair return" and will be influenced by the opportunity costs.

4. Many government agencies in the United States make economy studies using as the interest rate some assumed average cost of borrowed money. Historically there have been many economy studies that have, in effect, assumed a minimum attractive rate of return from 2½% to 3½%. In many other economy studies, particularly for highway agencies financing highway projects from current taxation, a 0% interest rate has been used. In the opinion of the authors, it usually is unsound public policy to select such very low values of $i^\star$ in the economic analysis of government projects; low rates in the 0%–3% range disregard opportunity costs (the rate of return the taxpayer could earn if he or she could invest the taxes paid) and tend to cause overdesign of public works projects. This topic is discussed in Chaper 19.

Differences Between Engineering and Accounting Viewpoints on the Time Value of Money

Engineering economy studies generally deal with *proposed* investments in machines or structures. As long as an investment is only proposed but not yet made, it is necessary to recognize interest in any calculations relative to the decision of whether or not to make it; there is always offered the alternative of an investment at interest. All of the consequences of the engineering economy decision lie in the future.

The engineer's usual viewpoint here is in contrast to that of the accountant. Accounting records deal generally with *past* investments, receipts, and disbursements. In calculations relative to past expenditures and disbursements, the time value of money may or may not be considered, depending upon the questions that it is desired to answer by means of the calculations. Many of the questions that the accounts of a business are called on to answer do not require the consideration of interest; thus, interest on ownership capital (i.e., where there is no actual interest payment to a creditor) is not generally considered as a cost in accounting, although there are some exceptions.

This difference between the engineer's viewpoint before the event and the accountant's viewpoint after the event often creates problems in trying to reconcile the calculations made for engineering economy studies with accounts of a concern; some of these problems will be discussed in Chapter 11, others are discussed in later chapters. Often controversies reflect a mutual misunderstanding of the legitimate objectives of procedures designed to serve different purposes. What is needed is a recognition by both engineers and accountants of the difference in the objectives of their calculations.

A Concluding Statement on Part II

In studies to determine a prospective rate of return (the subject matter of Chapter 8), the lowest rate of return deemed sufficient to justify a proposed investment may obviously be described as the minimum attractive rate of return. In annual cost comparisons (Chapter 6), present worth comparisons (Chapter 7), or benefit–cost comparisons (Chapter 9), the interest rate selected for use in equivalence calculations is—in effect—a minimum attractive rate of return regardless of whether or not it is so described.

The choice of a minimum attractive rate of return obviously has a great influence on decision making at all levels at which decisions are made between alternative investments in fixed assets. Proposed investments that look attractive at 3% appear to be undesirable at 7%; proposals that look good at 7% are properly vetoed at 15%. Further comments on the choice of a minimum attractive rate of return are made at various places in Part III. In the meantime the major point to keep in mind is that the controlling element in the choice of a minimum attractive return should ordinarily be either the return on the investment opportunity forgone or the overall cost of capital, all things considered. There is no figure for minimum attractive rate of return that is appropriate to all circumstances; it is reasonable that this figure should be much higher in some cases than in others.

PROBLEMS

10–1. A small, high-technology company has established a capital budget of $125,000 of equity funds to be invested in the next budget period. Proposed cost-reduction or profit-increasing investments are listed in the following tabulation:

Project	*Investment required*	*Estimated economic life (yrs.)*	*Estimated salvage value*	*Annual net positive cash flow*
K	$50,000	8	$5,000	$8,000
L	25,000	10	0	4,500
M	25,000	10	2,500	6,500
N	50,000	5	15,000	11,000
O	25,000	15	1,000	4,000
P	25,000	10	2,500	4,000
Q	50,000	12	3,000	6,800

(a) Calculate the prospective rate of return on each proposal and rank the proposals based on their prospective rates of return. Without consideration of any other matter except prospective rate of return, what should the company use as its after-tax minimum attractive rate of return, i^*, at this time?

(b) Since this is a very dynamic industry and both products and production equipment may become technologically obsolete in a very short time, the managers of the company would give preference to a short-lived project over a long-lived project by accepting a prospective rate of return 2% lower for each 5-year reduction in project life. How would this policy decision affect the minimum attractive rate of return proposed in part (a)? What new i^* should the company use? (*Ans.* (a) 10.69%; (b) changes the order but not the selection of projects. $i^* = 9.77\%$)

10–2. Examine the financial page of the daily newspaper or current issues of some financial journal to determine the current yields on United States government bonds and on representative municipal, public utility, and industrial bonds. What are current rates of interest on home loans and on loans on commercial property in your locality? Discuss possible reasons for the differences in these interest rates.

10–3. Find some index of the yield of high-grade corporate bonds that has been in existence for 25 years or more. Find a comparable index for medium-grade corporate bonds. Plot the two indexes on coordinate paper, showing the year-by-year figures for the past 25 years. Note the variations in the general level of interest rates and also the differences between the variations of the rates of high-grade and medium-grade bonds.

10–4. Mr. and Mrs. Howard Smith own a franchise for a limited-menu, family-style restaurant in a small city. They bought the franchise 5 years ago, making a down payment of $8,000 and agreeing to pay an additional $10,000 in 5 years, with interest at 9%, for the furnishings and equipment. The repayments were made in five equal annual year-end amounts of $2,571. They have just made the last payment.

Before acquiring this restaurant, Howard, who is 45 years old, had been the manager of a similar-style restaurant in the same city and his salary the last year there was $12,000. Now both of the Smiths devote full time, 50 or more hours per week each, to the management of their own restaurant. During the last year their total earnings (gross income less all expenses including depreciation but not including the debt repayment) before income taxes were $35,000. Since they did not incorporate, their income is taxed as individuals.

Recently Howard's former employer dropped into Howard's restaurant for a friendly visit, and in the process told Howard that he had just lost his restaurant manager. He asked Howard if he and his wife would be interested in coming back as manager and assistant manager. Although Howard did not indicate any interest in the offer, his former employer said, "I'll pay you $18,000 and your wife $15,000 a year and you will only have to work 40-hour weeks." Howard agreed to discuss it with his wife.

As Howard and Mrs. Smith thought about the offer, they also thought about their freedom, being their own bosses, their investment of $18,000, as well as their long hours, worry over the ups and downs of business, and meeting payrolls. They talked with some of their friends, and one friend offered to buy the franchise and the equipment for $15,000 cash.

(a) Suppose the Smiths did not own the restaurant franchise and were trying to decide between two alternative courses of action: (1) buy the franchise for $18,000; or (2) accept the two job offers and invest their $18,000 some other way. How would you estimate the prospective rate of return on their investment if they buy the franchise, and what before-tax i^* would you recommend they use in their decision process? Discuss your reasoning.

(b) Now suppose that the Smiths do own the restaurant franchise and are faced with the two alternatives: (1) continue to own and operate the restaurant; or (2) sell the franchise and equipment for $15,000 and accept the two job offers. Furthermore, assume that this occurs in your present location at the present time. Using data available locally, financial journals, and the *Wall Street Journal*, try to find reasonable investment opportunities for the Smiths' $15,000 if they decide to sell, and recommend a before-tax i^* for their use in considering the two alternatives. Discuss your reasoning.

10–5. Each year the Young Company's management requests each division manager to present cost-reduction or profit-increasing capital investment proposals for consideration by the board of directors. For the coming year the board of directors has allocated $200,000 from internally generated funds for capital investment projects. Any funds not invested in internal projects will be invested in short-term (3 to 6 months) money market certificates that will earn about 9.5% on an annual basis, so that the funds will be available whenever new attractive internal projects appear.

This year the following non-mutually exclusive proposals were presented:

Proposal	*Investment required*	*Estimated economic life (yrs.)*	*Estimated salvage value*	*Annual net positive cash flow after income taxes*
A	$50,000	10	$5,000	$11,000
B	70,000	7	4,000	19,100
C	20,000	4	0	7,150
D	40,000	6	2,000	10,900
E	50,000	8	3,000	11,300
F	60,000	14	6,000	11,850
G	80,000	20	10,000	15,600

(a) Without regard to anything except prospective rate of return, how should the company allocate its $200,000 for the next year? Arrange the projects in descending order based on prospective rate of return.

(b) Based on your answer to (a), what is the i^* for this company for next year?

(c) After reviewing the proposals and considering the differences in the estimated economic lives for the project, the board of directors decided that, considering the dynamic nature of the industry and the competition to develop new products, longer-lived projects ought to show at least 1% greater prospective rate of return for each additional 5 years of projected life. That is, a 20-year project would have to show a 2% greater prospective rate of return than a 10-year project. Using these guidelines, how would the board of directors allocate the $200,000 and what is the new minimum attractive rate of return?

10–6. Assume that you are a graduate student attempting to earn a Ph.D. in your chosen field. You have approximately 3 years to go to complete it. Fortunately you have just won $25,000 in a magazine sweepstake. After paying your income tax on it you have $20,000. Since you are married, you will have to pay income tax only on any income over $5,400 a year (assumes you do not itemize deductions). Assume your applicable combined state and federal income tax rate will be 16%. You want to have $7,500 available at the beginning of each of the next 3 years, if possible.

Investigate the current investment opportunities in your area and list these opportunities along with the effective annual prospective rates of return. Based on these returns and your needs, determine how you plan to invest the money over the next 3 years. Write a one-page explanation of your plan and your justification for its selection. What is your minimum attractive rate of return, i^*?

10–7. On page 172 we state, "A similar line of reasoning may be applied to government projects financed by current taxation; the collection of these taxes requires the taxpayer to forgo the opportunity to earn a return from personal investment of the money collected." It is rather obvious that investment opportunities for individual taxpayers vary as a result of differences in the amounts of money available to invest, short-term versus long-term cash needs, degrees of risk aversion, and so on. Thus, governmental decision makers need to consider the opportunities for the ordinary, average citizen in setting minimum attractive rates of return to be used in justifying the investment of public funds in long-range projects.

There are some guidelines that administrators might consider: (1) a large percentage of American families have home mortgages; taxpayers could reduce their debt faster if their taxes were reduced—a risk-free investment at the mortgage rate; (2) an even larger percentage of Americans owe money on car purchases; reduction of that debt also is a risk-free investment at the interest rate being paid; (3) most Americans have some kind of pension or retirement savings plans that earn interest each year (not social security); (4) many Americans have a regular savings plan into which they make monthly deposits. These are all "investment opportunities" into which more money could be invested by each person if taxes were reduced.

Write a one-page paper outlining to your Congressman and your Senator your views on what minimum attractive rate of return should be specified by the federal government in the justification of the proposed federal projects. This paper should be based on your research into the current rates of interest being paid and the current rates being earned on relatively secure investments by the ordinary citizen who has a limited amount of money to invest each month.

PART III

TECHNIQUES FOR ECONOMY STUDIES

11

SOME RELATIONSHIPS BETWEEN DEPRECIATION ACCOUNTING AND ENGINEERING ECONOMY

*A word is not a crystal, transparent and unchanged; it is the skin of a living thought, and may vary greatly in color and content according to the circumstances and the time in which it is used.—Justice Oliver Wendell Holmes**

Throughout the examples of economy studies in the prior chapters, estimates of the effect on cash flow of proposed decisions between alternatives have been stated, but no consideration of the effects of the decisions on the accounts of the enterprise have been mentioned. Accounting is the process by which the financial events of the enterprise are recorded in order to provide managers with information for managerial decisions and to comply with certain legal requirements. This chapter introduces the topic of the relationships between accounting and economy studies and deals specifically with one very important facet of accounting, namely, accounting for depreciable fixed assets (such as buildings, structures, machinery, and equipment.)

Why Should Students of Engineering Economy Learn About Depreciation Accounting?

Persons responsible for decisions on the acquisition and retirement of fixed assets need a general understanding of depreciation accounting for a variety of reasons. Four of these reasons are developed in the present chapter:

1. Often it is necessary to reconcile economy studies with the accounts of an enterprise. This need arises in various ways. Some data for economy studies come from the accounts and must be modified by conversion to cash flow before they are suitable for use in the economy studies. Frequently it is necessary that economy studies be related to the accounts for presentation to colleagues, to management personnel, or to the gen-

*In a 1918 United States Supreme Court decision, *Towne* v. *Eisner*, 245 U.S. 418.

eral public. The follow-up (if any) of decisions based on economy studies must be based in part on figures from the accounts.
2. Economy studies for private enterprise require estimates of the amounts and dates of the outlays for income taxes that will be affected by a choice among proposed alternatives. In most cases, such estimates involve consideration of the depreciation methods that will be used for tax purposes. Although some enterprises use different depreciation methods for internal accounting and for tax purposes, it is necessary to understand depreciation accounting if one is to understand the income tax treatment of depreciation. In fact, the authors believe that this relationship between depreciation accounting and income taxes makes it essential that persons responsible for economy studies have some understanding of depreciation accounting.
3. Several approximate methods used by some analysts for computing the annual cost of capital recovery are related to two methods of depreciation accounting, namely, the straight-line method and the sinking-fund method.
4. Several methods of making so-called rate of return studies that depend on depreciation accounts are in common use. The student of engineering economy needs to understand these methods in order to be aware of their deficiencies.

This book deals with depreciation accounting primarily in relation to engineering economy; there are many aspects of the subject that are not presented at all or that are discussed only briefly and superficially. Some other aspects of depreciation accounting referred to later in the book are as follows:

1. Chapters 16 and 17 point out certain common errors made in economy studies for retirements and replacements—errors that are related in part to a misunderstanding of depreciation accounting.
2. Chapter 18 shows the relationship between depreciation accounting and one source of funds for investment in fixed assets. It also discusses the relationship between depreciation accounting and the provision of funds for the replacement of assets.
3. Chapter 21 discusses a common misuse of depreciation accounting in computing so-called payoff periods.

A Classification of Reasons for Retirement

Property units are generally retired for one or more of the following reasons:

1. *The availability of improved machines or structures for performing the same service*. Research and development work by scientists and engineers is continually leading to new and more economical ways of doing existing jobs (technological obsolescence). If the prospective economies from new

methods are sufficient, it will pay to replace old assets with new ones or to relegate the old assets to standby purposes or other inferior uses.

2. *Changes in the amount and type of service requirements*. This includes such changes as increase or decrease in the amount of service required from the old asset, due to an increase or decrease in the demand for its product or service. It also includes changes in the product or service required. These changes frequently arise from competitive situations, either from competition among producers in a single industry or from competition with substitute products or services. They may also be caused by acts of public authority.
3. *Changes in the existing machines or structures themselves*. Machines and structures wear out, corrode, and decay as the result of age and use. Often this increases maintenance costs and decreases the quality and reliability of performance to the extent that it pays to replace assets that are still capable of continuing to render service. In some circumstances wearing out, corrosion, and decay may make retirement imperative rather than merely economical.
4. *Changes in public requirements regarding the machine or structure*. Public laws establishing new standards for occupational safety, health protection, and environmental pollution control may force the retirement of equipment because it cannot be economically retrofitted to meet the required standards.
5. *Casualties*.

These reasons for retirement are not mutually exclusive, but in most cases operate in combination with one another. Thus, an old machine might be replaced by a new one that (1) incorporated new automatic features that reduced unit labor costs, (2) provided increased capacity to meet an increased demand for the product, and (3) had the prospect of initial maintenance costs considerably lower than the current high maintenance costs of the old machine.

Obsolescence (1), inadequacy (2), wearing out (3), and so on, also are responsible for reductions in value of old assets long before the assets reach the point where immediate retirement is economical.

Mortality Dispersion for Physical Property Units

Not all human beings die at the same age. Some die young and others live to a ripe old age. Nevertheless, it is possible to analyze human mortality experience to determine curves and tables that give satisfactory estimates of average life and of the percentage of survivors at any age. Thus, a life insurance company can predict with confidence what percentage of 100,000 healthy white native males 20 years old will survive to the ages of 27 or 49 or 65, even though it is not possible to say with respect to any individual whether he will survive any given number of years.

Physical property units are like human beings in having a mortality dispersion. Of a number of property units that seem identical, some will be retired at an early age and others will serve many years before retirement. This fact of mortality dispersion is illustrated in Figure 11–1. This figure shows two curves that have been derived in the analysis of the retirement experience of telephone exchange underground cable.

The reversed curve starting in the upper left-hand corner of the diagram is called a *survivor curve*. It shows the percentage of this type of plant that, in the particular group of cable studied, may be expected to be in service at any given number of years after its installation. For instance, 70% will be in service after 15 years of life. The lower stepped curve is a *retirement frequency curve*. It shows the annual retirements expressed as a percentage of the original amount installed. Although the maximum number of retirements is at age 20, retirements are taking place from the 1st year of life to the 44th year.

These curves are mathematically fitted ideal curves derived from a certain

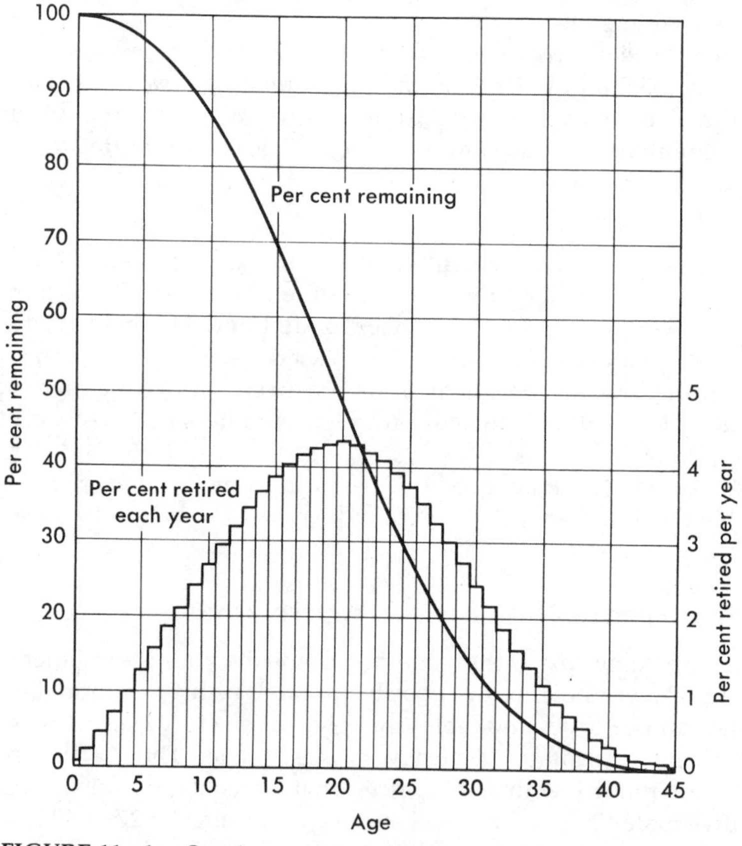

FIGURE 11–1. Survivor curve and retirement-frequency curve for certain telephone underground cable

group of actual data involving several million dollars' worth of exchange underground cable. Curves of this type have been developed in connection with the requirements of depreciation accounting.

Such curves may be developed from statistical analysis that uses data involving exposures to the risk of retirement and actual retirements of units of various ages. The curves therefore reflect retirements for all causes; they give weight to the results of technological progress and environmental changes as well as to the tendency of industrial assets to deteriorate as they get older. As might be expected, a series of such statistical analyses made over a period of years often indicates that average service lives are shortening for some classes of assets and lengthening for other classes.

Statistical analysis of the type used in obtaining curves such as Figure 11–1 can reflect only the ages at which retirements actually have been *made*. For various reasons mentioned in Chapter 17, it often happens that assets are kept in service beyond the date at which it would have paid to retire them. It follows that economic service lives often are shorter than the average service lives that will be determined from a statistical analysis of retirements.*

The Various Meanings of "Depreciation"

The meanings of words develop out of their use. Many words are used in a number of different meanings. Depreciation is such a word. In any use of the word, there needs to be a clear differentiation among these various meanings.

In his outstanding book, *Valuation of Property*, Professor J. C. Bonbright† points out that substantially all the different technical meanings attached to the word *depreciation* are variants of four basic concepts. These are:

1. *Decrease in Value*. This concept implies that the value of one asset is in some way computed at two different dates. The value at the later date subtracted from the value at the earlier date is the *depreciation* regardless of what combination of causes may have been responsible for the value change. When depreciation is used in everyday speech, this is the meaning generally implied; it is also implied by most dictionary definitions.

Like depreciation, *value* is a word with many meanings. As pointed out by Professor Bonbright, the two most important and useful of these meanings in the economic sense of the word are *market value* and *value to the owner*. Depreciation in the sense of decrease in value may apply to either of these two concepts of value. Values may be determined by actual market price, by appraisal, or in any other appropriate way.

*For a concise explanation of the various statistical methods of finding average service life, see E. L. Grant and P. T. Norton, Jr., *Depreciation* (New York: The Ronald Press Co., 1955), chap. v. See also Robley Winfrey, *Statistical Analysis of Industrial Property Retirements* (Ames, Iowa: Bulletin 125, Iowa Engineering Experiment Station, 1935).

†J. C. Bonbright, *Valuation of Property* (New York: McGraw-Hill Book Co., Inc., 1937), chap. x.

2. *Amortized Cost*. This is the accounting concept of depreciation, which is being discussed in the present chapter. From the viewpoint of accounting, the cost of an asset is a prepaid operating expense to be apportioned among the years of its life by some more or less systematic procedure. It should be emphasized that it is cost, not value, that is apportioned in orthodox accounting.

The accounting concept of depreciation is well described in a report of the Committee on Terminology of the American Institute of Certified Public Accountants as follows:*

> Depreciation accounting is a system of accounting which aims to distribute the cost or other basic value of tangible capital assets, less salvage (if any), over the estimated useful life of the unit (which may be a group of assets) in a systematic and rational manner. It is a process of allocation, not of valuation. Depreciation for the year is the portion of the total charge under such a system that is allocated to the year. Although the allocation may properly take into account occurrences during the year, it is not intended to be a measurement of the effect of all such occurrences.

Although we shall use the common phrase *book value* to describe the difference between the cost of an asset and the total of the depreciation charges made to date against the asset, this difference is more accurately described as *unamortized cost*.

3. *Difference in Value Between an Existing Old Asset and a Hypothetical New Asset Taken as a Standard of Comparison*. This is the appraisal concept of depreciation. Many appraisals of old assets are based on replacement cost. A replacement cost appraisal should answer the question "What could one afford to pay for this asset in comparison with the most economical new one?"

An upper limit on the value to its owner of an old asset may be determined by considering the cost of reproducing its service with the most economical new asset available for performing the same service. This most economical new substitute asset may have many advantages over an existing old asset, such as longer life expectancy, lower annual disbursements for operation and maintenance, increased receipts from sale of product or service. The deduction from the cost of the hypothetical new substitute asset should be a measure in money terms of all of these disadvantages of the existing old asset.

In the language of appraisal, this deduction is called *depreciation*. Appraisal depreciation, therefore, should mean the value inferiority at some particular date (the date of the appraisal) of one asset, the existing old one being appraised, to another asset, a hypothetical new one used as the basis of

**Accounting Terminology Bulletin No. 1: Review and Resumé* (para. 56), (American Institute of Certified Public Accountants, 1953).

valuation. This concept implies two assets and the measurement of their values at one date.

4. *Impaired Serviceableness.* As machines become older they are often unable to hold as close tolerances as when they were new. Similarly, the strength of structures may be impaired by the decay of timber members or the corrosion of metal members. Engineers have sometimes used the word *depreciation* to refer to such impaired functional efficiency.

It should be emphasized that this is not a value concept at all. Impaired serviceableness may result in decrease in value, but there are many other common reasons for decrease in value. Assets that are physically as good as new are not necessarily as valuable as when they were new. They may have higher operation and maintenance costs; they will nearly always have shorter life expectancy; service conditions may have changed; more economical alternative methods may have become available. As the use of depreciation in the sense of impaired serviceableness has generally led to confusion in valuation matters, the word is not used in this sense elsewhere in this book.

Concepts of Value*

As previously stated, the most useful economic concepts of value are *market value* and *value to the owner*.

Market value properly refers to the price at which a property could actually be sold. For certain types of valuation this is not an appropriate concept to use. Often only one owner is in a position to make effective use of a given item of property; although continued possession of this property may be of great monetary importance to its owner, the property might bring only a negligible price if sold to someone else.

Hence, the concept of value to a specific owner is of great importance in the valuation of property. Value to the owner may be defined as the money amount that would be just sufficient to compensate the owner if he were to be deprived of the property. Generally speaking, this value will not be greater than the money amount for which the owner could soon replace the property with the best available substitute, with due allowance for the superiority or inferiority of that substitute. And, generally speaking, value to the owner will not be less than the market price for which the property could be sold. The concept of value to the owner may properly be applied to a prospective owner as well as to a present owner.

The word *value* is also sometimes used in what might be called a neutral sense as any money amount that is associated with specific items of property for some given purpose. An example of this is the use of the phrase *book value*

*For an authoritative and thorough discussion of value concepts, see Bonbright, *op. cit.*, chaps. 3, 4, and 5.

to describe the unamortized cost of property as shown by the books of account.

The Balance Sheet and Profit and Loss Statement

Any presentation of the elements of accounting must focus attention on two important types of statements obtained from the accounts.

One of these, the *balance sheet*, describes the condition of an enterprise at a particular moment, for example, at the close of business on the final day of a fiscal year. The balance sheet shows what the enterprise owns (its assets), what it owes (its liabilities), and shows the "value" of the owners' equity in the enterprise as the excess of assets over liabilities. All balance sheet valuations are arrived at using the formal and systematic rules of accounting and sometimes differ greatly from market value of the same properties.

The other statement, the *profit and loss statement*, also called the *income statement*, gives the incomes and expenses of an enterprise as shown by the books of account for a period of time and states whether the enterprise has made a profit or suffered a loss and of how much. The longest period of time covered by the usual profit and loss statement is a fiscal year. (A business fiscal year does not always coincide with the calendar year, as it may start on some date other than January 1.) The figures in the profit and loss statement, like those in the balance sheet, are determined using the formal and systematic rules of accounting.

Example 11–1 illustrates the relationship between depreciation accounting and the balance sheet and profit and loss statement in an extremely simple case. Example 11–2 is adapted from an actual case that is relatively simple, illustrating a common method of estimating year-by-year return on investment from the accounts. Example 11–1 deals with only the first year of a business enterprise, but Problems 11–16 and 12–18 provide the operating results for the next 4 years, at which time the business was sold to another owner. Only at that time can the original owner determine the actual rate of return on the original investment. Example 11–2 deals with a terminated business which finally turned out to be much more profitable than it first seemed to be from an analysis of the accounts on a year-by-year basis.

EXAMPLE 11–1

Depreciation in the Accounts of an Equipment Rental Business

Facts of the Case

A man, whom we shall call Bill Black, moved into a small but rapidly growing suburban community and frequently needed to rent special equipment to use in improving his home and lot. In order to rent such tools as rototillers, ladders, and special plumbing tools, he had to drive 15 miles to another community. He realized that as many more families moved into that community there would be an increasing demand for rental equipment by the "do-it-yourself" homeowners.

After careful investigation and analysis of the equipment business, he decided to establish such an enterprise. He established Equipment Rentals with an initial investment of $50,000.

Bill then rented a suitable building and employed a full-time person to run the rental business and a part-time helper. He purchased a large array of tools for a total purchase price of $35,000, and opened for business.

Financial Statements for the First Year and Their Relationship to Depreciation Accounting

When Equipment Rentals opened for business its balance sheet was as follows:

Assets		Liabilities and Owner's Equity	
Cash	$15,000	B. Black, Invested Capital	$50,000
Tools	35,000		
	$50,000		$50,000

During the first year of operation, Equipment Rentals paid out $30,000 for wages, insurance, rent, and other business disbursements, but it received a total of $46,000 in rental fees for tools. Cash increased by $6,000, and receivables by $10,000. If Black had assumed that the tools were still worth their original cost, the balance sheet at the end of the year would have been:

Assets		Liabilities and Owner's Equity	
Cash	$21,000	B. Black, Invested Capital	$50,000
Accounts Receivable	10,000	B. Black, Retained Earnings	16,000
Tools	35,000		
	$66,000		$66,000

Black, of course, realized that the tools would not last indefinitely, and that he should set up a depreciation plan. He decided that the tools would have an average economic life of about 5 years and would have a zero net salvage value when disposed of. He recognized that some tools would probably have to be discarded before 5 years, but that some might last considerably longer. He also decided to use the straight-line depreciation method; that is, one-fifth of the original cost would be written off each year. After making the proper accounting entries to record the first year's depreciation, the balance sheet was:

Assets			Liabilities and Owner's Equity	
Cash		$21,000	B. Black, Invested Capital	$50,000
Accounts Receivable		10,000	B. Black, Retained Earnings	9,000
Tools	$35,000			
Less				
Allowance for				
Depreciation	7,000	28,000		
		$59,000		$59,000

The profit and loss statement for the year was:

Receipts from renting tools		$46,000
Less		
Operating Expenses		
Disbursements for rent, wages, etc.	$30,000	
Depreciation on assets	7,000	37,000
Profit from first year's operations		$ 9,000

Thus, the profit shown by the books of account represents 18% of Black's investment of $50,000, but that is before any consideration of the effects of his decision to invest in the equipment rental business on his income taxes for the year. Obviously, he will have to pay part of his "profit" to the government in the form of income taxes. Certain income tax aspects of this example are brought out in Problems 12–18, 12–19 and 12–20 following our discussion of income taxes in Chapter 12.

Comment on the Foregoing Financial Statements

This simple example illustrates several important points regarding depreciation accounting, as follows:

1. The operation of making the $7,000 depreciation charge on the books of account involved no cash flow. In effect, the $7,000 depreciation entry was an allocation to the first year of operation of a portion of a previous $35,000 cash flow that the accounts had viewed as causing the acquisition of assets rather than as the incurring of an expense.

(Later in this chapter, we shall see that even though the act of making a depreciation entry on the books does not in itself change the cash on hand, a depreciation entry in an income tax return influences the cash flow for income taxes.)

2. The profit as shown by the books of account depends on the depreciation charge and therefore is influenced by the estimated life and salvage value and by the depreciation accounting method selected. Thus, Black's use of straight-line depreciation and his assumption of a 5-year life with no salvage value gave him an income figure of $9,000. If he had assumed a 3-year

life and charged $11,667 depreciation for the year, the income would have been only $4,333. If he had assumed a 10-year life for the tools, the depreciation entry would have been only $3,500 and the income would have been $12,500.

Although, as the quoted definition said, depreciation accounting requires the writing off of cost "in a systematic and rational manner," there are many different systematic and rational ways to write off cost, and these different ways give different figures for profit.

3. The valuation of assets on the balance sheet is similarly influenced by the depreciation charge. The so-called book value of an asset is merely that portion of the cost of the asset that has not been charged off as depreciation expense. The book value of the tools was $28,000 at the end of the first year because $7,000 of depreciation had been charged against the year's operations and deducted from the $35,000 cost of the tools. If the depreciation charge had been less than $7,000, the book value would have been higher; if the depreciation charge had been more than $7,000, the book value would have been lower.

The quoted definition of depreciation accounting pointed out that "it is a process of allocation and not of valuation." The valuation shown for a depreciable asset on the books of account is not influenced by unpredicted fluctuations in market value. Thus, the $28,000 book value for the tools was not influenced by any current resale value or any other events, such as inflation, since their purchase.

EXAMPLE 11–2

Year-by-Year Estimates of Rates of Return From an Investment in Rental Property

Facts of the Case

Example 8–1 (page 118) described an investment in rental property. The cash flow series from this investment was shown in Table 8–1. The rate of return on the investment was 18.8% before income taxes, as computed in Table 8–2 by appropriate compound interest methods. Because a property costing $109,000 was sold for a net $209,000 after 7 years of ownership, the actual rate of return from the terminated investment was considerably higher than the rate of return had appeared to be during the period of ownership.

Table 11–1 shows the figures for annual profits as shown by the accounts of this enterprise for its first 6 years of operation. For purposes of accounting, the $109,000 investment was broken down as follows:

Land	$15,000
Buildings	80,000
Furniture	14,000

TABLE 11–1
Accounting figures for annual profits and current estimates of rate of return for data of Example 8–1

Year	Book value at start of year	Receipts for year	Repairs, taxes, etc.	Depreciation	Profits for year	Apparent rate of return on investment
1	$109,000	$15,000	$5,000	$3,400	$ 6,600	6.1%
2	105,600	18,000	5,500	3,400	9,100	8.6
3	102,200	18,000	5,700	3,400	8,900	8.7
4	98,800	18,000	4,500	3,400	10,100	10.2
5	95,400	18,000	3,600	3,400	11,000	11.5
6	92,000	18,000	4,300	3,400	10,300	11.2

In accordance with the conventions of accounting, no prospective decrease (or increase) in the value of land was considered in the accounts. For purposes of depreciation accounting, the buildings were assumed to have a 40-year life with no terminal salvage value, and the furniture was assumed to have a 10-year life with no terminal salvage value. Straight-line depreciation was charged in the accounts at $2,000 a year for the buildings and at $1,400 a year for furniture, a total of $3,400 a year. The profit figures for each year shown in the next-to-last column of Table 11–1 are the years' positive cash flows diminished by the $3,400 depreciation charge.

The table also shows the book value of the property at the start of each year. In the first year this was the investment of $109,000; each year it was diminished by the $3,400 depreciation charge made in the books of account. The final column of the table shows the profit for each year expressed as a percentage of the start-of-year book value. This latter figure may be thought of as a current estimate of rate of return on investment computed by a method that often is used in relation to the accounts of an enterprise.*

Tentative Character of Accounting Figures for Profit and for Rate of Return

Example 11–2 illustrates the point that intermediate judgments about profitability and about rates of return may be considerably in error. In this example, the rates of return for the first 3 years appeared to be less than 9% even though the overall rate of return on the terminated investment turned out to be approximately 19%.

Example 11–1 indicates an attractive 18% rate of return for the first year of

*An alternate method, also frequently used, is to express the year's profit as a percentage of the average of the book values at the start and finish of the year. For example, the fifth year's profit, $11,000, would be divided by $93,700 (the average of $95,400 and $92,000) to obtain an apparent return of 11.7%.

operation, but future years may show widely varying results. The assumption that such a rate will prevail throughout the life of the business venture could be seriously in error.

Of course the true overall rate of return obtained over the life of an investment cannot be known until ownership has been terminated. Nevertheless, year-by-year figures for profit are essential in the conduct of a business (and in the collection of income taxes); it is not practicable to wait until the termination of a business enterprise to draw conclusions regarding its profitability. It is natural to compute year-by-year rates of return by relating the profit figures taken from the accounts to some investment figure also taken from the accounts.

At this point in our discussion a clear distinction needs to be made between the problem of computing a single figure for rate of return on an investment based on the full period of consequences of the investment and the problem of estimating year-by-year figures for rate of return. As explained in Chapter 8, compound interest methods are required to find the correct figure for rate of return over the life of an investment. The foregoing statement applies both to the calculation of rate of return from a terminated past investment and to the calculation of estimated rate of return from estimates of cash flow for the entire period of service of a proposed investment.

It is rarely, if ever, of much importance to find the rate of return on a terminated past investment (such as the one in Example 11–2). Although the rate of return may be of historical interest, nothing can be done about terminated investments. The analysis of terminated investments can be helpful in economy studies in the future by providing guidance for the improvement of estimates of future events or consequences of proposed decisions.

In contrast, the estimation of rate of return on a proposed investment is a matter of great practical importance. An important criterion of attractiveness of a proposed investment is the single figure that expresses the prospective rate of return over the life of the investment. It needs to be emphasized that it is this overall rate of return that is significant for proposed investments, not year-by-year estimates of rates of return of the type illustrated in Table 11–1.

In the opinion of the authors of this book, there is seldom any valid reason for not using correct compound interest methods in analyzing *proposed* investments. A great deal of time and effort often goes into the making of estimates for economy studies relative to proposed investments. Only a few minutes more time are needed to apply methods of computing rate of return that are correct in principle than are needed to apply any of the various competing incorrect methods.

However, several incorrect methods of computing prospective rates of return on proposed investments are in common use in industry. Although these methods all have their origins in the type of calculations of rate of return on past investment illustrated in Example 11–2, the methods give results that differ greatly from one another. Sometimes these methods are used in the belief that they are good approximations to correct compound interest methods; sometimes they are used under the illusion that they are correct in principle. The weaknesses of several of these methods are illustrated in problems 11–9 through 11–15 at the end of this chapter.

Before we can examine these "approximate" methods, we must first discuss the most common methods of depreciation accounting.

General Comment on All Methods of Depreciation Accounting

As brought out in the quoted definition of depreciation accounting, it is the *cost* of tangible assets, less prospective salvage value, that is written off on the books of account. In effect, the cost of capital assets is viewed as a prepaid expense to be apportioned among the years of service of the assets "in a systematic and rational manner."

There are many different methods of writing off cost that obviously are "systematic." Moreover, methods that differ greatly from one another have been advocated as being "rational." In the United States the methods in common use have been influenced by changes in income tax laws and regulations, particularly by changes in 1934, 1954, 1962, 1971, and 1981.*

One way to classify depreciation accounting methods is as follows:

1. Methods that aim to give a greater write-off in the early years of life than in the final years of life
2. Methods that aim to give a uniform write-off throughout the entire service life
3. Methods that aim to give a smaller write-off in the early years of life than in the final years

In class (1) we shall discuss the declining-balance method, the sum-of-the-years-digits method, and certain multiple-straight-line methods. In class (2) we shall discuss the straight-line method. In class (3) we shall discuss the sinking-fund method. Because of its historical importance, the straight-line method is discussed first.

Straight-Line Depreciation Accounting

In the straight-line method, the full service life of the asset is estimated. The prospective net salvage value at the end of the life is also estimated and

*Three papers by one of the authors of this book that together give a historical picture of the income tax treatment of depreciation in the United States, particularly from 1962 through 1972, are:

"Life in a Tax-Conscious Society—Tax Depreciation Restudied," E. L. Grant, *The Engineering Economist* (Autumn, 1968), pp. 41–51.

"The ADR System and Other Related Income Tax Matters," E. L. Grant, *The Engineering Economist* (Spring, 1972), pp. 200–210.

"Some Lessons from Sixty Years of Treatment of Depreciation and Related Matters for Income Tax Purposes in the United States," E. L. Grant, *Technical Papers, American Institute of Industrial Engineers, Twenty-Fifth Anniversary Conference and Convention*, 1973, pp. 45–54.

expressed as a percentage of first cost. The annual depreciation rate to be applied to the first cost of the asset being written off is computed as follows:

$$\text{Straight-line rate} = \frac{\text{100\% minus estimated salvage percentage}}{\text{estimated service life in years}}$$

Consider a machine tool with a first cost of \$35,000, an estimated life of 20 years, and an estimated net salvage value of \$3,500. The salvage percentage is \$3,500 ÷ \$35,000 = 0.10 or 10%.

$$\text{Straight-line rate} = \frac{100\% - 10\%}{20} = \frac{90\%}{20} = 4.5\%$$

With this straight-line depreciation rate, the depreciation charge every year will be \$1,575 (i.e., 4.5% of the \$35,000 first cost).

The same \$1,575 may be computed without the use of the 4.5% figure, as follows:

$$\begin{aligned} &\text{Straight-line depreciation charge} \\ &= \frac{\text{first cost minus estimated salvage value}}{\text{estimated service life in years}} \\ &= \frac{\$35{,}000 - \$3{,}500}{20} = \frac{\$31{,}500}{20} = \$1{,}575 \end{aligned}$$

Formulas to assist in the computation of the depreciation charge for any year and the book value at the end of any year will be given for each of the depreciation methods described. The notation used in the formulas include:

P = first cost of the asset
S = the estimated salvage value at the end of the estimated service life
n = the estimated service life in years
D_r = the depreciation charge for the rth year
BV_r = the book value at the end of the rth year after the depreciation charge for the rth year has been made

The formulas for the straight-line depreciation method are:

$$D_r = (P - S)/n$$
$$BV_r = P - rD_r$$

Prior to 1934, it was common in the manufacturing industries of the United States for cost to be written off in a much shorter period than full service life. For example, the cost of machinery often was written off by a uniform annual charge during the first 10 years of its life. At the expiration of this period the machinery was carried on the books of account as "fully depreciated." It was not uncommon for the full service lives of machines so written off to be 20 or 25 years or even longer. Although in pre-1934 days, this method of write-off was described as "straight-line," we shall see that the method is more accurately described as a special case of multiple-straight-line depreciation accounting.

A change in policy of the U. S. Treasury Department in 1934 and thereafter was intended to force the writing off of cost for tax purposes over the full service life. Over the years this policy was gradually relaxed. It was definitely changed by legislation in 1971 and 1981.

Declining-Balance Depreciation Accounting

It is common for assets to be used for stand-by or other inferior uses during the final years of their lives. The contribution of assets to income often is much greater in the early years of life than in the final years. For these reasons and for other reasons brought out in Chapter 12, it usually is sensible to write off the cost of assets more rapidly in the early years of life than in the later years.

Several ways of making this more rapid write-off in the early years were authorized for income tax purposes in the United States in 1954. The use of these liberalized methods for tax purposes was restricted to assets having lives of three years or more that were acquired new by the taxpayers in 1954 or thereafter. One of these methods was the so-called double-rate declining-balance method.

In any declining-balance depreciation accounting, a given depreciation rate is applied each year to the remaining book value, that is, to that portion of the cost of an asset (or assets) that has not already been written off in a previous year. For example, if a 10% rate is applied to an asset that cost $35,000, the depreciation charge in the first year is 0.10($35,000) = $3,500. In the second year the charge is 0.10($35,000 − $3,500) = 0.10(31,500) = $3,150. In the third year it is 0.10($31,500 − $3,150) = 0.10($28,350) = $2,835. And so on.

In the double-rate declining-balance method authorized for income tax purposes in the United States in 1954, the depreciation rate is computed as 200% ÷ (estimated life in years). This rate is double the straight-line rate that would be allowed for an asset that has an estimated zero salvage value and the given estimated life. In computing the permissible declining-balance rate, any prospective terminal salvage value is disregarded.

Consider, for example, the $35,000 machine tool for which we computed the 4.5% straight-line rate. This had a 20-year estimated life and a $3,500 estimated salvage value. The permissible declining-balance rate for this asset is 200% ÷ 20 = 10%. The application of a 10% rate for 20 years will lead to a book value of $4,255 at the end of the 20th year.

Let f = the declining-balance rate expressed as a decimal, then:

$$
\begin{aligned}
f &= 2.0/n \\
BV_r &= P(1 - f)^r \\
D_r &= f(BV_{r-1}) = (BV_r - BV_{r-1})
\end{aligned}
$$

It is important to note that the salvage value does *not* enter into the computation of either the depreciation charge or the book value when using the double-rate declining-balance method.

At this point in our discussion, the so-called textbook method of computing a declining-balance rate requires a brief mention. Assume that a rate f is desired that will make the book value at the end of an n-year life exactly equal to an estimated terminal salvage value, S. Then

$$S = P(1 - f)^n$$

and

$$f = 1 - \sqrt[n]{\frac{S}{P}}$$

However, this method of setting a declining-balance rate is rarely if ever used. It cannot be used with zero salvage value. Small differences in estimated salvage value make a great difference in the computed rate. For example, consider two $1,000 assets each with a 20-year estimated life, both with small prospective salvage values. Asset X has an estimated salvage value of $50; asset Y has an estimated salvage value of $1. The declining-balance percentage computed for X is 13.91% and for Y is 29.20%.

In the actual use of the declining-balance method, it is better to select a depreciation rate that seems appropriate, all things considered, than to compute a rate from the textbook formula. The declining-balance rates permitted by the 1954 tax laws and regulations in the United States are intended to permit a write-off of about two-thirds of the cost of an asset in the first half of the estimated life.

Sum-of-Years-Digits Depreciation Accounting

This method, authorized in the United States by the 1954 tax law, apparently was never used in actual accounting practice before that date. The digits corresponding to the number of years of estimated life are added together. For example, consider our $35,000 machine tool with its estimated life of 20 years. The sum of the digits from 1 to 20 is 210. The depreciation charge for the first year is 20/210 of the depreciable cost (i.e., of the first cost minus the estimated salvage value). This is 20/210 ($35,000 − $3,500) = 20/210($31,500) = $3,000. In the second year the charge is 19/210($31,500) = $2,850. In the third year it is 18/210($31,500) = $2,700. And so on. The charge decreases by $150 (i.e., by 1/210 of $31,500) each year until it is $150 in the 20th year.

This method writes off about three-fourths of the depreciable cost in the first half of the estimated life. It may be formulated as follows:

$$SOYD = \text{sum of years digits} = (n)(n + 1)/2$$

$$\text{Let } C = (P - S)/SOYD$$

$$D_r = (n + 1 - r)C$$

$$BV_r = P - C\left[SOYD_n - \frac{(n - r)(n - r + 1)}{2}\right]$$

$$= P - C\,(SOYD_n - SOYD_{n-r})$$

Sinking-Fund Depreciation Accounting

This method visualizes an imaginary sinking fund established by uniform end-of-year annual deposits throughout the life of an asset. These deposits are assumed to draw interest at some stated rate, such as 6%, 4%, or 3%, and are just sufficient so that the fund will equal the cost of the asset minus its estimated salvage value at the end of its estimated life. The amount charged as depreciation expense in any year consists of the sinking-fund deposit plus the interest on the imaginary accumulated fund. The book value at any time is the first cost of the asset minus the amount accumulated in the imaginary fund up to date. The sinking-fund method is also known as the present-worth method; the book value at any time is equal to the present worth of the uniform annual cost of capital recovery for the remaining years of life plus the present worth of the prospective salvage value.

For example, assume that our $35,000 machine tool is to be depreciated by the 6% sinking-fund method. The annual sinking-fund deposit is ($35,000 − $3,500)($A/F$,6%,20) = ($31,500)(0.02718) = $856.2. This will also be the depreciation charge in the first year. In the second year the depreciation charge will be $856.2 + $856.2(0.06) = $907.6. In the third year it will be $856.2 + ($856.2 + $907.6)(0.06) = $962.0. And so on.

The book value at any age can be computed without year-by-year calculations by finding the amount in the imaginary sinking fund and subtracting this amount from the first cost. For instance, the sinking fund at the end of 12 years will amount to $856.2($F/A$,6%,12) = $856.2(16.870) = $14,444. The book value at this time is therefore $35,000 − $14,444 = $20,556. This explanation may be formulated as follows:

$$BV_r = P - (P - S)(A/F,i\%,n)(F/A,i\%,r)$$

Book values with the sinking-fund method are always greater than they would be with the straight-line method. The difference is greater for long-lived assets than for short-lived ones and is greater with high interest rates than with low ones. The straight-line method has sometimes been described as the limiting case of the sinking-fund method in which the interest rate has been assumed to be 0%.

Although the sinking-fund method was used in certain industries a few decades ago, it has ceased to be of much importance in actual accounting practice. It is explained here primarily because of the use of so-called sinking-fund depreciation in certain engineering economy studies in a manner explained later in this chapter.

An Illustration of Depreciation Charges and Book Values by Various Methods

Table 11–2 shows the year-by-year write-offs that would be made for our $35,000 machine tool by the four methods that have been explained. The table

TABLE 11–2
Comparison of depreciation charges and book values by four methods of depreciation accounting
(Asset has first cost of $35,000, estimated life of 20 years, and estimated salvage value of $3,500.)

	Depreciation charge for year				*End-of-year book value*			
Year	*Declining balance*	*Years digits*	*Straight line*	*6% sinking fund*	*Declining balance*	*Years digits*	*Straight line*	*6% sinking fund*
0					$35,000	$35,000	$35,000	$35,000
1	$3,500	$3,000	$1,575	$ 856	31,500	32,000	33,425	34,144
2	3,150	2,850	1,575	908	28,350	29,150	31,850	33,236
3	2,835	2,700	1,575	962	25,515	26,450	30,275	32,274
4	2,551	2,550	1,575	1,020	22,964	23,900	28,700	31,254
5	2,297	2,400	1,575	1,081	20,667	21,500	27,125	30,173
6	2,067	2,250	1,575	1,146	18,600	19,250	25,550	29,027
7	1,860	2,100	1,575	1,215	16,740	17,150	23,975	27,812
8	1,674	1,950	1,575	1,287	15,066	15,200	22,400	26,525
9	1,506	1,800	1,575	1,365	13,560	13,400	20,825	25,160
10	1,356	1,650	1,575	1,447	12,204	11,750	19,250	23,713
11	1,221	1,500	1,575	1,533	10,983	10,250	17,675	22,180
12	1,098	1,350	1,575	1,626	9,885	8,900	16,100	20,554
13	998	1,200	1,575	1,723	8,897	7,700	14,525	18,831
14	890	1,050	1,575	1,827	8,007	6,650	12,950	17,004
15	801	900	1,575	1,936	7,206	5,750	11,375	15,068
16	720	750	1,575	2,052	6,486	5,000	9,800	13,016
17	649	600	1,575	2,175	5,837	4,400	8,225	10,841
18	584	450	1,575	2,306	5,253	3,950	6,650	8,535
19	525	300	1,575	2,444	4,728	3,650	5,075	6,091
20	473	150	1,575	2,591	4,255	3,500	3,500	3,500

also shows end-of-year book values. In the declining-balance and the 6% sinking-fund methods, the book values have been rounded off to the nearest dollar for listing in the table; the figures shown in the table for yearly depreciation charges have been made consistent with the figures shown for book value. Figure 11–2 graphically displays the relative book values of the machine over its life by the different depreciation methods.

It is evident that "systematic" ways of writing off cost can differ greatly from one another. Comment on the rational basis of a choice among the different methods is deferred until later in the chapter.

Some Other Methods of Depreciation Accounting

Some types of capital assets can be identified with the production of specific units of output. For such assets, depreciation can be charged in proportion to units of production provided it is reasonable to estimate life in production units. The method has been used for assets associated with exhaustible natural resources where the factor limiting the life of the assets is the quantity of the natural resource in question. Thus, the cost of a sawmill might be depreciated at so much per thousand board feet sawed, or the cost of a coal mine tipple might be depreciated at so much per ton of coal mined. Frequently the unit-of-production method is used for motor vehicles with depreciation charged in proportion to miles of operation. The method is rarely used in diversified manufacturing because of the difficulty of finding any suitable production unit.

It is possible to use depreciation accounting methods where two or more straight lines are needed to show the decline in book value from first cost to estimated salvage value. Such a scheme may be described as a multiple-straight-line method. Under the United States 1954 tax law, multiple-straight-line methods may be used for tax purposes subject to the restriction that the amount written off at any time during the first two-thirds of life is not more than would have been permitted under the declining-balance method. For example, our $35,000 machine tool might be written off at $2,250 a year for the first 10 years and at $900 a year for the final 10 years.

A special case of the multiple-straight-line method is the one in which the entire first cost (less estimated salvage value, if any) is written off on a straight-line basis in some period shorter than the useful life. As previously stated, this was the common situation in the United States before 1934. When an asset with a 25-year life was written off at 10% a year for 10 years and 0% for the remaining 15 years, this was really a multiple-straight-line method with one of the straight-line rates as 0%.

Under the 1954 tax law in the United States, it is permissible to switch from the declining-balance method to the straight-line method at any time before an asset is retired. For example, Table 11–2 shows the book value of our machine tool by the declining-balance method to be $7,206 at the

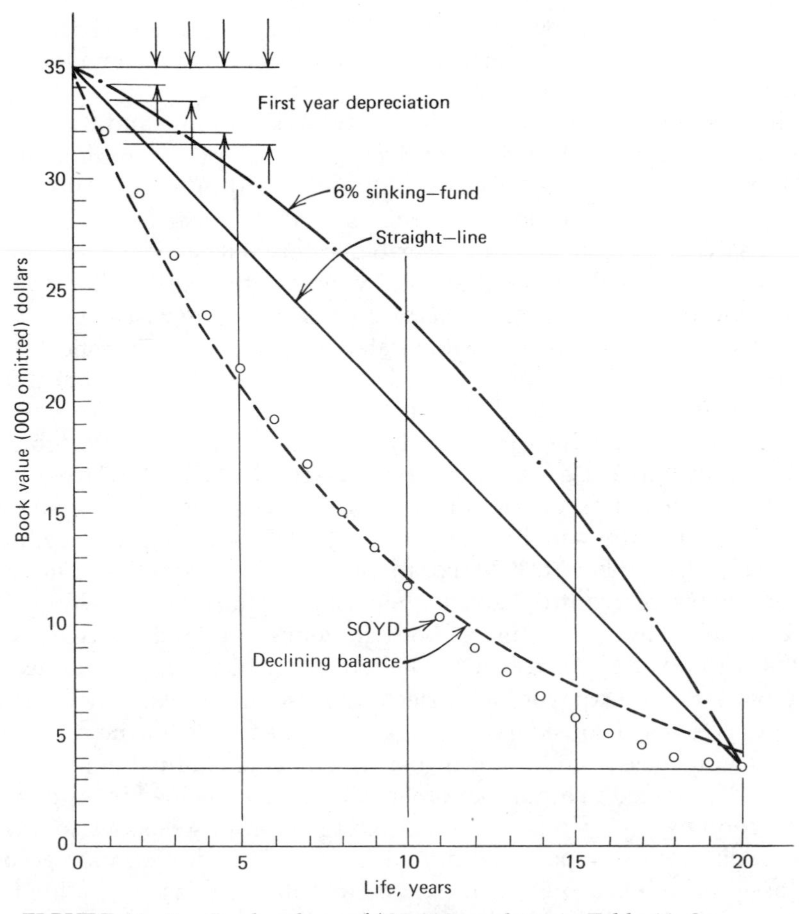

FIGURE 11–2. Book values of $35,000 machine in Table 11–2

end of 15 years. If the estimated remaining life on this date is still 5 years and the estimated salvage value is still $3,500, a straight-line write-off of $\frac{\$7{,}206 - \$3{,}500}{5} = \$741$ a year could be used for the final 5 years.

Estimates of Service Life and Salvage Value for Tax Returns, for Business Accounting, and for Economy Studies

Certain types of assets, such as automobiles, may have several owners before reaching the scrap heap. For each owner, the "life" for accounting purposes ordinarily is the expected period of service to the owner. In general, service life for accounting purposes is viewed as the number of years lapsing from an asset's acquisition to its final disposal, regardless of the different uses to which the asset may have been put during these years. Since 1934, the foregoing view of service life has governed the administration of income tax laws in the United States.

Estimates of terminal salvage value used for accounting and tax purposes should be consistent with estimated lives. For example, if automobiles are regularly purchased new and disposed of at the end of 3 years, the salvage percentage used in connection with the estimated 3-year life should be an estimated salvage percentage for 3-year-old automobiles.

Where sufficient data are available, various types of statistical studies may be made to estimate average service lives. In many cases such studies lead to survivor curves of the type illustrated in Figure 11–1. Statistical studies made year after year for a particular class of assets often indicate that the average realized life has been changing.

In 1962 the United States Treasury Department made available "guideline lives" for many broad classes of business assets. In 1971, the Treasury Department, with the subsequent approval of Congress, established an "asset depreciation range" (generally abbreviated to ADR) system for assets placed in service after December 31, 1970. For many classes of assets the ADR system permitted the use of lives from approximately 20% shorter to 20% longer than the 1962 guideline lives.* The use of both guideline lives and the ADR system was subject to a variety of restrictions about record keeping and other matters. In many cases, taxpayers who elected to use these relatively short lives for tax purposes used considerably longer estimated lives for the same assets in their own business accounting. In 1981 the income tax treatment of depreciation was liberalized in a number of ways that are explained in Appendix F.

An economy study relative to acquiring a proposed asset involves a somewhat different viewpoint from the keeping of the depreciation accounts for the asset once it is acquired. It may be reasonable to assume a life in the economy study that differs both from the life used for business accounting purposes and the life used for tax purposes. It is the economic life that usually is relevant in the economy study. Moreover, an economy study often relates only to the primary or initial type of service of an asset; possible stand-by or other inferior service during the final years of life is often disregarded on the grounds of having a negligible influence on the conclusions of the economy study.

In some cases an economy study may relate to an even shorter period than the expected economic lives of the various proposed assets; if so, it is necessary to estimate residual values at the end of the study period rather than terminal salvage values at the end of the economic lives. The use of such residual values in economy studies was illustrated in Example 9–1.

*The official document establishing the guideline lives was Revenue Procedure 62–21, 1962. The official documents setting up the ADR system were Treasury Decision 7128 and Revenue Procedure 71–25, both 1971.

Distinction Between Single-Asset and Multiple-Asset Depreciation Accounting

Assume a group of assets that have the mortality distribution shown in Figure 11–1. Although the average service life of these assets is 20 years, retirements occur all the way from the first to the 45th year. For the sake of simplicity in the following illustration, assume that the expected salvage value is zero and that each asset, when retired, actually has a zero salvage value. Assume that straight-line depreciation is to be used; the 20-year average service life and zero salvage value requires a 5% depreciation rate.

Now assume that each asset has its own account and that the depreciation on each asset is computed separately. (This is referred to as *single-asset* or *item* depreciation accounting.) Approximately half of the assets will be retired before they are 20 years old. Every such asset will have some book value when retired; with no realized salvage value, the books of account will show a "loss" for each retirement. That "loss" is referred to as "loss on disposal" and will affect the cash flow for income taxes in many cases. The half of the assets that survive for more than 20 years will be fully written off at age 20; no further depreciation charges will be made against them during their period of service from years 20 to 45.

In contrast, assume that one account is used to include the investment in all of these assets; each year the depreciation charge is made against the group of assets rather than against the individual asset and the book value at any time applies to the group with no identifiable separate figure applicable to each asset. With the assets considered in a group and with the fact of mortality dispersion recognized, it will be evident that the retirements at ages short of 20 years are not "premature" in the sense that they are inconsistent with the estimate of a 20-year *average* service life. If a set of unequal numbers are averaged, some of the numbers must be less than the average and others must be greater. The recognition of this point in multiple-asset depreciation accounting eliminates the "loss on disposal" entry under normal circumstances. However, in traditional multiple-asset accounting, depreciation charges continue beyond the average service life.

In the language used in publications of the U.S. Internal Revenue Service, a distinction is made among three general types of multiple-asset accounts, as follows:

1. *Group accounts*. Examples of such accounts would be passenger automobiles, punch presses, office desks.
2. *Classified accounts*. Some examples are transportation equipment, machinery, and furniture and fixtures.
3. *Composite accounts*. As an example, transportation equipment, machinery, and furniture and fixtures might be included in a single account.

Particularly in the manufacturing industries, the most common practice is to use classified accounts for income tax purposes. Often, the use of item accounts is limited to such assets as buildings and structures.

In this book the most important use of depreciation accounting is in connection with the calculation of cash flow for income taxes. In most of our examples and problems illustrating such calculations, our computations of the tax aspects of depreciation are made as if item accounts were to be used. This assumption of single-asset accounting is made chiefly for reasons of simplicity—to avoid the need to devote space to an explanation of the numerous technical details of multiple-asset accounting. In many studies, particularly those relative to proposed new assets, it makes no difference whether single-asset or multiple-asset accounting is assumed.

An economy study for a proposed retirement is one type of study in which the timing of cash flow for income taxes with single-asset accounting may differ greatly from the timing with multiple-asset accounting. Often the critical point in this type of study is whether a "loss on disposal" will be allowable for tax purposes. This topic is discussed in Chapter 17.

Certain Types of Assets Not Subject to Depreciation Charges in the Accounts

It is a convention of accounting that land is carried on the books at its original cost regardless of changes in its market value. No depreciation or appreciation on land is shown in the books of account. Of course, if land is finally sold at a price above or below its original cost, a "gain" or "loss" assigned to the year of the sale is shown on the books. But for accounting purposes, it is not necessary to make an estimate of the future sale price of land.

In the absence of any basis for a different forecast, some economy studies may follow the lead of orthodox accounting and assume, in effect, that investments in land will have 100% salvage values. But cases occur where this is not an appropriate assumption. Sometimes there is good reason to forecast that at the end of a study period land will be sold for much more or much less than its original cost. If so, this forecast is relevant in any analysis to guide the decision on whether or not to invest in the land; the estimated future sale price of the land is one element of cash flow to include in the economy study.

Many economy studies relate to the investment in plant to permit the sale of a new product or expanded sale of an existing product. In addition to the investment in physical plant, such a project usually requires an outlay for *working funds*. The working fund investment will usually include inventories of materials, work in process, and finished goods. It will also include the excess of accounts receivable associated with the project over accounts payable similarly associated. For certain types of product there may also be an inventory of returnable containers to be considered.

The working fund investment is not subject to depreciation on the books of account. For purposes of an economy study an outlay for working funds should ordinarily be treated as if it were an investment that had a prospective 100% salvage value at the end of the study period. That is, it usually is reasonable to expect that when a product is discontinued the accounts receivable will be collected and the inventories will be liquidated. This treatment was illustrated in Example 8–3. Comments on certain aspects of the estimation of working fund requirements are made in Chapter 16.

Economy-Study Estimates of Cash Flow for Income Taxes Based on Certain Simplified Assumptions

Many of the examples and problems in Chapters 6, 7, and 8 included estimates of the differences in cash flow for income taxes that would be caused by a choice among stated alternatives. In those studies applicable to competitive industry, these estimates were made under the following simplified assumptions:

1. The same estimated lives and salvage values used in the economy study will also be used for tax purposes and will be acceptable to the taxing authorities.
2. Straight-line item depreciation accounting will be used for tax purposes.
3. Throughout the period of the economy study, the applicable tax rate on an increment of taxable income will be 50%.
4. Except for capital items, receipts and disbursements before taxes in any year apply to the calculation of the taxable income for that year. (Capital items include first costs and salvage values of depreciable plant. They also include assets such as land and working capital that are not subject to depreciation charges in computing taxable income.)
5. Recovery of investments in nondepreciable assets such as land and working capital does not give rise to any tax consequences.

Both the absolute total amount and the timing of prospective income taxes are important in the after-tax evaluation of proposed investments. A reasonable question about the foregoing simple assumptions is whether they give results that are good enough for practical purposes in investment evaluation.

The answer is that sometimes these assumptions are fairly good ones in competitive industry in the United States and sometimes they are not good at all. In Chapter 12 we take a critical look at various matters that influence the amount and timing of the income tax consequences of decisions in competitive industry.

The treatment of prospective income taxes in economy studies for regulated public utilities in the United States is a specialized topic that is related to the way in which public utility rates are regulated. This topic is not discussed until Chapter 20.

Verification of Income Tax Differences Given in Examples 6–1 and 6–2

Consider the application of the foregoing assumptions to the data of Examples 6–1 and 6–2. The following estimates were made for the three plans compared in these two examples:

	Plan A	*Plan B*	*Plan C*
First cost	$0	$15,000	$25,000
Life	10 years	10 years	10 years
Terminal salvage value	$0	$0	$5,000
Annual before-tax disbursements	$9,200	$5,100	$4,050

These were alternatives with reference to production methods; the receipts from the sale of the product or service would not be influenced by the choice among the three plans. Therefore, the difference in estimated income taxes depends only on the differences in estimated tax deductions. The deductions from taxable income each year are as follows:

	Plan A	*Plan B*	*Plan C*
Straight-line depreciation	$0	$1,500	$2,000
Before-tax disbursements	9,200	5,100	4,050
Total deductions	$9,200	$6,600	$6,050

Because Plan B has $2,600 less deductions from taxable income than Plan A, it will involve higher annual income tax disbursements of $1,300, 50% of $2,600. This $1,300 was the income tax figure given in the statement of Example 6–1.

Because Plan C has $3,150 less deductions from taxable income than Plan A, it will involve higher annual income tax disbursements of $1,575, 50% of $3,150. This $1,575 was the income tax figure given in the statement of Example 6–2.

Explanation of the Method of Estimating Each Year's Income Taxes Given in Example 8–3

The third paragraph of Example 8–3 ended as follows:

> The specific estimates of net positive before-tax cash flow from operations year by year are: 1, $100,000; 2, $300,000; 3, $400,000; 4 through 11, $500,000; 12 and 13, $400,000; 14, $300,000; and 15, $200,000. These positive cash flows will be diminished by yearly disbursements for income taxes estimated to be 50% of the amount that each year's figure exceeds $100,000.

The depreciable investment in plant and equipment in this example was $1,500,000 made over a period of 2 years. The depreciation charge for accounting and tax purposes would not start until date 0 when the construction would be completed and the plant put in service. Because the estimated life was 15 years with no terminal salvage value, the annual straight-line depreciation is $1,500,000 ÷ 15 = $100,000. It is this $100,000 deduction from taxable income that was referred to in the second sentence quoted from the statement of Example 8–3.

The reader will note that Example 8–3 differs from Examples 6–1 and 6–2 in that the choice between the alternatives (accepting the project or rejecting it) involves differences in annual receipts. Moreover, because the estimated before-tax positive cash flow varies from year to year, the estimated disbursements for income taxes also vary from year to year.

According to the conventions of accounting and income taxation, the $300,000 outlay for land and the $200,000 outlay for working capital are non-depreciable and therefore have no influence on estimated disbursements for income taxes during the 15-year operating period of the project. Because it is assumed that these investments will be recovered without increase or decrease at the end of the 15-year operating period, the cash flow incident to their recovery at date 15 has no influence on income taxes in the 15th year, the terminal date of the project.

Some Economy-Study Methods That Appear to Tie Annual Capital Recovery Costs to the Depreciation Accounts

One of the topics presented in Chapter 6 was the conversion of a first cost, P, into an equivalent uniform annual figure over a life or study period. To make this conversion, it is necessary to have a stipulated minimum attractive rate of return or interest rate, $i^\star$, an estimated life or study period, n, and an estimated salvage value, S, at the end of the life or study period.

In Chapter 6, we gave the following equation to compute this annual cost of capital recovery with a return:

$$\text{CR} = P(A/P,i\%,n) - S(A/F,i\%,n)$$

This equation, which involves the use of the capital recovery factor and the sinking fund factor, obviously is independent of the depreciation accounts.

In some economy studies that use the method of annual costs, various combinations of depreciation figures and interest figures are used. The total of depreciation plus interest is intended to serve the same purpose as our annual cost of capital recovery with a return. Three methods that sometimes are used are as follows:

1. Sinking-fund depreciation plus interest on first cost
2. Straight-line depreciation plus interest on first cost
3. Straight-line depreciation plus average interest

These methods give satisfactory results in some cases and misleading results in others.

Sinking-Fund Depreciation Plus Interest on First Cost

This is a correct compound interest method that gives the same annual cost of capital recovery as our conventional method provided the assumed interest rate on the sinking fund is equal to the stipulated minimum attractive rate of return. Sometimes the method is referred to as *interest plus amortization*.

As explained in Chapter 4, the capital recovery factor is always equal to the sinking fund factor plus the interest rate. In Chapter 6 we showed two alternative ways of writing our basic equation for capital recovery at interest rate i^*. They were:

$$\text{CR} = (P - S)(A/P,i^*\%,n) + Si^*$$
$$\text{CR} = (P - S)(A/F,i^*\%,n) + Pi^*$$

In effect, the method of sinking-fund depreciation plus interest on first cost is based on this latter formula.

Let us assume an asset with a first cost of $22,000, an estimated life of 10 years, an estimated salvage value of $2,000, and a minimum attractive rate of return of 10%. Then, computation of the cost of capital recovery by the method of sinking-fund depreciation plus interest on first cost is:

Sinking-fund depreciation	$= (P - S)(A/F,i^*,n)$	
	= $20,000(0.06275)	= $1,255
Interest on first cost	$= Pi^* =$ $22,000(0.10)	= 2,200
Total annual cost of capital recovery with a 10% return		= $3,455

It should be noted that the figure for so-called sinking-fund depreciation in the foregoing calculation of capital recovery cost is not the full depreciation charge that would be made in the books if the sinking-fund method should be used; it is merely the *annuity* portion of the depreciation charge. If we should calculate year-by-year depreciation charges in the manner illustrated in Table 11–2, we would see that the sinking-fund depreciation charge for our $22,000 asset would vary from $1,255 in the first year to $2,959 in the 10th year; in our calculation of sinking-fund depreciation plus interest on first cost, only the $1,255 figure was used.

This annuity portion of the sinking-fund depreciation charge is sometimes referred to as *amortization*.* (The verb *amortize* comes from the same root as *mortal* and literally means "to make dead or destroy." It is applied to the

*For some reason this word is often mispronounced. The correct division into syllables is a-mor-ti-za-tion, with the second and fourth syllables accented. The common mispronunciation assumes the first syllable is *am* and accents the *am*.

provision for extinguishment of a debt or other obligation, particularly through periodic payments into a sinking fund.)

If the interest rate used on the amortization fund differs from the interest rate on first cost, the capital recovery cost obtained by this method obviously will differ from the one obtained by a conventional calculation using only one interest rate. For example, assume that a 10% rate is used to compute interest on first cost and a 3% rate is used in the sinking fund calculation. The capital recovery cost for our $22,000 asset will then appear to be:

Sinking-fund depreciation	= ($22,000 − $2,000)($A/F$,3%,10)	
	= $20,000(0.08723)	= $1,745
Interest on first cost	= $22,000(0.10)	= 2,200
Total		= $3,945

Some Pros and Cons on the Method of Sinking-Fund Depreciation Plus Interest on First Cost

This method was formerly widely used in economy studies for regulated public utilities and for government agencies, but it has largely been abandoned. In such studies it was usual to assume the same interest rate on first cost and on the sinking fund.

Where used in that manner with a single interest rate equal to the minimum attractive rate of return, i^*, the method gives exactly the same annual cost figure as the method employing the capital recovery factor, which is used throughout this book. The choice between these two methods must therefore be based on other grounds than correctness of results.

If sinking-fund depreciation accounting were actually to be used in the accounts, the use of sinking-fund depreciation plus interest on first cost for economy studies would help to reconcile the economy studies with the accounts of the enterprise. Sinking-fund depreciation accounting, however, is subject to the objection that it gives an unrealistically high proportion of the total write-off in the final years of the life of the asset.

Sinking-fund depreciation plus interest on first cost might tie an economy study to the financial policies of an enterprise if an actual sinking fund were to be established to recover an investment (less salvage, if any) in a fixed asset at the end of its life. But such sinking funds are rarely used.

In most cases, therefore, this method should be viewed as a compound interest conversion unrelated to the accounts or to the establishment of any actual fund. Viewed in this way, the method of sinking-fund depreciation plus interest on first cost does not seem to the authors to have any advantage over the conventional method of computing the annual cost of capital recovery with a return illustrated throughout this book. Often the method using sinking-fund depreciation is harder to explain than the conventional method because critics may make such comments as "There will be no actual sinking fund" or "We are not actually using this method of depreciation accounting." It is hard to give a satisfactory answer to these comments.

Valid objections may be raised to this method where two interest rates are used. A discussion of these objections follows the explanation of the method of straight-line depreciation plus interest on first cost because this latter method is really a special case of the analysis using two interest rates.

Straight-line Depreciation Plus Interest on First Cost

If this method were used, the annual capital recovery cost of our \$22,000 asset would be computed as follows:

$$\text{Straight-line depreciation} = \frac{P - S}{n}$$

$$= \frac{\$22{,}000 - \$2{,}000}{10} = \frac{\$20{,}000}{10} = \$2{,}000$$

$$\text{Interest on first cost} = Pi^{\star} = \$22{,}000(0.10) = 2{,}200$$

$$\text{Total} = \$4{,}200$$

Except in the special case of a salvage value of 100% (or more), this method invariably gives too high a figure for equivalent annual cost. The \$4,200 a year that the method indicates is needed to recover our proposed \$22,000 investment with a 10% return is in fact sufficient to permit recovery with a return of about 14.5%.

It has been pointed out that the straight-line method of depreciation accounting may be thought of as a limiting case of the sinking-fund method in which the interest rate on the sinking fund has been assumed to be 0%. Similarly, straight-line depreciation plus interest on first cost may be viewed as a special case of sinking-fund depreciation plus interest on first cost in which the sinking-fund interest rate is 0%.

Objection to Annual Cost Calculations Using Two Interest Rates

We have computed three different figures for the annual cost of capital recovery with a 10% return for our proposed \$22,000 investment in equipment. The conventional method would give us a figure of \$3,455; this same figure was obtained with 10% sinking-fund depreciation plus 10% interest on first cost. But 3% sinking-fund depreciation plus 10% interest on first cost gave us an annual capital recovery cost of \$3,945. And straight-line depreciation plus interest on first cost gave us \$4,200.

The usual objective of comparing equivalent annual costs in an economy study is to judge whether or not certain proposed investments will yield at least a minimum attractive rate of return. Assume that our proposed \$22,000 asset is expected to be responsible for increased receipts (or reduced disbursements) of \$3,800 a year for 10 years. This annual \$3,800 plus the final

$2,000 salvage value will in fact be sufficient to recover the $22,000 investment with interest at 12.1%.

If we should compare this annual figure of $3,800 with our computed annual capital recovery cost of $3,455, we would obtain the correct conclusion that the asset would yield more than a 10% return. But if we should compare the $3,800 with $3,945 or with $4,200, we would obtain the incorrect conclusion that the proposed investment was not justified because it would not yield 10%.

The reader may recall our discussion in Chapter 8 of the misleading conclusions obtainable from the calculation of an unknown rate of return by the Hoskold method, where reinvestment was assumed in an imaginary sinking fund bearing a low interest rate. The fallacy in the annual cost calculation using two interest rates (i.e., 3% and 10%, or 0% and 10%) is the same one that we examined in Chapter 8 in our discussion of the Hoskold method.

The error here is that the merits of *one* investment, the proposed $22,000 asset, are to be judged on the basis of the combined consequences of *two* investments, the asset and the 3% sinking fund. It is true that the asset, yielding 12.1%, and the sinking fund, yielding 3%, will not combine to give an overall rate of return of 10%. A total of $3,945 a year will be needed for the combined rate of return to be 10%; $3,800 is not enough. But unless there is to be an actual 3% sinking fund investment that *must* be made if the asset is acquired, this deficiency of the combined rate of return below 10% has no bearing on the merits of the proposed investment in the asset.

The foregoing paragraph may be altered so that it becomes a comment on the method of straight-line depreciation plus interest on first cost if 0% is substituted for 3% and $4,200 for $3,945.

Straight-Line Depreciation Plus Average Interest

This method of computing the annual cost of capital recovery with a return was widely used in economy studies in the United States in the 1930s and 1940s. Where straight-line depreciation was actually to be used in the accounts and where the assumed life and salvage value in the economy study were the same as used in the accounts, the method had the advantage of helping to reconcile economy studies with the accounts of an enterprise. The method is a close approximation to true equivalent annual cost in certain cases but a poor approximation in others.

With the decline in the application of straight-line depreciation accounting to new assets that started in 1954 in the United States, the use of this method no longer is justified on the basis of reconciling economy studies with the accounts except in a minority of cases.

A good way for the reader to compare this method with the conventional method is to look back at Table 3–1 (page 24) which compares different schemes of repaying (or recovering) $10,000 in 10 years with interest at 9%. Plan III in this table corresponds to the conventional viewpoint on computing the annual cost of capital recovery with a return. It is evident that the \$1,558.20 a year for 10 years will exactly repay \$10,000 with 9% interest.

Plan II is representative of another approach to the annual cost of capital recovery with interest. In this plan the annual payment on principal is uniform but the interest paid each year diminishes. With \$10,000 paid in 10 equal installments, each installment is \$1,000. Interest the first year is \$900, which is 9% of the full \$10,000, but diminishes by \$90 each year as the principal is repaid until in the 10th year it is only \$90. The average interest is \$495.

These interest payments form an arithmetic progression the first term of which is Pi and the final term of which is $(P/n)i$. The average interest is the average of the first and last terms or $\frac{Pi}{2}\left(\frac{n+1}{n}\right)$.

The method of straight-line depreciation plus average interest always computes an *average* (rather than an equivalent) annual figure for a repayment plan devised along the lines of Plan II in Table 3–1. In the general case with a first cost, P, an estimated terminal salvage value, S, an estimated life of n years, and an interest rate, i, straight-line depreciation is $\frac{P-S}{n}$ and average interest is $(P-S)\left(\frac{i}{2}\right)\left(\frac{n+1}{n}\right)+Si$. As applied to our \$22,000 asset, the calculation is:

$$\text{Straight-line depreciation} = \frac{\$22{,}000 - \$2{,}000}{10} = \$2{,}000$$

$$\text{Average interest} = \left[(\$22{,}000 - \$2{,}000)\left(\frac{0.10}{2}\right)\left(\frac{11}{10}\right)\right] + \$2{,}000(0.10) = \$1{,}100 + \$200 = \underline{1{,}300}$$

$$\$3{,}300$$

This is somewhat less than the exact equivalent uniform annual cost, which we computed to be \$3,455.

Limitations of Methods of Straight-Line Depreciation Plus Average Interest

Wherever the prospective terminal salvage is less than 100% of first cost, this method yields a figure for capital recovery cost that is too low. The method is an approximate one because it assumes the simple average of a diminishing series of payments to be the equivalent uniform annual payment. The error involved in its use increases with the length of the period of time considered and also increases with an increase in the interest rate.

Where annual cost methods are used to compare assets having approximately the same lives, where the lives are not too long, and where the minimum attractive rate of return is low, this method often gives satisfactory results. The greater the disparity in the lives of assets being compared and the higher the minimum attractive rate of return, the more likely it is that economy studies using this method will lead to misleading conclusions. The foregoing point is illustrated in Problems 11–7, 11–8, and 11–25.

Three Common Methods of Computing So-Called Rates of Return on Proposed Investments

There are many different ways in which prospective figures from the accounts are used to compute ratios that are alleged to be prospective rates of return. George Terborgh in the chapter entitled "Popular Rule-of-Thumb Tests of Investment Merit," in *Business Investment Policy** reports attending a conference where 14 companies reported 14 different methods of this type of calculation.

The numerous schemes that the authors of this book have observed all seem to be variants of the following three methods:

1. The ratio of prospective average annual profit after depreciation to original investment is alleged to be the prospective rate of return on investment. Often this is referred to as the "original book" method.

2. The ratio of prospective average annual profit after depreciation to the average book value is alleged to be the prospective rate of return on investment. Often this is referred to as the "average book" method. In many instances it gives a figure for rate of return that is double the figure given by the original book method.

3. The figures for prospective profit after depreciation for each year are divided by the prospective book value figures for the start of the respective year in the way that was illustrated in Table 11–1. This division gives a series of ratios that are alleged to be year-by-year prospective rates of return. Conceivably, these rates might be averaged to give an overall figure for rate of return.

These so-called rate of return methods are supposed to be substitutes for the correct rate of return method explained in Chapter 8. They could not be explained in Chapter 8 because they involve an understanding of depreciation methods and book value. Example 11–3 illustrates these so-called rate of return methods and calls attention to the kinds of errors that are encountered in their use as approximations for the true rate of return.

*George Terborgh, *Business Investment Policy* (Washington, D.C.: Machinery and Allied Products Institute, 1958), p. 33.

EXAMPLE 11–3

So-Called Rate of Return Calculations for Analysis of Proposed Investments

Facts of the Case

Among a number of proposed projects being considered by the management of a firm are two, an automatic palletizer and an energy control computer. Each of these units will have a total installed cost of $18,000, an estimated economic service life of 6 years, and an estimated zero salvage value. The palletizer will reduce operating disbursements by $8,000 the first year, $7,000 the second year, and so on to a net reduction of $3,000 in the sixth year. The energy control computer will save $3,500 in disbursements for fuel in the first year and the saving will increase by $1,000 each year.

The company requires a before-income-tax rate of return of 25% on all investments.

Using the correct rate of return methods explained in Chapter 8, the prospective rates of return on the two proposals are as follows:

Palletizer:

$$PW = 0 = -\$18{,}000 + \$8{,}000(P/A,i\%,6) - \$1{,}000(P/G,i\%,6)$$
$$i = 25.3\%$$

Computer:

$$PW = 0 = -\$18{,}000 + \$3{,}500(P/A,i\%,6) + \$1{,}000(P/G,i\%,6)$$
$$i = 20.4\%$$

Thus, the palletizer meets the company's criterion but the energy control computer does not.

If the so-called rate of return methods had been used to determine the attractiveness of these proposed investments, the results would have been quite different. Table 11–3, based on straight-line depreciation, shows the book value at the beginning of each year and the "profit" for the year. Then Columns E and I show the year-by-year rate-of-return based on the "profit" for the year divided by the beginning of year book value. The average rate-of-return on the palletizer is only 19.7%, while that for the computer is 55.1%. This so-called rate-of-return method completely reversed the decision, and made the computer appear to be extremely attractive.

The "original book" method requires that the average annual "profit" be divided by the original cost.

$$\text{Palletizer: } \$2{,}500/\$18{,}000 = 0.139 = 13.9\%$$
$$\text{Computer: } \$3{,}000/\$18{,}000 = 0.167 = 16.7\%$$

TABLE 11–3
Computation of book values, year-by-year profits, and year-by-year rate of return by so-called rate-of-return methods

	Palletizer				Computer			
Year A	*Cash flow before taxes* B	*Book value start of yr* C	*Profit* D	*Rate of return* E	*Cash flow before taxes* F	*Book value start of yr* G	*Profit* H	*Rate of return* I
0	−$18,000				−$18,000			
1	+8,000	$18,000	+5,000	27.8%	+3,500	$18,000	$ 500	2.8%
2	+7,000	15,000	+4,000	26.7	+4,500	15,000	1,500	10.00
3	+6,000	12,000	+3,000	25.0	+5,500	12,000	2,500	20.8
4	+5,000	9,000	+2,000	22.2	+6,500	9,000	3,500	38.9
5	+4,000	6,000	+1,000	16.7	+7,500	6,000	4,500	75.0
6	+3,000	3,000	0	0	+8,500	3,000	5,500	183.3
Σ	+15,000		+15,000	118.4	+18,000		+18,000	330.8
Average			+2,500	19.7%			+3,000	55.1%

By this method, neither of the proposed investments meets the company's criterion of 25% before income taxes.

The "average book" method requires that the average annual "profit" be divided by the average book value, $(P + S)/2$.

$$\text{Palletizer: } \$2{,}500/\$9{,}000 = 0.278 = 27.8\%$$
$$\text{Computer: } \$3{,}000/\$9{,}000 = 0.333 = 33.3\%$$

This method makes both proposals appear attractive, with the computer being more attractive than the palletizer.

Weaknesses in So-Called Rate of Return Methods

Problems 11–9 through 11–15 at the end of this chapter illustrate these three methods and demonstrate their unreliability. Although these particular problems, for simplicity, apply the methods to calculations of rates of return before income taxes, the methods are equally unreliable when used for calculations after income taxes. The problems are adapted with some modifications from an illustration used by Horace G. Hill, Jr.,* for many years Budget Director of The Atlantic Refining Company.

These seven problems illustrate a number of basic weaknesses of these

*Horace G. Hill, Jr., *A New Method of Computing Rate of Return on Capital Expenditures*, a pamphlet published privately by the author (Berwyn, Pa., 1953).

three types of method of computing so-called rates of return. Nevertheless, there is one other serious weakness that they do not illustrate. In a correct calculation of rate of return before income taxes by compound interest methods, it is immaterial whether a cash disbursement is to be capitalized on the books of account or whether it is to be charged as an expense in the year when it occurs. But the three methods that we have mentioned all use book value (original or average or the current year's) as the denominator in a fraction stated to be the rate of return. "Book value" ordinarily refers only to items that are capitalized on the books of account. Thus, two projects having identical cash flows might appear to have quite different rates of return if the two projects are to be treated differently in the accounts. This type of error may lead to incorrect ranking of projects and may also increase the inherent errors in the rates of return computed by these three methods.

Why Are Incorrect Methods of Computing Rates of Return in Common Use?

A reader of this book who is not already employed in an organization using one of these methods may well ask why methods of analysis that are so erratic and unreliable are so widely used. This is a good question to which the authors are unable to give a fully satisfactory answer. George Terborgh† refers to these and similar methods of analysis as "really industrial folklore, handed down from one generation of management to the next. They have no scientific rationale, no legitimate intellectual parentage."

In some instances it doubtless is true that the methods are used because the people who use them are not aware that any better methods are available. But there are many cases where analysts consciously reject compound interest methods of computing rates of return in favor of the original book or average book method. The choice of book methods usually is defended by the following arguments:

1. Book methods are alleged to be easier to apply.
2. Book methods are alleged to be easier to explain to others.

The greater ease of application of book methods is not a valid argument for their use. There are many economy studies in which compound interest methods can be applied as easily as book methods, once the compound interest methods are understood by the analyst. Moreover, even where compound interest methods require, say, an extra half-hour of an analyst's time, this extra time usually is trivial in relation to the many hours of time that have been spent on the gathering of data to be used in the economy study.

In many organizations it is true that the book methods are easier to explain. Unless managers and engineers have some understanding of compound interest, methods that seem to be tied to the books of account can be explained more readily than methods that use the mathematics of compound interest. The answer to this argument is that, as we have pointed out, book

†Terborgh, *op. cit.*, p. 28.

methods lead to erratic and unreliable answers that cannot be trusted as a basis for action; compound interest methods should be used in spite of the greater difficulty of explaining them.

What Meaning Shall Be Attached to the Phrase "Rate of Return" as Applied to Proposed Investments?

In our initial discussion of equivalence in Chapter 3, we examined a loan transaction in which an investor exchanged an initial cash disbursement of $10,000 for the prospect of end-of-year cash receipts of $1,558.20 a year for 10 years. It was shown (in the tabulation for Plan III, Table 3–1) that this series of receipts enables the investor to recover the investment with exactly 9% interest per annum. It might also be stated that the investor is scheduled to recover the investment with a 9% rate of return.

This is the only sense of rate of return that is a sound and consistent guide to action regarding proposed investments, and it is the only sense in which the phrase is used in this book. Where one of the methods based on the depreciation accounts is employed, we use the phrase "so-called rate of return."

(It may be of interest to apply the various methods based on the books of account to our simple case of the $10,000 investment recovered at 9% interest by $1,558.20 a year for 10 years. By the original book method, the rate of return appears to be 5.58%. By the average book method, it appears to be 11.16%. The year-by-year method depends on the scheme used for a time allotment of the investment among the 10 years; if a straight-line allotment is used, the return appears to vary from 5.58% in the first year to 55.82% in the 10th year.)

Arguments on the silly and unanswerable question "What does rate of return really mean?" have been so heated in some industrial and academic circles that some writers have tried to avoid the question by coining other phrases for rate of return obtained by correct compound interest methods. Thus, Reul uses the phrase *Profitability Index* or *PI* in this meaning. Weaver and Reilly have suggested the phrase *interest rate of return* to be used in the same sense.* Other terms frequently used are *solving rate* and *internal rate of return*.

Some Aspects of Depreciation Accounting Not Developed in This Book

We have already mentioned that the various types of statistical analysis of physical property mortality are not explained here, and that most of the technicalities of multiple-asset accounting have been avoided. Some other topics that we have omitted are as follows:

1. Assets generally are acquired during a fiscal year rather than exactly at the start of a year. For example, consider an asset acquired on March 31 in a business enterprise having the calendar year as its fiscal year. The question

*J. B. Weaver and R. J. Reilly, "Interest Rate of Return for Capital Expenditure Evaluation," *Chemical Engineering Progress*, LII, No. 10 (October, 1956), 405–12.

arises regarding the depreciation charge to be made for the remainder of the year. Because 9 months of the year remain, it might seem reasonable to charge 9/12 of the depreciation computed for the first full year of life under the depreciation accounting method selected. However, such a practice would require the application of a different fraction to each new asset, depending on its date of acquisition.

A more common practice is to use some type of *averaging convention*. The so-called half-year convention is a common one; all assets are given a half-year's depreciation charge during their acquisition year regardless of the time of the year when they were acquired. Another convention is to charge a full-year's depreciation for the acquisition year for the assets acquired during the first half of the year and no depreciation for those acquired during the second half.

In our examples we have, in effect, assumed that all assets are acquired at the start of a fiscal (and tax) year and that a full year's depreciation is charged during the year of acquisition.

2. There are various possible accounting treatments of gross amounts realized from salvage values and of costs of removing assets when retired. The way these matters are to be treated may influence the depreciation rate to be used and will determine the type of entry to be made on retirement.

3. There are differences between depreciation accounting where each year's acquisitions of a particular class of asset are kept in a separate account, sometimes called a *vintage* account, and where all the acquisitions of a particular class of asset over a period of years are merged together in a so-called open-end account.

4. It will rarely be true that average lives and salvage values will turn out exactly as estimated. Various problems arise associated with re-estimates of remaining lives and salvage values for existing assets. These problems differ with different depreciation accounting methods.

One purpose of mentioning the foregoing topics is to make it plain to our readers that our approach to depreciation accounting here is necessarily a simplified one—that there are many facets to the subject that we do not have space to explore. Another purpose is to suggest depreciation accounting and the tax aspects of depreciation as appropriate subjects for study by persons who are responsible for making or reviewing economy studies. Although an introduction to engineering economy can be made with simplified assumptions about depreciation accounting, a more sophisticated understanding of the subject will be helpful in certain types of economy studies.

PROBLEMS

11–1. An emergency gasoline-engine-driven pump has been acquired by a service company at an initial cost of $18,000. It is estimated to have an economic service life of 5 years with an estimated net salvage value at that time of $3,000. Find the first year depreciation charge based on: (a) straight-line depreciation; (b) double-rate declining-balance; (c) sum-of-years-digits; (d) sinking fund at 8%. (*Ans.* (a) $3,000; (b) $7,200; (c) $5,000; (d) $2,557)

11–2. Find the book value at the end of the third year (after the third year's depreciation has been charged) for the pump described in Problem 11–1. (*Ans.* (a) $9,000; (b) $3,888; (c) $6,000; (d) $9,700)

11–3. An asset with a first cost installed of $24,750 is estimated to have a 10-year economic service life with a zero salvage value at that time.

(a) What will be the depreciation charge for the 10th year by the SOYD method?

(b) What will be the depreciation charge for the 10th year by the double-rate declining-balance method and what will be the book value at the end of 10 years?

(c) If the asset is disposed of at the end of 10 years for a net salvage of zero, will there be any loss on disposal with the declining-balance method and, if so, how much? (*Ans.* (a) $450; (b) $664; (c) yes, loss on disposal will be $2,657.51)

11–4. A company purchased a commercial lot for $100,000 and built an office complex on it at a cost of $500,000. It is assumed that the building will be depreciated over 40 years with an estimated zero salvage value. What will be the book value of the property (land and building) at the end of the 10th and 20th years if (a) sum-of-years-digits is used; (b) double-rate declining-balance is used? (*Ans.* (a) $383,540; $228,050; (b) $399,368; $279,243)

11–5. An asset costing $26,000 is estimated to have an economic life of 15 years with a net salvage value of $2,000. Find the depreciation charge for the fourth year and the book value at the end of the fourth year by the (a) straight-line method; (b) sum-of-years-digits method; (c) double-rate declining-balance method; and (d) the sinking-fund method using an interest rate of 10%. (*Ans.* (a) $1,600; $19,000; (b) $2,400; $15,200; (c) $2,256; $14,671; (d) $1,005; $22,495)

11–6. Example 11–1 shows the results of the first year of operation for Equipment Rentals. The straight-line depreciation method was used to determine the profit of $9,000 before income taxes. What would the profit have been had the company (a) used the sum-of-years-digits method; (b) the double-rate declining-balance method of depreciation accounting? (*Ans.* $4,333; $2,000)

11–7. On page 217 the statement is made that the method of straight-line depreciation plus average interest understates the exact equivalent uniform annual cost of an investment and that the error increases as the interest rate and the life increase. Using an interest rate of 6%, compute the percentage by which the straight-line depreciation plus average interest understates the true equivalent uniform annual cost for lives of (a) 5 years; (b) 10 years; and (c) 20 years. (*Ans.* (a) 0.59%; (b) 2.11%; (c) 6.52%)

11–8. Change the interest rate of Problem 11–7 from 6% to 12% and compute the error by which the straight-line depreciation plus average interest understates the true equivalent uniform annual cost for the same lives. (*Ans.* (a) 1.95%; (b) 6.20%; (c) 15.60%)

11–9. Two investment proposals, a vacuum still and a product terminal, are competing for limited funds in an oil company. Each requires an immediate disbursement of $110,000, all of which will be capitalized on the books of account. In both cases the expected life is 10 years with zero terminal salvage value. The estimated positive cash flow before income taxes resulting from the vacuum still is $38,000 the first year and $34,000 the second, and will diminish by $4,000 each year until it is $2,000 in the 10th year. The estimated positive cash flow before income taxes resulting from the product terminal is $5,000 the first year and $9,000 the second, and will increase by $4,000 a year until it is $41,000 in the 10th year.

Find the approximate value of an interest rate that makes the present worth of these estimated cash flows just equal to zero for each of these competing proposals. (*Ans.* vacuum still, 19.2%; product terminal, 11.9%)

11–10. Apply the "original book" method to estimate the rates of return from the two competing projects in Problem 11–9. In this method the average annual estimated profit is divided by the estimated original investment. In applying this method assume straight-line depreciation, which will be $11,000 a year for each project. The profit for each year will be the net positive cash flow minus the depreciation. (*Ans.* vacuum still, 8.2%; product terminal, 10.9%)

11–11. Apply the "average book" method to estimate the rates of return from the two competing projects in Problem 11–9. In this method the average annual estimated profit is divided by the average investment. The average annual profit figures will be the same as in Problem 11–10, whereas the average investment will be one-half the figure in Problem 11–10. (*Ans.* vacuum still, 16.4%; product terminal, 21.8%)

11–12. Assuming straight-line depreciation, compute the apparent yearly rates of return for each of the 10 years of life of the alternatives in Problem 11–9 by dividing each year's profit (i.e., cash flow minus depreciation) by the book value at the start of the particular year. What is the arithmetic mean of the 10 computed rates of return for each project? (*Ans.* vacuum still, 1.7%; product terminal, 54.2%)

11–13. Solve Problem 11–12 using sum-of-years-digits depreciation instead of straight-line depreciation. (*Ans.* vacuum still, 22.2%; product terminal, 287.9%)

11–14. The comparison of vacuum still and product terminal in Problem 11–9 by correct compound interest methods showed an estimated rate of return of slightly over 19% for the vacuum still and slightly less than 12% for the product terminal. However, each of the "approximate" methods used in Problems 11–10 through 11–13 indicated that the product terminal would have a higher rate of return than the vacuum still. Do you think that these four so-called approximate methods will always give an incorrect ranking of projects competing for limited investment funds? If not, what aspects of the estimated cash flows for the vacuum still and product terminal caused all these methods to give the incorrect ranking?

11–15. The four methods of analysis used in Problems 11–10 through 11–13 gave widely different estimates of rate of return for each project. For example, the expected rate of return from the vacuum still seemed to be either 8.2%, 16.4%, 1.7%, or 22.2% depending on the method that was used. What differences among the so-called approximate methods seem to you to cause such extreme differences in the estimated rates of return?

11–16. Example 11–1 showed the results of the first year of operation for Equipment Rentals. At the end of the first year Black purchased another $10,000 worth of equipment and assumed the same life of 5 years with zero salvage value for it. At the end of 4 years Black was somewhat tired of running the business and, when he received an attractive offer, sold the business to a chain of rental stores. The cash flows for the 4 years are summarized in the following table:

Year	*Purchase of equipment*	*Cash disbursements for operating expenses*	*Cash receipts*
0	−$35,000		
1	− 10,000	−$30,000	+$46,000
2		− 36,000	+ 55,000
3		− 38,000	+ 57,000
4		− 41,000	+ 58,000

(a) Using the straight-line depreciation method, determine the before-income-tax profit each year.

(b) Using sum-of-years digits method of depreciation, determine the before-income-tax profit each year.

(c) None of the equipment had been retired at the end of 4 years. Determine the book value of the equipment by the straight-line method and by the sum-of-years-digits method.

11–17. Problem 11–16 gave the cash disbursements for the first 4 years of operation, but did not show the depreciation expenses. Assume that the estimated life of the equipment is changed to 10 years instead of 5. Determine the before-tax profit for each year based on (a) straight-line depreciation, (b) sum-of-years-digits depreciation, and (c) double-rate declining-balance depreciation.

11–18. A construction company is considering the purchase of a device to perform an operation that is quite costly as it is now being done. Two different devices are available. Each will cost $45,000 and is expected to last 5 years and have a zero salvage value at that time. The model "A" will reduce the operating disbursements by $9,000 the first year and $12,000 the second year, and the annual savings will increase by $3,000 each year. The other model, "B," will reduce the operating disbursements by $20,000 the first year, by $17,000 the second year, and by $3,000 less each year thereafter.

(a) Find the true prospective rate of return, before income taxes, for each model.

(b) Using the "original book" method illustrated in Example 11–3, find the so-called rate of return for each model.

(c) Using the "average book" method illustrated in Example 11–3, find the so-called rate of return for each model.

(d) Comment on the differences in the results obtained by using the three methods. Under what conditions might the original book method give a relatively good approximation?

Assume straight-line depreciation is being used.

11–19. Assuming straight-line depreciation, compute the apparent yearly rates of return for each of the 5 years of life of the two different alternative proposals in Problem 11–18. Follow Example 11–3 for the method, and compute the arithmetic mean of the returns for each proposal. Comment on the results obtained compared with the true rates of return found in Problem 11–18.

11–20. A company needs a storage building for its motor vehicles and is considering a prefabricated steel building versus a tilt-up concrete building. The steel building is estimated to cost $150,000 and to have a service life of 25 years. Annual maintenance and property taxes will be about $6,000 a year. The concrete building is estimated to cost about $200,000, but is estimated to have a useful service life of 50 years with an annual maintenance and property tax cost of about $4,000. Both buildings will have no realizable salvage value at the end of their service lives. The company uses a before tax i^* of 15%.

(a) Compute the equivalent uniform annual cost of both buildings using the exact method.

(b) Compute the approximate uniform annual cost of the two buildings using the straight-line depreciation plus average interest method.

(c) Explain the differences in your results.

(d) The company could rent suitable space for an annual rental of $32,000. Comment on the effects of the analysis method on the probable decision.

11–21. An engine lathe will cost $30,000 installed and is believed to have a reasonable service life of 25 years with a net realizable salvage value of $4,000. How much depreciation will be written off in the 10th year and what will be the book value at the end of the tenth year by the (a) straight-line depreciation method; (b) sum-of-years-digits method, and (c) the double-rate declining-balance method?

11–22. What declining-balance rate, f, will give a book value of (a) 2%, (b) 5%, and (c) 20% of the first cost of an asset at the end of 10 years?

11–23. A manufacturing firm uses a certain part in one of its products, but it does not have the necessary machines to make it. It can buy the part from other manufacturers or buy the necessary machines and produce the part. The following table gives the year-by-year cash flows for the two alternatives:

Year	*Buy parts*	*Make parts*
0	0	−$26,000
1	−$16,000	− 10,000
2	− 18,500	− 12,000
3	− 19,500	− 12,000
4	− 14,500	− 9,000
5	− 10,000	− 6,000
6	− 6,000	− 3,000
6		+ 2,000

(a) Assuming straight-line depreciation with a 6-year life, $2,000 estimated salvage value, and a 50% tax rate, compute the before-income-tax prospective rate of return on the investment of $26,000.

(b) Using the simplifying assumptions explained in this chapter, compute the after-income-tax prospective rate of return on the $26,000 investment.

11–24. A real estate company purchased an office building to be used as rental property. The total price was $700,000, of which $100,000 was the appraised value of the land. The building had an estimated remaining life of 20 years, with an estimated salvage value of zero. After having owned the property for 10 years, the company received an offer of $500,000 for it, of which $200,000 was the appraised value of the land.

Find the book value of the property at the end of 10 years using: (a) straight-line depreciation; (b) sum-of-years-digits depreciation; and (c) double-rate declining-balance depreciation.

The company had received net receipts after all disbursements for maintenance, property taxes, insurance, and so on, of $120,000 a year for the 10 years. Find the true rate of return (before income taxes) earned on the original investment if the property is now sold for $500,000.

11–25. Plot a curve showing the error introduced by using the straight-line depreciation plus average interest method instead of the exact equivalent uniform annual cost method as a function of years from 0 to 30, for (a) i = 10%; (b) i = 15%.

11–26. A warehouse has been acquired by a company at a total cost of $200,000, $50,000 of which was the appraised value of the land. The remaining useful life is estimated to be 15 years with a zero salvage value for the building. Of course, land is not depreciated and in the books of account is carried at its original cost. Determine the depreciation charge for the tenth year and the book value at the end of 10 years for the entire property if the building is depreciated by (a) the straight-line method, (b) the sum-of-years-digits method, (c) the double declining-balance method, and (d) the 6% sinking-fund method.

11–27. At the start of 1980 Kathy Brown, a certified public accountant, purchased a minicomputer for use in her private accounting practice. Its first cost was $25,000. She believed that it would have an economic life of 6 years with a net salvage value of $4,000. What depreciation expense would she have recorded for the third year and what would have been the book value at the end of the third year if she used the (a) straight-line method, (b) sum-of-years-digits method, (c) double declining-balance method, and (d) the 6% sinking-fund method of depreciation?

11–28. In 1980 the federal income tax laws of the United States permitted a taxpayer to deduct an "additional first year depreciation" of 20% on up to $10,000 of qualifying property. (See Problem 12–52.) Thus Kathy Brown (in Problem 11–27) could take 20% of $10,000 (or $2,000) in the first year plus regular depreciation on the remaining depreciable amount. Determine her depreciation write-off in the first and second years for her minicomputer with (a) the straight-line method, (b) the sum-of-years-digits method, and (c) the declining-balance method.

12

ESTIMATING INCOME-TAX CONSEQUENCES OF CERTAIN DECISIONS

*Our most important tax problem is to have tax systems at all levels of government that balance one another and are designed so that they promote business enterprise and permit our free-enterprise system to flourish.–H. A. Bullis**

Persons and business enterprises subject to income taxation need to consider prospective income taxes in relation to many types of choice among alternative courses of action. It often happens that the alternative that seems superior before income taxes seems clearly inferior after such taxes are considered. Moreover, in the special case of the choice among alternative investments in fixed assets in competitive industry, nearly *all* such investments are less attractive after income taxes than before.

In textbooks on engineering economy, examples and problems generally are more realistic when minimum attractive rates of return are stipulated as after-tax rates. A number of our examples and problems in Chapters 6, 7, and 8 did specify such rates. The differences among the alternatives in cash flow for income taxes were stated without any explanation of how such differences could be calculated. The simple assumptions used in estimating these differences were explained in Chapter 11.

Many economy studies in industry are made using similar simple assumptions about income taxes. But an analyst is not always safe in assuming that such assumptions are good enough for the purpose at hand. Often, it is a fairly complicated matter to predict the influence of a decision on cash flow for income taxes. The present chapter provides a more sophisticated look at certain aspects of income taxation that are related to investment decisions. Additional material about income taxes is introduced in a number of the following chapters and in Appendix F.

*H. A. Bullis, *Manifesto for Americans* (New York: McGraw-Hill Book Co., Inc., 1961), p. 84. Copyright © 1961 by Harry A. Bullis. Used with permission of McGraw-Hill Book Company.

Distinction Between Taxes on Net Income and Taxes on Gross Income

Where a flat tax is levied on gross receipts (so-called gross income), the estimates of future receipts and future tax rates supply an estimate of future tax payments. But taxes on gross receipts are subject to a number of serious objections. Therefore, income taxes are commonly levied on net income.

The concept of net income for tax purposes is similar to the concept of profit as measured by the accounts of a business enterprise. But legislative bodies are free to define net income in any way they see fit; in any specific case there may be important points of difference between the two concepts. The discussion here relates entirely to taxes on net income.

Widespread Applicability of the General Approach Presented Here

The graduated income tax in various forms is used in most industrialized countries and in many developing countries. Nevertheless, tax rates and the rules defining taxable income vary from country to country and change from time to time in any given country.

In an introduction to engineering economy, a general approach is needed that can be used with any set of tax laws, regulations, and rates. The authors have tried to develop such an approach in this chapter and in subsequent chapters. It is not our purpose to give detailed up-to-date information on income taxes in the United States or elsewhere. Detailed information about any income tax system would require much more space than is available.* Up-to-date information would call for annual revision.

A General Principle: Income Taxes Are Disbursements

Economy studies deal with prospective receipts and disbursements. In the simplified assumptions that we started to use in Chapter 6, it was recognized that prospective income taxes merely constitute another disbursement to add to those for operation, maintenance, property taxes, insurance, and so on.

In this chapter and in the following chapters, we continue to recognize that the inclusion of income taxes in an economy study requires the estimation of the amount and timing of this particular element of cash flow. But we shall see that the simplified assumptions that we used in Chapters 6 through 8 do not necessarily give us valid estimates either of amount or of timing. We shall start with a series of six examples that bring out the importance of the *timing* of prospective tax consequences of decisions.

We shall first consider an example that avoids the troublesome issue of how rapidly a proposed investment may be written off for tax purposes.

*For example, the *Federal Regulations on Income Tax* in the United States contain more than three times as many words as this entire book.

EXAMPLE 12–1

Rate of Return After Income Taxes for an Investment Expected to Have a 100% Salvage Value

Facts of the Case
An outlay of $110,000 is proposed for the purchase of land needed for the storage of certain equipment. Alternatively, the land may be rented for $25,000 a year. If the land is owned, $3,000 a year must be paid for property taxes and certain other matters incident to ownership. Except for income taxes, other receipts and disbursements will be the same whether the land is owned or rented. For the purpose of an economy study, it is estimated that the need for the equipment storage will continue for 10 years and that, if purchased, the land will be sold for a net $110,000 at the end of the period. It is estimated that any extra taxable income caused by this purchase will be subject to a tax rate of 49%.

Calculation of Rate of Return
Under the usual conventions both of accounting and of income tax laws, an investment in land will not be written off. (In this case we have estimated that the land will be resold at its original cost at the end of 10 years; therefore, no future gain or loss is expected for either accounting or income tax purposes.) Table 12–1 shows the derivation of the needed figures for after-tax cash flow.

Where the initial outlay, in this case $110,000, is expected to be recovered unchanged at the end of some stipulated period of years, and the annual positive cash flow is expected to be uniform throughout these years, the prospective rate of return may be computed merely by division. Here

TABLE 12–1
Estimation of cash flow after income taxes, Example 12–1

Year	*Cash flow before income taxes*	*Influence on taxable income*	*Influence of income taxes on cash flow* *−0.49B*	*Cash flow after income taxes* *(A + C)*
	A	B	C	D
0	−$110,000			−$110,000
1 to 10	+22,000 per year	+$22,000 per year	−$10,780 per year	+11,220 per year
10	+110,000			+110,000
Totals	+$220,000	+$220,000	−$107,800	+$112,200

our before-tax rate is $\frac{\$22,000}{\$110,000} = 0.20$ or 20%. The after-tax rate is $\frac{\$11,220}{\$110,000}$ $= 0.102$ or 10.2%.

In this simple case the 49% applicable tax rate has decreased the rate of return by 49%. In the next five examples we shall see that matters often are considerably more complex than this.

Although this example stipulated a 10-year study period with complete recovery of the $110,000 investment at the end of this period, the reader may note that neither the before-tax rate of 20% nor the after-tax rate of 10.2% really depend on the estimated life. For example, if the same annual cash flow before and after income taxes had been expected to continue from the 11th through the 20th years, and if the estimated terminal resale price of the land had continued to be $110,000, the two prospective rates of return would still have been 20% and 10.2%.

A General Principle: The Relationship Between Rates of Return Before and After Income Taxes Depends on the Rules Governing Write-Off for Tax Purposes

To illustrate the foregoing important principle, let us assume five proposed immediate disbursements of $110,000, each expected to reduce future disbursements (other than those for income taxes) by a net $26,300 a year for the next 10 years. At 20% interest the present worth of $26,300 a year for 10 years is $110,250. Therefore, the prospective rate of return before income taxes is a little less than 20.1%.

These five proposed investments yielding about 20% before income taxes are examined in Examples 12–2 through 12–6. Their rates of return after income taxes differ greatly. The sole cause of the difference (from 8.5% to 20.1%) is the difference in the way in which the initial outlay of $110,000 is permitted to be written off for income tax purposes.

The higher the tax rate, the greater the importance of the rules governing the speed of write-off. In Examples 12–2 through 12–6 we shall assume that a 49% tax rate is applicable throughout the entire period of each study.

EXAMPLE 12–2

Rate of Return After Income Taxes Assuming that the Initial Outlay is Written Off by the Years-Digits Method

Facts of the Case

An outlay of $110,000 for materials-handling equipment is proposed. It is estimated that this equipment will reduce disbursements for labor and labor extras by $32,000 a year for 10 years and will increase annual disbursements by $5,700 for maintenance, power, property taxes, and insurance. The estimated life of the equipment is 10 years with zero terminal salvage value; this estimate is acceptable to the taxing authorities. The taxpayer will use the sum-of-years-digits method in reporting depreciation for tax purposes.

Calculation of Rate of Return After Income Taxes

Table 12–2 shows the calculation of the prospective effect of the investment on cash flow after income taxes, assuming a 49% tax rate throughout the entire 10-year period. The rate of return after taxes may be determined by computing present worth of cash flow after taxes using rates of 12% and 13%.

$$
\begin{aligned}
\text{PW at 12\%} &= -\$110{,}000 + \$23{,}213(P/A,12\%,10) - \$980(P/G,12\%,10) \\
&= -\$110{,}000 + \$23{,}213(5.650) - \$980(20.254) \\
&= +\ \$1{,}300
\end{aligned}
$$

$$
\begin{aligned}
\text{PW at 13\%} &= -\$110{,}000 + \$23{,}213(5.426) - \$980(19.080) \\
&= -\$2{,}750
\end{aligned}
$$

Interpolation indicates a rate of return of about 12.3% after income taxes.

TABLE 12–2
Estimation of cash flow after income taxes, Example 12–2

Year	*Cash flow before income taxes*	*Write-off of initial outlay for tax purposes*	*Influence on taxable income*	*Influence of income taxes on cash flow*	*Cash flow after income taxes*
				−0.49C	(A + D)
	A	B	C	D	E
0	−$110,000				−$110,000
1	+26,300	−20,000	+$ 6,300	−$ 3,087	+23,213
2	+26,300	−18,000	+8,300	−4,067	+22,233
3	+26,300	−16,000	+10,300	−5,047	+21,253
4	+26,300	−14,000	+12,300	−6,027	+20,273
5	+26,300	−12,000	+14,300	−7,007	+19,293
6	+26,300	−10,000	+16,300	−7,987	+18,313
7	+26,300	−8,000	+18,300	−8,967	+17,333
8	+26,300	−6,000	+20,300	−9,947	+16,353
9	+26,300	−4,000	+22,300	−10,927	+15,373
10	+26,300	−2,000	+24,300	−11,907	+14,393
Totals	+$153,000	−$110,000	+$153,000	−$74,970	+$78,030

EXAMPLE 12–3

Rate of Return After Income Taxes Assuming that the Initial Outlay will be Written Off by the Straight-Line Method

Facts of the Case
This example deals with the same proposed investment in materials-handling equipment that was analyzed in Example 12–2. The only difference is that, if actually made, the investment will be written off for tax purposes by the straight-line method rather than by the sum-of-years-digits method.

Calculation of Rate of Return After Income Taxes
Table 12–3 shows the calculation of the prospective effect of the investment on cash flow after income taxes, assuming a 49% tax rate throughout the entire period. The rate of return after taxes may be determined by computing present worth of cash flow after taxes using rates of 11% and 12%.

$$\begin{aligned} \text{PW at } 11\% &= -\$110{,}000 + \$18{,}803(P/A,11\%,10) \\ &= -\$110{,}000 + \$18{,}803(5.889) = +\$730 \\ \text{PW at } 12\% &= -\$110{,}000 + \$18{,}803(5.650) = -\$3{,}760 \end{aligned}$$

Interpolation indicates a rate of return of about 11.2% after taxes.

EXAMPLE 12–4

Rate of Return After Income Taxes Assuming that the Initial Outlay is Written Off Over a Considerably Longer Time than its Period of Major Productivity

Facts of the Case
An outlay of $110,000 for certain machinery is proposed. It is anticipated that the period of primary service of this machinery will be 10 years. During these years it is estimated that the machinery will reduce annual disbursements for labor and labor extras by $36,300 and will increase annual disbursements for maintenance, power, property taxes, and insurance by $10,000.

Machinery of this type has an industry-wide average life of 25 years and has, in fact, had such an average life in the service of this particular taxpayer. Therefore, the taxing authorities insist that the write-off for tax purposes be based on a 25-year life. An estimate of zero salvage at the end of this life will be permitted. It is expected that the taxpayer's ownership of this machinery actually will continue for 25 years, with the machinery used for standby purposes during the final 15 years of its life. For purposes of the economy study, no net cash flow before income taxes will be assumed during this final 15 years. The straight-line method will be used in depreciation accounting and in reporting depreciation for tax purposes.

TABLE 12–3
Estimation of cash flow after income taxes, Example 12–3

Year	*Cash flow before income taxes*	*Write-off of initial outlay for tax purposes*	*Influence on taxable income*	*Influence of income taxes on cash flow* −0.49C	*Cash flow after income taxes* (A + D)
	A	B	C	D	E
0	−$110,000				−$110,000
1 to 10	+26,300 per year	−$11,000 per year	+$15,300 per year	−$7,497 per year	+18,803 per year
Totals	+$153,000	−$110,000	+$153,000	−$74,970	+$78,030

Calculation of Rate of Return After Income Taxes
Table 12–4 shows the calculation of the prospective effect of the investment on cash flow after income taxes. The prospective rate of return after income taxes may be determined by computing present worth of cash flow after taxes using rates of 8% and 9%.

$$\begin{aligned}\text{PW at } 8\% &= -\$110{,}000 + \$15{,}569(P/A,8\%,10) \\ &\qquad + \$2{,}156[(P/A,8\%,25) - (P/A,8\%,10)] \\ &= -\$110{,}000 + \$15{,}569(6.710) + \$2{,}156(10.675 - 6.710) \\ &= +\$3{,}020 \\ \text{PW at } 9\% &= -\$110{,}000 + \$15{,}569(6.418) + \$2{,}156(9.823 - 6.418) \\ &= -\$2{,}740\end{aligned}$$

Interpolation indicates a rate of return of about 8.5% after taxes.

TABLE 12–4
Estimation of cash flow after income taxes, Example 12–4

Year	*Cash flow before income taxes*	*Write-off of initial outlay for tax purposes*	*Influence on taxable income*	*Influence of income taxes on cash flow* −0.49C	*Cash flow after income taxes* (A + D)
	A	B	C	D	E
0	−$110,000				−$110,000
1 to 10	+26,300 per year	−$4,400 per year	+$21,900 per year	−$10,731 per year	+15,569 per year
11 to 25	0 per year	−4,400 per year	−4,400 per year	+2,156 per year	+2,156 per year
Totals	+$153,000	−$110,000	+$153,000	−$74,970	+$78,030

EXAMPLE 12–5

Rate of Return After Income Taxes Assuming that the Initial Outlay is Written Off Over a Considerably Shorter Period Than its Period of Major Productivity

Facts of the Case

The prospective cash flow before income taxes is the same as in Examples 12-2 and 12-3. But because the assets it is proposed to acquire meet certain specifications laid down in the tax laws, the cost of the assets may be written off for tax purposes at a uniform rate over a 5-year period.

Calculation of Rate of Return After Income Taxes

Table 12–5 shows the calculation of the prospective effect of the investment on cash flow after income taxes, assuming a 49% tax rate for the next 10 years. The rate of return after taxes may be determined by computing present worth of cash flow after taxes, using rates of 13% and 14%.

$$\begin{aligned}\text{PW at } 13\% &= -\$110{,}000 + \$24{,}193(3.517) + \$13{,}413(5.426 - 3.517)\\ &= -\$700\\ \text{PW at } 14\% &= -\$110{,}000 + \$24{,}193(3.433) + \$13{,}413(5.216 - 3.433)\\ &= -\$3{,}030\end{aligned}$$

Interpolation indicates a rate of return of about 13.2% after taxes.

EXAMPLE 12–6

Rate of Return After Income Taxes Assuming that the Initial Outlay is Written Off Against Current Income

Facts of the Case

A company's industrial engineering department has devised an improved plant layout for existing production equipment in an existing building. An immediate outlay of $110,000 will be required to rearrange the machinery. It is estimated that the new layout will reduce annual disbursements for materials handling by $26,300 for the next 10 years. Because the $110,000 outlay does not involve the acquisition of new assets or the extension of the lives of old ones, the entire amount will be treated as a current expense for accounting and income tax purposes in the year in which it is made (zero year on our time scale).

Calculation of Rate of Return After Income Taxes

Table 12–6 shows the calculation of the prospective effect of the $110,000 outlay on cash flow after income taxes, assuming a 49% tax rate from years 0

TABLE 12–5
Estimation of cash flow after income taxes, Example 12–5

Year	*Cash flow before income taxes*	*Write-off of initial outlay for tax purposes*	*Influence on taxable income*	*Influence of income taxes on cash flow* −0.49C	*Cash flow after income taxes* (A + D)
	A	B	C	D	E
0	−$110,000				−$110,000
1 to 5	+26,300 per year	−$22,000 per year	+$4,300 per year	−$2,107 per year	+24,193 per year
6 to 10	+26,300 per year	0 per year	+26,300 per year	−12,807 per year	+13,413 per year
Totals	+$153,000	−$110,000	+$153,000	−$74,970	+$78,030

to 10. It will be noted that in zero year the increased disbursement of $110,000 for machine rearrangement is partially offset by a decrease in the required disbursements to the tax collector amounting to $53,900. On the other hand, there is no subsequent depreciation deduction that results from the $110,000 outlay; the tax collector therefore will take $12,887 each year out of the $26,300 saving in materials-handling costs.

The following present worth calculation at 20% shows that the prospective rate of return after taxes is a little more than 20%, the same as the prospective rate before income taxes.

$$\text{PW at } 20\% = -\$56{,}100 + \$13{,}413(4.192) = +\$130$$

TABLE 12–6
Estimation of cash flow after income taxes, Example 12–6

Year	*Cash flow before income taxes*	*Write-off of initial outlay for tax purposes*	*Influence on taxable income*	*Influence of income taxes on cash flow* −0.49C	*Cash flow after income taxes* (A + D)
	A	B	C	D	E
0	−$110,000	−$110,000	−$110,000	+$53,900	−$56,100
1 to 10	+26,300 per year		+26,300 per year	−12,887 per year	+13,413 per year
Totals	+$153,000	−$110,000	+$153,000	−$74,970	+$78,030

The Need for After-Tax Analysis

If decisions in competitive industry are to be made from the viewpoint of the owners of business enterprises, it should be evident that comparisons of alternatives ought to be after taxes. Considered together, our six examples illustrated the point that an after-tax analysis of competing proposals may lead to quite different conclusions about their relative merits than would have been reached by a before-tax analysis. Our six proposed outlays of $110,000 all had prospective rates of return of about 20% before income taxes. But the after-tax rates differed greatly; they were, respectively, 10.2%, 12.3%, 11.2%, 8.5%, 13.2%, and 20.1%. If funds available for investment had been limited, and if these six proposals had been competing with other proposals and with each other for the limited funds, a sound choice could not have been made without an after-tax analysis.

This difference in after-tax attractiveness existed in spite of the fact that the applicable tax rate, 49%, was the same for all the proposals. In these examples, the source of the differences in after-tax rates of return was the difference in the speed of permissible write-off of an investment. We shall see in the remainder of this chapter and in some of the following chapters that there are many other ways in which income tax considerations can influence the relative attractiveness of proposed investments.

A General Approach to the Introduction of Income Tax Considerations into Economy Studies

Examples 12–1 through 12–6 each compare one alternative, which is continuing some present condition, with another alternative, which is making an immediate outlay of $110,000 to obtain some future advantage. The steps in an after-tax comparison that are illustrated in each example are as follows:

1. Estimates are made of the year-by-year prospective differences between the alternatives in cash flow before income taxes. (This is an essential step whether alternatives are to be compared before income taxes or after income taxes.)
2. The year-by-year differences between the alternatives in prospective taxable income are calculated. Usually these calculations involve some modification of the estimated differences in cash flow before income taxes. Examples 12–2 through 12–6 illustrated the use of different possible rules for write-off in estimating differences in taxable income. In Example 12–1, where no write-off was permitted, the estimated annual reduction in disbursements before taxes was also the estimated increase in taxable income.

   Sometimes the calculation of prospective differences in taxable income is much more complex than illustrated in these examples; an analyst may need a sophisticated understanding of the income tax laws and regulations applicable to the particular case.
3. The applicable income tax rates are estimated and applied to the esti-

mated differences in taxable income to compute the differences between the alternatives in cash flow for income taxes.

4. The year-by-year differences in cash flow before income taxes are combined with the year-by-year differences in cash flow for income taxes to obtain the estimated differences in cash flow after income taxes.
5. The cash flow after income taxes is then analyzed with reference to the selected criterion for decision making. The analysis in Examples 12–1 through 12–6 implied that the criterion in these examples was prospective rate of return after income taxes.

Comment on Certain Simplifying Assumptions in Examples 12–1 through 12–6

In these examples it was assumed that the cost reductions caused by the $110,000 initial outlay would continue at a uniform rate for a stated period of years and then abruptly cease. The tax payments for each year's taxable income were assumed to be concurrent with the other cash disbursements for the year. The end-of-year convention was assumed in all six examples. It was assumed that all taxable income throughout the period of the study would be taxed at the same rate, 49%. These assumptions were intended to simplify the calculations and to permit concentration of attention on the two main points at issue, namely, the general approach to comparisons made after income taxes and the importance of the speed of write-off in influencing the ratio between rates of return after income taxes and before such taxes.

The general method illustrated in the examples can, of course, be applied equally well when the appropriate assumptions are much more complex. Although many economy studies made after income taxes naturally have more complications than Examples 12–1 through 12–6, each example represents a type of case that has been common in modern industry. The reader will doubtless have observed that except for the use of the 49% tax rate instead of 50%, the tax assumptions in Example 12–3 were identical with the ones made in Chapters 6, 7, and 8 and explained in Chapter 11. Often such assumptions are good enough for practical purposes.

Nevertheless, Examples 12–4, 12–5, and 12–6 each illustrate a fairly common type of case where it would be misleading to assume that a proposed investment will be written off for tax purposes over the period of its expected favorable consequences. Example 12–4 illustrates the type of case in which the write-off for tax purposes must be over a much longer period. Example 12–5 illustrates the opposite type of case, where a considerably shorter write-off is permissible. Moreover, industry has occasional chances for productive outlays that, like the one in Example 12–6, may be charged off at once for income tax purposes.*

*Such an immediate write-off of the cost of business assets has been permissible for tax purposes in Great Britain since the early 1970s.

The Partnership of the Government in Productive Business Outlays

It is illuminating to examine the totals of the five columns in Tables 12–2 through 12–6 and to note that these totals are identical. In all five examples it is proposed to spend $110,000 at once to avoid spending $263,000 in the future. In each the prospective total addition to profits before income taxes is $153,000. The government will finally take 49% of this $153,000, leaving $78,030 for the owners of the business enterprise.

Although in all five cases the government's share of the $153,000 is $74,970, the dates when the government collects its share differ greatly. It is the difference in the timing of the collection of the $74,970 of taxes that makes the large difference among the after-tax rates of return to the taxpayer.

Only in the type of case illustrated in Example 12–6 is the taxpayer's rate of return undiminished by the government's partnership in the profits of the decision to make the $110,000 outlay. The effect of the tax treatment of this outlay in Example 12–6 is as if the taxpayer had made an investment of $56,100 from which there is received $13,413 a year for 10 years, a return of a little over 20%. Similarly, the government makes an investment of $53,900 from which it receives $12,887 a year for 10 years, also a return of slightly more than 20%.

Some Other Advantages of a Rapid Write-Off for Tax Purposes

Examples 12–2 through 12–6 illustrated the point that whenever the income tax rate is expected to remain constant, the taxpayer's rate of return after income taxes is improved by a rapid write-off for tax purposes. There are certain other advantages, which are:

1. Generally speaking, if matters turn out badly, they will not turn out so badly with a rapid write-off as they will with a slow write-off.
2. In the common case where enterprise funds are limited, more cash is made available for productive use at an early date by a rapid write-off than by a slow one.

The Partnership of the Government in Unproductive Business Disbursements

The government may be a partner in unproductive business outlays as well as in productive ones. For instance, if the new plant layout in Example 12–6 should fail to reduce future disbursements at all, the government will be a partner in the unproductive $110,000 outlay to the extent of $53,900. (That is, because of the outlay the government will forgo taxes of $53,900 that it would have collected without the outlay.) Or assume that the $110,000 investment in Example 12–3 turns out to be unproductive. If so, the taxpayer will have a $11,000 depreciation deduction for the next 10 years. This deduction will reduce taxable income and will reduce income taxes to be paid during the

10-year period; the government may be viewed as participating in the original investment to the extent of the present worth of the income taxes forgone.

Any partnership of the government in unproductive outlays depends on the taxpayer having taxable income from some other source. Otherwise there is no tax liability that can be reduced by the unproductive outlay. A future depreciation write-off (such as the $4,400 a year for 25 years in Example 12–4) will help only a taxpayer who continues to have taxable income.

In case an outlay turns out to be unproductive (or less productive than was forecast), the more rapid write-off for tax purposes has two advantages over the slower write-off. One advantage is that the tax saving comes sooner and therefore has a greater present worth. The other advantage is that a prospective tax saving in the near future seems more likely to be realized than a prospective tax saving in the distant future.

Influence of Rate of Write-Off on Available Cash

When enterprise funds are limited and have to be rationed among competing proposals, it is helpful to examine the immediate net cash requirements of various proposals and also to look at their net cash requirements in the near future. The cash requirements of proposals are greatly influenced by the rate of write-off for tax purposes. For instance, the proposal to spend $110,000 in Example 12–6 has an immediate net cash requirement of only $56,100,* whereas the proposals in the other five examples all require $110,000 at once.

It is pointed out in Chapter 21 that under certain circumstances it is helpful to compute a figure for crude "payback" after income taxes as a numerical index of the short-term cash aspects of various proposals. This payback is defined as the number of years required for net cash flow to equal zero (without consideration of interest) in a tabulation such as Tables 12–1 through 12–6. Other matters being equal, the more rapid the write-off for tax purposes, the shorter the payback period. This point is illustrated by the following comparison of payback periods in our examples:

Example	Write-off period	Payback period
12–6	at once	4.2 years
12–5	5 years (straight-line)	4.5 years
12–2	10 years (years-digits)	5.2 years
12–3	10 years (straight-line)	5.9 years
12–4	25 years (straight-line)	7.1 years
12–1	none	9.8 years

*For a qualification of this statement, see the discussion later in this chapter of the timing of income tax payments in relation to the timing of taxable income.

Choosing a Set of Income Tax Laws and Regulations for the Purpose of Exposition

Examples 12–1 through 12–6 avoided reference to the income tax laws and regulations of any particular country or other taxing authority. These six examples used an applicable tax rate of 49% throughout their respective study periods without any explanation of how the 49% figure was derived. Once the applicable tax rate was assumed, it was not necessary to refer to any particular system of income taxation to illustrate how the after-tax attractiveness of proposed investments is influenced by differences in allowable speed of write-off.

In contrast, one cannot explain how to decide on the applicable tax rate or rates without reference to some set of rules regarding the tax rates imposed on different classes of taxpayers and the circumstances under which various rates are applicable. It is conceivable that authors of books on engineering economy could devise an imaginary set of tax laws that could be used for purposes of illustration. But certain aspects of the subject can be explained to much better advantage if an actual set of tax laws and regulations is used. Moreover, if an imaginary system were to be made realistic, it might seem unbelievably complex to readers who were not familiar with modern income taxation. In this book, the authors have elected to illustrate a number of aspects of the subject using the laws and regulations for federal taxation of income that were in effect in the United States of America in 1980.*

A Simple Example of a Graduated Income Tax

In many systems of income taxation, the higher a taxpayer's taxable income the greater the percentage that must be paid as taxes. The lowest block of taxable income is taxed at some stipulated rate, the next higher block is taxed at a somewhat higher rate, and so on. This type of tax rate structure is illustrated in Table 12–7.

TABLE 12–7
1980 federal tax rate schedule on incomes of corporations in the United States

Taxable income	*Tax*
Not over $25,000	17%
Over $25,000 but not over $50,000	$4,250 plus 20% of excess over $25,000
Over $50,000 but not over $75,000	$9,250 plus 30% of excess over $50,000
Over $75,000 but not over $100,000	$16,750 plus 40% of excess over $75,000
Over $100,000	$26,750 plus 46% of excess over $100,000

*Two government pamphlets that are revised each year contain a great deal of helpful up-to-date information about federal tax laws and regulations in the United States. These are Internal Revenue Service Publication No. 17, *Your Federal Income Tax for Individuals*, and Internal Revenue Service Publication No. 334, *Tax Guide for Small Business*.

Consider a corporation with a taxable income of $130,000 that is subject to the foregoing tax schedule. Its income tax will be $26,750 + (0.46)($130,000 − $100,000) = $26,750 + $13,800 = $40,550.

The Need to Use Incremental Tax Rates in Economy Studies

In comparing alternatives the relevant figures are those for prospective differences in income taxes. This is one of the many applications of the basic concept in engineering economy that only the differences among alternatives are relevant in their comparison. (This concept was first mentioned in Chapter 1; some of its other applications are discussed in Chapter 16.)

Occasionally it happens that analysts who apply this concept elsewhere fail to use it when they are dealing with estimates of income taxes. For example, an analyst for the corporation just cited that had a tax of $40,550 on a taxable income of $130,000 might calculate that this tax is 31.2% of the taxable income. In comparing two alternatives the analyst might then multiply their estimated differences in taxable income by 31.2% to find a figure for difference in income taxes.

A look at the tax rate schedule in Table 12–7 should make it clear, however, that such a calculation would be incorrect. *All* taxable income above $100,000 will be taxed at 46%. If it is expected that total taxable income of the corporation will continue to be above $100,000 regardless of which of two alternatives is selected, it follows that the applicable tax rate should be 46% in any economy study comparing these alternatives.

A More Complex Example of a Graduated Income Tax

The federal government in the United States has taxed individual income since the ratification of the Sixteenth Amendment to the Constitution in 1913. In 1980 there were several tax rate schedules for individual income taxes; these had different brackets and different incremental tax rates applying to the various brackets. The schedule to be used depended on marital status and other matters. Table 12–8 is derived from the schedule for those single taxpayers who did not qualify under the tax law as either "unmarried heads of household" or "qualifying widows or widowers." Table 12–8 assumes that the single taxpayer to whom it applies is entitled to only one "personal exemption" of $1,000. It also assumes that the taxpayer does not itemize "nonbusiness deductions" and therefore uses the "standard deduction" of $2,300 available to single taxpayers. A further constraint on the use of this table, not evident in the table itself, was that income defined as "personal service income" could not be taxed at a rate of more than 50%.

It is evident that the tax rate table for individuals has many more brackets than the corresponding table for corporations. Nevertheless, from the point of view of applications to economy studies, Tables 12–7 and 12–8 are alike. In predicting how a proposed decision will influence the amount of money to be paid for income taxes, an analyst needs to use the incremental rates in those tax brackets that will be affected by the decision.

TABLE 12–8
1980 federal individual income tax rates on certain single taxpayers in the United States

Adjusted gross income	*Tax*
Not over $3,300	No tax
Over $3,300 but not over $4,400	14% of excess over $3,300
Over $4,400 but not over $5,400	$154 plus 16% of excess over $4,400
Over $5,400 but not over $7,500	$314 plus 18% of excess over $5,400
Over $7,500 but not over $9,500	$692 plus 19% of excess over $7,500
Over $9,500 but not over $11,800	$1,072 plus 21% of excess over $9,500
Over $11,800 but not over $13,900	$1,555 plus 24% of excess over $11,800
Over $13,900 but not over $16,000	$2,059 plus 26% of excess over $13,900
Over $16,000 but not over $19,200	$2,605 plus 30% of excess over $16,000
Over $19,200 but not over $24,500	$3,565 plus 34% of excess over $19,200
Over $24,500 but not over $29,800	$5,367 plus 39% of excess over $24,500
Over $29,800 but not over $35,100	$7,434 plus 44% of excess over $29,800
Over $35,100 but not over $42,500	$9,766 plus 49% of excess over $35,100
Over $42,500 but not over $56,300	$13,392 plus 55% of excess over $42,500
Over $56,300 but not over $82,800	$20,982 plus 63% of excess over $56,300
Over $82,800 but not over $109,300	$37,677 plus 68% of excess over $82,800
Over $109,300	$55,697 plus 70% of excess over $109,300

Relationship Between Corporate and Individual Income Taxes in the United States

One subject that has received much discussion by students of taxation is the double taxation of corporate income. If corporations are taxed on their profits, and if that part of the profits remaining after the corporate tax is paid is taxed again when received as dividends by the stockholders, this clearly is double taxation.

Over the years, the income tax laws of the United States have varied greatly on this matter of double taxation of corporate income. For 1980, individuals were permitted to exclude from gross income the first $100 of dividend income received during the year from domestic corporations. Otherwise, dividends were fully taxable as individual income.

Although corporate income in the United States is taxed as of the year in which it is earned, its second taxation as individual income does not take place until the dividends are received by the stockholders. The tax rates in the upper brackets on the individual income tax are higher than the maximum corporate rate. This leads to the following paradox.

In general, the owners of a business that is incorporated will, in the long run, pay a higher percentage of their profits as income taxes than they would pay if the same business were unincorporated. Nevertheless, under certain circumstances, the corporate form of organization has the possibility of being used as a device to reduce the immediate taxes on business profits. Assume that a corporation is controlled by stockholders whose incomes place them in the upper brackets. If, without penalty, these stockholders could elect to leave all the profits of this corporation in the business, the immediate income tax on these profits would be limited to the corporation tax. If the same profits could be taxed at the rates in the highest brackets of the individual income tax, the government would collect more taxes immediately.

To discourage the use of the corporate form of organization as a means of reducing the immediate payment of income taxes, the tax laws of the United States have imposed an additional tax on "corporations improperly accumulating surplus." This particular provision of the tax law was aimed particularly at corporations where a relatively few persons owned a large part of the stock. It exerted strong pressure on many such corporations to distribute a substantial part of their current earnings as dividends each year.

Ordinarily, all the business income of partnerships and sole proprietorships was subject to the individual income tax in the year earned regardless of whether any business profits were withdrawn by the owners. Under certain restrictions, corporations having 15 or fewer stockholders could elect to be taxed as partnerships.

What Viewpoint Toward Individual Income Taxes Should be Taken in Economy Studies for Corporations?

In any decision between alternatives made from the viewpoint of the owners of a business enterprise, all receipts and disbursements affecting the owners and influenced by the decision should properly be considered. For this reason the income tax aspects of an economy study for a corporation really involve both the corporation taxes and the individual income taxes to be paid by the stockholders on their dividends.

As a practical matter, prospective income taxes paid by the stockholders are likely to be disregarded in some corporations and considered in others somewhat as follows:

1. In corporations with many stockholders, in which most policy decisions are made by executives whose personal stock ownership is relatively small, the tendency is to consider only the corporate income tax. Most large corporations fall into this class. Where there are many stockholders whose personal incomes are unknown to the management, it is hardly practicable to consider stockholders' income taxes. Moreover, the performance of hired managers is judged by the profit showing on the corporate books.
2. In corporations with few stockholders, with those stockholders taking an active part in policy decisions, individual income taxes are much more likely to be considered. In such corporations the accumulated earnings tax often forces distribution of most of current earnings as dividends. Stockholders taking an active part in management are likely to be very conscious of the income taxes they pay on these dividends. Al-

though corporations in this class are generally smaller, they are more numerous.

Combining Income Tax Rates for Different Governmental Units

It often happens that the same income is subject to taxation by two or more governmental bodies. Sometimes income may be taxed by two countries, particularly where an enterprise incorporated in one country does business in another. In the United States it is common for the same income to be taxed by federal and state governments; many states have both corporation income taxes and individual income taxes. Some cities also have individual income taxes.

Economy studies are simplified when one applicable tax rate can be used to combine the incremental tax rates from the various governmental units that tax the same income. The appropriate rules for combining tax rates depend on the way in which the tax payments to each governmental unit influence the taxable income reported to the other governmental units.

In the United States, state income taxes imposed on corporations are deductible in computing the income to be taxed by the federal government. Individuals who do not take the standard nonbusiness deduction may include income taxes paid to a state among their itemized nonbusiness deductions on their federal tax returns.

A simple formula for combining state and federal incremental tax rates may be given for the common case where the state tax is deductible on the federal tax return but the federal tax is not deductible on the state return, as follows:

Let s represent the incremental state tax rate expressed as a decimal.
Let f represent the incremental federal tax rate expressed as a decimal.

$$\text{Combined incremental rate} = s + (1 - s)f$$

For example, assume that a $1,000 increment of income is subject to a 6% state tax rate and a 46% federal rate. The state tax on this increment of income is $60. The federal tax applies to the difference between the $1,000 income before taxes and the $60 state tax. Therefore, the federal tax is $432.40, 46% of $940. The total tax on the $1,000 increment of income is $60 + $432.40 = $492.40. This is 49.24% of the taxable income. Of course the same combined rate may be obtained from the foregoing formula as $0.06 + (1 - 0.06)(0.46) = 0.4924$.

Where there are two or more income taxes to be paid that are independent of one another, the tax rates may be combined by simple addition. As explained later in this chapter, individual taxpayers in the United States must either take a "standard deduction" or itemize their "nonbusiness deductions" on their federal returns. Unless a taxpayer elects to itemize, the amount owed for state income taxes does not influence the taxable income on the federal return. In this case the incremental rates on state and federal returns should be added to find the combined incremental rate.

The most difficult problems of combining tax rates arise where the tax due to each of the taxing units cannot be computed without somehow recognizing the past tax payments or current tax liabilities to the other taxing units. The laws and regulations under which such recognitions occur vary so greatly that no simple rules for combinations of tax rates can be given here. But it can be said that, given the specific laws and regulations, it usually is possible to compute an approximate combined incremental rate on an increment of income at any level.

The Concept of a Combined Applicable Tax Rate as a Net Incremental Rate

Even though income may be subject to two or more taxes with slightly different timing, it usually is good enough for practical purposes in an economy study to combine the taxes into a single rate. The phrase *applicable tax rate* used in examples and problems throughout the remainder of this book refers to a single incremental rate to use for a given year or group of years in a particular economy study. Each specified applicable tax rate is assumed to have been determined after consideration of all income taxes believed to be relevant, both individual and corporate and both federal and state, and after consideration of the interrelationships of these taxes and of the appropriate incremental brackets.

Sometimes the various governmental units taxing the same taxpayer have substantially different rules for determining taxable income. If so, it may not be advisable to simplify economy studies by the use of a single applicable tax rate.

Possible Changes in the Applicable Tax Rate During the Years Covered by an Economy Study

In Examples 12–1 through 12–6 we assumed that the applicable tax rate would be 49% throughout the entire study period. Presumably this rate was based on the existing tax laws and regulations and on the taxpayer's status at the time of the economy study. The adoption of *one* rate, rather than two or more rates applicable to different years, implies either that the analyst has no basis for estimating different rates for different years or that the expected changes from year to year are not large enough to have a significant influence on the results of the economy study.

In some instances it will be more reasonable to use two or more applicable rates throughout a study period. For example, it may be expected that a corporation with income in one of the lower corporate brackets will soon increase its income to where it will be taxed in the highest bracket. Or a high "excess profits" tax applicable to the final increment of a corporation's income may be scheduled to expire at a definite future date with the effect of dropping the corporation's incremental tax rate. Or an individual may expect considerably higher or lower income with a corresponding change in incremental rates. The same techniques of analysis that have been illustrated for a single applicable rate can be applied equally well if it is forecast that there will be two or more different tax rates applicable to different years of a study period. Several of the problems at the end of this chapter illustrate the assumption of two or more applicable tax rates.

Examples 12–1 through 12–6 illustrate the influence of the speed at which certain outlays are written off on the rate of return after taxes, assuming a uniform tax rate throughout the period of the economy study. The advantage of the rapid write-off is increased when it is expected that tax rates in the near future will be greater than in the more distant future. Conversely, the advantage of a rapid write-off is decreased or eliminated when tax rates in the near future are expected to be less than in the distant future.

Selecting Tax Rates for Use in Economy Studies for Corporations

A 50% applicable tax rate often is used in economy studies for corporations in the United States in the common case where it is expected that annual taxable income will continue to be greater than $100,000. This rate, which has the advantage of simplicity, allows for some state tax in addition to a 46% incremental federal tax.

But methods of introducing income tax considerations into economy studies can be shown more clearly with rates that do not divide taxable income equally between the taxpayer and the government. For this reason, none of the examples or problems in this chapter assume a 50% rate. (We have noted that our 49% rate is approximately correct for a combination of a 46% incremental federal rate and a 6% incremental state rate.)

In closely held corporations where consideration is given to all taxes on corporate income, both corporate taxes and personal taxes, the applicable tax rates may sometimes be extremely high. Suppose the incremental corporate tax rate is 46% and the stockholders' incremental rate on personal income is 55%. If the income remaining after corporate taxes should be paid out as dividends each year in order to avoid the penalty tax on "improper accumulation of surplus," each extra dollar of corporate income before taxes will lead to the payment of 46¢ in corporate taxes and 29.7¢ in personal taxes; this is a total tax rate of 75.7%

Selecting a Tax Rate in Economy Studies for Individuals

Because of the numerous brackets in both federal and state tax schedules in the United States, the selection of a single applicable tax rate is somewhat less appropriate for individuals than for corporations. In many decisions made by individuals where there are substantial differences in expected taxable income, the best thing to do may be to compute a total taxable income and a total income tax for each alternative under consideration.

Nevertheless, in many economy studies for individuals it is close enough for purposes of decision making to simplify matters by selecting a single rate applicable to the expected highest bracket of the individual's income. In the frequent case where the exact bracket is uncertain, a rate averaging two or more brackets may be used.

The Importance of the Applicable Tax Rate

It is evident that the greater the prospective tax rate on the increment of taxable income affected by an investment decision, the more important it will be to make the decision with at least one eye on the tax collector. This point is brought out by Table 12–9. This table shows the rates of return after income taxes as they would have been in Examples 12–1 through 12–6 with applicable tax rates of 30% and 15%. For comparison, the table shows the rates of return already calculated for tax rates of 49% and 0%. (The latter, of course, are also the rates before income taxes.) As might be expected, the lower the applicable tax rate, the less the differences among the proposals in attractiveness after income taxes. With a tax rate of 15% these differences are relatively unimportant.

Consideration of Differences in Income Tax Disbursements in Comparing Alternatives by Annual Cost or Present Worth

In Examples 12–1 through 12–6 we assumed that it was the prospective after-tax rate of return that was desired in order to guide investment decisions. In earlier chapters, using the simplified assumptions, we have also illustrated the introduction of outlays for income taxes in annual cost comparisons and in present worth comparisons.

If only two alternatives are to be compared, a year-by-year listing of their differences in taxable income should be made and the applicable tax rates applied to find the year-by-year differences in prospective income tax pay-

TABLE 12–9
After-tax rates of return in Examples 12–1 through 12–6 with different applicable tax rates

	Applicable tax rate			
Example	*49%*	*30%*	*15%*	*0%*
12–1	10.2%	14.0%	17.0%	20.0%
12–2	12.3%	15.6%	17.9%	20.1%
12–3	11.2%	14.8%	17.5%	20.1%
12–4	8.5%	12.9%	16.5%	20.1%
12–5	13.2%	16.2%	18.2%	20.1%
12–6	20.1%	20.1%	20.1%	20.1%

ments. This series of differences in income tax disbursements is then converted into an equivalent annual figure or into present worth, as the case may be, and added to the cost of the alternative requiring the higher income taxes. Problems 12–21 through 12–26 call for using the annual cost method to compare the alternatives in Examples 12–1 through 12–6; the stipulated after-tax i^* is 12%. Problems 12–27 through 12–32 involve comparing the same alternatives by the present worth method; the stipulated after-tax i^* is 11%. These problems all use the same applicable tax rate as the examples, namely, 49%. Answers are given to a number of these problems to allow the reader's calculations to be checked.

If three or more alternatives are to be compared, the alternative having the lowest prospective income tax payments should be selected as a base. The income taxes in excess of this base should be computed for each of the other alternatives. These should be converted to annual cost or present worth, whichever is required, and added to the cost totals for the respective alternatives. This technique was illustrated in Chapter 6 for annual costs.

The Timing of Income Tax Payments in Relation to the Timing of Taxable Income

In the United States the payment of federal income taxes by individuals and corporations is so organized that both pay most of their tax throughout the year in which the income is received. Employers are required to withhold income taxes from payments of wages and salaries. Corporations, and individuals having sources of income other than wages and salaries, are required to make declarations of estimated income taxes and to make quarterly payments based on these declarations.

The exact time lag, if any, between the cash receipts and disbursements affecting current taxable income and the quarterly payments of the various income taxes based on this taxable income may be an extremely important matter in the preparation of cash budgets for business enterprises. But in the usual economy studies where the minimum time unit is one year, it generally is good enough for practical purposes in the United States to disregard any short time lag between taxable income and the payment of income taxes. We have done so in Examples 12–1 through 12–6. The column showing cash flow after income taxes in Tables 12–1 through 12–6 was based on the assumption that tax payments would take place in the same year as the current cash flow influencing the year's taxable income.

At one time the laws of the United States were responsible for a common time lag of one year (or a little more or a little less) between the receipt of taxable income and the payment of the related income taxes. The investment analyst in a taxing jurisdiction other than the United States will need to consider the laws and regulations of his own jurisdiction before deciding whether or not to assume any such time lag. Even in the United States, special circumstances may arise in which it will be appropriate to assume a time lag in cash flow for income taxes. Economy studies that make this assumption are illustrated by certain problems at the end of this chapter.

Some of the Ways Prospective Income Taxes Influence the Relative Attractiveness of Proposed Alternatives

Every governmental unit that levies income taxes has its own laws and regulations that govern the way in which taxable income and tax liability are computed. These rules vary from one government to another and change from time to time in any given government. Generally speaking, legislative bodies are free to define taxable income in any way they see fit. Although the concept of taxable income corresponds in a general way to the concept of accounting income, there often are significant points of difference.

As already mentioned, the discussion in this book is intended to stress the general principles of introducing income tax considerations into economy studies. These principles are not limited to any particular set of tax laws and regulations. The steps outlined in the discussion that follows Example 12–6 are appropriate regardless of the laws and regulations that exist at the time and place of a particular economy study.

In our examples up to this point it was the rate of write-off for tax purposes of different proposed initial disbursements that influenced the relative attractiveness of different proposals. But we shall see that there can be many other ways in which tax laws and regulations can favor certain courses of action.

For example, one alternative may involve cash receipts that increase taxable income and a competing alternative may involve receipts that do not influence taxable income. This is illustrated in Example 12–7, which compares proposed investments in taxable and tax-exempt bonds.

Certain disbursements reduce taxable income, while competing prospective disbursements do not affect taxable income. This is illustrated in Examples 12–8 and 12–9, which deal with so-called nonbusiness deductions on income tax returns for individuals.

The prospective taxable income from one type of proposal is expected to be taxed at one tax rate, whereas the taxable income from a competing proposal is expected to be taxed at a different rate. This is illustrated in Example 12–10, which deals with the taxation of so-called long-term capital gains.

Certain proposed disbursements may cause a stipulated reduction in liability for income taxes (in contrast to a reduction in taxable income). This is illustrated in Example 12–11, which deals with the investment tax credit.

Although Examples 12–7 through 12–11 illustrate their points with reference to income tax laws and regulations in the United States of America, the points brought out are quite general with respect to the relationship of income taxes to decision making. Doubtless it would be possible to prepare examples illustrating these points using the laws and regulations of almost any taxing authority that levies income taxes.

EXAMPLE 12–7

Comparing a Proposed Taxable Investment with a Proposed Tax-Exempt Investment

Facts of the Case
Jane Doe, a resident of the United States, has decided to purchase a $5,000 bond. After considering a number of possibilities, she has narrowed her choice down to a bond of her local privately owned telephone company and a water revenue bond issued by a city in her home state. Either bond can be purchased for $5,000. In each case the borrower promises to pay the $5,000 principal of the bond at the end of 20 years and to pay interest semiannually during the 20 years. Jane believes the risk of default of interest or principal is very low with either bond; as far as she can tell by consulting the reports of bond rating services, there is no difference in risk between the two bonds. She also believes there is little or no difference in future marketability. However, there are two obvious and important differences, a difference in interest rate and a difference in the bond owner's liability for income taxes on the interest received.

The telephone bond has a nominal interest rate of 11%; interest payments of $275 are due every 6 months. The comparable rate on the municipal bond is 6.5%; interest payments of $162.50 are due every 6 months.

The $550 annual interest that Jane would receive if she should buy the telephone bond would be subject to both federal and state individual income taxes. In contrast, the $325 annual interest from the municipal bond would not be subject to any income taxes.

At present the highest increment of Jane's taxable income is subject to an 8% state income tax and a 39% federal income tax. Because she itemizes her nonbusiness deductions on her federal return, the tax rate on the highest increment of her taxable income is $0.08 + 0.92(0.39) = 0.4388$, or 43.88%.

Analysis and Action
When Jane multiplies the $550 annual interest on the telephone bond by her applicable tax rate of 43.88%, she finds that ownership of that bond will increase her annual taxes by $241.34; only $308.66 of the $550 will remain after taxes. This contrasts with the $325 annual interest that she will receive from the municipal bond, which will not be subject to income taxation.

All things considered, Jane believes it is likely that her applicable tax rate will go up a bit during the years she owns her bond. If so, the advantage of the municipal bond will increase as time goes on. Therefore, she decides in favor of the tax-exempt bond.

Some Comments About the Exemption of Certain Income from Income Taxation

For persons or organizations that pay income taxes, it usually is a good idea to compare proposed investments on the basis of performance *after* taxes. When proposals are competing for limited resources, such as Jane Doe's $5,000 in Example 12–7, tax-exempt investments have an obvious advantage over taxable ones. Depending on the applicable tax rate, the advantage of tax exemption may or may not be enough to overcome other advantages of a competing taxable proposal. The two bonds in Example 12–7 would have had equal after-tax cash flow with an applicable tax rate of 40.9%. Because Jane's applicable rate is about 43.9%, the tax-exempt bond gives her a somewhat better after-tax yield.

It often happens that *some* but not *all* of the favorable cash flow resulting from a particular investment will be exempt from income taxation. In the United States this is a common circumstance with respect to bonds issued by public agencies. Any capital gain that takes place when such bonds mature or are sold for more than their cost will be subject to income taxes. This complication in comparing taxable and tax-exempt bonds, which was not present in Example 12–7 because both bonds in that example were priced at $5,000, is illustrated in Problem 12–50.

Exemptions and Deductions on Individual Income Tax Returns in the United States

We have noted that positive cash flow associated with one proposed investment may increase taxable income, whereas positive cash flow associated with a competing investment may not affect taxable income at all. Similarly, one proposal may cause negative cash flow that will reduce taxable income and a competing proposal may cause negative cash flow that will have no influence on taxable income. Examples 12–8 and 12–9 illustrate this latter type of case.

Taxpayers who make individual income tax returns to the United States have *exemptions* and *deductions*. Both are subtracted from gross income to find the amount of income that is to be taxed. In 1980 each taxpayer had a $1,000 exemption for himself or herself. Taxpayers who supported persons classified under the law as *dependents* had a $1,000 exemption for each dependent. A taxpayer at least 65 years old had an extra $1,000 exemption, as did a taxpayer who was blind. Whereas each *exemption* was the same ($1,000 for 1980) and the number of permissible exemptions depended on the rules for eligibility, the amount of the *deductions* could depend on the particular taxpayer's outlay for certain stipulated purposes during the tax year.

In 1980 the personal expenditures that could be itemized as deductions included certain medical and dental expenses, taxes, charitable contributions, interest payments, and a few other miscellaneous items. For all these types of itemized deductions the tax law and regulations had specific rules about eligibility. Taxpayers had the option either of itemizing their deductions or of using a standard nonbusiness deduction (referred to in the tax forms as the "zero bracket amount"). In 1980 this standard deduction was $2,300 for single taxpayers and $3,400 for married taxpayers filing a joint return. It was to the taxpayer's advantage to use this standard deduction unless itemizing would give a total deduction of more than $2,300 or $3,400 as the case might be.

Our Table 12–8, which gave the 1980 tax rate schedule for certain single taxpayers, assumed an exemption of $1,000 and a deduction of $2,300. Taxpayers subject to this schedule were single persons with no eligible dependents; they were less than 65 years old and were not blind. Moreover, they elected to use the standard deduction rather than to itemize deductions.

The reader may recall that in Example 12–7 we computed an applicable tax rate of 43.88% for Jane Doe, based on incremental rates of 8% on her state return and 39% on her federal return. Our calculation was valid only because Jane itemized her nonbusiness deductions on her federal return. If she had used the standard deduction the amount of her state tax would not have influenced her federal tax. In this case her applicable tax rate would have been 47%, the sum of 8% and 39%.

The United States income tax on corporations has had no provision comparable to the exemptions allowed on individual income tax returns. But under certain restrictions, corporations have had one type of nonbusiness deduction. Charitable contributions have been deductible up to a stated percentage of taxable income. (This was 5% in 1980.)

EXAMPLE 12–8

Income Tax Consequences of a Charitable Gift

Facts of the Case

Henry Green, a resident of the United States, itemizes his nonbusiness deductions on both federal and state income tax returns. He believes that this year 34% is the highest tax bracket he will reach on his federal return. On his state return it will be 7%. He is considering whether or not to make a gift of $200 to a scholarship fund at his university. This gift, if he makes it, can be used as an itemized charitable deduction on both tax returns. Before deciding, Henry wants to know how much the gift would reduce his income tax payments.

Analysis

His state taxes will be reduced by $14, 7% of the $200 deduction. On his federal tax return his deduction for charitable contributions will be increased by $200, but his deduction for state income taxes will be reduced by $14. The net effect will be to give him an additional deduction of $186. This will reduce his federal tax by 34% of $186 = $63.24. His total tax reduction caused by the $200 gift will be $14 + $63.24, or about $77.

The same answer would have been obtained if we had computed his applicable tax rate as 0.07 + 0.93(0.34) = 0.3862 or 38.62%. This rate multiplied by his $200 deduction gives him a tax saving of $77.

When the income tax consequences are recognized, it is evident that Henry's overall negative cash flow caused by the gift will be only $123. He might reasonably ask himself whether he would get more satisfaction out of the $200 gift or out of some consumption expenditure (or, possibly, some investment) of $123.

EXAMPLE 12–9

Income Tax Consequences of Nonbusiness Real Property Ownership Financed in Part by Borrowing

Facts of the Case

A married couple, Alfred and Bernice Clark, make a joint income tax return. They itemize their nonbusiness deductions on both federal and state returns. Considering both federal and state taxes, their highest increment of taxable income is taxed at approximately 55%.

In past years the Clarks have rented a lakeshore cottage for their summer vacation. None of the expenditures in connection with their vacation rentals has been tax deductible.

Now the owner has offered to sell them this cottage for $36,000. They would pay $6,000 in cash and assume a mortgage loan of $30,000. The loan is for 15 years with $2,000 of principal to be repaid annually. Interest on the unpaid balance is payable at 9%. Annual ad valorem property taxes are approximately $800.

Analysis

One element in the Clark's decision is that owning this summer home will have definite income tax advantages over renting it. The property tax and the mortgage interest are nonbusiness deductions that can reduce their taxable income and therefore reduce their income tax payments. In their first year of ownership this income tax reduction would be approximately $1,925. (This is 55% of $3,500, which is the sum of the $800 deduction for property taxes and the $2,700 deduction for interest.) The tax saving will decrease each year as the loan is repaid.

Some Aspects of Interest as a Deduction from Taxable Income

Because the Clarks were taxpayers in the United States and because they itemized their nonbusiness deductions on their tax returns, the interest payments would reduce their taxable income even though these payments were unrelated to any business activity. In many other countries that levy income taxes, similar interest payments that are *consumption* expenditures are not deductible. On the other hand, interest payments for *business* purposes are nearly always deductible in determining taxable business income.

In the accounts of business enterprises, interest paid on a debt is an expense that reduces the accounting figure for profit (or increases the figure for loss). In contrast, a desired minimum attractive rate of return on money furnished by the owners of an enterprise is not deducted as an expense in the orthodox accounting determination of profit.

In this respect, as in so many others, the income tax concept of taxable income corresponds to the accounting view of profit. Hence, when an enterprise is financed entirely by ownership funds, the entire return on the investment is subject to income taxation. Where part of the financing is by borrowed money, the taxable income is reduced by the interest paid. This topic is discussed and illustrated in Chapter 18.

An assumption that a certain fraction of financing by long-term debt is to be associated with *all* proposed plant investment calls for some modification of the income tax analysis developed in the present chapter. The circumstances under which this assumption may or may not be reasonable are discussed briefly in Chapter 18. The assumption is applied in Chapter 20 in the development of a possible treatment of income taxes in economy studies for certain regulated public utilities in the United States.

Income Tax Treatment of the Depletion of Certain Natural Resources in the United States

Examples 12–8 and 12–9 describe proposed personal expenditures that were 100% deductible from personal taxable income. Presumably such expenditures often compete for limited funds with other types of personal expenditures that are not deductible at all. A slightly different but similar type of case can occur in industry when outlays for certain exhaustible natural resources are competing for limited funds against the usual type of proposed investments.

For accounting purposes the cost of an exhaustible natural resource (such as a mineral deposit) is written off over the estimated life of the resource. This write-off, usually on a unit-of-production basis, is described as *depletion* rather than as depreciation.

For income tax purposes, the laws of the United States have allowed taxpayers the option of making depletion charges for tax purposes by a method that may permit the total depletion deduction over the life of a resource to exceed its cost. This provision in the tax law recognizes the special risks associated with such investments and the public interest in having these risky investments made by someone. Under the law in 1980, taxpayers could make depletion deductions for certain stipulated exhaustible resources equal to a specified percentage of the gross income from the property during the year. These depletion percentages could be as high as 22% or as low as 5%

depending on the type of metal or mineral and on whether the deposit was in the United States. The depletion under this percentage method in any year was not permitted to exceed 50% of the net income of the taxpayer from the property in that year computed without the allowance for depletion.

In economy studies relative to the acquisition or disposal of any such natural resources, special consideration needs to be given to the income tax treatment of depletion.

Accounting Treatment of "Gains" and "Losses" on the Disposal of Certain Assets

It is a convention of accounting that certain business assets are carried on the books at their original cost regardless of changes in market value; some examples are land and certain securities (such as corporate stocks). It was explained in Chapter 11 that the cost of certain other assets is written off "in a systematic and rational manner." This systematic write-off applies not only to depreciable physical assets but also to certain intangible assets (such as patents).

Whenever an asset is disposed of for more or less than its book value, an increase or decrease occurs in the owners' equity as shown by the books of account. It is the convention of accounting to assign this so-called gain or loss on disposal entirely to the year in which the disposal takes place. For example, if land acquired for $10,000 in 1950 is sold for $100,000 in 1980, the $90,000 excess over the original cost increases the profit for 1980. Or if stock purchased for $50,000 in 1965 is sold for $30,000 in 1981, the $20,000 shortage reduces the profit for 1981.

Chapter 16 will point out that in an economy study dealing with the influence on future cash flow of a choice among alternatives, past outlays are irrelevant except as they are expected to influence future cash flows. From the viewpoint of economy studies, the chief importance of a prospective "gain" or "loss" on disposal consists of its influence on future cash flow for income taxes.

Income Tax Treatment of Such "Gains" and "Losses" in the United States

At one time in the United States such gains were included in taxable income and such losses were fully deductible from taxable income in the year in which disposal of the assets occurred. In the 1930s the law was changed in a way that subjected long-term gains to a lower tax rate than ordinary income and severely limited the extent to which capital losses could be deducted from taxable income except to offset similar gains. The exact rules on this matter have changed many times and have always been full of technicalities.

In an economy study in which any alternative involves prospective disposal of assets at more or less than their book values (either immediately or at some future date), it is necessary to consider the effect of the disposal on prospective cash flow for income taxes. To find this cash flow, one must determine the amount of the gain or loss that will be recognized for tax purposes and then determine the applicable tax rate or rates. This matter is discussed in Chapter 17 and illustrated in Examples 17–1 and 17–2.

For tax purposes, *long-term* capital gains and losses were defined as gains and losses on capital assets held for more than one year. Net long-term capital gains were taxed at lower rates than ordinary income. Capital losses could be used to offset capital gains; otherwise the extent to which they could influence current taxable income was limited.

Depreciable assets used in the trade or business were not "capital assets" as defined in the law. However, under certain complex restrictions, a part or all of a gain on the sale of certain depreciable assets could be taxed at capital gain rates, with the remainder of the gain, if any, taxed at the higher rates applicable to ordinary income. Under the technical rules of multiple-asset depreciation accounting, ordinary retirements of depreciable assets do not result in either a "loss" or a "gain." But where a loss on disposal of depreciable assets used in the trade or business was recognized for tax purposes (because of the use of item accounting or because of certain unusual circumstances), this loss was fully deductible from taxable income in the year of disposal.

EXAMPLE 12–10

An Economy Study Involving the Taxation of a Prospective Long-Term Capital Gain

Facts of the Case

George Spelvin, who is unmarried and has no dependents, has just received an inheritance. He is considering the investment of $7,500 in a certain piece of unimproved land. He believes that if he holds this real property for 5 years he will be able to sell it for $16,000. (This is the estimated net receipt to him after he pays a sales commission.) During his period of ownership he expects to pay $200 a year in property taxes. He calculates that if his estimates turn out to be correct, his rate of return before income taxes will be approximately 14.4%. He would like to have the corresponding figure for his after-tax rate of return.

George lives in one of the states of the United States that does not have an individual income tax law. The incremental tax rate on his federal return has been 39% and he expects that it will stay close to this figure. He itemizes his nonbusiness deductions and will continue to do so. If he has a long-term capital gain (such as the one he anticipates if he buys this unimproved property), only 40% of the gain will be taxable income.

Analysis

George's prospective before-tax cash flow from this investment is:

year 0	−$7,500
years 1 to 5	−200 per year
year 5	+16,000

Because he itemizes his nonbusiness deductions, his taxable income from years 1 to 5 will be reduced by $200 a year. This will decrease his income tax each year by $78, i.e., by 39% of $200. Therefore the net change in his cash flow will be −$122 per year.

When he sells the land after 5 years, he will have a capital gain of $8,500, the difference between his selling price of $16,000 and his cost of $7,500. Because this will be a long-term gain, only 40% of it will be added to his taxable income. When the $3,400 resulting increase in taxable income is taxed at 39%, the tax will be $1,326. George's net cash flow from the sale of the property will be $16,000 − $1,326 = $14,674. His prospective cash flow after income taxes will be:

year 0	−$7,500
years 1 to 5	−122 per year
year 5	+14,674

This will give him an after-tax rate of return of approximately 13.1%.

Some Comments on Economy Studies Involving Capital Gains or Losses in the United States

There were really two quite different incremental tax rates in Example 12–10, a 39% rate applicable to ordinary income and a 15.6% rate applicable to the long-term capital gain. Although the example stated that a 39% rate applied to 40% of the gain, this was—in effect—a 15.6% rate applied to the full gain. (If it had been estimated that the long-term gain would be large enough to move Spelvin's income into a higher tax bracket, both applicable rates would have been a bit higher for year 5.)

Certain states of the United States have income tax laws and regulations regarding long-term capital gains and losses that differ from the corresponding federal laws and regulations. (State laws also may differ from one another.) In cases where such differences are deemed to be significant, an economy study should have separate calculations to estimate how much state and federal taxes will be affected by a choice among alternatives. This type of complication was not illustrated in Example 12–10; it was indicated that Spelvin lived in a state without a personal income tax.

Reducing the Adverse Impact of Income Taxation on Capital Formation and Technological Progress

A high standard of living depends on the use of capital goods. Generally speaking, it is technological progress that makes it possible for the standard of living to be improved. Obstacles to capital formation and to technological progress are obstacles to improvements in the standard of living. They may also be obstacles to effective national defense.

Examples 12–2 through 12–6 brought out the point that the tax deterrent to capital investment is greatest when high income tax rates are combined with the requirement that depreciation be written off over a long period of years. Moreover, high income tax rates combined with low depreciation rates often create obstacles in the way of financing capital goods. It is of great social importance that the income tax laws be so drawn that this undesirable effect of income taxation be kept to a minimum.

Starting in 1962 in the United States, two different types of government actions have aimed to offset part of this particular undesirable consequence of high tax rates. One of these actions was to permit a larger fraction of the total depreciation to be charged in the early years by shortening the estimated lives acceptable for tax purposes. In 1962 the use of specified "guideline lives" was authorized for most classes of depreciable assets. For many classes of assets, these lives were appreciably shorter than the ones that previously had been required for most taxpayers. In 1971 an ADR (Asset Depreciation Range) system was authorized that allowed taxpayers to use lives as much as 20% shorter (or, at their option, as much as 20% longer) than had previously been permitted as guideline lives. To be entitled to use the ADR system, a taxpayer had to meet a number of fairly complex requirements about detailed fixed capital records and other matters.*

The other government action that reduced the tax deterrent to plant investment was the establishment of an "investment tax credit." Like the guideline lives, this started in 1962. But in contrast to the guideline lives and the ADR system, which merely affected the *timing* of taxable income (and therefore the timing of income tax payments), the investment tax credit reduced the total amount of income taxes to be paid.

EXAMPLE 12–11

Analysis of an Investment Proposal Eligible for an Investment Tax Credit

Facts of the Case

Example 12–2 analyzed a proposal to spend $110,000 for materials-handling equipment that was expected to reduce net cash disbursements before income taxes by $26,300 a year for 10 years. The equipment was estimated to have a net salvage value of zero at the end of the 10 years. These estimates led to a calculated before-tax rate of return of about 20.1%. With an applicable tax rate of 49%, and with years-digits depreciation used for tax purposes, the after-tax rate of return was shown to be about 12.3%.

In the present example we assume that this investment will be eligible for

*For a detailed explanation of this system, see Robert Feinschreiber, *Tax Depreciation Under the Class Life ADR System* (New York: AMACOM division of American Management Association, 1975).

an immediate investment tax credit of 10% of the cost of the new equipment. This will reduce income taxes due to the current year by $11,000. The net outlay caused by acquiring the equipment will be $99,000, the difference between the amount paid for the equipment and the tax saving.

Analysis
Except for the effect of the tax credit at year 0, the estimated after-tax cash flow will be the same as that shown in Table 12–2. The trial-and-error equations to find the after-tax rate of return are the same as those in Example 12–2 except that −$99,000 is substituted for −$110,000 and different interest rates are used.

$$\begin{aligned}\text{PW at } 15\% &= -\$99{,}000 + \$23{,}213(P/A,15\%,10) - \$980(P/G,15\%,10)\\ &= -\$99{,}000 + \$23{,}213(5.019) - \$980(16.979)\\ &= +\$870\\ \text{PW at } 16\% &= -\$99{,}000 + \$23{,}213(4.833) - \$980(16.040)\\ &= -\$2{,}530\end{aligned}$$

Interpolation indicates a rate of return of about 15.3% after income taxes. In this case the 10% investment tax credit has raised the prospective rate of return from 12.3% to 15.3%. But even with the 10% credit, the after-tax rate of return is considerably less than the 20.1% rate before income taxes.

Some Aspects of Investment Tax Credit Laws in the United States

The investment tax credit has been something of an "off again, on again" feature of tax legislation. Started in 1962, it was suspended for several months in 1966–67, repealed in early 1969, and reinstated in 1971. The rules for the credit were greatly liberalized starting in 1975. A somewhat simplified statement of certain features of the rather complex laws and regulations in effect for 1980 is as follows:

1. The tax credit applied to qualified investment in certain specified types of depreciable business property purchased and placed in service by the taxpayer and having an estimated useful life of at least 3 years. Qualified property generally was tangible property excluding buildings and their structural components. (Certain livestock was eligible property for farmers.) Land, which normally does not classify as depreciable property, was not eligible.

2. The qualified investment was equal to the sum of the cost (or other "basis" as defined in the tax laws and regulations) of eligible *new* property plus a limited amount of the cost of eligible *used* property acquired during the tax year (not over $100,000 in 1980).

3. The regular credit allowable was 10% of the eligible investment. In most cases if property had an estimated life of 7 years or more, the entire investment was eligible. For property with a useful life of 5 years or more but less than 7 years, only 2/3 of the investment was eligible. For property with a

useful life of 3 years or more but less than 5 years, only 1/3 was eligible. In effect, these limitations on eligibility permitted credits of 6⅔% and 3⅓%, respectively, on the full investment in these shorter-lived assets.

There was an additional 10% credit allowable for certain property designated as "energy property."

4. The credit reduced income tax liability for the year in which the eligible investment was made. In general, the allowable credit could not exceed the income tax liability. In 1980 the maximum credit was $25,000 plus 70% of the tax liability in excess of $25,000. (This percentage was to increase to 80% in 1981 and 90% in all subsequent years.) There was a provision for a carryback or carryover of any excess investment credit that could not be applied to the current year because of the foregoing limitations. Certain credits on energy property were subject to somewhat more liberal treatment.

5. Under certain restrictions, a lessor could elect to pass on to his lessee the right to use the investment tax credit on certain eligible property.

6. Part or all of a particular investment credit was subject to "recapture" by the government in certain cases where a taxpayer disposed of assets too soon.

The foregoing statement is intended to give the reader a general idea of certain aspects of a rather complex set of rules. It is not aimed to give enough information to enable the reader to compute what the tax credit would have been under the great variety of conditions that arise in practice.

The investment tax credit in the United States often has been referred to as a tax incentive for investment. But in cases where the applicable tax rate is high, the credit is more accurately referred to as a reduction of the tax deterrent to investment. This point is illustrated in Examples 12–2 and 12–11. These examples have the same before-tax cash flow, the same applicable tax rate (49%), and the same depreciation method used for tax purposes. Without the 10% credit (in Example 12–2) the after-tax rate of return was about 3/5 of the before-tax rate. With the credit (in Example 12–11) this fraction was increased to about 3/4.

Because the credit reduces the *tax* rather than merely reducing the *taxable income*, it can make a large improvement in the after-tax rate of return in cases where the applicable tax rate is low enough. In such cases it can be a strong incentive to choose those investments that are eligible for the credit. Problems 12–33 through 12–39 bring out some of the interrelationships among (1) applicable tax rate, (2) investment tax credit, and (3) the allowable speed of depreciation write-off for income tax purposes.

Carryback and Carryover Provisions of United States Income Tax Laws

It was pointed out in Chapters 8 and 11 that the profitability of an enterprise cannot really be measured until the enterprise has terminated; the accounting figures for profit or loss should properly be thought of as tentative judgments that may later turn out to have been either too favorable or not favorable enough. This viewpoint, in spite of its obvious soundness, is not a practicable one to adopt in the administration of income taxes. The laws, regulations, and practices governing income taxes are necessarily based on the fiction that there is some determinable correct figure for profit or loss for each year of the existence of an enterprise.

Adherence to this fiction is unfair to business enterprises that are of the feast-or-famine type unless the losses reported in tax returns for some years can be offset against profits reported for other years. Otherwise two enterprises that have equal net profits before taxes over a period of years may be taxed on quite different incomes. For example, Corporation A, producing consumer goods for which there is a fairly stable demand, may be contrasted with Corporation B, producing capital goods for which there are great fluctuations in demand. Over a 20-year period, each corporation has total net profits before income taxes of $6,000,000. The profits of Corporation A have never fallen below $150,000 or risen above $450,000. In contrast, Corporation B has had total profits of $10,500,000 for 14 profitable years and total losses of $4,500,000 for 6 unprofitable years.

In the absence of any provision for using loss years to offset profitable years in the determination of income tax liability, Corporation B will have to pay much more income tax than Corporation A. This clearly is inequitable.

The income tax laws of the United States have changed from time to time with respect to the use of loss years to offset years having before-tax profits. Neither a carryback nor a carryover (i.e., a carry forward) of losses was allowed from 1933 to 1939. Therefore, the business losses of a number of depression years were never used to offset the profits of earlier or later years in income tax returns. The law was liberalized by degrees, starting in 1940. In 1980 the rule in most cases allowed net operating losses to be carried back to the 3 preceding years or forward to the 7 succeeding years. For certain types of business under certain stipulated conditions, a slightly longer carryback or carryover was permitted.

In contrast to the recent laws of the United States, income tax laws of some of the states have not permitted either carryback or carryover of business losses.

Contrast Between Competitive Industry and Regulated Public Utilities (in the United States) with Respect to Economy Studies and Income Taxes

All of our examples and problems relating to competitive industry have assumed that the price of a competitive product or service will be established by market conditions unrelated to one producer's decision on alternative types of plant. We have assumed that economy studies for a competitive enterprise should be made from the viewpoint of the owner of the enterprise.

In Chapter 20 we shall see that the assumption that revenues are independent of the choice among alternative production methods is not necessarily valid for regulated privately owned public utilities in the United States. Under the rules that have evolved to govern the regulation of public utility rates starting in the 1890s, a utility's "revenue requirements" depend, among other matters, on the choice of production methods. Since the establishment of the federal income tax in 1913 (followed by state income taxes in later years), income tax requirements have been a necessary part of total revenue requirements.

It is brought out in Chapter 20 that under certain circumstances in the United States, decisions among alternate production methods for regulated public utilities should be made to minimize total revenue requirements as defined under the rules of rate regulation. In effect, decisions are made from the viewpoint of the public utility's customers. The implications of this for economy studies for regulated utilities are discussed in Chapter 20.

A Further Comment on the Question of How to Introduce Consideration of Income Taxes into Economy Studies in Competitive Industry

There is no room for debate on the point *that* income taxes need to be considered in choosing among alternatives in competitive industry; the only issue is *how* tax considerations can best be introduced into decision-making procedures. Two methods of considering income taxes in economy studies, both frequently used in industry, are the following:

1. Alternatives are compared *before* income taxes, using a minimum attractive rate of return before taxes that is high enough to yield a desired after-tax rate. A stipulated rate after taxes may be divided by one minus the applicable tax rate to find the minimum attractive rate before taxes. For example, if the minimum attractive rate after taxes were 10% and the tax rate were 49%, the before-tax rate would be 10% ÷ (1 − 0.49) = 19.6%.
2. The year-by-year differences among the alternatives in disbursements for income taxes are estimated. These estimates are part of an analysis that uses a minimum attractive rate of return *after* income taxes.

The first method has two major advantages. It is simpler in the sense of involving fewer calculations. Moreover, it can be used by persons who are not familiar with the relevant income tax laws and regulations. Nevertheless, the second method clearly is the one that is correct in principle. The first method is appropriate only when it is reasonable to expect that it will lead to the same *decisions* among alternatives that would be reached by applying the second method.

A rough generalization, subject to exceptions, is that the first method often is good enough for practical purposes in choosing among design alternatives but that it will rarely be satisfactory in comparing alternative projects on the level of capital budgeting.

In principle the interest rate used in the first method (the minimum attractive rate of return before income taxes) has the defect of discounting the future too greatly. Therefore, the first method places insufficient weight on the more distant consequences of alternatives.

When a number of investment proposals are competing for limited funds and it is desired to rank them in the order of their prospectives rates of return after income taxes, some circumstances requiring the use of the second method are as follows:

1. Alternatives with similar estimated lives differ greatly in the permissible speed of write-off for tax purposes. (This was illustrated in Examples 12–1 through 12–6.)
2. An investment tax credit applies to some proposals but not to others.
3. A substantial change in the applicable tax rate is anticipated during the period of the study.
4. The proposal is one that will make use of percentage depletion for income tax purposes. This will be the case in many mining and mineral enterprises in the United States. In such enterprises, after-tax analysis needs to be applied to *all* investment proposals. Generally speaking, some proposals will have the advantage of percentage depletion for tax purposes and others will not, and all proposals will be competing for limited funds.
5. One or more alternatives involve a prospective capital gain or loss or a gain or loss on disposal of real estate or depreciable assets used in the trade or business.
6. A crude payout after taxes is to be used as a supplementary criterion for investment decisions.

Income Taxation as a Source of Adverse Differential Price Change When Inflation is Expected

Much of Chapter 14, which deals with sensitivity studies, discusses the relationship between prospective price level changes and the making of decisions among investment alternatives. Our discussion there will point out that prospective price level changes would not be so troublesome to decision makers if *all* prices were expected to go up or down at the same rate. But decision making is more difficult in the usual case where various prices relevant to a given decision seem likely to change at different rates.

When price levels are rising, income taxation can be a source of adverse (and unfair) differential price change to many taxpayers. This has been true in the United States and in many other countries. Profits reported to the tax collector tend to be overstated in terms of purchasing power when depreciation and certain other expenses are required to be based on outlays that were made when price levels were lower.

A special kind of inequity occurs when there are graduated tax rate schedules such as those illustrated in Tables 12–7 and 12–8. If the tax rate schedules remain unchanged in the relevant monetary units (such as dollars) and if incomes and prices rise at approximately the same rate, a higher and higher proportion of taxpayers' purchasing power will go to pay income taxes.

To illustrate this point consider a numerical example using the tax rate schedule shown in Table 12–8. A taxpayer subject to this schedule who has an adjusted gross income of $16,000 will pay $2,605 in income taxes. Now assume that all prices double and that the taxpayer's adjusted gross income also doubles. If the tax rate schedule is not changed, the income tax will be $8,402 on the new income of $32,000. The tax will have increased by 223% even though price levels and the taxpayer's before-tax income increased by only 100%.

If this type of tax-related inequity is to be avoided in a period of inflation, the income figures applicable to each tax bracket need to be increased automatically each year to reflect the rise in price levels. If the income tax system has exemptions and standard deductions (such as the $1,000 and $2,300 that were used in preparing Table 12–8), these figures also should be increased automatically to compensate for the higher price levels. In fact, the income tax laws of many countries have included such automatic "indexing." But, as of 1980, such automatic adjustment for price level increases was not part of the federal income tax laws of the United States.

Some Other Income Tax Topics in the Remainder of This Book

Some of the problems at the end of this chapter use 1980 federal income tax laws of the United States to illustrate various points brought out in the text of the chapter. A by-product of this use is that a number of the problems contain information about features of income taxation in the United States that were not mentioned in the text.

Certain interrelationships among income taxes, inflation, and economy studies are examined in Chapter 14, which deals with sensitivity studies.

Chapter 17, which deals with economy studies for retirement and replacement, illustrates how certain retirements can influence income taxes.

Chapter 18 explores some interrelationships among business finance, economy studies, and income taxation.

As already mentioned, Chapter 20 includes an explanation of a traditional relationship between income taxation and economy studies for regulated public utilities in the United States.

We have noted that income tax laws and regulations seem to be altered from time to time wherever such taxes are levied. Significant changes in the federal tax laws of the United States were made in 1981. Some of these changes are described briefly in Appendix F. Certain legislation that influenced stipulated depreciation methods and depreciation rates is of special importance in relation to economy studies.

PROBLEMS

In the following problems assume 100% equity funds unless otherwise stated. Assume the end-of-year convention. Unless otherwise stated, assume that the cash flow for income taxes that results from each year's taxable income occurs at the end of the respective tax year. Do not assume an investment tax credit unless one is stated.

12–1. A $78,000 investment in machinery is proposed. It is anticipated that this investment will cause a reduction in net annual operating disbursements of $16,300 a year for 12 years. The investment will be depreciated for income tax purposes by the years-digits method using a 12-year life and zero salvage value. The forecast of zero salvage value is also to be used in the economy study. The applicable income tax rate is 48%. What are the prospective rates of return before and after income taxes? (*Ans*. 18.0%; 11.3%)

12–2. Compute the after-tax rate of return in Problem 12–1 using straight-line depreciation for tax purposes. (*Ans*. 10.3%)

12–3. Compute the after-tax rate of return in Problem 12–1 assuming a 10% investment tax credit taken at year zero. (*Ans*. 13.8%)

12–4. Compute the after-tax rate of return in Problem 12–1 assuming that the investment can be fully depreciated for tax purposes by a uniform write-off of $15,600 a year for the first 5 years. Assume that a 10% investment tax credit will be allowed at zero date. (*Ans*. 15.3%)

12–5. Compute the after-tax rate of return in Problem 12–1 under the stipulation that the investment must be written off for tax purposes by the straight-line method using a 20-year life and zero salvage value. This does not change the forecast that the operating saving will occur only during the first 12 years. Assume that the taxpayer will continue to have taxable income for the full 20 years; therefore the depreciation deduction can continue to be made. Assume that the applicable tax rate will continue to be 48%. (*Ans*. 9.0%)

12–6. Change the preceding problem so that it is now expected that the machinery will be discarded with zero net salvage value at the end of the 12th year. Assume that the so-called loss on disposal will be fully deductible from taxable income in the 12th year. (*Ans*. 9.3%)

12–7. A $126,000 investment in machinery is proposed. It is anticipated that this investment will cause a reduction in net operating disbursements of $34,000 a year for 8 years. The investment will be depreciated for income tax purposes by the years-digits method using an 8-year life and zero salvage value. The forecast of zero salvage value is also to be used in the economy study. The applicable income tax rate is 46%. What are the prospective rates of return before and after income taxes?

12–8. Compute the after-tax rate of return in Problem 12–7 using straight-line depreciation for tax purposes.

12–9. Compute the after-tax rate of return in Problem 12–7 assuming a 10% investment tax credit taken at year zero.

12–10. Compute the after-tax rate of return in Problem 12–7 assuming that the investment can be fully depreciated for tax purposes by years-digits depreciation using a 6-year life. The expected operating saving will continue for the full 8 years.

12–11. Modify the preceding problem by assuming that a 10% investment tax credit can be taken at year zero.

12–12. Compute the after-tax rate of return in Problem 12–7 under the stipulation that the investment must be written off for tax purposes by the straight-line method using a 20-year life and zero salvage value. This does not change the forecast that the operating saving will occur only during the first 8 years. Assume that the taxpayer will continue to have taxable income for the full 20 years; therefore, the depreciation deductions can continue to be made. Assume that the applicable tax rate will continue to be 46%.

12–13. Change the conditions of Example 12–6 so that there is a 60% applicable tax rate for year zero. Assume that this is an "excess profits" tax that is scheduled to expire at the end of year zero and that a 49% applicable tax rate is forecast for years 1 through 10. Under these assumptions what is the prospective after-tax rate of return?

12–14. At the time that the investment tax credit in the United States was increased from 7% to 10% for 1975, it was originally proposed that 10% of the cost of any asset eligible for the investment tax credit be subtracted from the cost of the asset before computing depreciation deductible for income tax purposes. For example, an eligible asset costing $20,000 would receive a $2,000 investment tax credit, but depreciation on the asset used in computing taxable income would have to assume its cost as $18,000 rather than $20,000.

Discuss the relative advantage to taxpayers of a 7% investment tax credit with no reduction in the depreciation base and a 10% investment tax credit with a 10% reduction in the depreciation base. How are the relative advantages of these two possible tax credits influenced by the applicable tax rate, by estimated lives of assets, by depreciation methods, and by the interest rate used to discount future tax savings?

12–15. A corporation is subject to the tax rate schedule shown in Table 12–7. In the coming year its estimated taxable income is $60,000. How much income tax will it owe? What percentage of its taxable income will be paid as income tax? What would you recommend as the appropriate applicable tax rate for its economy studies on the assumption that the tax rate schedule will not change and that its taxable income will continue at its present level?

12–16. Assume that in a certain state of the United States the rules for determining corporation taxable income for state and federal purposes are the same in all respects but one. The only difference is that the state income tax may be deducted in computing taxable income for federal purposes. The state tax rate is 6% on all taxable income. The federal rates are shown in Table 12–7.

Consider a corporation that has a taxable income for state purposes of $30,000. What will be the combined incremental rate on an additional dollar of taxable income for state purposes, considering both state and federal income taxes? What will this combined rate be for a corporation that has a taxable income for state purposes of $60,000? Of $90,000? Of $120,000?

12–17. An individual who does not itemize nonbusiness deductions is subject to the tax rate schedule shown in Table 12–8. Adjusted gross income varies from year to year but generally falls between $22,000 and $32,000. This range of incomes includes several incremental brackets in the tax rate schedule. Assume that this individual wants a single applicable income tax rate that will be approximately right for use in certain personal economy studies. What rate would you suggest? Explain your reasoning.

Note on 12–18, 12–19, and 12–20. Example 11–1 (pages 192 to 194) refers to the first year of a business enterprise, called "Equipment Rentals," that Bill Black established as a sole proprietorship. In general, a sole proprietor is subject to an individual income tax on all taxable income. Often such taxable income comes in part from the business enterprise and in part from other sources. Therefore, a sole proprietor's profits from a business enterprise (such as Equipment Rentals) are not subject to a separate tax on business income, as would be the case if the business were incorporated.

These three problems are intended to use Example 11–1 as a frame of reference for possible class discussion of certain aspects of the sole proprietorship as a form of business organization, with particular reference to the United States. All three problems assume that the events during the first year of operation of Equipment Rentals turned out to be as forecast by Black before he made his decision to enter the equipment-rental business. Problem 12–18 deals with analysis that might have been made if prospective income taxes had been disregarded. Problem 12–19 introduces income taxes into the analysis. Problem 12–20 deals with the choice of a depreciation method.

12–18. In Example 11–1 Black's total investment was $50,000. Only $35,000 of this was for equipment that he would rent. In effect, the remaining $15,000 was for working capital. (Presumably, this initial cash was necessary to finance accounts receivable and to meet payrolls and other current obligations.) Assume that he anticipated a $16,000 annual excess of receipts over disbursements, the sum that actually was obtained during the first year.

In a sole proprietorship a figure such as this $16,000 ought to be viewed as made up of two parts. One part is recovery of the investment with a return; in this case the investment was $50,000. The other part is compensation for the personal services of the sole proprietor. Often such services consist of full-time work by the proprietor. In the case of Equipment Rentals, Black employed a full-time manager but needed to give his enterprise general supervision. He estimated that had he not established Equipment Rentals, he would have been able to earn an annual $3,000 in the time that would otherwise be required for this supervision.

(a) What should Black have considered to be the prospective before-tax rate of return on his invested capital? Assume that before he made his investment he anticipated an annual excess of receipts over disbursements of $16,000 for 5 years. Assume that he expected to terminate this enterprise at the end of the 5 years. He looked forward to negligible terminal salvage value from his equipment but expected to recover his $15,000 of working capital undiminished. (*Ans*. 15.8%)

(b) What would Black's prospective before-tax rate of return on investment have appeared to be if his analysis had disregarded the annual $3,000 that he could have earned in the time needed to supervise his investment? (*Ans*. 23.3%)

12–19. Assume that before deciding whether or not to enter this new business, Black estimates that the added taxable income to be generated by Equipment Rentals will be taxed at a rate of approximately 40%. He expects that the $35,000 investment in equipment will make possible an investment tax credit of $6\frac{1}{3}\%$ of the investment. This credit may be taken at zero date.

Using the same estimates of before-tax receipts and disbursements that were made for Problem 12–18(a), find the prospective after-tax rate of return on the investment of $50,000. Assume that the $7,000 straight-line depreciation stated in Example 11–1 is also used for tax purposes. (Although there will be an income tax on the $3,000 a year for personal services, expected either with or without the equipment-rental business, such a tax is not relevant to this calculation.) (*Ans.* 11.3%)

12–20. Black could improve his prospective after-tax rate of return by using years-digits depreciation rather than straight-line depreciation for tax purposes. How much would this help?

Assume that he would have the option of using a 4-year life with years-digits depreciation, but that the choice of this shorter life would reduce his investment tax credit from $6\frac{2}{3}\%$ to $3\frac{1}{3}\%$. Would the tax advantage of the faster depreciation be enough to offset the disadvantage of the reduced investment tax credit?

12–21. Compare the alternatives in Example 12–1 by the method of equivalent uniform annual cost for 10 years using an after-tax i^* of 12%. (*Ans*. rental, $25,000; ownership, $26,980)

12–22. Compare the alternatives in Example 12–2 by the method of equivalent uniform annual costs for 10 years using an after-tax i^* of 12%. (*Ans.* reject the proposal, $32,000; accept it, $31,768)

12–23. Compare the alternatives in Example 12–3 by the method of equivalent uniform annual costs for 10 years using an after-tax i^* of 12%. (*Ans*. reject the proposal, $32,000; accept it, $32,665)

12–24. What special difficulty do you see in comparing the alternatives in Example 12–4 by the method of equivalent uniform annual costs? How would you suggest dealing with this difficulty? Make a comparison by the method you suggest using an after-tax $i^\star$ of 12%.

12–25. Compare the alternatives in Example 12–5 by the method of equivalent uniform annual costs for 10 years using an after-tax $i^\star$ of 12%.

12–26. Compare the alternatives in Example 12–6 by the method of equivalent uniform annual costs for 10 years using an after-tax $i^\star$ of 12%.

12–27. Compare the present worths of the net disbursements for the alternatives in Example 12–1 using an after-tax $i^\star$ of 11%. (*Ans.* rental, $147,220; ownership, $152,410)

12–28. Compare the present worths of the net disbursements for the alternatives in Example 12–2 using an after-tax $i^\star$ of 11%. (*Ans.* reject the proposal, $188,450; accept it, $182,720)

12–29. Compare the present worths of the net disbursements for the alternatives in Example 12–3 using an after-tax $i^\star$ of 11%. (*Ans.* reject the proposal, $188,450; accept it, $187,720)

12–30. Compare the present worths of the net disbursements for the alternatives in Example 12–4 using an after-tax $i^\star$ of 11%.

12–31. Compare the present worths of the net disbursements for the alternatives in Example 12–5 using an after-tax $i^\star$ of 11%.

12–32. Compare the present worths of the net disbursements for the alternatives in Example 12–6 using an after-tax $i^\star$ of 11%.

12–33. Examples 12–2 and 12–11 dealt with the same proposed investment of $110,000. The prospective before-tax rate of return was 20.1%. The applicable tax rate was 49%. With years-digits depreciation and no investment tax credit, the prospective after-tax rate of return was 12.3%. With a 10% investment tax credit taken at zero date, this was increased to 15.3%.

Alter the conditions of these two examples by changing the applicable tax rate to 25%. Calculate after-tax rates of return without and with the investment tax credit. (*Ans.* 16.4%; 19.5%)

12–34. Problems 12–1 and 12–3 dealt with the same proposed investment of $78,000. The prospective before-tax rate of return was 18.0%. The applicable tax rate was 48%. With years-digits depreciation and no investment tax credit, the prospective after-tax rate of return was 11.3%. With a 10% investment tax credit taken at zero date, this was increased to 13.8%.

Alter the conditions of these two problems by changing the applicable tax rate to 25%. Calculate after-tax rates of return without and with the investment tax credit.

12–35. In Problems 12–7 and 12–9 change the applicable tax rate from 46% to 30%. Calculate (a) the before-tax rate of return, (b) the after-tax rate of return without an investment tax credit, and (c) the after-tax rate of return with a 10% investment tax credit taken at zero date.

12–36. Solve the preceding problem changing the applicable tax rate to 20%.

12–37. A proposed investment in machinery of $63,000 will lead to estimated operating savings of $16,300 per year for 15 years. The machinery is expected to have zero salvage value at the end of the 15 years. The applicable income tax rate is 48%.

Compute the following: (a) prospective rate of return before income taxes, (b) prospective rate of return after income taxes using straight-line depreciation for tax purposes with a 15-year estimated life, (c) prospective rate of return after income taxes using years-digits depreciation for tax purposes with a 15-year estimated life, and (d) prospective rate of return after income taxes using years-digits depreciation for tax purposes with a 9-year estimated life and taking a 10% investment tax credit at zero date.

12–38. Solve Problem 12–37 changing the applicable tax rate to 40%.

12–39. Solve Problem 12–37 changing the applicable tax rate to 32%.

12–40. What would the after-tax rate of return have been in Example 12–11 if it had been necessary to put off the investment tax credit for 1 year? For 3 years?

12–41. Examples 12–2 and 12–3 illustrated an advantage of years-digits depreciation over straight-line depreciation for income tax purposes in the common case where no change is forecast in the applicable tax rate. The prospective after-tax rate of return was 12.3% with the years-digits method and 11.2% with the straight-line method.

Change the applicable tax rate to 32% for the first 3 years, 42% for years 4 and 5, and 49% thereafter. Under these circumstances what are the after-tax rates of return using years-digits and straight-line depreciation? Discuss the choice of a depreciation method for tax purposes in this type of case.

12–42. Consider the question of whether so-called income taxes on business enterprises ought to be levied on net income or on gross income. A critic of taxes on net income might point out that such taxes often are an obstacle to certain types of technological progress. For example, in Example 12–2 the 49% tax on net income reduced the prospective rate of return on cost-reducing machinery from about 20% to about 12%. With a tax on gross income such machinery designed to reduce costs would not have been subject to a tax penalty; presumably the cost reduction would not *increase* gross income.

But there are serious offsetting disadvantages to taxes levied on total business receipts (i.e., on so-called gross income). Discuss this topic, pointing out possible reasons for the common practice of taxing net income rather than gross income.

12–43. Compare crude payback periods after taxes for the six proposals for investment of $78,000 in Problems 12–1 through 12–6. Note that the only differences in cash flow among these proposals are the differences caused by income taxes.

12–44. Compare crude payback periods after taxes for the six proposals for investment of $126,000 in Problems 12–7 through 12–12. Note that the only differences in cash flow among these proposals are the differences caused by income taxes.

12–45. Susan Roe is subject to the tax rate schedule shown in Table 12–8. In her present employment her adjusted gross income is $28,000. She is considering an additional part-time selling job that she can do outside of her regular hours of employment. She hopes that this will add a net $6,000 to her adjusted gross income. How much of this $6,000 will she have left after paying her extra federal income taxes?

12–46. The discussion on page 248 of the selection of an applicable income tax rate to use in economy studies for a closely held corporation cites a case in which the corporate incremental rate is 46% and the incremental rate on the stockholders' personal income taxes is 55%. It is stated that under certain stipulated circumstances these figures lead to an applicable tax rate of 75.7%. Show the calculation needed to obtain this 75.7% figure.

12–47. Example 12–11 analyzed a case in which a proposed investment of $110,000 that had a before-tax rate of return of 20.1% turned out to have an after-tax rate of 15.3%. The applicable tax rate was 49%. The proposal had the advantage of years-digits depreciation and a 10% investment tax credit. What would the after-tax rate of return have been if the applicable tax rate had been the 75.7% calculated in the preceding problem?

12–48. Show the necessary calculations to verify the six after-tax rates of return given in Table 12–9 for an applicable tax rate of 30%.

12–49. Show the necessary calculations to verify the six after-tax rates of return given in Table 12–9 for an applicable tax rate of 15%.

12–50. Anna Allen is considering the investment of $15,000 in bonds. One possibility is to buy at par 3 $5,000 13.2% bonds of a new issue by her local public utility company. These mature in 8 years. Alternatively, she considers buying four $5,000 outstanding bonds of her home city that had been issued when interest rates were lower; these are now available for $3,750 apiece. The municipal bonds pay 4% interest on their face value; each will pay $5,000 at the end of 8 years. The annual interest on the public utility bonds is fully taxable. The annual interest on the municipal bonds is exempt from both federal and state income tax. However, it is expected that when these bonds mature at the end of 8 years, 30% of the long-term capital gain will have to be included in the buyer's taxable income. Allen expects her applicable tax rate to be 38%, considering both federal and state income taxes. Calculate the prospective after-tax rates of return to her on these competing investments.

12–51. David Davis, who lives in the United States, is undecided on whether or not to make a certain charitable gift of $500. He always itemizes his non-business deductions on federal and state income tax returns. This gift would be an eligible deduction. This year the highest tax rate on his federal return will be 44%; on his state return it will be 11%. His state income taxes are deductible on his federal return. What would be David's net cost of making this gift, considering its prospective saving in his income taxes?

12–52. Some years ago, as an aid to small business, the United States Congress passed a law allowing "additional first-year depreciation" on certain assets for income tax purposes. In 1980 this law permitted the deduction of 20% of the cost of qualifying property (determined without regard to salvage value) in addition to regular depreciation. The cost of property on which the additional allowance could be taken was limited to $10,000 on a separate return and $20,000 on a joint return. The allowance was applicable to equipment purchased second-hand as well as to new equipment.

Consider a sole proprietor in the construction industry, married, and submitting a joint return. In his depreciation accounting he uses an averaging convention under which assets acquired during the first half of the year have a full year's depreciation in the year of acquisition. His fiscal year is the calendar year.

At the start of January, 1980, he has a chance to purchase a used power shovel for $20,000. The estimated remaining useful life of this shovel is 10 years with no terminal salvage value. If he buys this shovel he can charge $5,600 of depreciation on it in his tax return for 1980. This $5,600 is a combination of $4,000 of "additional" depreciation (20% of the $20,000 first cost) and $1,600 of regular depreciation. This $1,600 is 1/10 of $16,000 (the first cost of $20,000 diminished by the $4,000 of additional depreciation). The total write-off during the 10-year life will be $20,000 just as in any other method. In 1980 he can take a 10% investment tax credit on this purchase.

In the past this contractor has rented a power shovel whenever one was needed for a job. He estimates that owning this shovel will reduce his rental disbursements by $8,000 a year for the next 10 years. This saving will be partially offset by extra disbursements of $2,400 a year for maintenance, storage, property taxes, and insurance.

His taxable income has fluctuated a bit from year to year. His highest tax bracket has varied between 35% and 55%. For purposes of the following questions, assume an applicable tax rate of 45% each year for the 10 years starting with 1980.

(a) Compute the prospective rate of return before income taxes on the power shovel investment.

(b) Compute the prospective rate of return after income taxes assuming that his right to use 20% additional first-year depreciation will be applied to this power shovel.

(c) Compute the prospective rate of return after income taxes assuming that ordinary straight-line depreciation based on a 10-year life and zero salvage value will be used.

(d) Assuming that money is worth 10% to him after income taxes, what is the present worth (on January 1, 1980) of this right to use additional first-year depreciation rather than ordinary straight-line depreciation?

12–53. Two $20,000 assets have been purchased, one with a 10-year estimated life and the other with a 20-year estimated life, both with zero salvage values. Is it better for a taxpayer to select the 10-year or the 20-year asset for the additional first-year depreciation described in the preceding problem? Or is the choice a matter of indifference? Show calculations to support your answer, using an applicable tax rate of 45%.

12–54. Two $20,000 used assets have been purchased. Although each has a 10-year estimated life, the salvage values for tax purposes are, respectively, zero and 30% of first cost. The additional first-year depreciation described in Problem 12–52 can be used for either but not for both. Which should be selected for the more rapid depreciation? Or is the choice a matter of indifference? Show calculations to support your answer, using an applicable tax rate of 40%. (The straight-line method of depreciation should be assumed for both assets.)

12–55. A small corporation is now subject to a 20% income tax rate on its highest increment of income. The owners expect that the taxable income will soon reach the higher brackets shown in Table 12–7 and that the income will reach the highest tax bracket of 46% within 5 years.

The corporation has various options in computing depreciation for tax purposes. For example, the estimated lives of certain assets might be 8, 10, or 12 years without being subject to challenge on audit. There is also a choice among straight-line, declining-balance, and years-digits methods. For certain assets the "additional first-year depreciation" described in Problem 12–52 is available. Discuss this corporation's problems in choosing among these depreciation options.

13

SOME ASPECTS OF THE ANALYSIS OF MULTIPLE ALTERNATIVES

The logical order of procedure in the case of any new enterprise—which is, first, to determine whether or not the project is a sound one, and to be carried out; and secondly, to make the necessary studies as to the manner of carrying it out—is not necessarily followed in order of time: often it cannot be, for the final decision as to the former often depends on the results of the latter, or on unknown future events. Nevertheless, although subsequent events may cause a revision of such assumptions, the mere initiation of the study of details implies a pro-forma conclusion, that the project as a whole is a wise one if wisely carried out, and can only fail by bad judgment in details. This premise must be from the beginning, therefore, under all circumstances, the basis of the engineer's action. From this it follows:

No increase of expenditure over the unavoidable minimum is expedient or justifiable, however great the probable profits and value of an enterprise as a whole, unless the increase can with reasonable certainty be counted on to be, in itself, a profitable investment. Conversely,

*No saving of expenditure is expedient or justifiable, however doubtful the future of the enterprise as a whole, when it can with certainty be counted on that the additional expenditure at least will, at the cost for the capital to make it, be in itself a paying investment.—A. M. Wellington**

A first step in learning how to reason clearly about choosing among many possible courses of action is to learn how to choose between *two* alternatives. Moreover, general principles of decision making often are clearer when only two alternatives are involved. For these reasons many of the examples and problems up to this point have dealt with alternatives in pairs.

Why Discuss Multiple Alternatives?

Obviously, many decisions call for a choice among a number of different proposals. Do these involve special problems or difficulties that differ in any way from the analysis of only two possibilities? The answer to this question is "Yes and no."

*A. M. Wellington, *The Economic Theory of Railway Location*, 2d ed. (New York: John Wiley & Sons, Inc., 1887), p. 15.

The answer is "No" in the sense that the same principles of analysis apply to multiple alternatives that apply to comparison of only two alternatives. Any multiple-alternative problem can be analyzed by considering alternatives in pairs.

But the answer is "Yes" because it is helpful to use various mathematical techniques to find optimal solutions for multiple-alternative types of problems and because it is desirable for analysts to be aware of certain pitfalls and limitations in the use of such techniques. Also the answer is "Yes" because certain types of errors in reasoning have been common when either the rate-of-return method or the benefit–cost ratio method have been used in the analysis of multiple-alternative problems.

A Classification of Multiple-Alternative Problems in Engineering Economy

For purposes of discussion, it is helpful to divide multiple-alternative problems into two classes:

1. Cases where only *one* of the alternatives will be selected, usually (although not always) because the alternatives are physically mutually exclusive. A convenient subclassification is:

(a) Cases where differences in estimated cash flows are entirely (or almost entirely) in disbursements. Many design alternatives and many cost-reduction alternatives are of this type. Examples 13–1 and 13–2 fall within this group.

Although Example 13–1 does not involve any special complications, it does illustrate a common state of affairs in which an increase in required investment in one place is accompanied by a decrease in required investment in another place. It also illustrates the treatment of income taxes in a multiple-alternative analysis that is based on equivalent uniform annual costs.

Example 13–2 deals with problems that are of great historical interest (as well as of continuing applied interest) in engineering economy. This example is simplified by using comparisons before income taxes. It serves here as an introduction to the use of mathematical models in dealing with multiple alternatives.

(b) Cases where there are also prospective differences in cash receipts (or in benefits). Decisions among alternate levels of development in a new project generally fall into this class. Examples 13–3 and 13–4 illustrate, respectively, rate-of-return analysis and benefit–cost ratio analysis as applied to this type of problem.

2. Cases in which proposals are sufficiently independent so that two or more can be accepted but some constraint (such as a limitation in finances or a limitation in manpower) controls the aggregate of the proposals that can be accepted. Some aspects of this common type of capital-budgeting problem were discussed in Chapter 10; other aspects are considered in subsequent chapters.

The present chapter is limited to consideration of the first class of case, namely, the case in which only one alternative is to be selected.

EXAMPLE 13–1

Preliminary Study to Determine the Economic Diameter of an Oil Pipeline

Facts of the Case
Table 13–1 shows preliminary estimates and calculations of annual costs in connection with a proposed pipeline. Four diameters of pipe are under consideration. The larger the diameter of the pipe, the lower the friction loss in the line. An increased size of pipe therefore reduces the necessary investment in pumping stations and reduces the amount of energy to overcome friction in the line.

In this preliminary study the estimated life of both the pipeline and the pumping stations is taken as 15 years with zero salvage value. The minimum attractive rate of return is 13% after income taxes. Income tax differences are computed assuming a 50% applicable tax rate and assuming straight-line depreciation. Estimated average annual property taxes and insurance are 3% of first cost. The comparison by equivalent uniform annual cash flow is shown in Table 13–1.

The table indicates that the 10-in. and 12-in. pipes have approximately equal annual costs. Both are clearly more economical than the 8-in. and 14-in. sizes. The choice between the 10-in. and 12-in. sizes should not be made without careful detailed cost estimates for both. If such estimates continue to

TABLE 13–1
Comparison of annual costs for different diameters of proposed oil pipeline (All figures in thousands of dollars)

Pipe diameter	*8-in.*	*10-in.*	*12-in.*	*14-in.*
A. First cost of pipeline	$ 9,600	$12,000	$14,300	$16,700
B. First cost of pumping stations	3,600	2,400	1,360	700
C. Total investment	$13,200	$14,400	$15,660	$17,400
D. Capital recovery cost ($i^\star$ = 13%)	$2,043	$2,228	$2,423	$2,693
E. Annual pipeline maintenance	314	350	350	390
F. Annual pumping station maintenance and attendance	340	225	108	70
G. Annual fuel cost for pumping	820	500	275	140
H. Annual property taxes and insurance	396	432	470	522
I. Extra annual income taxes over 8-in. line	0	142	252	234
J. Total equivalent annual cost	$3,913	$3,877	$3,878	$4,049

show the two sizes as having almost the same annual costs, the choice should be made on the basis of irreducible data. One important matter to consider would be the possibility of an increase or decrease in flow during the life of the pipeline.

This example illustrates a characteristic of many design problems involving alternative levels of investment. It will be noted that the larger the investment in the pipeline, the smaller the investment in the pumping plant. Moreover, certain operation and maintenance costs are increased by increased diameter whereas others are decreased.

It is a common condition that an increased investment in one place will reduce the necessary investment somewhere else. Whenever possible, this should be indicated directly in the cost estimates of total investment as has been done here.

Formulas for Minimum-Cost Point

Example 13–1 dealt with a case in which cost varied with a certain variable of design, namely, pipe diameter. Some elements of cost increased and others decreased with an increase in the value of the design variable. In this common type of case there is presumably some value of the design variable that makes the sum of all costs a minimum.

Wherever the variation of cost as a function of a design variable can be expressed by an algebraic equation, it is possible to use calculus to find the value of the design variable that results in minimum cost. Over the years an entire field of scientific literature has developed around the simple model illustrated in the following paragraphs. It is referred to as inventory control theory, but it might more properly be called economic inventory management.*

The simplest case is one in which one element of cost varies in direct proportion to the variable of design, a second element of cost varies inversely as the variable of design, and all other costs are independent of this variable. Although minimum-cost point formulas may, of course, be developed for situations much more complex than this, some of those that have traditionally been used by engineers did actually deal with situations of this type. A gen-

*The primary emphasis in this text is the presentation of the basic principles and techniques of capital expenditure analysis. Since inventory and lot-size decisions are very complex and technical operating decisions, it would be impossible to present a definitive exposition of the field here. However, since the basic principles discussed in Chapter 1 apply to both types of decisions, and, as Example 13–2 illustrates, since the general formulation may be applied to investment decisions of certain types, a brief discussion of both the advantages and disadvantages of such formulas is considered desirable at this point.

Most texts on production and inventory control contain extensive developments of economic production quantity and purchase quantity models. One excellent example is G. Hadley and T. M. Whitin, *Analysis of Inventory Systems* (Englewood Cliffs, N.J.: Prentice-Hall, 1963).

eral solution of the problem of finding the minimum-cost point in such circumstances is as follows:

Let y = total cost

and let x = the variable of design

The situation of cost variation just described may be expressed by the equation $y = ax + \frac{b}{x} + c$

Taking the first derivative, we find

$$\frac{dy}{dx} = a - \frac{b}{x^2}$$

Equating this to zero, and solving for x,

$$x = \sqrt{\frac{b}{a}}$$

This is the value of the design variable that makes cost a minimum.

When $x = \sqrt{\frac{b}{a}}$, the directly varying costs equal the inversely varying costs. This fact is illustrated in Figure 13–1 and may be demonstrated as follows:

$$ax = a\sqrt{\frac{b}{a}} = \sqrt{ab}$$

$$\frac{b}{x} = \frac{b}{\sqrt{\frac{b}{a}}} = \sqrt{ab}$$

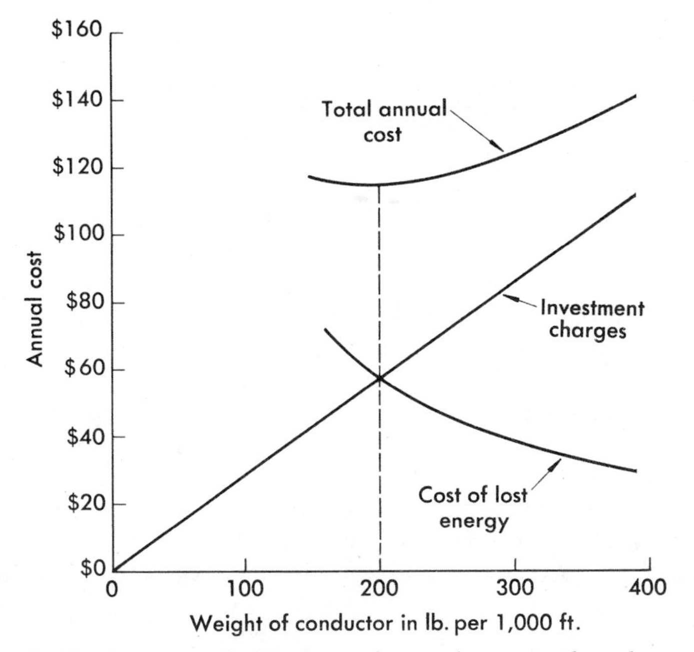

FIGURE 13–1. Comparison of annual costs in the selection of an electrical conductor

The formula, $x = \sqrt{\frac{b}{a}}$, can be applied to a number of different kinds of problems, but it should be obvious that certain precautions should be taken in its application. For example, the statement that the minimum-cost point occurs when the directly varying costs equal the inversely varying costs is not correct unless the line representing the directly varying costs goes through the origin. Also, the cost represented by ax must actually vary directly with x and the cost represented by $\frac{b}{x}$ must actually vary inversely. In the following example the cost per pound of wire must be the same for all different sizes of wire (which usually is not true), and the costs of energy losses must vary inversely with the wire size. A variable rate for electric energy or the existence of leakage loss and corona loss (such as occur in high-voltage transmission lines) interfere with the second assumption. Moreover, the analysis disregards any possible adverse consequences of voltage drop on the operation of electrical equipment. A lower limit on wire size may exist because of electrical code requirements.

EXAMPLE 13–2

Economical Size of an Electrical Conductor

Facts of the Case

The greater the diameter of an electrical conductor, the less the energy loss that will take place in it. (Power loss in watts is I^2R, where I = current in amperes and R = resistance in ohms. This may be converted to kilowatts by dividing by 1,000. Power loss in kw multiplied by the number of hours it occurs in a given period will give energy loss in kw-hr.) Thus, an increased investment in conductor metal will save an operating expense for electric energy.

Assume that a conductor is to be selected to carry 50 amperes for 4,200 hours per year, with the cost of wire at $1.75 per pound and electrical energy purchased at 3.4 cents per kw-hr. The life is estimated as 25 years with zero salvage value. The minimum attractive rate of return before income taxes is 14%, and average annual property taxes are estimated at 1.75% of first cost. These charges proportional to investment—namely, capital recovery cost of 14.55% and property taxes of 1.75%—are lumped together as investment charges of 16.3%.

The cross-sectional area of a copper conductor is expressed in circular mils, the weight of the conductor is directly proportional to the cross-sectional area, and the resistance to the flow of current is inversely proportional to the area. Therefore, let x represent the cross-sectional area in circular mils, and x_e represent the most economical size for the stated conditions. The resistance, R, for a conductor of 1,000 ft in length and 1 circular mil in cross-sectional area is approximately 10,580 ohms at 25°C, and the same conductor will weigh approximately 0.00302 pounds.

The investment in the conductor will be

$$\$1.75(0.00302)x$$

The annual cost will be $\$1.75(0.00302)(0.163)x$. Let

$$\$1.75(0.00302)(0.163) = a = \$0.000861$$

The annual cost of power loss is

$$\frac{I^2R(4{,}200)(\$0.034)}{1{,}000}$$

but

$$R = 10{,}580/x$$

Therefore, the cost of power loss is

$$\frac{(50^2)(4{,}200)(\$0.034)(10{,}580)}{1{,}000x}$$

Let

$$b = \frac{(50^2)(4{,}200)(\$0.034)(10{,}580)}{1{,}000} = \$3{,}777{,}060$$

From the formula developed in the previous article we know that the most economical value of x, x_e, occurs when

$$x = \sqrt{\frac{b}{a}}$$

$$x_e = \sqrt{\frac{\$3{,}777{,}060}{\$0.000861}} = 66{,}216 \text{ circular mils}$$

By examining a table of wire sizes (American Wire Gage, B & S) the closest available conductor is Gage No. 2, with 66,400 circular mils.

Comments on Example 13–2

In this example the size of the conductor was treated as a continuous variable. Actually, the conductors available are discrete sizes, increasing by a geometric progression at the rate of 1.123^2 for the cross-sectional area. The fact that there are specific sizes of wire available and that tables giving the resistance, weight, cross-sectional area, are also available makes another method of solution attractive.

In Table 13–2 five successive wire sizes that might be used for this application have been selected and the annual cost of each size was computed. The table shows that wire size No. 2 gives the lowest annual cost. This table also

TABLE 13–2
Comparison of annual costs of various wire sizes in the selection of an electrical conductor
(Investment in wire at \$1.75/lb; current of 50 amperes for 4,200 hrs/hr; all calculations based on 1,000 ft of copper wire)

A. Size of wire (AWG)	00	0	1	2	3
B. Weight of wire in lb	403	319	253	201	159
C. Investment in wire	\$705.25	\$558.25	\$442.75	\$351.75	\$278.25
D. Resistance in ohms	0.0795	0.100	0.126	0.159	0.201
E. Power loss—kw	0.1987	0.250	0.315	0.398	0.503
F. Annual energy loss in kw-hr	835	1,050	1,323	1,670	2.111
G. Investment charges at 16.3%	\$114.96	\$90.99	\$72.17	\$57.34	\$45.35
H. Cost of lost energy at 3.4¢ kw-hr	28.39	35.70	44.98	56.78	71.77
I. Total annual cost assumed to be variable with wire size	\$143.35	\$126.69	\$117.15	\$114.12	\$117.12

emphasizes the fact that one type of cost increases as the other decreases, as happens in many problems involving multiple alternatives. Furthermore, note that at the most economical alternative the investment charges approximately equal the cost of lost energy. This is characteristic of problems in which the total costs can be represented by the equation

$$y = ax + \frac{b}{x} + c$$

It was first pointed out by Lord Kelvin in 1881 that the economical size of conductor is that for which the annual investment charges just equal the annual cost of lost energy. This is well known in electrical engineering as Kelvin's Law. The costs in Figure 13–1 illustrate its application.

Minimum-Cost Point Formulas in Production and Inventory Control

One of the major operating problem areas faced by industry and commerce is that of determining purchase order quantities and manufacturing lot sizes. Possibly there has been more written about methods for determining appropriate economic values for these quantities than about any other single class of problems dealing with analysis for economy. This common problem arises whenever production, spacial, or financial resources may be shifted from one part or product to another.

Many solutions to lot-size problems employ some version of the minimum-cost point formula previously discussed wherein the decision variable, x, is the order or production quantity. In general, costs may be divided into three classes. The first includes those costs that vary directly with the total number of units required during a given period of time, usually a year, and are independent of the value of the decision variable. Such costs are illustrated by the direct labor, material, and/or parts costs required to produce or acquire the unit. The second class includes those that vary inversely with the value of the decision variable illustrated by costs associated with the ordering process or with preparations to manufacture—for instance, where annual expenditures for preparation will be cut in half if the sizes of lots are doubled.

Costs of the third class are those that vary directly with the number of units in the lot size and include such costs as storage, valuation taxes, insurance, and interest—for instance, where storage space requirements and inventory valuations will be doubled if lot size is doubled. It also means that the money outlay at the time a lot is procured or manufactured will be doubled, and thus more of the concern's funds will be used for the financing of inventories. If such extra investment of funds is to be justified, it must yield a return commensurate with the risk involved.

In applying the explanation of the simple minimum-cost point formula previously developed, the decision as to the value of the decision variable, x_e, the economic lot size, involves a trade-off between preparation costs and a combination of storage, tax, insurance, and interest costs. The appropriate interest cost in this class of decisions is the desired rate of return on funds invested in working capital.

As was the case with the analysis of Example 13–2, practical considerations may dictate the use of some lot size different from that found by application of the minimum-cost point formula. Thus, total costs for some range of values about the minimum-cost point may be of more value to the decision maker than simply the single value.

Some Comments About the Use of Mathematical Models in Engineering Economy

Many mathematical models have been developed to guide decisions among investment alternatives. Certain general statements may be made that are applicable to many of the complex models as well as to simple models such as the one leading to Kelvin's Law.

One generalization is that whenever a particular type of problem involving economic comparison occurs fairly often, a rule or formula is a timesaver in finding the most economic value of a variable of design. In such cases it has been fairly common to use a nomographic chart or special-purpose slide rule as a tool for timesaving. With many of the complex mathematical models developed by operations researchers and systems analysts, the modern high-speed electronic computer takes the place of the chart or slide rule that could be used to solve the simpler formulas.

The saving of time through the solution of a formula that gives a single "optimal" value for a variable sometimes has a hidden penalty associated with it. The single value has the limitation that it does not show the range of variation of the design variable through which cost (or some other chosen economic measure) will change very little. What often is needed is a minimum-cost *range* rather than a minimum-cost *point*. If this range is known, it is much easier to give weight to the irreducibles that enter into most economy studies. Thus, Table 13–2, which indicated a relatively small variation in cost between wire conductor gauge sizes 1, 2, and 3, gave more useful information than the solution of the formula in Example 13–2, which told us only that the minimum cost required a wire size of 66,216 circular mils.

The analyst who uses a formula to guide an economic decision ought to understand the assumptions underlying the mathematical model that generated the formula. Mathematical models describing economic matters generally require the model builder to make some compromises with reality. Sometimes these compromises are made so that the mathematics can be simplified; sometimes a mathematical treatment is impossible without a few compromises. These compromises are found in complex mathematical models as well as in simple ones.

Compromises may take the form of the omission of certain matters. For example, the derivation of Kelvin's Law disregarded the possible existence of leakage and corona losses and also disregarded any possible adverse consequences of voltage drop. Compromises may also take the form of a simplification of the relationship between certain relevant quantities that appear in the model. For example, in the derivation of Kelvin's Law it was assumed that the investment in an installed conductor is a linear function of the weight of conductor metal and that the appropriate rate of investment charges is the same for all wire sizes. These assumptions do not necessarily fit the facts of particular cases. It was also assumed that wire size is a continuous function, whereas in fact the function is discontinuous.

When an analyst decides whether or not to use a particular formula, the real issue that should be considered is whether or not the *decision* will be sensitive to the omissions and/or simplifications that exist in the mathematical model. The topic of tests for sensitivity is discussed in Chapter 14.

The attempt to devise a mathematical model to describe a type of problem calling for an economic decision can lead to a clearer understanding of the nature of the problem and the relevant matters that need to be considered. There have been many cases where this has occurred; one of them that is mentioned in Chapter 17 is concerned with mathematical models for equipment replacement.

Chapter 14 on sensitivity analysis deals with a topic of great importance both in the design of mathematical models and in the interpretation of formulas derived from them. The application of the mathematics of probability to investment-type decisions is the subject matter of Chapter 15; a number of the mathematical models developed by operations researchers and systems analysts involve probability mathematics.

Chapter 16 on increment costs and sunk costs develops the concept, first mentioned in this book near the end of Chapter 1, that it is only prospective *differences* among alternatives that are relevant in their comparison. Over the years, many published formulas for the solution of problems in engineering economy have given dangerously misleading guidance to decision makers because the authors of the formulas have not recognized this concept. Such misleading formulas have, generally speaking, made uncritical use of figures derived from the accounting systems of business enterprises.

Some Special Aspects of Comparing Mutually Exclusive Multiple Alternatives Using Rates of Return or Benefit–Cost Ratios

Certain examples in Chapters 8 and 9 brought out the point that in comparing three mutually exclusive proposals, each proposal should be able to sustain its economic challenge as compared to the best of the other two proposals. Specifically, it was pointed out that it is insufficient for each proposal for change to be compared only with the continuation of a present condition.

In using the rate of return method of analysis or the benefit–cost ratio method, there seems to be a special temptation for analysts to give undue weight to comparisons made with inappropriate alternatives. This temptation seems to be greatest when there are many alternative proposals. Example 13–3 illustrates the use and possible misuse of the rate of return method in examining a number of mutually exclusive proposals for private investment. Example 13–4 illustrates the use and possible misuse of the benefit–cost ratio method in examining a number of mutually exclusive proposals for a government project.

EXAMPLE 13–3

Rate of Return Analysis of a Set of Mutually Exclusive Alternatives

Facts and Estimates

Ten different plans are under consideration for the construction of a building for commercial rentals on a piece of city property. As shown on line A of Table 13–3, the investment in land will be $400,000 regardless of which plan is selected. Line B gives the required investment as estimated for each building design; the investments vary from $204,000 for a single-story building in Plan 1 to $3,000,000 for a luxurious multi-story building in Plan 10. The total investment shown in line C is the sum of the land and building investments.

The estimated annual receipts from rentals for each plan are shown in line D. The estimated total annual disbursements for operation and maintenance, property taxes, and insurance—in fact, all items except income taxes—are shown in line E. The estimated life of the project is 40 years with an estimated zero net terminal salvage value for the building. The income taxes shown in line F assume an applicable tax rate of 50%, straight-line depreciation with the same estimated life and salvage value used in the economy study, and financing entirely with equity funds. The annual net positive cash flow in line G is found by subtracting the total disbursements (lines E and F) from the receipts (line D).

TABLE 13–3
Prospective rates of return on total investment and on successive increments of investment for ten alternative plans for development of a commercial rental property
(All dollar figures in thousands)

	Plan 1	2	3	4	5	6	7	8	9	*10*
A. Investment in land	$400	$400	$400	$400	$400	$400	$400	$400	$400	$400
B. Investment in building	204	360	552	768	1,020	1,284	1,512	1,944	2,400	3,000
C. Total investment	$604	$760	$952	$1,168	$1,420	$1,684	$1,912	$2,344	$2,800	$3,400
D. Annual receipts	$250.7	$345.3	$474.5	$602.8	$757.9	$874.0	$979.3	$1,104.3	$1,214.0	$1,322.0
E. Annual disbursements other than income taxes	111.4	153.7	194.7	232.8	271.0	306.1	333.1	380.7	425.4	478.6
F. Annual disbursements for income taxes	67.1	91.3	133.0	175.4	230.7	267.9	304.2	337.5	364.3	384.2
G. Annual net positive cash flow	$72.2	$100.3	$146.8	$194.6	$256.2	$300.0	$342.0	$386.1	$424.3	459.2
H. After-tax rate of return on total investment	11.9%	13.2%	15.4%	16.6%	18.0%	17.8%	17.9%	16.4%	15.1%	13.4%
J. Increment of investment over plan with next lower investment		$156	$192	$216	$252	$264	$228	$432	$456	$600
K. Increment of annual net positive cash flow over plan with next lower investment		28.1	46.5	47.8	61.6	43.8	42.0	44.1	38.2	34.9
L. After-tax rate of return on increment of investment*		(18.0%)	(24.2%)	22.1%	24.4%	16.6%	18.4%	10.0%	(8.0%)	(5.0%)

*See the text for an explanation of why the rates in parentheses are not relevant in the choice among these alternatives.

Calculated Rates of Return

The rates of return shown in line H assume that the land will be sold for its original cost, namely, \$400,000, at the end of 40 years. This is a convenient simple assumption because the calculated rates of return are relatively insensitive to fairly large changes in the estimated terminal value of the land at such a distant future date.

The rates of return in line H are found by computing present worths and making interpolations in the manner that was explained in Chapter 8. For example, for Plan 6:

$$\text{PW at } 18\% = \$300{,}000(P/A,18\%,40) + \$400{,}000(P/F,18\%,40) = \$1{,}664{,}920$$
$$\text{PW at } 16\% = \$300{,}000(P/A,16\%,40) + \$400{,}000(P/F,16\%,40) = \$1{,}870{,}940$$

Interpolation using the \$1,684,000 investment indicates that the prospective rate of return is approximately 17.8%.

A rate of return analysis of multiple alternatives calls for consideration of incremental rates of return as well as total rates of return. Lines J and K give the cash flow figures needed to calculate these rates. Line L gives the prospective rate of return on each increment of investment. Because all the increments of investment apply only to the building and therefore have zero terminal salvage values, it is possible to compute these incremental rates by interpolating between capital recovery factors. For example, to compute the incremental rate of return of Plan 6 over Plan 5:

$$A/P = \$43{,}800/\$264{,}000 = 0.16591$$

An interpolation between the 40-year capital recovery factors for 18% and 16% (0.18024 and 0.16042, respectively) gives an approximate incremental rate of return of 16.6%.

We shall draw our conclusions from Table 13–3 on the assumption that the primary criterion for the choice among the alternatives is that the minimum attractive rate of return is 15% after income taxes. Of course this criterion should be applied not only to the total investment but also to each separable increment of investment.

However, none of the figures for rates of return in lines H and L depend on the stated i^* of 15%. Nevertheless, because the incremental rates for Plans 2, 3, 9, and 10 are not required in applying this i^* criterion, these rates have been shown in parentheses in line L.

The Challenger-Defender Viewpoint in Analyzing a Set of Mutually Exclusive Multiple Alternatives

In using Table 13–3 as a guide to decision making, it needs to be recognized that an untabulated alternative is to accept none of the plans. Plan 1 cannot sustain its challenge against this do-nothing alternative because its prospective 11.9% rate of return is less than the stipulated i^* of 15%. Neither can Plan 2 because $13.2\% < 15\%$.

The do-nothing alternative should be the defender against Plan 2 because Plan 1 was eliminated when it failed to sustain its challenge. Similarly, the elimination of Plan 2 makes the do-nothing alternative the defender against Plan 3. But Plan 3, with its 15.4% overall rate of return, eliminates the do-nothing alternative from the competition because 15.4% > 15%.

At this point in the analysis, incremental rates of return become relevant. Plan 3 now becomes the defender. Plan 4 is successful in its challenge to Plan 3 because the extra $216,000 will yield 22.1% and the overall rate of return is 16.6%; both of these rates exceed the stipulated 15%. Similarly, Plan 5 succeeds in its challenge to Plan 4 because the incremental rate of return of 24.4% and the overall rate of return of 18% both exceed 15%.

In spite of the fact that the overall rate of return decreases after Plan 5, Plan 6 succeeds in its challenge to Plan 5, and Plan 7 succeeds in its challenge to Plan 6. The extra investment of $264,000 in Plan 6 over Plan 5 has a prospective after-tax rate of return of 16.6%; because 16.6% exceeds the stipulated 15%, this extra investment is attractive. Similarly, the 18.4% on the extra investment of $228,000 in Plan 7 meets the given standard of attractiveness.

Beyond Plan 7, the proposed increments of investment are not attractive. Plan 8 does not succeed in its challenge to Plan 7 because the $432,000 extra investment yields only 10%. If we consider Plan 9 as a challenger to Plan 7, the last successful challenger, an extra investment of $888,000 is required to obtain an extra annual after-tax cash flow of $82,300 for 40 years; the approximate rate of return is 9.0%. (The 8% rate of return shown on line L for the extra investment in Plan 9 over Plan 8 is not relevant because Plan 8 was eliminated when it was compared with Plan 7.)

Plan 10 clearly fails to meet the 15% standard of attractiveness because its overall rate of return is only 13.4%.

The conclusion of the foregoing analysis is that the stipulation of an after-tax i^* of 15% leads to the selection of Plan 7.

A Diagram of the Steps in Making Comparisons Among Multiple Alternatives

The previous paragraphs described the challenger-defender approach to decisions involving mutually exclusive multiple alternatives. Example 13–4 extends this approach to the benefit–cost ratio method.

This approach stems from the fifth and sixth basic principles summarized at the conclusion of Chapter 2, namely:

5. *Only the differences among alternatives are relevant in their comparison.*
6. *Insofar as practicable, separable decisions should be made separately.*

In applying these principles, it is first necessary to array the alternatives in order of increasing investment magnitude and then insure that each increment of investment carries its own weight. This is accomplished by evaluating the differences between successive pairs of alternative investment levels, beginning with the do-nothing alternative, and concluding when the last alternative in the array is compared with the last *acceptable* alternative.

A diagram of the appropriate steps in the analysis applicable to either rate of return or benefit–cost ratio methods of analysis is shown in Figure 13–2.

Two Possible Types of Error in Interpreting Prospective Rates of Return on a Set of Mutually Exclusive Alternatives

If lines J, K, and L had been omitted from Table 13–3, the rates of return in Line H would not have been a sufficient guide for making a choice among the alternatives. There are two types of error that sometimes are made in drawing conclusions from such a curtailed table. Some persons who examine such a table will make the mistake of selecting the alternative that has the highest

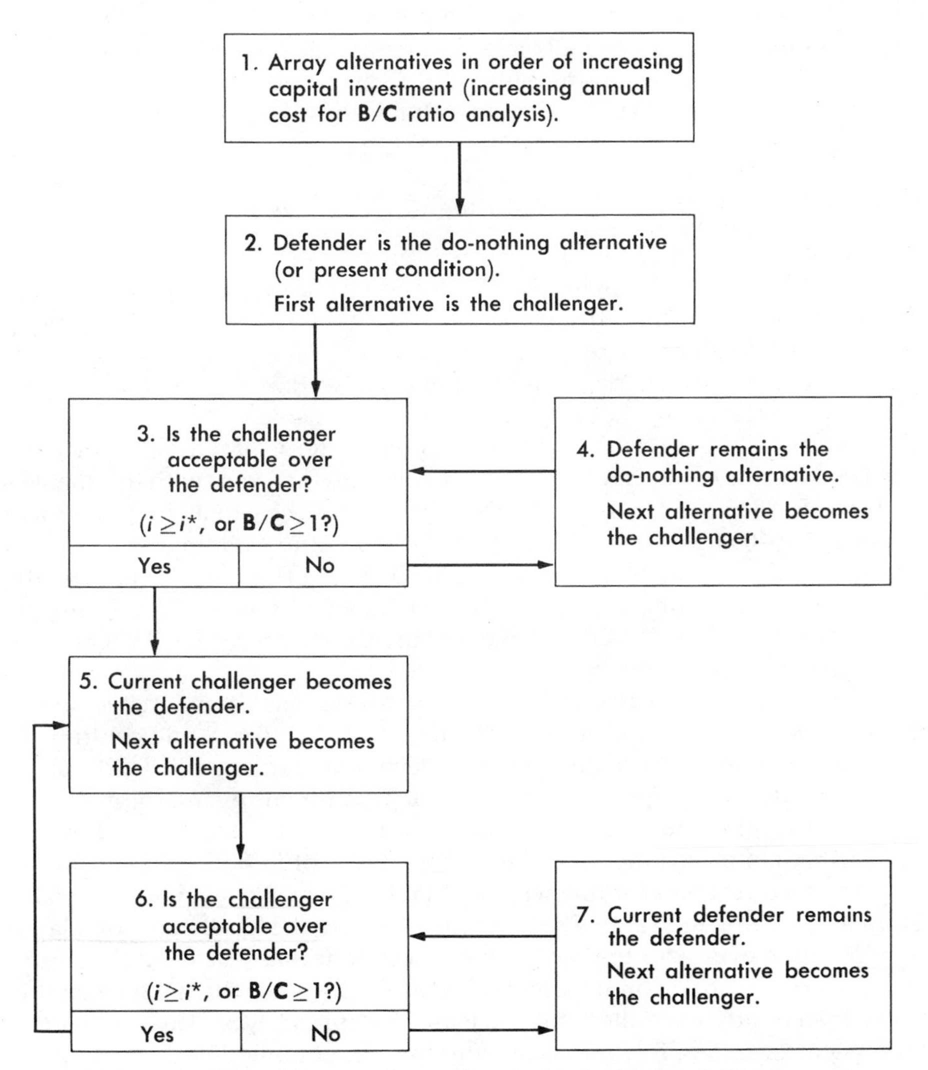

FIGURE 13–2. Flow diagram of procedure for analysis of multiple alternatives

total rate of return; this is Plan 5 in Table 13–3. Other persons will make the mistake of selecting the alternative with the highest investment that will yield at least the stipulated i^*; this is Plan 9 in Table 13–3.

If it is stipulated that the minimum attractive rate of return is 15%, presumably it is believed that the consequences of *rejecting* a proposed increment of investment will be to *accept* an unspecified investment elsewhere that will yield 15%. If Plan 5 should be selected in Example 13–3 on the grounds that it has the highest overall rate of return, the rejection of Plans 6 and 7 is, in effect, a choice of unspecified investments yielding 15% rather than specified ones yielding, respectively, 16.6% and 18.4%.

Also an i^* of 15% implies that the consequence of *accepting* a proposed increment of investment will be to *reject* an unspecified investment elsewhere that will yield 15%. If Plan 9 is chosen on the grounds that it has the highest total investment with an overall yield of at least 15%, an implied consequence of this choice is to reject an investment of $888,000 elsewhere that will yield 15% in favor of an avoidable extra investment in this project that will yield only 9%.

When a Calculated Incremental Rate of Return Is Irrelevant

The discussion of Example 13–3 pointed out that certain rates of return on increments of investment in line L of Table 13–3 were shown in parentheses to emphasize the point that these rates were not relevant in the choice among the stated alternatives. In general, if an alternative is rejected because it fails to meet the decision criteria, it should not be viewed as the defender against subsequent alternatives that require higher investments.

Occasionally there are irregularities in the patterns of rates of return that may lead to error in interpretation. As an example consistent with the analysis of Example 13–3, add a Plan 7a; this plan requires a total investment of $2,240,000 ($400,000 for land and $1,840,000 for the building) and has an estimated annual net positive cash flow of $359,700. The rate of return on the total investment will be approximately 16.0%, an acceptable figure. But the rate of return on the $328,000 increment of investment over Plan 7 will be only 4.5%, clearly an unacceptable figure.

Now imagine that Table 13–3 is changed to add Plan 7a and that in lines J, K, and L, Plan 8 is compared with the new Plan 7a. Plan 8 will continue to have a 16.4% rate of return on the total investment, and line L will now show a rate of return of 25.4% on the increment of investment. A superficial view of the matter might lead to the incorrect conclusion that Plan 8 is acceptable because both of these rates exceed the stipulated i^* of 15%. Such a conclusion would be incorrect because an incremental rate of return over an unacceptable alternative is not relevant. The defender against the challenger of Plan 8 should be the plan with the next lower investment that is *acceptable*, namely, Plan 7. The addition of an unacceptable Plan 7a does not change the fact that Plan 8 is an unattractive alternative to Plan 7 because its $432,000 increment of investment over Plan 7 promises an after-tax yield of only 10%.

EXAMPLE 13–4

Use of Benefit–Cost Ratios in Analysis of a Set of Mutually Exclusive Alternatives for a Public Works Project‡

Facts and Estimates

A certain section of highway is now in location A. A number of proposed designs at new locations and proposed improvements at the present location are to be compared with a continuation of the present condition at A. For purposes of analysis, continuing the present condition is designated as A-1.

Three possible new designs in the present location are referred to as A-2, A-3, and A-4. Two new locations B and C are also considered for this section of highway. There are five new designs to be analyzed at location B and four at location C. These 13 proposals, A-1 to A-4, B-1 to B-5, and C-1 to C-4, are mutually exclusive in the sense that only one proposal will be selected. Of course the various designs at each location contain a number of common elements.

Table 13–4 gives the investments and the estimated annual maintenance

TABLE 13–4
Estimates and annual cost comparison for certain mutually exclusive highway alternatives
(All figures in thousands of dollars)

Alternative	*First cost*	*Annual CR cost (i^* = 7%)*	*Annual maintenance cost*	*Annual highway costs*	*Annual road-user costs*	*Total annual costs influenced by choice*
A–1*	$ 0	$ 0	$60	$ 60	$2,200	$2,260
A–2	1,500	121	35	156	1,920	2,076
A–3	2,000	161	30	191	1,860	2,051
A–4	3,500	282	40	322	1,810	2,132
B–1	3,000	242	30	272	1,790	2,062
B–2	4,000	322	20	342	1,690	2,032
B–3	5,000	403	30	433	1,580	2,013†
B–4	6,000	484	40	524	1,510	2,034
B–5	7,000	564	45	609	1,480	2,089
C–1	5,500	443	40	483	1,620	2,103
C–2	8,000	645	30	675	1,470	2,145
C–3	9,000	725	40	765	1,400	2,165
C–4	11,000	886	50	936	1,340	2,276

*Continuation of the present condition.
†Minimum total annual cost.

‡This example is adapted from one given in a paper "Economy Studies for Highways," by E. L. Grant and C. H. Oglesby, published in Highway Research Board Bulletin 306, *Studies in Highway Engineering Economy* (Washington, D.C.: National Academy of Sciences—National Research Council, 1961).

costs for the various locations and designs. It also gives estimates of the annual costs to the road users for each alternative. It is assumed that there are no differences in other consequences that can be expressed in money terms.

In order to concentrate the reader's attention on the special problems involved in the comparison of multiple alternatives for a public works project, the facts of this example have been simplified in certain ways. It is assumed that road user costs will be uniform throughout a 30-year study period and that all alternatives will have zero terminal salvage values at the end of that period. (The reader may recall that when we used a highway example to introduce the benefit–cost ratio in Example 9–1, we assumed that annual traffic and road-user costs would increase and that certain alternatives would have substantial residual values at the end of the analysis period.) Our economic comparisons of the 13 alternatives are made assuming a minimum attractive rate of return of 7%.

Analysis Based on Minimum Annual Costs

The total annual costs of the 13 alternatives are compared in the final column of Table 13–4. For each alternative, this total is the sum of the capital recovery cost of the investment using an n of 30 and an i^* of 7%, the annual highway maintenance cost, and the annual costs to road users. The total annual cost is a minimum for design B-3, which requires an investment of $5,000,000.

Analysis Based on Benefit–Cost Ratios

In Table 13–5 the 12 alternatives to A-1, a continuation of the present condition, are arranged in order of increasing highway costs. Because each alternative first is to be compared to A-1, the highway costs shown are the annual costs in excess of the $60,000 maintenance cost for A-1; for example, the highway cost figure for A-2, shown at $156,000 in Table 13–4, is $96,000 in the second line of Table 13–5. The benefits shown in the first line of the table are the estimated reductions in annual road-user costs below the $2,200,000 that will continue if A-1 is selected; thus the benefit figure of $280,000 shown for A-2 is found by subtracting $1,920,000 from $2,200,000.

The third line of Table 13–5 gives the **B/C** ratio for each alternative as compared to A-1. All except the $11,000,000 C-4 design show **B/C** ratios greater than unity.

However, it is necessary to compare the 12 designs with one another as well as with the continuation of the present condition. The relevant incremental **B/C** ratios for this comparison are developed in the final four lines of Table 13–5. Of course, a consideration of these benefit–cost ratios leads to the same conclusion as the annual cost comparison in Table 13–4; B-3 turns out to be the best of all the alternatives.

It is of interest to note the challengers and defenders in the incremental comparison. A-2 is successful in its challenge to A-1 with an incremental **B/C** ratio of 2.92. Then A-3 eliminates A-2 with an incremental **B/C** ratio of 1.71. Neither B-1 nor A-4 can sustain their challenges to A-3; their respective incremental ratios are 0.86 and 0.38. Then B-2 eliminates A-3 and is in turn

TABLE 13–5
Benefit–cost ratios as compared to continuation of a present condition (A–1) and relevant incremental benefit–cost ratios for alternatives of Table 13–14
(Alternatives are listed in order of increasing annual highway costs. Annual benefit and cost figures are in thousands of dollars.)

Alternatives	*A–2*	*A–3*	*B–1*	*A–4*	*B–2*	*B–3*	*C–1*	*B–4*	*B–5*	*C–2*	*C–3*	*C–4*
Road user benefits as compared to A–1	$280	$340	$410	$390	$510	$620	$580	$690	$720	$730	$800	$860
Highway costs in excess of costs for A–1	$96	$131	$212	$262	$282	$373	$423	$464	$549	$615	$705	$876
B/C ratio as compared to A–1	2.92	2.60	1.93	1.49	1.81	1.66	1.37	1.49	1.31	1.19	1.13	0.98
Increment analysis compared to which defender	A–1	A–2	A–3	A–3	A–3	B–2	B–3	B–	B–3	B–3	B–3	B–3
Increment of benefits	$280	$60	$70	$50	$170	$110	–$40	$70	$100	$110	$180	$240
Increment of costs	$96	$35	$81	$131	$151	$91	$50	$91	$176	$242	$332	$503
Incremental **B/C** ratio	2.92	1.71	0.86	0.38	1.13	1.21	Negative	0.77	0.57	0.45	0.54	0.48
Decision in favor of	A–2	A–3	A–3	A–3	B–2	B–3	B–3	B–3	B–3	B–3	B–3	B–3

eliminated by B-3. In all of the subsequent comparisons, B-3 remains the defender; no design that has higher highway costs than B-3 can justify them by an increase in benefits that is greater than the increase in highway costs.

Some Comments on the Use of Rates of Return and Benefit–Cost Ratios in Comparing Mutually Exclusive Multiple Alternatives

Even though the techniques of analysis were different, the reader will doubtless have observed a similarity between the types of reasoning used in Examples 13–3 and 13–4. In the rate of return analysis in Example 13–3 it was not enough to compare each plan with a do-nothing alternative; the plans also needed to be compared with one another. And in using the **B/C** ratios in Example 13–4, it was not enough to compare each design with a continuation of a present condition; the comparison of the designs with each other was also essential. The same general type of reasoning applied to choosing the appropriate challengers and defenders in the incremental comparisons. The temptations to incorrect reasoning were similar in the two examples. The reader should recognize that the selection of a minimum attractive rate of return is of critical importance in analysis based on the benefit–cost ratio just as in the various other types of analysis.

Consideration of Multiple Irreducibles in the Analysis of Multiple Alternatives

Even with only two alternatives to be compared, it often happens that certain irreducible data favor one alternative whereas other irreducibles favor the other alternative. (This was illustrated in the story at the end of Chapter 2 about the choice between two proposed total energy systems.) When there are many alternatives to be compared and when there are a number of irreducibles with different impacts on the different alternatives, the problem of giving suitable weight to irreducibles may be particularly troublesome.

If there are multiple alternatives and if irreducibles need to be given weight in the final choice, it is helpful to examine alternatives in pairs. It is much easier to reach a conclusion about the impact of a variety of irreducibles, some favorable to one alternative and some favorable to another, when the decision maker looks at only two alternatives at a time. In Example 13–3, which applied the rate of return method, and in Example 13–4, which applied the **B/C** ratio method, we illustrated the technique of considering multiple alternatives in pairs, always viewing one as a defender and another as a challenger. Whenever a number of irreducibles are present, a similar technique can be applied with any method of analysis of multiple alternatives.

A systematic procedure for organizing and using data regarding multiple irreducibles is discussed at the end of Chapter 19. This is in connection with the "community factor profile" developed by C. H. Oglesby, A. B. Bishop, and G. E. Willeke to help with the problem of comparing alternative locations for urban freeways.

The System Viewpoint in a Choice Among Multiple Alternatives

In a choice among major alternatives, it is common for each alternative to have subalternatives, for each subalternative to have subsubalternatives, and so on. This point was stressed in the presentation of basic concepts in Chapters 1 and 2.

To reach sound decisions about the subalternatives, an analyst needs to recognize the interrelationships between the choices among subalternatives and the choices among major alternatives. Insofar as is practicable, it is desirable to take a system viewpoint in making decisions about subalternatives.

Where a choice among major systems must be made, one difficulty may be the existence of too many possible alternatives for an economical and convenient analysis. Often a complex problem of making a choice among possible systems can be reduced to workable dimensions by first considering the component subsystems. An instance where this was done is described in Examples 13–5 and 13–6.

EXAMPLE 13–5

Consideration of Relevant Subsystems in Planning an Economy Study for the Selection of a Materials-Handling System

Facts of the Case

One of the nation's largest independent drug distributors has eight distribution centers covering a major part of the country. It is now planning to remodel one of its warehouses that serves a population center of about 10 million people in an area about 200 miles wide by 500 miles long.

Each distribution center carries about 30,000 items in stock from which it fills several thousand orders per day. Most of the orders involve less-than-case lots of 10 to 50 different items. About 2,000 items account for 50% of its volume and another 2,000 items account for the next 30% of its volume. The company maintains its own fleet of trucks on which it delivers all its orders within a radius of 50 miles (about 6 million people) on a daily basis. Daily shipments are made to more distant points by way of commercial carriers. About 80% of its volume is delivered by its own trucks.

The company has experimented with a number of different materials-handling systems for filling orders and has warehouses in which the order selection is performed entirely by hand, others involving some mechanization, and one in which the most important 2,000 items are stored in a completely automatic, electronically controlled order-picking machine, with the other items being picked by hand and placed on a conveyor system for accumulating orders. The company is now trying to decide just what combination of methods to use in this installation.

The majority of the orders to this distribution center are telephoned in by the individual drugstores. A clerk types out a temporary invoice as she takes the order over the telephone. The items on the preliminary invoice must be picked from the stock in the warehouse and moved to the packing area. The items are checked against the invoice and any unfilled items are marked off. A final invoice in four copies is typed, with one copy serving as a shipping list, one as a receipt to be signed by the receiving clerk at the drugstore, one for the accounts receivable office for billing purposes, and the fourth for the inventory control section for use in maintaining stock levels.

It is important to visualize the interrelationships of the various activities. Prompt and complete shipments are essential for customer satisfaction. Errors, short shipments, and delays drive customers to competing distributors. Drugstores typically maintain only limited stocks of drug items and expect to be able to replace stock within 5 to 10 hours from local distributors.

This example deals primarily with the materials-handling system, but it was found that the paperwork systems were so closely tied to the materials-handling system that they must be considered simultaneously. Three major systems must be integrated: (1) the materials-handling (receiving, storing, order picking, and packing); (2) customer paperwork (taking order, preparation of invoice and shipping list, and end-of-month billing); and (3) inventory control (maintaining perpetual inventory, determination of order points, and writing purchase orders).

Identification of the alternatives available can best be done for the different subsystems individually. The physical handling system can be subdivided into three subsystems:

A. Receiving and storing
B. Order picking and assembly
C. Packing for shipment

Subsystem A. The products are always received in cases or in multi-case lots and must be transported to the proper storage area in the warehouse and stacked in the shelves and bins. The alternative methods that appear feasible are:

A1. A completely manual system, using four-wheeled shop trucks for transportation
A2. A pallet-fork truck system; manually palletize at the dock; small-volume items would require manual separation and stacking at point of storage
A3. Under floor tow chain towing four-wheeled trucks through the warehouse; requires manual loading, unloading, and stacking
A4. Powered belt or roller conveyors running from dock to storage area; requires manual loading, unloading, and stacking
A5. Completely automatic conveyorized system, employing live storage of cases on gravity roller conveyors and electronic dispatching and control

Subsystem B. A typical order consisting of from 1 to 12 bottles, jars, tubes, small boxes, etc., of 10 to 25 items. Most of the items are relatively small. The order picking and assembly can be performed in a number of ways:

B1. Completely manual system, employing four-wheeled shop trucks, with one person picking all the items for one order
B2. A belt conveyor system, nonautomatic, with manual picking and manual regulation of the placement of orders on the belt for transport to the packing tables
B3. Under floor tow chain with four-wheeled shop trucks, performing in same way as conveyors in B2
B4. Overhead chain conveyor system with independent, dispatchable carriers, but with manual picking
B5. Completely automatic order picking of the large-volume items and any of the other alternatives for the low-volume items

Subsystem C. The packing function is basically a manual job, allowing only minor variations in the arrangement of the work places to accommodate the order-picking system selected. The packing operation should include the final accuracy check on the shipment and the initiation of the customer paperwork.

Subsystem D. The records system dealing with the customer can be operated in a number of different ways, with a wide choice of actual equipment under each general alternative:

D1. Completely manual system
D2. Basically manual, but including a duplicating process to eliminate the retyping of invoices in multiple copies
D3. Basically manual, but including semiautomatic accounting machine to prepare invoices, shipping lists, and monthly bills
D4. An automatic system involving electronic data-processing equipment, punched cards, automatic printers, etc.

Subsystem E. The inventory control system must maintain a perpetual inventory for each item and see that an adequate supply is on hand at all times. Thus, it must not only have daily records of receipts and sales of each item, but must continuously analyze sales trends, market fluctuations, and prices in order to set the best purchasing policies. The available methods are similar to those dealing with the customer:

E1. Completely manual system
E2. Basically manual, but including some semiautomatic accounting machines to summarize daily sales and receipts, compute inventories, and compare balances with order points
E3. Fully automatic system, using electronic data-processing equipment to perform the operations

The complete system in this warehouse must contain one of the alternatives from each of the five subsystems. With 5 alternatives each for subsystems A and B, 1 for C, 4 for D, and 3 for E, there are a total of 300 possible combinations from which the most economical should be chosen. It is theoretically possible to estimate the expenses involved and the effects on receipts of each of the 300 possible alternatives, but it is doubtful that attempting to do so would be worth the effort.

In the first place, many of the 300 combinations can be eliminated as impracticable. For example, it would obviously be foolish to install subsystem A3, the under floor tow chain, for handling incoming materials and then use subsystem B2, the belt conveyor, for order picking. This would involve two separate systems covering the same floor area, two large investments in equipment, and would increase the complications of layout, because the conveyor would permanently occupy space and interfere with the layout of the tow chain system. Similarly, the choice of subsystem D3, the use of semiautomatic accounting machines for processing customers' orders and invoices, would make system E1 a foolish choice, because a great deal of the information needed by inventory control would be readily available from subsystem D3 in a form that could best be handled by semiautomatic machines rather than by hand. Furthermore, the selection of a fully automatic order-picking system for the large-volume items, subsystem A5, would practically demand the use of either subsystem D3 or D4 and either E2 or E3, because the information to the order-picking machine would have to be in a form that the machine could understand and be available at speeds that could not be possible with manual insertion.

Consequently, the selection of the system for this remodeled distribution center can best be made by making up a set of combinations for the materials handling problem and another set for the paperwork. Each alternative combination for the materials handling should be chosen so that the two subsystems will be compatible and so that the specific information needs (restrictions imposed by the handling system on the information system) will be known. Then the alternative combinations for the information and paperwork systems should be selected to make most economical use of whatever equipment is involved in that combination. One or more combinations should be devised to meet the information needs of each handling alternative. Thus, suboptimization can be employed to simplify the problem. Equivalent annual costs can be computed for each alternative combination, and the best pair of alternatives can be selected.

Comments on Example 13–5

This example illustrates a number of important concepts that must be understood by the person undertaking a complex economy study. The systems viewpoint is important to this case. If an attempt had been made at the beginning to identify only those complete systems that would be feasible, it is unlikely that all the alternatives would have been recognized. By breaking the whole system into subsystems and examining each individually, the analyst helped assure that some important alternative would not be overlooked. Furthermore, the analysis of the subsystems tends to bring the requirements of an acceptable whole system into proper focus.

The company had other distribution centers and many of the alternatives had been used in various forms. Cost data were available from the other centers, but the data could not be used in the existing form. Differences in such items as taxes, labor rates, insurance rates, and volumes to be handled among the different locations required adjustment of the cost data to suit the conditions at this location. None of the other distribution centers has a system that is considered to be ideal, and this remodeling presents another opportunity to try to develop the best possible system. Consequently, the new system should be a composite of many of the ideas from different plants in addition to some entirely new ideas. This is really a new problem and requires a complete analysis rather than dependence on past solutions. The final analysis of this problem is summarized in Example 13–6.

EXAMPLE 13–6

Economy Study to Select a Materials-Handling System

Estimates and Analysis

After study of the many possible combinations of subsystems described in Example 13–5, eight different combinations were chosen for preliminary economic analysis. Three of these combinations were eliminated by this initial analysis and the following five combinations were selected for detailed analysis:

Combination 1 consists of manually operated four-wheel shop trucks for both handling incoming shipments and order picking, along with a manual system for records keeping and inventory control. It includes a duplicating process for multiple copies.

Combination 2 consists of underfloor tow chain with four-wheel shop trucks for both handling operations and semiautomatic accounting equipment for all records keeping and inventory control.

Combination 3 employs an underfloor tow chain with four-wheel trucks for incoming materials, an overhead chain conveyor with independent dispatchable carriers for order picking, and an automatic electronic data-processing system for records keeping and inventory control.

Combination 4 employs fork-lift trucks for handling incoming materials, a completely automatic order-picking system for the 2,000 high-volume items and manual picking for all other items, and an automatic electronic data-processing system.

Combination 5 is the same as *combination 4* except that it has automatic order picking for 4,000 high-volume items.

The monetary comparison is shown in Table 13–6 using a before-tax i^* of 25%. This is a case where the choice among the alternatives does not affect the estimated revenues of the enterprise.

There are a number of irreducibles that need to be considered in the final decision. With combinations 3, 4, and 5, the invoices, bills, purchase orders, and reports will be available earlier than with a manual system. The combinations 4 and 5 will provide capacity to handle more orders per day with very little added expense, while all the other combinations can be expanded only by adding additional people (and increasing labor costs). The automatic systems will no doubt require a somewhat greater development and "debugging" period than the other systems. Also, a greater number of errors will probably be made in both order filling and billing during the initial period, but eventually the automatic systems should operate with fewer errors than the other combinations.

Table 13–6 reveals that combination 4 has the lowest prospective annual cost with a minimum attractive rate of return of 25% before income taxes. Combination 4 is only $5,633 better than combination 5, however, and consideration of such irreducibles as prospective increases in business, less de-

TABLE 13–6
Cost comparison of alternative materials-handling systems for wholesale drug distributor

Combination	*1*	*2*	*3*	*4*	*5*
A. Total first cost	$12,000	$32,500	$61,480	$326,000	$465,500
B. Estimated economic life, years	15	10	10	10	10
C. Estimated salvage value at end of economic life	0	$5,000	$8,000	$36,000	$60,000
D. Capital recovery factor, $i^* = 25\%$	0.25912	0.28007	0.28007	0.28007	0.28007
E. Equivalent annual cost of capital recovery	$3,109	$8,952	$16,978	$90,221	$128,569
F. Average annual labor cost	252,500	235,000	210,000	108,000	65,000
G. Average annual fuel and power costs	300	900	1,200	2,000	2,400
H. Average annual maintenance costs	1,000	1,500	2,200	4,600	5,800
I. Average annual taxes and insurance	360	975	1,844	9,780	13,965
J. Annual rental on data processing equipment	0	2,500	18,000	24,000	28,500
K. Total equivalent annual costs	$257,269	$249,827	$250,222	$238,601	$244,234

pendence on labor, and the prospect of fewer errors might lead the management to select combination 5.

Summary

This chapter has dealt with certain aspects of economic analysis in comparisons of multiple alternatives, with special reference to cases where only one alternative can be chosen. The first part of the chapter illustrated and discussed the use of simple mathematical models to obtain an optimal solution in this type of problem. The second part of the chapter illustrated and discussed certain dangers of misinterpretation of an analysis of multiple alternatives whenever rates of return or benefit–cost ratios are calculated. The final part of the chapter illustrated a complex problem in which it was desirable to recognize a variety of subsystems before selecting the particular combinations to be compared in the detailed economy study.

It has stressed the fundamental principle that each increment of investment in a proposed project should be expected to yield a return commensurate with the value of i^* stipulated by the decision maker or decision-making authority. When a decision is made based on the stipulated i^*, either the decision maker has another opportunity of equal risk at which any available funds can be invested at i^* or a conscious decision has been made that, for the risk involved in this set of proposals, no investment will be made that does not have the prospect of earning i^*.

PROBLEMS

13–1. Examine Table 13–3. With the stipulated after-tax i^* of 15%, our discussion of Example 13–3 indicated that Plan 7 should be selected. Which plan would you recommend if i^* were 17%? If it were 12%? If it were 20%? Explain your reasoning.

13–2. Make two changes in the facts of Example 13–3. Change the applicable income tax rate from 50% to 45%. Write off the building investment for income tax purposes at a straight-line rate of 5% a year for the first 20 years. Make the necessary calculations to find the prospective rate of return on the investment in Plan 7. Make the necessary calculations to determine whether the extra investment in Plan 8 should now be recommended assuming that the after-tax i^* continues to be 15%.

13–3. The analysis in Example 13–4 led to the recommendation that alternative B–3 should be chosen. The example assumed zero terminal salvage value for the highway facilities at the end of the 30-year study period. Make the necessary calculations to find the benefit–cost ratios for alternatives B–3 and B–4 assuming 100% terminal salvage values. Would this change be sufficient to justify the selection of B–4 rather than B–3? Explain your answer.

13–4. In a certain manufacturing company, decisions regarding approval of proposals for plant investment are based on the requirement of a 25% minimum attractive rate of return before income taxes. The mechanization of a certain costly hand operation has been proposed. Machines from six different manufacturers are under consideration. The estimated investment for each proposal and the estimated reduction in annual disbursements are as follows:

Machine	*Investment*	*Reduction in annual disbursements*
Onondaga	$54,000	$20,800
Oneida	61,200	22,700
Cayuga	72,000	26,700
Tuscarora	77,400	27,900
Seneca	91,800	32,100
Mohawk	108,000	37,300

Assume that each machine will have an 8-year life with zero terminal salvage value. Which one, if any, of these six mutually exclusive investments should be made? Show the calculations on which you base your recommendation.

13–5. Solve Problem 13–4 changing the stipulated minimum attractive rate of return to 15% after income taxes. Assume an applicable tax rate of 40%. Use straight-line depreciation assuming an 8-year life and zero terminal salvage value.

13–6. Solve Problem 13–4 changing the stipulated minimum attractive rate of return to 15% after income taxes. Assume an investment tax credit of 10% at zero date. Assume an applicable tax rate of 40%, with years-digits depreciation based on an 8-year life and zero terminal salvage value.

13–7. In a proposed flood control project, there are two possible sites for a dam and storage reservoir, designated as the Willow and Cottonwood sites. One or the other of these sites may be used but not both. A small hydroelectric power development may be added at the Willow site. Certain channel improvements are also considered. Seven alternate projects are set up for analysis and average annual damages due to floods under each plan estimated as follows:

Plan		**Damages**
A	Willow dam and reservoir alone	$100,000
B	Willow dam, reservoir, and power plant	120,000

C	Willow dam and reservoir, with channel improvement	40,000
D	Willow dam, reservoir, and power plant with channel improvement	60,000
E	Cottonwood dam and reservoir alone	180,000
F	Cottonwood dam and reservoir, with channel improvement	90,000
G	Channel improvement alone	330,000

With no flood control works at all, the average amount of flood damages is estimated as $680,000.

The estimated first cost of the Willow dam and reservoir is $5,000,000. The power plant will increase this first cost by $1,000,000. The estimated first cost of the Cottonwood dam and reservoir is $3,750,000. The estimated first cost of the channel improvement is $800,000. In the economic analysis a 100-year life with zero salvage value is to be used for the two dams and reservoirs, a 50-year life with zero salvage value is to be used for the power plant, and a 20-year life with $300,000 salvage value is to be used for the channel improvement. All equivalence calculations are to use an i^* of 5%.

On the basis of the cost of equal power from a steam electric plant, the "benefits" from the hydroelectric power are estimated to be $200,000 a year. Annual operation and maintenance costs will be:

Willow dam and reservoir	$60,000
Power plant	25,000
Cottonwood dam and reservoir	50,000
Channel improvement	70,000

Compute a benefit–cost ratio for each of the seven plans of development. Make any other calculations that you think are desirable to aid a choice among the different plans. Do you recommend that one of these plans be adopted? If so, which one? Why?

13–8. Solve Problem 13–7 changing i^* from 5% to 8%. Make no other changes.

13–9. An electrical conductor is to be selected to carry 40 amperes for 3,500 hours per year. The estimated cost of wire is $1.90 per lb. Electrical energy will be purchased at 7.2 cents per kw-hr. The minimum attractive rate of return before income taxes is 18%. Annual property taxes are estimated at 1.5% of first cost. The life of the wire is estimated to be 20 years with zero salvage value. Use the formula applied in Example 13–2 to find the economical value in circular mils for this wire cross section.

13–10. Prepare a table similar to Table 13–2 for the data of Problem 13–9. Use the same sizes of wire shown in Table 13–2. Which wire size has the least annual cost?

13–11. Example 13–2 and Table 13–2 dealt with an analysis using a before-tax i^* of 14%. Prepare a table similar to Table 13–2 for a comparison of these five wire sizes after income taxes. Assume an applicable tax rate of 50% and an after-tax i^* of 7%. Assume that straight-line depreciation will be used for income tax purposes with a 25-year estimated life and zero terminal salvage value. Make no other changes in the data of the example. Which wire size now appears to have the lowest annual cost?

(Note that a line needs to be added to the revised table to show for wire sizes 00 to 2 the extra annual income tax above the tax that would have been payable if size 3 had been chosen. Note also that the annual *disbursements* that will influence income tax liability include property taxes as well as payments for lost energy.)

13–12. Change the data of Problem 13–11 to recognize an investment tax credit of 10% of the investment in wire. Assume that the tax credit is taken at zero date. (The tax credit will not influence the depreciation deduction that enters into the determination of taxable income.)

13–13. Prepare a table similar to Table 13–2 comparing the annual costs of wire sizes 00 to 3, both inclusive, assuming the investment in copper wire at $2.25/lb, the price of electric energy at 4.2 cents/kw-hr, investment charges at 18.5%, for a current of 60 amperes flowing for 6,000 hr/year. Which size has the lowest annual cost? (*Ans*. Size 1)

13–14. In a certain manufacturing plant, schemes for cost-reducing machinery are judged on the basis of the "gross return," which is computed as the ratio of the annual saving in direct materials and labor to the investment. To justify the investment of funds in such projects, there must be a gross return of 30%. This 30% covers capital recovery (interest and depreciation), property taxes, insurance, and income taxes. Funds are available to finance any projects that meet this standard of attractiveness. On a certain operation, six alternative proposals for cost reduction are made. On the basis of the stated criterion, which one of these should be chosen? Why?

Proposal	**Required investment**	**Annual savings in direct materials and labor**
A	$ 6,000	$1,500
B	8,000	2,500
C	12,000	4,100
D	13,000	4,200
E	18,000	6,000
F	25,000	7,600

(*Ans*. Proposal E)

13–15. In a proposed flood control project, there are two possible sites, A and B, for a dam and storage reservoir. One or the other of these sites may be used but not both. Certain channel improvement is also considered; this will increase the capacity of the stream to carry flood discharge. Estimated first costs, lives, and annual operation and maintenance costs are as follows:

	Site A	*Site B*	*Channel improvement*
First cost	\$6,000,000	\$8,000,000	\$1,000,000
Life	75 years	75 years	25 years
Annual O & M	\$100,000	\$140,000	\$230,000

Annual capital recovery costs are to be computed using an i^* of 5½%. Assume zero salvage values at the end of the estimated lives.

The average annual amount of damages due to floods are estimated under various possible plans of development, as follows:

No flood control works at all	\$1,200,000
Development at Site A alone	380,000
Development at Site B alone	260,000
Channel improvement alone	520,000
Site A plus channel improvement	200,000
Site B plus channel improvement	120,000

Compute a benefit–cost ratio for each of the five plans of development as compared to the alternative of having no flood control. Assume that the annual costs of a dam and reservoir plus channel improvement will be the sum of the costs of the dam and reservoir alone and channel improvement alone. Compute any incremental **B/C** ratios that you believe are relevant. Which plan of development, if any, would you recommend? Why? (*Ans.* Development at Site A alone)

13–16. No construction actually took place on the flood control project analyzed in Problem 13–15. Now some years later matters have changed so that the average annual damages without flood control as well as the damages with each plan will be double those shown in the problem statement. Estimated first costs, lives, and annual operation and maintenance costs are unchanged from those stated in Problem 13–15. However, the appropriate i^* is now 8%.

With these changes in data calculate the **B/C** ratios that you believe are relevant for an analysis. Which plan of development, if any, would you now recommend?

13–17. The XYZ Manufacturing Co. has funds available for investment in machinery and tools to reduce direct manufacturing costs. All proposals for such investment are judged on the basis of "gross return." It has been decided that no avoidable investment is to be considered attractive that does not show a gross return of 25%. This is intended to cover capital recovery at the company's i^* over a stipulated service period plus property taxes, income taxes, and insurance.

A group of engineers has analyzed six different mutually exclusive proposals for investment in machinery and tooling to reduce costs on a certain operation. They have prepared the following table to provide a basis for comparing these proposals:

Proposal	*Investment*	*Annual saving in direct mfg. costs*	*Gross return on inv.*	*Extra inv.*	*Extra annual saving*	*Gross return on extra inv.*
A	$2,000	$ 200	10.0%			
B	8,000	2,320	29.0%	$6,000	$2,120	35.3%
C	16,000	4,800	30.0%	8,000	2,480	31.0%
D	18,000	5,340	29.7%	2,000	540	27.0%
E	24,000	5,760	24.0%	6,000	420	7.0%
F	40,000	10,240	25.6%	16,000	4,480	28.0%

At this point the engineers disagree on the question of which of these proposals is to be recommended to the management. Smith prefers C on the grounds that it has the greatest gross return on total investment. Jones chooses B because it has the greatest gross return on the extra investment. Johnson is inclined toward F as representing the maximum investment on which both the gross return on total investment and the gross return on the extra investment are over 25%.

Which of these six proposals would you recommend? Explain why.

13–18. Many valves are required in a certain pipeline that carries a corrosive chemical. In the past, cast iron valves have always been used and have required replacement every 2 years. Now valves of two corrosion-resistant alloys, A and B, are available. Estimates of installed first costs, lives, and salvage values for these three types of valve are as follows:

	Cast Iron	*Alloy A*	*Alloy B*
First cost	$3,500	$6,500	$11,000
Life	2 years	5 years	10 years
Salvage value	zero	zero	zero

The three types of valves differ only in first costs, lives, and annual disbursements for income taxes. Assume a 45% applicable tax rate and straight-line depreciation. Assume no investment tax credit for the cast iron valve, a 6⅔% tax credit on the dates of installation of Alloy A valves, and a 10% tax credit on the date of installation of the Alloy B valve. With an after-tax i^* of 10%, which valve would you recommend? Explain your analysis.

13–19. Five alternative proposals have been made for the development of a residential rental property. The required land is available on option at a price of $200,000. The following estimates have been made for the plans:

	Investment in building	*Annual receipts*	*Annual disbursements (not including income taxes)*
Plan A	$ 400,000	$176,000	$55,000
Plan B	560,000	260,000	70,000
Plan C	800,000	380,000	100,000
Plan D	1,080,000	467,000	125,000
Plan E	1,520,000	598,000	150,000

Assume that the choice among these alternatives is to be made using a before-tax i^* of 25%. Compute the rates of return before income taxes for each proposal assuming a 40-year life with zero salvage value for each building. Assume that the land will be sold for a net $200,000 at the end of the 40-year period. Compute any prospective rates of return on increments of investment that you believe are relevant for the decision. Which plan, if any, do you recommend? Explain your reasoning.

13–20. Solve Problem 13–19 changing the i^* to 15% after income taxes. Use an applicable tax rate of 40%. Assume that straight-line depreciation will be used for income tax purposes for any building. This will be based on a 40-year life and zero terminal salvage value.

13–21. Solve Problem 13–20 assuming that any building can be written off for tax purposes at a straight-line rate of 6⅔% a year during the first 15 years.

13–22. A certain section of mountain highway called Rocky Canyon Road is used only for through traffic and is in bad condition. Three proposals have been made for its improvement. One, designated R–1, involves resurfacing with no change in location. A second, R–2, involves resurfacing plus some minor relocation, particularly elimination of several dangerous curves. A third, R–3, calls for enough relocation to shorten the highway a bit. Alternatively, it has been proposed to eliminate this section of highway and replace it with one in an entirely new location. There are two designs for this Trail Creek Road, designated as T–1 and T–2. The following estimates have been made to compare these designs with one another and with a continuation of the present condition (designated as R–O):

Proposal	*Annual highway user costs*	*Portion of investment with 50-year estimated life*	*Portion of investment with 10-year estimated life*	*Annual public expenditures for highway upkeep*
R–0	$880,000	$0	$0	$100,000
R–1	760,000	0	700,000	70,000
R–2	740,000	500,000	700,000	70,000
R–3	700,000	1,400,000	600,000	60,000
T–1	630,000	3,000,000	550,000	55,000
T–2	580,000	4,000,000	500,000	50,000

Assuming zero salvage values for the different investments at the ends of their respective lives, make the necessary calculations to compare these alternatives using an i^* of 8%. Which proposal do you recommend? (*Ans.* R–3)

13–23. Solve Problem 13–22 using an i^* of 3%.

13–24. Solve Problem 13–22 using an i^* of 15%.

13–25. The subscription rates for a certain weekly publication are $25 for 1 year, $40 for 2 years, and $50 for 3 years, payable in advance in all cases. What is the rate of return on the extra investment in a 2-year subscription as compared to two 1-year subscriptions? On the extra investment in a 3-year subscription as compared to three 1-year subscriptions? Discuss the significance of these rates of return as a guide to a decision on the length of the subscription period. Compute any other rates of return that might reasonably be considered by someone who expects to continue to subscribe for many years and who wishes to choose among these three available subscription periods. What irreducibles or other matters do you think might reasonably influence such a subscriber's decision on his subscription period? (Refer to Appendix B.)

14

PROSPECTIVE INFLATION AND SENSITIVITY ANALYSIS

What's to come is still unsure.—Shakespeare

The examples and problems in the preceding chapters have involved choosing among alternatives with a single set of forecasts given. Presumably in each case the stated forecasts represent someone's best judgment on the extent to which future receipts and disbursements and other matters will be influenced by the choice among the stated alternatives.

There is always uncertainty about the future; it will rarely if ever turn out that events occur exactly as forecast. This uncertainty in itself is not a reason for not making the best forecasts that we can and then being governed by our analysis of these forecasts.

The Concept of Sensitivity

Different types of forecasts may enter into an economy study that compares proposed alternatives. Disbursements need to be estimated for various elements of cost. Many studies call for estimates of receipts. Nearly all such studies need decisions about the length of the study period and estimates of lives and salvage values. Studies in competitive industry usually need estimates of differences in income taxes.

The usual purpose of an economy study is to reach a decision or a recommendation for a decision. Often it helps the decision-making process to consider the consequences of moderate changes in certain forecasts. There will be some elements in a given economy study for which it will be possible to make substantial changes in forecasts without altering the conclusion of the study. In contrast, there may be other elements for which a relatively small change in forecast will change the recommended decision among the alternatives.

As the word is used in this book, *sensitivity* refers to the relationship between the relative change in forecast of some element of an economy study and the measure of attractiveness of an alternative. If one particular element can be varied over a wide range of values without affecting the recommended decision, the decision in question is insensitive to uncertainties about that element. On the other hand, if a small change in the estimate of some element will alter the recommended decision, the decision is said to be sensitive to uncertainties about that element.

Because all estimates are subject to some uncertainty, the sensitivity approach is helpful in decision making. The application of the sensitivity concept can be an intermediate step between the numerical analysis based on the best estimates for the various elements and the final choice among the alternatives. Any desired element can be tested to find out whether the decision will be sensitive to moderate changes in an estimated value. The results of the sensitivity analysis can be given weight in the final decision-making process.

Much of the discussion in this chapter develops the subject of sensitivity analysis by examples that deal with prospective changes in specific prices and in price levels. Application of sensitivity analysis to forecasts of matters other than price is discussed briefly in the text at the end of the chapter and illustrated in a number of the problems that follow the chapter.

Using an Index to Measure Price Level Change

There are two quite different aspects of changes in prices. One is a change in the general level of prices; this really is a change in the purchasing power of the monetary unit. The other is differential price change; during any given period there will be some prices that change at a different rate from the change in the general price level.

A price index is a ratio of the price of some stipulated "market basket" of commodities (usually both goods and services) at one date to the prices of the same market basket at some other date. The index is generally expressed as a percentage even though the percent sign is omitted. Three widely used price indexes prepared by agencies of the United States government are the Consumer Price Index (CPI), the Wholesale Price Index, and the Implicit Price Index for the Gross National Product. Our discussion in the remainder of this chapter uses the CPI as a measure of price level change. The list of items recently priced for this index included some 400 different goods and services.*

Because governmental fiscal and monetary policies (a major influence on price level change) and other matters vary so much from country to country, there may be considerable differences in the rates of price level change in different countries. For example, assuming the 1967 index as 100 in all the listed countries, the Bureau of International Economic Policy and Research of

*For a brief explanation of the CPI, see *Handbook of Labor Statistics 1978* (Washington, D.C.: U.S. Government Printing Office, 1979), pp. 9–10.

the United States Department of Commerce reported the following 1979 indexes of consumer prices:*

United States	217.4
Canada	221.0
Japan	261.3
France	258.5
Germany	166.6
Italy	328.5
United Kingdom	359.1

Indexes of Cost of Living Are Necessarily Imperfect

There are no solutions that are entirely satisfactory for some of the problems involved in preparing an index of cost of living for the entire population of a country. Certain troublesome questions arise in choosing the market basket of prices to be included and the weights, if any, to be given to each price. Difficulties exist in the collection of valid data. Special problems are present in the usual case where there is technological progress and an index is to be interpreted as reflecting changes in living costs over a period of years.

It should be evident that different market baskets of goods and services are appropriate for different income groups. Usually market baskets for urban and rural populations ought to differ. Any market basket chosen must be a compromise that can give a somewhat misleading estimate of the change in costs of living to part of the population.

In collecting data to use in price indexes, it is not always easy to find the true average prices being paid for certain goods and services. Frequently, identical products are sold at quite different prices by different vendors. Moreover, there may be a difference between quoted prices and the prices actually paid. In a period of shortages actual prices may exceed quoted prices; in a recession they may be less than quoted prices.

It is difficult for price indexes to give weight to the favorable effect of technological progress in causing large quality improvements.† For this reason indexes tend to overstate the long-run increase in the cost of living.

Nevertheless, in spite of the difficulties of measurement, discussion of certain consequences of price change is helped by the assumption that there is such a thing as a price "level" and that changes in this level are measurable.

*See *Economic Indicators, August 1980* (Washington, D.C.: U.S. Government Printing Office, 1980).

†One could fill a book with instances of this. As one example, consider the increase in the quality of automobile tires over several decades. The senior author of this book recalls the time when an automobile trip of a few hundred miles without at least one flat tire was the exception rather than the rule. More recently he drove for more than ten years with no flat tires at all.

Lacking an ideal (but unobtainable) index of price level change, the best that can be done is to use an index that is available.

Some History of Price Level Changes in the United States

A rise in general price levels is described as *inflation*, a fall as *deflation*. In the 1980s, when nearly continuous inflation has been experienced in the United States for some 40 years, it may seem hard to believe that over the years of record since 1800, there were also extended periods of deflation. Throughout the 1800s, with the exception of two periods of wartime inflation (1812–1814 and 1862–1865), the general trend of price levels was downward. Although the CPI did not start until 1913, a study linking this modern index to earlier price indexes shows that the price level in 1900 was half of that in 1800.* Table 14–1 gives values of the CPI from 1913 through 1980.

TABLE 14–1
United States Consumer Price Index, 1913–1980

Year	*CPI*	*Year*	*CPI*	*Year*	*CPI*
		1936	41.5	1961	89.6
		1937	43.0	1962	90.6
1913	29.7	1938	42.2	1963	91.7
1914	30.1	1939	41.6	1964	92.9
1915	30.4	1940	42.0	1965	94.5
1916	32.7	1941	44.1	1966	97.2
1917	38.4	1942	48.8	1967	100.0
1918	45.1	1943	51.8	1968	104.2
1919	51.8	1944	52.7	1969	109.8
1920	60.0	1945	53.9	1970	116.3
1921	53.6	1946	58.5	1971	121.3
1922	50.2	1947	66.9	1972	125.3
1923	51.1	1948	72.1	1973	133.1
1924	51.2	1949	71.4	1974	147.7
1925	52.5	1950	72.1	1975	161.2
1926	50.0	1951	77.8	1976	170.5
1927	52.0	1952	79.5	1977	181.5
1928	51.3	1953	80.1	1978	195.4
1929	51.3	1954	80.5	1979	217.4
1930	50.0	1955	80.2	1980	247.7
1931	45.6	1956	81.4		
1932	40.9	1957	84.3		
1933	38.8	1958	86.6		
1934	40.1	1959	87.3		
1935	41.1	1960	88.7		

*Table 116, The Consumer Price Index, 1800–1977, on p. 397 of *Handbook of Labor Statistics 1978*, op. cit., gives the 1800 index as 51 and the 1900 index as 25. These particular values are based on the selection of 1967 as the base year for which the index was 100. But the conclusion that the purchasing power of a U.S. dollar doubled during the nineteenth century is unrelated to the selection of the base year.

Figure 14–1 shows the annual percentage of change in the CPI for the 30-year period from 1950 to 1980.* Although this was a period of inflation, it is interesting to note that for the 13-year period from 1952 to 1965 the annual rate of inflation was only 1⅓%. (The CPI was 79.5 for 1952 and 94.5 for 1965; the change over this period would have occurred if the CPI had increased at 1.34% compounded annually.) In contrast, the change from 1973 to 1980 reflected an annual rate of inflation of more than 9%, compounded annually.

The Concept of Analysis of Investments in Terms of Money Units of Constant Purchasing Power

In Chapters 1 and 2 of this book we pointed out that decision making is simplified when the anticipated consequences of a choice among alternatives can be made commensurable with one another. To be commensurable, consequences need to be expressed in numbers with the same units applicable to all the numbers. In economic decision making, money units are the only ones that meet this specification. But money units at different points in time are not commensurable without calculations that somehow reflect the fact that money has a time value.

When price levels rise, the purchasing power of the monetary unit goes down. Therefore, prospective inflation can be an additional source of lack of commensurability of cash flows at different dates. A possible solution of this difficulty is to apply the relevant price indexes to convert all cash flows to units of constant purchasing power. Our first illustration of this technique is Example 14–1, where the technique is used to analyze a past investment. The second illustration is Example 14–2, which deals with the analysis of a proposed investment.

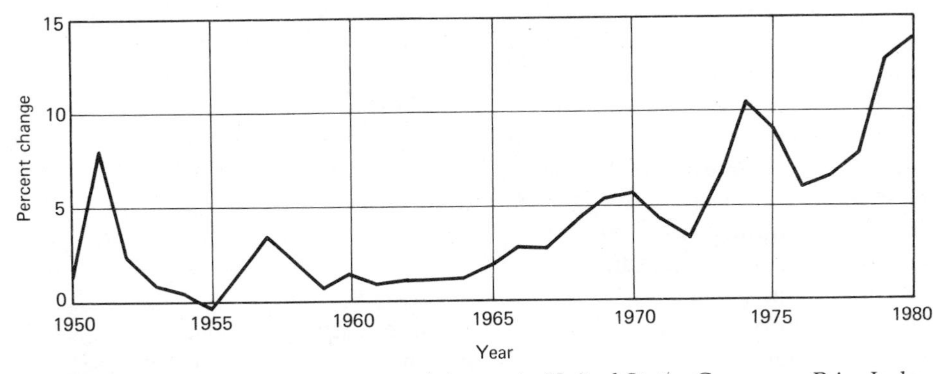

FIGURE 14–1. Annual percentage of change in United States Consumer Price Index

*Each number in Table 14–1 is the *average* CPI for its respective year. The percentages of change are derived from these averages. For example, the figure plotted for 1951 is

$$\left(\frac{77.8 - 72.1}{72.1}\right) 100\% = 7.9\%.$$

EXAMPLE 14–1

Use of Units of Constant Purchasing Power to Analyze a Past Investment

Facts of the Case

A $10,000 15-year 5% tax-exempt bond of a certain water district in the United States was purchased for $10,000 on July 1, 1964, and held until its maturity on July 1, 1979. Table 14–2 shows the results of this investment expressed in units of constant purchasing power. Measured in this way, it is evident that the investor's yield in purchasing power was negative.

Each cash flow is converted into 1964 dollars by multiplying it by the ratio of the 1964 CPI to the current year's CPI. For example, for 1975 the interest of $500 received by the investor had the same purchasing power that $288 would have had in 1964. That is, $500 (92.9/161.2) = $288. The table has been simplified a bit by making calculations as if interest had been payable annually rather than semiannually.

EXAMPLE 14–2

Use of Units of Constant Purchasing Power to Analyze a Proposed Taxable Investment that will be Adversely Affected by Inflation

Facts of the Case

In a period when inflation is expected, an investor is considering whether or not to purchase a $10,000 12% 10-year public utility bond. The investor wishes to analyze the proposed investment in units of constant purchasing power assuming annual inflation rates of 4%, 6%, and 8%. Because the tax rate on the highest bracket of the investor's income is expected to be in the neighborhood of 30%, the analysis is to be made assuming 30% as the applicable tax rate.

In our examination of a terminated investment in Example 14–1, it was possible to use known past values of a price index to express the cash flows of different dates in terms of purchasing power at some one chosen date (which was 1964 in Table 14–2). But in examining a proposed future investment, the future price indexes are unknown. Where some inflation is expected, a sensitivity study assuming different inflation rates may be a helpful guide to decision making.

Even though it is unlikely that the rate of price level change will be constant, there will rarely be any valid basis for forecasting the ups and downs of future inflation rates. In fact, variability in annual inflation rates (such as that shown in Figure 14–1 for the United States) seems to be the common experience. The most practical approach for an economy study usu-

TABLE 14–2
Analysis of a terminated bond investment

Year	*Cash flow in current dollars*	*Consumer Price Index*	*Cash flow converted to 1964 dollars*
1964	−$10,000	92.9	−$10,000
1965	+500	94.5	+492
1966	+500	97.2	+478
1967	+500	100.0	+464
1968	+500	104.2	+446
1969	+500	109.4	+425
1970	+500	116.3	+399
1971	+500	121.3	+383
1972	+500	125.3	+371
1973	+500	133.1	+349
1974	+500	147.7	+314
1975	+500	161.2	+288
1976	+500	170.5	+272
1977	+500	181.5	+256
1978	+500	195.4	+238
1979	+10,500	217.4	+4,487
	+$7,500		−$338

ally is to assume a uniform rate of inflation (or, in a sensitivity study, two or more different inflation rates).

In Table 14–3 the after-tax cash flows from the proposed investment in the 12% bond have been converted into dollars that have the purchasing power of a dollar at zero date. Price indexes for each year have been computed assuming 6% annual inflation. In such calculations it is convenient to assume that the price index is 100 at zero date; indexes for later dates may be obtained from the single payment compound amount factors in the interest table that corresponds to the inflation rate. (Table D–11, Appendix D, provides our figures for Table 14–3.) The final column of Table 14–3 shows all cash flows converted to units of purchasing power at zero date.

With 6% inflation our investor will do only slightly better than recovering the purchasing power given up at zero date. Present worth calculations indicate that the prospective rate of return is about 2.3% when cash flow is converted to units of constant purchasing power. Similar analysis assuming 8% annual inflation indicates that the rate of return will be only 0.4%. At 4% annual inflation the rate of return is 4.2%.

These contrast with an apparent after-tax rate of return of 8.4%. We can see that 4% annual inflation will cut this in half if an analysis is made in units of constant purchasing power. Inflation at 8% will practically eliminate the return.

TABLE 14–3
Analysis of a proposed investment in a 12% bond assuming 30% income taxes and a 6% annual inflation rate

Year	*Before-tax cash flow in current dollars*	*Cash flow for 30% income tax*	*After-tax cash flow in current dollars*	*Estimated price index assuming 100 for zero date*	*After-tax cash flow in zero-date dollars*
0	−$10,000		−$10,000	100.00	−$10,000
1	+1,200	−$360	+840	106.00	+792
2	+1,200	−360	+840	112.36	+748
3	+1,200	−360	+840	119.10	+705
4	+1,200	−360	+840	126.25	+665
5	+1,200	−360	+840	133.82	+628
6	+1,200	−360	+840	141.85	+592
7	+1,200	−360	+840	150.36	+559
8	+1,200	−360	+840	159.38	+527
9	+1,200	−360	+840	168.95	+497
10	+11,200	−360	+10,840	179.08	+6,053
	+$12,000	−$3,600	+$8,400		+$1,766

Illustration of One Type of Differential Price Change in Examples 14–1 and 14–2

If, on the average, prices rise 10% (in dollars or other relevant currency units) in a particular year, it can be expected that some prices will rise more than 10% and others less than 10%. There has been a *differential price change* in all the prices that have not changed by 10%.

A price that remains constant in currency units is nevertheless subject to a *differential* price change (i.e., a change in purchasing power) whenever the price level changes. Interest received by a lender of money has a differential price change adverse to the lender whenever there is inflation. The repayment of principal also is subject to a differential price change adverse to the lender. These were illustrated in Example 14–1, which showed how past inflation affected an investment during the years 1964–1979. (Although this example described one actual investment, it obviously is representative of a great many such investments made during these years.) Example 14–2 showed how future inflation could cause adverse consequences to a present investor.

Economy Studies When Price Level Change Is Expected Without a Net Differential Price Change

Examples 6–1 and 8–5 compared the same two alternatives, referred to in both examples as Plans A and B. Estimated cash flows, presumably based on prices expected to be in effect at zero date, were as follows:

Year	Plan A	Plan B
0	$ 0	−$15,000
1 to 10	−9,200 per year	−6,400 per year

Let us assume that now it is desired to compare these alternatives assuming that the general price level will rise at 6% per year during the 10-year study period. The estimated annual disbursements for Plan A were entirely for labor and labor extras. The comparable disbursements for Plan B included labor and labor extras ($3,300), power ($400), maintenance ($1,100), property taxes and insurance ($300), and extra income taxes ($1,300).

The analyst forecasts that the prospective inflation will cause total annual disbursements to increase 6% per year for each plan. This is the same rate of increase that is expected for the general price level. (In other words, no differential price change is anticipated.) The estimated cash flows in current dollars will be:

Year (n)	Plan A	Plan B
0	$ 0	−$15,000
1	$-9{,}200(1.06)$	$-6{,}400(1.06)$
2	$-9{,}200(1.06)^2$	$-6{,}400(1.06)^2$
3	$-9{,}200(1.06)^3$	$-6{,}400(1.06)^3$
	and so forth	

A multiplier of $(1.06)^n$ has been applied to each original estimate of cash flow made in dollars that had the purchasing power of a dollar at zero date. The result of each multiplication is estimated cash flow in current dollars for the nth year.

These estimated future cash flows expressed in current dollars of their respective dates all need to be expressed in dollars that have the purchasing power of zero date if an economy study is to be made in money units of constant purchasing power. In general, the estimated actual cash flow for the nth year should be multiplied by the ratio of the estimated price level index for zero date to the estimated price level index for the nth year. If the index for zero date is designated as unity and the estimated annual rate of general price inflation is 6%, the multiplier for the nth year is $1/(1.06)^n$. The table of cash flows expressed in dollars of constant purchasing power then becomes:

n	Plan A	Plan B
0	$ 0	−$15,000
1	$-9{,}200\left(\frac{1.06}{1.06}\right) = -\$9{,}200$	$-6{,}400\left(\frac{1.06}{1.06}\right) = -\$6{,}400$
2	$-9{,}200\left(\frac{1.06^2}{1.06^2}\right) = -9{,}200$	$-6{,}400\left(\frac{1.06^2}{1.06^2}\right) = -6{,}400$
	and so forth	

The point illustrated by the foregoing tabulation is that when it is desired to make an analysis in terms of money units of constant purchasing power and when no net differential price change is forecast, it is sufficient to base all estimated cash flows on the prices at zero date. In other words, the conclusions of an economy study should not be influenced by the prospect of inflation or deflation as long as no net differential price change is expected in the sum of the items included in the estimated year-by-year cash flows.

It should be mentioned that no *net* differential price change may be forecast even though it is believed that not all prices will change at the same rate. For instance, it might be forecast that power costs in Plan B will increase at a more rapid rate than 6% a year but that this will be offset by a less rapid rate of increase for property taxes and insurance.

Prospective inflation (or deflation) causes difficulties in economy studies chiefly because of the expectation of differential price change. Later in this chapter we shall make the point that expected differential price change is relevant in an economy study even though *no* change is expected in the general price level.

Our illustration based on Examples 6–1 and 8–5 involved one item that might reasonably have been expected to be a source of differential price change. This was the extra $1,300 for annual income taxes expected with Plan B. The reasons that income taxes often are a source of prospective differential price changes are discussed later in this chapter. Problem 14–29 illustrates one possible reason for expecting a differential price change in income taxes in comparing Plans A and B in Example 8–5.

A Word of Caution

An economy study aims to guide a choice among alternatives. Sometimes another purpose is to estimate the funds that will be needed to finance a project. If the project is not to start at once, or if its completion will extend over a period of rising prices, short-run forecasts of price changes are necessary to determine how much money will be required. This need exists regardless of the presence or absence of prospective differential price change during the years covered by the economy study. For this purpose there is need for forecasts of changes that will take place in the near future in specific prices of the different elements of a project.

The Relationship Between Interest Rates and the Expectation of Inflation

Borrowed money is a common source of differential price change in periods of inflation. In the frequent case where payments of interest and principal are fixed in currency units (such as dollars), inflation can reduce or eliminate the rate of return to lenders if the analysis is made in units of constant purchasing power. This point was illustrated in Examples 14–1 and 14–2. The effect of inflation on borrowers naturally is opposite to the effect on lenders; a cost of money may be very low or even negative when cash flow is converted to units of constant purchasing power. (Some illustrations of this point are given in Chapter 18.)

Our 1964 investor in Example 14–1 did so badly because the considerable inflation that followed 1964 was not anticipated; for many years preceding

1964, prices had risen very slowly. In part, it is lenders' collective *expectation* of inflation that raises the interest rates that must be paid when money is borrowed. The rising rate of inflation during the late 1970s in the United States caused an expectation of further considerable inflation. This expectation naturally increased interest rates. The United States government paid nearly 14% on one 30-year bond issue financed early in 1981.

The Relationship Between High Interest Rates Caused by Expected Inflation and the Choice of a Minimum Attractive Rate of Return

How should such high interest rates influence economy studies that compare engineering alternatives? Generally speaking, this influence should be indirect rather than direct. That is, the influence—if any—should be on i^* rather than on the cash flow series. Estimates of cash flow for engineering economy studies rarely need to include either disbursements associated with borrowing or receipts associated with lending. As we first pointed out in Chapter 1 and continue to emphasize throughout this book, a good rule for decision making is that separable decisions should be made separately. It will be stressed in Chapter 18 that acquisition decisions and financing decisions generally are separable.

At first impression it may appear to be self-evident that when interest rates go up because of prospective inflation there ought to be a corresponding increase in the minimum rate of return that is deemed sufficient to justify a proposed investment in physical property. Nevertheless, the matter is not as simple as it might seem.

In considering how i^* ought to be influenced by the high interest rates associated with prospective inflation, it is helpful to consider two cases separately. The first case is where the total funds for the capital budget are generated within the business enterprise. The second is where consideration is given to supplementing the funds generated internally by funds acquired from sources outside the enterprise, such as long-term borrowing, long-term leasing, or securing new ownership funds. In the first case, interest may need to be considered from the viewpoint of a possible *lender*. In the second, it may need to be considered from the viewpoint of a possible *borrower*.

The first type of case was illustrated in Table 10–1 (page 167). That table referred to circumstances in which all the $90,000 available for the capital budget was generated internally. Proposals for capital expenditures totaled $207,000. Because the $90,000 available would be exhausted by projects that had estimated after-tax rates of return of 15% or more, the minimum attractive rate of return, in effect, was 15%.

Conceivably some or all of the available $90,000 might have been loaned at interest. To have been competitive with the poorest acceptable internal investment, such a loan would have had to yield at least 15% after taxes.

But if the interest rate available on such a loan had been high because of the prospect of inflation, a direct comparison with the rate of return expected on an internal investment would have been misleading. Unless an adverse differential price change is forecast, the 15% (from Project S in Table 10–1) presumably represents an after-tax rate of return in money units of constant purchasing power. The yield from the loan needs to be analyzed using money units of constant purchasing power as illustrated in Example 14–2. The authors believe that it rarely will happen that a loan so analyzed will be competitive when the total capital budget is limited to funds generated internally.

(As pointed out in Chapter 10, part or all of a limited capital budgeting fund may be used for a short-term loan because it is believed that good opportunities for internal investment will be available in the near future. Here it is not the yield on the loan that is relevant. The real competitor for the limited funds is some future unspecified internal investment expected to have a better yield than the marginal acceptable proposal currently available, such as Project S in Table 10–1.)

High interest rates caused by the expectation of inflation rarely should have much weight in the selection of a minimum attractive rate of return in competitive private industry in our second type of case in which funds for the capital budget come in part from outside sources. Examples 14–1 and 14–2 illustrated the point that inflation reduces rates of return received by lenders when analysis is made in units of constant purchasing power. Obviously, the same inflation that will reduce rates of return to lenders will also reduce costs of money to borrowers. Moreover, income taxes that reduce lenders' after-tax rates of return also reduce borrowers' after-tax costs of money whenever interest paid is deductible from taxable income. Therefore, the apparent high interest rates paid by borrowers during periods of inflation may not really be high when all relevant matters are considered. This topic is discussed further in Chapter 18.

Indexing Payments of Interest and Principal on Loans

We saw in Examples 14–1 and 14–2 how inflation could cause lenders to receive rates of return that were very low or even negative when measurement was made in money units of constant purchasing power. After an extended period of inflation, lenders naturally expect that inflation will continue. This expectation causes interest rates to rise. The greater the rate of inflation and the more the uncertainty about future inflation rates, the greater the difficulty in establishing satisfactory relationships between lenders and borrowers.

Some of these difficulties can be reduced by laws that require the "indexing" of payments of principal and interest on loans. In fact, many countries that have experienced rapid inflation have made such laws. The greater the rate of inflation, the more important it becomes to have such legislation. Otherwise, there may be very serious obstacles to any lending of money. Such obstacles interfere with the functioning of an industrial society.

With indexing, the repayment of a loan is intended to have the same purchasing power as the money originally borrowed. Also the interest payments are intended to be, in effect, in units of the original purchasing power. Obviously, these objectives are not easily accomplished. Indexing may be viewed as an imperfect solution of a troublesome problem that does not have a perfect solution.

If the law requires that the amount of certain types of money payment must depend on the ratio of the most recent value of a stipulated price index to the value of the same index at some stipulated earlier date, the index becomes a very important figure. It clearly is desirable that there be widespread confidence in the integrity of the person or persons finally responsible for computing the index. (The authors know of one country in which one individual has had this responsibility for many years, through a number of administrations that had extremely different political orientations.) The index must be issued at frequent intervals. All persons whose actions might depend on a new index figure should have access to that figure at the same moment.

As mentioned earlier in this chapter, costs of living tend to change at different rates for different segments of a population. Because an index of cost of living never can be based on a "market basket" of goods and services that are appropriate for everyone, any such index that influences the size of certain stipulated types of payments is bound to seem unfair to some of the persons affected by it.

Under such indexing the interest rate on a loan transaction does not need to contain an element to compensate the lender for the prospect of receiving payments of principal and interest in currency units of decreased purchasing power. Therefore, stated interest rates tend to be lower with indexing than without it.

As far as the authors of this book are aware, through the year 1980 no political leaders in the United States had proposed laws indexing the payments of principal and interest on loans.

Income Taxation as a Possible Source of Differential Price Change

When prices go up, income taxes may cause differential price change in two ways. One way relates to the calculation of taxable income. The other relates to the determination of the tax rates that apply to that income.

Inflation may influence taxable income in ways that are adverse to taxpayers because, in most instances, the rules for calculating taxable income are the same as the rules for determining profit or loss from the books of account. Generally speaking, conventional accounting does not reflect changes in the purchasing power of the relevant monetary unit. Example 14–3 illustrates how a taxpayer may be penalized in a period of inflation when the allowable depreciation deduction fails to reflect the change in the value of the dollar. Taxpayers also are penalized in periods of inflation when conventional accounting rules are used for calculating so-called gains or losses on disposal of assets; this point is illustrated in Problems 14–10, 14–11, and 14–12.

Inflation may gradually increase applicable income tax rates. The result may be a differential price change in income taxes; that is, income taxes may rise faster than the general price level. This topic is discussed following our comments on Example 14–3.

EXAMPLE 14–3

How Prospective Rates of Return in Periods of Inflation are Reduced by Differential Price Changes in Income Taxes Related to Allowable Depreciation Deductions

Facts of the Case

The examples at the start of Chapter 12 brought out the point that, generally speaking, it is unfavorable to a taxpayer to be required to put off the deduction of an expense that decreases his taxable income. In a period of inflation it is even worse to put off such a deduction unless the size of the deduction can be increased enough to compensate for the rise in price level. (For reasons explained in Chapter 12 the foregoing statements assume a constant or declining applicable tax rate.) We may illustrate the effect of inflation by assuming that the facts of Example 12–2 apply to a period for which inflation is forecast.

In that example an outlay of $110,000 for materials-handling equipment was expected to reduce cash disbursements (other than disbursements for income taxes) by $26,300 a year for 10 years. Zero salvage was estimated at the end of the 10 years. The applicable income tax rate was 49%. Years-digits depreciation was to be used for tax purposes. Example 12–2 showed that under these circumstances a prospective before-tax rate of return of 20.1% was reduced to a prospective after-tax rate of 12.3%.

Now let us change the conditions of this example by assuming annual inflation of 6%. Let us also assume that the annual depreciation deductions from taxable income are based on the $110,000 outlay at zero date with no adjustment for change in price levels. And let us make our analysis in money units of constant purchasing power in the manner illustrated in Examples 14–1 and 14–2. Table 14–4 provides such an analysis. Conventional calculations of present worths (not shown here) will tell us that the prospective rate of return now is 10.8%. This contrasts with the 12.3% figure found in Example 12–2. If no such differential price change had been forecast, the figures in Column H would have been identical with the figures in the final column (E) of Table 12–2; this statement is true for any assumed rate of inflation. The one source of differential price change in Table 14–4 is the change in income taxes. But here there is no change in the expected applicable income tax *rate*, which is 49% throughout the 10 years. The differential price change in income taxes has only one cause. This cause is the basing of the allowable depreciation deduction on a dollar investment that was made when the dollar had more purchasing power.

Consider the numbers on the line for year 5. Because the annual saving that would be $26,300 at zero-year prices is assumed to increase at the same

TABLE 14–4
Estimation of cash flow after income taxes in zero-date dollars, Example 14–3

Year	Cash flow before income taxes in zero-date dollars	Estimated price index assuming 100 for zero date	Estimated before-tax cash flow in current dollars	Write-off of initial outlay for tax purposes	Taxable income from project	Cash flow for income taxes in current dollars	Cash flow after income taxes in current dollars	Cash flow after income taxes in zero-date dollars
			BA/100		(C + D)	−0.49E	(C + F)	100G/B
	A	B	C	D	E	F	G	H
0	−$110,000	100.00	−$110,000				−$110,000	−$110,000
1	+26,300	106.00	+27,878	−$20,000	+$7,878	−$3,860	+24,018	+22,658
2	+26,300	112.36	+29,551	−18,000	+11,551	−5,660	+23,891	+21,263
3	+26,300	119.10	+31,323	−16,000	+15,323	−7,508	+23,815	+19,996
4	+26,300	126.25	+33,204	−14,000	+19,204	−9,410	+23,794	+18,847
5	+26,300	133.82	+35,195	−12,000	+23,195	−11,366	+23,829	+17,807
6	+26,300	141.35	+37,307	−10,000	+27,307	−13,380	+23,927	+16,868
7	+26,300	150.36	+39,545	−8,000	+31,545	−15,457	+24,088	+16,020
8	+26,300	159.38	+41,917	−6,000	+35,917	−17,599	+24,318	+15,258
9	+26,300	168.95	+44,434	−4,000	+40,434	−19,813	+24,621	+14,573
10	+26,300	179.08	+47,098	−2,000	+45,098	−22,098	+25,000	+13,960
	+$153,000		+$257,452	−$110,000	+$257,452	−$126,151	+$131,301	+$67,250

rate as the general price level, it is *multiplied* by 1.3382 to become $35,195 in year 5 (as shown in Column C). If it had been desired to evaluate the proposal *before* income taxes, the $35,195 could then have been *divided* by 1.3382 to show that the fifth year saving is $26,300 when expressed in dollars that have the zero-date purchasing power. The differential price change in income taxes occurs because the $12,000 depreciation deduction from $35,195 is based on zero-date dollars. If the 12,000 zero-date dollars had been converted to 16,058 fifth-year dollars by multiplication by 1.3382 (the ratio of the fifth-year price index to the zero-year price index), the purchasing power of the income tax would not have been increased by inflation. If the deductible depreciation had been adjusted in this way each year to reflect the change in the purchasing power of the dollar, there would have been no differential price change in income taxes. With such an adjustment the rate of return using dollars of constant purchasing power would have been 12.3%, the same as the rate that was calculated in Example 12–2.

Sensitivity to Assumed Inflation Rate and to Stipulated Depreciation Method
Examples 12–2 through 12–6 all dealt with the same prospective before-tax cash flows. Therefore, all had the same prospective before-tax rate of return (20.1%). Moreover, all had the same applicable income tax rate (49%) throughout their entire study periods. The only differences among the examples were in the rules governing the write-off of the $110,000 investment for tax purposes. Because of the differences in these rules, prospective after-tax rates of return in these examples varied from 8.5% to 20.1%.

We can use the data from these examples to examine the extent to which inflation reduces the rate of return with various depreciation methods and with various inflation rates. Table 14–5 gives the results of applying the same type of analysis used in Table 14–4 with inflation rates of 6% and 12%. In all cases rates of return have been calculated after expressing all cash flows in dollars of constant purchasing power.

As might be expected, Table 14–5 shows that the fastest write-off is the most advantageous and the slowest the least advantageous, either with or without inflation. For any particular depreciation method, the reduction in rate of return is slightly less between inflation rates of 6% and 12% than between rates of 0% and 6%.

It should be emphasized that the rates of return in Table 14–5 are based on the assumption of a constant applicable tax rate (in this case, 49%) throughout the study period. As brought out in Chapter 12, the advantage of more rapid depreciation for tax purposes may be reduced or eliminated when applicable tax rates are expected to increase during a study period.

Tables 14–4 and 14–5, like Tables 12–2 through 12–6, assume the end-of-year convention. If all of these analyses had been made using continuous compounding of interest and the uniform-flow convention (discussed in Appendix A), the prospective rates of return all would have been a bit higher. But the general conclusions would have been unchanged about the relative

TABLE 14–5
Effect of differential price change in income taxes caused by failure to adjust depreciation deductions for changes in purchasing power of the monetary unit

Investment of $110,000 from	*Rate of return with depreciation deduction adjusted for inflation*	*Rate of return with unadjusted depreciation deduction (6% inflation)*	*Rate of return with unadjusted depreciation deduction (12% inflation)*
Example 12–2 (years-digits depreciation; 10-year life)	12.3%	10.8%	9.7%
Example 12–3 (straight-line depreciation; 10-year life)	11.2%	9.6%	8.4%
Example 12–4 (straight-line depreciation; 25-year life)	8.5%	6.9%	6.0%
Example 12–5 (straight-line depreciation; 5-year life)	13.2%	11.7%	10.6%
Example 12–6 (immediate write-off)	20.1%	20.1%	20.1%

advantages of different depreciation methods for tax purposes and about the adverse effect of inflation on rate of return when depreciation deductions for income tax purposes are not allowed to reflect changes in the value of the monetary unit.

Consideration in Economy Studies of Prospective Differential Price Change in Income Taxes Caused by the Required Use of Conventional Depreciation Deductions During Periods of Inflation

Earlier in this chapter we pointed out that the prospect of inflation (or deflation) does not necessarily call for economy studies that make use of estimates of price level change. Under certain conditions it is satisfactory to make all price estimates using the prices expected at zero date. Such estimates meet the reasonable objective of making an analysis using monetary units of constant purchasing power whenever no differential price change is expected.

Example 14–3 brought out a general point that is applicable whenever taxes are levied on income caused by investment in depreciable assets during periods of inflation and the depreciation deductions from taxable income are not adjusted for change in purchasing power of the monetary unit. This point is that a differential price change in income taxes will take place that will be adverse to the taxpayer. In the United States and in many other jurisdictions that levy income taxes, no such adjustments of depreciation deductions have been allowed.

A comparison of the column totals in Tables 12–2 and 14–4 illustrates the way in which taxpayers are penalized. These tables deal with identical before-tax cash flows and identical depreciation methods. In both tables a tax rate of 49% is applied to the taxable income in each of the 10 years. In both tables the total of Column A is $153,000; this is the net cash flow before income taxes expressed in zero-date dollars. In Table 12–2, with no inflation assumed, the 49% tax rate takes away 49% of this $153,000; the taxpayer is left with $78,030, which is 51%. In contrast, in Table 14–4 the 49% tax rate takes away $85,750 expressed in zero-date dollars; this is 56% of the before-tax cash flow. The taxpayer is left with $67,250 expressed in zero-date dollars as shown in the total of Column H; this is only 44% of the before-tax purchasing power generated by the investment. In effect, a distortion of the calculation of taxable income is responsible for an actual average tax rate (56%) that is considerably higher than the stated rate (49%).

It is evident that during periods of inflation a large proportion of economy studies can be expected to involve some prospective differential price change in income taxes that will be adverse to the taxpayer. In principle, any differential price change that is expected ought to be included in an economy study. But as we have mentioned in various places in this book, the analysis in an economy study is an intermediate step in reaching a decision or a recommendation for a decision. It is a good *decision* that should be the objective; good analysis is important because it is reasonable to assume that good analysis will lead to better decisions than would be made with poor analysis or with no analysis at all.

The problem of whether or not the type of complication illustrated in Example 14–3 should be introduced into economy studies made in periods of inflation is similar to the common problem of whether to make economy studies before income taxes or after income taxes. An economy study before income taxes is always simpler. As we noted in Chapter 12, such taxes may be introduced—after a fashion—into before-tax studies by an appropriate increase in the minimum attractive rate of return, i^*. Is this adjustment all that is needed to reach the same decisions as those that would result from a more specific after-tax analysis? Sometimes the answer is "Yes"; sometimes it is "No." To identify those cases where the answer is "No," an analyst needs a fairly sophisticated understanding of the rules of income taxation.

Similarly, in periods of inflation economy studies will be simpler if they avoid the detailed type of analysis involving the prospective differential price change in income taxes that is illustrated in Example 14–3. Conceivably, matters could be simplified by making economy studies with a somewhat higher after-tax i^* than the figure really deemed to be the *minimum* attractive rate of return. Depreciation deductions illustrated in Examples 12–2 and 12–5 are representative of those permitted under many income tax laws. We may note from Table 14–5 that for these examples the failure to adjust depreciation deductions for price level change reduces rate of return by about 12% with 6% inflation and by about 20% with 12% inflation. But it is not safe to assume that a simple modification of i^* is appropriate in all cases.

The special type of case illustrated by Example 12–6 is of particular interest. In that example the $110,000 outlay could be written off at once for income-tax purposes. The before-tax and after-tax rates of return both were 20.1%; the 49% income tax rate did not reduce the investor's rate of return at all. Table 14–5 tells us that this 20.1% will not be reduced even when depreciation deductions are not modified to reflect changes in purchasing power of the dollar; when the deduction is made at once, the purchasing power has not had time to change.

Several of the problems at the end of this chapter illustrate the point brought out in Example 14–3. But this special aspect of the interrelationship between inflation and income taxes is not pursued throughout the remainder of this book. Generally speaking, the examples and problems in the remaining chapters do not deal with the special difficulties in economy studies that are introduced by the prospect of inflation.

Differential Price Changes in Income Tax Rates During Periods of Inflation

To see how *price level* changes can influence tax *rates*, it is helpful to look at a tax rate schedule such as Table 12–8 (page 244). This table gives the United States individual income tax rates for certain single taxpayers for the year 1980. Like the other tax rate schedules for individuals in the United States at that time, this had many brackets, with rates starting at 14% and going up to 70%.

Consider a taxpayer subject to this schedule who has an adjusted gross income of $18,000. Table 12–8 shows that the incremental rate that applies to an extra dollar of income is 30%. Now assume that the price level increases by 50% and that the taxpayer's income (in dollars of reduced purchasing power) also increases by 50%. If the tax rate schedule remains unchanged, Table 12–8 shows that the incremental tax rate will be 39%. If price levels should double and the taxpayer's dollar income should also double, the incremental rate would go up to 49%.

If an economy study is being made for this taxpayer during a period of inflation, and if it is forecast that inflation will continue and that the tax rate schedule will remain unchanged, the prospective adverse differential price change in cash flow for income taxes needs to be recognized in the study.

Indexing Income Taxes

The Advisory Commission on Intergovernmental Relations (ACIR) in the United States is a permanent national bipartisan body, created by Congress in 1959 to recommend improvements in the American federal system. In 1980 the ACIR issued a report that dealt with the relationship between inflation and personal income taxes.* Two quotations from this report are:

> When the Advisory Commission on Intergovernmental Relations first studied the effect of inflation on income tax burdens in 1976, little consideration had been given to the matter in this country because the United States historically had not suffered from prolonged high rates of inflation. The Commission, however, found what it felt was a serious and growing problem. Namely, inflation automatically interacts with the progressive income tax systems of the federal government and of most states to increase personal income tax burdens at a faster rate than inflation. This not only makes it difficult for taxpayers to keep up with inflation, but it allows the government to receive windfall revenue gains without the Congress or state legislature overtly voting a tax increase. For several reasons, including prospects for continued rapid inflation, the Commission recommended that the federal and state governments index their personal income taxes for inflation—i.e., annually adjust the fixed-dollar features of the tax code, such as the personal exemptions, standard deduction, and income brackets, by the rate of inflation—to prevent the automatic, unlegislated "inflation tax" increases that would otherwise result.
>
> Indexing the individual income tax would promote the goal of tax equity in two ways. By neutralizing the effects of inflation on tax burdens, it preserves the tax burden distribution as approved by Congress or the state legislature so that legislative intent and existing equity are maintained despite inflation. Second, indexing will, in effect, move state and federal income taxes toward true equity—i.e., based on ability to pay—because it shifts the tax base toward real income or real purchasing power. The latter is a better measure of ability to pay than money income, which becomes bloated by inflation with no increase in purchasing power.

The ACIR report pointed out that other countries that had been indexing their personal income taxes for several years included Australia, Brazil, Canada, Chile, Denmark, and The Netherlands. Several of the states of the United States also had adopted laws for automatic indexing of their personal income taxes.

Where indexing is not adopted, tax rate schedules on individual income and related matters such as personal exemptions and standard deductions (where relevant) have been changed from time to time during extended periods of inflation. This has been true in the United States and elsewhere. Often these changes are intended to offset the effect of inflation on income taxes, at least in part. Although such changes are generally described as tax

*_The Inflation Tax: The Case for Indexing Federal and State Income Taxes_, Advisory Commission on Intergovernmental Relations (Washington, D.C.: U.S. Government Printing Office, 1980).

reductions, they may fall short of complete adjustment for the "inflation tax" mentioned in the ACIR report. Problems 14–33 and 14–34 deal with this subject.

The 1980 federal tax rate schedule on incomes of corporations in the United States was given in Table 12–7 (page 242). This schedule contains five brackets, with tax rates of 17%, 20%, 30%, 40%, and 46%. All taxable income over $100,000 was taxed at 46%. It follows that inflation did not change the incremental tax rate for those numerous corporations that continued to have taxable incomes over $100,000. In contrast, inflation could make substantial changes in the incremental tax rate for corporations with lower incomes, particularly those in the 20% and 30% tax brackets. Problems 14–13 and 14–14 illustrate this point. As far as the authors have been able to discover, no serious proposals have been made for *automatic* indexing of the brackets of the corporate income tax in the United States.

How is an analyst to decide whether a particular economy study ought to include an estimate of a differential price change in the applicable income tax rate that will be caused by inflation? Among other matters, the answer depends on whether the economy study is being made for a corporation or for an individual. It also depends on whether the relevant income tax is expected to be indexed for inflation, either automatically, or, in effect, through frequent changes in the tax law. In general, no such change in applicable tax rate to be caused by inflation will need to be included where the income tax law is indexed. Moreover, inflation will not cause a change in the applicable tax rate for taxpayers expecting to continue to be in the highest tax bracket. For one reason or another there are many economy studies in periods of inflation in which increase in applicable income tax rates should not be viewed as a source of a differential price change.

In this connection it should be noted that there may be prospective changes in applicable income tax rates that are unrelated to a forecast of inflation and that need to be recognized in economy studies. This point was mentioned in Chapter 12 and is illustrated in several of the problems at the end of that chapter.

Differential Price Changes in the Costs of Goods and Services

A long-sustained differential price change in any particular group of prices seems to be the exception rather than the rule. This is brought out in Table 14–6. The 400 or so prices that made up the CPI in the United States from 1935 to 1977 were divided by the government into eight groups as shown in the table. For purpose of comparison of the different rates of growth, the years from 1935 to 1977 are here divided into four periods, each of which had a growth of approximately 45% in the index of all prices. The table shows the percentage growth in each component index during each period. The reader may observe that not one of the eight component indexes grew faster than the overall CPI during all four periods. Neither did any component index grow more slowly than the CPI during all the periods.

TABLE 14–6
Growth of the different groups of prices that compose the United States Consumer Price Index during four periods in which the index of all items grew approximately 45%

	Percentage of increase from			
	1935 to 1946	*1946 to 1958*	*1958 to 1972*	*1972 to 1977*
All items	*42*	*48*	*45*	*45*
Food	**59**	**52**	*40*	**56**
Housing	*23*	*45*	**47**	**47**
Apparel and upkeep	**65**	*30*	*40*	*26*
Transportation	*18*	**71**	*39*	**48**
Medical care	*23*	**65**	**81**	**53**
Personal care	**60**	*47*	*38*	*43*
Reading and recreation	**54**	*30*	**46**	*29*
Other goods and services	*32*	*44*	**49**	*27*

Percentages were calculated from index values given in Table 117, "The Consumer Price Index and Major Groups, 1935–1977," *Handbook of Labor Statistics 1978* (Washington, D.C.: U.S. Government Printing Office), p. 398. Where prices in a group grew faster than the CPI, figures are in **boldface**; where they grew more slowly, figures are in *italics*.

Generally speaking, there may be good reasons for an analyst to make a specific forecast of a differential price change in certain goods or services in the near future. Also, generally speaking, analysts ought to be cautious in forecasting that such a change will continue indefinitely at the same rate.

Past experience indicates that some differential price changes always seem to occur. But in many cases in a period of inflation there may be no valid basis for forecasting *which* specific prices will rise faster than the general price level and *which* will rise more slowly. It is only those differential price changes that are forecast that enter into an economy study.

In a period of inflation a forecast that a certain price will remain constant or will increase more slowly than the general price level may be reasonable because the price is fixed by contract over a period of years. The bond investment in Example 14–1 illustrated this. Government price controls may also put constraints on the increase of certain prices. When artificial price controls cause certain prices to increase more slowly than the general price level for a while, it is likely to happen that as soon as the controls are removed these prices will rise more rapidly than prices in general.

There are two classes of prices that for many years have tended to rise relative to the general price level. One is the price of land. The other is the price of labor, particularly in industrialized countries. These two types of differential price change have dissimilar causes. As the amount of land is fixed and as the number of people has increased, land has become a scarcer and scarcer resource relative to the total population. On the other hand, increased productivity of labor over the years has caused its price to go up. This increased productivity started with the industrial revolution some 200 years or so ago; it has been a consequence of technological progress and the increased use of capital goods.

It does not follow that any particular economy study ought to contain a specific forecast of a differential price change for land or labor. Each economy study involves estimates of the prices of certain particular relevant items; the future prices of these items will not necessarily correspond to an expected general trend. For instance, it may be estimated that a certain parcel of land is greatly overpriced even though one alternative in an economy study requires the purchase of this land. No favorable differential price change can be expected for this particular land even though land prices in general are expected to go up faster than the general price level. Moreover, a differential price change that will not affect cash flow for a number of years has a relatively low present value; the price at which it is believed land will be sold at some distant date may have little impact on the conclusions of a study. (With an i^* of 12%, a dollar 40 years hence has a present worth of one cent.) An expected upward differential price change in the price of labor may be so slow that it will not have much influence on the conclusions of a study. Often the sensible thing to do about such possible differential price changes is to view them as irreducibles to be taken into account in the final choice among alternatives rather than as something calling for sensitivity studies. Some of the matters mentioned in this paragraph are illustrated in Problems 14–19 through 14–23.

Certain groups of prices may increase less rapidly than the general level of prices for a number of years and then suddenly start to increase much more rapidly. This has been true of many prices associated with the production and sale of energy. For a long period before the early 1970s, these prices had a negative differential price change, caused in large measure by steady technological progress. Then matters changed (in part because monopoly pricing of petroleum was made possible by worldwide growth of demand) so that these prices had a large positive differential price change. Example 14–4 uses a forecast of an upward differential price change in the cost of electric energy to illustrate sensitivity analysis of differential price change.

EXAMPLE 14–4

Sensitivity of Economic Choice of Size of a Certain Electrical Conductor to a Prospective Differential Price Change in the Price of Electricity

Facts of the Case

Example 13–2 and the subsequent discussion deals with a before-tax economic analysis of a choice among several possible wire sizes for an electrical conductor. The comparison of the different sizes was made using equivalent uniform annual costs. There were two general classes of cost. One class, the investment charges, was directly proportional to the area of wire cross section. The other, the cost of lost energy, was proportional to the electrical resistance of the wire and therefore was inversely proportional to the area of wire cross section.

The electrical conductor to be selected was expected to carry 50 amperes for 4,200 hours per year. The cost of wire was $1.75/lb and electrical energy was to be purchased at 3.4 cents/kw-hr. The life of the wire was estimated to be 25 years with zero net terminal salvage value. The before tax i^* was 14%. Annual property taxes were 1.75% of first cost. Under these conditions the most economic wire size was shown to be Gage No. 2 (American Wire Gage, B & S).

In a period in which the price of purchased electric energy is expected to rise faster than the general price level, a decision on wire size ought to give weight to the estimated amount and duration of the prospective differential price change. The present example applies sensitivity analysis to the data of Example 13–2, making various assumptions about future differential price changes.

Initial Analysis

Table 13–2 (page 284) showed a comparison of the before-tax annual costs for 1,000 feet of copper wire. The monetary figures in that table have been reorganized in Table 14–7. The reader who compares the two tables will note that the investment charges taken from Table 13–2 are now broken down into capital recovery costs and property taxes.

For reasons explained earlier in this chapter, the prospect of a price level increase with no differential price change in any of the items estimated for an economy study should not alter the conclusions of the study. This assumes the usual case in which it is deemed appropriate to make an analysis in money units of constant purchasing power. It follows that a sensitivity analysis of the effect of a possible differential price change will give correct conclusions when

TABLE 14–7
Comparison of annual costs before income taxes of various wire sizes for an electrical conductor
(Data of Table 13–2)

A. Size of wire	00	0	1	2	3
B. Investment in wire	$705.25	$558.25	$442.75	$351.75	$278.25
C. Capital recovery cost (14.55%)	102.61	81.23	64.42	51.18	40.49
D. Property taxes (1.75%)	12.34	9.77	7.75	6.16	4.87
E. Cost of lost energy @ 3.4¢/kw-hr	28.39	35.70	44.98	56.78	71.77
F. Total annual cost influenced by wire size	$143.34	$126.70	$117.15	$114.12	$117.13

all cash flows are stated in monetary units that have the same purchasing power expected for a monetary unit at zero date.

In the present example, some upward differential price change is expected in the price of purchased electric energy. Table 14–8 illustrates a possible first step in examining the relationship between the *decision* and the extent of this upward differential price change.

This table shows the annual costs (in dollars having the zero-date purchasing power) as they would have been with different prices assumed for electric energy throughout the entire 25-year study period. In addition to the actual initial unit price of 3.4¢, the assumed unit prices are 1.5, 2, 2.5, and 3 times this figure. On each line of Table 14–8 the lowest annual cost is shown in italics. It is evident that size 1 would have been the economic choice for a price up to 6.8¢ and that size 0 should then have been chosen up to something more than 10.2¢.

Sensitivity of the Choice to the Prospective Amount and Timing of Future Differential Price Change

In effect, Table 14–8 told us that not much prospective differential price change is needed in the unit price of purchased electric energy to justify choosing size 1 instead of size 2. But a further increase in conductor cross section will not be justified unless it is believed that the price of this energy will increase more than twice as fast as the increase in the general price level and unless it is believed that this increase will take place fairly soon. A sensitivity analysis can give some guidance on how soon this differential price change needs to be expected to occur and how far it needs to be expected to go for the selection of size 0 to be justified.

As an example, let us assume that the unit price per kilowatt-hour (stated in monetary units that have the purchasing power of zero date) will increase from 3.4¢ to 10.2¢ by uniform annual amounts throughout the 25-year study period. The annual cost of lost energy for each wire size will finally rise to a figure 3 times the one shown in line E of Table 14–7. The *average* figure will be 2 times the corresponding figure from line E. But with conversion using an i of 14% each equivalent uniform annual figure will be considerably less than this average.

TABLE 14–8
Influence of different unit costs per kilowatt-hour of electrical energy on annual costs of the electrical conductor of Example 13–2

A. Size of wire	00	0	1	2	3
B. Annual cost @ 3.4¢	$143.34	$126.70	$117.15	*$114.12*	$117.13
C. Annual cost @ 5.1¢	157.54	144.55	*139.64*	142.51	153.02
D. Annual cost @ 6.8¢	171.73	162.40	*162.13*	170.90	188.90
E. Annual cost @ 8.5¢	185.92	*180.25*	184.62	199.29	224.78
F. Annual cost @ 10.2¢	200.12	*198.10*	207.11	227.68	260.67

Using the appropriate gradient factors, we can calculate that the equivalent uniform annual cost of lost energy over the 25-year life would be $54.03, $68.07, and $85.93, respectively, for wire sizes 0, 1, and 2. (The calculation to obtain these figures is asked for in Problem 14–24.) If these numbers are substituted in line E of Table 14–7, the total annual costs in line F will become $145.03 for size 0, $140.24 for size 1, and $143.27 for size 2. It follows that the prospect of such a gradual differential price change in the cost of electric energy is not sufficient to justify the extra investment needed to install wire size 0 rather than size 1.

Table 14–8 already indicated that if the unit price of electric energy is expected to triple immediately and to continue at that higher figure relative to the general price level, size 0 will be the economic choice. The solution of Problem 14–25 shows that the annual costs of sizes 0 and 1 will be nearly equal with a uniform differential price change that triples the relative price of the purchased electricity in 10 years with no subsequent change relative to the general price level. Problem 14–26 deals with the circumstances under which size 0 should be selected when it is believed that the unit price will go up only to 2.5 times its present figure.

In general, our sensitivity studies indicate that size 1 should be selected rather than size 0 unless it is believed that the price of this electricity will increase to at least 2.5 to 3 times its present figure measured in currency units of constant purchasing power, and unless it is believed that this substantial increase will occur fairly soon, say within 5 or 10 years.

Some Comments on Topics Related to Example 14–4

The statement of the facts in Example 14–4 gave no forecast of the amount or timing of any change in the general price level. In this example the prospective differential price change in electric energy was the relevant matter. But the reader should note what is meant when we estimate that the price of this electricity will triple relative to the general price level in, say, 25 years. If the general level of prices should also triple in that period, the electric energy that sold for 3.4¢ at zero date will cost 30.6¢ in the currency units of 25 years later. Our economy study expresses this prospective 30.6¢ as 10.2¢ only because we aim to have all money figures commensurable by stating them in units of constant purchasing power.

Example 14–4 was derived from Example 13–2. In that example the facts were simplified in various ways to illustrate Kelvin's Law, which possibly is the earliest decision rule in engineering economy that is based on a mathematical model. One type of simplification was the omission of certain electrical matters that sometimes are important in economic choice of conductor size. This type of simplification has been continued in Example 14–4.

Another type of simplification in Example 13–2 was the stipulation of a minimum attractive rate of return *before* income taxes. This avoided the need to estimate prospective differences in disbursements for income taxes. (There were *no* income taxes in Lord Kelvin's day.) Problems 13–11 and 13–12 added certain income tax considerations to the facts of Example 13–2. We have continued to use the stipulated before-tax i^* in Example 14–4 in order to concentrate attention on a single source of differential price change, namely, the price of energy. We have noted earlier in this chapter that income taxes are also a possible source of differential price change; this point is illustrated in Problems 14–27 and 14–28 using the data of Example 14–4.

Our sensitivity studies in Example 14–4 assumed that the relative price of the electric energy would increase according to an arithmetic progression rather than according to a geometric progression. This assumption simplified the calculations a bit. It may be observed that there is no particular assumption about future differential price change that is correct in principle. Given an estimated upper limit on the relative price increase (such as an estimate that the energy price ultimately will triple relative to the general level of prices) and an estimated date for reaching that limit (such as 10 years from zero date), the assumed prices between zero date and the limiting date will be somewhat higher under the assumption of arithmetic growth.

The reader should note that a sensitivity analysis of a possible differential price change is appropriate whether or not general inflation (or, for that matter, deflation) is anticipated. Thus, our analysis in Example 14–4 was not contingent on the prospect of general inflation.

In economy studies where both general inflation and differential price change are forecast, analysts may conceivably be dealing with three different geometric growth rates. One, of course, is compound interest. The other two possibilities relate to general inflation and to differential price change; it may be forecast that one or both of these will occur at a geometric rate throughout the entire economy study period. An analyst may be tempted to combine these geometric rates for purposes of mathematical analysis. The authors advise against any such combination. It should always be kept in mind that the purpose of an economy study is to reach a *decision* among alternatives. Generally speaking, decisions are likely to be made more intelligently if the implications of any assumptions about future price changes are not concealed by the method of analysis that has been used.

An Overview of the Discussion of Inflation in This Book

Inflation is a many-sided subject. It is treated here chiefly with reference to its impact on engineering economy. We have not discussed inflation's numerous adverse consequences to society (even though we have explained its adverse consequences to certain types of investors). Neither have we attempted to introduce the complex and controversial topic of what national governments ought to do to prevent inflation or to reduce inflation once it has started.

In relation to the impact on economy studies of prospective inflation (or deflation) or of prospective differential price changes, we have, in effect, considered four types of cases, namely:

1. The case in which there is no reason to forecast any differential price change even though inflation (or deflation) is anticipated
2. The case in which some differential price change is expected in one or more elements of future cash flow and this change will influence the actual money amounts of this cash flow
3. The special case in which a differential price change is expected to occur because an element of future cash flow is fixed in money amount whereas the purchasing power of the money unit is expected to change
4. The special case in which a differential price change in income taxes will accompany inflation because the depreciation deduction from taxable income will not be adjusted for the change in purchasing power of the relevant monetary unit.

In our first type of case no special problems for economy studies are caused by the prospect of either inflation or deflation. If we start with the reasonable premise that it is desired to make an analysis using an assumed monetary unit that has a constant purchasing power, it is sufficient to make all estimates of future cash flow using prices that are in effect at zero date. In other words, no modification of the conventional economy study is required because of the prospect of either inflation or deflation.

The second type of case was illustrated in Example 14–4. The reader will have noted that, just as in the first type, no estimate is needed of the prospective rate of change in the general price level. It is the expected differential price change that is the important matter. In most instances a good approach to this type of case is to make a sensitivity analysis using different assumptions about the amount and timing of the future differential price changes.

The third type of case, in which the actual money amount of some set of receipts or disbursements is fixed regardless of changes in purchasing power of the monetary unit, was illustrated in Example 14–2. Here it is the amount and timing of the change in the *general* price level that must be estimated. A suggested approach, used in that example, is to make a sensitivity study with two or more different assumptions about the change in the general price level.

Chapter 18 of this book deals with the influence on economy studies of sources of investment funds. In that chapter we examine this third type of case from the point of view of a prospective borrower.

The fourth type of case was illustrated in Example 14–3. Just as in the third type, the size of the differential price change depends on the amount and timing of the change in the general price level. Again, a sensitivity study is appropriate.

The Concept of the Break-Even Point

Often we have a choice between two alternatives where one of them may be more economical under one set of conditions and the other may be more economical under another set of conditions. By altering the value of some one of the variables in the situation, holding all of the other points of difference between the two alternatives constant, it is possible to find a value for the variable that makes the two alternatives equally economical. This value may be described as the break-even point.

It frequently happens that a knowledge of the approximate value of the break-even point for some variable of design in the comparison of two alternatives is a considerable help in preliminary engineering studies and designs.

The term "break-even point" is also used in management literature to describe the percentage of capacity operation of a manufacturing plant at which income will just cover expenses.

The literature of engineering economy contains many formulas to determine break-even points, although such formulas are not necessarily described by this name.

Break-even-point formulas deal with such matters as the investment justified by a prospective cost saving, annual hours of operation necessary before a proposed extra investment is profitable, the period of time in which a proposed investment will "pay for itself" (i.e., the life to break even). In general, although such formulas may appear to be complicated, the only mathematics involved in their preparation is elementary algebra. The formulas are merely expressions of cost situations in symbols rather than in figures; the apparent complexity is the result of a large number of symbols.

Break-even formulas do not really save much time as compared to direct calculations for the break-even point. Moreover, the mistake is often made of using such formulas as substitutes for direct comparisons of the costs of specific alternatives under specific circumstances. When they are so used—a choice between alternatives being based on the relation between a calculated break-even point and the hours per year (or life or investment or whatever the break-even point has been calculated for) actually expected—the analyst is less apt to give intelligent weight to irreducibles than if there had been a direct comparison of alternative costs. Just as in the minimum-cost-point situations described in Chapter 13, the use of the formulas as a time-saver may be at the sacrifice of desirable information. Where a choice is to be made between specific alternatives, a direct comparison of the expected costs for each is likely to be more illuminating than any break-even-point calculation.

The break-even-point calculation may be particularly useful in the situation where a decision is very sensitive to a certain variable. If the break-even point for that variable can be calculated, it may be possible to estimate on which side of the break-even point the operations may fall, even though there may be considerable uncertainty regarding the exact value of the variable. Even in this use, however, it is desirable to investigate the range of values of the variable that would permit that alternative to be attractive, and to estimate the consequences of its occurring outside that range. Several of the problems at the end of this chapter illustrate the calculation of different types of break-even points.

It was brought out in Chapter 10 that the choice of a value of $i^\star$ is not necessarily a simple, straightforward matter, subject to no legitimate difference of opinion. Any calculation of an unknown rate of return may be viewed as a calculation of a break-even point. Where two alternatives are being compared at a stipulated i^*, and where the monetary calculation (by equivalent uniform annual cost or by net present worth) favors one of them and the irreducibles favor the other, a rate of return calculation may be thought of as a sensitivity test that can be useful to the decision maker.

Some General Comments Regarding the Sensitivity Point of View

In any particular economy study, there will usually be certain estimates for which a moderate change will have a relatively small influence on the conclusions of the study, and there will be other estimates for which a moderate change will have a relatively large influence. When an analyst is aware of this relationship, maximum effort can be put on the estimates that are of the greatest importance. It follows that it is helpful to have an awareness of sensitivity in the initial stages of any economy study.

In the final or decision-making stage of a study, the sensitivity viewpoint continues to be desirable. The first two chapters of this book stressed the importance of defining alternatives, estimating their consequences, expressing the consequences in money terms, and analyzing the monetary figures with reference to a chosen decision criterion. But it was pointed out that such an analysis does not necessarily settle the problem of choice. Finally, someone must make a selection among the alternatives giving due consideration to matters that for one reason or another were left out of the formal economic analysis. One aspect of the decision maker's assignment is to recognize that "What's to come is still unsure" (as pointed out in the Shakespearean quotation that started this chapter). Sensitivity analysis may help in making a sound decision. Various types of calculation using the mathematics of probability may also be useful in contemplating an unsure future; this topic is discussed in Chapter 15.

At the capital budgeting level in a business enterprise, it is possible to establish certain formal secondary decision criteria based on the concept of sensitivity; such criteria will be discussed briefly in Chapter 21.

In Chapter 13 we made certain comments about the use of mathematical models for making or influencing economic decisions. It was pointed out that when a general class of economic decisions is described by a mathematical formula, it often is necessary to make certain omissions and simplifications. In making new mathematical formulations, it is desirable to restrict omissions and simplifications so that decisions will be relatively insensitive to the matters omitted or simplified. Moreover, in deciding whether or not to apply an existing mathematical model to a particular decision, it is a good idea to evaluate the model in part from the viewpoint of sensitivity.

PROBLEMS

14–1. On July 1, 1960, Ambrose Ames, a resident of the United States, paid $9,600 for a 10-year 4% $10,000 bond issued by a city in his home state. He collected $400 interest each year and the $10,000 principal on July 1, 1970.

Neither state nor federal income tax was payable on his annual interest. However, he paid a $100 tax on his so-called capital gain when the bond matured in 1970. If no consideration is given either to inflation or to his capital gains tax, what seems to be his rate of return? With an analysis using dollars of constant purchasing power along the lines of Example 14–1, what was his rate of return after taxes? Use the appropriate CPI values from Table 14–1. (As in Example 14–1, simplify matters by making calculations as if interest had been payable annually rather than semiannually.) (*Ans.* 4.5%; 1.8%)

14–2. In Example 14–2 it is stated that "similar analysis assuming 8% annual inflation indicates that the rate of return will be only 0.4%." Make the necessary calculations to verify this 0.4%.

14–3. In Example 14–2 it is stated that "at 4% annual inflation the rate of return is 4.2%." Make the necessary calculations to verify this 4.2%.

14–4. Table 14–5 gives 8.4% as the prospective after-tax rate of return for the $110,000 investment in Example 12–3 assuming annual inflation of 12%. This figure was based on an analysis in money units of constant purchasing power. It assumes only one type of differential price change, the change caused by the use of a depreciation deduction from taxable income that is not adjusted for changes in purchasing power of the dollar. Make the necessary calculations to verify this 8.4% figure.

14–5. Make the necessary calculations to verify the 6.9% figure given in Table 14–5 as the prospective rate of return in Example 12–4 assuming 6% annual inflation.

14–6. Make the necessary calculations to verify the 10.6% figure given in Table 14–5 as the prospective rate of return in Example 12–5 assuming 12% annual inflation.

14–7. Assume that the economy study described in Example 9–1 (page 149) is made at a time when a rapid upward differential price change is expected in road user costs. No such change is anticipated either in residual values or in maintenance costs. Discuss the effect of this prospective differential price change on the choice between locations J and K. Make your analysis along the lines indicated in Example 14–4.

14–8. Assume that the economy study asked for in Problem 13–7 (page 304) takes place at a time when there is a rapid upward differential change in the value of electric energy such as would be produced under Plan B. No such change is anticipated in other items. Discuss the effect of this prospective differential price change on the choice among the alternative plans. Make your analysis along the lines indicated in Example 14–4.

14–9. In Chapters 1 and 2 there is a true story about the need to expand a "total energy system." First there was a hunch decision that turned out to be unsound. Subsequently there was an economy study to examine all of the promising alternatives and to choose among them on grounds of long-run economy. In Chapter 2 it was stated that these events took place in the United

States before the 1970s; the economy study was not influenced either by the prospect of rapid inflation or by the prospect of an upward differential price change in certain items associated with the production of energy. Suppose the economy study had been made at a later date when both rapid inflation and the differential price change were expected. In what ways, if any, do you think the economy study ought to have been influenced by these expectations?

14–10. Certain unimproved land was purchased for $100,000. It was held for 10 years and then sold for $320,000. The sales commission was $20,000 leaving a net $300,000. During the 10 years the land was rented for just enough to pay property taxes. Price levels doubled during the 10 years. When the land was sold the $200,000 so-called profit was taxed as ordinary income at a 45% rate.

(a) If the change in the purchasing power of the dollar is not recognized, what seems to be the rate of return before income taxes?

(b) If the change in the purchasing power of the dollar is not recognized, what seems to be the rate of return after taxes?

(c) If an analysis is made in dollars of constant purchasing power, what is the rate of return after income taxes?

14–11. Solve Problem 14–10 making only one change in the data. Assume that the so-called $200,000 profit on the sale is taxed at only 15% on the grounds that it is a "long-term capital gain."

14–12. Solve Problems 14–10(c) and 14–11(c) on the assumption that the taxable gain on the sale is deemed to be $100,000 because the cost of the land is measured in dollars having the purchasing power of a dollar on the date of the sale. (That is, $100,000 at zero date had the same purchasing power as $200,000 10 years later.)

14–13. In Example 9–1 (page 149), residual values were estimated as $300,000 for location J and $550,000 for location K, both at the end of the 20-year analysis period. The **B/C** ratio of J as compared with continuing the present location at H was computed as 1.94; the incremental **B/C** ratio of K as compared to J was 0.57. To test the sensitivity of these **B/C** ratios to the estimated residual values, recompute the two ratios (a) increasing each residual value by 50%, and (b) decreasing each residual value by 50%. (*Ans.* (a) 2.17; 0.61; (b) 1.75; 0.54)

14–14. Tanks to hold a certain chemical are now being made of Material A. The first cost of a tank is $35,000 and the life is 6 years. When a tank is 3 years old it must be relined at a cost of $12,000. It has been suggested that it might be preferable to make the tanks from Material B. Accelerated tests give the estimate that the life using Material B will be 15 years and that no relining will be needed. The first cost of the tanks using Material B is not yet known. If the minimum attractive rate of return is 25% before income taxes, what is the greatest amount that it would be justifiable to spend for a tank constructed of Material B? Assume zero salvage value for tanks made of either material.

14–15. Wilma Wood can buy a 14% $10,000 public utility bond at par. She expects an applicable income tax rate of 25% and an annual inflation rate of 5%. Use the method illustrated in Example 14–2 to calculate her prospective after-tax rate of return in zero-date dollars (a) if the bond will mature in 10 years, and (b) if it will mature in 20 years.

14–16. A hotel owner in a developing country depends on tourism for most of his business, and in order to maintain a high occupancy rate, advertises "Pure Drinking Water" on all of his bill boards, letterheads, and business cards. At present he is providing bottled water for his guests at a cost of #1 per gallon, but occasionally some guest drinks the tap water and becomes ill. (The symbol # represents the *adler*, the currency unit of the country.) Because he believes that every such illness is bad for his business, he is seeking an economical means of providing pure water for all guest consumption. There is an ample supply of water available in the town through a reasonably reliable pipe system, and at very low rates. That system is now being used for all sanitary and guest purposes except for drinking.

A visiting civil engineer has proposed to him that he install an automatic filtration and chlorination plant and process all the water he uses. After careful investigation he finds that he can buy such a unit in either of two sizes, 2,500 or 5,000 gallons per day. The 2,500 unit will be adequate for his immediate needs and for the next 5 years. After 5 years he expects to double the size of the hotel and will need to double the purification capacity.

The smaller unit will cost about #32,000 installed and will require maintenance and repairs costing about #500 a year. The larger unit will cost about #48,000, and annual maintenance and repairs will cost #750. Both units have estimated lives of 15 years with zero salvage value. The owner has decided to purchase one or the other of these units.

Chlorine is available in steel cylinders for #350 per cylinder. Each cylinder will treat 150,000 gallons of water. Diatomaceous earth, used in the filter, costs #60 per 100 pounds. A hundred pounds will filter 75,000 gallons of water. The foregoing rates apply to both units. However, electric energy to pump the water through the units will cost #20 per 100,000 gallons for the smaller unit and #15 per 100,000 gallons for the larger.

The owner wishes to select the unit under the assumption that he will need 1,000 gallons per day next year and that the need will increase by 350 gallons per day each year for the following 10 years. Thereafter it will remain constant at 4,500 gallons per day. The owner wants an i^* of 15% on any investment that can be deferred. There will be no income taxes.

Use the present-worth method to determine whether he should buy a small unit now and another one in 5 years or buy a large unit now. Make some type of sensitivity analysis to judge how sensitive your conclusion is to the i^* used and to the estimated rate of growth in demand for water.

14–17. Over a 7-year period a small corporation was subject to the tax rate schedule shown in Table 12–7 (page 242). In year zero it had a taxable income of $45,000. This led to a tax of $8,250, which is 18⅓% of the taxable income. The tax rate on the highest increment of income was 20%.

During the next 6 years the general price level rose at 15% a year. The corporation's income before taxes also rose at 15% a year; that is, the before-tax income just maintained its purchasing power. However, the after-tax income lost purchasing power each year because the tax rate schedule was not adjusted for inflation.

Tabulate for years 0 to 6 the percentage of taxable income paid out in income taxes and the tax rate on the highest increment of income.

14–18. A small corporation subject to the tax rate schedule shown in Table 12–7 has a taxable income of $40,000. Obviously the tax rate on its highest increment of income is 20%. If the taxable income increases at the same rate as the rate of inflation, what will be the tax rate on the highest increment of income with a 50% rise in the price level? With a rise of 100%? Of 150%?

14–19. Modify the facts of Example 12–1 (page 231) incorporating a forecast of 12% annual inflation with no differential price change in cash flow before income taxes. That is, assume that if the land is not purchased its rental price for the nth year will be $\$25{,}000(1.12)^n$, that if the land is purchased the disbursements incident to ownership for the nth year will be $\$3{,}000(1.12)^n$, and that at the end of 10 years the land will have a net resale price before taxes of $341,600, which is $\$110{,}000(1.12)^{10}$.

(a) Make an analysis in dollars of constant purchasing power to find the prospective rate of return before income taxes.

(b) Assume a 49% applicable income tax rate throughout the entire 10 years. Assume that if the land is purchased the difference between the $341,600 net selling price at the end of 10 years and the $110,000 original cost will be taxable income and will be taxed at 49%. Make an analysis in dollars of constant purchasing power to find the prospective rate of return after income taxes.

(c) In part (b) change the tax rate on the so-called profit on the resale of the land from 49% to 20%. Make no other change. Make an analysis in dollars of constant purchasing power to find the prospective rate of return after income taxes.

14–20. Change the conditions of Problem 14–19 by introducing a prospective favorable differential change in the price of the land. Assume that its estimated selling price at the end of 10 years will be $550,000 rather than $341,600. Now find the prospective rates of return asked for in parts (a), (b), and (c).

14–21. Change the conditions of Problem 14–19 by introducing a prospective unfavorable differential change in the price of the land. Assume that its estimated selling price at the end of 10 years will be $220,000 rather than $341,600. Now find the prospective rates of return asked for in parts (a), (b), and (c).

14–22. Example 12–1 examined a proposed investment in nondepreciable property before and after income taxes with no inflation assumed. Problems 14–19, 14–20, and 14–21 introduced inflation with different assumptions about taxes and about the expected resale price of the nondepreciable property. What general points seem to you to be illustrated by your solutions of these three problems considered in relation to the conclusions of Example 12–1?

14–23. Change the conditions of Example 12–2 by assuming that wage rates will increase a bit during the 10-year period. The reduction in cash flow before income taxes caused by the materials-handling equipment will be $26,300 in the first year, $27,300 in the second, and $28,300 in the third and will continue to increase by $1,000 each year.

(a) What will now be the prospective rates of return before and after income taxes? Assume no change in general price levels. How much has this prospective differential price change improved the prospective rates of return before and after income taxes?

(b) How would your answer in (a) be changed, if at all, by the expectation of inflation during the 10-year life of the materials-handling equipment? Discuss this without making any actual calculations.

14–24. In Example 14–4 it is stated that "using the appropriate gradient factors we can calculate that the equivalent uniform annual cost of lost energy over the 25-year life would be $54.03, $68.07, and $85.93, respectively, for wire sizes 0, 1, and 2." Make the calculations necessary to verify these stated cost figures.

14–25. In Example 14–4 assume that the unit cost of electricity is 3.4¢ per kilowatt-hour at zero date and increases by 0.68¢ each year to become 10.2¢ in the tenth year. Thereafter, the unit cost remains unchanged at 10.2¢. (These costs are expressed in cents that have the purchasing power of zero date.)

What will be the equivalent uniform annual cost of lost energy over the 25-year life for wire sizes 0, 1, and 2? What will be the total annual cost for each of these three wire sizes? Continue to use the before-tax i^* of 14% that was used in Examples 13–2 and 13–4.

14–26. Assume that for Example 14–4 it is believed that the price per kilowatt-hour will not go beyond 8.5¢, expressed in cents that have the purchasing power of zero date. Table 14–8 indicates that if this differential price change from 3.4¢ to 8.5¢ should occur at once, size 0 would have a slightly lower annual cost than size 1. Explain how you would judge how soon this differential price change would need to occur for the equivalent uniform annual costs to be equal for sizes 0 and 1.

14–27. Example 13–2 dealt with a before-tax study to select economic wire size; the stipulated i^* was 14% before income taxes. Problem 13–11 (page 305) asked for an after-tax analysis using a 50% applicable tax rate and an after-tax i^* of 7%.

Without making actual calculations, discuss how you believe the choice among the competing wire sizes ought to be influenced by a forecast of inflation in the near future. Assume no forecast of any differential price change in the cost of electric energy. Assume that depreciation deductions to determine taxable income will be based on the dollar investment at zero date with no adjustment for changes in the purchasing power of the dollar.

14–28. Without making actual calculations, discuss how you believe the choice among the competing wire sizes in the preceding problem ought to be influenced by each of the following:

(a) a 10% investment tax credit available at zero date

(b) depreciation of the investment for income tax purposes allowed at 20% a year for the first 5 years of life

(c) a forecast of an upward differential price change in the cost of electric energy during years 1 to 5

(d) a forecast of a downward differential price change in the cost of copper wire during years 1 to 5

14–29. We used the facts of Examples 6–1 and 8–5 to make the point that the prospect of inflation ought not to change the conclusions of an economy study unless a differential price change is expected in some item included in the study. For these examples it was mentioned that the $1,300 extra annual income taxes estimated for Plan B were a possible source of such a differential change if inflation should occur.

This $1,300 figure was based on using straight-line depreciation for income tax purposes with a 10-year life and zero terminal salvage value; the assumed applicable tax rate was 50%. If there should be annual inflation of 6% and if the depreciation deduction from taxable income should not be adjusted for the change in the purchasing power of the dollar, there will be a differential price change in income taxes. Make an analysis along the lines of Example 14–3 to find out how much the 13.3% prospective after-tax rate of return computed in Example 8–5 will be reduced by this differential price change. Continue to assume straight-line depreciation for tax purposes just as in Examples 6–1 and 8–5.

14–30. A maintenance operation on a plating tank consists of installing a new lining. The type of lining now being used costs $3,000 installed and has an average life of 3 years. A new lining material has been developed that is more resistant to the corrosive effects of the plating liquid. Its estimated cost installed is $5,800. With a stipulate i^* of 12%, how long must the new type of lining last for it to be equally economical with the present type? (In this case the before-tax and after-tax rates of return are the same; the circumstances under which this is true were illustrated in Example 12–6.)

14–31. In the design of a single-story county office building, the question arises as to whether certain structural provisions should be made for the addition of another story at a later date. The architect prepares two designs;

the one that provides for this possible expansion will cost $810,000 to build; the one without it will cost $720,000.

With the more costly initial design, it is estimated that the second story will cost $400,000 to build; with the less costly design, $540,000. The estimated life of the building is 40 years from the date of construction of the first story. Maintenance costs will not be influenced by this structural provision for expansion. How soon must the expansion be needed in order to justify the more costly initial design? Assume an i^* of 8%. Because this is a government building there will be no property tax or income tax.

14–32. Assume that the choice between the two designs in the preceding problem is for a privately owned competitive business rather than for a county government. The estimated cash flows for construction are the same as before, but prospective property tax differences and income tax differences will now enter into the analysis. Assume a building life of 40 years from the date of completion of the first story just as in Problem 14–31. Also assume an after-tax i^* of 8%.

Annual property taxes throughout the life are estimated as 2% of the investment already made. Assume that the depreciation deduction for income tax purposes will be based on a 5% write-off of any investment during the first 20 years after the date of the investment. If it is believed that the second story will be needed in 5 years, should the initial investment be $810,000 or $720,000? Explain your analysis.

14–33. In year zero Sam Smith and Jennifer Jones are subject to the income tax schedule shown in Table 12–8 (page 244). Assume that for year 1 a "tax reduction" bill alters this schedule by multiplying all dollar figures in the table by 1.05. The exemption is increased from $1,000 to $1,050; the nonbusiness deduction is increased from $2,300 to $2,415; the width of each bracket is increased by 5%. In year 1 the general price level is 10% higher than in year 0. In year 1 Sam and Jennifer were each subject to the revised schedule.

(a) Sam's adjusted gross income was $16,000 for year 0 and his income tax was $2,605. For year 1 his adjusted gross income was $17,600; it increased by the same 10% as the general price level. By what percentage did his income tax increase? (*Ans.* 14.2%)

(b) Jennifer's adjusted gross income was $30,000 for year 0 and her income tax was $7,522. For year 1 her adjusted gross income was $33,000; it also increased by the same 10% as the general price level. By what percentage did her income tax increase? (*Ans.* 13.8%)

14–34. In the preceding problem Sam and Jennifer were fortunate in having 10% increases in their before-tax incomes in a year that had 10% inflation. Nevertheless, it is evident that their income taxes increased more rapidly than the general price level in spite of the "tax reduction" bill. What type of adjustment in the tax rate schedule seems to you to be desirable in a period of inflation if the burden of income taxes is not to be increased by the inflation?

15

USE OF THE MATHEMATICS OF PROBABILITY IN ECONOMY STUDIES

*What can we say about the future behavior of a phenomenon acting under the influence of unknown or chance causes? I doubt, that, in general, we can say anything. For example, let me ask: "What will be the price of your favorite stock thirty years from today?" Are you willing to gamble much on your powers of prediction in such a case? Probably not. However, if I ask: "Suppose you were to toss a penny one hundred times, thirty years from today, what proportion of heads would you expect to find?," your willingness to gamble on your powers of prediction would be of an entirely different order than in the previous case.—W. A. Shewhart**

The statement that tomorrow will probably be a rainy day is perfectly clear and understandable, whether you agree with it or not. So also is the statement that Smith is more likely than Jones to receive a promotion. In general, the word *probability* and its derivative and related words such as *probable, probably, likelihood, likely,* and *chance* are used regularly in everyday speech in a qualitative sense and there is no difficulty in their interpetation.

But consider a statement that the probability is 0.01 that there will be a runoff of 500 cu ft/sec or more next year at a certain highway drainage crossing. Or a statement that if a 13-card hand is selected at random from a standard deck of 52 playing cards, the probability is 0.00264 that it will contain all four aces. In such statements, *probability* is used in its quantitative or mathematical sense. It is evident that some special explanation of the meaning of *probability* is necessary before these statements can be understood. A critical consideration will show that the two statements not only call for more explanation but that they need somewhat different explanations.

*W. A. Shewhart, *Economic Control of Quality of Manufactured Product,* p. 8. Copyright 1931, by Bell Telephone Laboratories. Reprinted by permission. This book was originally published by the Van Nostrand Reinhold Co., New York.

Two different traditional definitions of *probability* in its mathematical sense may be given. One may be described as the *frequency definition*, the other as the *classical definition*.

Definitions of Probability*

Probability may be thought of as relative frequency in the long run. This may be phrased somewhat more precisely as follows:

Assume that if a large number of trials are made under the same essential conditions, the ratio of the number of trials in which a certain event happens to the total number of trials will approach a limit as the total number of trials is indefinitely increased. This limit is called the probability that the event will happen under these conditions.

It may be noted that this limit is always a fraction (or decimal fraction), which may vary from 0 to 1. A probability of 0 corresponds to an event that never happens under the described conditions; a probability of 1 corresponds to an event that always happens.

It is because *probability* describes relative frequency in the long run that the concept is so useful in practical affairs. But its use would be severely limited if the only way to estimate any probability were by a long series of experiments. Most mathematical manipulations of probabilities are based on another definition, which may be stated as follows:

If an event may happen in **a** *ways and fail to happen in* **b** *ways, and all of these ways are mutually exclusive and equally likely to occur, the probability of the event happening is* **a/(a + b)**, *the ratio of the number of ways favorable to the event to the total number of ways.*

This is called the *classical definition*. It represents the approach to the subject developed by the classical writers on the mathematics of probability, many of whom wrote particularly about probabilities associated with games of chance. Experience shows that where properly used, this definition permits the successful forecasting of relative frequency in the long run without the necessity of a long series of trials prior to each forecast.

Our statement about the probability of a runoff of at least 500 cu ft/sec implies knowledge about past runoff at this or comparable sites; it would be impossible to enumerate a number of equally likely ways in which the runoff

*With a few minor changes, our discussion of definitions is taken from pages 171–172 of the fifth edition of *Statistical Quality Control* by E. L. Grant and R. S. Leavenworth (New York: McGraw-Hill Co., 1980). The definitions are included here because it is necessary to make specific reference to them in our discussion of various probability applications in economy studies. However, limitations of space in this book make it impracticable for us to explain the various theorems of the mathematics of probability such as those that are applied in the examples. Many of our readers will already have been exposed to these theorems elsewhere in the study of algebra, probability, or statistics. If not, the reader can take our simple probability calculations on faith for the time being until the opportunity presents itself to study the subject in greater detail.

could be above or below 500 cu ft/sec. In contrast, the statement about the probability of the hand with all four aces is based on a counting of equally likely ways in which the hand might contain all four aces or less than four aces; even though not based on the evidence of actual trials, a statement of this sort may be made with strong confidence that the stated probability is really the relative frequency to be expected in the long run.

It is possible to raise philosophical and practical objections to both of the traditional definitions of probability. For example, in the frequency definition how long a series of trials should one have to estimate relative frequency in the long run? In the classical definition, how can one tell which ways are equally likely? Such types of objections have led many of the modern writers on probability to view probability merely as a branch of abstract mathematics developed from certain axioms or assertions. The establishment of any relationship between actual phenomena in the real world and the laws of probability developed from the axioms is viewed as an entirely separate matter from the mathematical manipulations leading to the probability theorems.* The numerical values of probability sometimes are established as a matter of judgment; in such cases they are merely educated guesses. We shall have more to say on the use of *subjective* or *intuitive* probabilities later in this chapter.

The mathematical theorems that deal with the interrelationships among probabilities are the same regardless of the definition that is used. Generally speaking, persons who use probability calculations in economy studies will find it satisfactory to think of probability as meaning relative frequency in the long run.

Probabilities of Extreme Events

One important application of probability theory in engineering economy is in estimating the *expected value* of certain kinds of extreme events. We shall use this type of application to introduce the subject.

First, we shall examine the calculation of expected value in a certain game of chance. Here we can use the classical definition of probability to compute numerical values of extreme probabilities. A study of the gambling game of Keno will help to clarify certain principles even though this is not an engineering economy application. Examples 15–1 and 15–2 refer to this game.

Then in Example 15–3 we shall look at calculations of the expected value of the adverse consequences of extreme natural phenomena such as floods. Here we shall find it necessary to use the frequency definition of probability. We shall see that often it is necessary to use estimates of probability values

*For a clear discussion of the relationship of the axiomatic definition of probability to the two traditional definitions, see G. A. Wadsworth and J. G. Bryan, *Applications of Probability and Random Variables*, 2 ed. (New York: McGraw-Hill Book Co., Inc., 1974), pp. 2–9.

that clearly are imperfect. Finally, in the same example we shall illustrate the use of such calculated expected values in the analysis of a problem in engineering economy.

The Gambling Game of Keno

This is a variant of a game of Chinese origin. Our description applies to the version of the game played in a Nevada gambling house which we shall call the XYZ Club.

In each game, a mechanical device makes a random drawing of 20 plastic balls from a set of 80 balls that are numbered from 1 to 80. Prior to a drawing, each player chooses certain numbers to have marked on a card for which 60¢, or some multiple thereof, is paid. The chosen numbers are called "spots." The player may elect to choose from 1 to 15 spots. In effect, he or she is betting that enough spots will be the same as some of the 20 numbers selected by the mechanical device to entitle the player to a payment by the XYZ Club. The size of the payoff, if any, depends on how many spots agree with some of the chosen 20. (A spot that agrees is described as a "winning spot.") The stipulated amounts of payoff for players who elect to choose either 2 or 7 spots are given in Examples 15–1 and 15–2. (Certain other payoff schedules are given in problems at the end of this chapter.)

The number of different sets of 20 numbers that may be selected out of 80 numbers is the number of combinations of 80 things taken 20 at a time. In general, the symbol C_r^n may be used to represent the number of combinations of n things taken r at a time. It is shown in algebra that $C_r^n = \dfrac{n!}{r!(n-r)!}$.

Here the expression $n!$, read as "factorial n" or "n factorial," is used for the product of the first n integers. By definition, $0! = 1$. $C_{20}^{80} = \dfrac{80!}{20!60!}$ $= 3{,}534{,}816{,}142{,}212{,}174{,}320$. Presumably, if the mechanical device operates in a way so that the 80 plastic balls have equal chances to be selected, each of this large number of combinations is equally likely.

Suppose a player selects 7 spots. To find the probability that he or she will have exactly 5 winning spots, one first must compute how many of the foregoing combinations will contain exactly 5 out of a set of 7 specified numbers. This computation calls for multiplying C_5^7 by C_{15}^{73}. The classical definition of probability may then be applied to compute the desired probability, as follows:

$$\frac{C_5^7 C_{15}^{73}}{C_{20}^{80}} = \frac{7!73!20!60!}{5!2!15!58!80!} = \frac{18{,}054}{2{,}089{,}945} = 0.008638505$$

EXAMPLE 15–1

Calculation of Expected Value With a Single Payoff Figure

Calculations for Two-Spot Version of Keno

Suppose a player buys a Keno ticket for 60¢ and elects to have only two spots marked on it. If both of the spots are winning spots, the house will pay $7.50; otherwise it will pay nothing. The probabilities of the three possible mutually exclusive results are as follows:

$$\text{Prob. of 0 winning spots} = \frac{C_0^2 C_{20}^{78}}{C_{20}^{80}} = \frac{177}{316}$$

$$\text{Prob. of 1 winning spot} = \frac{C_1^2 C_{19}^{78}}{C_{20}^{80}} = \frac{120}{316}$$

$$\text{Prob. of 2 winning spots} = \frac{C_2^2 C_{18}^{78}}{C_{20}^{80}} = \frac{19}{316}$$

The sum of all of the mutually exclusive probabilities is 316/316 = 1. (The necessity for the sum of all of the mutually exclusive probabilities to be 1 often supplies a useful check on probability calculations.) These probabilities can be combined with the payoff schedule to obtain the expected value of the payoff on a 60¢ ticket, as follows:

Winning spots	Probability	Payoff	Expected value
0	177/316	$0.00	$0.00000
1	120/316	0.00	0.00000
2	19/316	7.50	0.45095
Totals	316/316		$0.45095

EXAMPLE 15–2

Calculation of Expected Value With Several Different Payoff Figures

Calculations for Seven-Spot Version of Keno

For a player who elects to have 7 spots marked on a 60¢ Keno ticket, the XYZ Club will pay nothing if the ticket has 0 to 3 winning spots, 60¢ for exactly 4 winning spots, $12 for exactly 5, $245 for exactly 6, and $5,500 for exactly 7. To determine the expected value of a player's winnings, it is necessary to add the various expected values for the different possible numbers of winning spots, as follows:

Winning Spots	Probability	Payoff	Expected value
0 to 3	0.938414048	$ 0.00	$0.0000
4	0.052190967	0.60	0.0313

5	0.008638505	12.00	0.1037
6	0.000732077	245.00	0.1794
7	0.000024403	5,500.00	0.1342
Totals	1.000000000		$0.4486

Comment on the Significance of an Expected Value as Illustrated by Examples 15–1 and 15–2

In the long run, Keno players at the XYZ Club get back as winnings approximately 45¢ per 60¢ ticket purchased. (This is true not only for the 2-spot and 7-spot versions of the game but for all versions from 1-spot to 11-spot as well.) From the viewpoint of the XYZ Club, about 75% of the money taken in on 60¢ tickets goes for payment of players' winnings and the other 25% goes for miscellaneous expenses, including taxes, and for profits, if any.

From the viewpoint of the player who buys *one* 60¢ ticket, it is impossible to get back exactly 45¢. If, say, the 2-spot version is played, the player either will win $7.50 or nothing. Our probability calculations show that if a great many such games are played, a player will win nothing in about 94% of them and will win $7.50 in about 6% of them. The 6% of the games in which $7.50 is won will give the player 45¢ for each 60¢ bet. This is the sense in which $0.45 is the expected value of the game.

Although the arithmetic is somewhat more complicated in analyzing the 7-spot version of Keno, the meaning of the calculated expected value is the same. Nevertheless, there are some important differences between the 2-spot version and the 7-spot version from the viewpoint of a prospective gambler. Any reader interested in gambling will doubtless have noted that in any single 7-spot game a player has less than 1 chance in 100 of doing better than getting back 60¢. But the low probabilities of having 6 or 7 winning spots are offset by the higher payoffs. We shall have occasion to comment further on this difference between Examples 15–1 and 15–2 after we have examined calculations of expected values of flood damages and used such calculations in an economy study.

Incidentally, it should be noted by the reader that the phrase *expected value* is a mathematicians' phrase referring to the product of a probability and an associated monetary figure and carries no implication about desirability or lack of it. Although a $5,500 payoff for a Keno player who is fortunate enough to have 7 winning spots on a 60¢ ticket will be viewed quite differently by the player and by the XYZ Club, the expected value of this occurrence is the same for the player and the club. In this usage, the word *value* is applied in a neutral sense. Thus, one can speak of the expected value of something that is viewed as undesirable, such as the adverse consequences of a flood, as readily as something viewed as desirable, such as a successful gamble. Example 15–3 presents a simplified valuation of adverse consequences employing the relative frequency definition of probability.

EXAMPLE 15–3

Using Expected Values in the Comparison of Alternative Investments

Determining the Economic Capacity of a Spillway

The following illustration is adapted from one prepared by the late Allen Hazen:*

A large public utility company has acquired a recently constructed small hydroelectric plant in connection with the purchase of a small utility property. It seemed probable to the engineers of the large company that the spillway capacity of 1,500 cu ft/sec provided by the dam at the plant was inadequate; if a flow occurred that exceeded the capacity of the spillway, the stream was likely to cut a channel around the power plant. A rough estimate of the cost of the necessary repairs if this should occur was $250,000.

In order to estimate what, if any, increase in spillway capacity was justified, the engineers estimated the costs of increasing the spillway capacity by various amounts. They also estimated from the available records of stream flow the probabilities of occurrence of floods of various magnitudes in any one year.

Flood flow in cu ft/sec	Probability of greater flood occurring in any one year	Required investment to enlarge spillway to provide for this flood
1,500	0.10	No cost
1,700	0.05	$24,000
1,900	0.02	34,000
2,100	0.01	46,000
2,300	0.005	62,000
2,500	0.002	82,000
2,700	0.001	104,000
3,100	0.0005	130,000

Annual investment charges on the spillway enlargement were assumed as 10.5%. This figure was the sum of annual capital recovery cost of 7.5% at an i^* of 7% over the estimated remaining life of 40 years [$(A/P,7\%,40) = 0.07501$], property taxes of 1.3%, and income taxes of 1.7%. For each spillway size, the expected value of annual flood damages was the product of the probability that the spillway capacity would be exceeded and the $250,000 estimated damage if this event should occur. The following tabulation shows

*Allen Hazen, *Flood Flows* (New York: John Wiley & Sons, Inc., 1930).

that the sum of these costs is a minimum if the spillway is designed with a capacity of 2,100 cu ft/sec. This will take care of a flood such as would be expected, on the average, 1 year in 100.

Spillway capacity	Annual investment charges	Expected value of annual flood damages	Sum of annual costs
1,500	$ 0	$25,000	$25,000
1,700	2,520	12,500	15,020
1,900	3,570	5,000	8,570
2,100	4,830	2,500	7,330
2,300	6,510	1,250	7,760
2,500	8,610	500	9,110
2,700	10,920	250	11,170
3,100	13,650	125	13,775

Certain Implications of the Use of Expected Value of Annual Damages in Design Against an Extreme Event

In Example 15–3 the remaining life of the power plant and spillway was estimated to be 40 years. Nevertheless, the economy study indicated that it was desirable to design against a flood that would be expected on the average only once in 100 years.

It is reasonable for such an analysis to give weight to the hazard of events that are so extreme that they may not happen at all during the study period. In effect, the calculations in Example 15–3 recognize that the flood is expected on the average only once in 100 years is as likely to occur next year (or in any other specified year) as to occur 100 years from now.

It is because it is impossible to predict the *time* at which such extreme events will occur that it is desirable to predict their relative frequency in the long run. One way to look at the solution in Example 15–3 is to say that the analysis leads to the conclusion that it is a better gamble to design against the 100-year flood than against some more frequent or less frequent flood. Another way to interpret the expected values of damages used in the economy study is to say that they would represent a fair charge for flood insurance of $250,000 if such insurance were available at the bare "cost of the risk."

Needless to say, this example has been simplified greatly. It assumes the damages from a flood are a constant dollar amount regardless of the flood flow. Additionally, it ignores the possibility of more than one flood flow of a given magnitude in any one year. Different assumptions with respect to either of these factors will increase the number of events to be considered in arriving at the most economical design alternative.

Computing Expected Values When Different Expected Damages Are Associated with Different Probabilities of Extreme Events

In Example 15–3, the damage from a flood exceeding the spillway capacity was estimated to be $250,000 regardless of whether the flood was slightly more than capacity or a great deal more. Under this simple assumption, the expected value (i.e., the prospective average annual damage in the long run) was the product of $250,000 and the probability that the spillway capacity would be exceeded in any year.

Often the damage from an extreme event will vary with the magnitude of the event. For example, in analyzing most flood control projects, it is necessary to recognize that, generally speaking, the larger the flood, the greater the prospective damage.

We may illustrate the way to compute expected values under such circumstances by a slight modification of the data of Example 15–3. Assume that it is estimated the damage will be $0 for any flood less than the spillway capacity, $250,000 if the flood exceeds the spillway capacity by not more than 200 sec ft, $300,000 if the flood exceeds capacity by from 200 to 400 sec ft, $400,000 if the flood exceeds capacity by more than 400 sec ft. The required calculations for a spillway capacity of 2,100 sec ft are as follows:

Damages $F(x)$	**Probability of stated damage** $P(x)$	$E(x) = F(x)P(x)$
$ 0	1.00 − 0.01 = 0.99	$ 0
250,000	0.01 − 0.005 = 0.005	1,250
300,000	0.005 − 0.002 = 0.003	900
400,000	0.002 − 0.000 = 0.002	800
Expected value of annual damage		$2,950

Similar calculations for spillway capacities of 1,900 and 2,300 sec ft will yield figures for average annual damages of $6,000 and $1,450, respectively. The reader will doubtless recognize that the analysis here is similar to the analysis of the 7-spot Keno game in Example 15–2.

A more general case is one in which the expected damage $F(x)$ varies continuously with the magnitude of the flood flow. In such a case, $F(x)$ can be plotted on rectangular coordinate paper as a function of $P(x)$; the area under the curve will give the expected value of annual damages.* In Example 9–2 and in our various problems dealing with the evaluation of proposed flood control projects, it is assumed that this method has been used.

If the relationship between $F(x)$ and $P(x)$ can be described by a mathematical function, the expected value may be determined by integration.

*For a good illustration of this method, see J. B. Franzini, "Flood Control—Average Annual Benefits," *Consulting Engineer* (May, 1961), pp. 107–109.

General Comments on Economic Aspects of Engineering Design to Reduce Risk of Some Undesired Event

Example 15–3 is representative of a type of problem occurring in practically every field of engineering. The selection of a safety factor in structural design implies the balancing of the greater risk of structural failure with a lower safety factor against the larger investment required by a higher safety factor; the design of a sewer that carries storm water implies balancing the damages that will occur if sewer capacities are inadequate in severe storms against the extra cost of building sewers with larger diameters; a similar problem exists in the selection of the waterway area for any highway or railway drainage structure. Similarly, a program of acceptance inspection of manufactured product may require the balancing of the costs associated with the risk of passing defective parts or defective final product against increased inspection costs if more product is inspected; the allotment of public funds to grade-crossing-elimination projects implies balancing the prospective accidents at any crossing against the cost of eliminating that crossing; the design of an electric generating station and distribution system implies weighing the prospective damages from service outages against the cost of reducing the chance of such outages. Illustrations of this sort might be multiplied indefinitely.

If such problems are to be solved quantitatively rather than by someone's guess, estimates must be made of:

1. The expected frequencies of the undesired events under various alternative plans of design and operation.
2. The expected consequences if the undesired events occur, expressed in money terms insofar as is practicable.
3. The money outlays (both immediate investment and subsequent disbursements) to make various degrees of reduction in the risk.
4. Any other future differences—either measurable in money terms or otherwise—associated with alternative designs or operating policies to be compared.

The practical difficulty lies in evaluating these items, particularly the first two. It is generally easy to recognize cases in which it clearly pays to reduce a risk because the cost of reducing the risk is small and the risk itself and the prospective damages if the undesired event occurs are large. Similarly, it is easy to recognize cases at the other extreme where the cost of reducing the risk is high, and the risk of the undesired event occurring is very slight.

But often it is not possible to make a quantitative approach to those many troublesome cases in which the answer is not obvious, simply because of the absence of any reliable information as to the frequencies of the events against which it is desired to protect, and as to the amount of the damages that will occur if the events take place. Often this information does not exist merely because of the absence of any systematic effort to get it. In some cases, however, even after such systematic effort is started, the securing of useful data may require the passage of a considerable period of time.

Even though records are available as to the frequencies of undesired events and as to the kinds of damages that have occurred from them, it may be difficult or impossible to place a money valuation on all of the damages. This is particularly true in estimates relative to those types of catastrophe that involve human suffering. Thus, irreducibles may play an important part in economy studies of this type, even though the necessary facts are available to take decisions out of the "hunch" class.

Some Difficulties in Estimating Probabilities of Extreme Events

The prediction of an extreme event such as a Keno player having 7 winning spots on a 7-spot card is a fairly simple and straightforward matter. All that is required is to do some arithmetic and to apply the classical definition of probability.

In contrast, if an economy study calls for estimates of the probabilities of infrequent events and the frequency definition of probability must be used, the problem usually is more troublesome. For instance, in our flood spillway example, flood probabilities were estimated as low as 0.0005, that is, 1 year in 2,000. On most streams in the United States, the available records on which such estimates must be based vary from none at all to records extending for approximately 80 years!

In estimating flood frequencies and other extreme hydrologic phenomena, two types of extrapolation may be required. One is an extrapolation in space; in the common case where the stream flow records are not available at the particular site in question, it is necessary to draw conclusions from records at some other site on the same stream, or possibly from records on other streams deemed to be comparable. The other is an extrapolation in time, which obviously is required if, say, a 500-year flood is to be estimated from a 25-year record. Different extrapolators can and do reach widely different conclusions about the probabilities of extreme events.

In other engineering fields there are other types of difficulties in estimating the hazard of extreme events. For instance, proposals for changes in highway design often are based on the assertion that a particular location involves more than normal accident hazard and that this hazard can be reduced by a proposed design change. Some practical difficulties in the collection and interpretation of highway accident data are indicated in the following paragraph from a letter written by a distinguished highway engineer:*

> The identification of an accident-prone location is the bone of contention. Police

*This letter, dated July 25, 1966, from Robert W. Sweet, Chief Engineer of the New York State Department of Transportation, to A. E. Johnson, Executive Secretary of the American Association of State Highway Officials, was printed in *Highway Research News*, Highway Research Board, Washington, D.C. (Winter 1967), pp. 46–49, under the title "Traffic Safety and the Highway Engineer."

reports may, at best, provide a guide. They are never complete or conclusive. Seldom do they accurately pinpoint the location and often they err as to cause. A long, straight stretch of road may develop a lot of accidents and thus be identified as an accident-prone location. Actually, the location has nothing to do with it. Poor speed law enforcement and lack of adequate police patrolling may be the cause. I'll bet you that wouldn't show up in the report. There are many dangerous situations where the necessary coincidence of conditions and events has not occurred, and, therefore, no accident record exists. This does not change the hazardous nature. The Angel of Death has merely withheld his hand. I cite the infamous Dansville Hill on our Route 36 as an example. Here is a long steep hill terminating in the main business section of a moderate sized village. To my personal knowledge, at least 20 large trucks have lost their brakes on that hill and roared down into the Village. So far no one has been hit or killed! It took four long years to finally convince the statistics boys that here was a hazardous situation. We now have a relocation under construction. There are many other potential killers of a similar nature.

The Concept of a Constant System of Chance Causes

The following quotations present important ideas that should be helpful in an analyst's decision on whether or not to use probability calculations in any particular economy study.

Wadsworth and Bryan give this clear explanation of the concept of a random variable:†

> On an abstract level, ordinary mathematics is concerned with independent variables, the values of which may be chosen arbitrarily, and dependent variables, which are determined by the values assigned to the former. In the concrete domain, science aims at the discovery of laws whereby natural phenomena are interrelated, and the value of a particular variable can be determined when pertinent conditions are prescribed. Nevertheless, there exist enormous areas of objective reality characterized by changes which do not seem to follow any definite pattern or have any connection with recognizable antecedents. We do not mean to suggest an absence of causality. However, from the viewpoint of an observer who cannot look behind the scenes, a variable produced by the interplay of a complex system of causes exhibits irregular (though not necessarily discontinuous) variations which are, to all intents and purposes, random. Broadly speaking, a variable which eludes predictability in assuming its different possible values is called a *random variable*, or synonymously a *variate*. More precisely, a random variable must have a specific range or set of possible values and a definable probability associated with each value.

In writing about manufacturing quality control, Dr. W. E. Deming made the following observations that have widespread general application:*

> There is no such thing as a constancy in real life. There is, however, such a thing

†G. A. Wadsworth and J. G. Bryan, *op. cit.*, p. 40.

*W. E. Deming, "Some Principles of the Shewhart Methods of Quality Control," *Mechanical Engineering*, Vol 66, (March 1944), pp. 173–77.

as a *constant-cause system*. The results produced by a constant-cause system vary, and in fact may vary over a wide band or a narrow band. They vary, but they exhibit an important feature called *stability*. Why apply the terms *constant* and *stability* to a cause system that produces results that vary? Because the same percentage of these varying results continues to fall between any given pair of limits hour after hour, day after day, so long as the constant-cause system continues to operate. It is the *distribution* of results that is constant or stable. When a manufacturing process behaves like a constant-cause system, producing inspection results that exhibit stability, it is said to be in *statistical control*. The control chart will tell you whether your process is in statistical control.

The frequency definition of probability given near the start of this chapter began with the clause "Assume that if a large number of trials are made under the same essential conditions." The phrase "constant system of chance causes" coined by Walter Shewhart in his pioneer writings on quality control deals with the issue of whether or not the same essential conditions have existed in the past and are expected to continue to exist in the future. The Shewhart quotation that opened this chapter emphasized the point that some quantities (such as the number of heads in 100 tosses of a penny) seem to behave over long periods of time as if they were random variables, whereas other quantities (such as the price of a favorite stock) do not behave in this way.

Presumably, when an economy study includes expected values obtained by multiplying one or more probabilities by one or more associated money amounts, the analyst who makes the multiplication believes that he or she is dealing with a random variable and that a constant system of chance causes will continue to be present throughout the analysis period. Or, at least, the analyst believes that prospective changes in the system of chance causes will be so small that the conclusions reached in the economy study will be relatively insensitive to these changes.

EXAMPLE 15–4

Using Decision Trees in the Evaluation of Expected Values

A Choice Among Expansion Plans

A large chain of retail convenience stores is considering the expansion of one of its outlets. Results of a market study indicate that a continuation of the current rate of growth will require expanding the hours of operation and resorting to overtime pay if substantial business is not to be lost.

Alternately, the company is considering enlarging the store. This expansion may be accomplished in either of two ways. The least costly plan is to relayout the store and make a small addition on one side. This small expansion is believed to be sufficient for 5 years, the usual planning period for the company, and will cost $12,000. It is considered to have no residual value at the end of that time. Because of the possibility of substantial growth in the neighborhood, an alternate plan is to remodel the entire store and nearly double its floor space. This plan will cost $41,000 initially. It is assigned a residual value at the end of 5 years of $15,000.

A preliminary analysis before income taxes is made, based on the market study already mentioned, using an i^* of 20%. The present worths of estimated net cash flows (receipts minus disbursements) before income taxes affected by this decision are as follows for each of three assumed classes of business conditions:

Business condition	Estimated probability of occurrence	PW of net cash flow before income taxes under each plan		
		No expansion	Expand–small	Expand–large
Good	0.25	$115,000	$140,000	$190,000
Moderate	0.60	100,000	120,000	145,000
Poor	0.15	10,000	15,000	10,000

Figure 15–1 shows the resulting decision tree. The dollar figures given at the right-hand end of the branches are the net present worths, before taxes, from adoption of each alternative and having the indicated future business condition result. In the no expansion case, since there is no investment required, the values on the decision tree branches are the same as those previously tabulated for each future condition. The expand—large case requires

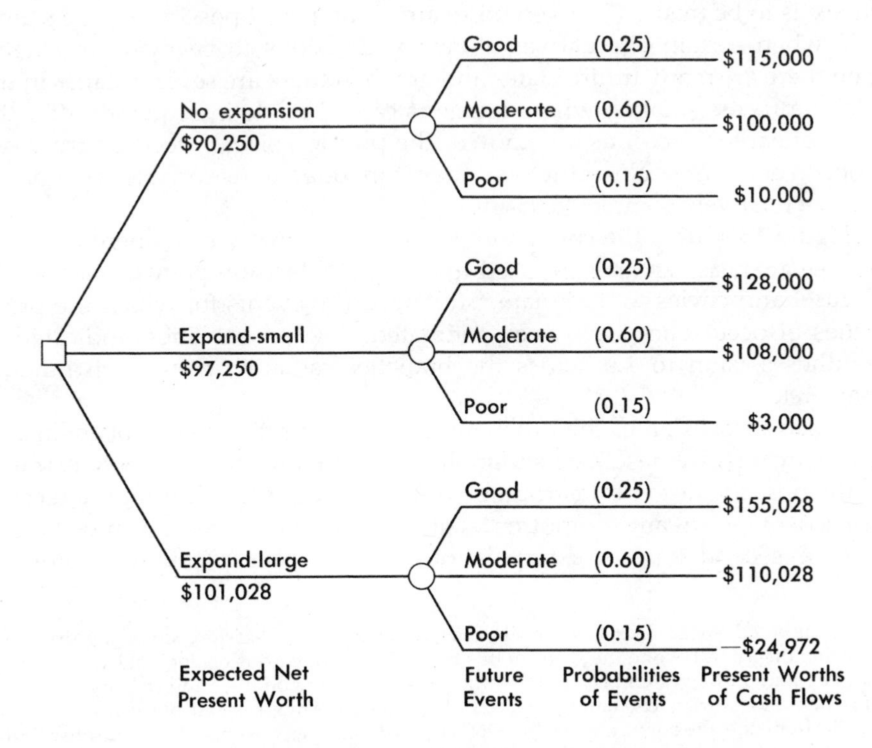

FIGURE 15–1. Decision tree describing Example 15–4

reduction of the present worths of future cash flows by the outlay for remodeling less its associated residual value, i.e., \$41,000 − \$15,000 (*P/F*, 20%,5).

The expected present worths of cash flows before taxes are indicated for each alternative beneath the name designation for that alternative. For each case, the expected present worth is found by multiplying the probability associated with the result of (1) choosing that alternative and (2) having the designated future business condition occur times the associated net present worth dollar figure; and summing the results for each alternative algebraically. The result of the company's analysis indicates that it should choose the large remodeling expansion since that alternative has the largest expected net present worth.

Discussion of Example 15–4

Decision trees have become very popular in recent years in the literature of decision theory and financial decision making.* This example presents a very simple form of decision tree in which there are three alternatives to be considered and only three possible future states, i.e., business conditions, to be considered. Decision trees are usually used when problems are far more complex: (1) when there is a large number of alternative courses among which a choice is to be made, (2) when there are a number of possible future states, or (3) when certain alternatives allow for decisions to be made in stages. When there are many future states and/or when there are several stages in the decision process, that is, when decisions can be made in sequence after the results of earlier decisions are known, the physical size of decision trees can become very large. Nevertheless, they can offer a reasonable method to visualize relatively complex decisions.

Figure 15–1 uses the conventions adopted by many proponents of decision tree analysis; squares are used to designate decision points, only one in this case, and circles to designate random future events for which the probabilities of occurrence have been estimated. The probabilities, indicated in parentheses, sum to 1.0 across the branches radiating from each random event circle.

In our discussion of the problems involved in estimating probabilities of extreme events, we described situations in which it was necessary to extrapolate the probabilities of catastrophic events, such as a flood, from the records of one river or stream to another. This was referred to as extrapolation in space. A second type of extrapolation, in time, was required in order to

*A more advanced discussion of risk analysis, or complex decisions involving the use of probability theory, is beyond the scope of this book. The more serious reader will want to see the articles by John F. Magee and David B. Hertz concerning decision trees and risk analysis in the *Capital Investment Series*, containing reprints of articles from the *Harvard Business Review*, and available from the President and Fellows of Harvard College, Boston, Massachusetts, 02163. Howard Raiffa's book *Decision Analysis: Introductory Lectures on Choices Under Uncertainty* (Reading, Massachusetts: Addison-Wesley, 1968) provides an excellent introduction to the subject for those with even a modest introduction to probability theory.

associate a 1/100 or 1/500 chance with an annual event based on possibly a record of only 25 years. The same types of extrapolation, at the very least, are necessary if explicit probabilities and descriptions of future states are to be used for analysis of problems of the type described in Example 15–4. By their very nature, market studies must rely on similarities between locations and extrapolate the results obtained from one or more similar locations to the location under study. Similarly, they must extrapolate in time in order to predict the future events, with respect both to outcome and to probability. This implies the assumption that there has been a constant system of chance causes that will continue to operate.

The use of decision trees is limited to those cases in which future events occur in a discrete fashion or where it is reasonable to combine random fluctuations in such a way that future events can be treated as discrete occurrences. This was the approach taken by the market analysts in Example 15–4. Computer simulation may be used to analyze such problems when it is desired to treat random fluctuations as continuous distributions.

Matrix Formulation of Example 15–4

Because of the symmetry of the random outcomes of the decision tree in Example 15–4, the decision problem may be presented in decision matrix form. We shall use the following symbols:

A_i = symbol used to designate each of the mutually exclusive alternatives being considered, $i = 1, 2, \ldots, n$

S_j = symbol used to designate each of the mutually exclusive future states that may occur, $j = 1, 2, \ldots, m$

$P(S_j)$ = probability associated with the occurrence of possible future state S_j. The sum of these probabilities must equal 1

Θ_{ij} = physically definable outcome of selecting alternative A_i and having future state S_j occur

$V(\Theta_{ij})$ = value, usually a monetary figure, associated with each Θ_{ij} outcome. These may be present worths or equivalent uniform annual figures. As explained in Chapter 13, they should not be rates of return or benefit–cost ratios because of the necessity to consider the differences between pairs of mutually exclusive alternatives when these methods are used.

By convention, the alternatives (A_is) are assigned to the rows of decision matrix and the future states (S_js) are assigned to the columns. Each $V(\Theta_{ij})$ is an element in the matrix. The problem in Example 15–4 appears as follows:

	Future states		
	Good	**Moderate**	**Poor**
Alternative	S_1	S_2	S_3
No expansion A_1	$115,000	$100,000	$10,000
Expand—small A_2	$128,000	$108,000	$ 3,000
Expand—large A_3	$155,028	$110,028	−$24,972
$P(S_j)$	0.25	0.60	0.15

The $V(\Theta_{ij})$s in this matrix are the present worths derived for the decision tree of Figure 15–1. The general formula for the expected present worth (EPW) of each alternative is:

$$\text{EPW}(A_i) = \sum_{j=1}^{m} V(\Theta_{ij})P(S_j)$$

For the data of the decision matrix:

$$\begin{aligned}
\text{EPW}(A_1) &= \$115{,}000(0.25) + \$100{,}000(0.60) + \$10{,}000(0.15) = \$\ 90{,}250.\\
\text{EPW}(A_2) &= \$128{,}000(0.25) + \$108{,}000(0.60) + \$3{,}000(0.15) = \$\ 97{,}250.\\
\text{EPW}(A_3) &= \$155{,}028(0.25) + \$110{,}028(0.60) - \$24{,}972(0.15) = \$101{,}028.
\end{aligned}$$

Naturally, the results of analysis of this matrix are identical to those of the decision tree analysis because the mathematical operations are identical.

Expected Values May Be Misleading Guides to Certain Types of Decisions

The use of expected values as a basis for decision making may be questioned on two counts. First, the estimates of both the outcomes of events and their respective probabilities must be assumed to be accurate. There were three possible future states described in Example 15–4. The formulation and analysis of the situation assumes that one of these states will result, and that, when the market analysts add the outcome of this decision to their storehouse of information, it will not alter the character of the accumulated data; that is, a constant system of chance causes was and is in operation.

Second, even though a constant system of chance causes is anticipated, the estimates of the relevant probabilities are deemed to be entirely satisfactory, and the monetary evaluations of the associated events are believed to be correct, it does not follow that it is always reasonable to base economic decisions on expected values. Such values are based on relative frequencies in the long run. Not all decision makers are in a position to take advantage of what is anticipated in the long run. Only one decision may be involved, or, at most, only a relatively few decisions.

Vogt and Hyman* illustrate this point by citing the case of a man who is given the choice between two favorable alternatives (a) the certain receipt of $1,000, and (b) the receipt of $10,000 only if the result of a single toss of a fair coin is heads. The expected value of (a) is $1,000 and of (b) $5,000; it therefore appears that (b) is clearly the economic choice. But this man needs exactly $1,000 for an immediate operation to restore his vision and has no other

*E. Z. Vogt and Ray Hyman, *Water Witching, U.S.A.* (Chicago: University of Chicago Press, 1959), p. 196.

source of this money. His choice therefore is between (a) getting his sight back, and (b) a 0.5 probability of getting his sight back plus an additional $9,000. With this translation of monetary consequences into resulting non-monetary consequences, it is clear that the sensible choice is (a) in spite of its lower expected money value.

Consider also the case of an automobile owner considering the purchase of liability insurance or a homeowner deciding whether or not to take fire insurance. It is clear that only a part of the total premiums that an insurance company receives can be used to pay damage claims. (This is approximately half in the case of many fire insurance companies in the United States.) The remainder must go for sales commissions and operating expenses. If the insurance company has made a good evaluation of the risks associated with each policy, the expected value of the recovery by the insured is bound to be considerably less than the amount paid by the insured.

But the monetary figures used in computing expected values may be an imperfect measure of the favorable or adverse consequences of a particular event. For an individual, an insurance policy can give protection against an extreme event that might mean financial ruin or severe financial hardship if the event should actually occur. Thus, although large geographically diversified enterprises may find it economical to act as self-insurers, individuals and small business enterprises are well advised to carry fire, casualty, and liability insurance even though a decision to do so does not minimize the expected value of a prospective negative cash flow.* Similarly, the public utility company in Example 15–3 might well conclude to design, say, against the 200-year flood rather than against the 100-year flood.

In the case of our large chain of retail convenience stores, a decision based on expected monetary value may be reasonable. The risk is spread across many very similar, if not identical, investments. It cannot be anticipated that each market survey leading to the expected value calculation will yield identical numbers. Nevertheless, where the required investment is small in relation to the total capital and when many similar productive investments are made, the averaging effect will tend to produce a result that is some weighted average of the individual expected values. This statement assumes that the individual investments are statistically independent. That is, the failure of one investment does not increase the probability of failure of any of the others. In many business enterprises this is not the case. Cyclical effects and adverse developments in the economy may drive all investments in the same direction at the same time. Decisions based on expected value calculations are particularly suspect when the investment under consideration is unique and very large and there is a reasonable probability of failure.

*The authors once knew a brilliant mathematician who refused to buy insurance of any sort on the grounds that the expected value of the recovery by the insured was necessarily less than the cost of the insurance. He seemed to us to be deficient in judgment about practical affairs in spite of his undoubted genius in his special field.

Intuitive Probabilities

We started this chapter with the classical and frequency definitions of probability. The application of either of these definitions implies the existence of some event for which it is possible to make many trials "under the same essential conditions."

Obviously, not all events in the uncertain future are of this type. Sometimes it is clear that the concept of relative frequency in the long run does not apply because only one trial of a future event will be made under a particular set of conditions. Moreover, it may be evident that there has been no constant system of chance causes operating in the past that can be deemed to apply to the future event in question.

Our first two examples dealt with games of chance in which it is possible to utilize either the classical or relative frequency definitions of probability to arrive at satisfactory numerical values. In the third example, the relative frequency definition was used. However, as previously discussed, it is often necessary to apply judgment, or subjective logic, to estimates of the tails of distributions. The fourth example utilized a market study to provide outcomes and probabilities. To the extent that subjective logic rather than objective knowledge of a set of conditions influences the evaluation of outcomes and probabilities, both become the result of intuition. The company in Example 15–4 may have had considerable experience with market analysis. Thus, intuition may not play a large role in its analyses. Nevertheless, since it is doubtful that it made many investments in localities where prior analysis indicated unfavorable results, its storehouse of data is likely to be biased.

In other circumstances, where there is little objective knowledge on which to base an analysis, subjective evaluation frequently is advocated. For example, a proposal is made in a manufacturing company to manufacture and market a product that no one has ever made before. Certain modern writers on operations research and statistical decision theory advocate the use of probabilities obtained purely by intuition.* A whole body of mathematical procedures intended to guide economic decisions has been based on the use of such probabilities. Various forms of decision trees frequently are employed. To the extent to which many such intuitive probabilities seem to be similar to crystal ball gazing, the validity of these procedures is currently a matter of controversy.

In the extensive literature pro and con about the use of intuitive probabilities, there is one point adverse to their use that we have never seen mentioned. Consider any formal analysis to be submitted to guide managerial decisions about an event where there will be only one trial. Where there are pressures on analysts to support certain predetermined conclusions, the use

*For a clear exposition of the case for the use of such "personal" probabilities, see Robert Schlaifer, *Probability and Statistics for Business Decisions* (New York: McGraw-Hill Book Co., Inc., 1959).

of intuitive probabilities may give the analysts an easy way to cheat.† It is pointed out in Chapter 21 that there is a chance to post-audit many of the estimates that analysts must make in connection with economy studies. But there can never be a chance to post-audit an intuitive probability where only a single trial is to be made.

An Alternate Decision Criterion: Maximum Security Level

An alternate or possibly secondary decision rule that may be useful in situations like the ones described in the previous paragraphs is a strategy that may be referred to as maximizing the decision maker's security level.* We shall explain this criterion by using the data from Example 15–4.

The company in Example 15–4 is quite large and owns a great many outlet stores similar to the one discussed. Using an expected value approach appears rational because the company must make many such decisions and is therefore able to spread its risk. In addition, it has a large marketing organization both experienced and capable in the area of market analysis.

Without changing the data, assume that this decision problem is faced by the sole proprietor of a small business. The proprietor is able, perhaps with the aid of a marketing consultant, to arrive at the same set of estimates. Not satisfied with the three expected net present worth calculations, a table of the present worths from the right-hand side of Figure 15–1 is prepared as follows (all figures in thousands of dollars):

	Future states		
	Good	**Moderate**	**Poor**
Alternative	S_1	S_2	S_3
No expansion A_1	115	100	10
Expand—small A_2	128	108	3
Expand—large A_3	155	110	−25

†The authors have observed what seemed to them to be cheating even when the conventional concept of mathematical probability was applicable but it was necessary to extrapolate to estimate probabilities of extreme events. On one proposed flood control project, distinguished hydrologists using the generally preferred method of analysis estimated a certain flood, much larger than any that had occurred during the period of record, to be a 1,000-year flood (i.e., to have a probability of 0.001 of taking place in any one year). However, political pressures in favor of the project were very strong. The hydrologists making the official analysis for a certain government agency used an extrapolation technique that contained a "fudge factor" that could be manipulated to give the conclusion that this same flood was a 100-year flood. The expected value of the damages that would be caused by this flood was of course 10 times as great with a probability of 0.01 rather than 0.001. The consequence of the use of the fudge factor in the probability estimate was to change an unfavorable **B/C** ratio to a favorable one.

*The reader interested in decision strategies of this type should read R. Duncan Luce and Howard Raiffa, *Games and Decisions: Introduction and Critical Survey* (New York: John Wiley & Sons, Inc., 1957). In game theory, maximizing the decision maker's security level is called a minimax strategy if the figures are presented as costs to the decision maker; it is called a maximin strategy if the figures are presented as positive returns. In our presentation, all monetary figures are expressed algebraically (including the proper sign); therefore, only the maximin strategy needs to be presented.

Immediately the troublesome problem, the potential $25,000 net present value loss if alternative A_3 (the expand—large alternative) is chosen, becomes apparent. In discussing this problem with a business associate, our proprietor learns of another criterion, called a *maximum security level strategy*, which may be summarized as follows:

When favorable or unfavorable money amounts are to be compared, prepare a decision matrix similar to that shown previously but excluding the probabilities;

1. List the possible outcomes across the top of a page.
2. List the alternatives on the left side of the page and fill in the money amounts (+ or −) associated with each outcome for each alternative.
3. Identify the minimum money amounts for each alternative and list them in a column to the right of the table.
4. Select the alternative which has the maximum of these minimum money amounts.

These rules may be applied as written to problems stated in terms of Equivalent Uniform Annual Cost or the PW of all cash disbursements. However, it is necessary to include the minus (−) sign preceding the money amounts because the decision rule requires that money amounts be considered algebraically.

Following the prescribed steps, our proprietor lists the minimum net present worths as follows (figures in thousands of dollars):

No expansion	A_1	10
Expand—small	A_2	3
Expand—large	A_3	−25

Unlike many problems of this type, all of the minimums come from the same column, future state, or event, S_3. The decision criterion leads to the choice of alternative A_1, the "no expansion" alternative. This is a completely different choice than that made by applying the expected value criterion.

The reader should note that the probabilities associated with each future business condition played no part in the decision. Had the probability associated with a "poor" business prospect been 0.5 or 0.001 rather than 0.15, the same result would have been obtained. Likewise, the same conclusion would have been reached if the proprietor had had no idea of the probabilities, or little confidence in their accuracy.

The two decision criteria discussed so far in this chapter, expected value and maximum security level, afford a means for a decision maker to arrive at a decision. Whether or not either one is rational depends on the context of the decision, the circumstances in which the decision is to be made. It is not possible to say categorically that one criterion or the other *should* be used by any rational decision maker when risk plays an important part in the results. The company with the experienced marketing group in Example 15–4 uses the expected value criterion, and possibly *should*, because the marketing group has performed reliably in the past and because it can spread the risk over many similar decisions. Our sole proprietor, however, may make this decision only once.

Differences in Personal Attitudes Toward Risk Taking

Jones pays $250 for a homeowner's insurance policy hoping that no disaster will occur permitting collection of any insurance proceeds. However, considering the probabilities associated with various home disasters and the payments that will be made by the insurance company if a disaster occurs, the expected value of the recovery is only $125. Jones' decision to take the insurance policy reflects a personal aversion to a particular type of risk.

Smith pays 60¢ for a 7-spot Keno card at the XYZ Club, hoping that the card will have 7 winning spots so that the club will pay $5,500. However, as we saw in Example 15–2, the expected value of this recovery is about 45¢. Our calculations in that example showed that in the long run, Smith could expect to win the $5,500 in only 1 game in approximately 41,000 and that winnings in excess of the 60¢ price of a ticket would occur in only 1 game in approximately 105. Smith's decision to buy the 7-spot Keno card reflects a preference for a particular type of risk.

Johnson, who pays 60¢ for a 2-spot Keno card which also has a 45¢ expected value of recovery, shows still a different attitude toward risk. The XYZ Club has a Keno game every few minutes; if Johnson spends several evenings playing 2-spot Keno, there is a good chance for an occasional win and that a reasonable fraction of the investment in Keno tickets (say, at least half) will be returned. In fact, Johnson may win back as much as the tickets cost, thus paying nothing for the pleasure of gambling.* On the other hand, there is only a slight chance of any substantial winning over a long series of games.

The observed fact that different persons have quite different attitudes toward risk has led certain theorists to hunt for ways to quantify such attitudes. *Cardinal utility theory*, originally proposed by von Neumann and Morgenstern, provides one approach.†

Simply stated, cardinal utility theory attempts to quantify an individual's preference among alternatives in risk situations. Utility is measured on an arbitrary scale of units called "utiles." The relationship between utility and dollars may be determined for an individual by asking an appropriately designed set of questions. A plot of the individual's responses to these questions is the utility function for that individual.

*If 4 games out of 100 are won, half of the $60 bet will be returned; if 8 games are won, all of the $60 will be returned. A student of probability may apply the binomial distribution to find that in 100 games the probability of winning at least 4 is approximately 0.25; that of winning more than 10 games is only 0.04.

†John von Neumann and Oskar Morgenstern, *Theory of Games and Economic Behavior* (Princeton, N.J.: Princeton University Press, 1947).

In any choice between risky alternatives, the theory assumes that an individual decision maker will choose the alternative that maximizes utility. Thus, given knowledge of the utility function, the probabilities associated with possible outcomes, and the monetary consequences of each outcome, an analyst should be able to predict the decision maker's choice.*

Consider again the decision to be made by our sole proprietor based on the data of Example 15–4. The factor that led to seeking another decision rule in preference to choosing the alternative with maximum expected net present worth was the possible $25,000 net present worth loss. Suppose our proprietor is not satisfied with the choice of the "no expansion" alternative either, which resulted from application of the maximum security level strategy criterion. The final decision, based on intangible factors and a *subjective* view of the quantifiable factors, is alternative A_2, the "expand—small" alternative. Advocates of cardinal utility theory would argue that this decision indicates the decision maker's preference among various alternatives in risk situations. Whether the proprietor *should* make this choice is a question that is nearly impossible to answer. To this extent, the application of utility theory is *descriptive* in nature rather than *prescriptive*. It may be useful to predict what a decision maker *would* do in particular risk situations; it cannot tell what *should* be done.

It seems to the authors of this book that utility theory may at some future time make useful contributions to engineering economy but that it has not yet reached the point where it is able to do so. Personal attitudes toward risk depend greatly on the surrounding circumstances and are not necessarily fixed for any given individual. Moreover, there are practical difficulties in measuring attitudes toward risk; although an individual may answer questions that define what he or she *says* will be done in risk situations described in terms of known probabilities and expected values, it rarely is possible to measure the risks and expected values associated with actual decisions. Finally, it is not clear what the appropriate use of a utility function would be as a guide to rational decision making, even though such a function were a valid description of a personal attitude and could be accurately defined. Some writers contend that their main value may be in policing internal decision making inconsistencies.

Some Special-Purpose Definitions that May Lead to Semantic Confusion

Certain writers on operations research and statistical decision theory classify decisions into three groups, described as (1) decisions under certainty, (2) decisions under risk, and (3) decisions under uncertainty. Where only one set of estimates of the outcome is made for each alternative, the decision is described as being made under "certainty." Where two or more different

*For our readers who have not been exposed elsewhere to the subject of cardinal utility theory, we recommend the following article: Ralph O. Swalm, "Utility Theory—Insights Into Risk Taking," *Harvard Business Review* (November–December 1966) pp. 123–136.

mutually exclusive estimates of outcomes are made for an alternative, and probabilities (which must add to 1) are somehow assigned to each estimate, the decision is described as being made under "risk." (In many instances, such an assignment of probabilities can be made only by intuition.) Where two or more different mutually exclusive estimates are made for an alternative and these estimates are presumed to include all possible outcomes, and where no probabilities are assigned, the decision is described as being made under "uncertainty." Based on these definitions, the maximum expected net present worth criterion of Example 15–4 is a risk decision rule; the maximum security level strategy criterion is an uncertainty decision rule.

To anyone who is not one of the limited number of professional specialists who have adopted the foregoing terminology, any discussion that assumes these meanings can be extremely misleading because these meanings do not fit the common usage of the words *certainty, uncertainty,* and *risk*. The consequences of a choice among alternatives are rarely, if ever, *certain*; few, if any, decisions of the type we examine in this book can be made under conditions of certainty if the word *certainty* is interpreted as having its usual meaning. The distinction that the special-purpose terminology makes between risk and uncertainty is arbitrary and does not fit any of the common meanings of these two words. Furthermore, the degree of subjectivity used in determining probabilities may make any given decision maker shy away from a choice among alternatives based on their use.

Another possible source of confusion in the use of words exists because of the difference between the meaning of the phrase *expected value* when used in the sense illustrated in Examples 15–1 and 15–2 and its meaning when used in its ordinary sense. The word *expected* used in its popular sense can be applied to any estimate about the future. Therefore, any estimate of future cash flow made for an economy study might conceivably be called an expected value. However, an estimate of future cash flow is not an expected value in the probabilistic sense unless it has been obtained by two or more different estimates of some future cash flow, unless each of the estimates of this future cash flow has been multiplied by its estimated probability and the probabilities add to 1, and unless these products have been added together.

The Troublesome Problem of Making Investment Decisions with Imperfect Estimates of the Future

Our discussion in Chapters 1 and 2 outlined a suggested approach to an engineering economy study. Alternatives should be defined. The differences in their consequences should be estimated and expressed insofar as practicable in commensurable units (i.e., in money units). A primary criterion for decision making should be applied to the monetary figures; this criterion should be chosen with the objective of making the best use of the limited available resources. In some cases, also, secondary decision making criteria should be used; such secondary criteria may be related to the inevitable lack of certainty associated with all estimates of the future. Finally, someone must make a decision among the alternatives. The decision maker should consider the analysis as related to the primary criterion. The analysis should be considered as related to the secondary criteria if any such have been established. Recognition also should be given to any prospective differences in consequences that were not expressed in terms of money (i.e., to what we have called the irreducible data).

Where should uncertainties about the future be introduced into the foregoing sequence of procedures? In the view of the authors of this book, there is no one best way to do this under all circumstances.

Moreover, there seems to be no standard practice in this matter in either industry or government. In many types of economy studies there is no formal recognition of uncertainty. But where there is a systematic policy of allowing for uncertainty, the allowance sometimes is made in the estimating, sometimes in the selection of the minimum attractive rate of return where this rate is used as the primary criterion for investment decisions, sometimes through sensitivity analysis (possibly related to formal secondary decision criteria), and sometimes at the final decision-making level through viewing uncertainty as one of the irreducibles.

In many economy studies there is a good case for doing nothing at all about the observed fact that an estimate of future cash flow nearly always seems to miss its mark on one side or the other. If experience indicates that the favorable misses tend to be as frequent and as large as the unfavorable misses, and if there is no way to judge in advance which particular estimates will have the favorable misses and which the unfavorable ones, there is nothing useful that can be done about the matter. Of course there should be a sufficient number of estimates involved to give the favorable and unfavorable misses a chance to offset one another. Such a condition might exist, say, with reference to estimates for cost reduction projects in a large manufacturing company or with reference to estimates to guide decisions on alternative designs in an expanding public utility company.

Sometimes uncertainty is allowed for in estimates for economy studies through the practice of shading estimates to be "conservative" by estimating positive cash flows on the low side and negative cash flows on the high side. Sometimes there is a formal allowance for uncertainty made in the estimates by using probabilities in connection with associated monetary figures; if so, some of the estimates entering into the economy study will be *expected values* in the sense that this phrase is used in Examples 15–1 through 15–4 and in a number of the problems at the end of this chapter. As pointed out earlier in the chapter, a distinction needs to be made between uncertainties that can legitimately be expressed as probabilities and other types of uncertainties.

Gaming strategies, such as the maximum security level strategy, may prove useful in analyzing problems with potentially catastrophic results. They may be used in a primary capacity in making a choice among mutually exclusive alternatives or in a secondary capacity in eliminating certain alternatives involving extremely adverse consequences.

Where dissimilar types of proposed investment compete for limited funds, and where experience indicates that some types have a considerably greater risk of loss than other types, it is common to require a higher prospective rate of return from the proposals deemed to be riskier. As explained in Chapter 21, such a requirement often is applied at the capital budgeting level without a stipulation in advance of the minimum attractive rate of return for any particular project. However, a case of an advance stipulation was mentioned in Chapter 10; a certain oil company required an after-tax i^* of 10% for projects in marketing and transportation, 14% for projects in refining, and 18% for projects in production.

As pointed out in Chapter 14, sensitivity studies may be a great help to decision makers, particularly where certain constraints put upper and lower bounds on the extent to which actual values will depart from estimated values. Chapter 21 describes a case of the systematic use of secondary decision criteria based on the concept of sensitivity.

A useful piece of advice is not to make allowances for the *same* uncertainties at a number of different stages in the process of analysis and decision making. For instance, do not shorten estimated lives, then shade estimates of cash flow by arbitrary decreases in estimates of cash receipts and increases in estimates of cash disbursements, then require an extremely high i^* because of risk, and then also give weight to the risk at the final step of making the decision. Such a policy of quadruple counting of a recognized hazard may cause a failure to approve many proposed investments that actually would be very productive.

PROBLEMS

15–1. In a factory with fire insurance based on a value of $2,400,000, an engineer suggests the installation of an improved automatic sprinkler system costing $85,600. This will reduce the annual fire insurance cost from 1.32% to 0.54%. On the hypothesis that insurance rates need to be set high enough so that only half of the total premiums collected by an insurance company will be available to pay for fire damages, the management estimates that the expected value of the damages that would be compensated by insurance will be one-half the insurance premium. It also estimates that on the average the total adverse consequences of a fire will be 3 times the losses on which recovery can be made. The annual cost of operation and maintenance of the sprinkler system is estimated as $2,500.

Assuming a before-tax i^* of 30% and assuming a 20-year life for the sprinkler system with zero salvage value, make calculations to compare the alternatives: (a) continue without sprinkler system (b) install sprinkler system. Of course a $2,400,000 fire insurance policy will be continued either with or without the sprinkler system.

What matters not reflected by the use of expected values do you think might reasonably enter into management's choice between these alternatives? (*Ans*. sprinkler, $54,236; no sprinkler, $63,360)

15–2. Two sites are under consideration for a proposed warehouse building. At either site a $1,500,000 fire insurance policy will be taken on the building and its contents. At site A, the annual insurance rate will be 0.65%; at site B it will be 0.90%. It is estimated that the expected value of the damage that would be compensated by insurance is one-half of the insurance premium. It also is estimated that on the average the total adverse consequences of a fire will be 2.5 times the losses on which recovery can be made from the insurance company. Sites A and B differ only in the cost of land and in risk of fire damage. The estimated life of the warehouse is 40 years with zero salvage value.

Considering the differences in insurance costs and in the expected value of the adverse consequences of a fire that will not be compensated by insurance, and using a before-tax i^* of 20%, how much extra could be paid for site A? Assume that there will be no change in land values over the life of the warehouse. (*Ans*. $32,810)

15–3. In the selection of the spillway capacity of a proposed dam, estimates are made of the spillway cost to provide for various flows, and of the probabilities of the flows being exceeded, as follows:

Design	Flow in cu ft/sec	Probability of greater flow in any one year	Investment in spillway to provide for this flow
A	8,500	0.08	$200,000
B	10,000	0.05	225,000
C	13,000	0.02	300,000
D	14,500	0.01	350,000
E	17,500	0.005	392,000
F	19,000	0.002	480,000

Investment charges on the spillway are to be calculated on the basis of a 75-year life with zero salvage value, an i^* of 7%, and no taxes (as this is a public works project). Operation and maintenance costs will be unaffected by the spillway capacity chosen. The estimated damages are approximately $500,000 if the flow exceeds the spillway capacity.

What spillway capacity makes the sum of investment charges and expected value of damages a minimum? Discuss the sensitivity of this minimum cost capacity to the chosen i^*, to the estimated damages if spillway capacity is exceeded, and to the estimated probabilities of the extreme floods. (*Ans*. Design D, expected annual cost = $29,654)

15–4. In the 7-spot version of Keno at the XYZ Club, a $6 ticket pays off according to the following schedule:

Winning Spots	Payoff
0 to 3	None
4	$6

5	120
6	2,450
7	25,000

Use the probabilities tabulated in Example 15–2 to compute the expected value of the winnings of a player who buys such a $6 ticket. (*Ans*. $3.754)

15–5. Consider the solutions in Example 15–2 and Problem 15–4. Discuss possible reasons why the XYZ Club gives the buyers of $6 tickets relatively less favorable treatment than the buyers of 60¢ tickets. (The club has a policy, announced to all players, of never paying more than $25,000 to the aggregate of the winners in any game.) Do you see any similarity between what you think is the reasoning of the XYZ Club on this matter and the type of reasoning appropriate in decisions about designs intended to reduce the adverse consequences of extreme events such as floods, fires, earthquakes, and tidal waves?

15–6. On page 356, in the discussion of a change in the data relative to estimated damages from floods greater than the spillway capacity in Example 15–3, it is stated: "Similar calculations for spillway capacities of 1,900 and 2,300 sec ft will yield figures for average annual damages of $6,000 and $1,450, respectively." Show your calculations to check these two figures.

15–7. There is 1 chance in 10 that a certain event will occur in any given year; if it occurs it will require an expenditure of $50,000. With an i^* of 15%, what is the justifiable present expenditure to eliminate the risk for 10 years? What is the justifiable present expenditure to reduce the risk from 1 in 10 to 1 in 50 for the same 10 years? Assume that the expected value of the outlay if the event occurs is deemed to be an appropriate measure of the annual cost of the risk.

15–8. Compare the sums of annual investment charges and expected values of damage for the 6 spillway capacities of Problem 15–3 assuming the damage will be $350,000 if the spillway capacity is exceeded by not more than 1,500 cu ft/sec, and $750,000 if it is exceeded by more than 1,500 cu ft/sec. Estimated probabilities that certain other flood magnitudes will occur in any one year are as follows:

Flow (cu ft/sec)	**Probability**
11,500	0.03
16,000	0.0075
20,500	0.001

15–9. For the data of Problem 15–3, calculate the appropriate benefit–cost ratios to determine the most economical design. Assume that Design A is the base situation. Which design is most economical?

15–10. The owner of an orange grove in a relatively frost-free area is considering the question of whether or not to purchase and operate smudge pots as a protection against frost. An investigation of the weather records in this area shows that during the past 50 years there have been 4 years with freezes sufficiently bad to injure the fruit. It is estimated that on the average such a damaging freeze will reduce the cash receipts from the sale of the crop by about 75%. Average annual gross receipts from this grove during a year without frost are estimated as $125,000.

The initial investment in smudge pots to protect this grove will be $15,000. The life of the pots is estimated as 10 years. Annual labor cost for setting out pots in the autumn, removing in the spring, and cleaning is estimated as $1,500. Average annual fuel cost will be $500. Assuming that pots will give complete protection against loss due to frost, show calculations to provide a basis for judgment on whether it is desirable to purchase and operate the pots. Assume a before-tax i^* of 25%.

What matters not reflected by the use of expected values do you think might reasonably enter into the owner's decision on this matter? Discuss the sensitivity of the conclusions of your analysis to changes in the various parameters, for example, changes in i^*, in the estimated probability of a damaging freeze, in the price of oranges, and in the labor cost associated with using the smudge pots.

15–11. (This is adapted from an example in a paper by C. H. Oglesby and E. L. Grant published in Volume 37 of *Highway Research Board Proceedings*.)

It is desired to select a size for a box culvert for a rural highway in central Illinois. The drainage area has 400 acres of mixed cover with slopes greater than 2%. The culvert will be 200 ft long. As the headroom is critical, the culvert can be only 4 ft high. If the water rises more than 5 ft above the streambed, the road will be overtopped. Damage to highway and adjacent property for each overtopping will be $30,000.

If the project is to be built at all, the minimum acceptable culvert for this location is a simple box 10 × 4 ft, which will be overtopped, on the average, once in 5 years. Four possible designs, with associated initial costs, capacities, and probabilities of overtopping, are as follows:

	First cost	*Capacity cu ft/sec*	*Probability of overtopping in any one year*
A. Single culvert 10 × 4 ft	$32,500	300	0.20
B. Double culvert 8 × 4 ft	$42,000	400	0.10
C. Double culvert 10 × 4 ft	$50,000	500	0.04
D. Triple culvert 8 × 4 ft	$62,000	600	0.02

Which design gives the lowest sum of the annual cost of capital recovery of the investment and the expected value of the annual damage from overtopping? Assume an i^* of 10%. Assume culverts will have lives of 50 years with zero salvage value.

15–12. For the data of Problem 15–11, calculate the appropriate benefit–cost ratios to determine the most economical design. Assume that Design A, the single 10 × 4 ft culvert, is the base situation.

15–13. Ewing Oil Company must decide to drill on an option lease property or sell the lease option. The lease can be sold for $100,000. Drilling will cost $750,000 and may yield an oil well, a gas well, a combination oil/gas producer, or a dry hole. To date, 65 wells have been put down in this field. The success record is as follows:

Dry	15
Gas	12
Combination	20
Oil	18

Successful wells in this field are currently expected to last 15 years. The present worth of the net cash flows (annual receipts minus disbursements and royalties) discounted at a minimum attractive rate of return before income taxes of 40% are:

Gas	$850,000
Combination	1,200,000
Oil	1,500,000

Draw a decision tree to describe the alternatives and probable events in this case. If the company makes its decision based on maximum expected value, which alternative will be chosen? Discuss the various ramifications of the outcomes, risks, and probabilities associated with this decision.

15–14. Set up the decision matrix for Problem 15–13 as described in the text. Find the maximum security level strategy. What are the implications of a strategy of this type applied to decisions of the type faced by a small independent oil drilling company such as Ewing Oil? Discuss this question in relationship to relative frequency probabilities, as used in Problem 15–13, and intuitive probabilities.

15–15. The following decision tree results from a company's analysis of the expansion of one of its distribution centers over a 5-year period and is expressed in terms of the present worth of receipts minus disbursements over the period. The alternatives are A_1, no expansion; A_2, expand the building and re-equip the entire facility; and A_3, expand the building now but delay re-equipping the facility for 2 years to see whether the expected growth develops.

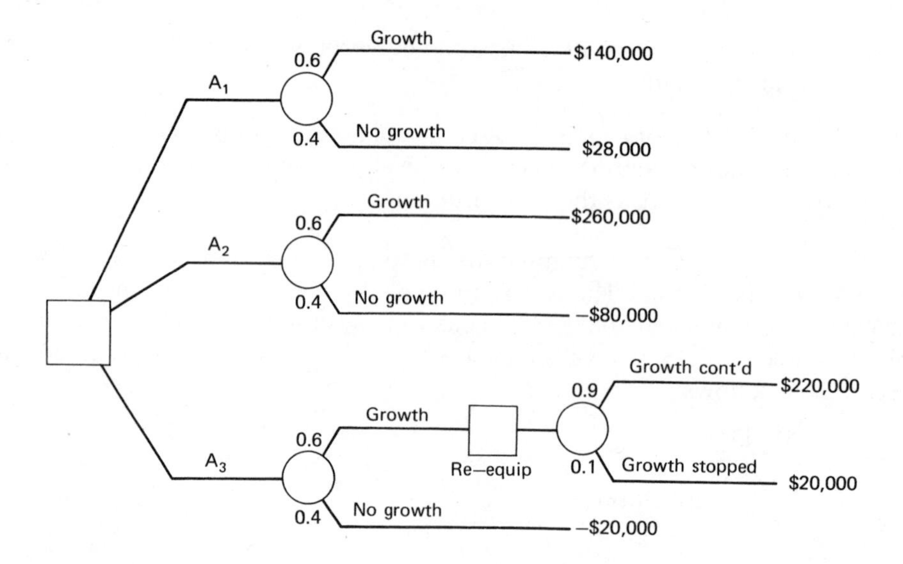

15–16. The following decision matrix presents the analysis of three mutually exclusive alternatives in terms of annual cost over a 10-year period (figures in thousands of dollars):

	S_1	S_2	S_3	S_4
A_1	−26	−30	−24	−36
A_2	−17	−32	−30	−28
A_3	−35	−20	−22	−30

(a) Apply the maximum security level decision rule to choose among these alternatives.

(b) Assume the probabilities of each future state are 0.4, 0.2, 0.3, and 0.1 for states S_1 through S_4, respectively. Which alternative has the minimum expected annual cost?

15–17. A large chain of motels is planning to locate a new 150-unit motel near the intersection of two interstate highways. Land cost is $400,000. The building will cost $2,800,000 and furnishings, which must be replaced every 5 years, will cost $475,000. It is expected that the motel will be sold at the end of 15 years for $2,000,000. Annual disbursements for operation and maintenance are expected to be $150,000.

The average daily room rate is expected to be $30 with the motel operating 365 days per year. Market study indicates that there is a 30% chance the motel will have an 80% annual occupancy rate, a 60% chance of 60% occupancy rate, and a 10% chance of only a 35% occupancy rate.

Determine the expected net present worth before income taxes of this investment opportunity using an i^* of 20%. What considerations other than the expected net present worth might the company want to investigate before making the decision to proceed with this project?

16

INCREMENT COSTS AND SUNK COSTS

*"What is a unit of business?" . . . The most important unit is a single business decision.—J. M. Clark**

Throughout this book it has been emphasized that it is always prospective *differences* between alternatives that are relevant in making a choice. In a going concern all past receipts and disbursements and many future ones will be unaffected by a particular decision. It often happens that average costs per unit (sometimes called "unit costs") are misleading guides to choosing between given alternatives.

Frequently an economy study ought to be based on a critical look at the effect of a proposed course of action on prospective receipts and disbursements. Examples 16–1 and 16–2 illustrate this point in relatively simple cases. These examples serve the incidental purpose of illustrating two common types of electric rates.

EXAMPLE 16–1

Increment Cost of Electricity for Household Purposes

Facts of the Case

A family purchases electric energy under the following monthly rate:

Customer charge	$1.75/month
plus	
Energy charge	
First 240 kw-hr	@ 3.7¢/kw-hr
Next 240 kw-hr	@ 5.9¢/kw-hr
All over 480 kw-hr	@ 8.1¢/kw-hr

*J. M. Clark, *Studies in the Economics of Overhead Costs* (Chicago: University of Chicago Press, 1923), p. 213.

The family's monthly consumption of electricity is about 500 kw-hr. This leads to a monthly bill of $26.41. The average cost per kilowatt hour is 5.282¢.

The family is considering the possibility of improving the illumination in certain rooms. It is estimated that the proposed changes will increase the monthly consumption by 100 kw-hr. An examination of the rate shows that the change will add $8.10 to the monthly bill; each extra kilowatt hour will cost 8.1¢. The cost of 600 kw-hr per month will be $34.51.

A superficial but incorrect view of the matter would have been to assume that because the previous unit cost had been 5.282¢/kw-hr, 100 extra kilowatt hours would add $5.28 to the bill.

EXAMPLE 16–2

Increment Cost of Electricity for Industrial Purposes

Facts of the Case

A small manufacturing company buys electricity under the following rate:

A monthly demand charge is computed as follows:

First 40 kw of billing demand	$91.00
Next 260 kw of billing demand	1.99/kw
All over 300 kw of billing demand	1.82/kw
plus	
An energy charge of 5.1¢/kw-hr	

In a representative month when the company used 80,000 kw-hr with a billing demand of 400 kw, its bill was computed as follows:

Demand charge:	
First 40 kw	$91.00
Next 260 kw @ $1.99	517.40
Last 100 kw @ $1.82	182.00
	$790.40
Energy charge:	
80,000 kw-hr @ 5.1¢	$4,080.00
	$4,870.40

The average cost per kilowatt hour in this month was $4,870.40 ÷ 80,000 = $0.06088 or 6.088¢.

The company considers installing equipment that will reduce labor cost on certain special jobs. This new equipment would add 100 kw to the billing demand. Because its use will be infrequent it will add only 1,000 kw-hr of energy per month. The increase in the monthly bill can be computed from the rate to be $233. This is the sum of an extra $182 of demand charge and an extra $51 of energy charge.

An uncritical application of the unit cost of 6.088¢/kw-hr would have led to the incorrect conclusion that the extra 1,000 kw-hr would add only $60.88 to the monthly bill. This $60.88 is only a little over one-fourth of the $233 that actually would be added. Here is a case in which it would have been extremely misleading to have made the assumption that additional energy could be purchased at the average cost per kilowatt hour of the energy already purchased. Nevertheless, this type of assumption seems to be a common one in industrial economy studies.

Demand Charges in Electric Rates

Because electric energy cannot be stored economically, an electric utility needs to have enough generating station capacity to meet its expected maximum load. Also, the required capacities of electric transmission and distribution systems must be enough for the expected maximum loads on these systems. It follows that there are a number of electric utility costs that depend largely on a utility's need to be ready to serve its customers at the time of maximum load.

Demand charges such as the one illustrated in Example 16–2 are a common feature of electric rates. In a general way, such charges aim to charge customers their fair shares of costs of readiness to serve.

Load Factor and Capacity Factor

Two useful phrases originating in the generation of electricity are *load factor* and *capacity factor*. Load factor is defined as the ratio of average load to maximum load. Average and maximum loads may be taken for any desired period of time. Thus, a power company will have a daily load factor each day that is the ratio of the average load for that day to the day's maximum load, and it will have an annual load factor that is the ratio of the average load for the year to the maximum load occurring during the year. A company with a typical daily load factor of 70% might conceivably have an annual load factor of 55% because of seasonal fluctuations in load.

Capacity factor is defined as the ratio of average load to maximum capacity. In comparing entire power systems it is more common to use load factor than capacity factor. This is chiefly because of the difficulty of securing a satisfactory uniform measure of power system capacity. It is appropriate, however, to speak of the capacity factor of an individual generating unit or of a generating station.

Although the terms "load factor" and "capacity factor" have a definite quantitative meaning in the electric power field, they are often used loosely in a qualitative sense in other fields of production. Thus, one might say that a factory with a seasonal demand for its product might improve its load factor by taking on another product to be produced in the off season.

Relation of Economy Studies to the Accounts of an Enterprise

The tendency to use "unit cost" figures that are readily available, rather than to take a critical look at differences, seems to be even greater when data for economy studies are drawn from the accounts of an enterprise. Many economy studies will combine information obtained from the accounting system with information obtained from other sources, such as time studies or other types of performance tests. The usefulness of accounting information will depend upon the detail of the classification of accounts, the skill with which it has been drawn, and the care with which actual expenditures have been charged to the appropriate accounts.

However, even the best accounting systems may give misleading conclusions if the figures shown by the books of account are uncritically used. The point of view of an estimator for an economy study is necessarily different from the point of view of the accountant. The economy study is concerned with prospective differences between future alternatives. The accounts of the enterprise are a record of past receipts, disbursements, and certain other matters. They generally involve apportionments of past costs against future periods of time, and apportionment of joint costs among various services or products; such apportionments are sometimes misleading to estimators who are making economy studies.

Illustrations of Incorrect Inferences from Accounting Apportionments

The following two cases relate to the experience of a city that owned its electrical distribution system, retailing electric energy. Although most of its power was purchased at wholesale rates from a large power system, a small part was generated at peak load periods by its diesel engines (of relatively small capacity) in its own generating plant. In the accounts of this city the expenses of the electric utility were carefully distinguished from the cost of carrying on governmental functions and from the cost of running the municipally owned water utility. The expenses of joint departments were prorated between the city government and the electric and water utilities on an equitable basis; the governmental departments and the water utility were charged by the electric utility for electricity used at rates such as might reasonably have been charged by a privately owned corporation. Two situations arose in which this well-organized plan of accounting served to block an understanding of the true differences between the alternatives which it was desired to compare.

In the first situation some of the council members examined the municipal report, which showed the cost of purchased energy at less than half of the cost of generated energy. They concluded from these figures that a substantial saving would take place if the diesel plant were shut down and all electricity were purchased.

It turned out that no such conclusion as this was justified by the facts. Many of the charges included in the unit cost of generation were allocated charges that would not have been reduced if the generating station had been shut down; they would simply have been allotted to some other account than "Power Generation." The crew that operated the diesel engines also operated the substation and the pumping plant for the water department; the labor cost was, therefore, divided uniformly among those three activities. But as it still would have been necessary to run the substation and waterworks, no reduction in labor cost could have been made by discontinuing the operation of the diesels. Another charge allotted against the "Power Generation" account was the depreciation of the diesel engines. This, however, was simply a time allotment against the current year of an expenditure that had been made many years before; no part of that past expenditure could be eliminated or recovered by shutting down the diesels.

Thus, the generating costs that were really relevant to the question "Shall we continue to operate the diesels, or shut them down and purchase all power?" appeared to be merely fuel, lubricants, and maintenance; these were less than the average cost of purchased energy. Because energy was purchased on a rate that included a substantial charge for maximum demand, the diesels had been operated only at periods of peak load in order to reduce the demand charge. If the diesels were shut down, the extra cost of purchasing the peak load energy would be considerably greater than the average cost for the base load energy. When the members of the council finally recognized this as a problem in determining differences between alternatives, they were able to see that it was clearly economical to continue to generate peak load energy, despite the apparent showing of the accounts that generated energy cost much more than purchased energy.

In the second situation the question arose regarding the economy of the city's building its own generating station rather than continuing to purchase power. Here, in estimating the expense of the proposed generating station, the engineer's report included merely direct generating station costs and made no allowances for increased costs in other departments. In justification of this it was argued that overhead costs were apportioned in the accounting system in proportion to direct costs and that as direct costs would not be increased with the proposed plant there would, therefore, be no increase in indirect costs as a result of the plant.

A more realistic examination of this situation indicated that the engineer's reasoning was in the same class with assuming that the increment cost of extra miles on your automobile was merely the out-of-pocket expense for gasoline; that is, this reasoning took a too short-run viewpoint. It seemed reasonable to believe that doubling the investment in the electric utility, as this would do, would increase the responsibility of the manager of that utility. In the long run the manager would be better paid and would require an assistant sooner with the generating station than without it. It also seemed reasonable to believe that the operation of a generating station would involve more of such services as engineering, accounting, purchasing, and storekeeping than would the paying of a single power bill once a month. These conclusions were reinforced by an examination of the experience of other cities, which seemed to indicate quite definitely that expenses of these service departments always tended to move upward with any increase in activity. From such a study, it was possible to make a judgment as to what this long-run increase in expense in indirect departments would be. This item of cost had considerable weight in throwing the final decision against the generating plant which, on the basis of preliminary study, had appeared as if it might be economical.

The "With or Without" Viewpoint and the Concept of Cash Flow as Aids in Decision Making

The two cases just described had one characteristic in common with one another and with Examples 16–1 and 16–2. In each instance one alternative was to make some proposed change from the existing way of doing things and the other alternative was to continue this existing way. A search for the differences between such alternatives calls for a prediction of what will happen *with* the proposed change and *without* it. The phrase "with or without" will sometimes help to clarify the issues in such cases.

Throughout this book we have emphasized that, wherever practicable, the prospective physical differences between alternatives should be converted into prospective differences in cash flow. Although our concentration of attention on cash flow has been aimed particularly at the securing of data permitting the calculation of rates of return on proposed investments, the cash flow viewpoint is also helpful in avoiding the types of estimating errors that are discussed in the present chapter. An analyst who adopts the cash flow viewpoint is likely to take a critical look at all figures based on allocations.

Treatment of Unused Capacity in Economy Studies

In any organization it is likely that various kinds of unused facilities will exist from time to time. Extra space may exist in office or warehouse or factory; extra capacity may exist for various service facilities (such as water, steam, or compressed air). Under such circumstances, a proposal for a new activity needing such facilities (e.g., expanded output, a new product, the production of items previously purchased) may not require any immediate investment to secure the facilities.

A similar, but slightly different, condition exists when it is possible to carry out the proposed new activities using existing facilities at some kind of a cost penalty until the time comes that the need for still further capacity makes it practicable to install a new unit of economic size. For example, the storage of additional material in an existing warehouse, already crowded to its economic capacity, might cause a disproportionately large increase in materials-handling costs.

In an expanding organization most unused capacity that now exists is likely to be only temporary. For this reason it usually is incorrect for economy studies to assume that proposed new activities that will use existing unused facilities will never be responsible for investments in new facilities of the types in question. On the other hand, because an immediate investment in such facilities will not be required, it obviously is incorrect to make the economy studies assuming an investment before it will need to be made. Frequently, it should be recognized that the question at issue is the *timing* of an investment that will be required eventually.

Analysts should guard against assuming that such an investment will be deferred into the distant future. Ray Reul has suggested that the assumed period of deferment should never be more than five years; for projects with lives of less than 15 years, it should not exceed one-third of the project life.*

Burden Rates are Seldom Adapted for Use in Economy Studies

Cost accounting in the manufacturing industries classifies production expense into direct labor, direct material, and indirect manufacturing expense. This last class may include a wide variety of items (such as salaries and wages of supervisors, inspectors, clerical employees, crane operators, and storekeepers; operating supplies; packing and unpacking; shop losses due to defective material or workmanship; purchasing; receiving; shipping; heat, light, and power; plant maintenance; taxes, insurance, and rentals; and depreciation).

Indirect manufacturing expense is called by various shorter names, the most common of which are "burden," "overhead," and "expense." Although there is great variation in cost accounting systems, in many of them each item included in burden is allotted among the various departments (or production centers) on some basis that appears reasonable. Then the total burden allotted against a department is charged against the product of that department in proportion to some "burden vehicle." This burden vehicle will be something that is readily measurable regarding the product (such as its direct labor cost, its total direct cost, its labor hours). More often than not the burden rate is "predetermined"; that is, it is based on the allotment of estimated indirect manufacturing expenses and on estimated direct labor cost (or other burden vehicle) reflecting estimated production.

In a department in which the vehicle is direct labor cost and the burden rate is 130%, each product is charged with $1.30 for every dollar of direct labor cost. Such burden allocation in cost accounting serves many useful purposes.

It does not follow, however, that a change in production methods that reduces direct labor cost by $1,000 will reduce indirect manufacturing expense

*R. I. Reul, "Profitability Index for Investments," *Harvard Business Review*, Vol. 35, No. 4 (July–August, 1957), p. 122.

by $1,300. Or that if one of two alternative new machines involves $5,000 less direct labor cost than the other it will also involve $6,500 less indirect manufacturing expenses. The only way to judge the relative effect of two alternatives on burden is to consider their probable effect on each of the individual items of indirect manufacturing expense that have been combined together in the burden rates.

Although the error of assuming that any saving in direct cost will be accompanied by a proportionate saving in indirect cost should be obvious, it appears that this error (which has appeared in several published formulas for determining economy) is often made in industry.

Allocations in Accounting and in Economy Studies

Thus, it is necessary that the engineer look beneath the surface of an accounting figure before using it in an economy study.

For instance, an engineer for a railway company was called upon to compare the cost of increasing the generating capacity of the power plant serving one of the railway company's shops with the cost of purchasing power. In making the comparison, it was necessary to recognize that the accounting charges to shop power that the railway made under its accounting routines did not reflect all of the costs that were pertinent to the comparison. For instance, one of the major savings that would result from purchased power was in the cost of coal, and a considerable portion of the coal cost was the cost of its transportation from the mine. But this transportation cost was not allocated to the account showing shop power expense; the railway accounts considered it merely as part of the cost of conducting transportation. Another major advantage of purchasing power as compared with increasing generating capacity was in the lower investment involved, with corresponding lower investment costs of interest, depreciation, and taxes. But none of these investment costs was allocated against shop power expense in the railway's accounts.

Treatment in Economy Studies of an Investment-Type Disbursement Charged as a Current Expense in the Accounts

In tabulating the effect of a decision on cash flow before income taxes, a disbursement is negative cash flow regardless of whether it is capitalized on the books of account or treated as a current expense. It often happens that certain nonrecurrent disbursements that will be "expensed" in the accounts are associated with the acquisition or replacement of physical assets. For example, in considering the replacement of a railway bridge, the cost of handling traffic during the replacement period may be different between alternative replacement structures. This cost of handling traffic will be considered as an operating expense in the railway accounts. Nevertheless, its effect on cash flow before income taxes is just the same as if it were capitalized in the accounts.

However, in any analysis to determine cash flow after income taxes, it is essential to differentiate between disbursements that are to be capitalized and those that are to be expensed. This point was brought out in Example 12–6.

Cost Information That Accounting Records Will Not Give

Where the economy of some new process or machine is involved, estimates of operating costs must be obtained by experimental studies rather than from accounting records. Time studies or laboratory studies regarding the characteristics of new machinery may be combined with known wage rates and material costs and with an analysis of indirect costs in order to arrive at cost figures.

Frequently, an economy study requires the consideration of some cost that, by its very nature, cannot be isolated by accounting charges. For instance, in considering the economy of automatic block signal systems for railway trains, an important saving will be in the elimination of train stops. But what is the cost of stopping a train? The railway accounts cannot isolate this cost. It must be determined on the basis of fuel saved from an analysis of locomotive performance curves, and on the basis of the economies resulting from time saved.

In designing any cost accounting system, the question always arises as to the detail to which accounting records ought to be carried. To what extent may approximations be used in allocating costs in place of more precise methods of cost allocation that are more expensive? There is always the conflict between the expense of getting better accounting records and the value of the information obtainable from such records. No cost accounting system is justified that will not pay its way by giving information that is worth more than the cost of getting it. Where cost systems are planned with the idea of simplifying determinations of economy, the question is likely to arise whether it will be more economical to maintain regular continuous records or to make an occasional analysis when a particular sort of cost information is required. There will be, in many organizations, cost information useful to engineers, that must be obtained by analysis, rather than from the accounting records, simply because it does not pay to keep accounting records in such detail as would be required to furnish this information.

Allocation of a Previously Incurred Loss

It is not practicable for accountants to go back into past records and revise past figures when an error is discovered.

Once an industrial engineer made some suggestions of methods aimed to reduce the operating cost of the power plant in a factory. In the course of the investigation it was also discovered that the amount of coal in the coal pile was much less than the amount shown by the inventory figure in the books of account.

The engineer was greatly surprised some months later when the factory superintendent reported that the power plant operating costs had gone up rather than down. On investigation the engineer discovered that the accountant had decided to spread the cost of the fuel shortage uniformly over the next 6 months after it was discovered, in order not to make it seem as if expenses had been very high in the month in which the shortage was discovered. This avoided distorting the comparative operating cost statements for various months. But the fuel shortage cost charged subsequent to its discovery more than neutralized the economies resulting from the engineer's suggestions! On superficial examination the situation appeared as if the fuel costs had not been reduced by the suggestions.

Increment Cost

The phrase *increment cost* is used in this chapter to refer to a prospective difference in cost in certain "with or without" situations. This is not a phrase that can be given a precise definition in general terms; it is a phrase that is useful chiefly in reference to specific alternatives. Other phrases sometimes used in the same meaning are *incremental cost* and *differential cost*.

These phrases are also used in economic literature in connection with discussions of determination of pricing policies. The topic of pricing policy is a complex one that is beyond the scope of this book.

Difficulty in Estimating Increment Costs

Assume that an automobile owner is uncertain whether to use public transportation or a private automobile for a certain 600-mile trip. Before making a decision, the owner wants to estimate the extra outlays in connection with the automobile if it is used for this 600 miles.

Two extremes, neither valid, are sometimes observed when estimates of this type are made. One involves calculating a "unit cost" per mile of operation by dividing estimated total costs over the entire period of ownership by estimated total miles during this period; this unit figure is then multiplied by the mileage of a projected trip to find the cost of the trip. If our hypothetical car owner estimates total costs over the period of ownership to be 30¢ per mile and uses this type of reasoning, he or she will conclude that his 600-mile trip will involve automobile costs of $180.

Another extreme is to estimate merely the out-of-pocket expense for gasoline during the trip and to view this as the trip's cost for decision-making purposes. If our owner takes this extreme short-run view and estimates gasoline costs as 10¢ per mile, he or she will conclude that the 600-mile trip will cost $60.

The reader will recognize that neither type of extreme estimate recognizes the "with-or-without" aspect of the car owner's decision. The total costs of car ownership include many items not affected by the decision to drive the car 600 extra miles. On the other hand, it is unlikely that the $60 for motor fuel covers all the extra long-run outlays that will be caused by the trip. Increased mileage tends to cause increased outlays for lubricants, tires, repairs and maintenance, and possibly accidents, even though no such outlays may happen to occur *during* a particular trip. If our car owner is to base the decision in part on cost, some estimates are needed of how these items are influenced by extra miles. Our traveler should recognize that an estimate of increment costs should be the objective, even though it cannot be checked by the best possible cost and performance records on a single automobile.

Similarly, even though increment costs are not precisely determinable in industry, they need to be estimated wherever decisions are to be based on cost. As has been pointed out, questions of relative economy of technical alternatives are often complicated by the difference between the short-run and long-run viewpoints. This difference is even more troublesome in cases of increment cost pricing.

Cost Allocations in Accounting in Relation to Economy Studies

Illustrations have been given in situations in which accounting allocations of cost seemed to block a clear recognition of differences between alternatives. This suggests an important problem arising in the management of every business organization having more than one department, that is, "What should be the basis of joint cost allocations and interdepartmental charges?"

Of course, this question cannot be answered in general terms. Interdepartmental charges, for instance, may be established on an average cost basis or on an increment cost basis. (Either permits considerable room for controversy; in fact, the question of interdepartmental charges may be a source of bitter argument between department heads within an organization.) The point to be emphasized here is that no matter what basis is used, there will be some types of decisions in which it will be misleading to use the interdepartmental charges and joint cost allocations as established by the accounting system; no one answer can serve all purposes. No scheme is satisfactory without managerial understanding of its limitations.

The Concept of a Sunk Cost

Once the principle is recognized that it is the *difference* between alternatives that is relevant in their comparison, it follows that the only possible differences between alternatives for the future are differences in the future. The consequences of any decision regarding a course of action for the future cannot start before the moment of decision. Whatever has happened up to date has already happened and cannot be changed by any choice among alternatives for the future. This applies to past receipts and disbursements as well as to other matters in the past.

From the viewpoint of an economy study, a past cost should be thought of as a *sunk cost*, irrelevant in the study except as its magnitude may somehow influence future receipts or disbursements or other future matters. Although this principle that a decision made now necessarily deals with the future seems simple enough, many people have difficulty in accepting the logical implications of the principle when they make decisions between alternatives. This seems particularly true when sunk costs are involved. Although some of the failures to recognize the irrelevance of sunk costs involve a misuse of accounting figures, these mental obstacles to clear reasoning are by no means restricted to people who have had contact with the principles and methods of accounting.

This concept of the irrelevance of past costs is illustrated in a simple way in Example 16–3.

EXAMPLE 16–3

Irrelevance of a Past Outlay

Facts of the Case

Brown and Green are partners in a small engineering consulting firm. One year ago the firm purchased a new electronic calculator to replace the old one it had used for 10 years. Rapid advances in calculator technology had resulted in the production of units capable of being programmed with short but highly sophisticated routines. The newer units contained many automatic functions that were not available at any price at the time the firm purchased the old electronic calculator. These features would eliminate the necessity to constantly refer to tables of logarithms, trigonometric functions, and the like.

Before the purchase actually was made, the choice had been narrowed to a diode display unit and one that also provided a paper-tape printout of data inputs and solutions. The prices were $400 and $850, respectively. Although the paper-tape printout had the advantage of greater ease of verification of results, this advantage was not believed to be sufficient to justify the required extra investment of $450. Thus, the diode display unit was selected and a 10-year life for the unit was assumed for accounting purposes.

Now, one year later, the company that made the paper-tape printout unit has considerably improved its product. Among other improvements are (a) greatly increased speed, memory storage capacity, and mathematical capability and (b) expanded programming capability. Moreover, the price of the improved unit has been reduced to $795. The year-old diode display unit can be sold for approximately $125.

Experience with the present unit over the past several months has demonstrated that the paper-tape feature would have saved a fair amount of labor time in verifying results and hunting for errors. In addition, it is evident that improvements in the paper-tape unit now available should save some labor time in making calculations. Although Brown and Green's accounting system does not provide any basis for a formal measurement of the labor costs that would be saved by replacing the present unit, Brown estimates that they will be at least $300 per year.

Brown also investigated the possibility of replacing the calculator with a small microprocessor system. Although these systems were faster than available calculators, they were not as portable and would require extensive programming that could not be afforded at the time. On this basis, Brown favors buying the new paper-tape calculator and selling the year-old unit on the second-hand market.

Green accepts Brown's $300-a-year savings figure but opposes making the change. Green points out that the book value of the present unit is $360 (its $400 cost minus the $40 depreciation charge for one year based on the straight-line method and a 10-year estimated life). Thus, there will be a $235 "loss" if it is sold for $125. Green reasons that this $235 loss should be added to the $795 price of the paper-tape printout calculator to find the "real" investment required by the replacement unit. With this $235 added to the $795 price Green contends that the paper-tape unit is not economically justified.

Green's attitude illustrates the type of mental obstacle that so often interferes with correct decisions. The firm paid $400 for its present calculator and has received one year's service from it. Its early retirement not only leads to a write-off on the books of account, but it also seems to admit a past error of judgment that would not be admitted in the absence of the replacement.

Finally, Brown was able to demonstrate that Green's initial reasoning was incorrect. Brown prepared a table as follows:

Year n	Do not replace diode unit	Replace with paper-tape unit
-1	$-\$400$	$-\$400$
0	0	$-\$795 + \$125 = -\$670$
1–10	0	$+\$300$ per year
Totals	$-\$400$	$+\$1,930$

The realistic view regarding this past outlay of $400 is, of course, that the money has been spent regardless of which alternative is selected for the future. Because the past outlay is the same regardless of the alternative selected, it should not influence the choice among alternatives.

(Aspects of this matter that neither Brown nor Green recognized are the appropriate treatment of present salvage values in replacement studies, aspects of estimated lives of equipment, and the possibility of an income tax advantage from early retirement of the diode display unit. These topics are discussed in Chapter 17.)

Retirements and Replacements

A decision on a *retirement* is a decision regarding whether or not to continue to own some fixed asset, for example, some machine or structure. In some cases an asset retired may be scrapped, with no salvage value (or even with a negative salvage value if the cost of removal and disposal exceeds any receipts from disposal). In other cases there may be substantial net resale value, sometimes even higher than the original cost.

An asset retired may or may not be replaced. In modern industry, with its frequent improvements in design and changes in service requirements, it is common for the replacement machine or structure to differ in various ways from the machine or structure being retired. Often the new machine or structure may serve other functions in addition to providing the same services that were given by the machine or structure retired.

In some cases a new asset may be acquired to replace the services of an old asset, but the old asset will not be retired. Sometimes the old asset may be used for another purpose, as when an old main-line railroad locomotive was relegated to branch-line service. In other instances the old asset may be continued in the same general type of service but used less frequently than before; for example, an old steam power plant originally used to carry base load might be used only a few hours a year for peak load purposes.

Thus, a retirement may be made either with or without a replacement. And the services of an existing old asset may be replaced or augmented by the services of a new one either with or without the retirement of the old asset.

The Question of the Cost of Extending the Service of an Asset Already Owned

In any economy study involving a prospective retirement, it is necessary to consider the money difference between disposing of the old asset at once and disposing of it at some future date. In determining this figure, the past investment in the asset is irrelevant. The current book value is a result of the past investment and the past depreciation charges made in the accounts; this also is irrelevant. So also are the future depreciation charges to be made in the books if the asset is continued in service.

The relevant estimates include the prospective net receipts, if any, from disposal of the asset (1) on the assumption that it is retired immediately, and (2) on the assumption that it is continued in service for the immediate future and retired at some specified later date. Under assumption (2) it is necessary to consider all prospective receipts and disbursements that will take place if ownership is continued but will not take place if the asset is retired immediately. In many economy studies regarding retirements, it may be desirable to consider several different specified future dates on which the asset might be retired if continued in service for the time being. For example, it may be appropriate to consider the cost of keeping the old asset in service for 1 year, for 2 years, for 3 years, etc.

Example 16–4 illustrates the estimates and calculations that are needed relative to the capital recovery costs of extending the service of an asset. This example provides essential background for the discussion in Chapter 17 of practical problems of judging the economy of proposed retirements and replacements. The reader should therefore examine Example 16–4 carefully before starting Chapter 17.

EXAMPLE 16–4

Computing Capital Recovery Costs of Extending the Service of an Asset for Various Periods

(a) Capital Recovery Cost of Extending a Service for One Year

A 2-year-old piece of construction machinery had a first cost of $25,000 and has been depreciated on the books of its owner by the straight-line method at 20% a year. Its present book value is therefore $15,000. Its present net resale value in a secondhand market is $13,000. It is estimated that this resale value will decrease to $10,000 if the machine is held for another year, to $7,500 if it is held for 2 years more, and to $5,500 if held for 3 years more. The question arises whether to dispose of this machine immediately for $13,000 and to rent a similar machine if one should be needed, or to continue it in service for a year or more. Interest (minimum attractive rate of return) is 15% before income taxes.

As a first step in finding the money differences between disposing of the machine at once and disposing of it at some later date, it should be noted that at the present moment it is possible to have either the $13,000 or the machine but not both. As far as immediate money receipts and disbursements are concerned, the differences are as follows:

Keep machine	**Dispose of machine**
No receipts or disbursements	Receive $13,000

The immediate money difference is just the same as if the question at issue were the purchase of a secondhand machine for $13,000. In this case the immediate receipts and disbursements would be:

Buy machine	**Do not buy machine**
Disburse $13,000	No receipts or disbursements

Whether it is a question of keeping a machine that can be sold for $13,000 or acquiring one that can be purchased for $13,000, we have $13,000 more without the machine than with it. If all future estimates (i.e., annual receipts and disbursements, future salvage values, and irreducible data) are the same,* an economy study comparing the alternatives of (a) continuing an

*In most cases there would be some differences between the cost estimates that would seem appropriate for a used asset already owned and an apparently identical used asset to be acquired. For example, the net realizable salvage value of a machine already installed in the plant is the secondhand price *minus* the cost of removing it, transporting it to the market, and selling it. In contrast, the installed cost of a purchased secondhand machine is the secondhand price *plus* the costs of buying it, transporting it to the plant, and installing it. Moreover, the appropriate estimates of future repair costs might differ. There might well be a great deal more known about the maintenance history of a used machine already owned than about that of an apparently identical used machine purchased in the secondhand market. Hence, the factor of uncertainty might well lead to a higher estimate of repair costs for a purchased machine. The point made in Example 16–4 does not bear on these matters but is simply that a decision to continue an asset in service is, in principle, identical with a decision to acquire the same asset at an outlay equal to the present net realizable value.

asset in service, and (b) disposing of the asset at once, is identical with an economy study comparing the alternatives of (a) acquiring the same asset at a price equal to its present net realizable value if disposed of, and (b) not acquiring it. This general principle may be applied in all economy studies regarding proposed retirements.

In all such economy studies the capital recovery costs on any asset already owned should be based on its present net realizable value if disposed of (i.e., on the amount of capital that could be recovered from its disposal). With interest at 15% the capital recovery cost of extending the service of the 2-year-old asset for one more year is:

$$\begin{aligned} CR &= (\$13{,}000 - \$10{,}000)(A/P,15\%,1) + \$10{,}000(0.15) \\ &= (\$13{,}000 - \$10{,}000)(1.15) + \$10{,}000(0.15) = \$4{,}950 \end{aligned}$$

Another way to express the capital recovery cost of extending the service for one year is:

$$\begin{aligned} \text{Depreciation} &= \$13{,}000 - \$10{,}000 = \$3{,}000 \\ \text{Interest} &= \$13{,}000(0.15) = \underline{\quad 1{,}950} \\ \text{Total capital recovery cost} &= \$4{,}950 \end{aligned}$$

This is mathematically identical with the preceding calculation. The depreciation figure used here is depreciation in the popular sense of decrease in value; the value figures are market values. In effect this calculation says that by extending the service of the asset one year more, we receive $3,000 less for the asset and we lose the use of $13,000 in cash for a one-year period. With interest at 15%, this use is valued at $1,950. The total is $4,950 as of a date one year hence.

(b) Equivalent Annual Capital Recovery Cost of Extending a Service for Two or More Years

Now consider the question of the capital recovery cost of extending the service of the asset for 3 years more. The net realizable value 3 years hence is $5,500.

$$\begin{aligned} CR &= (\$13{,}000 - \$5{,}500)(A/P,15\%,3) + \$5{,}500(0.15) \\ &= \$7{,}500(0.43798) + \$5{,}500(0.15) = \$4{,}110 \end{aligned}$$

It is of interest to relate this equivalent uniform annual cost of 3 years of service to the separate capital recovery costs of extending the service for each of the next 3 years.

$$\begin{aligned} \text{CR cost next year} &= (\$13{,}000 - \$10{,}000) + \$13{,}000(0.15) \\ &= \$3{,}000 + \$1{,}950 = \$4{,}950 \end{aligned}$$

CR cost 2nd year = (\$10,000 − \$7,500) + \$10,000(0.15)
= \$2,500 + \$1,500 = \$4,000
CR cost 3rd year = (\$7,500 − \$5,500) + \$7,500(0.15)
= \$2,000 + \$1,125 = \$3,125

These separate costs of extending service year-by-year may be converted into equivalent annual cost for 3 years by finding their present worths and multiplying the sum of the present worths by the capital recovery factor, as follows:

PW of \$4,950 = \$4,950(0.8696) = \$4,305
PW of \$4,000 = \$4,000(0.7561) = 3,024
PW of \$3,125 = \$3,125(0.6575) = 2,055
Sum of the present worths = \$9,384
CR = \$9,384(0.43798) = \$4,110

This agrees with the \$4,110 figure previously obtained by considering only the \$13,000 present realizable value and the \$5,500 realizable value 3 years hence. The year-by-year capital recovery costs of extending the service of an asset may always be converted into an equivalent annual cost that is equal to the capital recovery cost of extending the service for the entire period of years under study. This is mathematically true regardless of the pattern of year-by-year reductions in salvage value.

(c) Year-by-Year Capital Recovery Costs Throughout the Life of an Asset
Assume that the asset in question had a net realizable value of \$16,000 at the end of its first year of life and will have the net realizable values shown in column B of Table 16–1 at the end of each year thereafter. The resulting capital recovery costs of extending service for each year are shown in column E. The first year's cost, \$12,750, the sum of \$9,000 depreciation and \$3,750 interest, is the capital recovery cost of one year's service to a *prospective purchaser* of the machine for \$25,000 on the assumption that the machine is disposed of for \$16,000 at the end of its first year of life. The succeeding figures in column E are the capital recovery costs of each successive year's extension of service to a *present owner* of the machine in question.

Column F gives the present worth at zero date of each year's capital recovery cost of extending service. Column G gives the sum of the present worths of these costs for n years. It will be noted that when the net realizable value reaches zero, the sum of these present worths must equal the first cost, \$25,000. Column H shows the annual cost of capital recovery for n years' service computed by multiplying the figure from column G by the appropriate capital recovery factor. (Present worth factors and capital recovery factors are taken from the 15% table, Table D–20.) The main purpose of including this calculation is to demonstrate the identity of the capital recovery cost computed

TABLE 16–1
Year-by-year capital recovery costs of extending service of a $25,000 machine and equivalent annual costs if held for *n* years, with interest at 15%

Year n (A)	*Net realizable value at year end (B)*	*Decrease in realizable value during nth year (C)*	*Interest on realizable value at start of year (D)*	*Capital recovery cost of extending service through nth year (E)*	*Present worth of capital recovery cost for nth year (F)*	*Present worth of capital recovery costs for n years (G)*	*Equivalent uniform annual capital recovery cost if retired after n years (H)*
1	$16,000	$9,000	$3,750	$12,750	$11,087	$11,087	$12,750
2	13,000	3,000	2,400	5,400	4,083	15,170	9,332
3	10,000	3,000	1,950	4,950	3,255	18,425	8,070
4	7,500	2,500	1,500	4,000	2,287	20,712	7,255
5	5,500	2,000	1,125	3,125	1,554	22,266	6,642
6	3,500	2,000	825	2,825	1,221	23,487	6,206
7	2,000	1,500	525	2,025	761	24,248	5,828
8	1,000	1,000	300	1,300	425	24,673	5,498
9	0	1,000	150	1,150	327	25,000	5,239
10	0	0	0	0	0	25,000	4,981

in this way with the capital recovery cost computed in the conventional way from first cost and the salvage value at the end of the life.

For example, if the machine is retired at the end of 3 years with a $10,000 salvage value, the conventional calculation is:

$$\begin{aligned} CR &= (\$25{,}000 - \$10{,}000)(A/P,15\%,3) + \$10{,}000(0.15) \\ &= \$15{,}000(0.43798) + \$1{,}500 = \$8{,}070 \end{aligned}$$

Or if retired after 6 years with $3,500 salvage value, it is:

$$\begin{aligned} CR &= (\$25{,}000 - \$3{,}500)(A/P,15\%,6) + \$3{,}500(0.15) \\ &= \$21{,}500(0.26424) + \$525 = \$6{,}206 \end{aligned}$$

Irrelevance of Book Value and Current Depreciation Accounting Charges in a Before-Tax Analysis to Guide a Decision on a Proposed Retirement

In Example 16–4(a) it was stated that the 2-year-old asset in question had been depreciated by the straight-line method at 20% ($5,000) a year and had a current book value of $15,000. But these figures were given no weight in calculating the cost of extending the service for one or more years. This neglect of the book value and the current annual depreciation charge in the accounts was entirely proper for this particular purpose.

The original $25,000 purchase price of this asset was spent 2 years ago. This money has already been paid out regardless of whether it is decided to retire the asset at once or to continue it in service. No future decision regarding the disposal or retention of the asset can alter the fact of this past $25,000 disbursement.

As brought out in Chapter 11, a depreciation charge in the accounts is simply a time allotment of a past disbursement, which, when made, was considered in the accounts to be a prepaid expense of service for a number of years. The book value of an asset or group of assets is simply that portion of the cost that has not yet been written off in the accounts as depreciation expense. Regardless of the date of the retirement and regardless of the method of depreciation accounting in use, eventually the first cost less salvage will be written off on the books. Nevertheless, as explained in Chapter 11, the entry made on the books to record a retirement under the single-asset (item) method of depreciation accounting differs from that under multiple-asset methods (group, classified, or composite).

The straight-line item method of depreciation accounting was in common use in the United States at one time, particularly in the manufacturing industries. The accounting entries to record retirements under the item method led to great confusion of thought on the part of many engineers and industrialists regarding the economic aspects of proposed replacements.

A Common Error in Reasoning in Economy Studies Involving Prospective Replacements

If the 2-year-old asset in Example 16–4 was depreciated by the straight-line item method at $5,000 a year, $10,000 of the original $25,000 investment has been written off and the current book value (unamortized cost) is $15,000. If the asset is now disposed of for a net $13,000, the item method requires that $2,000, the difference between the book value and the net salvage value, be written off at once. This $2,000 might be charged to an account with some such title as "Loss on Disposal of Fixed Assets."

This "loss on disposal" type of entry once common in the United States proved to be an obstacle to clear thinking on matters of replacement economy. Much of the early literature of this subject involved formulas or other methods of analysis in which the excess of book value over net salvage value of the old asset was considered as an addition to the first cost of the proposed new asset. For example, if an economy study were to be made to determine whether to replace the 2-year-old asset in Example 16–4 with a new asset costing $30,000, these writers on replacement economy would consider the first cost of the new asset to be $32,000, the sum of the $30,000 purchase price and the $2,000 "loss on disposal."

The preceding discussion of sunk costs and of Example 16–4 has shown the fallacy of this idea. In further consideration of the unsoundness of this view, it should be pointed out that the loss on disposal entry related only to the timing of the write-off of a prepaid expense. The money spent for an asset already owned has been spent whether or not the asset is to be retired immediately. This past outlay cannot be altered by the timing of an accounting write-off. Under straight-line item accounting the need for a loss on disposal entry resulted from the past use of a depreciation rate that turned out to be insufficient to write off the difference between first cost and actual salvage value during the actual realized life.

Need To Consider Book Value and Current Depreciation Accounting Charges in Estimating the Influence of a Proposed Retirement on Cash Flow for Income Taxes

Before World War II, income tax rates in the United States were low enough for most economy studies to be made without examining the income tax consequences of proposed decisions. Like other literature of that period, the incorrect formulas we have mentioned did not consider income taxes.

Because the concept of taxable income corresponds in most respects to the concept of accounting income, matters that affect the accounts usually influence income tax payments. Thus, the retirement of an asset under circumstances where the accounts show a "gain" or "loss" will generally have a tax consequence. If the retirement of an asset eliminates a depreciation charge that would continue if the asset were not retired, this also will affect cash flow for income taxes. A discussion of the foregoing aspects of economy studies for retirements is deferred until Chapter 17, where we examine the general problem of estimating the income tax consequences of retirement decisions.

Some Suggested Readings on Topics Introduced in This Chapter

Clark's classic work on overhead costs, published in 1923, is desirable background reading for the subject matter of this chapter. Modern writings by Dean, Goetz, and Reul are excellent. Norton's writings on engineering economy contain helpful examples along these lines. One of the authors of the present book, in collaboration with Norton, has discussed at length the relationship between depreciation accounting and the viewpoint developed in this chapter.*

*Detailed references to the writings cited are as follows: J. M. Clark, *Studies in the Economics of Overhead Costs* (Chicago: University of Chicago Press, 1923). Joel Dean, *Managerial Economics* (Englewood Cliffs, N.J.: Prentice-Hall, Inc., 1951). B. E. Goetz, *Management Planning and Control* (New York: McGraw-Hill Book Co., Inc., 1949). E. L. Grant and P. T. Norton, Jr., *Depreciation* (New York: The Ronald Press Co., 1955), particularly chap. 15. W. G. Ireson and E. L. Grant (eds.), *Handbook of Industrial Engineering and Management*, 2d ed. (Englewood Cliffs, N.J.: Prentice-Hall, Inc., 1971), see particularly the sections by Joel Dean on "Managerial Economics," by P. T. Norton, Jr., on "Engineering Economy," and by R. I. Reul on "Capital Budgeting."

Increment Cost Aspects of the Estimation of Working Capital Requirements for an Investment Proposal*

Most of the discussion in this book deals with the economic analysis of proposals involving the flow of business funds for so-called fixed assets (land, buildings and structures, machinery, transportation equipment, furniture and fixtures, and so forth). Many proposals for fixed assets also influence cash flow associated with such matters as accounts receivable and payable and inventories of raw materials, work in process, and finished goods. It is as important for an analyst to recognize the cash flow associated with working capital, wherever relevant, as to recognize the cash flow associated with proposed investments in physical plant.

It was pointed out in Chapter 11 that economy studies usually treat proposed investments in working capital as if they will have 100% salvage values at the end of the life of a project (or possibly at the end of an assumed study period). Because working capital investments are not depreciated for accounting or income tax purposes, such treatment is consistent with the books of account. Nevertheless, the analyst making an economy study that involves working capital requirements will not find adequate guidance in the standard accounting definition of net working capital as the excess of current assets (chiefly cash, receivables, and inventories) over current liabilities (usually obligations payable within one year). It is essential to apply the "with or without" viewpoint to the influence of a proposal on prospective cash flow in order to make a rational estimate of working capital requirements.

For example, consider the prospective cash flow involved in the financing of additional accounts receivable in a manufacturing business. A company's books will always show accounts receivable at the full selling price. However, the commitment of cash necessary to finance, say, 30 days' accounts receivable is considerably less than the selling price of the product sold in 30 days. The accounts will "value" accounts receivable at the selling price. But this price normally includes such items as allowances for profit and depreciation that have not involved current cash outlays by the manufacturer. Similarly, book values for inventories of finished goods and work in process include depreciation and possibly other noncash items that should be excluded in the estimation of working capital requirements for purposes of an economy study.

*Our exposition of this topic is largely influenced by a paper by J. B. Weaver, Director, Development Appraisal Dept. of Atlas Chemical Industries, given at a conference sponsored by the Engineering Economy Division of the American Society for Engineering Education at Pittsburgh, Pa., in June, 1959. For a more complete presentation of Weaver's viewpoint on this topic, the reader is referred to his articles in the June and August, 1959, issues of *Industrial and Engineering Chemistry*.

The kinds of proposals that obviously call for the estimates of working capital requirements are proposals for a new product or for expanded production of an existing product. But the common use of simple rules of thumb to apply to *all* such proposals in a business organization (e.g., 30 day's accounts receivable; 60 days' inventory) is likely to disregard important differences between different products and between different material sources. For instance, although natural gas and fuel oil are interchangeable for many purposes, they may involve quite different working capital requirements. A user of natural gas generally receives it via a pipeline and therefore carries no inventory; in fact, there may be a negative element in the working capital requirement because of the time lag between the use of gas and the payment for it. In contrast, in many cases the economical way to purchase oil is in tankers or barges, and it may be necessary to carry a considerable inventory at all times to ensure against running out.

Summary

The following points brought out in this chapter may be restated for emphasis as follows:

Average costs per unit, whether generated from accounting records or elsewhere, should not be used uncritically as guides to decision making. It should always be remembered that it is prospective *differences* between alternatives that are relevant to their comparison.

It is particularly important to keep in mind that all economy studies start from the moment of decision. The only possible differences between alternatives for the future are future differences. Past receipts and disbursements and other past events are irrelevant in economy studies except as they may influence future receipts or disbursements or other future events.

For purposes of an economy study to decide whether or not to dispose of assets already owned, capital recovery costs on these assets should be based on the present net realizable value if disposed of rather than on original cost.

PROBLEMS

16–1. The family in Example 16–1 concluded that any additional electric energy purchased each month would add 8.1¢/kw-hr to its electric bill. Assume that the proposal to add 100 kw-hr to each month's use has been rejected on the grounds of being too expensive. Now various decreases in use of electricity are proposed that would combine to reduce monthly consumption from 500 kw-hr to 400. Will these changes reduce the monthly bill by 8.1¢/kw-hr? If not, what is the average saving per kilowatt hour that can be expected from these economies? (*Ans.* 6.34¢) Do you consider this to be a unit cost that is significant for the purpose of any decision?

16–2. Consider the proposed new equipment in Example 16–2 that would use 1,000 kw-hr/month and would increase the monthly electric bill by $233. When the extra cost of $233 is divided by the extra number of kilowatt hours used, the quotient is 23.3¢/kw-hr. What is the significance, if any, of this apparent unit cost of extra energy?

16–3. A certain type of machine has a first cost of $10,000. Prospective end-of-year salvage values are as follows:

Year	Salvage values
1	$5,300
2	2,900
3	2,100
4	1,600
5	1,200
6	900

Assuming interest at 15%, what is the capital recovery cost of extending service for each year of life? (*Ans.* (1) $6,200; (2) $3,195; (3) $1,235; (4) $815; (5) $640; (6) $480)

16–4. Using only the first cost of $10,000 and final salvage value of $900, find the equivalent uniform annual cost of capital recovery for a machine in Problem 16–3 purchased new and disposed of at the end of 6 years. Use interest at 15%. Also find the uniform series for 6 years equivalent to the irregular series of end-of-year capital recovery costs obtained in the solution to Problem 16–3. These two figures should be the same. (*Ans.* $2,540)

16–5. A year-old machine of the type referred to in Problem 16–3 is acquired for $5,300. It is disposed of for $1,200 when it is 5 years old. Compute the uniform annual cost of capital recovery with interest at 15%. Also find the uniform series for 4 years equivalent to the year-by-year capital recovery costs of extending service for the second, third, fourth, and fifth years obtained in the solution to Problem 16–3. These two figures should be the same. (*Ans.* $1,616)

16–6. Assume that you are selling goods on a commission basis. On the sale of a given article you will earn a commission of $200. To date you have spent $140 promoting a certain sale that you have not yet made. You are confident that this sale will be assured by an added expenditure of some undetermined amount. What is the maximum amount (over and above what you have already spent) that you should be willing to spend to assure the sale?

16–7. What differences can you see in the reasoning that ought to influence the decisions of Frank and James under the following circumstances?

Frank wishes to raise $20,000 immediately to make a down payment on the purchase of a home. He has only two possible sources of funds. (1) He may sell 500 shares of XY Co. that is currently paying annual dividends of $4 a share. He bought this stock a few years ago for $65 a share; its present market price is $40 a share. Or (2) he may borrow $20,000 at 10% interest from his life insurance company with his life insurance policy as security.

James also wants to raise $20,000 immediately to make a down payment on the purchase of a home. He also has only two possible sources of funds. (1) He may sell 500 shares of XY Co. stock that is currently paying annual dividends of $4 a share. He bought his stock for $25 a share several years before Frank made his purchase; its present market price is $40 a share. Or (2) he may borrow $20,000 at 10% interest from his life insurance company with his life insurance policy as security.

16–8. In a period of rising prices a merchant attempted to maintain his stock of goods at a constant physical volume. He had purchased his stock of one item some time ago at $15 per unit. He sold these items at $25 per unit (applying his usual markup) and immediately replaced them by identical ones purchased at the new wholesale price of $30 per unit. What do you think of the profitableness of this transaction?

16–9. A student, arrested for speeding, was given his choice between a $20 fine and a day in jail. He elected the latter. On emerging from jail, he wrote a story about his experience there, which he sold to a newspaper for $25. Commenting on this incident, one of his friends remarked, "Steve made $5 by going to jail, the $25 from the newspaper minus the $20 fine." "No," said another, "he made $25 by the decision to go to jail as he didn't pay the fine." Which friend do you agree with, if either? Explain.

16–10. A small company buys electric energy under the rate described in Example 16–2. In a typical month it buys 55,000 kw-hr with a billing demand of 280 kw. What will the electric bill be for such a month? What is the average price per kilowatt hour?

New machinery is proposed that will add 50 kw to the billing demand and 2,000 kw-hr to the amount of energy purchased. How much will this add to the utility bill? What will the average cost be per kilowatt hour added by this additional energy?

16–11. The small company in Problem 16–10 is considering a rescheduling of certain activities in a way that will reduce its billing demand by 30 kw and its monthly energy consumption by 1,000 kw-hr. How much will this change reduce its monthly electric bill? What will be the average saving per kilowatt hour?

16–12. A paperboard manufacturing company has two plants, one in Georgia and one in Florida, producing equivalent grades of "cardboard." The Georgia plant has been operating at 75% capacity, producing 2,700 tons per month at a total cost per ton of $115. The Florida plant has been operating at 60% capacity, producing 3,600 tons per month at a total cost per ton of $126.

Included in the total cost per ton is the cost of waste paper, the major raw material. For each 100 tons of product, 80 tons of waste paper are required. At the Georgia plant the local waste paper costs $28 per ton (of waste paper), but the supply is limited to 1,440 tons per month. At the Florida plant, local waste paper costs $30.50 per ton and is limited to 3,200 tons per month. Additional waste paper must be purchased through brokers at $36.50 per ton (delivered at either plant). The direct cost to the company of transshipment of waste paper from one plant to the other is $8 per ton.

Of the total monthly costs at the Georgia plant, $94,050 is estimated not to vary with the production level. The remainder of the costs, with the exception of the cost of waste paper, are expected to vary in direct proportion to output. The comparable figure for the Florida plant is $162,000.

(a) If the total production of both plants is to be continued at the present rate of 6,300 tons per month, would there be any apparent advantage to shifting part of the scheduled production from one of the plants to the other? If so, which plant's production should be increased and by how much? Why?

(b) If production requirements increased to 9,000 tons per month, how much would you recommend be produced at each plant? What would the total cost be per month for each plant in this case?

16–13. Jones and Smith are engineers from the United States employed in a foreign country. Both expect to stay on this assignment for another year or more. Each has just converted $1,000 (U.S.) to the currency of this country at the existing rate of 8 to 1, thus receiving 8,000 currency units in exchange. Suddenly, to the surprise of Jones and Smith, the exchange rate changes to 10 to 1. Now 8,000 currency units changed back into U.S. dollars will bring only $800. A few days later Jones and Smith learn that they are to be transferred back to the United States in 2 weeks. Each is confronted with the question of what purchases, if any, he will make of various products of the country to take back to the United States. The two men take different attitudes toward this. Jones says that he can get so few dollars for his currency units that he is going to spend them all before going back to the United States. In contrast to this, Smith says that each of his currency units cost him 12½ cents and that he will not spend one unless he believes he is getting 12½ cents worth of goods for it; otherwise he will convert his currency units back into dollars.

Which of these views seems reasonable to you? Or is there some other point of view, not expressed by either Jones or Smith, that seems more sensible than either? If so, what is it? Explain your reasoning.

16–14. A public surveyor owns two transits that he purchased from the Surveyor's Service Co. a year ago for $750 each. He is currently renting a third one from this company at $12 per month. One of his transits becomes damaged in a way not covered by insurance. The S.S. Co. representative estimates the repair cost to be $120. He suggests that the surveyor sell him the two transits "as is" for $720 and rent two more transits at $12 a month. He reasons that his company can rent a transit for less than it costs the surveyor to own one "because we get them wholesale and have our own setup for repairing and adjusting them." He presents the following cost comparison to the surveyor:

Cost of continuing to own two transits		Cost of renting two transits	
Depreciation = $1,500/10	= $150	Rental cost 12($24)	= $288
Taxes and insurance = 3% of $1,500	= 45	Less:	
Repair cost	120	Depreciation saved by sale of transits $720/9	= 80
Cleaning and adjusting	40		
Net cost	$355		$208

Criticize the salesman's analysis.

16–15. Burden rates in a certain factory are on a machine-hour basis. They are established by first apportioning all of the various expected indirect manufacturing expenses of the factory at normal output among all of the machines in the factory; the total estimated indirect manufacturing expense apportioned to any given machine is then divided by the expected normal hours of operation of this machine in order to arrive at a rate per machine-hour. Thus, if the estimated indirect manufacturing expense apportioned to a certain milling machine is $4,820, and its expected normal hours of operation are 1,800 per year, the burden rate per hour of operation is $2.70.

Most of the product of this factory is manufactured to buyers' specifications on contract jobs. In planning many of the operations carried on in this production, the question arises as to whether they shall be done on general-purpose machines or special-purpose machines. This involves a comparison of the cost of machine setups and the direct labor and material costs for each given operation on alternative machines. In such comparisons the problem arises as to what use, if any, should be made of the machine burden rates.

Discuss this problem, considering as separate cases (a) the factory operating at about the expected normal hours of operation, (b) the factory operating at greatly curtailed output (such as 30% of normal in a period of business depression), and (c) the factory operating at more than normal output with a large volume of unfilled orders.

16–16. A writer on management subjects discussed the topic of the use of rate of return as a criterion for investment decisions somewhat as follows:

"Prospective rate of return is of limited usefulness as a guide to decision making. For instance, this criterion would be of no use in the following case:

"At the end of his freshman year at the XYZ College, John Doe is offered the campus concession for a certain soft drink. He must pay $900 for the concession and certain equipment. He estimates that the concession will bring him $450 a year for the next 3 years in addition to a reasonable payment for his labor, and that the concession and equipment will be salable for $900 at the end of the 3-year period. He therefore expects a 50% rate of return on his investment.

"However, Richard Roe, who for many years has been concessionaire for a competing soft drink, would prefer less energetic competition than he expects to receive from Doe. Roe therefore offers Doe an outright immediate payment of $100 if Doe will refrain from purchasing this concession. Because Doe will make no investment at all if he accepts Roe's offer, it is evident that his rate of return will be infinite.

"Doe's problem of decision making between the purchase of the concession and the acceptance of Roe's offer illustrates the weakness of the rate-of-return technique. Doe appears to have the choice between a 50% rate of return and an infinite rate of return. But this would also appear to be his choice if Roe had offered him only $1 or if Roe had offered him $1,000. Thus, the rate-of-return technique does not permit Doe to give any weight whatsoever to the size of Roe's offer. No matter how small Roe's offer is, it appears to have a better rate of return than the 50% expected from the purchase of the concession."

Do you agree with the writer that the rate-of-return technique cannot be used by Doe to help him choose between these two alternatives? Or do you see any way to apply the rate-of-return technique to this particular case? Explain your answers fully.

16–17. A large chemical company has recently acquired two plants that manufacture a certain chemical. These plants use different production processes, although their products are identical. During the first 6 months of operation neither plant is operated at capacity.

The Los Trancos plant has produced 400 tons per month of output at an average cost of $160 per ton. Of the total monthly costs it is estimated that $23,200 will remain fixed regardless of substantial variations in output either upward or downward, and that the remainder of costs will vary in direct proportion to output. The San Francisquito plant has produced 320 tons per month at an average cost of $148 per ton. Of the total monthly costs it is estimated that $12,480 will remain fixed regardless of substantial variations in output either upward or downward and that the remainder of the costs will vary in direct proportion to output.

(a) If the total amount produced at the two plants is to continue at the present figure of 720 tons per month, does there appear to be any advantage in increasing the production at one of the plants and making an equal decrease at the other? If so, at which plant would you increase production?

(b) Assume that for reasons of policy it is not desired to make the production shift indicated in (a). The total required production increases from 720 to 810 tons per month. At which plant would it be more economical to produce the extra 90 tons?

16–18. In Example 16–3 technological progress played a major role in causing consideration of an early replacement. Originally, the advantage of the paper-tape unit had not seemed sufficient to justify its required extra investment of $450.

Assume that when Green finally accepts Brown's reasoning about the irrelevance of the past outlay, he still has another objection to Brown's analysis. Because the diode display unit turned out to be obsolete in one year, he is unwilling to accept Brown's estimate of a 10-year life for the paper-tape unit.

Compare the two calculators before income taxes using a 4-year remaining life for the diode display unit and a 4-year life for the paper-tape unit. Assume that either unit will have zero salvage value at the end of the 4 years. Use a before-tax i^* of 20%.

16–19. An industrial company buys electricity under the following monthly rate:

Demand charge:

$9,700 for the first 5,000 kw of maximum demand or less

Next 5,000 kw of maximum demand at $1.67/kw

All over 10,000 kw of maximum demand at $1.48/kw

Energy charge:

A block equal to 500,000 kw-hr plus a number of kilowatt hours that is the product of 300 and the maximum demand, all at 4.3¢/kw-hr

All additional energy at 3.9¢/kw-hr

In a representative month the company's maximum demand is 9,500 kw and its energy consumption is 2,000,000 kw-hr.

(a) What is the average cost per kilowatt hour?

(b) What will be the incremental cost per month of an additional load that adds 1,000 kw to the maximum demand and 350,000 kw-hr to the monthly energy consumption? What will be the average cost per kilowatt hour of this extra load?

16–20. The company in the preceding problem closes for a summer vacation of 2 weeks, with no activity except certain deferred maintenance. During the month when the vacation occurs energy consumption is 60% of normal even though the maximum demand is unchanged. What is the electric bill for the vacation month? What is the average cost per kilowatt hour for this month?

16–21. The company in Problem 16–19 plans to go to two-shift operation. Although the maximum demand will be unchanged, the monthly energy consumption will be increased by 75%. What will the average cost be per kilowatt hour of the additional energy required by the second shift?

16–22. A certain machine used in the construction industry has a first cost of $15,000. End-of-year salvage values are $10,000 at the end of the first year, $6,000 at the end of the second, $3,000 at the end of the third, and $1,000 at the end of the fourth. Using an interest rate of 18%, compute the capital recovery cost of extending the service for each year of life.

16–23. Using only the first cost of $15,000, the $1,000 final salvage value, and the i of 18% from the preceding problem, calculate the equivalent uniform annual cost of capital recovery for an asset with a 4-year life. Also calculate the uniform series for 4 years equivalent to the irregular series of year-by-year capital recovery costs obtained in the preceding problem. These two figures should be the same.

17

ECONOMY STUDIES FOR RETIREMENT AND REPLACEMENT

The hand of time lies heavy on the works of man, whether ancient or modern.

This is a fact, obviously, of the most practical consequence. It confronts the owners of these nominally "durable" but nevertheless ephemeral goods with two problems. The first is to distinguish the quick from the dead; in other words, to tell whether goods not yet physically exhausted have outlived their economic usefulness, either generally or for the particular function they now perform. The second is to make financial provision against the wastage of durable assets over their service life. The one involves replacement, or reequipment policy; the other, depreciation policy.—George Terborgh*

Chapter 11 listed various types of causes of property retirement, namely: (1) improved alternatives, (2) changes in service requirements, (3) changes in the old assets themselves, (4) changes in public requirements, and (5) casualties. It was pointed out that these causes are not mutually exclusive; for instance, an asset may be retired partly because of obsolescence (cause 1), partly because of inadequacy (an example of cause 2), and partly because of increasing annual disbursements for repairs and maintenance (an example of cause 3).

In modern industry the usual experience is that assets are retired when they are still physically capable of continuing to render service. Someone must make a *decision* to make such retirements. Generally speaking, such decisions should be made on grounds of economy. This chapter discusses the kinds of analysis needed to guide such economic decisions.

Distinction Between Retirement and Replacement

The disposal of an asset by its owner is referred to as a retirement. Not all retirements involve the actual scrapping of the asset retired; many assets, retired by their present owners, may be used by one or more other owners before reaching the scrap heap.

*George Terborgh, *Dynamic Equipment Policy* (Washington, D.C., copyrighted 1949, Machinery and Allied Products Institute), p. 1. Reprinted by permission.

If an asset (or group of assets) is retired, and another asset (or group of assets) is acquired to perform the same service, this is a replacement.

It frequently happens that new assets are acquired to perform the services of existing assets, with the existing assets not retired but merely transferred to some other use—frequently an "inferior use" (such as standby service). In such cases, the acquisition of the new assets sometimes is also described as a replacement.

Two Words That are Useful in Discussing Replacement Economy

In Chapter 13 we used the words *defender* and *challenger* with special meanings that were helpful in discussing the problem of choice among multiple alternatives that are physically mutually exclusive. In the present chapter, we use these words with the meanings originally assigned to them by George Terborgh in *Dynamic Equipment Policy* for the discussion of replacement economy. An existing old asset, considered as a possible candidate for replacement, is called the *defender*. The proposed new replacement asset is called the *challenger*.

Some Characteristics of Economy Studies for Retirements and Replacements

This chapter concentrates attention on certain special aspects of studies for retirement and replacement, as follows:

1. Capital recovery costs for extending the services of assets already owned (e.g., defenders in replacement economy studies) are computed differently from capital recovery costs for assets yet to be acquired (e.g., all proposed assets in economy studies discussed in previous chapters, and challengers in replacement economy studies). The reasons for this difference were explained in Chapter 16. Various aspects of this topic are illustrated in all of the examples in the present chapter.
2. There are special difficulties in estimating the effect of a decision regarding retirement or replacement on cash flow for income taxes. Some aspects are illustrated in Examples 17–1 and 17–2.
3. In many replacement studies the appropriate assumption regarding the defender is that, if retained in service at all, it will be kept for a relatively short time, often only one year. In contrast, the appropriate assumption regarding the challenger may be that, if acquired, it will be kept for its full economic life. Although we have already given some consideration (particularly in Chapters 6, 7, and 14) to the interpretation of economy studies in which the alternatives have different services lives, there are special aspects to this problem in replacement economy. Some of these aspects are discussed in connection with Examples 17–3, 17–4, and 17–5.

In order to start our discussion by concentrating attention solely on points (1) and (2), Examples 17–1 and 17–2 deal with cases where the expected remaining service life of the defender is the same as the expected full service life of the challenger.

These two examples will first be analyzed by an annual cost comparison before income taxes. The effect on prospective disbursements for income taxes of the decision on retirement or replacement will then be computed and an annual cost comparison will then be made after income taxes. Finally, prospective rates of return before and after income taxes will be calculated.

EXAMPLE 17–1

Analysis of a Proposed Retirement

Facts of the Case

A manufacturing company owns a warehouse in a city some distance from its main plant. The warehouse is used by the branch sales office in the area to make delivery of certain products from stock. An equal amount of storage space is available in a new commercial warehouse. As a favorable offer for the old warehouse has been received, consideration is given to its sale and the rental of the needed space.

The property was not new when it was purchased 10 years ago for \$100,000. For accounting and income tax purposes, this was divided into \$80,000 for the warehouse building and \$20,000 for land. Straight-line item depreciation on the building has been used for accounting and tax purposes, assuming a 40-year remaining life and zero salvage value. The warehouse can now be sold for a net \$150,000 after payment of selling expenses.

In recent years annual warehouse expenses have averaged \$6,440 for operation and maintenance, \$2,240 for property taxes on the land and building, \$420 for fire insurance on the building, and \$1,400 for fire insurance on the stock. It is expected that costs will continue at approximately these figures if the warehouse is not sold.

Equal space in the commercial warehouse can be rented for \$31,600 a year. Estimated annual disbursements for operation and upkeep of this space are \$2,800. Fire insurance on the stock will be reduced to \$900 a year.

If the decision is made not to accept the favorable offer for the warehouse, it is estimated that it will not be sold for another 10 years. The estimated net selling price 10 years hence is \$90,000.

In before-tax studies in this company, a minimum attractive rate of return of 20% is used; in after-tax studies, 10%. An applicable tax rate of 50% on ordinary income is assumed in all economy studies. It is assumed that a "gain" on the sale of the property either now or later will be taxed at 28%.

Before-Tax Comparison of Annual Costs

Using the stipulated i^* of 20%, the comparative annual costs for a 10-year study period are as follows:

Comparative annual cost—continued ownership of warehouse

CR = \$150,000(*A*/*P*,20%,10) − \$90,000(*A*/*F*,20%,10)	\$32,310
Operation and maintenance	6,440

Property taxes	2,240
Fire insurance on building and stock	1,820
Total	$42,810

Comparative annual cost—rental of equal space

Rent	$31,600
Operation and upkeep	2,800
Fire insurance on stock	900
Total	$35,300

The foregoing comparison favors disposal of the warehouse and rental of equal space.

After-Tax Comparison of Annual Costs

The decision to sell the warehouse at once will cause an immediate disbursement for the tax on the gain on disposal. As the property has been depreciated for tax purposes at $2,000 a year for the past 10 years, its present book value is $80,000, the difference between the $100,000 original cost and the $20,000 of depreciation charged against the warehouse to date. There will be a taxable gain of $70,000, the difference between the $150,000 net selling price and the $80,000 book value.

Because this gain will be taxed at 28%, the tax will be $19,600. The effect of this tax will be to reduce the net amount received from the sale from $150,000 to $130,400. In effect, $130,400 is the net realizable salvage value after taxes.

A similar analysis will show that if the warehouse is sold for $90,000 in 10 years, the gain will be $30,000, the tax will be $8,400, and the net amount realized after taxes will be $81,600.

Annual taxes on ordinary income will also be influenced by the decision on retirement. If the warehouse is continued in service, annual deductions from taxable income will be the $2,000 annual depreciation plus current disbursements of $6,440, $2,240, and $1,820, a total of $12,500. If space is rented, annual deductions from taxable income will be $35,300. With continued ownership, annual taxable income will therefore be $22,800 (i.e., $35,300–$12,500) higher than with rental. Using the applicable tax rate of 50%, annual disbursements for income taxes will be $11,400 more with ownership than with rental.

Using the stipulated after-tax i^* of 10%, an after-tax comparison of equivalent annual costs is as follows:

Comparative annual cost—continued ownership of warehouse

CR = $130,400($A/P$,10%,10) − $81,600($A/F$,10%,10)	$16,100
Operation and maintenance	6,440
Property taxes	2,240
Fire insurance on building and stock	1,820
Extra income taxes above those under rental	11,400
Total	$38,000

Comparative annual cost—rental of equal space

Rent	$31,600
Operation and upkeep	2,800
Fire insurance on stock	900
Total	$35,300

The margin favoring rental is considerably less in the after-tax analysis than in the before-tax analysis in spite of the added item of $11,400 for extra income taxes. If a 50% tax rate rather than 28% should be applicable to the gain on the present sale, the two alternatives would have approximately equal annual costs.

Calculation of Rates of Return

The prospective differences between cash flows under continued ownership and under rental (i.e., continued ownership *minus* rental) may be tabulated as follows:

Year	*Difference before taxes*	*Difference in income taxes*	*Difference after taxes*
0	−$150,000	+$19,600	−$130,400
1 to 10	+24,800 per year	−11,400 per year	+13,400 per year
10	+90,000	−8,400	+81,600

Trial-and-error calculations of the type that have been illustrated many times in the preceding chapters give the conclusion that the present worth of the before-tax series is zero with an interest rate of approximately 14.4%, and the present worth of the after-tax series is zero with an interest rate of approximately 7.6%.

Interpretation of Analysis in Example 17–1

Most of the economy studies discussed in previous chapters have dealt with proposals to make investments in fixed assets—to convert money into capital goods. In contrast, Example 17–1 might be described as a proposal for a disinvestment; the question at issue is whether it is desirable to convert capital goods into money.

The description of the facts of the example did not tell what would be done with the $130,400 (after taxes) realizable from the sale of the warehouse. However, the statement that the after-tax i^* was 10% implied that this money could be invested in the business in a way that would provide an after-yield of at least 10%. Our analysis indicated that a decision to keep the warehouse would be, in effect, a decision to invest $130,400 at a yield of 7.6% after taxes. If, all things considered (including irreducibles not mentioned in the "facts of the case"), a 7.6% after-tax rate of return is considered to be unsatisfactory in this instance, the analysis favors the sale of the warehouse.

Just as in our previous analyses by comparative annual costs and by rate of return, the two methods of comparison lead to the same decision between the alternatives as long as the minimum attractive rate of return is used as the interest rate in compound interest conversions in the annual cost method.

Comment on Economy Studies Where Retirements Involve Prospective Purchase of Product or Service of Asset Retired

The general type of situation illustrated in Example 17–1 is a common one. It often is necessary to consider the question "Can we buy this product or this service at a lower figure than our cost of continuing to produce it?" Many such economy studies require determination of increment costs and recognition of sunk costs. The book figure for present costs of production may contain certain allocated costs that would not be eliminated if production were stopped. And the depreciation charges in the books are, of course, based on the original cost, a figure that is irrelevant for purposes of the retirement economy study except for its influence on future income taxes.

The higher the present realizable value of the assets that are being considered for retirement, the more attractive is the proposal to dispose of the assets and purchase the goods or services that they have been producing. This point is generally recognized. If the warehouse in Example 17–1 had only a small resale price, the economy study would favor continued ownership.

Another important element in such comparisons, perhaps not so generally recognized as significant, is the estimated prospective future realizable value. The greater the future resale price, the more attractive is continued ownership, and vice versa. This may be illustrated in Example 17–1 by changing the $90,000 estimate of future net resale price before taxes to a substantially higher and to a substantially lower figure. Of course, the nearer the date of prospective future disposal, the greater the importance of the future resale price.

In many economy studies of this type a third alternative may be to purchase the goods or services in question from the outside source, but to keep the old assets for standby and emergency purposes. Conditions favorable to this third alternative are as follows: (1) a low present net realizable value; (2) relatively low annual disbursements required for continuing the ownership of the assets; and (3) the likelihood that continued ownership with its threat of resumption of production will strengthen bargaining power in the establishment of the price for the purchased goods or services.

EXAMPLE 17–2

Reviewing an Error in Judgment in Equipment Selection

Facts of the Case

A year ago, an irrigator purchased a pump and motor which cost $1,925 installed. This person knew nothing of pump selection, and—making the choice on the recommendation of a clerk in a hardware store—selected a pump that was unsuited to the requirements of head and discharge under which it was to operate. As a result, the year's power bill for its operation was $900; this was much higher than if a suitable pump had been purchased.

Just at the start of the current irrigation season, a pump company representative offers the irrigator another pump and motor, suited to the requirements. This will cost $1,650 installed and is guaranteed to reduce power requirements to a point where the electric energy for pumping the same amount of water as before will cost only $500. The original pump and motor can be sold to a neighbor for $375.

For purposes of this economy study, we shall assume a 10-year study period with estimated zero salvage value for both pumps at the end of this period. It will be assumed that other expenditures (such as property taxes, insurance, and upkeep) will not be affected by the type of pump being used for this service. The irrigator's applicable tax rate on ordinary income is 40%. A minimum attractive rate of return of 15% is to be used in the before-tax analysis and a rate of 9% is to be used in the after-tax analysis.

The irrigator has used years-digits item depreciation for tax purposes on the present pump, assuming a 10-year life and zero salvage value. If the new pump is substituted,the same depreciation method, life, and salvage value will be used.

Before-Tax Comparison of Annual Costs

Using the stipulated 15% minimum attractive rate of return before taxes, the comparative annual costs for a 10-year study period are as follows:

Comparative annual cost—present pump

CR = $375(*A*/*P*,15%,10)	$75
Electric energy for pumping	900
Total	$975

Comparative annual cost—proposed new pump

CR = $1,650(*A*/*P*,15%,10)	$329
Electric energy for pumping	500
Total	$829

The foregoing comparison favors the replacement.

After-Tax Comparison of Annual Costs

Under the years-digits method the past year's depreciation charge on the defender was 10/55 of $1,925 = $350. The defender's present book value is therefore $1,925 − $350 = $1,575. If the defender is sold for $375, the books will show a $1,200 "loss" on this sale.

On the assumption that this transaction occurs in the United States (and subject to certain limitations discussed later in this chapter), this $1,200 is viewed for tax purposes as a loss on disposal of depreciable property used in the trade or business. As such, the sale will reduce current taxable income by $1,200; with a 40% tax rate, it will reduce current taxes by $480. Considering the retirement separately from the acquisition of the replacement asset, a consequence of the retirement will be an immediate positive cash flow of $855, the sum of the $375 receipt from the buyer and the $480 reduction in disbursements to the tax collector. The after-tax net realizable salvage value of the defender is therefore $855.

If the defender is kept in service, the next-year deductions from taxable income that are relevant in our economy study will be the depreciation charge of $315 (9/55 of $1,925) and the $900 outlay for the electric energy, a total of $1,215. Because years-digits depreciation is being used, this total will decrease by $35 a year to $900 in the 10th year. (For simplicity in this example, we are assuming the depreciation charge on it will continue for only 9 years more.)

If the challenger is acquired, its relevant next-year tax deductions will be $300 depreciation (10/55 of $1,650) and $500 for electric energy, a total of $800. This total will decrease by $30 a year to $530 in the 10th year.

The next-year difference in taxable income between challenger and defender will therefore be $415 (i.e., $1,215 − $800). This difference will decrease by $5 a year to $370 in the 10th year.

With a tax rate of 40%, the next-year difference in disbursements for income taxes will be $166 (i.e., 40% of $415). This difference will decrease by $2 a year to $148 in the 10th year.

Using the stipulated 9% minimum attractive rate of return after taxes, an after-tax comparison of equivalent annual costs is as follows:

Comparative annual cost—present pump

CR = $855(*A/P*,9%,10)	$ 133
Electric energy for pumping	900
Total	$1,033

Comparative annual cost—proposed new pump

CR = $1,650(*A/P*,9%,10)	$ 257
Electric energy for pumping	500
Extra income tax = $166 − $2(*A/G*,9%10)	158
Total	$ 915

This after-tax comparison also favors replacement.

Calculation of Rates of Return

The prospective differences in cash flow between defender and challenger may be tabulated as follows:

Year	*Difference before taxes*	*Difference in income taxes*	*Difference after taxes*
0	−$1,275	+$480	−$795
1	+400	−166	+234
2	+400	−164	+236
3	+400	−162	+238
4	+400	−160	+240
5	+400	−158	+242
6	+400	−156	+244
7	+400	−154	+246
8	+400	−152	+248
9	+400	−150	+250
10	+400	−148	+252
Totals	+$2,725	−$1,090	+$1,635

Trial-and-error calculations indicate that the present worth of the before-tax series is zero with an interest rate of about 29% and the present worth of the after-tax series is zero with an interest rate of a little less than 28%.

The Relationship Between the Past Investment in a Defender and Decision Making About its Proposed Retirement

Examples 17–1 and 17–2 provide numerical illustrations of a point that has often been emphasized in this book. This point is that in choosing between alternatives, only the differences between the alternatives are relevant. No past occurrences can be changed by a decision between alternatives for the future. Nevertheless, past events may influence the future in different ways, depending on decisions that are made for the future.

In Example 17–1, the prospective difference in cash flow before income taxes between disposal and continued ownership was unaffected by the past investment in the warehouse. And in Example 17–2, the prospective difference in cash flow before income taxes between the defending pump and its challenger was unaffected by the past investment in the defender. In general, in economy studies for retirements and replacements, differences in cash flow *before* taxes will not be influenced by past investments in assets considered as candidates for retirement. Past investments should normally be viewed as irrelevant in before-tax studies.

(As explained in Chapter 16, there may be mental blocks to applying the foregoing rule in particular cases. Thus, our irrigator in Example 17–2 might be reluctant to admit the past mistake in judgment in equipment selection; human nature might lead this person to say "I cannot afford to replace the pump until I get my money out of it.")

In contrast, in economy studies involving proposed retirements, differences in cash flow for income taxes usually will be influenced by past investments. Example 17–1 illustrated prospective taxes on a gain on disposal of an asset. Example 17–2 illustrated how a disposal at less than book value could reduce immediate income taxes. Both the gain and the loss on disposal were influenced by the past investments and by the past method of making depreciation charges for tax purposes. In both examples, the depreciation deductions from taxable income if the existing asset were continued in service depended on the past investment in the asset. Generally speaking, past investments in defenders need to be considered in after-tax studies for retirements and replacements.

In fact, the rational consideration of the income tax aspects of retirements may lead to conclusions exactly opposite from the ones that many people would reach by intuition. Although intuition often favors disposal at an apparent profit, a taxable gain reduces the net cash realized from such a disposal. Although intuition often is opposed to disposal at an apparent loss, a loss deductible from taxable income increases the net cash realized from such a disposal.

Example 17–2 illustrated the tax advantage of such a deductible loss. The prospective rate of return before income taxes on the purchase of the challenger was 29%. Ordinarily, one would expect a 40% tax rate to make a nearly proportionate reduction in the rate of return. However, the after-tax rate of return was nearly 28%, a negligible reduction.

It should be noted that in Example 17–2, the entire $1,925 investment in the defender, less salvage value, was scheduled to be deducted from taxable income at some time regardless of whether the decision favored the challenger or the defender. The selection of the challenger merely changed the *timing* of the deduction in a way that favored the challenger.

Some Comments on Income Tax Aspects of Disposal of Fixed Assets in the United States

The tax aspects of proposed retirements often are more complex than the reader might assume from studying Examples 17–1 and 17–2. As related to retirements in the United States, an adequate discussion of this topic would not be possible without a fairly detailed consideration of various types of multiple-asset depreciation accounting and of income tax regulations. Such a discussion would require more space than would be appropriate in this book.

A number of separate issues arise in finding out whether the retirement of an asset would have an immediate tax consequence, either a favorable one (i.e., a reduction of taxes) such as was assumed in Example 17–2, or an unfavorable one (i.e., an increase in taxes) such as was assumed in Example 17–1. Still other issues arise in finding out whether the "gain" or "loss" will influence taxes at the rate applicable to ordinary income (such as the 50% in Example 17–1 and the 40% in Example 17–2) or whether some lower tax rate should be applied to these special items (such as the 28% in Example 17–1).

Both Examples 17–1 and 17–2 assumed item accounting. However, where there are many assets, multiple-asset accounting is sound in principle and often is used. Ordinary retirements in multiple-asset accounting do not cause the recording of so-called gains or losses in the books of account.

Multiple-asset accounting reflects the existence of mortality dispersion, a topic mentioned in Chapter 11. Figure 11–1 (page 188) shows a representative survivor curve for a group of assets that had an average service life of 20 years; this average resulted from retirements that occurred all the way from age 1 to age 45. In the usual case where it is reasonable to expect mortality dispersion, retirements short of the estimated average life are not "premature" in the sense that they are inconsistent with the estimated average; it is to be expected that there will also be assets that survive considerably longer than the estimated average life. The foregoing viewpoint is recognized in multiple-asset depreciation accounting by the continuation of depreciation charges after the expiration of the estimated average life and by the prohibition, under ordinary circumstances, of entries for loss on disposal when assets are retired before they have reached the estimated average life. Moreover, the policy of the U.S. Treasury Department has been, in effect, to apply the rules of multiple-asset accounting even when item accounting is used, wherever there are a number of assets of a similar type and it is reasonable to believe that the life estimate applies to the *average* life of the assets. For this reason, a retirement at low salvage value short of the estimated life will not necessarily lead to a tax-deductible loss in the United States even when item accounting is used.

The $1,200 prospective loss on disposal of the defender in Example 17–2 illustrates the conventional treatment in item accounting. Where the taxpayer has only one asset of the type being retired, there is no question of the appropriateness of item accounting for tax purposes.

In the United States, the trade-in of certain assets on new ones leads to "nontaxable exchanges" that transfer to the new asset the remaining tax basis of the old asset that was traded in. In effect, the tax consequences of a retirement are deferred, conceivably for a very long time. Our examples and problems in this book do not illustrate the tax aspects of such exchanges.

The issue of whether the applicable tax rate is the rate on ordinary income or some lower rate depends in part on rather complicated laws and regulations that are different for different classes of assets. These laws and regulations have changed a number of times. The applicable rate also depends in part on the rules for the offsetting of gains and losses. It should be observed that whereas it is favorable to the taxpayer to have a *lower* rate apply to gains, it is also favorable to have a *higher* rate apply to losses. Certain aspects of these matters are illustrated in several of the problems at the end of this chapter.

Still another question arising in an after-tax analysis of a proposed retirement is the effect of the retirement on future depreciation deductions from taxable income. This matter is closely related to the topic of entries for gain or loss on disposal. Examples 17–1 and 17–2 illustrated the required calculations under item accounting. In both examples, there were year-by-year depreciation deductions that would occur if the defender remained in service and would be eliminated by its retirement.

In multiple-asset accounting, the rules on this matter differ from those illustrated for item accounting. For example, in declining-balance group accounting using a so-called open end account, a retirement at zero salvage value will have no influence on future depreciation charges. Moreover, different types of multiple-asset accounting have different rules.

The examples in this book that deal with economy studies for retirements and replacements are based on the assumption of item accounting. The chief purpose in using this assumption is to avoid the necessity of devoting considerable space to an explanation of the technical aspects of a number of different types of multiple-asset depreciation accounting. Incidentally, the assumption provides uniformity in the examples and problems. It should go without saying that after-tax studies in industry should be based on the particular type of depreciation accounting that is being used for the assets that are candidates for retirement.

Annual Cost versus Rate of Return as a Method of Analysis in Replacement Economy

In Examples 17–1 and 17–2, the stipulated minimum attractive rate of return was used as the interest rate in the annual cost comparisons. Where this is done, the alternative favored by an annual cost analysis will be the same as the one favored by an analysis based on rate of return. In this respect economy studies for retirements and replacements are no different from other economy studies.

In Examples 17–1 and 17–2 the assumed remaining service lives were the same for all alternatives. In this characteristic these examples are not typical of the majority of replacement studies. As already pointed out, we have introduced the subject with examples of this type in order to concentrate attention on certain aspects of replacement economy while avoiding the special complications often introduced by the difference between a relatively short assumed remaining service life for the defender and a relatively long one for the challenger.

Where two alternatives have the same service life, as in Examples 17–1 and 17–2, an unknown rate of return can be computed by finding the interest rate that makes the present worth of the difference in cash flow equal to zero. Where the lives are different, this method cannot be used without making specific assumptions about differences in cash flow after the expiration of the life of the shorter-lived alternative.

In Examples 17–3 and 17–4 the assumed remaining service life of the defender is considerably shorter than the assumed service life of the challenger. In these examples our presentation will be simplified by making our comparisons solely on the basis of annual costs.

Where To Introduce Present Defender Net Salvage Value in Annual Cost Comparisons

Chapter 16 presented the concept that in making decisions on whether or not to continue to own an asset, the capital costs of extending its service should be based on its present net salvage value and on its prospective future net salvage value at the end of a study period. (See Example 16–4.) This concept, usually called the *opportunity cost* concept, was applied in computing capital recovery costs of existing assets in the annual cost comparisons of Examples 17–1 and 17–2.

An alternate method sometimes advocated for replacement economy studies is to assume zero capital costs for the defender and to subtract the present net defender salvage value from the challenger's first cost before computing capital costs of the challenger. In this method the capital recovery costs of the challenger are based on the new money required rather than on the total investment.

In a certain limited group of cases this alternate method gives a difference in annual costs identical with the difference obtained by the method illustrated in Examples 17–1 and 17–2. For instance, consider the use of this alternate method for the after-tax comparison of annual costs in Example 17–2 as follows:

Comparative annual cost—present pump

Electric energy for pumping	$900

Comparative annual cost—proposed new pump

CR = ($1,650 − $855)(*A*/*P*,9%,10)	$124
Electric energy for pumping	500
Extra income tax	158
Total	$782

The difference in annual costs favors the challenger by $118, just as in our solution in Example 17–2.

Nevertheless, this alternate method has certain weaknesses that lead the authors to recommend against its use in annual cost comparisons. The method clearly is not applicable where the defender has an estimated future salvage value. It is troublesome to interpret in multiple-alternative studies. And it can lead to misleading conclusions in the common case where the remaining life assumed for the defender differs from the assumed service life of the challenger. We shall have further opportunity to study these weaknesses when we discuss Example 17–4.

An error that sometimes occurs in replacement studies is to base capital costs of the defender on present net salvage value and *also* to subtract this salvage from challenger first cost before computing the capital costs of the challenger. It should be evident that this procedure counts defender salvage value twice—as a basis for a plus item in defender annual costs and as a basis for a minus item in challenger annual costs. Such double counting has the effect of biasing the analysis in favor of the challenger.

EXAMPLE 17–3

Analysis to Determine Whether Certain Gas Mains Should Be Replaced

Facts of the Case

A city is engaged in the distribution of natural gas, purchased at a flat rate of $2.50 per MCF (thousand cubic feet). As a result of corrosion, its gas mains eventually develop small leaks which increase as time goes on. The question arises of how much gas should be lost before it is economical to replace any given section of main.

Mains have no salvage when retired. New mains cost about $20,000 per mile installed. Because of improved protective coatings and better methods of installation, it is believed that new mains will be tight for a longer period than was the case for the mains in the original distribution system. The engineer for the city's gas department estimates that under average conditions, a new main will lose no gas during its first 15 years of service; starting with the 16th year, gas losses will increase by about 100 MCF per mile of main per year. The only operation cost that is variable with the age or condition of the main is the cost of lost gas.

In this municipally owned utility, economy studies are made using an interest rate (minimum attractive rate of return) of 7%.

Calculation of Annual Cost of Challenger

In this comparison any section of main that is losing gas is a possible defender in a replacement economy study. In all cases the challenger is a new main that has an estimated first cost of $20,000, zero salvage value at all times, and operating costs that start at $250 in the 16th year and increase by $250 a year thereafter. Table 17–1 provides a basis for estimating challenger life and equivalent annual cost.

The actual compound interest calculations are not shown in the table. A sample calculation assuming retirement after 25 years is as follows:

CR = $20,000 ($A/P$,7%,25) = $20,000(0.08581)	= $1,716
Equivalent annual cost of gas lost	
= $250($P/G$,7%,11)($P/F$,7%,14)($A/P$,7%,25)	
= $250(32.466)(0.3878)0.08581)	= 270
Total equivalent annual cost	= $1,986

In this particular case, it is evident that the equivalent costs of the challenger are relatively insensitive to differences in assumptions regarding challenger service life over a considerable range of possible lives. Although the

TABLE 17–1
Equivalent annual costs for a mile of $20,000 gas main retired at various ages, assuming a minimum attractive rate of return of 7%

		Equivalent uniform annual costs if retired after n years		
Year n	*Cost of gas lost during year*	*Capital recovery of $20,000*	*Equivalent uniform annual cost of gas lost*	*Total equivalent uniform annual cost*
A	B	C	D	E
3	$ 0	$7,621	$ 0	$7,621
6	0	4,196	0	4,196
9	0	3,070	0	3,070
12	0	2,518	0	2,518
15	0	2,196	0	2,196
16	250	2,117	9	2,126
17	500	2,049	25	2,074
18	750	1,988	46	2,034
19	1,000	1,935	72	2,007
20	1,250	1,888	100	1,988
21	1,500	1,846	132	1,978
22	1,750	1,808	165	1,973
23	2,000	1,774	199	1,973
24	2,250	1,744	234	1,978
25	2,500	1,716	270	1,986
26	2,750	1,691	306	1,997
27	3,000	1,669	342	2,011
28	3,250	1,648	378	2,026
29	3,500	1,629	414	2,043
30	3,750	1,612	449	2,061

lowest equivalent annual cost, $1,973, is obtained at 22 years, this figure is changed by less than 3% within the range from 19 to 28 years.*

Calculation of Annual Cost of Extending Service of Defenders

It has been stated that the gas mains in this example have zero salvage values at all times. Therefore, there are no capital costs associated with extending the service of any defenders. It has also been stated that there is no difference in maintenance cost between old and new mains.

It follows that the costs of gas lost are the only defender costs that are relevant in comparison with challenger costs. For any particular section of main that is being considered for replacement, it is necessary to estimate the

*The annual cost for 23 years' service is $1,973 also. For reasons explained later in this chapter, however, 22 years should be assumed as the economic life.

MCF of gas that will be lost next year, to express this quantity as a rate per mile of main, and to multiply the rate per mile by the unit price of $2.50 per MCF.

For instance, consider section X of main losing gas at 600 MCF per mile; this has a next-year cost of $1,500 per mile. Or consider section Y losing gas at 1,200 MCF per mile; its next-year cost is $3,000 per mile.

Because gas losses tend to increase from year to year, it may be presumed that the annual costs in subsequent years for any section of main will be greater than those next year. Therefore, it merely is necessary to estimate next-year defender costs for purposes of comparison with challenger annual costs.

Comparison of Defender and Challenger

Consider the economy study relative to section X of main:

Comparative annual cost of defender = $1,500 per mile (for next year)
Comparative annual cost of challenger = at least $1,973 per mile (for 22 years, more or less)

It is evident that section X should be continued in service.

It is worth emphasizing that the use of next-year costs for the defender carries no implication that, if the decision favors the defender, it will be kept in service *only* one year more. It is merely that, for purposes of comparison, we are using the period most favorable to the defender, the cost figure that is most advantageous to the defender. The challenger also has the benefit of the period most favorable to it, namely, approximately 22 years, that is, the cost figure that is most advantageous to the challenger.

In contrast to the foregoing study, consider the comparison relative to section Y of main:

Comparative annual cost of defender = $3,000 per mile (for next year)
Comparative annual cost of challenger = at least $1,973 per mile (for 22 years, more or less)

This comparison seems to favor the challenger. However, in a case of this type, it is always pertinent to raise the question of what is expected to happen after the expiration of the one-year period assumed as the remaining life of the defender. This question is discussed later in the present chapter.

At the present stage of our discussion, we can say that if the next-year cost of extending the service of an existing main is not at least $1,973 per mile, it is doubtful whether replacement is justified. In other words, the sections of main where replacement should be considered are those expected to lose at least 790 MCF per mile per year (i.e., $1,973 per mile divided by the price of $2.50 per MCF).

EXAMPLE 17–4

Analysis of Proposed Replacement of Certain Construction Equipment

Facts of the Case

The first cost of a machine used to perform a certain service in construction is $25,000. It is expected that the secondhand value of this machine will be $16,000 after its first year, $13,000 after 2 years, and $10,000 after 3 years, and will continue to decline as shown in Table 16–1, page 398. It is believed that the disbursements in connection with ownership and operation of this machine will be $8,000 for the first 3 years and that disbursements will increase each year as shown in column C of Table 17–2.

The owner of a machine performing this service has a chance to sell it for a favorable price of $12,500. The initial cost was $26,000 one year ago. However, the first cost of new machines of this type has declined and the economy of performance has been improved. A decision must be made between continuing this machine (defender) in service and replacing it with a new machine (challenger) that is expected to have costs for various lengths of service life as shown in Table 17–2. All studies are made before taxes with an i^* of 15%.

The estimated future salvage values of the defender are $8,000 next year, $5,000 2 years hence, and $2,200 3 years hence. If the defender is continued in service, annual disbursements are estimated at $8,900 next year, $10,500 the following year, and $12,500 the year after that.

Calculation of Annual Cost of Challenger

It is evident from column G of Table 17–2 that there is considerable range of possible challenger lives for which the equivalent uniform annual cost is very close to $15,200. However, as a 6-year life yields slightly lower annual costs than any other life, this is the assumed life that is most favorable to the challenger. Thus, $15,143 should be used as the challenger advantage cost (CAC) for comparison purposes.

Comparative annual cost of challenger (6-year life)

CR = $25,000(*A/P*,15%,6) − $3,500(*A/F*,15%,6)	
= $25,000(0.26424) − $3,500(0.11424)	$6,206
Equivalent annual disbursements [$8,000(*P/A*,15%,3)	
+ $9,000(*P/F*,15%,4) + $10,500(*P/F*,15%,5)	
+ $12,000(*P/F*,15%,6)](*A/P*,15%,6)	8,937
Total (challenger advantage cost)	$15,143

Calculation of Annual Cost of Defender

The first impulse of anyone making this economy study might be to compare

TABLE 17–2
Equivalent annual costs of $25,000 machine of Table 16-1 for assumed lives of 1–10 years interest at 15%

	Cost of extending service during the nth year			*Equivalent uniform annual costs if retired after n years*		
Year n	*Capital recovery cost (column E of Table 16–1)*	*Disbursements during year*	*Total marginal cost (B + C)*	*Equivalent uniform annual capital recovery cost (column H of Table 16–1)*	*Equivalent uniform annual disbursements*	*Total equivalent uniform annual cost (E + F)*
A	*B*	*C*	*D*	*E*	*F*	*G*
1	$12,750	$ 8,000	$20,750	$12,750	$ 8,000	$20,750
2	5,400	8,000	13,400	9,332	8,000	17,332
3	4,950	8,000	12,950	8,070	8,000	16,070
4	4,000	9,000	13,000	7,255	8,200	15,455
5	3,125	10,500	13,625	6,642	8,542	15,184
6	2,825	12,000	14,825	6,206	8,937	15,143
7	2,025	14,000	16,025	5,828	9,394	15,222
8	1,300	15,000	16,300	5,498	9,802	15,300
9	1,150	15,500	16,650	5,239	10,142	15,381
10	0	16,800	16,800	4,981	10,470	15,451

next year's costs for the defender with the foregoing challenger advantage cost.

Comparative annual cost of defender (1 year more)

CR = $12,500(A/P,15%,1) − $8,000(A/F,15%,1)	
= $12,500(1.15) − $8,000(1.0)	$6,375
Operating disbursements	8,900
	$15,275

It is evident that $12,500(1.15) − $8,000(1.0) may also be written as ($12,500 − $8,000)(1.0) + $12,500(0.15). In general, where the assumed remaining life of a defender is 1 year, it is convenient to think of the capital recovery cost as made up of the prospective decline in salvage values (depreciation in the popular sense) and interest on the present salvage value. An alternate statement of next-year defender costs is therefore:

Decline in salvage value = $12,500 − $8,000	$4,500
Interest on present salvage value $12,500(0.15)	1,875
Operating disbursements	8,900
Total	$15,275

The challenger is favored slightly by a comparison of this $15,275 next-year defender cost with the $15,143 challenger advantage cost. However, in this case the decline in defender salvage value forecast for next year ($4,500) is much greater than the decline forecast for the following year ($3,000). A comparison favorable to the defender by a slight margin will be obtained if defender costs are considered over the next 2 years.

Comparative annual cost of defender (2 years more)

CR = $12,500(A/P,15%,2) − $5,000(A/F,15%,2)	$5,363
Equivalent annual disbursements [$8,900(P/F,15%,1) + $10,500(P/F,15%,2)](A/P,15%,2)	9,644
Total (defender advantage cost)	$15,007

A similar calculation assuming 3 years more for the defender yields an equivalent annual cost of $15,308. Thus, the most advantageous annual cost to be used for the defender in a comparison with the best challenger, the defender advantage cost (DAC), is $15,007.

One purpose of including this example has been to emphasize the point that in any declining salvage situation, it may be advisable to consider defender costs for several different assumed remaining lives. The practical conclusion of this particular study should be that the comparative cost figures are so close that the decision between challenger and defender ought to be made on so-called irreducibles—that is, on considerations that have not been reduced to money terms in the study.

The $26,000 original cost of the defender is, of course, irrelevant in any before-tax study.

Some Aspects of Tables Such as 17–1 and 17–2

The figures given in column E of Table 17–1 and column G of Table 17–2 are valid equivalent annual costs for the respective challengers, given the assumed interest rate and first cost, and the predictions of the patterns of salvage values and operating costs. Nevertheless, even though the various disbursements turn out exactly as predicted, the tables do not tell us that it will necessarily be economical to replace the challenger of Table 17–1 when it is 22 years old or the challenger of Table 17–2 when it is 6 years old. The economic replacement date may be sooner or later, depending on the costs and prospective performance of challengers that may be available in the future, and on other matters.

Calculations such as those illustrated in Tables 17–1 and 17–2 may be viewed as yielding "economic lives" only under very restrictive assumptions. All future challengers must have the same first cost as the present challenger, the same expected salvage values at each age, and the same expected operating costs for each year. Moreover, it must be assumed that the need for the particular service continues indefinitely and that the minimum attractive rate of return remains unchanged.

In general, replacement decisions should not be based on such assumptions, which obviously are unrealistic in our modern world of changing technology and changing prices. The question in a replacement decision should not be "How old is the defender?" The issue should always be viewed as an economic choice between extending the service of the defender, regardless of its age, and its retirement in favor of the best available challenger. But a temporary view of Tables 17–1 and 17–2 as "economic life" calculations may be helpful as an aid to the exposition of certain theoretical aspects of replacement economy and as a basis for pointing out a useful check on the arithmetic and the analysis in any tables or formulas of this type.

For instance, consider Table 17–1 as applicable to future challengers as well as to the present one. As long as the cost of extending the service of the present challenger is less than $1,973, the annual cost of a future challenger, it will pay to keep the present challenger in service. Because salvage values are zero at all times, the only cost of extending service is the operating disbursement (cost of lost gas) in column B. Consider the start of the 22nd year; as the column B figure is $1,750, it will not pay to replace. Now consider the start of the 23rd year; as the prospective cost of extending service another year is $2,000 and as $2,000 exceeds $1,973, replacement is justified after 22 years. The various elements in the analysis are shown graphically in Figure 17–1. The minimum point on the curve of total equivalent annual cost occurs where the curve is intersected by the dashed line for cost of lost gas during the year.

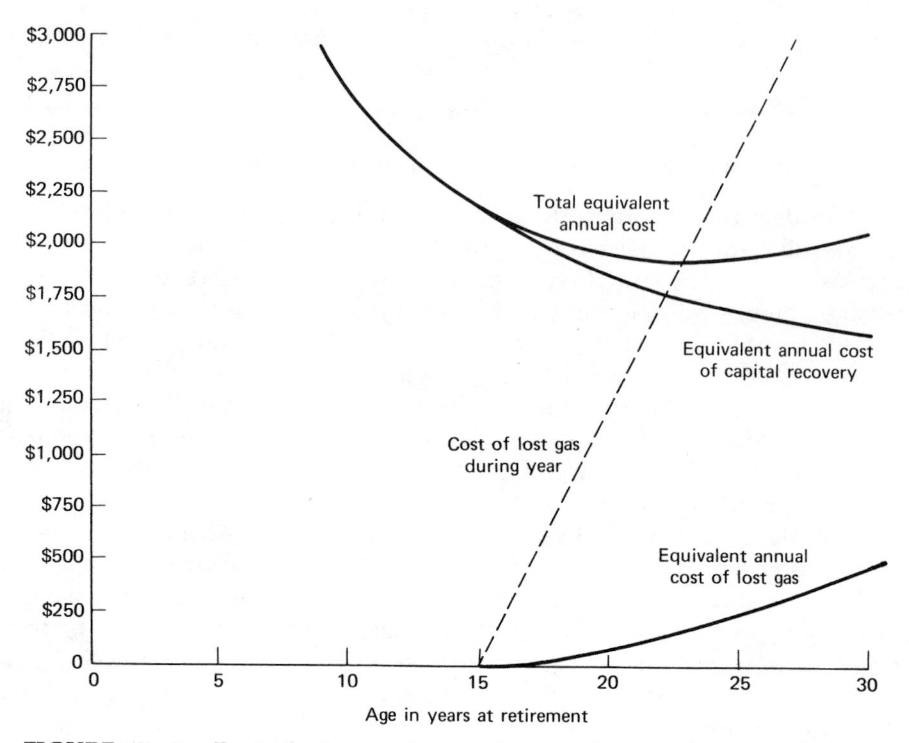

FIGURE 17–1. Equivalent annual cost of gas main as a function of age at retirement—data of Table 17–1

Figure 17–2 shows the same type of relationship for the more complicated facts of Table 17–2. Because salvage values are not zero and are declining, the cost of extending service in any year is the sum of operating disbursements, interest on salvage value at the start of the year, and decline in salvage during the year. But the same principle applies that the minimum annual cost is reached when the marginal cost of extending service next year is greater than the equivalent annual cost to date.

If the curves of total equivalent annual cost in Figures 17–1 and 17–2 should be interpreted as the basis for finding the economic life, they would suggest the conclusion that there is a wide range of possible lives of these assets that are almost equally economical. Although there are minimum cost points, these curves are relatively flat through a considerable range. Mathematical models developed in connection with replacement economy often indicate a substantial minimum cost range as well as a minimum cost point.

If one could imagine a static society in which prices and service requirements never changed and in which challengers always repeated the cost history of their defenders, the exact timing of replacement in such a society doubtless would be of little consequence. In our dynamic modern industrial society where challengers often differ greatly from their defenders, the timing

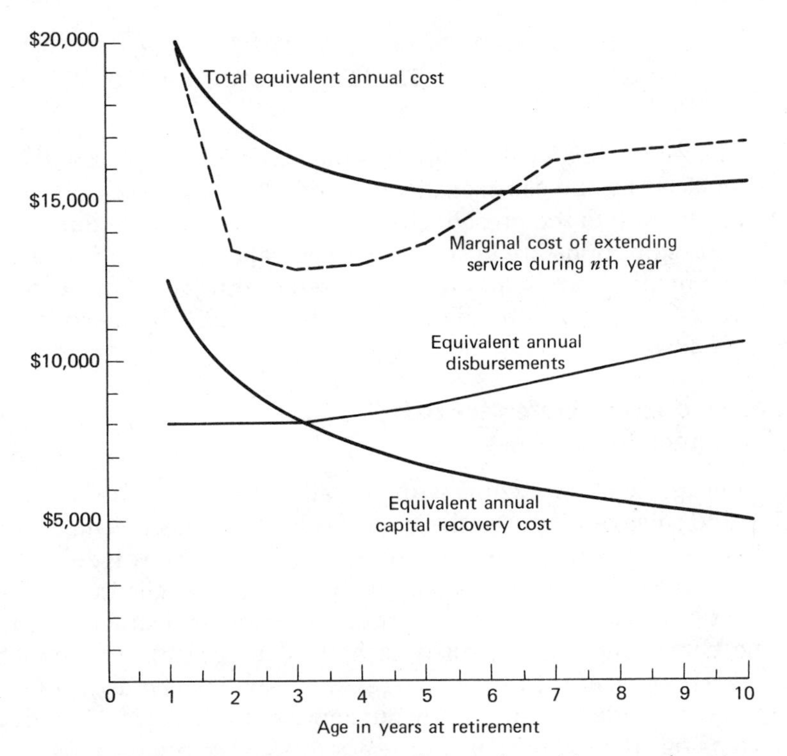

FIGURE 17–2. Equivalent annual cost of machine as a function of age at retirement—data Table 17–2

of replacements is likely to be much more important than would be suggested by any mathematical model.

Economic Life of the Challenger and Its Successors

The implications of an economic life from a study such as that illustrated in Tables 17–1 and 17–2 were discussed in the previous section. They may be illustrated further by using a capitalized cost approach to the replacement problem of Example 17–4.

If replacement is made this year, the capitalized cost of the challenger and all of its replacements will be:

$$\text{CC} = \$15{,}143/0.15 = \$100{,}953$$

At each 6-year interval, the machine will be replaced by a machine with the same operating disbursement and salvage value time profiles.

If the defender is held for the 2-year period for which the defender advantage cost was $15,007, the capitalized cost is:

$$\begin{aligned} CC &= \$15{,}007(P/A,15\%,2) + \$15{,}143(P/F,15\%,2)/0.15 \\ &= \$15{,}007(1.626) + \$15{,}143(0.7561)/0.15 = \$100{,}732 \end{aligned}$$

This solution assumes explicitly that when the defender is replaced in 2 years, it will be replaced by the same challenger and all of its successors, also assumed to be identical to the present challenger, over an infinite time horizon. Where these assumptions are known not to be valid, other explicit assumptions must be made. Corresponding capitalized cost figures for the defender for 1 or 3 years' service will be higher than the $100,953 figure for the challenger.

Impact of Deducting Defender Salvage Value from Challenger Initial Cost

Let us assume an existing machine with operating disbursements as shown in Table 17–2 and salvage values as shown in Table 16–1. This machine is now 6 years old. It is suggested that the present salvage value of the defender should be subtracted from the initial cost of the challenger in making the study so that only the "new money" required for the investment is charged against the challenger (the so-called cash flow approach). The previously determined economic life of the challenger of 6 years is to be used.

Since the estimated operating disbursements from Table 17–2 are constantly increasing after the third year, the appropriate comparison figure to be used for the defender is the next-year operating disbursement.

Defender next-year operating disbursement = $14,000

In developing a cost figure for the challenger, the initial cost of $25,000 is reduced by the current salvage value of the defender from Table 16–1, $3,500. Thus, the equivalent uniform annual cost for the challenger is:

$$\begin{aligned} &\textbf{CR} = \$21{,}500(A/P,15\%,6) - \$3{,}500(A/F,15\%,6) && = \$5{,}281 \\ &\textit{Equivalent annual operating disbursements} = [\$8{,}000(P/A,15\%,3) \\ &\quad + \$9{,}000(P/F,15\%,4) + \$10{,}500(P/F,15\%,5) \\ &\quad + \$12{,}000(P/F,15\%,6)](A/P,15\%,6) && = \underline{8{,}936} \\ &\textit{Total equivalent annual cost} && = \$14{,}217 \end{aligned}$$

These figures differ significantly from those in Table 17–2, which show that the next-year marginal cost for the defender (the DAC) is $16,025, which should be compared to the challenger advantage cost of $15,143. Not only are they significantly different but the decision has been reversed. If our analysis in Table 17–2 is correct, replacement should be made every 6 years.

The error in the foregoing analysis which led to lower seventh-year cost for the defender than the equivalent annual cost of the challenger at its economic life is in the treatment of the $3,500 salvage value of the defender.

Looking again at Table 16–1, it can be seen that the \$3,500 defender salvage value at the end of year 6 is reduced to zero by the end of year 9; it is extinguished in 3 years. However, we have, in effect, allocated it over the 6-year expected life of the challenger in this analysis. That is:

$$\textit{Challenger CR} = (\$25{,}000 - \$3{,}500)(A/P,15\%,6) - \$3{,}500(A/F,15\%.6) = \$5{,}281$$

If \$3,500(A/P,15%,6) = \$926 is added to the previous \$14,217 challenger annual cost, the figure is brought back to that of Table 17–2, \$15,143. However, the corresponding amount should not be added to the defender's next-year cost figure because the actual CR cost for the seventh year is \$2,025, not \$926.

We have shown that, when the remaining life of a defender differs from the economic life of a potential challenger, reducing the challenger initial cost by the salvage value of the defender may lead to an incorrect conclusion. Invariably the potential error will be to retain the defender beyond the time at which it should be replaced. Our analysis was based on the assumption of a completely steady-state economic situation. Nevertheless, a methodology which can be demonstrated to be incorrect in a steady-state environment cannot be correct in the dynamic environment of the real world. In making replacement decisions, defender salvage values should be treated as opportunity costs of the defender. The choice should be based on a comparison between the most advantageous defender annual cost and the most advantageous equivalent annual cost of the best available challenger.

What Shall Be Assumed as Remaining Defender Service Life in a Replacement Economy Study?

Examples 17–3 and 17–4 are typical of many replacement studies in which the assumed remaining defender life is shorter than the assumed challenger life. In such studies, one problem is to select the defender remaining life that is most favorable to the defender.

When a major outlay for defender alteration or overhaul is needed, the remaining defender life that will yield the least annual cost is likely to be the period that will elapse before the next major alteration or overhaul will be needed.

When there is no defender salvage value now or later (and no outlay for alteration or overhaul), and when defender operating disbursements are expected to increase annually, as in Example 17–3, the remaining life that will yield the least annual cost will be 1 year (or possibly less).

When salvage values are expected to decline from year to year, as in Example 17–4, it may be necessary to try several remaining service lives before selecting the one to be used in the comparison.

It should be emphasized that the selection of a remaining defender service life for use in a replacement study does not imply a commitment that, if the decision favors the defender, it will be retired at the end of this period. The problem is merely to find the period that yields the lowest annual cost of continuing the service of the defender.

How Prospective Occurrences After the Expiration of the Assumed Remaining Life of the Defender Should Influence the Choice Between Defender and Challenger

Consider the comparison of annual costs in Example 17–3 relative to section Y of main:

Defender—$3,000 per mile for next year
Present Challenger—$1,973 per mile for 22 years

Such a comparison appears to favor the present challenger. Nevertheless, it does not necessarily follow from such a comparison that a decision in favor of the present challenger will lead to maximum economy in the long run. If the replacement is put off for a while, a better challenger than the present one may be available. If the future challenger is enough better than the present one, a long-run comparison may favor continuing the defender in service for the time being in order to realize the advantages obtainable from the future challenger.

As a simple numerical illustration of this point, assume that it is expected that the first cost of new gas mains will decline a little during the coming year and that protective coatings will be somewhat improved. Let us assume that a monetary allowance for these changes leads to a figure of $1,850 a year as the annual cost per mile for 21 years of service from next year's challenger. Now our comparison is:

Defender and Next Year's Challenger—$3,000 next year followed by $1,850 a year for 21 years.
Present Challenger—$1,973 a year for 22 years

The defender alternative may be converted (at 7%) to an equivalent uniform annual cost of $1,947 for 22 years. If our assumption about next year's challenger turns out to be correct, it is better to make our present decision in favor of the defender.

More often than not, an analyst is likely to feel that no basis exists for such a specific numerical estimate of the difference in annual cost between a known present challenger and an unknown future one. But coming events may cast their shadows sufficiently well for the matter to be considered on a qualitative basis. In one instance it may be known that greatly improved designs of certain machines are in the development stage and that new commercial models may be expected in the near future. In another case new models containing drastic improvements may have recently appeared, and it may be thought unlikely that machines with further great improvements will be available soon. Sometimes there may be good reason to believe that future challengers will have higher first costs than the present one (perhaps because of their greater complexity or because of rising prices); sometimes it may be expected that future challengers will be lower priced (perhaps because of economies realizable from their production in greater quantities).

Any forecast about differences between the present challenger and future challengers is relevant in decision making about a proposed replacement—particularly so in borderline cases. A forecast that future challengers will be superior to the present one or that they will have lower first costs is favorable to postponing replacement. A forecast that future challengers will be inferior* or that they will have higher first costs is favorable to immediate replacement. If an analyst desires to consider such forecasts quantitatively but is unwilling to make a single monetary estimate, the sensitivity of the replacement decision to forecasts about future challengers may be tested by making an analysis of several different estimates.

Since 1949 in the United States there has been a considerable vogue for the use of mathematical models as a guide to decisions on replacements. A first step in the understanding of such models is an understanding of calculations of minimum annual cost such as the ones illustrated in Tables 17–1 and 17–2. A second step is an understanding of the problem of giving weight to the expectation that if replacement is deferred for a while a better challenger will be available. Several of the more popular models involve a formal monetary allowance (as a charge against the challenger) for the expectation that if replacement is deferred a better challenger will be available 1 year hence.

What Shall Be Assumed as Challenger Service Life in a Replacement Economy Study?

In Examples 17–3 and 17–4 we assumed the challenger life that would yield the lowest annual cost for the challenger. (We noted that in Examples 17–3 and 17–4, because of consideration of the prospective increase of annual disbursements with age, the challenger annual cost was not sensitive to a moderate change in assumed challenger life.)

In these examples we did not shorten the assumed life of the present challenger because of the expectation that better challengers would be available in the near future. However, decisions on replacements in industry

*Of course the normal expectation in modern industry is that technological progress will cause future challengers to be superior. However, it sometimes happens that style change or technological change causes new machines to be less well adapted to certain services. Moreover, periods occur in history when it seems likely that the quality of new assets available in the future will be inferior to those presently available. For example, the period 1939–1941 in the United States, prior to the entry of the country into World War II, was a time when an analyst making a replacement study might have forecast that challengers in the near future, if any, would be inferior to present ones.

sometimes are made by methods of analysis that in effect make an excessive shortening of challenger service life for this reason.

The use of a very short payout period for a challenger that has a relatively long prospective life should not be regarded as "conservative." Using a 1- or 2- or 3-year payout period for such a challenger has the effect of continuing the service of defenders that are obsolete and uneconomical.*

Nevertheless, *some* shortening of assumed challenger service life may be appropriate in lieu of a formalized allowance for the prospect that if replacement is deferred, a superior challenger will be available.

The Problem of Securing Adequate Cost Data for Certain Types of Replacement Economy Studies

It is a common condition for operation and repair costs of machinery and other assets to increase with age.† For adequate replacement studies, it is not enough to be aware of these matters in a qualitative way; specific quantitative information is needed applicable to each study.

In the frequent case where an organization operates many similar machines of different ages, there is an opportunity to secure such information. Usually the accounting system does not furnish data in the form needed by an analyst studying replacement policy. For example, the accounts may report the total repair costs for all machines of a given type but not give information on individual machines; an analyst may need to examine and classify individual repair work orders. The evaluation of some matters, such as the adverse consequences of breakdowns or other unsatisfactory performance, may require considerable information gathering and analysis that is entirely apart from the accounting system.

Limitation on Uses of "Comparative Annual Costs" Computed for Replacement Studies

It should be emphasized that "costs" such as have been calculated in the examples of replacement studies in this chapter are useful solely for the comparison of the specific alternatives under consideration. Such "comparative annual costs" are properly thought of merely as convenient measures of the differences between the proposed alternatives. Obviously no inferences should be drawn from them except inferences as to the relative merits of the alternatives being compared. This statement applies with particular force to the drawing of inferences about pricing policies.

*George Terborgh has written forcefully on this point. See *Dynamic Equipment Policy, op. cit.*, chap. xii.

†See Terborgh, *Dynamic Equipment Policy, op. cit.*, pp. 70–71, for an example of graphs showing the relation between age and repair cost per unit of service for certain metal-working equipment, textile machinery, locomotives, farm implements, light trucks, passenger automobiles, intercity buses, and local buses. In all of these graphs repair costs per unit of service increase with age. In some instances the relationship is linear; in others the increase per year is less in the later years than in the earlier years.

The Need to Find the Best Challenger

In replacement studies, just as in all other economy studies, it is essential that all promising alternatives be considered. It is not good enough for a particular challenger to be more economical than the defender; it also needs to be more economical than other possible challengers.

Several of the problems at the end of this chapter illustrate the comparison of several possible challengers with one another as well as with a single defender.

Some Aspects of the Various MAPI Systems

The following section contains a brief discussion of certain features of several related mathematical models for investment analysis developed and modified over a period of nearly 20 years. These models were described and their applications were illustrated in a number of publications written by George Terborgh, Research Director for the Machinery and Allied Products Institute (MAPI), a trade association of producers of capital goods.

Terborgh's four major books that deal with the various "MAPI systems" for investment analysis are:

Dynamic Equipment Policy (1949)
MAPI Replacement Manual (1950)
Business Investment Policy (1958)
Business Investment Management (1967)

All are now published by Machinery and Allied Products Institute, Washington, D.C. (*Dynamic Equipment Policy* was originally published by McGraw-Hill Book Company, Inc.)

The first two books deal chiefly with decisions about the *replacement* of machinery and equipment. Although the mathematical models used in the two later volumes were greatly influenced by certain aspects of the problem of making replacement decisions, these books deal more broadly with matters related to all kinds of investment in physical plant.

A present challenger may have various advantages over a defender that will influence prospective year-by-year cash flow. These advantages may lead to lower disbursements to provide a given service, higher receipts from the sale of that service, or both. Similarly, future challengers may have such advantages over the present challenger. The models developed by Terborgh assume that this advantage will increase linearly year by year; that is, that technological progress will occur at a uniform rate each year. Therefore, a penalty cost is charged against this year's challenger amounting to the present worth in perpetuity of the operating inferiority of the present challenger to the challenger that will be available next year. The models in the earlier two books gave what amounted to an equivalent annual cost for the challenger to be compared with a defender next-year operating cost.

For the sake of brevity, we shall refer to the model used in *Business Investment Policy* as the 1958 model and the one used in *Business Investment Management* as the 1967 model.

Both models broadened the problem to an evaluation of *any* proposed investments in plant and equipment. Nevertheless, many of the features of the earlier models were maintained. Anyone who wishes to understand the later models should first study *Dynamic Equipment Policy*, which is the only one of the four volumes that contains a step-by-step exposition of underlying theory.

Just as in the earlier books, a formal allowance was made for the prospect that future challengers will be superior to present ones. And this prospect of superiority of future challengers was, in effect, converted to a single equivalent monetary amount to permit a next-year comparison between the alternatives of (1) selecting the present challenger at once, and (2) continuing with the defender for one more year and then acquiring the next-year challenger. The 1958 and 1967 models were like the 1950 model of *MAPI Replacement Manual* in basing their assumptions regarding the superiority of future challengers on the estimated life and salvage value of the present one. The 1950, 1958, and 1967 models were alike in making the conclusions of a study quite sensitive to the estimated salvage value percentage of a challenger.

However, the 1958 and 1967 MAPI methods had a number of points of difference from the earlier ones. Some important differences were:

1. The previous models had involved comparisons before income taxes; the 1958 and 1967 techniques involved after-tax comparisons.

2. Whereas the previous models had given a next-year difference in annual cost between defender and challenger, the 1958 and 1967 models gave a next-year rate of return on the extra investment needed to acquire the challenger.

3. In the earlier methods, the gradual increase of operation and maintenance costs (and/or decrease in revenues) with the age of an asset was also assumed to follow a uniform gradient. In *Business Investment Policy*, such a uniform increase is described as the "standard" projection pattern. But other projection patterns also were made available in the 1958 volume; it was possible to assume that annual disbursements increase at either a decreasing rate or an increasing rate.*

4. The most complicated aspects of the 1958 and 1967 models came from the introduction of income taxes. The earlier models had discounted the infinite chain of estimated differences in cash flow between the present challenger and its successors and the next-year challenger and its successors. In making a similar calculation that reflected income taxes, it became necessary, in effect, to discount an infinite chain of differences in cash flow for income

*An earlier mathematical model assuming that annual disbursements for operation and maintenance increase at a decreasing rate was described in A. A. Alchian's *Economic Replacement Policy*, Publication R-224 (Santa Monica, Calif.: The Rand Corporation, 1952).

taxes. The 1967 volume supplied the analyst with charts applicable to straight-line, years-digits, and declining-balance depreciation methods. (In 1958, it had been stipulated that a years-digits chart would be used whenever the declining-balance method was to be adopted.) In all cases it was necessary to assume that the challenger life and salvage value adopted for the MAPI evaluation would also be used for income tax purposes. (The reader may recall that we made a similar simplification in our calculation of income tax differences in Chapters 6 through 11.)

In computing the future chain of income tax differences, it was assumed that the tax rate would be 50%, and that 25% of any project would be financed by debt bearing an interest rate of 3%.

5. In the discounting of the infinite chain of cash flow differences in the earlier models, the discount rate had been the stipulated value of *i* (in effect, the before-tax minimum attractive rate of return). But the 1958 and 1967 volumes dealt with rate-of-return models, not annual-cost models.

In conventional rate-of-return calculations such as those we have illustrated starting with Chapter 8, the purpose of the analysis is to find an unknown value of *i* applicable to the entire study period; there is no figure for *i* stipulated in advance to be used in compound interest conversions. But the 1958 and 1967 MAPI models aimed to find a one-year rate of return. Therefore, it was necessary to adopt a value of *i* to apply to the infinite chain of estimated cash flow differences after date 1. For this purpose it was assumed that there would be 75% of equity capital earning an average after-tax return of 10% and 25% of borrowed capital at an average interest cost of 3%. The weighted average of these gave an *i* of 8.25% to be used in the discounting. Comments regarding project evaluation based on the assumption of a standardized mixture of debt and equity capital are deferred to Chapter 18. However, we should point out here that one of the dangers in using forms, graphs, and procedures such as those required in applying the 1958 and 1967 models is that assumptions as to capital structures and interest rates rarely fit specific companies at fixed points in time. It would be hard to imagine, for instance, a company able to borrow capital at an interest rate of 3% or attract equity capital at 10% in the early 1980s.

The 1958 and 1967 models, which gave as their answer a one-year rate of return, therefore constituted an analysis using two interest rates, namely, the discount rate applied to all cash flow differences after the first year, and the next-year rate of return found at the conclusion of the analysis. To the extent that the former is unrealistically low, the latter will be misleadingly high.

6. Beginning with the first MAPI methods in 1949, all MAPI systems have been geared to a next-year comparison as the desirable one. But all have provided approximate methods for applying the MAPI philosophy to comparison periods of more than one year in cases where a one-year period was not appropriate for the defender. The 1967 book provides a special auxiliary chart to be used in such cases.

Both the 1958 and 1967 volumes are useful reading for business managers and analysts engaged in project evaluation even though it is not planned to use any of the various MAPI systems. Most of the topics discussed by Terborgh are of general application in investment analysis, and his comments are of interest apart from the particular MAPI system that is being explained.

The Issue of How to Consider the Prospect of Future Technological Change in Making Present Replacement Decisions

Although the various MAPI methods have differed from one another, one element that they have had in common has been the use of a formal monetary figure—usually implied rather than explicitly stated—to reflect the likelihood that future challengers will be superior to present ones. In fact, this formal allowance has been the new element that these mathematical models have introduced into replacement economy.

In the common case where replacement decisions are made without the use of a mathematical model of this type, one possibility is to make the required monetary comparisons with no special allowance for the prospect that there will be a better challenger in the near future. Then, whenever deemed relevant, this prospect can be recognized as a nonmonetary element in making the decision. In effect, this prospect is viewed as part of the irreducible data for decision making.

An alternate possibility is to require challengers in replacement studies to meet more severe criteria than are imposed on other proposals for new assets. For example, somewhat higher minimum attractive rates of return might be used in replacement studies than in other studies. More commonly, this device of using severe criteria takes the form of shortening the estimated service life of the challenger.

An extreme shortening of challenger service life in replacement studies constitutes an irrational obstacle to technological progress. If old assets are kept in service until new ones will "pay for themselves" in one or two years, the old assets will generally be kept for much too long a time. The adverse social consequences to be expected from a widespread use of extreme short payoff periods were stressed by George Terborgh in *Dynamic Equipment Policy*. On the other hand, *some* shortening of the estimated life of a challenger in its initial or primary service has an effect not unlike the one obtained from using certain MAPI models.

Some General Comments on the Use of Complex Mathematical Models as Aids in Making Economic Decisions

If the use of a mathematical model is under consideration as a possible basis for decision making—either on a proposed replacement or on any other matter—certain questions need to be examined. One question is which of the available models seems to be best adapted to the particular circumstances. (It sometimes happens that although no available model is suitable, a new one can be devised that is more nearly adequate.) Another question is whether or not the formalized assumptions of the best model are close enough to the facts. In general, these questions cannot be answered in a satisfactory way except by someone who understands both the facts and the assumptions underlying the various models. Judgment on these questions is more difficult for complex mathematical models than for simple ones.

It is not a valid objection to a particular model to point out some way in which the model fails to fit the exact circumstances of a case. Assumptions that can be incorporated into mathematical formulas will seldom fit economic facts perfectly. In judging the suitability of a certain model to a particular case, the important matter is to judge whether the *decision* based on the use of the model is likely to be sensitive to the various ways in which the assumptions depart from the facts of the case.

The foregoing comment about the importance of sensitivity may be illustrated with reference to several of the assumptions made in all of the MAPI models. Consider the assumptions (1) that the service period is perpetual, (2) that all future challengers will have the same first costs, salvage values, and lives as the present one and that technological progress will occur at a uniform rate, and (3) that the appropriate comparison is between replacement at once with the present challenger and continuing the defender in service one year more and then replacing it with the next-year challenger.

1. In general, with the interest rates of 8% or more commonly used in replacement studies, the present worth of the assumed distant consequences of present decisions is quite small. If a service is likely to be needed for a fairly long time, no serious error in decision making will be caused by using a model that assumes perpetual service. But such a model clearly is inappropriate whenever it is expected that the service will terminate at a foreseeable date in the near future.

2. Following the discussion of Example 17–4, it was pointed out that there sometimes is a reasonable basis for making specific forecasts about the challengers to be expected in the next few years. Different forecasts have different implications regarding the present replacement decision. For example, the expectation that challengers in the near future will have considerably higher first costs than the present challenger is favorable to immediate replacement; the expectation of substantial technological improvement in the near future is favorable to putting off replacement. Where a basis exists for such specific forecasts, better decisions are likely to be made by an analysis incorporating these forecasts than by an analysis using a mathematical model that depends on generalized forecasts about future challengers.

3. The question of selecting the defender remaining life that is most favorable to the defender also was discussed. It was pointed out that under certain circumstances this life is one year; under other circumstances it may be considerably longer. In those circumstances where a defender life of more than one year should be used in a replacement study, a next-year analysis may lead to an incorrect decision favorable to the challenger.

In general, it should be emphasized that a prerequisite to the use of *any* mathematical model to guide an economic decision should be a clear understanding of the assumptions of the particular model by the person who makes (or recommends) the decision.

Often the Source of a Product or Service Is Replaced Without a Retirement

The examples and problems in this chapter generally imply that if a challenger is acquired the defender will be retired. Often the circumstances of the case make it reasonable that the old asset be retired as soon as the new one is available.

But frequently new assets are acquired for a given service in a way that displaces the old assets but does not involve their retirement. The old assets are simply relegated to a different use—one that might generally be described as an inferior use. Assignment for peak load or standby purposes is an example of this. Another familiar example is the assignment of a displaced main-line railroad locomotive to branch-line service. It may be many years before an asset displaced in this way is finally retired; in fact, there may be several successive different inferior uses before retirement.

In periods of expanding demand for a product or service (generally but not necessarily, corresponding to periods of general business prosperity), the tendency is to acquire new assets of improved design but not to retire the older less economical assets. In periods when there is apparent overcapacity (generally, but not necessarily, in business depressions), the older assets are retired. Much replacement of industrial assets occurs with this type of time lag. Where this takes place, the economy studies in periods of expanding demand involve the displacement of the sources of a service and those in overcapacity periods involve retirements, but there are no replacement economy studies as such.

Common Errors in Replacement Studies

Observations of the practice of industrialists in such studies and of the published literature of the subject indicates four errors which it seems are often made in dealing with replacement economy:

1. Considering the excess of present book value over the net realizable value of the old asset as an addition to the investment in the new asset. This error increases the apparent cost associated with the new asset, and thus tends to prevent replacements that are really economical.
2. Calculating depreciation and interest (i.e., capital recovery) on the old asset on the basis of its original cost rather than its present net realizable value. This usually increases the apparent costs associated with the old asset, and thus tends to favor replacements that are really uneconomical.
3. Where indirect costs (burden) are allotted in the cost accounting system in proportion to direct costs (usually in proportion to direct labor cost),

assuming without investigation that a reduction of direct expenditures will effect a corresponding saving in indirect expenditures. This error usually makes the apparent saving from proposed replacements greater than the saving that is actually possible to realize, and thus tends to favor replacements that are really uneconomical.

4. In cases where the proposed new asset provides more capacity than the old asset, comparing calculated unit costs realizable only with full-capacity operation, rather than comparing the actual costs realizable with the expected output. Where such excess of capacity is not likely to be used, this unit cost comparison tends to favor the asset with the surplus capacity, and is therefore favorable to replacements that are really uneconomical.

The first two of these errors cited result from a failure to recognize the true nature of depreciation accounting as a time allotment against future dates of money already spent. The third results from a failure to understand clearly the nature of cost accounting allocations. The fourth is merely an unrealistic use of unit costs.

EXAMPLE 17–5

Influence of Budgetary Considerations on the Timing of Replacement Expenditures

Facts of the Case

As a result of a change from the purchase of manufactured gas to natural gas, a city that had its publicly owned gas distribution system was faced with a great increase in leakage of gas from its mains. (The moist manufactured gas contains bituminous compounds that tend to plug up any small holes caused by corrosion of the mains; the dry natural gas dries these plugs, with the result that the gas escapes through the holes.) An analysis of the cost of lost gas and the cost of main replacement indicated that it would pay to replace a number of the oldest mains and services. There were other old mains and services that had not quite reached the point where replacement was economically justified even though considerable gas was being lost from them.

If all the gas mains that justified replacement were to be replaced at once, the only possible source of funds would have been a bond issue; this would have required an affirmative vote of two-thirds of the voters at a bond election. As gas main replacement was not considered well adapted to contract, a program of immediate replacement would have necessitated the hiring of a number of temporary employees for the gas department.

For a number of years this gas department had operated with a pay-as-you-go policy for financing capital improvements, with all capital expenditures budgeted out of current earnings from the sale of gas. If this policy were to be adopted for the replacement program, only enough funds could be budgeted annually to keep one crew busy on main replacement. As a result it would take several years to replace all of the mains requiring replacement.

After careful consideration, it was decided to continue this pay-as-you-go policy. It was believed to be desirable to avoid bond issues except for major expenditures for plant expansion. Moreover, it was felt that a single crew working full time on main and service replacement would become more efficient and, therefore, would operate at a lower cost than several temporary crews working a shorter period on the same type of work. Considerations of personnel policy were also against the hiring of temporary crews.

After several years, all of the mains that were losing enough gas to justify immediate replacement had been replaced. It was then decided to continue to budget funds for replacement for another year or two. These funds would be used to replace those mains that were losing considerable gas even though the losses had not quite reached the point where replacement was justified strictly on economic grounds. It was clearly possible to finance these replacements at once whereas the possibilities of financing at a later date were less certain.

Thus, because of budgetary considerations and other reasons, some mains were replaced a few years later than the date indicated by economy studies and others were replaced a few years sooner than the apparent economic date. Considerations such as these quite properly enter into decisions on the timing of expenditures for replacements.

Need for a System Viewpoint in Certain Types of Replacement Studies

Care must be taken in the preparation of many repair and replacement economy studies to ensure that the same savings is not counted more than once. Consider, for example, the maintenance of equipment in a small machining job shop. Each major area of the shop is treated as a separate profit center. If, say, a stamping machine in the stamping shop breaks down and is not repaired or replaced, the $20,000 annual profit of that shop will be lost. If it costs $16,000 to repair the machine, clearly an excellent return is made on the investment since $20,000 per year would be saved.

Suppose that another machine in the same shop breaks down the next year. This time it costs only $12,000 to save $20,000 per year in profit. This process of saving $20,000 per year could be repeated *ad infinitum*. The trouble is that it is the same $20,000 per year that is being saved each time.

In situations like that described, the analyst needs to apply a system viewpoint. The equipment in question should be described in terms of functions performed and considered in relation to all other functionally associated equipment. Estimates should be made of repairs and replacements of all associated items for the anticipated life of the need for the function, or at least for some reasonable planning horizon period.

In the case of the machine shop, a forecast should be made of the period over which the $20,000 profit is expected to extend. Then all anticipated expenditures for repairs, replacements, additions, operations, and so forth, should be made covering the same period.

Summary

The following points brought out in this chapter may be restated for emphasis:

1. It is helpful to use the word *defender* to apply to an existing asset being considered for replacement and to use the word *challenger* to apply to a proposed new replacement asset.
2. Capital recovery costs for extending the service of a defender should be based on its present net realizable value.
3. In evaluating the effect of a proposed replacement decision on cash flow for income taxes, it is necessary to consider the past investment in the defender as well as the depreciation accounting method that has been used for the defender in past tax returns. It may also be necessary to consider whether the old asset will actually be retired in determining the after-tax consequences due to differences in depreciation deduction amounts.
4. As in all comparisons of alternatives, prospective receipts and disbursements that are unaffected by the choice may be omitted from a replacement study.
5. Generally speaking, the annual cost of extending the service of a defender should be based on the number of years of additional service that results in the lowest figure for this annual cost.
6. Where the remaining life assumed for the defender is shorter than the life expected for the challenger, it is appropriate to recognize that if replacement is deferred, a different challenger from the present one is likely to be available. This prospect may be given weight in the economy study in a number of different ways.

PROBLEMS

17–1. Two years ago a clay-pipe manufacturing company designed, built, and installed a gas-fired drying oven at a total cost of $27,000. A firm manufacturing drying ovens of a new and radically different design now offers a new oven for $58,000. The maker of the new oven guarantees a fuel saving of $9,000 per year compared to the present oven. Repairs and maintenance are estimated at $2,000 per year.

For accounting and income tax purposes, depreciation has been charged on the present oven at $2,250 a year, using the straight-line method with a 12-year life and zero salvage value. This depreciation rate will be continued if the oven is not retired. If the new oven is purchased, its depreciation charges will be based on a 10-year life, zero salvage value, and the straight-line method. Item depreciation accounting is used and is appropriate because only one oven is owned.

Repairs and maintenance costs on the present oven are about $1,500 a year and it is estimated they will continue at this figure. All other costs except fuel in connection with the drying operation will be unaffected by the type of oven used. Either oven, once installed, has value only as scrap if removed; the scrap value is about equal to the cost of dismantling and removal.

The company's applicable tax rate on ordinary income is 50%. If the present oven is retired now, it is believed there will be no offsetting gains during the current year that will prevent securing the full advantage of the tax loss.

Assume that if continued in service the defender will be kept 10 years more, and that the challenger life is 10 years. If the company's minimum attractive rate of return is 14% after income taxes, which oven should be selected? Explain your answer. (*Ans*. defender, $12,657; challenger, $15,594)

17–2. Five years ago, at a cost of $12,600, a film-processing concern designed and installed the necessary equipment to develop semiautomatically a certain size of film. A salesman now proposes that this equipment be discarded and that a new machine that his firm has just put on the market should be installed to do this work. His estimate of a $3,000 annual saving in the cost of labor and supplies seems to be reasonable. It is believed that there will be no other difference in annual disbursements between the old and new equipment. The price of the new machine is $22,500. The salesman has found a buyer willing to pay $1,000 for the old equipment.

The present equipment is fully depreciated on the books of account and for tax purposes, as the assumed life was 5 years with zero salvage value. Item accounting has been used. However, this equipment is still serviceable and could be used for several years more at approximately the present level of annual disbursements. Unless sold at once, it is expected to have zero salvage value. If acquired, the new machine will be depreciated by the years-digits method, assuming a 5-year life and zero salvage value; this estimated life and salvage value will also be used in the economy study to guide the decision on this suggested purchase.

The company's applicable tax rate is 50% on ordinary income and 28% on capital gains. If the company acquires the new equipment, it will be able to take full advantage of the 10% investment tax credit at the time of the acquisition. If the minimum attractive rate of return is 12% after income taxes, would you recommend purchase of the new equipment? Explain any assumptions that you make in your analysis. (*Ans*. defender, $4,725; challenger, $5,618)

17–3. The manager of the film-processing firm described in Problem 17–2 would like to have the new equipment. How much would he have to receive from the sale of the old equipment in order to justify its replacement on economy grounds? Assume a 2-year remaining life for the old equipment in service to the company. (*Ans*. at least $3,988)

17–4. A high-speed, special-purpose, automatic strip-feed punch press costing $45,000 has been proposed to replace three hand-fed presses now in use. The life of this automatic press has been estimated to be 10 years, with no salvage value at the end of that time. Expenditures for labor, maintenance, etc., have been estimated to be $8,000 per year.

The general-purpose, hand-fed punch presses cost $8,000 each 10 years ago and were estimated to have a life of 20 years, with no salvage value at the end of that time. Their present net realizable value is $1,000 each. This net value is expected to decrease at a rate of about $100 per press per year. Operating expenditures for labor, etc., will be about $6,500 per year per machine.

It is expected that the required service will continue for only 5 years more and that the salvage value of the challenger will be $15,000 at the end of this period.

(a) Make an analysis before income taxes to provide a basis for the decision on whether or not to make this replacement. Assume a before-tax minimum attractive rate of return of 30%.

(b) Make a similar analysis after income taxes using a minimum attractive rate of return of 15%. Assume a tax rate of 52%. Although the expected service period of the challenger is only 5 years, it will be necessary to use a 10-year life (with zero salvage value) for depreciation used in tax returns; years-digits depreciation will be used. Straight-line depreciation has been used for the defenders. Assume that item depreciation is used in this company and that any "losses" on disposal will be fully deductible from taxable income in the year of disposal. However, any "gain" on disposal will be taxed at a capital gains rate of 28%. The company will be able to take full advantage of a 10% investment tax credit on the acquisition of this new equipment. (*Ans.* (a) defender, $20,566; challenger, $24,817: (b) defender, $21,222; challenger, $21,052)

17–5. A municipal water department has as part of its system of delivery a flume that is 60 years old. Maintenance costs on this flume have averaged $6,800 a year during the last 10 years. Gagings made at both ends of the flume indicate an average loss of water in its length of about 15%.

This water department serves its customers in part with surface water brought in the aqueduct of which this flume is a part; the remainder of its water supply is pumped from wells located in the city. The loss of water in the flume thus requires additional pumping from wells. The average amount of water delivered at the inlet end of the flume is 10 million gallons per day. The increment cost of pumping from wells is $30.75 per million gallons.

It is estimated that a standard metal flume to replace the existing one would cost $185,000. It is believed that this would reduce the average loss of water in the flume to not more than 3%, and that the maintenance cost of such a flume would not exceed $2,000 a year. For the purposes of an economy study, the life of such a replacement flume is to be assumed as 20 years, although it might actually be much longer.

Irreducible factors favorable to the change are the lessened danger of breaks in the aqueduct line, and the reduced drain on the underground water supply. One of the more important elements of the increment pumping cost is the cost of electric energy which has steadily increased over the past several years.

Using an i^* of 8%, determine whether you would consider it economical to make the replacement. (*Ans*. old, $20,269; new, $20,842)

17–6. A grain elevator requires additional capacity for the grinding of feed. Two plans for securing this capacity are to be compared.

Plan A. Replace the existing centrifugal bar mill with a patent feed grinding mill manufactured by the XYZ Company. This XYZ mill will provide 3 times the present capacity.

Plan B. Continue the present centrifugal bar mill in service and supplement it by another similar mill of like capacity. This will provide double the present capacity, which appears to be adequate for all needs.

The existing centrifugal bar mill cost $4,000 5 years ago. Straight-line depreciation has been charged on the books against it based on an estimated life of 15 years. It has a present net realizable value of $1,000. A new similar mill will now cost $5,100. The XYZ mill has a first cost installed of $9,500.

It is estimated that the maintenance cost of the existing centrifugal bar mill will average at least $250 a year in the future. A new centrifugal bar mill with somewhat improved design can be maintained for about $175 a year, it is believed. The XYZ mill is expected to have an annual maintenance cost of $200.

Energy requirements for a centrifugal bar mill of this capacity are 23 kw-hr per hour of operation. It is expected that if two such mills are operated, each will operate about 1,200 hours per year. Because of its greater capacity, the XYZ mill will grind the same total amount of feed as the other two in 800 hours per year. Its energy requirements are 41 kw-hr per hour of operation. Energy is purchased for 4.3 cents per kw-hr. Labor costs and taxes will be practically unaffected by the choice of plan.

Make a before-tax comparison of equivalent annual costs for Plans A and B assuming a 10-year remaining life for the present mill and a 10-year life for both new mills with zero terminal salvage values for all. Use a minimum attractive rate of return of 25%. (*Ans*. Plan A, $4,271; Plan B, $4,507)

17–7. Solve Problem 17–1 assuming that the defender will be used for only 6 years more. Assume that depreciation will be taken as stated in the problem and that the loss on disposal at the end of 6 years will be fully deductible from ordinary income at that time. Should replacement be made?

17–8. In the circumstances described in Problem 17–4, the company intended to use years-digits depreciation for the new machine. Repeat the after-tax analysis required in Problem 17–4 (b) assuming straight-line depreciation will be used for the new machine. How does this change of method of accounting for depreciation for tax purposes affect the analysis and the decision to replace the three hand-fed presses?

17–9. A study of the costs of operating a 4-year-old piece of construction equipment indicates that the only costs variable with age are repair costs and losses due to lost time. These costs, and the corresponding projected costs for the next two years, are as follows:

Year	Repair cost	Cost of lost time
1	$176	$100
2	264	200
3	830	400
4	1,280	700
	Projected costs for next two years	
5	$2,150	$900
6	3,750	$1,500

The present realizable salvage value of this piece of equipment is $4,400. An estimate of the net realizable salvage value at the end of the third year of life is $5,800; at the end of the fifth year it is $3,200, and the sixth, $2,200.

The first cost of a new unit is $15,000. Due to certain improvements in design, it is believed that the new equipment will reduce annual costs for fuel and supplies by $300; however, it is expected that repair costs and lost time will behave about as before. A before-tax analysis is to be performed using a minimum attractive rate of return of 20%. Assume the same salvage values apply to the old and new assets.

Compare the annual cost of extending the service of the old equipment one year more with the most advantageous annual cost for the proposed new equipment.

17–10. Solve Problem 17–6 on an after-tax basis using an i^* of 13%. Assume an applicable 50% income tax rate and that mills purchased now will be depreciated by the years-digits method and that the 10% investment tax credit will be taken immediately. Should replacement be made now?

17–11. It is stated in the discussion following Example 17–1 that a substantial increase in the future resale price of the land and warehouse will make continued ownership more attractive. Assume that this future resale figure is increased to $150,000, the present resale value. Does this changed future resale value change the conclusion reached in Example 17–1? Base your conclusion on an analysis after income taxes. At what future resale value is the choice a matter of indifference between retaining the warehouse and selling it?

17–12. Assume a machine with operating disbursement and salvage value time profiles as shown in Table 17–2 is now one year old (a defender). Calculate the appropriate defender advantage cost to be compared with the challenger advantage cost. Is replacement indicated? Is there a defender remaining economic life implied by your analysis? Why or why not?

17–13. Answer the questions in Problem 17–12: (a) assuming the defender is 5 years old; (b) assuming it is 6 years old.

17–14. On page 430 it is stated that "Corresponding capitalized cost figures for the defender for 1 or 3 years' service will be higher than the $100,953 figure for the challenger." Make the necessary calculations to verify the accuracy of this statement. Find the marginal cost for the third year for the defender and comment on the appropriateness of its use in a choice between challenger and defender.

17–15. A corporation engaged in large-scale farming operates a fleet of tractors. Major overhauls generally have been needed when tractors are 4 and 8 years old. In the past most tractors have been retired after 8 years, just before the second major overhaul. However, a few tractors have been retired after 4 years, just before the first overhaul. Some others have been overhauled when 8 years old and kept in service until they were scrapped at the end of 12 years. In reviewing past replacement policy, an analyst estimates typical year-by-year annual disbursements that appear to have been influenced by the age of a tractor (chiefly routine maintenance, fuel, and lubricants) as follows:

Year	*Disbursements*	*Year*	*Disbursements*	*Year*	*Disbursements*
1	$6,000	5	$8,000	9	$13,000
2	6,500	6	8,500	10	13,000
3	7,000	7	9,000	11	13,000
4	7,500	8	9,500	12	13,000

The first cost of a tractor is $46,000. A typical salvage value before overhaul at the end of 4 years is $20,000. A typical cost of overhaul at the end of 4 years is $8,500. A typical salvage value before overhaul at the end of 8 years is $10,000. A typical cost of overhaul at the end of 8 years is $10,000. Salvage value at the end of 12 years is about $5,000.

Assuming a minimum attractive rate of return of 18% before income taxes, compare the before-tax equivalent annual costs of service for lives of 4, 8, and 12 years, using the foregoing "typical" figures for the relevant costs.

17–16. In Problem 17–15, consider the decision of whether a typical 8-year-old tractor (a defender) should be overhauled and kept in service for 4 years more or replaced with a new tractor (a challenger). Using an i^* of 18%, make a before-tax comparison of equivalent annual costs of (a) extending the service of the defender for 4 years or (b) acquiring a new tractor to be retired after 8 years.

17–17. In Problem 17–15, consider the decision of whether a typical 4-year-old tractor (a defender) should be overhauled and kept in service for 4 years more or replaced with a new tractor (a challenger). Using an i^* of 18%, make a before-tax comparison of equivalent annual costs of (a) extending the service of the defender for 4 years or (b) acquiring a new tractor to be retired after 8 years. How much must the salvage value be for the 4-year-old tractor in order to make it attractive to replace it with a new tractor?

17–18. The GSC company manufactures sportswear on contract to several large retail stores. It uses a large number of special-purpose and general-purpose sewing machines on which it keeps cost records. The cost experience on the average double-needle Crooner machine, used at 50 work stations in the plant, is as shown in the table below. A minimum attractive rate of return before income taxes of 20% has been used in developing the figures. A new machine costs $4,800. Annual disbursements are primarily for maintenance and lost time.

Year	*Salvage value*	*Disbursements for year*	*Total cost for year n*	*EUAC for n years*
1	$2,600	$250	$3,410	$3,410
2	1,600	320	1,840	2,696
3	1,000	480	1,400	2,340
4	600	730	1,330	2,152
5	300	1,050	1,470	2,060
6	0	1,580	1,940	2,048
7	0	2,160	2,160	2,057
8	0	3,120	3,120	2,121

Determine whether replacement should be made under the circumstances described in the following questions. Assume that, in general, the stated conditions will repeat indefinitely.

(a) A particular machine has proved unusually well made, and disbursements have averaged 30% less than expected. The machine is now 6 years old.

(b) The machine now in use was purchased used 3 years ago from another plant for $1,000. It was 2 years old at the time. Disbursements during the last 3 years have been $850, $1,427, and $2,120, in that order.

(c) A general inflation has increased new machine prices and normal disbursements by 30% over the past 2 years. The machine is now 5 years old.

(d) The company is now able to buy, at a bankruptcy sale, four 2-year-old machines, all in good condition, for $900 each. Four of the present machines are now 5 years old or older.

17–19. A furniture manufacturer is considering the installation of an automatic boring machine to replace two machines that now provide the same total capacity. The proposed machine will cost $27,500 ready to operate. If acquired, it will be depreciated for accounting and tax purposes by the years-digits item method assuming a 10-year life and zero salvage value. Annual disbursements for its operation are estimated as follows:

Direct labor will be one operator at $400 per week for 48 weeks per year.

Labor extras are 30% of direct labor cost.

Power will be 4,000 kw-hr per year at 5¢ per kw-hr.

Annual repairs and supplies are estimated at $600.

Property taxes and insurance total 3% of first cost.

The two present machines cost $8,000 each and are 5 years old. Straight-line item depreciation has been used for accounting and tax purposes assuming a 15-year life and zero terminal salvage value. Their present realizable value is $2,000 each. Annual expenditures are:

Direct labor is two operators at $250 per week each for 48 weeks per year.

Labor extras are 30% of direct labor cost.

Power is 2,500 kw-hr per machine per year at 5¢ per kw-hr.

Annual repairs and supplies are $400 per machine.

Property taxes and insurance total 3% of first cost.

Assuming a 20% minimum attractive rate of return before income taxes, compute the before-tax annual cost of extending service of the defenders for 10 more years. Compare this with the annual cost of the challenger over a 10-year service life. Assume zero salvage values for both defender and challenger at the end of the 10 years.

17–20. Make an after-tax comparison of annual costs in Problem 17–19 assuming a 10% minimum attractive rate of return. Use a tax rate of 50%. Assume the company will take advantage of the 10% investment tax credit if the challenger is acquired. Consider extra annual income taxes as an additional annual cost of the challenger. Assume that any "loss" on disposal of the defenders will be fully deductible from ordinary income in the year of retirement; consider the resulting tax saving as an addition to the net salvage value of the defenders.

17–21. The supervisor in the shop considering replacement of the boring mills in Problem 17–20 (and 17–19) believes that advances in technology will result in much improved models of automatic boring mills in the future. The model available next year, for example, should save $150 per year after taxes more than the model available this year. That available the following year should offer another $150 annual savings, etc. The supervisor argues that this $150 inferiority gradient should be converted to an equivalent uniform annual cost and the resulting penalty cost should be added to the annual cost of the challenger. Find the next-year annual cost of this technological inferiority of today's challenger to those expected to be available in the future (a) for a 20-year period, and (b) in perpetuity. How does this additional charge to the challenger affect the decision in Problem 17–20?

17–22. A manufacturer must double the plant's press brake capacity for making certain sheet metal parts. Two plans are set up for comparison:

A. Continue the present XY–10 press brake in service, and supplement it by an XY–12 (a newer model) of equal capacity. The XY–10 cost $15,000 5 years ago. It has been depreciated for tax purposes by the straight-line item method assuming a 15-year life and zero salvage value. A new XY–12 model will cost $24,000. If purchased, it will be depreciated for tax purposes by the years-digits item method assuming a 15-year life and zero salvage value.

B. Sell the XY–10 for $4,500. Buy an XZ–20 model press brake that has double the capacity of the XY–10. The first cost of the XZ–20 is $36,000. If purchased, it will be depreciated for tax purposes by the years-digits item method assuming a 15-year life and zero salvage value.

It is desired to compare these two plans on the assumption that the need for this particular operation will terminate after 5 years. Expected salvage values 5 years hence are $3,000 for the present XY–10, $7,150 for the XY–12, and $13,750 for the XZ–20. Average annual operating disbursements for the three press brakes are estimated as follows:

	XY–10	*XY–12*	*XZ–20*
Direct labor	$9,000	$7,800	$14,000
Labor extras	2,250	1,950	3,500
Maintenance	1,800	1,200	2,000
Power	300	225	400
Taxes and insurance	750	1,200	1,800

The XZ–20 has the same space requirements as the total of XY–10 and XY–12. Receipts from the sale of product are not expected to be influenced by the choice between A and B. The manufacturer's applicable tax rate is 50% on ordinary income and 28% on capital gains.

(a) For years 0 to 5, both inclusive, tabulate the difference in prospective cash flow before income taxes. What interest rate makes plans A and B equivalent to one another?

(b) For years 0 to 5, both inclusive, tabulate the difference in prospective cash flow after income taxes. Assume that any "loss" on disposal can be treated as a deduction from taxable income in the year of disposal. What interest rate makes plans A and B equivalent to one another?

(c) Discuss the interpretation of your computed interest rates in (a) and (b) as a basis for the choice between the two plans.

17–23. The following problem deals with a decision that it was necessary to make in the RST Company in November of 1977.

Early in 1976, the maintenance department of this company ordered a specially designed machine, Machine X, costing $120,000. Nearly 2 years was needed to secure delivery of Machine X. It was delivered and paid for in mid-November 1977. However, it could not be used until it was installed. The installation period would require another month and the estimated cost of installation was $24,000.

The maintenance department had two objectives in mind when Machine X was originally ordered. The primary objective was to improve the diagnosis of certain types of troubles that had occurred in equipment subject to periodic overhaul. A secondary objective was to effect a moderate reduction in the labor, fuel, and power costs in the testing of overhauled equipment. This secondary objective alone was not sufficient to have justified the purchase of Machine X.

When Machine X was ordered, the management of the RST Company knew that the equipment that Machine X was designed to service would gradually be superseded by a different type of equipment by the end of 1985. It was believed that after 1985, Machine X could no longer be used and that it would have a negligible salvage value at that time.

In the period of nearly 2 years that had elapsed between the order date and the delivery date of Machine X, other methods had been developed for diagnosing the troubles in the equipment Machine X was designed to service. It was believed that these other methods were satisfactory enough that Machine X was not actually needed for purpose of diagnosis.

Mr. A, the controller of the RST Company, was therefore opposed to spending the $24,000 to install Machine X. He said: "We are always short of funds for plant investment. Why send good money after bad?" He favored disposing of Machine X at the best available price. A prospective buyer was found who was willing to pay $36,000 for this machine. The market for specialized machines of this type was quite limited, and it seemed unlikely that Machine X could be sold for more than this figure.

Mr. B, the superintendent of the maintenance department, favored keeping Machine X. He pointed out that there would be substantial savings in disbursements for the testing of overhauled equipment if Machine X were installed. The industrial engineering department made careful estimates of these savings, as follows:

1978	$32,000	1981	$19,000	1984	$10,000
1979	25,000	1982	16,000	1985	7,000
1980	22,000	1983	13,000		

The following questions are to be answered as if it were November of 1977. They deal with various aspects of the choice between the alternatives, (1) to sell Machine X at once for $36,000, and (2) to install Machine X and operate it for the next 8 years. Where necessary to make assumptions about the timing of receipts and disbursements, assume that money payments near the end of 1977 are made on January 1, 1978 (i.e., at zero date on your time scale). Assume the end-of-year convention with respect to the savings in annual disbursements. In answering questions that involve income taxes, assume an end-of-year convention with regard to tax payments. For example, assume that anything that affects 1978 taxable income influences a tax payment at the end of 1978, etc.

(a) Assume alternative (2) is selected. At what rate of return will the *total* investment in the purchase and installation of Machine X be recovered?

(b) Discuss the relevance of the rate of return computed in your answer to (a) in the decision between alternatives (1) and (2).

(c) Determine the year-by-year differences between alternatives (1) and (2) in cash flow before income taxes.

(d) Use compound interest calculations to make an analysis of the cash flow series in your answer to (c) in a way that would provide a rational basis for choice between alternatives (1) and (2) if the RST Company were not subject to income taxes.

(e) Analyze the effect of the choice between (1) and (2) on cash flow for income taxes. Assume an applicable tax rate of 50% throughout the period of the study. Assume that if Machine X is installed, it will be depreciated for tax purposes by the years-digits method using a life of 8 years with zero salvage value. Assume that depreciation charges will start with the year 1978 and that a full-year's depreciation will be charged in 1978. Assume that if Machine X is sold at once, any loss sustained will be deductible from 1977 taxable income. Tabulate the differences between (1) and (2) in cash flow after income taxes.

(f) Use compound interest calculations to make an analysis of the final cash flow series in your answer to (e) in a way that provides a rational basis for choice between (1) and (2).

(g) How would you recommend that your analysis in your answer to (f) be used by the management of the RST Company in choosing between alternatives (1) and (2)?

18

THE INFLUENCE ON ECONOMY STUDIES OF SOURCES OF INVESTMENT FUNDS

*The long-run profitability of the enterprise hinges on the solution of two problems of management of corporate capital: (1) sourcing (acquisition) of capital funds and (2) rationing (investment) of that capital. They should be quite separate. Investment proposals should compete for corporate funds on the basis of financial merit (the productivity of capital), independent of the source or cost of funds for that particular project. Investable funds of the corporation should be treated as a common pool, not compartmented puddles. Similarly, the problem of acquiring capital should be solved independent of its rationing and also on the basis of merit (the comparative costs and risks of alternative patterns of sourcing).—Joel Dean**

Engineering economy impinges on many other fields. We have seen some relationships between engineering economy and accounting and income taxation. In this chapter we shall examine some relationships between engineering economy and certain aspects of business finance and government finance. Some additional aspects of government finance are discussed in the next chapter.

We shall see how certain types of confused reasoning may be introduced into economy studies where calculations appropriate for judging long-run economy of proposed investments in physical assets are combined with calculations related to the financing of these assets. We shall also see how certain techniques of engineering economy, particularly cash flow analysis and rate of return calculations, can be used to help in comparing the merits of alternative financing plans.

Possible Sources of Funds for Financing of Fixed Assets

Plant and equipment may be financed by (1) ownership funds or by (2) borrowed funds or by a combination of the two. In many cases another possible source of funds is (3) long-term leasing.

*Joel Dean, Sec. 2, Managerial Economics, in *Handbook of Industrial Engineering and Management*, 2d ed., W. G. Ireson and E. L. Grant, eds. (Englewood Cliffs, N.J.: Prentice-Hall, Inc., copyright 1971), p. 108.

As the name implies, ownership funds are those furnished by the owners of an enterprise. In the case of an individual this is limited by personal resources. A partnership may secure ownership funds from the existing partners, or it may possibly reorganize by taking another partner—a new owner. Similarly, a corporation may secure ownership funds in various ways from its existing stockholders, or it may acquire some more owners by selling stock to new stockholders. The ownership funds of governmental bodies are those supplied by current taxation and other current revenues.

Before we can discuss the relationship between borrowing and economy studies, some general background material is needed.

Borrowing

For funds to be borrowed, some security is usually required by lenders. This may be specific property that is put up as collateral. Thus, an individual may finance the purchase of a home by giving the lender a mortgage—a deed that conveys title to the lender (after appropriate legal action) in case the borrower fails to pay interest or principal as promised. Similarly, by means of somewhat more elaborate legal devices, a corporation may give a mortgage on specific physical property as security for a bond issue. Often, however, borrowing takes place without specific collateral, the security being the general credit of the borrower; this is the case with nearly all governmental loans.

Whatever the security, borrowed money must ultimately be repaid. It follows that if a proposed investment is to be financed by borrowing, two questions must be answered regarding it:

1. Is it economical in the long run as compared to other possible alternatives?
2. Do the conditions of repayment make it advisable in the short run?

Various Possible Plans for Repayment of Borrowed Money

Table 3–1 (page 24) describes four different ways in which money borrowed for a period of several years or more might be repaid. These are representative of the ways of repaying personal loans, business loans (including corporate borrowings through bond issues), and government loans.

In Plan I, interest was paid each year.* No annual payments at all were made on the principal of the debt that was to be repaid in a lump sum at the end of a stipulated number of years. This is a common plan of repayment for corporate bond issues as well as for secured loans for business and personal purposes.

Where money is borrowed to be repaid in a lump sum, some plan is necessary for securing the lump sum to make the repayment. There are two possibilities here: (1) refunding, or (2) the accumulation of the necessary funds through the use of a sinking fund.

*As brought out in Chapter 7, the almost universal practice in bond issues of private corporations is for interest to be payable semiannually. This is also a common practice in other types of long-term loans.

Refunding means that the money for repayment is secured through a new borrowing. This is common in corporate bond issues, particularly those of public utilities (including railroads) where the public regulation of rates generally makes it impracticable to accumulate sufficient funds to pay off bonds out of earnings. Although individual bonds carry a definite maturity date and there is an obligation to pay the bondholders on that date, it is merely a case of borrowing from Peter to pay Paul. When the refunding of a bond issue is contemplated, there are no short-run burdensome repayment obligations to influence decisions between engineering alternatives. From the viewpoint of the engineer at the time the money is borrowed, the debt may be thought of as perpetual. Whenever refunding is contemplated, the managerial problem is to maintain the company's assets and earning power so that the company's credit position will permit a new borrowing at the time refunding is necessary.

Plans II and III in Table 3–1 called for annual payments on the principal of the debt. In Plan II the debt repayment was uniform from year to year with a resulting steady decrease in the annual interest payment. In Plan III the debt repayment increased from year to year in a way that permitted the sum of the principal and interest payments to be constant.

Where a bond issue provides for *serial maturities*, repayment plans similar to Plans II and III are used. For example, a $2,500,000, 20-year bond issue might consist of 500 bonds, each of $5,000 denomination. Of these, 25 might mature at the end of the first year, 25 more at the end of the second, and so on until the 20th year when the final 25 bonds would mature. Each separate bond with its definite maturity date would be comparable to Plan I. However, the bond issue as a whole would be comparable to Plan II. This type of uniform serial maturities is characteristic of bond issues by states, counties, cities, school districts, and the many special types of local improvement districts. In such public borrowing, the use of serial maturities often is required by law in order to prevent the diversion of sinking funds to other purposes than debt repayment.

If uniform annual payments of principal plus interest are desired in a bond issue, it is possible to schedule serial maturities so that these payments are roughly uniform, approximating the equivalent uniform annual cost of capital recovery. This is sometimes done in real estate bond issues and in bond issues by local governments. The Plan III type of uniform payment is also characteristic of real estate loans to individuals.

In Plan IV no current interest was paid during the life of the loan and the final lump-sum payment covered both principal and compound interest. Series EE United States Savings Bonds are of this type. However, this scheme of payment is seldom used by business.

The different plans of debt repayment may sometimes be combined. For instance, a bond issue might run "flat" without repayment for 5 years, after which time uniform serial maturities start. Or half the bonds might mature serially, the other half coming due in a lump sum at the end of 20 years, with refunding contemplated.

Serial Maturities Versus Sinking Funds

Where a bond issue is to be paid off in a lump sum and refunding is not contemplated, it is necessary to establish a sinking fund to provide the lump sum. This involves the use of current receipts to amortize the debt just as is necessary when serial maturities are used. A sinking fund to pay off a bond issue might involve uniform annual deposits such as were discussed in the chapters on compound interest. However, it is also possible for the sinking fund deposits to be variable, possibly depending on net earnings each year.

One difficulty with sinking funds is that if they are conservatively invested with a view to maximum safety, the interest rate is likely to be considerably less than that which is being paid on the outstanding debt. For instance, a corporation with 12% bonds outstanding might be able to invest with safety at only 7% after taxes. For this reason the best investment for current revenues intended for ultimate debt repayment is to pay off some of that debt immediately. This means that a borrowing corporation or public body may best invest its sinking fund in its own outstanding bonds, provided they can be purchased without paying too high a premium above par. If a company purchases its own bonds at par, this amounts to an investment of its sinking fund at the coupon rate of the bond issue.

The adoption of a plan calling for serial maturities has the same effect as if a regular sinking fund investment were made each year in the borrower's bonds at par. The serial maturities insure that such bonds will be available, and thus avoid all possibility of loss of interest due to the difference between what the borrower pays and what can be obtained from a conservative sinking fund.

From the standpoint of keeping to a minimum the total disbursements on account of the bond issue, the serial maturity plan would thus seem to have a definite advantage over the plan of establishing a sinking fund to provide for lump-sum repayment. In cases where either plan might be used, the choice between the two plans should depend somewhat on the prospective relative difficulty of meeting serial repayments and of maintaining a sinking fund. In the case of an industrial borrower anticipating wide fluctuations in annual earnings, a fixed annual repayment obligation may be a severe burden in bad years, and a sinking fund plan in which deposits (if possible in the form of purchases of outstanding bonds) vary with earnings may be advantageous. On the other hand, the tax revenues of local governments may be expected to be reasonably stable; here serial maturities on public bond issues prevent the possibility of pressure on public officials to appropriate existing sinking funds to current needs. Thus, serial maturities are advantageous in bond issues by cities, counties, states, and special improvement districts.

Many enterprises, if they are to finance at all, must do so on a serial repayment basis in order to satisfy the lender's requirements that the margin of security behind the loan be maintained or increased. This is particularly true if the loan is secured by a single property (such as an office building subject to depreciation). It is also likely to be true of hazardous enterprises or relatively small enterprises; in such cases the required repayment period is likely to be much shorter than the estimated life of the property being financed.

The Relationship Between Debt Repayment and Economy Studies

Many economy studies relate to the proposed acquisition of fixed assets that must be financed by borrowing. In such studies, in addition to considering the question "Will it pay in the long run?" it is also necessary to consider the question "Can the required repayment obligation be met?" If short-run repayment obligations for a proposed alternative are such as to prevent ever reaching the "long run," it is impractical to select that alternative even though it may show theoretical long-run economy.

However, the two questions of long-run economy and practicability of debt repayment should not be confused. It is only in those rare cases in which a proposed plant is to be financed entirely by borrowing, with the debt repayment spread fairly uniformly over its entire estimated life, that the debt charges may be correctly substituted for the annual cost of capital recovery (depreciation plus interest) in a study of long-run economy. This situation occurs in connection with some public improvements made by cities and other local governmental units, but seldom happens in private enterprise.

The mistake is sometimes made of including both debt repayment and depreciation as "costs" in an economy study. It should be obvious that the estimator who does this is using badly confused reasoning, and that the calculations serve to answer neither the question of long-run economy nor the one of practicability of debt repayment.

A General Principle—Debt Charges Are Equal to the Annual Cost of Capital Recovery Where the Life of the Debt Is Equal to the Capital Recovery Period and the Interest Rate on the Debt Is Used as the Minimum Attractive Rate of Return

A city owns its electric distribution system, generating part of the electric energy that it distributes and purchasing the rest. It is proposed to increase the capacity of the generating station to eliminate the necessity of purchasing energy. An economy study is to be made to determine whether or not this will be economical.

The estimated cost of the new generating facilities is $2,000,000. If constructed, they will be financed entirely by long-term borrowing repayable from the revenues of the electric system. The proposed bonds will mature serially, with the entire issue to be paid off at the end of 20 years. Consultation with investment bankers indicates that such a $2,000,000 bond issue will need to have an interest rate of 7%. Serial maturities are to be scheduled in a way so that the sum of interest and bond repayment will be as nearly uniform as possible throughout the 20-year period. This annual payment of interest plus principal will, of course, be approximately $2,000,000(*A/P*,7%,20) = $2,000,000 (0.10185) = $203,700.

If the estimated life of the generating plant is 20 years, with zero terminal salvage value, and if 7% is used as the minimum attractive rate of return, this annual outlay for debt service is exactly equal to the annual cost of capital recovery that should be used in the economy study. In an engineer's presentation to city officials of an analysis of the proposed investment, it may be more understandable to show this annual charge for debt service instead of an annual figure for capital recovery with a 7% return (or an annual figure for depreciation plus 7% interest). But it would be double counting to include *both* a figure for bond repayment with interest and a figure for depreciation plus interest. This would be charging the proposed generation of power with the investment costs of two generating stations during the life of one.

However, in this connection we should mention that in Chapter 10 we explained that the minimum attractive rate of return used in economy studies should ordinarily be greater than the bare cost of borrowed money. Further comments on this point, with particular reference to governmental projects, are made in Chapter 19.

Advantages and Hazards of Doing Business on Borrowed Money

Consider a business enterprise financed entirely by ownership funds and earning a 16% return after taxes. Assume that this business expands by borrowing money at 10% interest and that the after-tax cost of this borrowed money is 5%. If the new funds are as productive as the old ones, the owners will earn 11% (i.e., 16% − 5%) on the borrowed money without increasing their personal investments in the enterprise.

Using an ownership investment as a basis for borrowing is referred to as "trading on the equity." In effect, the owners' equity in an enterprise provides a margin of security that makes it possible to secure a loan. It is a general principle that trading on the equity makes good business better and bad business worse. The owners who succeed in earning more on the borrowed funds than the cost of the borrowed money will, of course, increase the rate of return on their own investments. On the other hand, if such an enterprise does not continue to be prosperous, all of its earnings may have to go to pay interest and principal on the debt, and the owners will have no return at all; in the absence of sufficient earnings to meet the debt charges, foreclosure on the part of the lenders may result in the owners losing their entire investment.

The foregoing comments about financing by borrowing are also applicable—with some modifications—to financing by long-term leases of fixed assets. The question of whether any particular business enterprise ought to finance in part by long-term loans and/or long-term leases cannot be analyzed without considering many aspects of business finance that are outside the scope of this book. However, it should be evident that long-term obligations are less suitable to enterprises of the feast-or-famine type than to enterprises with stable earning power. An enterprise financed entirely by ownership funds may weather a few bad years that could be fatal to an enterprise that has substantial fixed obligations.

When rate-of-return studies are made assuming financing by borrowing or by leasing, there is danger of an incorrect interpretation of computed rates of return. However, a rate-of-return-type analysis of alternate possible financing plans is an extremely useful tool to aid decisions on methods of financing. The foregoing ideas can be explained to best advantage with the help of numerical examples. Examples 18–1 and 18–2 are designed for this purpose.

EXAMPLE 18–1

Analysis of Before-Tax and After-Tax Effects of Equipment Purchase Using Debt Financing

Facts of the Case

It is proposed to acquire certain equipment having a first cost of $20,000, an estimated life of 8 years, and an estimated zero terminal salvage value. (These estimates of life and salvage value will be used for income tax purposes as well as in the economy study.) It is estimated that this equipment will reduce disbursements for labor and related costs by $7,000 a year. This saving in annual disbursements will be partially offset by increased expenditures of $800 a year for maintenance and property taxes. The net reduction in estimated annual disbursements (disregarding capital costs and income taxes) is therefore $6,200. It is believed this saving will continue throughout the 8-year life.

Straight-line depreciation will be used for accounting and income tax purposes. An applicable income tax rate of 50% is to be assumed throughout the 8-year period.

This equipment might be purchased from equity funds. An alternative proposal calls for buying the equipment on time. The initial payment will be $8,000, with the remaining $12,000 to be paid off over the 8-year period. At the end of each year, $1,500 will be paid on the principal of the debt plus 10% interest on the unpaid balance. Thus, the payments will decline uniformly from $2,700 at the end of the first year to $1,650 at the end of the eighth year.

Rate of Return Assuming Purchase from Equity Funds

Interpolation indicates a rate of return before income taxes of approximately 26.2%. (A/P = $6,200 ÷ $20,000 = 0.310).

The investment will increase annual taxable income by $6,200 − $20,000/8 = $3,700.

Annual disbursements for income taxes will therefore be 0.50($3,700) = $1,850 and annual cash flow after income taxes will be $6,200 − $1,850 = $4,350.

Interpolation indicates a rate of return after taxes of approximately 14.3% (A/P = 0.2175).

Rate of Return Assuming Purchase on Time

To compute the rate of return before income taxes, the cash flow of column A, Table 18–1, must be combined with the cash flow associated with the debt service. The result is shown in column D. A trial-and-error calculation with the aid of the table of gradient present worth factors is:

$$\text{PW at } 40\% = -\$8{,}000 + \$3{,}500(P/A,40\%,8) + \$150(P/G,40\%,8) = +\$829$$
$$\text{PW at } 45\% = -\$8{,}000 + \$3{,}500(P/A,45\%,8) + \$150(P/G,45\%,8) = -\$52$$

The approximate before-tax rate of return on equity capital is 44.7%.

To compute the rate of return after income taxes, the analysis proceeds as shown in Table 18–1. The two segments of the annual payments to the lender, principal repayment in column B and interest payment in column C, have been shown separately because, under U.S. tax law, interest is a deductible operating expense for income tax purposes. Thus, the amount of taxable income shown in column F is comprised of the column A cash flow before debt service and taxes less the interest paid during the year, column C, and less the allowable depreciation in column E. The column G cash flow for taxes is subtracted from the cash flow after debt service, column D, to yield cash flow after debt service and taxes, column H. In each year from 1 through 8 there will be a negative cash flow for debt repayment with interest when a portion of the investment is financed through borrowing. This will be partially offset because disbursements for income taxes will be reduced by 50% of the interest. For example, in year 1, the cash flow after income taxes is +\$4,350 − \$2,700 + 0.50(\$1,200) = \$2,250.

Trial-and-error calculations are:

$$\text{PW at } 25\% = -\$8{,}000 + \$2{,}250(P/A,25\%,8) + \$75(P/G,25\%,8) = +\$86$$
$$\text{PW at } 30\% = -\$8{,}000 + \$2{,}250(P/A,30\%,8) + \$75(P/G,30\%,8) = -\$933$$

The approximate rate of return on equity capital is 25.4%.

After-Tax Cost of Borrowed Money

It is of interest to compare the after-tax cash flow under complete equity financing with that under partial debt financing, as follows:

Year	*After-tax cash flow, equity financing*	*After-tax cash flow, purchase on time*	*Difference in after-tax cash flow*
0	−\$20,000	−\$8,000	+\$12,000
1	+4,350	+2,250	−2,100
2	+4,350	+2,325	−2,025
3	+4,350	+2,400	−1,950
4	+4,350	+2,475	−1,875
5	+4,350	+2,550	−1,800
6	+4,350	+2,625	−1,725
7	+4,350	+2,700	−1,650
8	+4,350	+2,775	−1,575

TABLE 18–1
Calculations to determine cash flow after income taxes, Example 18–1 assuming financing of $12,000 of the initial first cost of $20,000 by a 10% loan

Year	*Cash flow before debt service and taxes*	*Cash flow for debt repayment*	*Cash flow for interest on debt*	*Cash flow after debt service*	*Depreciation*	*Taxable income*	*Cash flow for taxes*	*Cash flow after taxes*
	A	*B*	*C*	*D*	*E*	*F*	*G*	*H*
0	{ −$20,000 +12,000			−$8,000				−$8,000
1	+6,200	−$1,500	−$1,200	+3,500	−$2,500	+$2,500	−$1,250	+2,250
2	+6,200	−1,500	−1,050	+3,650	−2,500	+2,650	−1,325	+2,325
3	+6,200	−1,500	−900	+3,800	−2,500	+2,800	−1,400	+2,400
4	+6,200	−1,500	−750	+3,950	−2,500	+2,950	−1,475	+2,475
5	+6,200	−1,500	−600	+4,100	−2,500	+3,100	−1,550	+2,550
6	+6,200	−1,500	−450	+4,250	−2,500	+3,250	−1,625	+2,625
7	+6,200	−1,500	−300	+4,400	−2,500	+3,400	−1,700	+2,700
8	+6,200	−1,500	−150	+4,550	−2,500	+3,550	−1,775	+2,775
	+$41,600	−$12,000	−$5,400	+$32,200	−$20,000	+$24,200	−$12,100	+$12,100

No trial-and-error present worth calculations are needed here to see that 5% is the interest rate that will make the present worth of the final column equal to zero. That is, the series of disbursements from years 1 through 8 would repay $1,500 of principal of a $12,000 debt each year with 5% interest on the unpaid balance, so that the debt would be completely repaid with 5% interest at the end of 8 years.

It should also be obvious that our final column of differences in cash flow is unrelated to the merits of the equipment that it is proposed to finance. The figures in the column depend only on the amount borrowed, the repayment schedule, the interest rate on the loan, and the applicable income tax rate.

Our tabulation of difference in after-tax cash flow between equity financing and purchase on time was really unnecessary in this simple case. If we borrow money at 10% and if our applicable income tax rate is 50%, the after-tax cost of our borrowed money obviously is 5%. However, we shall see in Example 18–2 that a similar tabulation will be a useful tool in the more complex circumstances that exist in many leasing agreements.

EXAMPLE 18–2

Analysis of Before-Tax and After-Tax Effects of Proposed Acquisition of Equipment By Leasing

Facts of the Case

The equipment in Example 18–1 can also be acquired under an 8-year lease. The proposed lease contract calls for an initial deposit of $2,000, which will be returned at the end of the period of the lease. Rental charges will be $6,000 at the *beginning* of each of the first 3 years and $2,000 at the *beginning* of the remaining 5 years. Under this contract the lessee will pay all maintenance, insurance, and property taxes just as if the equipment were owned.

Rate of Return Before Income Taxes

Table 18–2 shows before-tax cash flow and calculation of present worths at 25% and 30%. Interpolation indicates a rate of return of approximately 27.9%.

Rate of Return After Income Taxes

Table 18–3 develops the year-by-year effect on taxable income of the proposal to lease the equipment. The taxable income each year will be increased by the $6,200 operating saving and decreased by the rental payment applicable to the particular year. It should be noted that because each year's rental is *prepaid*, the effect on taxable income applies to the year following the payment date; the influence on cash flow for income taxes of each item of cash flow for rental occurs one year after the rental payment. (Under our end-of-year convention, we assign cash flow for income taxes to the end of the tax year.) Of course the $2,000 negative cash flow for the deposit at zero date and the $2,000 positive cash flow for the refund at date 8 have no effect on taxable income. As the equipment is not owned, no depreciation is recognized in computing taxable income. (Depreciation will enter into the *lessor*'s, taxable income but not the lessee's.)

The present worth of the cash flow series in column F of Table 18–3 may be computed to be +$412 at 16% and −$287 at 18%. The approximate after-tax rate of return on equity funds is 17.2%.

TABLE 18–2
Before-tax analysis of a proposal to lease certain equipment

Year	*Effect on cash flow of use of equipment*	*Cash flow for payments related to lease*	*Combined cash flow*	*Present worth at 25%*	*Present worth at 30%*
0		−$8,000	−$8,000	−$8,000	−$8,000
1	+$6,200	−6,000	+200	+160	+154
2	+6,200	−6,000	+200	+128	+118
3	+6,200	−2,000	+4,200	+2,150	+1,912
4	+6,200	−2,000	+4,200	+1,720	+1,471
5	+6,200	−2,000	+4,200	+1,376	+1,131
6	+6,200	−2,000	+4,200	+1,101	+870
7	+6,200	−2,000	+4,200	+881	+669
8	+6,200	+2,000	+8,200	+1,376	+1,005
Totals	+$49,600	−$28,000	+$21,600	+$892	−$670

Cost of Money Provided by the Lessor

The following tabulation shows the differences in cash flow before and after income taxes between financing by the proposed lease agreement and financing entirely from ownership funds.

Year	***Before-tax difference in cash flow between equity financing and leasing***	***After-tax difference in cash flow between equity financing and leasing***
0	+$12,000	+$12,000
1	−6,000	−4,250
2	−6,000	−4,250
3	−2,000	− 250
4	−2,000	−2,250
5	−2,000	−2,250
6	−2,000	−2,250
7	−2,000	−2,250
8	+2,000	+1,750
Totals	−$8,000	−$4,000

The present worth of the before-tax series is −$855 at 20% and +$225 at 25%. In effect, the funds made available by the lessor will cost the owners approximately 23.9% before taxes. The present worth of the after-tax series is −$106 at 10% and +$481 at 12%. Considered after income taxes, the cost of money provided by the lessor is approximately 10.4%.

It will be noted that, although $20,000 worth of equipment is being leased, the effect of the leasing agreement is to make only $12,000 of cash

TABLE 18–3
After-tax analysis of a proposal to lease certain equipment

Year	Effect of proposal on cash flow before income taxes	Effect of use of equipment on taxable income	Effect of rental payment on taxable income	Combined effect on taxable income (B + C)	Effect on cash flow for income taxes (−0.5D)	Effect on cash flow after income taxes (A + E)
A	B	C	D	E	F	G
0	−$8,000					−$8,000
1	+200	+$6,200	−$6,000	+$200	−$100	+100
2	+200	+6,200	−6,000	+200	−100	+100
3	+4,200	+6,200	−6,000	+200	−100	+4,100
4	+4,200	+6,200	−2,000	+4,200	−2,100	+2,100
5	+4,200	+6,200	−2,000	+4,200	−2,100	+2,100
6	+4,200	+6,200	−2,000	+4,200	−2,100	+2,100
7	+4,200	+6,200	−2,000	+4,200	−2,100	+2,100
8	+8,200	+6,200	−2,000	+4,200	−2,100	+6,100
Totals	+$21,600	+$49,600	−$28,000	+$21,600	−$10,800	+$10,800

available to the lessee. Because of the requirement of a $2,000 deposit and a prepayment of $6,000 for the first year's rent, an initial outlay of $8,000 of ownership funds is required under the leasing agreement.

It will also be noted that Example 18–2 is like Example 18–1 in that the difference in cash flow between equity financing and outside financing is unrelated to the merits of the equipment that it is proposed to acquire. The before-tax and after-tax series of cash flow differences can be derived solely from the terms of the leasing agreement, the allowable depreciation if the equipment is to be owned, and the applicable tax rate. However, unlike the case of debt financing in Example 18–1, the after-tax cost of money with a 50% tax rate is not exactly half of the before-tax cost. The favorable timing of the tax consequences of this particular leasing agreement is such that the after-tax cost of the money (10.4%) is considerably *less* than half the before-tax cost (23.9%).

Desirability of Separating Decision Making on Physical Plant from Decision Making on Methods of Financing

Throughout this book, the general principle has been stressed that separable decisions should be made separately. Some cases occur where a decision to acquire the use of certain fixed assets is inseparably linked with one particular plan for financing the assets. In such instances it is rational to apply a single analysis to the proposed acquisition and financing. But in the more common case where decisions among alternative fixed assets are clearly separable from decisions regarding policies on methods of financing, both types of decisions are likely to be made more intelligently if the analyses are separated.

Difficulties in Judging Merits of Proposed Plant Investments on the Basis of Prospective Rate of Return to Equity Capital Where It Is Assumed that a Portion of Financing Will Be by Borrowing or Leasing

In Examples 18–1 and 18–2 we have computed rates of return to that portion of the proposed investment to be made from ownership funds, 25.4% to the $8,000 equity investment in Example 18–1 and 17.2% in Example 18–2 after taxes. Calculations such as these are often made in industry. Given the assumptions regarding income taxation and method of financing, these are unquestionably valid as exercises in the mathematics of compound interest.

Nevertheless, the authors recommend against using such computed rates of return on equity investments as guides to decisions about the acquisition of fixed assets. Their objections to such use are as follows:

1. Because such computed rates of return depend on the fraction of equity funds assumed, they are not appropriate to determine the relative merits of types of plant that are to be financed in different ways.
2. In the common case where the cost of outside money (secured by borrowing or leasing) is less than the return earned by the equipment itself (as computed under the assumption of equity financing), a computed rate of return on a fractional equity investment tends to give an unduly favorable impression of the productivity of the equipment.

The foregoing objections apply to studies made either before or after income taxes. A numerical example may help to clarify the reasoning underlying the objections. We shall use the after-tax analysis from Example 18–1.

It will be recalled that with complete equity financing, the after-tax return was 14.3%. With 60% of the cost of the equipment borrowed at an after-tax cost of 5%, the return on the remaining 40% of equity capital was 25.4%.

Now let us change the leverage exerted by the low after-tax cost of borrowed money by assuming that 80% ($16,000) can be borrowed at 10% before taxes. A calculation similar to that made in Example 18–1 will show that the after-tax return on the 20% ($4,000) of equity capital is approximately 40.7%. If we assume 90% can be borrowed, the return on the remaining 10% equity capital can be computed to be approximately 66.5%. If we assume that 100% can be borrowed, the absurdity of the rate of return analysis is evident because the computed rate of return is infinite. (That is, there is money return even though no equity investment is attributed to the proposed equipment.)

A point made earlier in this chapter with regard to borrowing to finance equipment should be restated. That is the point that borrowing may make good business better and it may make bad business worse. Referring again to Example 18–1, should the cash flow before debt service and taxes, column A of Table 18–1, be reduced to $3,175 for each of years 1 through 8, the rate of return on the $8,000 equity capital would be reduced to 0%. That is, the after-tax cash flow would be adequate to return the initial $8,000 equity investment but without interest. This same annual cash flow would yield a rate of return after taxes of approximately 2.9% with 100% equity financing. Thus, rates of return on equity investment are much more sensitive to random fluctuations in annual cash flows, or errors in their estimates, when a substantial portion of the investment is financed through borrowing. Again this is a characteristic of the method of financing, not of the merit of the proposed investment.

Shall Economy Studies Assume a Stipulated Percentage of Debt Financing in the Analysis of *All* Investment Proposals?

The quotation from Joel Dean at the start of this chapter stated that in the evaluation of investment proposals "investable funds of the corporation should be treated as a common pool, not compartmented puddles." Our discussion in which we assumed various percentages of debt for the data of Example 18–1 illustrated the rationale of Dean's colorful statement. If proposed alternatives are compared on the basis of prospective rate of return on equity investments, and if it is assumed that each alternative will be financed from its own "compartmented puddle," and if the different puddles involve different percentages of debt or different interest rates on the debt or different types of repayment schedules, the comparison is likely to give a badly distorted view of the actual merits of the different alternatives.

The distortion of relative merits can be avoided if it is assumed that *all* proposals are to be financed from a "common pool" that contains a stipulated percentage of debt financing at some stipulated average interest rate. A number of mathematical models for investment analysis have been developed that have made such an assumption. One of these models was mentioned in Chapter 17.

Certainly it is better to assume constant percentages of debt and equity financing for all proposals than to distort comparisons by changing the assumed percentages from proposal to proposal. Nevertheless, in economy studies for competitive industry, it is questionable whether any advantage is gained by having all analyses based on the assumption of partial debt financing. And there are some practical disadvantages in departing from the practice of making analyses that assume 100% equity financing. A minor disadvantage is a somewhat greater difficulty in computing the required figures for cash flow, particularly in an after-tax analysis. A more serious disadvantage arises in the danger of misinterpretation of the results of the economy studies, particularly when these results are expressed in terms of rate of return to equity capital. Although, generally speaking, the relative attractiveness of the various investment proposals is not changed by assuming a stipulated percentage of debt financing, all the acceptable proposals seem better because the return is computed on equity rather than on the total investment. Moreover, the leverage exerted by the debt financing has a disproportionate influence on the proposals that have the higher prospective rates of return. Also the assumption of debt financing makes the computed rates of return much more sensitive to moderate changes in estimated cash flows.

For reasons that are explained in Chapter 20, there are special reasons for assuming stipulated percentages of debt and equity financing in calculations to introduce income taxes into economy studies for regulated public utility companies under the rules of regulation that have developed in the United States.

The Usefulness of Computed After-Tax Costs of Money

Throughout this book we have stressed the point that interest rates are not always what they seem. In this connection we strongly recommend calculation of the after-tax cost of money to be borrowed or secured through leasing agreements, with such costs expressed as an interest rate in the manner illustrated at the ends of Examples 18–1 and 18–2. In these two examples it is evident that leasing at an after-tax cost of 10.4% is unattractive if we can borrow at an after-tax cost of 5.0%. In general such calculations are a great help in comparing the merits of alternate plans of financing. Such calculations employ techniques that are similar to the ones we use in engineering economy, since they start with estimates of differences in cash flow and apply compound interest mathematics to these differences to find unknown interest rates.

The calculation of the after-tax cost of borrowed money is not always as simple as in Example 18–1. Some sources of complications in finding the true cost of borrowed money were illustrated in Chapter 8. These included the difference between a cash price and a price on terms, initial disbursements incident to securing a loan, and required annual disbursements throughout the life of a loan. Although the illustrations in Chapter 8 dealt with the before-tax cost of borrowed money, the same complications arise in computing after-tax costs.

Some Aspects of the Analysis of Proposed Leasing Agreements

There is great variety in the types of leasing agreements made in industry. In many cases both the lessor and lessee recognize that the lease may continue only for a relatively short term; such leases are terminable by the lessee on short notice.

The more interesting problems for the student of engineering economy arise where the lease is used as a device for long-term financing. In such cases it often is true that the same assets would be acquired with leasing or outright ownership and that the assets would be continued in service for the same number of years whether leased or owned. From the viewpoint of the lessee, the long-term lease is a fixed obligation not unlike an obligation for debt interest and repayment.

Frequently the analysis of difference in cash flow before income taxes between leasing and ownership is more complicated than in Example 18–2. Certain disbursements necessary if an asset were owned (e.g., major maintenance overhauls, property taxes, insurance) may be made by the lessor—not by the lessee—as stipulated in Example 18–2. There may be substantial prospective salvage values, realizable under ownership but not under leasing.

An analysis of the difference in cash flow for income taxes requires a tabulation of deductions from taxable income under ownership and leasing. The after-tax analysis made at the end of Example 18–2 could also have been obtained by means of the following tabulation:

Year	*Deduction for depreciation*	*Deduction for rental*	*Difference in taxable income*	*Difference in cash flow for income taxes*
1	$2,500	$6,000	−$3,500	+$1,750
2	2,500	6,000	−3,500	+1,750
3	2,500	6,000	−3,500	+1,750
4	2,500	2,000	+500	−250
5	2,500	2,000	+500	−250
6	2,500	2,000	+500	−250
7	2,500	2,000	+500	−250
8	2,500	2,000	+500	−250
Totals	$20,000	$28,000	−$8,000	+$4,000

Long-term leases of machinery and equipment generally are like Example 18–2 in requiring the highest rental charges during the early years. As illustrated in the foregoing tabulation, such an arrangement leads to a favorable timing of the tax consequences of leasing. The large difference between the before-tax cost and after-tax cost of money secured by leasing in Example 18–2 was due to the fact that a tax credit in the near future is more valuable than one in the more distant future. The longer the expected life of the equipment to be leased, the more favorable to leasing may be the tax consequences of the difference in timing of tax deductions under leasing and ownership.

When Is an Agreement That Purports to Be a Lease Not Recognized as a Lease for Income Tax Purposes?

In order to secure the advantage of larger tax deductions in the early years of life, some agreements that are, in effect, sales on time are drawn in a way that makes them appear to be leases. Such agreements are viewed critically by taxing authorities in the United States. A Treasury publication contained the following advice to taxpayers on this topic:*

> *Lease or purchase*. To determine if you may deduct rent payments, you must first determine if your agreement is a lease or a conditional sales contract. If under the agreement you acquired, or will acquire, title to or equity in the property, the agreement should be treated as a conditional sales contract. Payments made under a conditional sales contract are not deductible as rent.
>
> Whether the agreement is a lease or a conditional sales contract depends upon the

**Tax Guide for Small Business, 1980 Edition*, Publication No. 334 (Washington, D.C.: Department of the Treasury, Internal Revenue Service), p. 36.

intent of the parties. Intent is generally determined by the agreement together with the facts and circumstances existing at the time the agreement was made.

Determine the intent. An agreement may be considered a conditional sales contract rather than a lease if:

1. The agreement applies part of each "rent" payment toward an equity interest that you will acquire.
2. You are given the right to buy the property after making all the required payments.
3. You are required to pay, over a short period of time, an amount that is a large part of what you would pay to buy the property.
4. You pay rent that is much more than the current fair rental value for the property.
5. You have an option to buy the property at a price that is small compared to the value of the property at the time you may take advantage of the option. Determine this value at the time of the agreement.
6. You have an option to buy the property at a price that is small compared to the total amount you are required to pay under the lease.
7. The lease designates some part of the "rent" payments as interest, or part of the "rent" payments are easy to recognize as interest.

In the problems at the end of this chapter requiring the after-tax analysis of various leasing agreements, it will be assumed that the agreements described are legitimate leases and not conditional sales contracts.

A Classification of Sources of Equity Funds in Business Enterprise

Although the issues involved in choosing between equity financing, on the one hand, and long-term obligations involving borrowing or leasing, on the other hand, are beyond the scope of this book, it is helpful to classify sources of equity funds as follows:

1. New equity money (e.g., the sale of new stock by a corporation)
2. Profits retained in the business
3. Capital recovered through the depreciation charge

The relationship between depreciation charges and the financing of fixed assets is not always clear to persons who have not studied accounting. Example 18–3 presents this topic in a simplified form.

EXAMPLE 18–3

Financing Fixed Assets with Capital Recovered Through the Depreciation Charge*

*For a more complete version of this same example and a more detailed discussion of this subject, see E. L. Grant and P. T. Norton, Jr., *Depreciation* (rev. prtg., New York: The Ronald Press Co., 1955), Chap. 14.

Facts of the Case

The accounting expense of depreciation was discussed in Chapter 11. One of the reasons for recognizing depreciation in the accounts is to include it in the selling price of the product or service. A concern that includes an adequate amount for depreciation in its selling price year after year will at least succeed in recovering the invested capital that is being used up through the decrease in value of its plant and machinery. In such a concern a part (or all) of the excess of income currently received over expenses currently paid out (i.e., expenses not including depreciation) will not be profit but will be recovered capital.

Assume that on January 1 of some specified year, a manufacturing company has the following balance sheet:

Assets

Current Assets			
Cash		$ 300,000	
Accounts Receivable		500,000	
Inventories		600,000	$1,400,000
Fixed Assets			
Land		150,000	
Plant and Machinery	$2,000,000		
Less			
Allowance for Depreciation	800,000	1,200,000	1,350,000
			$2,750,000

Liabilities and Owners' Equity

Current Liabilities		
Accounts Payable	$ 300,000	
Notes Payable	200,000	$ 500,000
Fixed Liabilities		
Mortgage Bonds Outstanding		500,000
Owners' Equity		
Capital Stock	1,500,000	
Surplus	250,000	1,750,000
		$2,750,000

Suppose there is no profit or loss for the year, the company just clearing its expenses including estimated depreciation. A condensed form of its profit and loss statement for the year might be as follows:

Sales	$1,200,000
Less	
Cost of Goods Sold (including $100,000 Depreciation Expense)	850,000
Gross Profit on Sales	$ 350,000

Less	
Selling, Administrative, and Financial Expense (including Bond Interest)	350,000
Net Profit for Year	—

The cost of goods sold includes $750,000 currently paid out and $100,000 of depreciation expense, prepaid in previous years, which is balanced by a $100,000 increase in allowance for depreciation. Thus, total expenses currently paid out are $1,100,000 ($750,000 + $350,000); with revenues of $1,200,000 there remains $100,000 of additional assets in the business to compensate for the estimated decrease in value of plant and machinery.

For the capital invested in the business to be conserved, this $100,000 must be left in the enterprise in some form. Conceivably it might all be an addition to cash. Or it might increase other current assets through financing additional accounts receivable, or financing a larger investment in inventories. Or the cash might be applied to decrease liabilities, reducing accounts or notes payable or paying off some of the mortgage bonds. In certain circumstances one of these uses or some combination of them might be imperative. But it is generally recognized that the primary use for such recovered capital should be the financing of fixed assets. This includes both those assets needed to replace those retired and other fixed assets acquired to reduce operating costs or increase revenues. If $30,000 of the $100,000 should be devoted to the replacement of assets having an original cost of $20,000 that were retired during the year, $50,000 to the acquisition of other new plant and machinery, $15,000 to financing increased inventories, and $5,000 to an increase of cash, and if the receivables and payables should be the same at the year end as at the start of the year, the year-end balance sheet will be as follows:

Assets

Current Assets			
Cash		$ 305,000	
Accounts Receivable		500,000	
Inventories		615,000	$1,420,000
Fixed Assets			
Land		150,000	
Plant and Machinery	$2,060,000		
Less			
Allowance for Depreciation	880,000	1,180,000	1,330,000
			$2,750,000

Liabilities and Owners' Equity

Current Liabilities		
Accounts Payable	$ 300,000	
Notes Payable	200,000	$ 500,000

Fixed Liabilities		
Mortgage Bonds Outstanding		500,000
Owners' Equity		
Capital Stock	1,500,000	
Surplus	250,000	1,750,000
		$2,750,000

This corporation has made no profit during the year. If, as may well be the case, it is impossible to borrow more money or to sell more stock, the capital recovered through the depreciation charge is the only possible source of ownership funds to finance the purchase of plant and machinery. These funds are therefore of great importance in the continuation of the business.

As pointed out in Chapter 17, the factors that make replacements economical do not usually make them imperative. Machines and structures do not collapse like the "one hoss shay"; it is possible to continue them in service for a long time after it has become economical to replace them.

A business cannot continue indefinitely to operate profitably with an uneconomical plant; if its plant is worn out or obsolete, the concern is likely to have difficulty competing with rival concerns that have lower operating costs because their plants are new and modern. A manufacturing company that year after year diverts its recovered capital to uses other than the financing of fixed assets may ultimately find itself in this unfavorable competitive position. The difficulty of securing funds for other capital requirements (such as financing receivables and inventories or reducing debt) and the ease of putting off plant modernization may combine to bring about such diversion of funds.

Deferring replacements beyond the date when they are economical not only results in an uneconomical plant; it also has the hazard of causing depreciation rates that are much too low. Such low rates cause an overstatement of profits; the possible unfortunate consequences of such overstatement are obvious.

Some Ways in Which Proposed Investments May Be Vetoed Because of Considerations of Financing

If it is impossible to finance proposed physical assets in some way—from either equity funds, borrowing, or leasing—it is merely of academic interest that they would earn a good rate of return. And it is not only the proposals that *cannot* be financed that are eliminated due to considerations related to their financing. It often happens that one or more ways exist in which assets might be financed, but that for one reason or another these ways are unattractive from the viewpoint of the decision maker. This point is illustrated in Example 18–4. It also was illustrated in Examples 10–1 and 10–2.

EXAMPLE 18–4

Effect on an Investment Decision of a Requirement for Rapid Debt Repayment

Facts of the Case
A family living in a rented house was offered a chance to purchase this house at a favorable price. However, because the house was an old one, only 50% of the purchase price could be borrowed on a long-term first mortgage. The family had funds for only a 10% down payment. The present owner agreed to take a second mortgage for the remaining 40%. But he insisted that one-fifth of the principal of this second mortgage be repaid each year. No better source of money could be found for this 40% of the purchase price.

An analysis of the long-run economy of home ownership and renting was favorable to home ownership. However, because of the high repayment requirements on the second mortgage, the money outlays to finance home ownership would be much higher than rent. These outlays included principal and interest payments on both mortgages, maintenance, property taxes, and insurance. When these outlays were considered with relation to the prospective family income, it was evident that not enough money would be left for the other necessary family living expenses during the next 5 years. Therefore, it was advisable to continue to rent.

The Possible Influence of Project Financing on the Choice of Decision Criteria

There are exceptions to the rule that an investment proposal ought to be evaluated apart from its "compartmented puddle" of financing. An exception sometimes exists when a project must be financed partly or entirely by borrowing, when the borrowed funds must be repaid too rapidly in relation to the useful life of the project, and when there is too great a limitation on the financial resources of the prospective investor.

In effect, where the financial resources of an individual or family or business enterprise are too small in relation to a proposed investment, the requirement for too rapid debt repayment may change the appropriate primary criterion for the investment decision. It may be desirable to give short-run cash flow greater weight than long-run economy. A similar relationship between repayment requirements and decision criteria may exist with reference to certain types of local governmental works for which 100% debt financing is common. Further comments regarding financial aspects of local governmental projects are made in Chapter 19.

Some Comments on Relationships of Prospective Inflation to Business Finance and to Engineering Economy

Certain aspects of price level change were discussed in Chapter 14. Our discussion there pointed out that in choosing among alternative investment proposals, it is desirable in principle to make analyses in units of constant purchasing power. If *all* prices are expected to rise or fall by the same percentage, a comparison of alternatives based on the assumption of a continuation of present prices will, in effect, be a comparison in monetary units of constant purchasing power. But to the extent that a differential price change can be forecast, such differential change is relevant and needs to be given weight in the choice among proposed investments.

Whenever a series of prospective cash flows is fixed in dollars (or in whatever other monetary units are relevant), any change in the purchasing power of the monetary unit causes a differential price change with respect to the particular set of cash flows. This category of future cash flows fixed in money amounts includes interest and principal payments on nearly all long-term borrowings as well as rental payments on many long-term leases.

Examples 14–1 and 14–2 analyzed two loans, one past and one prospective, from the viewpoint of the lender. In both examples the payments of interest and principal to be received were fixed in the relevant monetary units (which were dollars in these cases) regardless of changes in purchasing power. These examples made the point that when, in a period of inflation, a loan is examined in monetary units of constant purchasing power, the rate of return to the lender may be very low indeed. It may even be negative, as was illustrated in Example 14–1.

Generally speaking, inflation, which hurts the lender, helps the borrower. That is, when the rate of return to a lender is very low measured in monetary units of constant purchasing power, the cost of money to a borrower is also very low when measured in the same way.

The foregoing discussion might lead the reader to conclude that a forecast of inflation favors financing by long-term borrowing. But this matter is not quite as simple as it might seem at first glance. An expectation of inflation by *lenders* naturally tends to increase the general level of interest rates to compensate, at least in part, for the expected decrease in the value of money. When very rapid inflation is expected, it may be impossible to borrow for long terms; even short-term loans may carry extremely high interest rates. We noted in Chapter 14 that some countries with high inflation rates have indexed the payments of interest and principal on loans in order to eliminate or at least reduce this obstacle to lending. In this book we do not attempt to give an exposition of the complex subject of business finance; a particularly troublesome segment of this important subject deals with the relationship between business finance and past or prospective price level change.

Some Complications in the Financing of Fixed Assets Created by a Rise in Price Levels

Example 18–3 brought out the point that replacement assets tend to be financed in part through internally generated funds, with the amount of such funds related to the depreciation charge made in the accounts. When inflation occurs, the conventions of accounting and the rules of income taxation create special difficulties in the way of financing replacement assets.

As pointed out in Chapters 11 and 14, the depreciation charge in the accounts is a writing off of *cost*. In the years immediately following a price-level rise, the depreciation charges in the accounts apply in large measure to assets acquired before price levels had risen. Hence, the funds that appear to have been made available through the depreciation charge tend to remain nearly constant whereas the funds needed to finance the replacement of any asset are greatly increased. At the same time, rising price levels create an increased demand for funds to finance inventories and accounts receivable. In such a period the funds needed for investment in fixed assets often cannot be secured without either a substantial diversion of profits to this purpose or the issuance of new securities or both. The rules of income taxation, which generally use cost as the basis for the depreciation deduction from taxable income, complicate the problem of plowing back "profits" into the business enterprise to finance replacement assets.

Summary

Some of the points brought out in this chapter may be restated for emphasis as follows:

1. In most instances where it is expected that the major part of the cost of fixed assets will be financed by borrowing or leasing, the prospective rate of return on a small increment of equity capital is not a sound guide to investment decisions.
2. In comparing different possible schemes of financing through borrowing or leasing, it is desirable to compute an after-tax cost of money for each scheme. A tabulation should be made of prospective differences in cash flow after taxes between equity financing and each other proposed scheme of financing. Appropriate compound interest calculations should then be made to find the unknown interest rate in each case.
3. Whenever it is proposed to acquire the ownership of property through the borrowing of money, consideration should always be given to the question of whether or not it is practicable to meet the repayment obligations.
4. The question of the practicability of debt repayment should not be confused with the question of long-run economy; repayment obligations and depreciation should not appear as "costs" in the same study.
5. The problem of meeting short-run repayment obligations does not arise

in connection with a bond issue on which refunding is contemplated; it does arise whenever bonds mature serially.

6. Doing business on borrowed money has the tendency to make good business better and bad business worse.
7. Available funds for plant expansion and replacement are often limited; if so, the problem is likely to be to select, from a number of desirable uses for these funds, those which seem likely to yield the highest returns.
8. Where for some reason it is impossible or undesirable to secure funds by borrowing or to secure new ownership funds, the sources of financing for new fixed assets are limited to profits and to capital recovered through the depreciation charge. It often happens that such funds are practically limited to the latter source by company policies or by income tax considerations influencing the use of profits.

PROBLEMS

Unless otherwise stated, the calculation of prospective interest rates and prospective rates of return should assume annual compounding. A date stated as the *beginning* of any year should be treated as identical with the *end* of the preceding year. Payments *during* each year should be treated as if they took place at the end of that year.

18–1. A machine-tool builder will rent new machine tools for 3 years under the following rental contract:

The purchase price of a tool is designated as 100N. The lessee must deposit 10N at the start of the rental period; when the tool is returned to the lessor in good condition at the end of 3 years, this deposit is refunded to the lessee. Annual rental payments are 24N, payable at the start of each year. Lessee pays all operation and maintenance costs, including property taxes and insurance.

A user of machine tools has determined that he intends to use a certain machine tool for a 3-year period. He is undecided whether to purchase the machine and sell it in the secondhand market at the end of 3 years or to rent it. Designate purchase as Plan P and rental as Plan R. Show the year-by-year differences in cash flow before income taxes for these two plans. Assume that the machine can be sold for a net figure of 55N after 3 years of use. Find the interest rate that makes the two plans have equal present worths of net cash flow before income taxes. (This rate might be interpreted as a basis for decision making in an organization not subject to income taxes—such as a nonprofit organization.) (*Ans*. (P − R) cash flow; 0, −66N; 1, +24N; 2, +24N; 3, +45N; interest rate, 17.1%)

18–2. In Problem 18–1 assume an applicable income tax rate of 45%. If Plan P is selected, years-digits depreciation will be used for tax purposes assuming a 3-year life and 55% salvage value. If Plan R is selected, assume that rental paid at the start of any year will affect income taxes for that year. What is the after-tax cost of the money made available by the lease agreement? (*Ans.* 9.9%)

18–3. Consider a prospective lessee's analysis of the leasing agreement described in Problem 18–1 and analyzed after taxes in Problem 18–2. These problems were solved assuming a net realizable salvage value of 55N at the end of 3 years. The analyst for a certain prospective lessee subject to income taxation believes that the actual salvage value will turn out to be somewhat less than 55N. Discuss the question of whether an expectation of a lower salvage value increases or decreases the attractiveness to the prospective lessee of leasing in this case.

18–4. Discuss the sensitivity of the computed after-tax cost of money in Problem 18–2 to the applicable income tax rate. Other things being equal, do you believe that high applicable income tax rates tend to encourage or to discourage financing by leasing? Explain your answer.

18–5. (a) It is estimated that a proposed equipment investment of $60,000 to be financed from equity funds will cause an excess of receipts over disbursements of $19,400 a year for 5 years. The equipment will have zero salvage value at the end of 5 years. What is the prospective rate of return before income taxes? (*Ans.* 18.5%)

(b) Assume that this equipment can be financed by a $10,000 down payment from equity funds. The remainder will be paid at $10,000 a year plus 12% interest on the unpaid balance. What is the prospective before-tax rate of return on the equity investment? (*Ans.* 41.3%)

18–6. Find the after-tax rates of return on equity investment in both parts of Problem 18–5 assuming a 45% applicable income tax rate and years-digits depreciation based on a 5-year life and zero salvage value.

18–7. Consider the circumstances described in Problem 18–1 as applied to a nonprofit organization not subject to income taxes. The analyst for this organization anticipates moderate inflation of perhaps 4% a year for the next 3 years. Should this prospect of inflation favor a decision for outright ownership or a decision to lease, or is it irrelevant in this case? Explain your answer.

18–8. Answer the question in the preceding problem for the organization in Problem 18–2 that is subject to the 45% applicable tax rate. Assume that the inflation will not change either the tax rate or the size of the depreciation allowance for tax purposes.

18–9. (a) It is estimated that a certain equipment investment of $90,000 will cause an excess of receipts over disbursements of $22,000 a year (before income taxes) for 9 years. The equipment will have zero salvage value at the end of the 9 years. It will be financed entirely from equity funds. What is the prospective rate of return before income taxes?

(b) Assume that this equipment can be financed using a down payment of $27,000 from equity funds. The remaining $63,000 will require payments of $7,000 at the end of each year for 9 years plus 12% interest on the unpaid balance. What is the prospective before-tax rate of return on the equity portion of the investment?

18–10. Find the after-tax rates of return on equity investment in both parts of the preceding problem with a 45% applicable tax rate. Assume years-digits depreciation based on a 9-year life and zero terminal salvage value.

18–11. For the circumstances described in Problems 18–9 and 18–10 it is expected that there will be an inflation rate of 6% a year, more or less, for the next few years. Assume that, whatever the rate of inflation may be, the before-tax excess of receipts over disbursements will increase in proportion to the increase in the general price level; that is, no differential price change in this figure is expected. Discuss whether the prospect of inflation (a) favors equity financing, (b) favors financing by a combination of debt and equity, or (c) is not relevant to the decision about financing.

18–12. In the discussion of Example 18–1 it is stated that "A calculation similar to that made in Example 18–1 will show that the after-tax return on the 20% ($4,000) of equity capital is approximately 40.7%." Make the necessary calculations to verify this statement.

18–13. A writer on the subject of the determination of the costs of public hydroelectric power projects included the following items as costs: (1) interest on the first cost of the project; (2) depreciation by the straight-line method based on the estimated life of the project; (3) an annual deposit in an amortization sinking fund sufficient to amount to the first cost of the project at the end of 50 years (or at the end of the life of the project if that should be less than 50 years); (4) where money is borrowed, the annual disbursements for bond interest and bond repayment; (5) all actual annual disbursements for operation and maintenance of the project.

Do you believe that annual cost should properly be considered as the sum of these items? Explain your answer.

18–14. A proposal is made for a city-owned public garage to be built in the business district of a certain city. The purchase price of the required land is $700,000 and the estimated first cost of the garage building is $2,500,000. The proponents of this garage project suggest that it be financed by $3,200,000 of general obligation bonds of the city. These bonds would mature at $160,000 a year for 20 years and would pay 7% interest per annum. The estimated life of the garage building is 50 years.

A committee from a local garage owners' association has submitted a report on this project. This report, given to the city council, seems to show that the expenses of the project will exceed its revenues. The tabulated annual expenses include the following items:

Depreciation on building ($2,500,000 ÷ 50)	$ 50,000
Bond repayment	160,000
Bond interest	224,000
Sinking fund at 5% for building replacement	11,950

Comment on the relevance of these items in the economic evaluation of this project. Discuss also the question of the different types of decision criteria that it might be desirable to apply to this project and the relevance of these four items in applying each suggested criterion.

18–15. Company X is a lessor of capital assets. This company holds title to office buildings, hotels, warehouses, stores, machine tools, construction equipment, slot machines, and various other types of assets intended to produce income for their lessees. All of the assets are leased to the respective operators under an agreement that requires the lessee to pay all upkeep, property taxes, and insurance. Mr. A, one of the officials of Company X, is very articulate in pointing out why he believes it is to the advantage of the operators of capital assets to acquire the use of such assets by means of leasing rather than by outright ownership. He states that working capital usually is the most productive capital invested in a business enterprise and that leasing agreements (such as those made by Company X) have the effect of making more working capital available to the lessee. He uses the following example to compare a 10-year lease of a fixed asset with outright ownership of the same asset:

An asset having a first cost of $110,000, a life of 10 years, and zero salvage value may be purchased or leased. If leased, the lessor requires an initial deposit of $10,000 to be refunded at the end of the term of the lease. The rental charge will be $2,500 a month for the first 5 years and $500 a month over the final 5 years. If purchased, the asset will be depreciated for tax purposes by the years-digits method. If leased, the rental charge will be deductible from taxable income for the year in which it is made. For the sake of simplicity in Mr. A's example, an applicable tax rate of 50% is used. He assumes that any funds made available by choosing the lease rather than outright purchase will serve to increase working capital. Because Mr. A contends that working capital generally earns at least 20% after income taxes, his example assumes that all funds made available by the choice of the lease agreement will earn a rate of 20%. His conclusion from the following table is that the decision to lease this $110,000 asset rather than to own it outright will yield the company leasing the asset an additional $149,300 after income taxes in 10 years. Therefore, he asserts, it is better to lease than to own.

Year	*Difference in cash flow before income taxes between leasing and ownership*	*Difference in cash flow for income taxes*	*Difference in cash flow after income taxes*	*Cash accumulation from differences after allowing for 20% return each year*	
0	+$100,000		+$100,000		$100,000
1	−30,000	+$5,000	−25,000	1.2(100,000) − 25,000 =	95,000
2	−30,000	+6,000	−24,000	1.2(95,000) − 24,000 =	90,000
3	−30,000	+7,000	−23,000	1.2(90,000) − 23,000 =	85,000
4	−30,000	+8,000	−22,000	1.2(85,000) − 22,000 =	80,000
5	−30,000	+9,000	−21,000	1.2(80,000) − 21,000 =	75,000
6	−6,000	−2,000	−8,000	1.2(75,000) − 8,000 =	82,000
7	−6,000	−1,000	−7,000	1.2(82,000) − 7,000 =	91,400
8	−6,000	0	−6,000	1.2(91,400) − 6,000 =	103,680
9	−6,000	+1,000	−5,000	1.2(103,680) − 5,000 =	119,416
10	+4,000	+2,000	+6,000	1.2(119,416) + 6,000 =	149,300

(a) Has Mr. A made a correct analysis of the prospective differences in cash flow after income taxes? Given his figures for differences in cash flow, is his figure of a difference of $149,300 in compound amount at the end of 10 years consistent with his assumptions? Explain your answers. (It will be noted that he has used an end-of-year convention with respect to monthly rental payments to be made during a year and with respect to the payment of income taxes on each year's income. In your analysis of this problem, assume that this end-of-year convention is satisfactory.)

(b) The leasing agreement may be viewed as providing a source of investment capital for the lessee. Disregarding the effect of the leasing agreement on income taxes, express the cost of this capital as an interest rate.

(c) Considering the effect on income taxes of the difference between leasing and outright purchase, express the cost of this capital as an interest rate. Accept the assumptions of years-digits depreciation and an effective tax rate of 50% throughout the 10-year period.

(d) Discuss the relevance of Mr. A's analysis and of the two interest rates that you computed in your answers to (b) and (c) as a basis for a decision between leasing and outright ownership to be made by the prospective lessee. Assume that equity funds can be made available for the purchase of this $110,000 asset if it is decided that outright purchase is more advantageous than leasing. What information other than leasing terms, tax rate, and depreciation method, if any, would you need before deciding on your recommendation between leasing and ownership? How would this additional information influence your recommendation?

18–16. A manufacturer of machine tools offers the following plan for the rental of his product. The outright price of a tool is designated as 100C. Upon rental a deposit of 18C is required, to be returned upon termination of the rental. For the first 4 years, the rental charge, payable at the *start* of each year, is 20C. For the next 3 years, the rental charge, payable at the start of each year, is 16C. Thereafter, the annual prepaid rental is 10C. The lessee may terminate the agreement at the end of the fourth year, at the end of the seventh year, or at the end of any year thereafter.

The GH Company uses many of these tools that have been purchased outright. Typically these have been retired when 10 years old with zero salvage value. The company's applicable tax rate is 45%. Consideration is being given to a change in policy to rental for a 10-year period. Payments for operation, maintenance, and property taxes will be the same under rental as under outright ownership. If purchased, tools are depreciated for tax purposes by the years-digits method. If tools should be leased, what will be the after-tax cost of the money made available by the leasing agreement?

18–17. How would your answer in Problem 18–16 be changed if the GH Company should be entitled to a 10% investment tax credit at zero date under outright ownership but not under leasing? How would it be changed if the 10% credit should be available either with outright ownership or with leasing? Explain your answer.

18–18. In Problem 18–16 should GH Company's attitude toward leasing be influenced by an expectation of moderate inflation (say 4% to 6%)? Assume that the manufacturer's leasing plan described in the problem was established when the price level was not changing and that this plan has been continued unchanged in spite of some inflation.

18–19. A manufacturer has established a leasing plan such as the one described in Problem 18–16 during a period of fairly level prices. Will the plan be as advantageous to the manufacturer during a period of inflation as it was during the period of level prices? Explain your answer.

18–20. The annual report of a municipally owned waterworks showed the following figures for January 1, 1981:

A. Total investment in plant now in service	$959,435
B. Estimated depreciation to date on this plant	340,200
C. Net present value (A − B)	619,235
D. Bonds outstanding against the plant	400,000

The accounts of the plant were kept on a "cash" basis, so that this figure for net present value was merely a statistical figure built up from past records of the cost of plant additions and the city engineer's estimates of the amount of depreciation chargeable each year. The cash receipts and disbursements from the waterworks fund for the year 1981 were as follows:

Receipts		**Disbursements**	
Sale of water to private customers	$193,300	Labor	$ 37,500
Sale of water to city departments including hydrant rental	22,600	Operating supplies	3,400
		New well	13,300
		Power for pumping	27,100
		Pump house and pump at new well	9,600
		Repairs to pumps	4,200
		Maintenance of distribution system	10,900
		New water mains	12,800
		Interest on bonds	28,000
		Repayment of bonds	25,000
		Transfers to other city departments (used for police department and street lighting)	44,100
	$215,900		$215,900

The city engineer's figure for depreciation chargeable in 1981 was $45,270.

A committee of citizens, criticizing the financial policy of the city in running the waterworks, points out that no sinking fund has ever been maintained in connection with the depreciation of the plant. Therefore, they contend, the city has been negligent in failing to make desirable financial provision for depreciation of its waterworks property.

Discuss the topic of whether this criticism is sound, giving specific answers to the following questions:

(a) What outlays, if any, during 1981 do you think should be viewed as financial provision for depreciation? If the city engineer's figure for 1981 depreciation is accepted, do you think this financial provision has been adequate? Why or why not?

(b) It is known that the original construction of this plant was financed entirely by a bond issue. If operation and maintenance, debt service, and plant additions have been financed solely from the sale of water, and if the city engineer's overall depreciation figure is accepted as reasonable, what can you say about the overall financial provision for depreciation up to the end of 1981?

18–21. In Problem 18–1 it was stipulated that the lessee must make payments of 24N at the *start* of each year for 3 years. Change matters so that instead of prepaying rent for each full year, the lessee is required to pay 2N at the start of each month. This changed requirement is somewhat more favorable to the lessee. Therefore, the before-tax cost of the money that the leasing agreement makes available to the lessee really is somewhat less than the 17.1% computed under the stipulations of Problem 18–1.

To obtain an approximate figure for this lower cost of money, assume that the rental payments are made uniformly throughout the year (rather than monthly) and calculate the cost of money as an effective interest rate assuming continuous compounding. To solve this problem it is necessary to make use of Table D–31. To understand the use of this continuous compounding table it is desirable to read Appendix A.

Note that it is only the annual rental payment that is treated as if it took place uniformly throughout the year. The lessee's deposit at zero date and its return at date 3 are unaffected. Also, under outright ownership the disbursement to buy the machine tool and the receipt from its ultimate sale each take place at a particular moment in time. (*Ans.* 14.0%)

19

SOME ASPECTS OF ECONOMY STUDIES FOR GOVERNMENTAL ACTIVITIES

*Very little can be accomplished by theory and procedure without data. Similarly, a jumble of data without model relationships and procedures is equally powerless. There can be no other point of departure in a systems cost-effectiveness study than to develop an understanding of the physical characteristics of the system, of the equipment within the system, of the ground support equipment, of the inter-relation between man and machine, and of the effect of enemy action and technology.**

The words *benefits* and *costs* as applied to the economic evaluation of proposed public works projects were introduced in Chapter 9. It was pointed out that benefit–cost analysis generally requires calculations of present worths or equivalent uniform annual money amounts, and that such calculations need to be preceded by the choice of a minimum attractive rate of return. Certain pitfalls in the use of the *ratio* of benefits to costs (which we abbreviated to **B/C**) were noted both in Chapters 9 and 13.

Chapter 9 pointed out that conceptually the evaluation of proposed government projects often is more complex than the evaluation of projects in private enterprise. Moreover, it was stated that the application of concepts generally is more difficult when an analyst deals with government projects. The present chapter expands the foregoing statements.

The Case for Economy Studies in Public Works

Sometimes it is contended that because governments are not organized to make a profit, there is no need to make any economic analysis of proposed government projects. In the opinion of the authors of this book, such a contention is unsound. Two factors that make economy studies particularly de-

*Weapons System Effectiveness Industry Advisory Committee, "Final Report of Task Group IV, Cost Effectiveness Optimization (Summary, Conclusions, and Recommendations)", AFSCR-TR-65-4, Vol. 1 (January 1965).

sirable for proposed public works are the limitation of available resources and the diversity of citizen viewpoints regarding any given project.

Whether one looks at national governments, state and regional governments, or local governments, one is likely to find that the government has limited resources and that citizens are making many demands on these resources that it is not practicable to meet.

Let us contrast private business with public works. In private business there is a unified owner interest—the interest in the securing of profits—and ownership control of decisions on technical matters, at least in the case of large corporations, is comparatively remote. (That is, the individual stockholder is not likely to be greatly concerned about technical decisions under the jurisdiction of the chief engineer; even if concerned, a stockholder is not in a position to do much about it.) In public works there may be a wide variety of owner interests (assuming that the citizens of a community bear a relation to its government similar to the relation of the stockholders to their corporation) that often are in conflict with one another, individual owners frequently are much more articulate about decisions on technical matters, and their influence on such decisions can be fairly direct, particularly in local government.

Moreover, the customers of a private business are generally in the position of making a voluntary purchase of goods or services for which they make payment in the form of a price that is presumably less than their valuations of the benefits they receive. The beneficiaries of public works, on the other hand, generally receive their services without any specific voluntary purchase; payment that is made for these services in the form of taxes does not bear any necessary relation to benefits received by the individual taxpayer. This tends to intensify the diversity of owner interests that we have noted. Because the beneficiaries of public works do not pay for their benefits directly in the form of a price, there is a constant pressure on public officials to undertake projects and activities that are decidedly uneconomical from any commonsense viewpoint. This pressure for an unreasonably high standard of service is much less likely to exist from the customers of a private enterprise. On the other hand, because there are nearly always a number of taxpayers who do not receive benefits that are in excess of their payments for particular public works projects, there is frequently determined opposition by influential taxpaying groups to projects that are sound and economical.

It sometimes is stated that an economy study with reference to proposed public works makes it possible to examine alternatives from the viewpoint of the silent majority rather than from the viewpoints of various vocal minorities.

A Classification of Consequences to Whomsoever They May Accrue

The famous passage from the United States Flood Control Act of 1936, which we quoted at the start of Chapter 9, used the words "if the benefits to whomsoever they may accrue are in excess of the estimated costs. . . ."

A reasonable interpretation of what Congress presumably had in mind in 1936 is that the favorable consequences deemed to be relevant ought to be greater than the unfavorable consequences deemed to be relevant. The classification of possible consequences that we shall adopt for purposes of our discussion in this chapter is as follows:

1. Favorable consequences to the general public apart from consequences to any governmental organization. We illustrated such consequences in Chapter 9 by savings in costs to highway users and by reductions in expected damages due to floods.

2. Unfavorable consequences to the general public apart from consequences to any governmental organization. We illustrated such consequences in Example 9–2 by damages to anadromous fisheries and by a loss of land for agricultural purposes.

3. Consequences to governmental organizations. Usually consequences of this type can be estimated as net disbursements by governments over a period of years. And usually, such disbursements must ultimately be financed by taxation. Thus, once removed, these may be viewed as unfavorable consequences to the taxpayers, i.e., to the general public.

In our examples in Chapters 9 and 13, we described the first two of these categories as *benefits* (which we summarized algebraically in Example 9–2 when we viewed category 2 as "disbenefits"). The consequences to governments we described as *costs*. But we noted in Chapter 9 that our classification there was entirely arbitrary and that some analysts classify certain adverse consequences to the general public as costs, whereas other analysts classify certain maintenance costs paid by government as disbenefits. We also noted that the excess of favorable over unfavorable consequences (i.e., **B** − **C**) was not changed by decisions on classification, whereas the ratio **B/C** sometimes was greatly influenced by arbitrary decisions on classification.

Estimating "Costs" for Purposes of Comparing Alternatives in Public Works

Within the foregoing classification of possible consequences, our third category is the simplest to identify conceptually and the easiest to evaluate practically. If it were necessary only to estimate consequences to governmental organizations, we could eliminate many of the troublesome and controversial aspects of making economy studies for public works. Nevertheless, certain controversial matters would remain.

One of these matters is the selection of the i^* to be used. This is a basic and critical question even though (as will be illustrated in Example 19–1) the alternatives involve no differences to the general public except those in connection with governmental disbursements.

Another matter is the consideration, if any, to be given to taxes forgone. Although this question does not arise as an issue in most governmental economy studies, it may be critical in certain types of studies.

The Controversial Question of the Treatment of Interest in Economy Studies for Public Works

Engineers have not always agreed on the point of view that should be taken toward the treatment of interest in judging the soundness of proposed public works expenditures. Some different viewpoints on this subject have been as follows:

1. Costs should, in effect, be computed at zero interest rate. The advocates of this viewpoint have generally restricted its application to those public works that were financed out of current taxation rather than by borrowing.
2. Costs should be computed using an interest rate equal to the rate paid on borrowings by the particular unit of government in question. If the proposed public works are to be financed by borrowing, the probable cost of the borrowed money should be used. Otherwise the average cost of money for long-term borrowings should be used.
3. Just as in private enterprise, the question of the interest rate to be used in an economy study is essentially the question of what is a minimum attractive rate of return under the circumstances. Although the cost of borrowed money is one appropriate element in determining the minimum attractive rate of return, it is not the sole element to be considered. In most instances the appropriate minimum attractive rate of return should be somewhat higher than the cost of borrowed money.

Our discussions in Chapters 9 and 10 made it clear that the authors of this book favor the view stated under heading 3. Some further aspects of the case supporting this view are developed following Example 19–1.

EXAMPLE 19–1

The Effect of the Selection of the Minimum Attractive Rate of Return on a Comparison of Highway Bridge Types

Facts of the Case

In a certain location near the Pacific Ocean, two alternative types of highway bridge are under consideration for the replacement of an existing timber trestle bridge on a state highway in a rural area. The first cost of a steel bridge will be $850,000; the first cost of a concrete arch bridge will be $930,000. Maintenance costs for the steel bridge consist chiefly of painting; the average annual figure is estimated to be $5,000. Maintenance costs on the concrete arch bridge are assumed to be negligible over the life of the bridge. Either bridge has an estimated life of 50 years. The two bridges have no differences in their prospective services to the highway users.

It is evident that in this instance the choice between the two types depends on the assumed interest rate or minimum attractive rate of return. A tabulation of annual costs with various interest rates is as follows:

	Annual cost		*Difference in annual cost*	
Interest rate	*Steel*	*Concrete*	*Favoring steel*	*Favoring concrete*
0%	$22,000	$18,600		$3,400
2%	32,047	29,593		2,454
4%	44,568	43,292		1,276
5%	51,563	50,945		618
6%	58,924	58,999	$75	
8%	74,479	76,018	1,539	
10%	90,731	93,800	3,069	
12%	107,357	111,991	4,634	

If i^* is below 5.9%, the concrete bridge is more economical for this location. If above 5.9%, the steel bridge is more economical.

The Need for Some Minimum Attractive Rate of Return in Economy Studies for Public Works

Examples 9–1 and 19–1 both dealt with economy studies for state highway projects. In general, such projects in the United States have been financed chiefly by current highway-user taxes and have involved little or no public borrowing. This is the field in which the advocates of the 0% interest rate in public works projects were most articulate. It is also a field in which the funds available in any year have been limited by current tax collections, and in which there often have been many desirable projects that could not be constructed because of the limitation on current funds.

Example 19–1 represents a type of decision that usually is made on the level of engineering design rather than on the policy level of determining the order of priority of projects competing for funds. If each authorized project is to be designed to best advantage, it is essential that economy studies be made to compare the various alternative features in the design. If such studies were made at 0% interest, and if the conclusions of the studies were accepted in determining the design, many extra investments would be made that would yield relatively small returns (such as 1% or 2%). These extra investments in the projects actually undertaken would absorb funds that might otherwise have been used for additional highway projects. If the additional projects put off by a shortage of funds should be ones where the benefits to highway users represented a return of, say, 15% on the highway investment, it is clearly not in the overall interest of highway users to have invested funds earning a return of only 2%. In other words, where available funds are limited, the selection of an appropriate minimum attractive rate of return calls for consideration of the prospective returns obtainable from alternative investments. This is as sound a principle in public works as it is in private enterprise.

If the time should ever be reached when economy studies indicate that all the highway funds currently available cannot be used without undertaking a number of highway investments yielding very low returns (such as 2%), a fair conclusion would be that highway-user taxes should be lowered. In such a case the alternative investments would be those that might be made by individual taxpayers if taxes should be reduced. Money has a time value to the taxpayers; this is a fact that should be recognized in the use of funds collected from taxpayers.

Opportunity Cost Versus Cost of Borrowed Money in Selecting i^* for Economy Studies for Public Works*

Chapter 9 emphasized the point that the question of what ought to be the minimum attractive rate of return, all things considered, is still present when decision making is based on comparisons of annual costs, comparisons of present worths, or an analysis of benefits and costs. These methods all require the use of an interest rate for conversion of nonuniform money series to equivalent uniform annual figures or to present worths. The operational effect of using a particular interest rate in calculations of annual costs, present worths, or benefits and costs, is to adopt that interest rate as the minimum attractive rate of return. Moreover, the foregoing statement is true regardless of what the interest rate may be called; various names such as discount rate, vestcharge, or imputed interest rate sometimes are used.

Historically, government agencies in the United States that have made economy studies to evaluate proposed public works have inclined toward the selection of interest rates based on costs of borrowed money. But whenever i^* is chosen to reflect the costs of borrowing by a particular unit of government, the question arises whether the rate ought to represent an average interest cost on long-term debt already outstanding (sometimes called the *imbedded cost* of borrowed money) or the prospective cost of new borrowings.† Under the rules stipulated in the early 1950s by the Bureau of the Budget for the economic evaluation of federal water projects in the United States, the interest rate was based on the average rate that was being paid on long-term bonds of the United States. Under these rules, rates in the range from 2½% to 3¼% were used for a number of years when outstanding long-term U.S. bonds were selling to yield 4½% to 5% and *no* new long-term federal borrowing was

*For an expansion of the ideas presented here followed by a discussion that reflects a variety of opinions held by government engineers and economists, see "Interest and the Rate of Return on Investments" by E. L. Grant in *Highway Research Board Special Report 56, Economic Analysis in Highway Programming, Location and Design*, pp. 82–90. Special Report 56 contains the proceedings of a "workshop conference" held in September 1959 and is published by the Transportation Research Board, Washington, D.C. (The name of the Highway Research Board was changed to Transportation Research Board in 1974.)

†Over the years, there seems to have been a tendency always to select the *lower* of these two figures. When new public borrowings were at very low interest rates, these low rates were used even though outstanding debt had a considerably higher interest cost. When the cost of new borrowing rose above the cost of outstanding debt, the imbedded cost of debt was used.

possible because of a legal limitation that placed an interest ceiling of 4¼% on debt maturing in more than 5 years. In the 1960s and 1970s, the stipulated interest rates for benefit–cost studies on proposed new federal water projects were raised by degrees. As mentioned in Chapter 14, in 1981 inflation caused very high yields (nearly 14%) for certain long-term bonds of the United States government that were sold under competitive bidding.

Two analyses made in 1962 gave a good picture of the practices of highway agencies in the United States at that time with respect to the choice of interest rate in economy studies.* Under the chairmanship of Evan Gardner, a subcommittee of the Committee on Highway Engineering Economy of the Highway Research Board used a mailed questionnaire to survey the practices of the highway agencies of the 50 states plus the District of Columbia and Puerto Rico with respect to various aspects of their use of engineering economic analysis. Replies were received from 50 of the 52 agencies. One of the questions dealt with the interest rate used in economy studies by the various organizational units in each agency. The answers to this question showed that different organizational units in any one agency often used quite different interest rates. On the average, 45% of the organizational units that made economy studies used an interest rate of 0%, 22% used rates from 2% to 3¾%, and 33% used rates from 4% to 7%.

In the same year, Charles Dale, then Research Engineer of the U.S. Bureau of Public Roads, analyzed 130 economy studies prepared by state highway departments or their consultants. Projects examined in these studies included (1) alternate highway locations, (2) alternate river crossing schemes, (3) grade-separation studies, and (4) surface-type determinations. Of those reports that stated the interest rate used in the analysis, 20% used 0%, 22% used rates from 0.1% to 3.9%, 45% used rates from 4% to 5.9%, and 13% used rates from 6% to 7%.

The diverse practices of federal government agencies in the United States in 1968 are described and criticized in published testimony before a congressional committee.† As might have been expected, the hearings brought out considerable difference of opinion regarding the appropriate concept that should govern the selection of an interest rate for benefit–cost studies. However, Otto Eckstein and Arnold C. Harberger, the two distinguished economists who testified, favored the concept of opportunity cost; under 1968 conditions, this concept led them to suggest, respectively, rates of about 8% and about 12%. More recent studies by the Federal Highway Administration

*See the paper by David Glancy entitled "Utilization of Economic Analysis by State Highway Departments" included in *Highway Research Record Number 77, Engineering Economy 1963*, pp. 121–32. This is publication 1262 of the Highway Research Board, Washington, D.C. The results of the analysis by Charles Dale were reported in C. H. Oglesby's discussion of the Glancy paper.

†*Economic Analysis of Public Investment Decisions: Interest Rate Policy and Discounting Analysis,* Hearings before the Subcommittee on Economy in Government of the Joint Economic Committee, Congress of the United States, Ninetieth Congress, Second Session (Washington, D.C.: Government Printing Office, 1968).

have used a range of interest rates, frquently 5%, 10%, and 15%. In this way, the sensitivity of a decision to the choice of an interest rate is made clear.

In the opinion of the authors of this book, there are a number of reasons why the i^* used in economy studies for government projects usually ought to be greater than the bare cost of borrowed money. These reasons may be summarized as follows:

1. The opportunity cost within the particular government agency may be quite high in the sense that there are many good investment projects that the agency cannot finance because of the limited amount of funds available to it.* The acceptance of a proposed investment that has a low prospective rate of return will inevitably cause the rejection or postponement of some other proposed investment with a much higher rate of return.

2. The opportunity cost outside the government agency may be high in relation to the cost of borrowed money to the government. This may be because of desirable investment opportunities that are being foregone either by other government agencies or in the private sector of the economy.†

3. Whenever an analyst attempts the difficult task of placing money valuations on consequences to whomsoever they may accrue, there are obvious risks that the estimates will turn out to be incorrect. If risk has not been allowed for in estimates of benefits and costs, risk can be allowed for by increasing i^* just as decision makers often do in private enterprise.

4. When an economic analysis of proposed public works is made on the basis of relevant consequences to whomsoever they may accrue, it often is evident that such consequences are distributed quite unevenly among the population. Certain persons may be affected extremely favorably. Other persons will not be affected at all. Still other persons will be adversely affected. This uneven distribution of consequences of certain public works is a reason why such works should not be deemed to be justified unless their prospective rate of return is appreciably higher than the bare cost of money to the government.

5. Interest rates at which governmental units can borrow money do not always fully reflect the adverse consequences of borrowing. For example,

*A noteworthy illustration of this point was given in a Highway Research Board paper entitled "Sufficiency Rating by Investment Opportunity," by E. H. Gardner and J. B. Chiles. This paper described a computerized method for determining the order of priority over a period of years for proposed highway improvements over an entire state highway system. For the particular state in question, the computer derived a minimum attractive rate of return of 20%. This high rate, in effect, was a cutoff rate similar to the 15% in our Table 10–1 and reflected the opportunity cost of capital within the state highway system considering the limitation of highway funds and the many productive projects competing for these limited funds.

†Consider, for instance, certain investment opportunities that are being forgone by the highway users who pay the taxes that finance highway improvements. For the many taxpayers who have to borrow money for one purpose or another, a gilt-edge risk-free investment is to borrow less money or to reduce the amount of an outstanding loan. For those taxpayers who borrow to finance homes, this risk-free investment will yield, say, from 8% to 13%. For those numerous taxpayers (all of them highway users) who borrow to finance automobiles, such a risk-free investment might yield from 12% to 24%.

increased federal borrowing may contribute to inflation. State and municipal borrowings in the United States have had a concealed subsidy because of the exemption of the interest on the debt from federal income taxes and from certain state income taxes. A large issue of general obligation bonds of a state may have the effect of increasing future interest rates to be paid by cities, school districts, and other civil subdivisions of the state.

It is easier to accept the general proposition that the value of i^* used in government economy studies ought to be greater than the bare cost of borrowed money, either imbedded or current, than to defend any single value for i^* against all possible challengers. One of the co-authors of this book has suggested a value of 7% for highway economy studies.* In the examples and problems in this book for which it has been necessary to stipulate i^* for economy studies for public works, we have generally used values in the range from 5% to 10%. However, higher and lower values of i^* have sometimes been introduced to illustrate sensitivity.

In a 1968 report of a United States Congressional subcommittee that had held extensive hearings on this matter, the general conclusions seem to be summarized in the titles of the final five chapters, as follows:†

III. The discounting procedure must be used if good public investment decisions are to be made.

IV. Current discounting practices in the Federal agencies are neither adequate nor consistent.

V. The appropriate interest rate concept is the opportunity cost of displaced private spending.

VI. The current risk-free interest rate which should be used for evaluating public investments is at least 5%.

VII. All Federal agencies should establish consistent and appropriate discounting procedures utilizing an appropriate base interest rate computed and published on a continuing basis.

Taxes Forgone as a "Cost" in Certain Types of Economic Comparisons for Governments

Taxes are levied to meet the costs of government. When a governmental decision cuts off some sources of taxes without a corresponding reduction in the cost of conducting government, presumably one of the long-run consequences of the decision will be higher tax rates than would otherwise have been required. Such a consequence ought to be recognized in an economy study that examines all relevant consequences "to whomsoever they may accrue."

*In *Highway Research Board Special Report 56, op. cit.*, p. 86.

†*Economic Analysis of Public Investment Decisions: Interest Rate Policy and Discounting Analysis, A* Report of the Subcommittee on Economy in Government in the Joint Economic Committee, Congress of the United States, together with Separate and Supplementary Views (Washington, D.C.: U.S. Government Printing Office, 1968).

The foregoing statement bears on any economy study to evaluate a proposal that a government produce goods or services that it would normally buy from private industry. Part of the purchase price of the goods or services acquired from private industry ordinarily will be returned to government in the form of taxes. These particular taxes will be denied to government if the decision is made in favor of government production of the goods or services. In principle, the comparison of the two alternatives should be somewhat as follows:

I. Cost of purchase from private industry:	
Purchase price of goods or services	A
Less: Portion of purchase price returned as taxes	B
Net cost to government	A − B
II. Cost of production by government	C

The foregoing indicates that (A − B) should be compared with C. However, the same difference between the alternatives will be obtained if the taxes forgone are viewed as one of the "costs" of production by government, and A is compared with (C + B).

This problem of considering taxes forgone does not arise in the usual economy studies relative to public works alternatives, either for project formulation or project evaluation.

Whose Point of View?

It is possible to consider the economy of a public works proposal from several viewpoints:

1. That of the particular governmental body (or governmental department) concerned
2. That of all the people of a particular area (such as a state, county, city, or special district)
3. That of all of the people in the country

It is necessary to have clearly in mind whose viewpoint is being taken before it is possible to proceed with such an economy study.

In many cases the first impulse of the engineer will be to take viewpoint 1, considering only the prospective receipts and disbursements by the governmental body—or, in some cases, merely the particular governmental department—concerned. This appears to be comparable to an economy study for a private corporation in which the relevant matters are the prospective receipts and disbursements of that corporation. It should be clear that this viewpoint is a sound one in public works economy studies only when the alternatives being compared provide identical services to the people whom the government is organized to serve. For instance, this viewpoint might be correct in the choice between centrifugal pumps and reciprocating pumps for a municipal water works if it appeared that the services received by the water users would be equally satisfactory with either; the differences between the alternatives would then merely be differences in costs to the city's water department.

But where there are differences in the service provided by two alternatives, it is necessary to recognize the broader viewpoint that what the government does is simply something done collectively by all the people. If the ideal of democratic government that it is the objective of government to "promote the general welfare" is to be followed, it is necessary to consider the probable effects of alternative governmental policies on all of the people, not merely on the income and expenditures of a particular governmental unit.

Ideally, perhaps, it should be viewpoint 3 rather than 2 that should be considered in the public works policies of cities, counties, and states. For example, in comparing alternative plans for sewage treatment and disposal, a city should give consideration to the differences in their effects on downstream communities that take their water supply from the stream into which the sewage is discharged. Practically, however, experience indicates that the public officials and people of a community look at matters from the standpoint of what they consider to be the self-interest of their own community. Usually, the most that can be hoped for in economy studies for local governmental units is viewpoint 2 rather than viewpoint 1. If the broader question of the effect of one community's action on other communities is to be considered, it must be by the governmental authority of an area that includes both. Thus, a state board of health may regulate the sewage treatment policies of individual communities from the standpoint of the interests of all of the people of a state.

Viewpoint 3 definitely seems to be the correct one in all federally financed public works. Nevertheless, because the direct effects of most works of this character (e.g., navigation, flood control, reclamation) seem to be concentrated in a particular locality, there may be difficulty in applying this broad viewpoint even here. All of the effects on the people of a nation of a particular public improvement may be hard to trace, and doubly hard to evaluate quantitatively, even though the prospective local effects are fairly clear.

Analyses that are unsound from the broad viewpoint of the best use of limited national or state resources are fairly common when a national or state government has a policy of paying a substantial part of the cost of certain types of projects for which the chief benefits are local. Although Example 19–2 describes an actual case (with certain data changed a bit to disguise the source), the authors have observed many cases where the same type of reasoning was used. Too often, local governments take the attitude that money to be obtained from a higher level of government will be costless.

EXAMPLE 19–2

An Unsound Analysis of a Public Works Proposal

Facts of the Case

A certain consulting firm was employed to make a benefit–cost analysis of a proposed county expressway project. The conclusion was that annual benefits would be $400,000 and annual costs would be $1,000,000 yielding a **B/C** ratio of 0.4. On this basis, the project did not appear to be justified.

However, the members of the firm had what seemed to them to be a bright idea to suggest to the county supervisors. They proposed that the supervisors make an effort to have this expressway incorporated into the interstate highway system. If this could be done, 90% of the cost would be paid by the federal government. They advised the supervisors that this would reduce the local annual costs to $100,000 and that the **B/C** ratio would then be $400,000 ÷ $100,000 = 4.0.

Some Aspects of Estimating Benefits of Public Works Expenditures

A number of the problems at the end of this chapter illustrate the monetary evaluation of benefits and disbenefits. The reader will observe that some types of data that are desirable for the economic planning of public works cannot be secured on the spur of the moment when the decision is made to consider the merits of some proposed expenditure. Systematic fact finding on a continuing basis is needed to permit valid economy studies to be made for many types of projects.

For example, one of the objectives in many highway investments is the improvement of highway safety. The influence of alternative proposals on highway accidents cannot be judged without a complete and accurate system of reporting highway accidents whenever they occur. Regular traffic counts carried out on a routine basis throughout a highway system are essential for this purpose as well as for other aspects of the economic planning of highways.

Another good example is the recognized need for systematic collection of information regarding flood damages. This information is difficult to secure with reasonable accuracy except immediately following a flood; memories of specific aspects of a flood become dimmed after a very few years.

Double-Counting in Economy Studies for Public Works

Without careful reasoning there is danger that the same benefits from a public works project may be measured in different ways and added together.

For example, the common approach to estimating the benefits of flood protection works lies in making estimates of the flood damages that they will eliminate. An alternative approach, possible only in those circumstances where an extreme flood has followed a long period without any floods, is to determine the reduction in property values that has taken place as a result of the flood. In this approach it is reasoned that property values before the flood were based on the assumption that no destructive flood would ever occur; property values shortly after the flood (assuming reconstruction has restored property to something like its preflood condition) discount the expected damages from future floods.

It should be clear that these two approaches are alternative ways of trying to measure the same thing. Because both are imperfect measures they will not agree; nevertheless, the reduction in property values represents a collective estimate of the present worth of the costs of future flood damages. However, in some studies of flood protection economy these two measures were added together to determine the prospective benefits of flood protection!

A Classification of Consequences of Highway Improvements

Some of the difficulties of placing money valuations on benefits and disbenefits to whomsoever they may accrue can be illustrated by a brief look at the field of highway economy.* One suggested classification of data for a highway economy study is as follows:†

I. *Expenditures of the highway agency*

A. Capital outlay, including the costs of rights-of-way, final design, construction, and field engineering

B. Annual expense for maintenance, operation, etc.

II. *Consequences to highway users*

A. Market consequences (i.e., those where the market provides a basis for money valuations). These include:
(a) Motor vehicle operating costs
(b) Time costs to commercial vehicles
(c) Direct costs of motor vehicle accidents (including the overhead costs of insurance that can be demonstrably influenced by highway improvements)

B. Extra-market consequences (i.e., those where the market does not provide a basis for money valuations)

1. Consequences where a basis may be found for a somewhat arbitrary assignment of money valuations. These include:
(a) Deaths and permanently disabling injuries from highway accidents
(b) Time saving to non-commercial vehicles
(c) Increase or decrease in the value of parks, recreational facilities, and cultural and historical areas where the principal gain or loss is to highway users

*For an excellent and comprehensive discussion of this subject, see Robley Winfrey's *Economic Analysis for Highways* (Scranton, Pa.: International Textbook Co., 1969).

†This is quoted from a paper by C. H. Oglesby and E. L. Grant, "Economic Analysis—The Fundamental Approach to Decisions in Highway Planning and Design," *Highway Research Board Proceedings 37th Annual Meeting* (Washington, D.C.: Highway Research Board, 1958), pp. 45–57.

2. Consequences to which (at least at present) money values cannot be assigned
   (a) Sightseeing and driving for pleasure

III. *Consequences to other than highway users*

A. Market consequences

(a) Costs or cost reductions to public services (i.e., public transit, police and fire departments, school bus operations, etc.)
(b) Damages or savings from increased or decreased hazards created by the improvement (e.g., flooding of property)
(c) Increases in land values or in the value of crops or natural resources (but not both) where areas are made more readily accessible
(d) Changes in the value of land and improvements or changes in business activity (but not both) where these changes can be clearly attributed to the highway improvements

B. Extra-market consequences

(a) Overall impact of motor vehicle use, highway expenditures, and the character and location of the highways themselves on the economic and social well-being
(b) Increase or decrease in the value of parks, recreational facilities, and cultural and historical areas where the principal gain or loss is to other than highway users

Some of the troublesome aspects of placing money values on a number of the foregoing types of consequences will be discussed following Example 19–3.

EXAMPLE 19–3

An Economy Study Comparing Alternate Highway Locations

Facts and Estimates

A certain rural highway is to be relocated. Two locations are under consideration.

Location A involves a distance of 8.6 miles with a first cost for right-of-way, grading and subsurface preparation, and structures of $1,850,000. The surface will cost $731,000 and will require renewal after 15 years. Annual maintenance cost is estimated as $2,000 per mile.

Location B involves a distance of 7.1 miles. Because of the much heavier grading required to shorten the distance, the first cost of right-of-way, grading and subsurface preparation, and structures will be $2,900,000. The surface will cost $610,000 and will require renewal after 15 years. Annual maintenance cost is estimated as $4,000 per mile; this is higher than the estimate for Location A because of deeper cuts and the consequent possibility of earth slides.

In comparing the economy of the two locations, an i^* of 7% is to be used. Considering the likelihood of obsolescence of location, the assumed life for either location is 30 years with zero salvage value.

The present average traffic is 2,120 vehicles per day, made up of 1,800 passenger cars, 215 light trucks (under 5 tons), and 105 heavy trucks. It is estimated that this volume of traffic will remain fairly constant throughout the 30-year analysis period. Average figures for the increment costs per mile of operation of these three types of vehicles are estimated to be 13.0¢, 25.2¢, and 82.5¢, respectively; these figures will be the same at both locations. Therefore the difference of 1.5 miles in distance between Locations A and B involves an annual difference in cost of vehicle operation of:

Passenger cars—1,800(365)(1.5)($0.13)	$128,115
Light trucks—215(365)(1.5)($0.252)	29,664
Heavy trucks—105(365)(1.5)($0.825)	47,427
Total annual difference	$205,206

Because of the shorter distance there will be a time saving to all traffic. The average value of this saving is estimated as 25¢/vehicle-minute for all commercial traffic. It is estimated that 20% of the passenger cars are commercial. Passenger cars and light trucks are estimated to travel on this road at an average speed of 40 miles/hour, and will thus save 2.5 minutes each if Location B is selected. Heavy trucks are estimated to travel at an average speed of 30 miles/hour and will thus save 3 minutes each with Location B. The total money value of the annual time saving by commercial vehicles is:

Passenger cars—1,800(0.20)(365)(2.25)($0.25)	$73,913
Light trucks—215(365)(2.25)($0.25)	44,142
Heavy trucks—105(365)(3.0)($0.25)	28,744
Total annual value	$146,799

It is estimated that costs influenced by grades, curvature, and stops will be approximately equal for the two locations and that there will be no difference in accident hazard.

Comparison of Alternatives

With an i^* of 7% and a 30-year life with zero terminal salvage values, annual highway costs at the two locations are:

Location A

CR—Right-of-way, grading, and structures = $1,850,000(A/P,7%,30) =	$149,092
CR—Surface = $731,000(A/P,7%,15) =	80,256
Maintenance cost = 8.6($2,000) =	17,200
Total	$246,548

Location B

CR—Right-of-way, grading, structures = $2,900,000(A/P,7%,30) =	$233,711
CR—Surface = $610,000(A/P,7%,15) =	66,972

Maintenance cost = 7.1($4,000) =	28,400
Total	$329,083

A comparison of the relevant annual costs of the two locations considering the total highway costs and the differences in highway-user costs is as follows:

	Location A	**Location B**
Highway costs	$246,548	$329,083
Extra user costs due to differences in vehicle operating costs	205,206	
Extra user costs due to difference in time	146,799	
Totals	$598,553	$329,083

If comparison by the incremental **B/C** ratio is desired, the annual benefits from the shorter location would be the savings in highway-user costs of $205,206 + $146,799 = $352,005. This figure is to be compared with the extra highway costs of Location B, namely, $329,083 − $246,548 = $82,535. The incremental **B/C** ratio therefore is $352,005 ÷ $82,535 = 4.26.

One element omitted from the preceding analysis is the saving in time by noncommercial traffic. If this time saving is evaluated at a figure of, say, 10¢/vehicle-minute, the resulting annual figure is 1,800(0.80)(365)(2.25)($0.10) = $118,260. By adding this to the benefits, the **B/C** ratio is increased from 4.26 to 5.70.

Comments on Example 19–3

In our examples that compared highway alternatives in earlier chapters (Examples 9–1 and 13–4), the prospective annual benefits due to savings in road user costs were given in total dollars without any indication as to how they were derived. Example 19–3 brings out the point that to find road user benefits it is necessary to estimate the amount and character of the motor vehicle traffic throughout the study period. To find relevant motor vehicle operating costs for alternative designs, it is necessary to estimate how much the incremental costs of operation will be affected by differences in design. To secure a monetary figure for the benefit of a prospective saving in time to road users, it is necessary to estimate average speeds for various classes of vehicles and to assign a value per unit of time saved by each class.

Certain common complications that were illustrated in the earlier highway examples were omitted from Example 19–3. Example 9–1 illustrated estimated growth of traffic during the initial years of the study period. Example 19–3 assumes constant traffic. Example 9–1 illustrated the use of estimated residual values at the end of the study period whereas Example 19–3 assumes zero terminal salvage values. Both Examples 9–1 and 13–4 illustrated the common problems arising in the analysis of more than two alternatives.

Consider the 13¢/mile used as the increment cost of operation for passenger cars. Presumably this is an estimated average figure that is deemed to apply to the mixture of many makes and models of cars that will use this highway. Presumably, also, this average cost of vehicle operation has been built up from an analysis of such components as fuel, lubricants, tires, repairs and maintenance, and depreciation. And presumably it is deemed to be applicable to the particular conditions of speed, surface, grade, curvature, and other conditions of operation applicable to the particular highway.

Obviously a fair amount of fact finding by someone is required to find an appropriate unit value to be placed on prospective time saving by commercial vehicles. The assignment of an average value of 25¢/vehicle-minute in Example 19–3 implies that such fact finding has taken place.

Although prospective time to be saved by pleasure traffic is an extra-market consequence of a proposed highway improvement, analysts often wish to place a money valuation on such a time saving. Usually the unit figure per vehicle-minute, such as the 10¢ in Example 19–3, is chosen quite arbitrarily. However, some researchers have studied the division of traffic between free roads and competing toll roads to obtain a figure for the value that pleasure motorists seem to be placing on their time.* Because many factors other than time saving can enter into motorists' choices in such cases, research studies of this type tend to be somewhat inconclusive.

If monetary values are assigned to extra-market benefits, it always is desirable to follow the policy illustrated in Example 19–3 of stating the relationship between benefits and costs both without and with such benefits included. When **B** − **C** or **B/C** is stated only *with* the extra-market benefits included, decision makers often are unaware of the values that were somewhat arbitrarily assigned to the extra-market benefits.

Where no monetary valuations are assigned to certain extra-market consequences (such as time savings to pleasure traffic), such consequences should be considered as part of the irreducible data of the economy study.

Irreducibles in Economy Studies for Public Works

Although measurements of benefits of public works in money terms make it possible to take many decisions out of the "hunch" class, such measurements have obvious limitations. These limitations lie not only in the difficulties of measurement, but also in the much greater importance of irreducibles in public works than in private enterprise.

*For instance, see P. J. Claffey, "Characteristics of Passenger Car Travel on Toll Roads and Comparable Free Roads," *Studies in Highway Engineering Economy, Highway Research Board Bulletin 306* (Washington, D.C.: Highway Research Board, 1961), pp. 1–22.

In many respects the irreducibles in public works projects create problems of judgment similar to those that arise in personal economy studies. The best that an individual can do in dealing with personal problems of economy may be to note the satisfactions that will come from particular expenditures, and to consider them in the light of their long-run costs and in the light of the individual's capacity to pay.

A similar analysis may be applied to such a public works project as one for park improvement. Even though the services provided by the park are not expressible in money terms, it is pertinent to estimate how many people will use the proposed facilities, and in what ways. It is also pertinent to estimate the long-run cost of the park improvement, considering not only the immediate investment (translated into annual cost in the conventional manner) but also the necessary annual expenditures for upkeep. The annual cost of the service of the proposed park improvement may even be expressed as so much money per unit of use in order to permit its comparison with other similar improvement proposals. Finally, any such proposed expenditure must always be considered in relation to the capacity of the community to pay for it, particularly in the light of other possible uses for available public funds.

Use of a Money-Based Index Where Major Consequences Are Not Reducible to Money Amounts

Assume that the expected favorable consequences from proposed governmental expenditures that are competing for limited funds can be stated in some units that are deemed appropriate in relation to the objectives of the expenditures. Assume also that there is no satisfactory way to convert these units into money units. Therefore, an analysis cannot be made to determine whether benefits exceed costs.

In such cases, it still is possible to secure a money-based index that is helpful in comparing alternatives that are intended to reach the same general type of objective. Such an index might be computed as follows:

$$\text{Cost Effectiveness Index} = \frac{\text{Units that somehow measure net favorable consequences}}{\text{Costs to government}}$$

If the outlays by government are for public works that will be used over a period of years, it is desirable in principle that the denominator of the index fraction be an equivalent uniform annual cost that includes operation and maintenance costs and uses an appropriate value of i^* in finding the capital costs. But in cases where the annual costs for all the alternatives that are being compared will be approximately the same percentage of first cost, it is good enough for practical purposes to use first cost as the denominator.

Using a Cost Effectiveness Index to Establish Priorities on Rural Road Construction in a "Developing" Country

Problems 19–11 and 19–12 illustrate two such indexes of cost effectiveness as used in Mexico and as described in a research paper by Henry M. Steiner.* One index applied to proposed "social roads" and the other applied to proposed "economic penetration roads."

Social roads are those that will reach isolated population groups where there are no natural resources that are likely to make possible substantial new production as a consequence of better transportation. Steiner comments about such roads in part as follows:

> A social road undoubtedly will bring about a decisive change in the way of life of the inhabitants of the affected region. It will facilitate the establishment of schools, both for the children and for adults (over one million Mexicans speak no other language than their own tribal dialect; thus some Mexicans need interpreters to speak to other Mexicans). Sanitary and welfare services will be established. There will be an opportunity to engage in commerce with the rest of the country, that is, a market economy will replace barter.

The priority criterion for proposed social roads is the ratio of investment cost to number of inhabitants served. In effect, this is an inverse index of cost effectiveness; the lower the value of the index, the higher the priority. The denominator of this index, inhabitants served, is a rough measure of the expected favorable consequences. The numerator, the first cost of the project to the government, is money-based.

Economic penetration roads also serve isolated population groups but in areas where it is believed that natural resources exist to make possible a considerable increase in economic activity, particularly in agriculture. In these areas, economic development has been held back because of the absence of satisfactory roads. However, it would not be possible to show an excess of benefits over costs for proposed roads classified as economic penetration roads.

The numerator of an index used to compare proposed economic penetration roads is the estimated gross value of the agricultural output from the area served in the fifth year of service of the road. The denominator is the esti-

*These problems are adapted from H. M. Steiner's *Criteria for Planning Rural Roads in a Developing Country: the Case of Mexico, Report EEP-17* (Stanford, Calif.: Program in Engineering-Economic Planning, Stanford University, 1965). At the time of the Steiner study, Mexico had 0.024 miles of rural road per square mile of area whereas the United States had 0.868 miles of rural road per square mile. The United States therefore had approximately 37 times as dense a rural road network as Mexico in spite of having almost the same population density, 50.6 persons/sq. mile in U.S. as compared to 46.2 in Mexico. Thus, where the rural road problem in the United States was how best to use limited resources in the *improvement* of existing roads, the comparable Mexican problem was how best to use limited resources to construct *new* rural roads.

mated first cost of the project to the government. This ratio is a direct index of cost effectiveness, so that projects with higher ratios have higher priorities.

A Limitation on the Use of Economic Measures to Guide Governmental Decisions

At the time of the Steiner research, proposals for improvement in the existing highway network in Mexico were analyzed by conventional benefit–cost comparisons. These were carried out along the lines of the better studies of this type made by highway agencies in the United States; the chief difference was that in Mexico a minimum attractive rate of return of 12% was used. Such benefit–cost studies could give decision makers guidance about the best use of the limited funds available to improve existing highways.

Similarly, the inverse cost effectiveness index chosen for social roads could be used to compare projects competing for the limited funds available to build such roads. And the direct cost effectiveness index adopted could be used to compare competing proposals for economic penetration roads.

Nevertheless, it is evident that none of these three economic measures could provide a basis for dividing the total highway funds among the three general categories of social roads, economic penetration roads, and improvements to the existing highway network. This division had to be made on the basis of other criteria, some of which would be hard to quantify.

We have mentioned earlier the importance of irreducibles in many government decisions and the similarity of a number of resource allocation decisions in government to resource allocation decisions in a family. The decision in Mexico to allocate certain highway funds to social roads and economic penetration roads is similar to many other decisions by governments and by families that are not made primarily on economic grounds. But once such overall decisions on resource allocation are made, it is a desirable objective to make the best possible use of the resources to be devoted to each major purpose. The use of some measure of cost effectiveness to compare competing proposals for the use of the limited resources allotted to each major purpose can help in achieving this desirable objective.

Who Gets the Benefits and Who is Responsible for the Costs?

As has been suggested, certain public works may be a benefit to some people, a matter of indifference to others, and possibly a detriment to others. This raises the question not only of what are the benefits, but who gets them.

The effort is frequently made in public works to allocate taxes or other charges according to a price principle that recognizes benefits and responsibility for costs. This is likely to involve somewhat arbitrary allocations of joint benefits and joint costs; in some instances, however, it is necessary to recognize increment costs in such studies.

Consider, for example, the case of a trunk-line sewer that, in addition to collecting the sewage of the buildings on its own street, also carries the sewage from a large tributary area. Here, the only charge that may be made legitimately in the form of a special assessment against the abutting property is a charge sufficient to build a sewer of the size necessary to serve that property. The difference between the actual cost of the trunk sewer and the estimated cost of a sewer adequate to serve the local needs should be financed in some other way, perhaps by considering it as a general benefit to the entire community and paying for it out of general taxation, or perhaps by considering it as a special benefit to the entire area tributary to the trunk sewer and distributing it over the property in that area in the form of a special assessment.

Some General Aspects of Cost Allocations

Because cost allocation problems arise so often in connection with public works, it is desirable to make certain general observations about all kinds of cost allocations, as follows:

1. A proposal that certain costs should be "allocated" suggests that there is no unique "correct" method of dividing the costs among the various purposes involved. This fundamentally indeterminate nature of most cost allocations tends to make many such allocations extremely controversial. (However, it is rare for persons who make cost allocations to concede—at least, in public—that they believe their allocation problem has no correct answer.)

2. In spite of the somewhat arbitrary aspect of many cost allocations, the allocations can often be extremely important because of their consequences. Therefore, in making a cost allocation or in judging the merits of cost allocations made by someone else, it is desirable to consider what types of decision (if any) are likely to be influenced by the particular allocation. Here it may be necessary to look beyond the stated purpose of a particular allocation. (In public utilities and public works, the stated purpose may be to make or to influence—or possibly to justify—some sort of pricing decision.) In this connection, one also should look at the possible consequences of making no allocations at all.

3. An important matter in any cost allocation is the choice of the "purposes" among which costs are to be allocated. This choice needs to be related to the intended uses of the allocation. In some cases, it may be desirable to have two or more classifications of "purposes" among which certain costs are to be allocated; for example, a manufacturer with a variety of products might allocate distribution costs among products, among geographical areas, and among different marketing channels.

4. Wherever it is proposed to make any use of a cost allocation, there are usually different groups of persons with conflicting interests in the results of the allocations. This is particularly true where an allocation influences budgets, pricing decisions, taxes, or the evaluation of the performance of persons or organizations.

5. The word "cost" is capable of many different definitions. It, therefore, is important to start any allocation with a set of rules for establishing the

"costs" that are to be allocated. In considering the interests of various persons and organizations in certain allocations, the choice of a definition of cost often is of greater importance than the choice of a method to make the particular allocation.

6. A first step in any cost allocation ought to be to identify all costs that are separable (or incremental) among the various chosen purposes. By so doing, the size of the job of allocation is reduced by limiting it only to the fraction of the total cost that is residual, that is nonseparable. Nevertheless, the identification of separable costs is not necessarily a straightforward and noncontroversial matter. (Consider, for example, the troublesome problem of allocating highway costs among different classes of motor vehicles.) In judging the merits of any particular allocation, existing or proposed, one matter to examine is the method used for identifying the separable costs.

7. One cannot judge the merits of any plan for allocating nonseparable costs without some criterion—or, preferably, some set of criteria—for judging merit. (The rather awkward word *nonseparable* is used here because of special meanings that have been attached to *joint costs, common costs,* and *residual costs*.)

8. The concept of sensitivity, so useful in engineering economy, also is useful in any study of cost allocation. It is of interest to note the sensitivity of the results of various cost allocations to the definition of cost, to the method used to identify separable costs, and to the method used for allocating nonseparable costs.

9. Because, as stressed throughout this book and as emphasized particularly in Chapter 16, the analysis of problems in engineering economy calls for examining *differences* between alternatives, cost allocation is not a useful tool for solving problems in engineering economy.

An important field of cost allocation of public works has been in connection with multiple-purpose water projects. The importance has arisen out of the common practice of having the costs allocated to certain purposes repaid by the beneficiaries and having the costs allocated to other purposes paid for out of general taxation. Example 19–4 illustrates a method of making such allocations that has been favored by many writers and government agencies.

EXAMPLE 19–4

An Illustration of Cost Allocation for a Multipurpose Water Resource Project

The Separable Costs–Remaining Benefits Method of Cost Allocation

Table 19–1 is reproduced from a United Nations manual on water resource projects.* It illustrates the use of the separable costs–remaining benefits

*Division of Water Resources Development of Economic Commission for Asia and the Far East, *Manual of Standards and Criteria for Planning Water Resource Projects*, Water Resources Series No. 26, United Nations Publications Sales Number: 64, II. F, 12 (New York: United Nations, 1964), p. 53.

TABLE 19–1
Summary of cost allocation by separable costs–remaining benefits method (Unit: thousands of dollars)

Item	*Flood control*	*Irrigation*	*Power*	*Domestic and industrial water*	*Fisheries*	*Navigation*	*Totals*
1. Costs to be allocated							46,853
a. Construction costs							(39,500)
b. Op., maint., and repl. costs (capitalized)							(7,353)
2. Benefits (capitalized)	10,975	54,875	12,622	16,462	1,975	2,195	99,104
3. Alternative costs	20,000	28,800	12,622	15,600	—	—	—
4. Justifiable expenditure	10,975	28,800	12,622	15,600	1,975	2,195	72,167
5. Separable costs	8,110	9,434	9,992	1,929	—	—	29,465
a. Construction costs	(8,000)	(6,800)	(6,700)	(1,600)	—	—	(23,100)
b. Op., maint., repl. costs (capitalized)	(110)	(2,634)	(3,292)	(329)	—	—	(6,365)
6. Remaining justifiable expenditure	2,865	19,366	2,630	13,671	1,975	2,195	42,702
7. Percent distribution	6.7	45.4	6.2	32.0	4.6	5.1	100.0
8. Remaining joint costs	1,165	7,894	1,078	5,564	800	887	17,388
a. Construction costs	(1,099)	(7,446)	(1,017)	(5,248)	(754)	(838)	(16,400)
b. Op., maint., repl. costs (capitalized)	(66)	(448)	(61)	(316)	(46)	(51)	(988)
9. Total allocated cost	9,275	17,328	11,070	7,493	800	887	46,853
a. Construction cost	(9,099)	(14,246)	(7,717)	(6,848)	(754)	(836)	(39,500)
b. Op., maint., repl. costs (capitalized)	(176)	(3,082)	(3,353)	(645)	(46)	(51)	(7,353)
10. Annual operation, maintenance and replacement costs	8	141	153	29	2	2	335

method of allocating costs as applied to a project that serves six purposes, namely, flood control, irrigation, hydro-power, domestic and industrial water supply, fisheries, and navigation. The project includes a dam and reservoir, irrigation canals and distribution works, a power plant and transmission facilities, and a domestic-industrial pipeline. The total first cost is $39,500,000. For purposes of the cost allocation, the project life is assumed to be 100 years with no terminal salvage value. The capitalized figures in Table 19–1 (in lines 1b, 2, 5b, 8b, and 9b) are present worth of estimated annual figures for a 100-year period computed with an i of 4½%.

This method of cost allocation requires estimates of costs for single-purpose alternates that would produce equivalent single-purpose benefits. Such estimates, shown in line 3, involved a single purpose reservoir for flood control, irrigation, and domestic and industrial water. For power, the alternate was a thermal plant with the same capability and production as the project hydro plant. Because the power benefits from the water project were measured by the cost of power from the most economical alternate source (thermal power), lines 2 and 3 under "Power" contain identical figures. No alternates were deemed possible for fisheries or for navigation.

The justifiable expenditure given in line 4 for each purpose is the lesser of the figures in lines 2 and 3. That is, the justifiable expenditure clearly should not exceed the benefits; neither should it exceed the costs of a single-purpose project that will have the same consequences.

To obtain the separable costs given in line 4 for each purpose, it is necessary to estimate the construction costs and the annual operation, maintenance, and replacement costs for a multipurpose project from which the particular purpose is excluded. The excess of the total project costs over the costs with the purpose excluded gives the separable costs. For instance, if flood control were omitted, the construction costs of the project would be $31,500,000 as contrasted with $39,500,000 with flood control included; the difference of $8,000,000 is the separable construction cost of flood control. Without flood control, the annual cost for operation, etc., would be $330,000 as contrasted with $335,000 with flood control included; the separable annual cost of flood control for operation, etc., is $5,000.

The separable costs do not need to be allocated; the allocation procedure applies only to the excess of total costs over the sum of the separable costs. Thus, since total construction cost is $39,500,000 and the sum of separable costs of construction is $23,100,000, only $16,400,000 of construction cost needs to be allocated.

The annual separable operation, maintenance, and replacements costs are $5,000 for flood control; $120,000 for irrigation, $150,000 for power; and $15,000 for domestic and industrial water. As the sum of these separable costs is $290,000, only $45,000 out of the annual total of $335,000 needs to be allocated.

The allocation of nonseparable costs is in proportion to the ratio of the remaining justifiable expenditure (line 6) for each purpose to the sum of the remaining justifiable expenditures for all purposes. For example, for flood control $2,865,000 ÷ $42,702,000 = 0.067 or 6.7%.

Special Problems of Public Borrowing

When cities, counties, or special improvement districts (such as water districts, sanitary districts, irrigation and drainage districts and bridge districts) wish to construct public works of any substantial magnitude, it is often necessary to finance such works by borrowing 100% of their cost. In many instances only minor works can be financed out of current taxes and revenues.

In order to prevent excessive borrowing and assure systematic repayment, the states of the United States, either through constitutional provisions or legislative enactments, have put various restrictions on borrowing by local governments. These restrictions are commonly of three types:

1. The total borrowing power of a local government is limited to a specified percentage of the assessed valuation of the property within its area.
2. No bonds may be issued without the approval of the voters at an election; with the exception of "revenue bonds" in certain jurisdictions, it often is specified that this approval must be by a two-thirds affirmative vote.
3. Repayment must be within a specified number of years, and must be in accordance with a specified plan. This plan often requires uniform serial maturities over the life of a bond issue.

In considering any proposal for local public works to be financed by bond issues, the effect of these legal restrictions must be considered. Will borrowing power, considering probable future needs, be impaired by the proposed issue? Are the chances good for a two-thirds favorable vote? Can required obligations for interest and repayment of principal be met, particularly in the earlier years when they may be the greatest?

The necessity for 100% borrowing to finance many local public works, and the requirement that the bonds be completely paid off within a limited period, usually 40 years or less, makes the financial background quite different from that found in private enterprise. Private enterprises seldom have the opportunity for 100% borrowing; on the other hand, they are not confronted with the necessity for retiring all of their capital obligations within a limited period.

Borrowings by states are generally subject to restrictions similar to those enforced on cities, counties, and special districts. However, state highway improvements, which are the major state-financed public works, are now generally financed on a pay-as-you-go plan through the proceeds of gasoline taxes and other forms of motor vehicle taxation. Thus, the problems of debt limit, bond elections, and debt repayment, which are so common in connection with municipal public works, arise less frequently in connection with state works.

The situation with respect to borrowing by the federal government is entirely different from that existing in other government units. Whereas each bond issue by a state, city, county, or special district is for a definitely specified purpose, borrowings by the federal government are for the general purpose of supplying funds to take care of the excess of current disbursements over current receipts.

The relationship between the financing of federal deficits, the banking and credit mechanism, and the general price structure, is much too complex for discussion here. Let it suffice to point out that federal borrowing (except that engaged in for purposes of debt refunding) serves to create purchasing power and is thus a stimulant to business activity that may be used for the purpose of promoting recovery from periods of business recession. Federal borrowing also, if carried on in large amounts, has the tendency to cause a great rise in price levels with all of the ills that the experience of the world has demonstrated to be attendant upon inflation. Both of these effects—one good, the other bad—are decidedly relevant in connection with any federal public works proposal made when the federal government is operating at or near a deficit. They are, however, a long way removed from matters of engineering technology.

Shadow Pricing in Economy Studies for Proposed Public Works

Sometimes it may be felt that market prices do not provide a suitable measure of the opportunity costs associated with certain elements of project input and output when such costs are looked at from the national viewpoint. If there is substantial unemployment of unskilled labor that would be likely to be used in a proposed project, or if a proposed project will make appreciable demands on limited resources of foreign exchange, it may be a good idea to make a supplementary economic analysis that substitutes "shadow prices" for market prices. For example, in one research study, W. W. Shaner assumed that opportunity costs were reflected by shadow prices for unskilled labor, skilled labor and domestic materials, and foreign exchange which were 50%, 100%, and 120% of their respective market values.* The need for such a supplementary analysis arises chiefly in developing countries.

A Proposed Technique for Improving Decision-Making Procedures Where Many Irreducibles are Present

In any economy study that involves irreducible data, a first step should be to identify those consequences that are to be given consideration as irreducibles in the final choice among the alternatives. By definition, what we have called "irreducibles" are expected consequences of a decision that it is not practica-

*W. W. Shaner, *Economic Evaluation of Investments in Agricultural Penetration Roads in Developing Countries: A Case Study of the Tingo Maria-Tocache Project in Peru, Report EEP-22* (Stanford, Calif.: Program in Engineering-Economic Planning, Stanford University, 1966).

ble to express in units of money for purposes of the particular economy study. Nevertheless, even though these matters are not to be quantified in monetary units, a desirable step is to quantify each irreducible in some manner; this involves finding an appropriate unit to measure the favorable or unfavorable consequences that are deemed to be relevant. Finally, as we pointed out in Chapter 13, whenever there are multiple alternatives and multiple irreducibles, it is helpful to compare alternatives in pairs.

Oglesby, Bishop, and Willeke have proposed a systematic procedure for dealing with multiple irreducibles in one particularly troublesome field of public works decision making.* The choice among several proposals for a location of an urban freeway often is a controversial matter. Different locations have different expected impacts on the community. Generally speaking, it is not practicable to express the various community impacts in money terms so that they can be counted as benefits or disbenefits in a benefit–cost analysis. A common complication is that different community impacts favor different locations, and that the locations that look best if only irreducibles are considered are not the same locations that look best if only estimated benefits and costs are considered.

These writers list many different possible types of community impacts and suggest one or more different possible units that might be applied to each type. For example, under the general category *Neighborhood Impact*, one community impact is *family units displaced*; this unfavorable consequence can be measured in numbers of living units. Under the general category *Community Planning*, one community impact is *developable land to which freeway provides excellent access*; this favorable consequence could be measured in acres.

Oglesby, Bishop, and Willeke make the following pertinent remarks about each measured community consequence:

> The time period over which the consequences of the various factors are evaluated is also important. Otherwise short run consequences might be given more weight in the decision as compared to the long run effects, or vice versa. An example might be the community concern that elderly people would be displaced from their homes in a given area. At the same time, the community master plan may indicate that the area is suitable for high-density apartments and a survey show that the transition is already under way. In this instance, an appreciation of the time factor is extremely important to a rational appraisal of the possible alternatives.

They suggest plotting the various quantified irreducibles for all of the alternative proposed freeway locations on a "community factor profile," which they illustrate. (See Figure 19–1.) They describe this "profile" in part as follows:

*C. H. Oglesby, A. B. Bishop, and G. E. Willeke, "A Method for Decisions Among Freeway Location Alternatives Based on User and Community Consequences," *Highway Research Record No. 305* (Washington, D.C.: Highway Research Board, 1970), pp. 1–14.

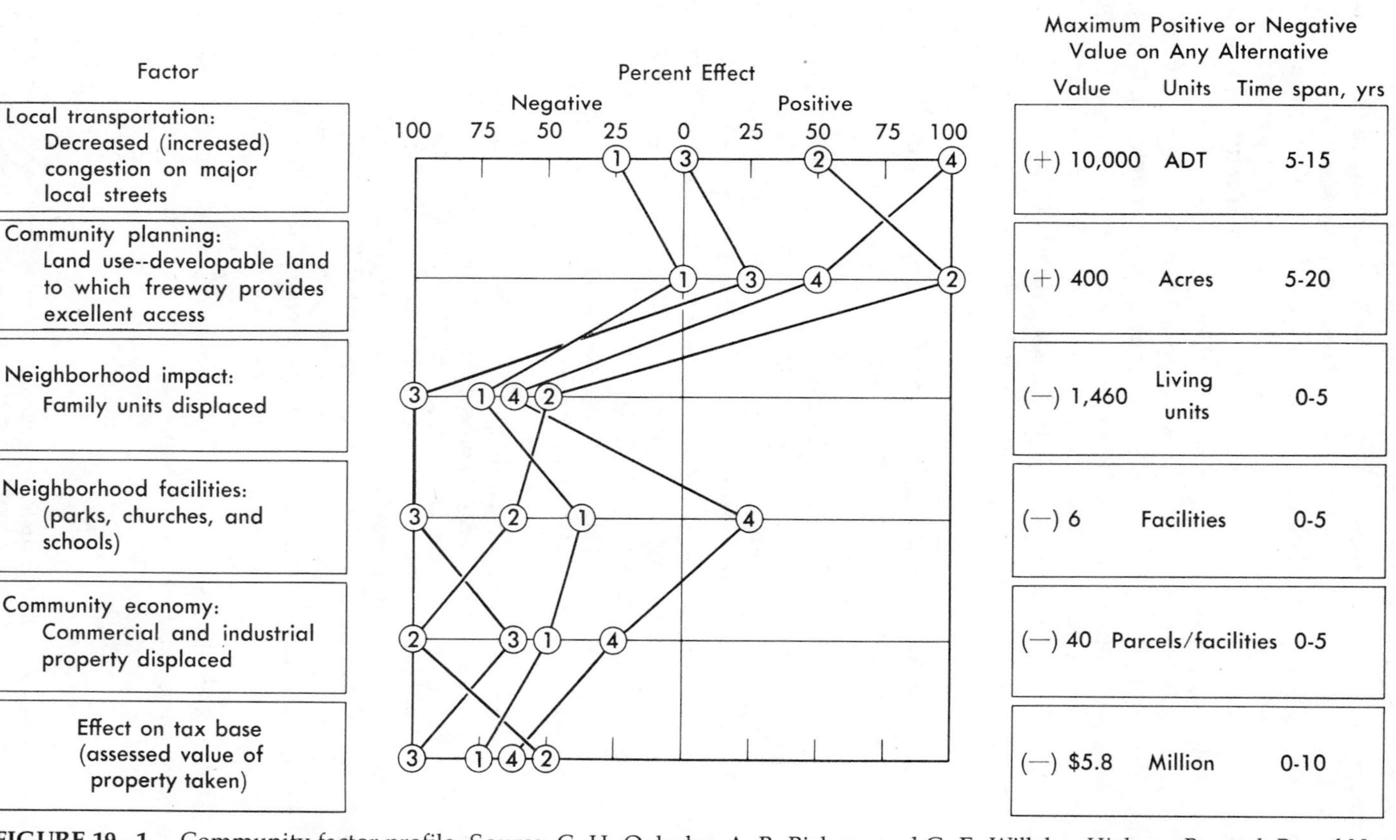

FIGURE 19–1. Community factor profile. Source: C. H. Oglesby, A. B. Bishop, and G. E. Willeke, *Highway Research Record No. 305*, 1970

> The community factor profile is a graphical description . . . of the effects of each proposed freeway location alternative. . . . On this figure, each profile scale is on a percentage base, ranging from a negative to a positive 100%. One hundred either negative or positive is the maximum absolute value of the measure that is adopted for each factor. Reduction to the percentage base simplifies scaling and plotting the profiles. The maximum positive or negative value of the measure, the units, and the time span are indicated on the righthand side of the profile for reference. For each alternative, the positive or negative value for any factor is calculated as a percent of the maximum absolute value over all alternatives, and is plotted on the appropriate abscissa.

An important aspect of the system of analysis proposed by these writers is the use of paired comparisons of multiple alternatives, somewhat along the lines that we mentioned in Chapter 13. Figure 19–2 shows their flow diagram for a systematic procedure that employs such paired comparisons. They comment on this point in part as follows:

> A highly simplified example to illustrate the paired comparison approach is given by the question: "Is it preferable to save $50,000 per year to local residents in vehicle operating costs by adopting a shorter route or to retain a commercial enterprise employing ten people and paying $20,000 per year in property taxes? It is estimated that a substitute enterprise will develop in five years." It is admitted that this example is far simpler than those of the real world where the factor profile would include several elements. Even so, such comparisons make clear the actual points at issue and may greatly reduce the number of irrational arguments that accompany most controversial decisions.

Environmental Impact of Proposed Public Works

As pointed out near the start of this chapter, there is often a diversity of citizen viewpoints regarding a proposed government project. Generally speaking, although not always, these differences in viewpoint reflect differences in the economic interests of different groups of citizens.

The "environmental impact statement" required for many proposed federal public works projects in the United States provides one approach to the review of the diverse consequences of certain public works. M. L. Manheim, et al., comment on this matter in part as follows:*

> The evaluation method is designed to produce periodic reports documenting the planning process. These reports can serve as the basis for the environmental impact statement (EIS) called for by Section 102(2)(C) of the National Environ-

*M. L. Manheim, J. H. Suhrbier, E. D. Bennett, L. A. Neumann, F. C. Colcord, Jr., and A. T. Reno, Jr., "Transportation Decision Making—A Guide to Social and Environmental Considerations," *National Cooperative Highway Research Program Report 156* (Washington, D.C.: Transportation Research Board, 1975), p. 123.

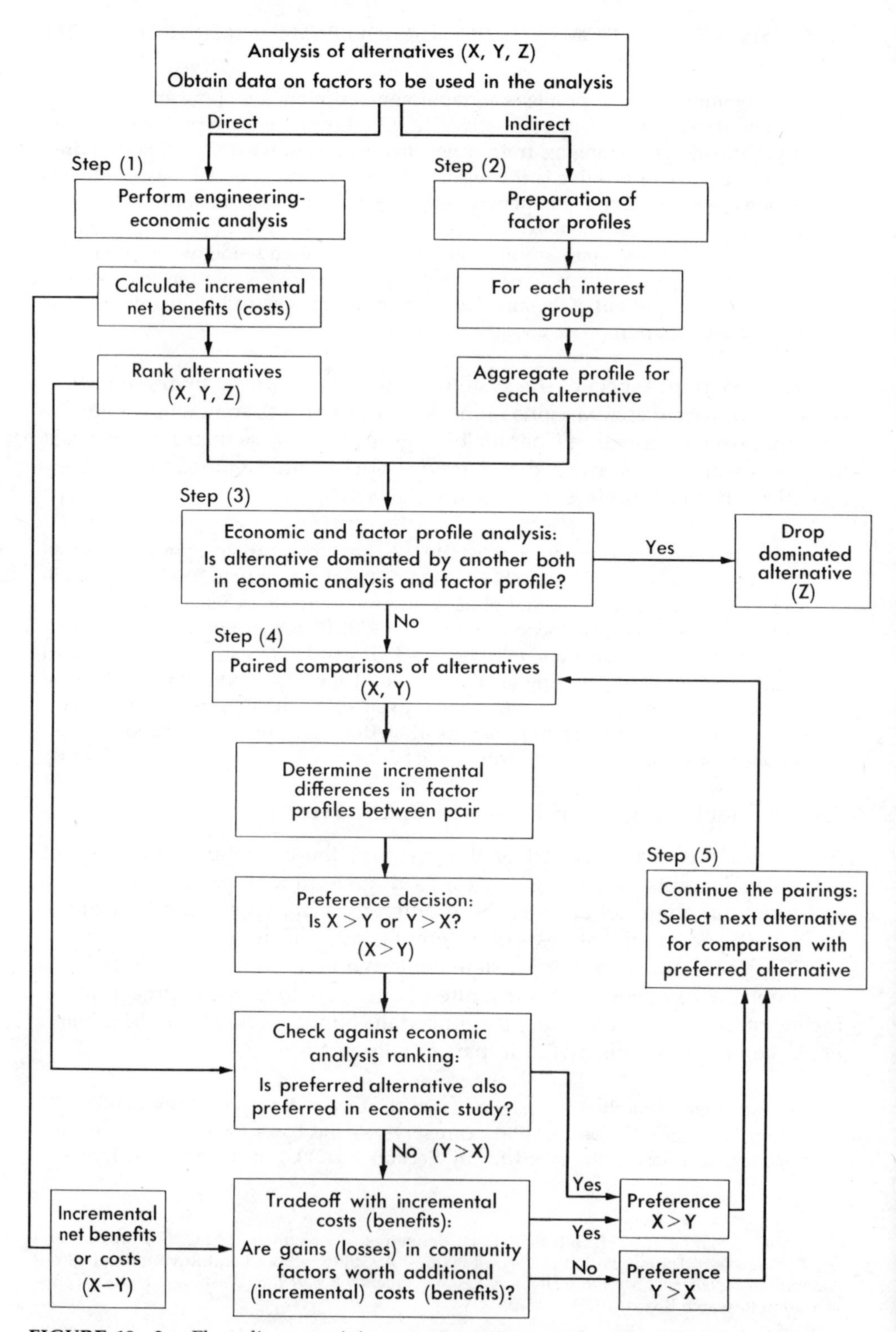

FIGURE 19–2. Flow diagram giving a systematic procedure for evaluating alternative freeway routes. Source: C. H. Oglesby, A. B. Bishop, and G. E. Willeke, *Highway Research Record No. 305*, 1970

mental Policy Act of 1969, which requires that before decisions are made on major federal actions, a statement be circulated describing

(i) The environmental impact of the proposed action,

(ii) Any adverse environmental effects which cannot be avoided should the proposal be implemented,

(iii) Alternatives to the proposed action,

(iv) The relationship between local short-term uses of man's environment and the maintenance and enhancement of long-term productivity, and

(v) Any irreversible and irretrievable commitments of resources which would be involved in the proposed action should it be implemented.

The environmental impact statement must satisfy two conflicting but essential considerations. It must serve as a public disclosure mechanism, making information on the proposed project readily available to interested parties; and it must demonstrate that the responsible agency has performed careful investigations of the potential consequences of its proposed actions. To inform people, an impact statement should be brief and readable, and people should not have to wade through hundreds of pages to extract the information of interest to them. Yet for many projects demonstrating that a thorough analysis has been carried out may require a massive statement.

The way out of this dilemma may be to have one part of an environmental impact statement provide a cogent summary with other sections filling in details.

Diverse Impacts of Some Public Regulations

Examination of expected consequences from a variety of viewpoints is desirable not only for proposals for public works but also for certain types of proposals for the public regulation of various actions by individuals and business enterprises. It often happens that laws and regulations are adopted that lead to favorable results if only one type of consequence is considered, but lead to very unfavorable results if all consequences are recognized "to whomsoever they may accrue." Example 19–5 gives a brief description of an actual case where this occurred. Although this example is described without the use of any money amounts, Problems 19–22, 19–23, and 19–24 at the end of this chapter use money figures to illustrate some possible kinds of economic analysis of the particular type of public regulation illustrated by the example.

EXAMPLE 19–5

Public Regulation of Certain Urban Motor-Vehicle Traffic

Facts of the Case

The residential area of a moderate-sized city in the United States contained a fairly high percentage of street intersections where drivers had limited visibility of traffic on cross streets, often because their visibility was impeded by hedges, trees, and other vegetation on corner lots. Many drivers drove faster than the legal speed limit. As a result of the limited visibility and the excessive speeds, there were occasional collisions between automobiles at intersections. A number of the collisions resulted in property damage to one or both of the colliding vehicles. Once in a while an accident caused personal injury to a driver or a passenger.

The city manager decided that the best course of action was to slow traffic down by the installation of stop signs. On each street a stop sign was installed every second block. Thus, *every* intersection was a two-way stop and no car could be driven on any given street for more than two blocks without being required to make a full stop. Over the years, often in response to suggestions made by residents living near the particular intersection in question, a number of the intersections were changed from two-way stops to four-way stops.

Initially the change had the effect of reducing the number of collisions at street intersections, although such collisions were not entirely eliminated. As time went on, more drivers tended to disregard the stop signs, so accident rates increased a bit.

Apparently the city officials gave no weight in their decision making to various undesirable effects of a great increase in the number of motor vehicle stops. Some of these effects were as follows:

1. Every motor vehicle stop involves incremental costs for motor fuel and oil. Moreover, in the long run stops shorten the lives of tires and brake linings and otherwise tend to cause higher maintenance costs. They also use the time of drivers and passengers.
2. Each stop tends to increase atmospheric pollution.
3. Any regulation in the United States that tends to increase motor-fuel consumption has the adverse effect of increasing the dependence of the United States on foreign sources for energy.

PROBLEMS

19–1. Two alternate locations for a new rural highway are to be compared. Location X involves a distance of 11.3 miles. Total first cost is estimated to be $6,776,000. The location and grading are assumed to be permanent. Resurfacing and reconstruction of the base will be required every 15 years at an estimated cost of $160,000 per mile. In addition, annual maintenance cost will be $4,000 per mile.

Location Y involves a distance of 13.5 miles. Total first cost is estimated to be $4,640,000. Costs per mile for resurfacing and annual maintenance are the same as for X. Location and grading are assumed to be permanent.

The estimated average traffic over this highway is 700 vehicles per day, of which about 15% will be trucks and an additional 10% will be commercial passenger cars. The increment cost per mile of vehicle operation is assumed as 21.6¢ for passenger cars and 80¢ for trucks. Traffic will travel at an average speed of 40 miles/hour. The money value of time saving to commercial traffic is estimated as 24¢ per vehicle-minute.

Assuming an i^* of 8%, compute a **B/C** ratio applicable to the extra investment of Location X. (*Ans.* 1.48)

19–2. For the data of Problem 19–1, find the approximate value of i^* above which the incremental **B/C** for the extra investment required by Location X will be less than unity. (*Ans.* 10.8%)

19–3. For the multipurpose water project of Example 19–4, what is the overall **B/C** ratio assuming an i^* of 4½% (i.e., assuming the same i that was used in computing all the capitalized values of Table 19–1)? What are the respective incremental values of **B/C** for flood control, irrigation, power, and domestic and industrial water? (*Ans.* 2.12; 1.35; 5.82; 1.26; 8.53)

19–4. Using the data of Example 19–4 and an i^* of 4½%, what would have been the respective values of **B/C** for single-purpose projects for flood control, irrigation, and domestic–industrial water? Assume that the costs of alternates given in line 3 of Table 19–1 include capitalized operation, maintenance, and replacement costs as well as construction costs. (*Ans.* 0.55; 1.91; 1.06)

19–5. In the construction of an aqueduct to serve a city with water, a tunnel is necessary. In order to determine whether it will pay to build this tunnel to the ultimate capacity of the aqueduct, the engineers have forecast the growth of the demand for water in terms of tunnel capacity as follows.

A one-third capacity tunnel will be adequate for 10 years.

A one-half capacity tunnel will be adequate for 20 years.

A two-thirds capacity tunnel will be adequate for 35 years.

Estimated construction costs are as follows:

One-third capacity	$2,000,000
One-half capacity	2,400,000
Two-thirds capacity	2,700,000
Full capacity	3,400,000

Extra pumping costs for the smaller size tunnels above costs for the full-capacity tunnel are estimated as follows:

One-third capacity tunnel, running full, $11,000 a year. Thus, for an ultimate development of 3 such tunnels the extra pumping costs would be $33,000 a year.

One-half capacity tunnel, running full, $10,000 a year. Thus, for an ultimate development of 2 such tunnels the extra pumping costs would be $20,000 a year.

Two-thirds capacity tunnel, running full, $8,000 a year. Thus, for an ultimate development of 1 two-thirds and 1 one-third capacity the extra pumping costs would be $19,000 a year.

Set up the four plans of development suggested by these estimates. Compare them on the basis of the capitalized cost of perpetual service, using an $i^\star$ of 9% and assuming the continuance of present price levels. In calculating excess pumping costs, assume that the differences in costs given apply from the date a tunnel is put in service. What size tunnel would you recommend for present construction?

19–6. For a certain proposed government project, annual capital costs to the government are $300,000 and annual operation and maintenance costs to the government are $100,000. Annual favorable consequences to the general public of $800,000 are offset by certain annual adverse consequences of $300,000 to a portion of the general public. What is the **B/C** ratio if all consequences to any of the general public are counted in the numerator of the ratio and all consequences to the government are counted in the denominator? What will it be if the classification of the $300,000 adverse consequences to the general public is changed from a disbenefit to a cost? If the $100,000 operation and maintenance cost to the government is changed from a cost to a disbenefit? If both changes are made? Regardless of the classification of these items, what is the excess of benefits over costs?

19–7. Two alternate storage flood control projects are proposed for the Crow River. A low dam (LD) project has an estimated first cost of $2,000,000 and annual operation and maintenance costs of $28,000. A high dam (HD) project has an estimated first cost of $3,300,000 and annual operation and maintenance costs of $47,000. The "expected value" of the annual cost of flood damages is $410,000 with a continuation of the present condition of no flood control, $155,000 with LD, and $20,000 with HD. The HD project has annual disbenefits of $15,000 due to a loss of land for agricultural purposes in the reservoir site. Find the relevant **B/C** ratios to evaluate the merits of these projects using an $i^\star$ of 6%, a 50-year life for the projects, and zero terminal salvage values.

19–8. Solve Problem 19–7 changing the $i^\star$ to 10%.

19–9. Solve Problem 19–7 changing the estimated terminal salvage values to 75% of the first costs.

19–10. Two counties in the Sierra foothills are considering the construction of a dam to provide flood control and water for domestic and irrigation uses. The site has been selected and the only question remaining is how high a dam should be built. It has been decided that a minimum attractive rate of return of 8% and an economic life of 50 years will be used in making a benefit–cost study. The engineers' estimates of relevant costs and income from the sale of water are as follows:

Project	*Investment*	*Salvage value*	*Annual O & M disbursements*	*Expected annual flood damages*	*Revenue from sale of water*
No flood control	—	—	—	$1,700,000	—
Dam A	$12,000,000	$1,000,000	$30,000	900,000	$40,000
Dam B	15,000,000	2,500,000	40,000	600,000	60,000
Dam C	18,000,000	3,500,000	50,000	300,000	120,000

Find the relevant **B/C** ratios to evaluate the merits of these projects and select the alternative with the greatest economy.

19–11. Rank the following proposed Mexican social roads according to the criterion of cost per inhabitant served.

Proposed road	*Length in kilometers*	*Cost per kilometer in pesos*	*Population served*
1. Las Norias–Cruillas	32	110,000	1,999
2. El Capulin–Bustamente	32	109,000	1,692
3. Palmillas–Miquihana	43	112,000	2,853
4. Mendez–Entronque	45	119,000	3,037

19–12. Rank the following proposed Mexican economic penetration roads according to the criterion of the ratio of the estimated gross value of agricultural product in the fifth year of operation to the construction cost of the road.

Proposed road	*Principal crops*	*Estimated hectares of area in production, fifth year*	*Estimated fifth year average value of output per hectare in pesos*	*Construction cost of road in pesos*
1. Hidalgo–La Mesa	corn	1,600	1,400	2,500,000
2. Altamira–Aldama	tomatoes	4,500	2,100	6,000,000
3. El Barretal–Santa Engracia	corn	1,300	1,400	1,000,000
4. Casas–Soto La Marina–La Pesca	corn, cotton	15,500	1,400	10,300,000
5. Limon–Ocampo	citrus, fruit, cane	3,400	1,800	5,500,000
6. Llera–Gonzalez	cane, corn	8,900	1,500	6,600,000
7. Ebano–Manuel	corn, cane	4,700	1,900	7,400,000
8. El Barretal–Padilla	corn, cotton, fruit	3,400	2,300	3,900,000
9. Jimenez–Abasolo	cotton, fruit, corn	2,100	2,500	3,500,000
10. Mendez–Burgos	corn, beans, cotton	5,300	1,400	3,720,000

19–13. A city of 40,000 population uses an average of 150 gallons of water per capita per day. Its water supply has a total hardness of 320 parts per million (ppm). A municipal water softening plant is proposed to reduce this to 70 ppm.

The plant capacity must be double the average daily consumption; the plant will cost $35,000/million gallons per day (mgd) of capacity. The plant would be financed by 20-year, 7% serial bonds, with a uniform number of bonds maturing each year. Chemicals are estimated to cost 15¢/mg per ppm of hardness removed. Plant labor costs at the water-treatment plant will be increased by $12,000 a year. Pumping in the softening plant will cost $2.40/mg pumped. Average annual maintenance cost is estimated as 3% of investment. The life of the plant is estimated as 20 years with a negligible salvage value. The city will raise water rates sufficiently to cover the extra operating costs for water softening plus first year's bond interest and repayment.

Assume a saving in annual per capita soap consumption from 38.5 lb to 30.8 lb as a result of the water softening, with an average retail soap price of 45¢/lb. Assume a saving in cost of chemicals to customers, already softening their water, of 29¢/mg per ppm of hardness removed; this applies to 110 mg/year. It is estimated that the life will be doubled for 4,000 storage water heaters having an average life of 8 years under present conditions; the average investment per heater is $140.

Estimate the required increase in water rate per 1,000 gallons. Make an analysis to determine whether the foregoing estimated monetary savings are sufficient to justify water softening.

19–14. The following question is adapted from one used a number of years ago in a state examination for registration as a professional engineer:

Estimates are made for the first costs of various elements of a new highway assuming designs for different numbers of lanes, as follows:

	4 Lanes	**6 Lanes**
Right of way	$240,000	$320,000
Pavement	960,000	1,440,000
All other elements of first cost	1,200,000	1,440,000
Total	$2,400,000	$3,200,000

It is estimated that 4 traffic lanes will be sufficient for the next 10 years; after that, 6 lanes will be required. The interest rate is 4%. How much money, in addition to the cost of a 4-lane highway, can be spent economically at once? How should your recommended total be divided among the foregoing elements of first cost? Explain your answers.

Give the solution that you think was expected by the examiner.

Assume that you are confronted by this type of problem in an actual case. What additional data and estimates, if any, would you want before arriving at your recommendations? How would you use the additional data in your analysis?

19–15. In the cost allocation in Table 19–1, 77.4% of the nonseparable costs were allocated to two of the six purposes of the water project, namely, irrigation and domestic–industrial water. None of the other four purposes had an allocation of more than 6.7% of the nonseparable costs.

In the solution to Problem 19–3, the incremental **B/C** ratios for irrigation (5.82) and for domestic–industrial water (8.53) were much higher than for flood control (1.35) or power (1.26). Does it seem likely to you that the separable costs–remaining benefits method of cost allocation will ordinarily allocate higher percentages of the nonseparable costs to the purposes that have the highest incremental **B/C** ratios? Use your answer to this question as a starting point for a general discussion of what seems to you to be the rational foundation that underlies the separable costs–remaining benefits method.

19–16. Annual benefits in Example 19–4 (which were used to compute the present worth figures in line 2 of Table 19–1 but which were not explicitly stated in the text of the example) are as follows:

Flood control	$500,000
Irrigation	2,500,000
Power	575,000
Domestic–industrial water	750,000
Fisheries	90,000
Navigation	100,000
Total	$4,515,000

Compute the various values of **B/C** asked for in Problem 19–3 assuming an i^* of 7% instead of the 4½% used in that problem.

19–17. Recompute the cost allocation of Table 19–1 using an i of 7%. (Annual figures for benefits are given in Problem 19–16, and the required annual figures for the separable costs of operation, etc., are given in the text of Example 19–4. Because the source of the example does not break down the figures for alternate costs in line 3 into the components of construction costs and operation costs, etc., it will be necessary to assume that these figures are also appropriate for an i of 7%).

Discuss the sensitivity of the results of this cost allocation to this change in the chosen value of i.

19–18. What is the break-even value of i^* above which it would not have paid to include flood control as one of the purposes of the water project of Example 19–4? Answer the same question for the hydro-power features of the project. (See Problem 19–16 for certain necessary data not given in the example.)

19–19. In Example 19–4 it was assumed that the water project would have a life of 100 years with a zero terminal salvage value. In Problem 19–3, the value of **B/C** was computed to be 2.12 with an i^* of 4½%. To examine the sensitivity of **B/C** to certain assumptions, compute this ratio assuming a perpetual life. Compute **B/C** assuming a life of 50 years with zero terminal salvage. Using the life and salvage of Example 19–4, compute **B/C** first with an i^* of 4% and then with an i^* of 5%.

19–20. The monetary figures in Table 19–1 are present worths computed for a 100-year period with an i^* of 4½%. How would the allocation have differed if equivalent uniform annual figures rather than present worths had been used for the various benefits and costs shown in the table?

19–21. A critic of highway economy studies such as the one illustrated in Example 19–3 points out that savings to highway users because of lower vehicle operating costs are offset by lost revenues to service stations, oil companies, tire dealers, repair shops, and others. If consequences "to whomsoever they may accrue" are to be examined, the critic contends that the favorable consequences to highway users are counterbalanced by unfavorable consequences to others and therefore should be given no weight in the economic evaluation of proposed highway improvements. Discuss this contention, explaining whether or not you agree with this critic. If so, why? If not, why not?

19–22. In the residential district of the city described in Example 19–5, the intersection of Doe Street and Roe Avenue is now a 4-way stop. A group of citizens proposes that a vehicle- and pedestrian-actuated traffic signal be installed at this intersection. This would require an investment of $29,500 by the city. It is estimated that the signal will have a 12-year life with negligible net salvage value when retired. The estimated annual operation and maintenance cost to the city is $1,600. There is no evidence that there will be any difference in police department costs or other municipal disbursements.

The differences in road user consequences that the proponents of the traffic signal deem to be relevant in an economic analysis are:

1. Extra costs of motor vehicle operation due to required motor vehicle stops at the intersection.
2. Extra costs of the time of commercial vehicle traffic due to vehicles stopping or slowing down at the intersection.
3. Costs associated with accidents.

For the expected volume of traffic at this intersection and for actual vehicle speeds, and considering the hourly traffic distribution throughout the day, the division of the traffic among passenger vehicles, single-unit trucks, and combination trucks, the division of traffic between commercial and noncommercial vehicles, the division of traffic between Doe Street and Roe Avenue and the proportion of the various types of turning movements, the following estimates are made:

1. Average daily extra motor vehicle operating costs due to slowing down

and stopping will be $41.60 with the present 4-way stop and $29.10 with the proposed traffic signal.

2. Average daily costs of time delays will be $33.40 with the present system and $19.40 with the proposed system. This estimate considers only time delays for commercial vehicles and disregards time delays for noncommercial vehicles.
3. A study of accident records in this and nearby cities leads to an estimate of average daily accident costs of $6.80 with the 4-way stop and $10.30 with the traffic signal.

Assuming a 365-day year and a 12-year study period and using an i^* of 10%, make an economic analysis of the benefits and costs of the proposed change.

19–23. For the next two years, more urgent projects make it impracticable for the city to budget construction funds for the proposed traffic signal evaluated in Problem 19–22. The group of citizens who originally proposed the traffic signal now suggest that the intersection of Doe Street and Roe Avenue be changed from a 4-way stop to a 2-way stop for the immediate future. They point out that at least 75% of the traffic is on Doe Street and that, generally speaking, this traffic will be benefited by the change to a 2-way stop. No costs to the city will be involved; the salvage value of the two stop signs removed will be equal to the cost of removal, and no differences are expected in police department costs or other municipal disbursements.

(a) Make an economic analysis of this proposal assuming that the change to a 2-way stop will reduce the average daily extra motor vehicle operating costs due to slowing down and stopping from $41.60 to $33.60, that it will reduce the average daily costs of time delays to commercial vehicles from $33.40 to $21.70, and that it will increase the average daily accident costs from $6.80 to $11.50.

(b) Make an economic comparison between the 2-way stop and the traffic signal system described in Problem 19–22.

19–24. The economic analysis in Problems 19–22 and 19–23 placed money values on the following estimated differences in road user consequences: (1) motor vehicle operating costs, (2) time of commercial traffic, and (3) accidents. No weight was given either to the road user consequence of time of movement of noncommercial vehicles or to the community consequence of the effect of vehicle stops on atmospheric pollution. If weight should be given to each of these consequences, do you think it would favor the 2-way stop, the 4-way stop, or the traffic signal system? Explain your reasons. What other types of consequences, if any, do you think it would be appropriate to consider in choosing among these three types of traffic control systems?

19–25. Our examples and problems related to highway engineering economy have not broken down incremental highway user costs per mile of vehicle operation into such components as fuel, lubricants, tires, maintenance and repairs, and depreciation. However, it is obvious that each component needs to be analyzed and priced in order to determine relevant highway user costs in any actual case.

Assume that the average retail price of gasoline is $1.35 per gallon, and that this $1.35 figure is made up of 11¢ of state and federal highway-user taxes and $1.24 paid for the gasoline. Some analysts would use the $1.35 figure and others would use the $1.24 figure in computing the benefits to highway users from a reduction in vehicle operating costs. Which would you use? Explain your answer.

19–26. A city has its own municipal utility district which supplies electricity, gas, water, solid waste disposal, and sewer services for the residences and business establishments. The electric service needs to expand its generating capacity and is considering the construction of a hydroelectric generation station on a large stream in a mountain park owned by the city. Thus, there is no cost for the land and there will be certain secondary benefits from building a dam and creating a lake in the park. Preliminary engineering estimates indicate that there are three feasible heights for a dam, 173 feet, 194 feet, and 211 feet. The generating capacity in kilowatts (kw) increases with increasing height of the dam (hydraulic head). For simplicity, the city places a wholesale value of $37.50/kw/year on generating capacity from this project. The following tabulation gives a summary of the information available for the decision:

		Dam height in feet	
	174	***194***	***211***
First cost of dam	$4,800,000	$5,300,000	$6,100,000
First cost of buildings	180,000	180,000	180,000
First cost of equipment	885,700	987,700	1,074,400
Total investment	$5,865,700	$6,467,700	$7,354,400
Annual operation & maintenance costs	$124,000	$142,000	$169,000
Kilowatts developed	19,450	21,675	23,600

It is assumed that the economic life of the dam, buildings, and equipment is 40 years, with zero net realizable salvage value at that time. The municipal utility district does not pay federal or state income taxes. The city uses an i^* of 9% for all capital investments. Calculate the appropriate **B/C** ratios to determine which dam height, if any, should be built. What is the annual cost per kilowatt of generating capacity for the best alternative?

20

SOME ASPECTS OF ECONOMY STUDIES FOR REGULATED BUSINESSES

"American society experienced a virtual explosion in Government regulation during the past decades. Between 1970 and 1979, expenditures for the major regulatory agencies quadrupled, the number of pages published annually in the Federal Register nearly tripled and the number of pages in the Code of Federal Regulations increased by nearly two-thirds.

"The result has been higher prices, higher unemployment and lower productivity growth. Over-regulation causes small and independent businessmen and women, as well as large businesses, to defer or terminate plans for expansion and, since they are responsible for most of our new jobs, those jobs aren't created.

"We have no intention of dismantling the regulatory agencies—especially those necessary to protect the environment and to assure public health and safety. However, we must come to grips with inefficient and burdensome regulations—eliminate those we can and reform those we must keep."*

As the foregoing quotation implies, it was not too many years ago that mention of "regulation" automatically triggered the assumption that regulation applied only to the public utilities, such as telephone, telegraph, electricity, gas, water, and transportation. Now every business enterprise in the United States, regardless of how small, is subject to rules and regulations by numerous local, state, or national agencies. Much of this regulation stems from rapidly increasing social awareness in a long period of affluence and has such objectives as protection of the public and employees from health hazards, improvement of the environment, and so on. There is no doubt that all of this regulation adds to the cost of doing business and that the added costs must eventually be paid for in the form of higher prices for goods and services. It is equally obvious that the justification for such controls and regulation *should* be based on the benefits to be derived by the citizens of the locality, state, or nation. It is, however, still the prerogative of the managers of businesses to

*President Reagan, State of the Nation Address to the Congress, February 18, 1981.

choose methods of complying with the regulations that will minimize the extra costs. Selection of the most favorable method is the subject to be considered here.

This chapter will address two aspects of regulation: (1) the general nature of regulation and types of economy studies for both private and governmental activities, and (2) the regulation of public utilities in the United States. In a great many instances the general regulations apply equally to ordinary businesses (manufacturing, mining, service, etc.) and to public utilities. Examples of such regulations include those promulgated by the Environmental Protection Agency, Occupational Safety and Health Administration, Internal Revenue Service, Department of Labor (Equal Employment Opportunity legislation), and others. An oversimplified statement of the methods employed by these regulatory agencies is that they set minimum standards with which individual businesses must comply, but the choice of methods by which to comply is left up to the business managers.

Regulated public utilities, especially telephone, telegraph, electricity, gas, and water, are subject to additional regulations and regulatory processes because they tend to be more efficient when allowed to have a monopoly with respect to a service area. The privilege of operating a monopoly to provide service for a certain area is granted by the political subdivision with a regulatory agency having the responsibility to protect consumers by regulating such matters as general level of rates, specific rate structures, standards of service, accounting methods, and issuance of securities. Regulation is supposed to replace competition as a means of assuring good service to the public at a fair price.

The transportation industry has been an exception to the general concept of a utility monopoly because there are many competing modes of transportation vying for the customers' business. One of the principal regulatory functions relative to transportation has been to assure uniform rates for commodities and to prevent unfair competition by the granting of special rates or services to preferred customers. In recent years there has been much discussion of "deregulation" of transportation services and regulation of the airline industry has been relaxed considerably. One of the arguments for deregulation of transportation, especially of trucking companies, is that the minimum rate structure tends to keep marginal companies in business and tends to maintain freight rates that would otherwise be reduced. Legislative action to reduce the regulation of both railroads and truck lines is under consideration by the Congress of the United States at the time of this writing. Special regulatory matters concerning public utilities will be discussed in detail in the latter part of this chapter.

Types of Economic Problems Related to General Regulation

The typical situation encountered by most businesses is that a regulatory agency in its role of carrying out a mandate of the legislative body establishes certain minimum or maximum standards of performance by all businesses subject to the legislative action. The legislation usually authorizes the agency to require that each business prepare certain reports initially and then periodically to demonstrate its compliance with regulatory standards. If the business is not in compliance at the time of the publication of the standard, there is usually some time period within which it must comply. In addition, the regulatory agency may be empowered to make on-site inspections of the business in order to determine that the enterprise has actually complied or to determine that the proposed actions will probably bring the condition into compliance by the specified time. In certain cases the standards may be progressive. That is, the standards will change over time, as was the case with the EPA's mandated miles per gallon of gasoline for automobiles sold in the United States. There are many other examples of progressive tightening of regulatory standards.

The publication of the standard and the requirement for status reporting brings the problem to the attention of management. If the firm is not in compliance, action must be taken. From that point on the solution of the problem is a completely typical engineering economy problem. The problem must be fully identified and defined. The alternative technical "fixes" must be identified and all the technical and engineering data for each alternative method assembled. Especially, all of the consequences of choosing each of the alternatives must be identified and translated into cash flows at specific times in the future. Then the usual time-value of money analysis is done to determine which method will minimize the cost of complying with the regulatory standards.

Although the description of the process is very simple and straightforward, the execution of the process is frequently very difficult. There are several factors that tend to increase the difficulty. First, some regulatory commissions or agencies have been charged with the responsibility to see that "the best available control technology (BACT) that is consistent with economic reasonableness" be adopted. That charge poses two problems: what is the "best available technology" and what is "economic reasonableness"? In practically all situations there are two or more different technologies *and* several levels of application of the technology to obtain greater and greater levels of attainment of the desired goal. Each additional level of attainment normally requires an incremental investment and may increase annual operation and maintenance expenditures. At the same time, increased attainment of one goal may cause deterioration of other factors.

"Economic reasonableness" requires that benefits be evaluated and compared with costs. An analysis based entirely on the minimization of cost for some factor, such as cost per pound of emission removed by an emission-control device, ignores the benefits of the removal of emission.* If the benefits, computed at an appropriate rate of return, i^*, exceed the incremental cost of obtaining an additional pound of emission removal, then a benefit–cost analysis would indicate that the incremental expenditure should be made.

*See C. D. Zinn and W. G. Lesso, "Putting the Economics in Best Available Control Technology or Pollutants in the Air and Dollars to the Wind," *Interfaces*, Vol. 9, No. 1 (November 1978), pp. 68–71.

The difficulty, however, is that it is nearly impossible to obtain agreement on the benefits to be derived from the incremental expenditure. Such agencies as OSHA and EPA are constantly trying to find means of putting a price tag on the effects of hazardous conditions, air pollution, and other contributors to disease and ill health. Some of the results are directly measurable, such as medical costs and costs of lost time because of accidents. The value of discomfort, potential for later illness, and suffering are much more controversial.

These comments only emphasize the difficulty encountered in attempting to apply BACT. The BACT must be determined on a case-by-case basis, taking energy, environment, and economic impacts and other costs into consideration.

The following example, while not solving a problem, indicates some of the complications faced by both private business and the regulatory agency.

EXAMPLE 20–1

How Should the Thermal Electric Utilities of Ohio Meet EPA Environmental Standards and Who Shall Pay for the Clean-up?*

Ohio, a highly industrialized state, has vast coal deposits and uses coal for a large percentage of its electric generation. The use of coal has been encouraged by the national government as a means of conserving oil and natural gas and reducing the dependence of the nation on imported fuel. In recent years Ohio has mined about 47 million tons of coal a year, of which about 70% was used to produce electricity.

Ohio coal, unfortunately, is rather high in sulfur, varying around 3.5%, while so-called western coal is usually less than 1%. If the utilities continue to use Ohio coal, some method must be used to reduce the SO_2 and other emissions to a level prescribed by the 1977 amendment to the Clean Air Act of 1970. Basically, the utility companies have two alternatives: install flue-gas desulfurization equipment ("scrubbers") or use low-sulfur coal from the western part of the United States.

From the utilities' viewpoint the use of "scrubbers" means a very large capital investment as well as significant annual operating and maintenance expenses, while buying low-sulfur coal from outside the state will only increase the operating costs by increasing the delivered-fuel cost. The low-sulfur western coal would be blended with the high-sulfur Ohio coal to meet the required emission standards, but that would result in a reduction of about one-third in the demand for Ohio coal. That in turn would result in the elimination of about 13,000 jobs in Ohio.

*Most of the information for this example was gleaned from an article by Mark N. Dodosh, "Ohio Utilities, Coal Industry Battles EPA over Clean-Air Bill and Who Must Pay It," *Wall Street Journal* (February 15, 1979), p. 40.

The utilities estimate that the residential customers will face an increase in electric bills of only 3% to 12% if outside coal is used, while they will face increases of 10% to 28% if "scrubbers" must be installed.

The EPA is empowered by the 1970 Clean Air Act (Section 125) to use "regionally available coal" if "significant local or regional disruption or unemployment" occurs as a result of buying outside coal. At the same time EPA must "take into account the final cost to the consumer."

Differences in the age, design, and operation of generating plants by the many different companies means that EPA faces a wide spectrum of conditions and either must make a series of company-by-company decisions or a blanket decision for all large coal users. It has several alternatives to consider: (1) allow each utility company to make its own decision as to how to meet the emission standards, (2) require all large coal users to install flue-gas desulfurization equipment, (3) allow Ohio coal to be used if "washing" of the coal will permit the standards to be met, or (4) require some combination of outside coal blended with Ohio coal and any additional treatment necessary.

While this case revolves primarily around the matter of environmental regulations, the Department of Energy is also deeply concerned because a decision not to insist on "regionally available coal" could seriously disrupt the production of coal in other states such as Kentucky, where over 80% of its steam coal is shipped out of the state.

Comments on Example 20–1

There are several relevant points illustrated by the example. First, this air pollution emission standard with which the industries will be required to comply does not depend upon the health and other benefits that will accrue to the citizens from the reduction of SO_2 in their environment. The Act does, however, recognize that the economics of the area, especially employment, must be considered in the administration of the Act's mandates without setting any specific guidelines as to how the economic effects are to be evaluated. The utilities' response is, in general, in accord with the usual regulatory agencies' requirement that regulated public utilities make their investment decisions so as to minimize revenue requirements. (That matter will be treated more fully later in this chapter.) Assuming the data presented are correct, the use of outside, low-sulfur coal will minimize the cost of electric energy to consumers, but the burden of unemployment and welfare payments may be increased, resulting in higher social costs and taxes for these same consumers. Thus, secondary and tertiary economic effects tend to be neglected.

Presumably the citizens of the state desire clean air and are willing to pay something for it, but they are unlikely to associate an increase in their electric bills with the benefits of clean air. The consumers are not consulted in this matter and are likely to protest any electric rate increases granted to provide for environmental improvement.

This example illustrates the conflict of interests among the participants in a large-scale public decision. The coal miners want local coal to be used and local politicians fear the consequences of a major increase in unemployment. All sectors of the economy would be adversely affected. The utility companies want to provide the necessary electricity at as low a unit price as possible while complying with the Clean Air Act's requirements. The capital requirements for investment in "scrubbers" is very large and would have to be added to the rate base for rate-setting purposes, which probably would be discouraged by the state utility regulatory commission. In addition, the investments would have to be made at a time of very high interest rates, saddling the companies with long-term debt, which in turn will make it more difficult to raise additional capital for plant expansion or renewals. EPA is required to see that the standards are met and would like to avoid political in-fighting among the interest groups. The general public wants clean air but also wants low electric rates. It is obvious that no decision can satisfy all the interest groups.

How Large Is the Cost of Regulation

Previously we have noted the difficulty of estimating the benefits to be derived from compliance with specific regulations. On a nationwide basis it is almost as difficult to estimate the costs to the citizens. Every report, document, or record adds to the clerical and overhead costs of doing business. Huge, almost unbelievable quantities of paper containing the required information are prepared each month for national, state, and local governments. Thousands of persons are employed each day in collecting, analyzing, and preparing data for these documents. Similarly thousands of government employees spend their time reviewing these documents, yet this is only one aspect of the total cost of regulation. One of the announced goals of President Carter's administration (1977–81) was to drastically reduce bureaucracy and "red tape" by deregulation. Progress was being made, but there was a long way to go.

In 1979 the Transportation Department's Federal Railroad Administration proposed the elimination of "unnecessary and burdensome maintenance inspection requirements" on freight cars and estimated that this, along with some other changes in regulations, could save the railroad industry $100 million a year.* Mr. Willard C. Butcher, President of the Chase Manhattan Bank, told the Commonwealth Club of California on September 16, 1978, that based on an economic analysis conducted by the bank, regulation cost the United States $100 *billion* in 1977. He placed the cost of compliance with pollution and abatement controls at $32 billion, general compliance costs at $53 billion, and auto safety and pollution equipment costs at $7.5 billion. The U.S. Department of Commerce's *Current Industrial Reports*: "Manufacturers' Pollution Abatement Capital Expenditures and Operating Costs,"

**Wall Street Journal* (September 4, 1979), page 1.

Advance Report, 1978, issued in November 1979, gives much lower comparable figures for these pollution-abatement costs. These astronomical numbers illustrate why both business and government ought to examine very carefully both the costs and the benefits of imposing any specific set of regulations *before* reaching a decision to do so.

Public Utilities Require Special Consideration

Regulated public utilities are required to have the capital equipment necessary to provide the desired service in place and operating before the actual demand occurs. In order to do that, in a capital-intensive industry, the utility companies require a very large investment in plant and equipment in comparison to their annual revenues and annual operating costs. This relatively high plant investment is the most distinguishing characteristic of the public utility field. A large proportion of public utility costs are so-called fixed costs such as taxes, depreciation, and cost of capital. For this reason, a monopoly in the public utility field is highly economical. For example, two competing telephone companies serving the same community must duplicate their physical plant with the result that the total investment is much higher than if all the service were provided by one company. This means that in order to provide a fair rate of return the total revenues, and hence the rates charged the users, must be higher than if there were only one supplier. There are additional nuisance factors to be considered, such as duplicate pole lines in the streets.

Relationship Between Regulatory Policies and the Rapid Growth of the Need for Public Utility Service

The general experience in the United States and other industrialized countries has been one of continuous increase in demand for utility service. Since 1900 this has been particularly true of electric, gas, water, and telephone utilities. There are high social costs caused by inadequate capacity of utility plant. The public interest therefore requires that utility companies be in a position to finance necessary expansion and modernization of plant. If new capital to finance expansion is to be secured, a utility must have the prospect of earnings sufficient to attract such capital.

Rates, therefore, should be high enough to attract capital. At the same time they should be low enough so that the charges to the public are reasonable, all things considered. A common method used by commissions to accomplish these two ends has been to set rates at a level so that the prospective total revenues received by a utility from the sale of its public utility services will equal its prospective operating expenses including an allowance for depreciation, plus income taxes, plus a "fair return" on the investment in the property employed in the utility service.

The appropriate rate of return to be allowed in any given instance should reflect the overall cost of money to the utility being regulated, considering both the interest that must be paid for borrowed money and the dividend rate needed to attract new equity capital. The return will properly be higher for a relatively risky utility enterprise (such as an urban transportation company) than for a relatively secure one (such as a large electric light and power company). The rate of return necessary to attract capital will also be influenced by the size of the utility and the extent to which it is well known to investors. The appropriate fair return will change from time to time with changes in the general level of interest rates.

During the 1970s interest rates rose very rapidly and tended to fluctuate at levels considerably higher than had been common in the 1960s. For example, examination of the bond price listings in financial journals reveals that many public utility companies have outstanding bonds with coupon rates of as low as 3.5% for bonds issued in the 1950s and as high as 13% for bonds issued in 1980. All borrowers have experienced sharp increases in the interest rates.

Most regulatory agencies recognized the difficulty faced by public utilities in raising capital for necessary facilities to meet the ever-increasing demand for service. During the same decade of the 1970s, regulatory agencies gradually raised the allowed fair rates of return to correspond with increasing bond rates and the need for higher return on equity capital. While fair rates of 6.0% to 7.5% were common in the 1960s, fair rates of return for telephone companies in some states were as high as 9.5% in the 1970s. Rates for certain other public utilities were even higher. In many rate cases the commissions took into consideration the need for prospective rate of return on equity capital to be substantially higher than the fair rate of return and typically used from 11.5% to as much as 14% as the rate necessary to raise private capital.* Unfortunately there usually was a "regulatory lag" of a year or more in taking these actions.

As interest rates rose the market value of shares of common stock of public utilities tended to decline. Shares which commonly sold at between $35 and $40 in the 1960s in many instances declined to $18 to $21 by the late 1970s. Thus, if a company wished to raise a given amount of equity capital it had to sell almost twice as many shares in 1979 as in 1965. Inflation caused the cost of utility equipment and construction to rise very rapidly, with the effect that many companies had to increase their capital budgets by many millions of dollars each year in order to be able to maintain their service capability. Many found that interest rates on their new bond issues were as great as or greater than the currently allowed fair rate of return. It is easy to understand why the public utilities had to seek higher fair rates of return in order to attract the necessary capital.

*The rules governing rate regulation have been laid down in a series of decisions by the United States Supreme Court and other courts. These decisions have taken place over a period of years. The viewpoint of the Supreme Court on underlying principles has changed from time to time. Later court decisions permitted regulatory bodies to exercise much more latitude than was allowed in earlier years. See particularly *Federal Power Commission* v. *Hope Natural Gas Co.*, 320 U.S. 591 (1944).

Occasionally one hears the statement that public utilities are *guaranteed* a return on their investments. Any such statement is incorrect. The most that a regulatory commission can do is to allow a rate schedule that gives the utility a chance to earn a fair return *if it can get it*. Certain utilities (such as some urban transportation companies) find it impossible to earn an adequate return with any schedule of rates.

Authority of Regulatory Commissions

In the United States, any public utility company engaging in interstate commerce is regulated by two or more commissions. Interstate activities are regulated by a federal commission, such as the Federal Power Commission, the Federal Communications Commission, or the Interstate Commerce Commission. The intrastate activities usually are regulated by a state public utility commission. Although federal and state commissions consult with one another, there is still the possibility that their decisions on certain matters will not be consistent.

The complexities of the actual regulation of utility rates and services are beyond the scope of this book; we shall not attempt to examine all the numerous controversial issues involved in this interesting and important subject. Nevertheless, it is desirable to examine certain aspects of the authority of regulatory commissions and to note the possible effects of commissions' decisions on a regulated utility's engineering economy studies.

Commissions generally have the authority to:

1. Set the "fair rate of return" that the company may earn on its rate base.
2. Determine what the rate base includes and how it is to be computed.
3. Determine what expenditures are allowed to be recovered as allowable expenses, including approval of salaries to be paid to executives.
4. Prescribe the permissible ways of applying depreciation, tax credits, and tax incentives.
5. Approve or disapprove the rate schedules presented by the company.
6. Prescribe a uniform classification of accounts and stipulate certain accounting methods to be used by the company.
7. Require periodic reports on all matters under commission jurisdiction.

It is possible that the "fair return" may be adjusted downward on the initiative of the regulatory commission as a means of penalizing a company for inadequate or poor service. Such action may occur because of complaints by customers. On the other hand, if a utility's earnings are higher than the stipulated "fair return," the commission either can ignore the matter, or it can require that new rate schedules be presented that will bring the net earnings down to the specified level. Because all rate schedules are based upon predictions of future conditions and such predictions of growth in demand, labor costs, and fuel costs are frequently in error, there has been a trend to specify a range of acceptable returns rather than a single rate.

The fair rate of return frequently is computed by regulatory commissions on the basis of the estimated cost of capital where total capital is assumed to be composed of a certain percentage of debt at the "embedded cost of debt," a certain percentage of preferred stock at the specified dividend rate and the remainder of the capital from equity at a rate that will attract additional private investments. The "embedded cost of debt" is the average interest rate being paid on all outstanding bonds, including old bonds nearing their redemption dates and bearing very low interest rates (3.5% to 5%) as well as the most recent issues, which may be at very high rates (8.5% to 12.0%). Thus, if a commission decides that 13% is a fair return on equity capital, and if the utility is paying an average of 7.5% on debt and 8.2% on preferred stock, the fair rate may be computed as follows:

Source	*Percent of total capital*	*Cost (%)*	*Contribution to fair rate of return*
Debt	45	7.5	3.375
Preferred stock	15	8.2	1.230
Common stock	40	13.0	5.20
Fair rate of return			9.805%

Many utility economists argue that using the "embedded cost of debt" is unfair to investors and subsidizes current customers; the current cost of new debt should be used in the computation of the fair rate of return. During the early 1980s, most utility companies were finding it necessary to offer bonds at 11.0% to 13.0% and preferred stock, if it could be sold at all, at rates and prices to yield 12% to 14%. Using these current costs of capital, the new fair rate of return becomes 0.45 (12%) + 0.15 (13%) + 0.40 (13%) = 12.55%. In a period of deflation and declining earnings rates, the fair rate of return computed on current capital costs could fall below those based on embedded cost of capital.

The rate base presents many special problems. Many commissions in the United States define the rate base as the depreciated book value of the assets used in providing service. They include working capital as well as fixed assets (land and plant). But assets acquired in advance of immediate needs, such as land on which a utility expects to build new facilities 10 years hence, sometimes are excluded from the rate base. Under certain circumstances, some plant may be included in the rate base but at a lower price than actually paid by the utility.

Plans to issue new stock or to sell bonds must be approved by the state commission. Generally, such plans are accompanied by a statement of the intended use of the funds. Commissions are interested in these matters for two major reasons. First, the plan for raising new capital will affect the cost of money to the utility, and the commission wants to see that capital is obtained in the most advantageous way. Second, although part of the new capital may be to refund maturing bonds, some of it doubtless will be used to acquire assets that will increase the rate base. The commission wants to know what development, improvement, or service extension plans are to be funded so that it can judge the appropriateness of the plans.

The foregoing discussion indicates some of the complexities of the regulatory problem. Utility managements have the responsibility of maximizing the long-run value of the stockholders' investments, and commissions have the responsibility of protecting the public, collectively, from monopolistic pricing. At the same time, commissions need to encourage the development of the desired services at the lowest possible cost to the consumer. This latter objective provides a basis for judging the attractiveness of competing plans for the provision of a service. Generally, both commissions and utilities should make decisions among alternative plans on the basis of minimizing revenue requirements (minimizing the long-run cost to customers).

Some Differences Between Economy Studies in Regulated Public Utilities and in Competitive Industry

Some contrasts between public utilities and competitive industry that bear on the central theme of this chapter are as follows:

1. As already pointed out, the funds available for investment in new fixed assets are often limited in enterprises engaged in competitive industry. In contrast a public utility is likely to be raising new capital for expansion at frequent intervals and consideration of capital rationing due to limitation of funds does not enter into the determination of the minimum attractive rate of return. The i^* for such expansion projects is likely to be at or only slightly higher than the estimated cost of capital. On the other hand, all utility companies are under pressure to improve efficiency and reduce costs as prices of labor, materials, and equipment rise, and cost reduction proposals are likely to be subject to capital rationing concepts, with the i^*s for such proposals usually determined by the opportunity costs.

Although most utility regulatory agencies assume that the supply of capital is unlimited, and therefore require that projects be selected that minimize revenue requirements when calculated with i^* equal to the fair rate of return, this condition does not always prevail. Such a selection from a set of mutually exclusive alternative proposals will frequently require so much capital that discretionary investments at much higher prospective rates of return must be foregone. Sosinski, et al., demonstrated that the selection of discretionary investment proposals on the basis of opportunity costs will result in lower total (corporate-wide) revenue requirements.*

2. Just as in competitive industry, the overall, or composite, cost of money to a public utility, considering both borrowed money and equity capital, tends to establish an appropriate *lower limit* for the minimum attractive return to be used in economy studies. Many regulatory agencies insist that the *upper limit* should seldom be much greater than the cost of capital, commonly the fair rate of return allowed; when they do so their studies are, in

*J. H. Sosinski, P. Gose, B. C. Huntzinger, and W. G. Ireson, "Incorporating Investment Opportunities into Economic Evaluations to Minimize Corporate Revenue Requirements," *Transactions on Power Apparatus and Systems*, (New York: IEEE, January 30–February 4, 1977).

effect, being made from the viewpoint of the utility's customers. They believe that decisions based on such an i^* will minimize the rates that will be charged for the utility's service; but, as pointed out in previous paragraphs, this is not always true, especially if the fair rate of return and the i^* are based on the embedded cost of capital.

A policy of making engineering economy studies to minimize a utility's revenue requirements not only takes the viewpoint of the customers but also implies that the interests of the customers and stockholders are identical. This may be true in the long run in many utilities, but it is not true in some and may not be true in the short run in many cases, particularly when the major consequences of a decision will occur between rate cases. Whether customers' and stockholders' interests are really identical depends on a number of factors, foremost of which is the matter of how the regulating commission treats investments for rate base purposes and what expenditures it allows as operating expenses. It also depends on whether or not the "fair return" is realistic relative to the cost of capital at the time of a decision on rates. If the permitted rate of return is too low, this tends to cause utility companies to put off investments in new facilities to provide better service. Some of the social costs of inferior or inadequate utility service are discussed later on in this chapter.

Thus it can be argued that the principle of basing engineering decisions on minimizing revenue requirements, while promoted as the best policy from the customers' viewpoint, may occasionally be inappropriate for both customers and stockholders.

3. It has been pointed out that the common rate-making formula for utilities is designed to permit the utility to earn a "fair return" *after* income taxes. In effect the rate-making authorities view prospective income taxes as an element of expense to be included in the rates charged for utility service.

It is convenient, when calculating the revenue requirements for a proposed investment, to be able to express the income taxes as a percentage of that investment. The appropriate percentage will depend on the income tax rate, the life and percentage salvage value of the fixed asset, the depreciation method being used, the average interest rate being paid on long-term borrowing, and the proportion of the capital raised by borrowing. One method of calculating the percentage to use for income taxes will be explained later in this chapter.

Special Aspects of Treatment of Income Taxes in Economy Studies for Regulated Public Utilities in the United States

Under common rules of regulation in many parts of the United States, a utility's revenue requirements are made up of current operating disbursements (not including interest on debt), an allowance for depreciation, income taxes, and a "fair return" on a rate base that usually is approximately equal to depreciated book value. The income taxes paid, of course, depend on the taxable income; as explained in Chapter 12 and 18, interest on debt is a deductible expense in computing taxable income. The income subject to taxation will depend on the interest rate paid on borrowed money, on the debt/equity ratio, on the "fair return," and on the rate base. The following three equations show the relationships between the permitted revenues and the income taxes:

I. Permitted revenue − (Current operating disbursements + Depreciation + Interest paid on debt) = Taxable income
II. Taxable income × Applicable tax rate = Income taxes
III. Permitted revenue = Current operating disbursements + Depreciation + Income taxes + (Fair return × Rate base)

We have noted that economy studies for regulated utilities usually aim to minimize a utility's revenue requirements. Given the rules under which utility rates are to be regulated, it usually is possible to express the income tax element in revenue requirements as a percentage of the first cost of the plant under consideration. The appropriate formulas for this percentage will be dependent on the expected rules of rate regulation.

Examples of Formulas for the Ratio of Income Tax Requirements to First Cost of Plant for Regulated Utilities

The following formulas assume that:

1. Straight-line depreciation is used both in the determination of regulated rates and in the calculation of taxable income.
2. Fair return is allowed on a rate base equal to the depreciated book value.
3. The fair rate of return is used as the i to convert a diminishing series of income tax requirements to an equivalent uniform annual series.
4. The debt/equity ratio will remain constant throughout the life of the asset.
5. The fair rate of return and the interest rate on borrowed money will remain constant throughout the life of the asset.

The presentation of the formulas is simplified by using the functional representation of the gradient factor rather than the algebraic symbols for this factor.

Let a = rate of return ("fair return") on depreciated investment
b = interest rate paid on borrowed funds
c = fraction of plant investment financed by borrowing
e = applicable income tax rate
n = life of plant
s = ratio of terminal salvage value to first cost of plant
t = ratio of equivalent annual income taxes to first cost of plant (with equivalence calculated at rate a)

The general formula for t is

$$t = \frac{e}{1-e}(a - bc)\left[s + (1-s)\left(1 - \frac{(A/G,a\%,n)}{n}\right)\right] \quad (1)$$

For the special case where the salvage value is zero, the formula becomes

$$t = \frac{e}{1-e}(a - bc)\left(1 - \frac{(A/G,a\%,n)}{n}\right) \quad (2)$$

For the special case of 100% salvage value, the formula becomes

$$t = \frac{e}{1-e}(a - bc) \quad (3)$$

For the special case of 100% salvage value and 100% equity financing, it is

$$t = \frac{e}{1-e}a \quad (4)$$

Explanation of the Basis of the Foregoing Formulas

In explaining the assumptions on which these formulas are based, it is helpful to use several numerical examples and to start with the simplest case, gradually adding various complicating factors to the examples. In all of the following examples, the applicable income tax rate e is 0.49 (i.e., 49%), and the permitted rate of return a is 0.10 (i.e., 10%).

First assume $1,000 of investment in an asset assumed to have 100% salvage value (such as land) and assume that the utility is financed entirely from equity funds. The utility will be permitted to earn $100 after income taxes on this investment, i.e., 10% of $1,000. The before-tax earnings, subject to 49% income tax, must be high enough to cover the tax and leave $100 remaining after taxes. Let T represent the income tax. Then

$$T = 0.49(\$100 + T)$$

$$T - 0.49T = 0.49(\$100)$$

$$T = \frac{0.49}{0.51}(\$100) = \$96.08 \text{ or } 9.61\% \text{ of the investment}$$

The foregoing reasoning is the basis of formula (4), which we may apply as follows:

$$t = \frac{e}{1 - e}\, a = \frac{0.49}{1 - 0.49}\,(0.10) = 0.0961$$

Now change the conditions of the example by assuming that this \$1,000 asset is financed half by equity funds and half by money borrowed at 7% interest. In the terminology of our formulas, $b = 0.07$ and $c = 0.50$. The permitted earnings, before interest but after income taxes, will still be \$100. However, \$35 of this will go for interest on debt (7% of the \$500 borrowed), and this \$35 will be a deduction from taxable income. Thus

$$T = 0.49(\$100 + T - \$35)$$
$$T - 0.49T = 0.49(\$100 - \$35)$$
$$T = \$62.45 \text{ or } 6.25\% \text{ of the investment}$$

The foregoing reasoning is the basis of formula (3), which we may apply as follows:

$$t = \frac{e}{1 - e}\,(a - bc) = \frac{0.49}{0.51}\,[0.10 - 0.07(0.5)] = 0.0625$$

Now make a further change in the conditions of the example by assuming that the \$1,000 asset has an estimated life of 10 years with zero salvage value. In the first year of life of the asset the 10% earnings permitted will apply to the full \$1,000 of investment, and the tax deduction for interest will apply to 7% of \$500, just as in the previous example; it follows that the income tax will be \$62.45 just as before. But in the second year, the 10% will apply to the depreciated book value of \$900, and the 7% interest is applicable to a debt of \$450. And so on. Year-by-year figures are shown in Table 20–1.

The income tax made necessary by the \$1,000 investment starts at \$62.45 in the first year and reduces each year by \$6.245. At 10% interest, the equivalent uniform annual figure is \$62.45 − \$6.245 $(A/G,10\%,10)$ = \$62.45 − \$6.245(3.73) = \$39.16 or 3.92% of the investment.

The foregoing reasoning is the basis of formula (2), which we may apply as follows:

$$t = \frac{e}{1 - e}\,(a - bc)\left(1 - \frac{(A/G,a\%,n)}{n}\right)$$
$$= \frac{0.49}{0.51}\,[0.10 - 0.07(0.5)]\left(1 - \frac{3.73}{10}\right) = 0.03916$$

TABLE 20–1
Year-by-year income tax payments caused by $1,000 investment in utility plant having 10-year life and zero salvage value
Applicable tax rate assumed as 49%. "Fair return" on depreciated book value assumed as 10%. Straight-line depreciation assumed for both rate regulation and income tax purposes. Half of utility financing by debt with interest at 7%.

Year	*Required earnings to cover depreciation, "fair return," and income taxes*	*Depreciation deduction from taxable income*	*Interest deduction from taxable income*	*Taxable income*	*Income taxes*
1	$200 + T_1	$100	35	$65 + T_1	$62.45
2	190 + T_2	100	31.50	58.50 + T_2	55.73
3	180 + T_3	100	28.00	52.00 + T_3	49.96
4	170 + T_4	100	24.50	45.50 + T_4	43.72
5	160 + T_5	100	21.00	39.00 + T_5	37.47
6	150 + T_6	100	17.50	32.50 + T_6	31.23
7	140 + T_7	100	14.00	26.00 + T_7	24.98
8	130 + T_8	100	10.50	19.50 + T_8	18.74
9	120 + T_9	100	7.00	13.00 + T_9	12.49
10	110 + T_{10}	100	3.50	6.50 + T_{10}	6.25

Now make a still further change in the conditions of the example by assuming that the $1,000 asset has a 10-year life with a prospective 50% salvage value at the end of the life. This may be thought of as a $500 asset with 100% salvage value, responsible for a tax of half of $62.45 or $31.23 and another $500 asset with a zero salvage value, responsible for a tax of half of $39.16 or $19.58. The total tax will be the sum of these two figures, $50.51, midway between the figure for zero and 100% salvage values. In general, the income tax percentage for any salvage value above zero and less than 100% can be found by a linear interpolation between the figures for salvage percentages of 0 and 100. Formula (1), our general formula, may be viewed as giving a linear interpolation between the results computed from formulas (2) and (3).

Contrast Between Regulated Public Utilities and Competitive Industry with Respect to Economy Studies and Income Taxes

Our discussion has assumed that, in a regulated utility, income taxes will be allowed by the regulatory authority as one of the components of the price of the utility's product or service, and the economy studies comparing alternative types of plant should, in effect, be made from the viewpoint of the utility's customers. In contrast, our discussion in Chapter 12 implied that the price of a competitive product or service will be established by market conditions unrelated to an individual producer's decision on alternative types of plant and that economy studies for a competitive enterprise should be made from the viewpoint of the owners of the enterprise.

Some important aspects of this difference in viewpoint may be clearer if we examine a specific numerical example. Consider Example 12–3 in which an investment of $110,000 for materials-handling equipment reduced the annual disbursements by $26,300 and increased the annual depreciation charge by $11,000 giving a net increase in taxable income of $15,300. If this equipment were installed in a manufacturer's plant (non-regulated industry), his income tax would be increased by 49% of the net increase of $15,300, or $7,497. The net after-tax income would be $18,803.

In contrast, if this were a proposed investment in the plant of a regulated public utility the reduction of $26,300 in disbursements would be taken into consideration in the determination of the revenue requirements. The influence of the proposed investment on income taxes would depend on the size of the investment itself and on the company's overall cost of money, reflected in its permitted fair rate of return. The higher the after-tax return permitted on the $110,000, the greater the before-tax return must be and the greater the element in income taxes caused by the investment.

A Further Comment on Income Tax Considerations in Economy Studies for Regulated Utilities

The formulas given in this chapter for ratio of equivalent annual income taxes to first cost of plant in a regulated utility were based on certain stated assumptions about the rules of regulation and the method of depreciation used in computing taxable income. They have been presented here to illustrate how formulas may be developed to fit a particular set of assumptions on these matters. It is not intended to suggest that these are the only such formulas that are appropriate. In many public utilities other assumptions regarding the rules of rate regulation and methods of tax depreciation may be closer to the facts. Various other formulas have been published based on a number of different assumptions.*

Two Approaches to a Special Utility Engineering Economy Problem

A common problem in public utilities is caused by the need to add service capability at frequent intervals to meet the growing demand of an area. Expected service lives of units added at different dates may be approximately the same. Moreover, some units may be retired to be replaced by larger units. Often there is no "natural" study period for an economy study to compare different plans to meet expected growth in demand.

*American Telephone and Telegraph Co., *Engineering Economy*, 2d ed. (New York: McGraw-Hill Book Company, 1977).

This type of problem was first introduced in Examples 7–1 and 7–2. The question arises of how best to formulate such a problem in regulated public utilities. Two methods are commonly used. One is called the "coterminated plant" plan; the other is the "repeated plant" plan.

In the coterminated plant plan, the assumption is made that there is some terminal date at which all the units then in service will end their lives. This arbitrary assumption is made only for convenience of calculation, and requires that a "residual value" be estimated to reflect the remaining service life of the units that actually will remain in place and in use at the end of the arbitrarily chosen study period. A simple way to implement this assumption is to calculate the present worth of the costs for the portion of the life that will take place up to the end of the study period. Another way is to compute the present worth of the unrecovered investment at the end of the study period. A third way is simply to use the book value at the end of the study period.

The repeated plant type of analysis was first discussed in Chapter 7 in connection with present worth comparisons. This method makes the assumption that each replacement of an asset will repeat the cost history of the first one. If a unit added to a system to expand capacity is identical with the first unit, then its equivalent annual cost, once it has been installed, is assumed to be the same as for the first unit. The economy study needs to reflect the fact of the deferment of an investment and its related costs until the need actually arises. The equivalent annual cost of each added unit starts at a different time and the assumption is made that this specific annual cost will be repeated indefinitely. Consequently, the capitalized cost of infinite service can be computed for the date of installation and brought back to zero date by using the appropriate single payment present worth factor.

There are at least three different ways by which the effects of the coterminated assumption can be analyzed in economy studies, but the authors prefer the method illustrated in the following example. Two other methods are described in Problems 20–6 and 20–7 at the end of this chapter.

EXAMPLE 20–2

Comparison of Two Proposed Development Plans For Utility Service to a New Residential Community

Facts of the Case

A real estate development company owns a tract of land that it has held for several years in order to develop it for middle-income homes. Available housing in the adjacent city has become scarce and the prices for apartments and private homes have been increasing rapidly. The company has decided that it is time to start the actual development. Since the company does its own contracting the management is well aware of the necessity to build homes at a rate that will enable it to sell all the houses as they become available and still not depress the housing market. Therefore, it has reached a final plan that will require about 8 or more years for completion. These plans have been presented to the local utility company in order that the utility company can start its engineering designs and be ready to install underground conduits, etc., as the streets are being built, and to be able to supply energy to the early home buyers.

The electric company's engineers have arrived at two different plans to meet the demand, based on the projected number of homes to be occupied each year and the gradual growth in the utilization of electrical energy by the homeowners. The first plan, called A, will provide full capacity for 20 years with the initial installation. The second plan, called B, will be a two-step development, with the second step being supplied in 8 years.

The "fair return" allowed by the state public utility commission is 10%, but the company is now borrowing money at 9%. It is the company's policy to maintain about a 50/50 ratio between debt and equity capital at all times. The applicable tax rate for both federal and state income taxes is 54%.

Plan A calls for an initial investment of $200,000 with a prospective salvage value of 20% of first cost 20 years hence. Operation and maintenance disbursements are estimated to be $15,000 a year and ad valorem taxes will be 2% of the first cost.

Plan B calls for an immediate investment of $140,000 and a second investment of $160,000 8 years later. The O & M disbursements will be $9,000 a year for the initial installation, and $8,000 a year for the second installation, making a total of $17,000 a year after 8 years. It is believed that each installation will have an economic life of 20 years with a prospective salvage of 20% of the initial investments. Ad valorem taxes will be 2% of the first cost of all plant in place at any time.

The utility company needs to determine which plan is better, based upon the concept of minimization of revenue requirements.

Comparison of Revenue Requirements
Using the methods previously described:

$$t = \frac{0.54}{(1 - 0.54)}(0.10 - 0.09 \times 0.50)\left[0.20 + (1 - 0.20)\left(1 - \frac{(A/G,10\%,20)}{20}\right)\right]$$

and substituting 6.51 for $(A/G,10\%,20)$, we find that

$$t = 0.0478$$

Plan A

Capital recovery = $160,000(A/P,10%,20) + $40,000(0.10)	= $22,794
O & M disbursements	15,000

Ad valorem taxes = (0.02)\$200,000	=	4,000
Equivalent annual income tax disbursement = 0.0478(\$200,000)	=	9,560
Total equivalent annual revenue required		\$51,354

Plan B

First installation:

Capital recovery = \$112,000(A/P,10%,20) + \$28,000(0.10)	=	\$15,955
O & M disbursements		9,000
Ad valorem taxes = (0.02)(\$140,000)	=	2,800
Equivalent annual income tax disbursement = (0.0478)(\$140,000)	=	6,692
Total equivalent annual revenue required		\$34,447

Second installation:

Capital recovery = \$128,000(A/P,10%,20) + \$32,000(0.10)	=	\$18,235
O & M disbursements		8,000
Ad valorem taxes (0.02)(\$160,000)	=	3,200
Equivalent annual income tax disbursement = 0.0478(\$160,000)	=	7,648
Total equivalent annual revenue required		\$37,083

The second installation is made at year 8, so only 12 years of that service is "chargeable" to the first 20 years of service. Therefore, the equivalent annual revenue required for Plan B is:

$$
\begin{aligned}
AR &= \$34{,}447 + \$37{,}083(P/A,10\%,12)(P/F,10\%,8)(A/P,10\%,20) \\
&= \$34{,}447 + \$37{,}083(6.814)(0.46651)(0.11746) = \$48{,}292
\end{aligned}
$$

This calculation indicates that Plan B requires \$3,062 less annual revenue than does Plan A.

The previous method of analysis clearly "charges" to the 20-year study period the prorated share of the revenue requirements incurred by the second installation and spreads that increment of revenue requirements over the entire 20 years by a logical method.

The Reason for Using Debt/Equity Ratio and Cost of Debt in Computing Public Utility Revenue Requirements

In Chapter 18 we gave a number of reasons why, generally speaking, it is undesirable to make comparisons of alternatives on the basis of return to equity capital. The reader should note that we are not disregarding our own advice when we explain the use of the revenue requirements approach in economy studies for public utilities. The "fair return" adopted as i^* applies to total capital, not merely to equity capital. The debt/equity ratio and the cost of borrowed money are used in the economy study *only* as data that are required to calculate the income tax portion of revenue requirements. In making this calculation for regulated utilities, we, in effect, assume that all items of new plant will be financed by the same proportion of debt and equity capital with all debt carrying the same interest rate. With respect to the debt/equity ratio, this simplifying assumption is a reasonable one for the many regulated utility companies in the United States that endeavor to maintain a fairly constant ratio of debt to equity capital in spite of the need for frequent new financing.

"Fair Return" Is Not Guaranteed for Regulated Public Utilities

We have noted that a "fair return" authorized by a regulatory body for a utility company merely grants permission to earn that rate of return on the rate base *if it can*. Since 1973 some utilities costs have skyrocketed, and there may be no reason to believe that the costs of materials, labor, services, and supplies will decrease in the future. After a rate schedule has been approved by the regulatory commission, a company is likely to be under pressure to effect economies in order to continue to earn something near the stipulated "fair return." The rate schedule usually is based on the company's performance in a given "test year" and the test year is likely to be more favorable to the company than any future year.

The difference between the allowed or fair rate of return and the actual rate earned by a large public utility is shown in Fig. 20–1. Pacific Gas and Electric Company, one of the largest utility companies in the United States, has found it impossible to earn as much as the utility commission allowed for 5 years in spite of massive efforts to reduce costs and improve efficiency. The earning curve clearly shows the results of increasing costs of labor, interest, materials, and equipment and how they have prevented the company from earning the allowed rate of return.

Many economy studies are made by engineers for regulated utilities to find the least-cost method of meeting the service demands over a long period of time. Frequently these studies are, in effect, cost-reduction studies involving the selection of equipment, the design of service extension programs (full capacity now versus stepped development), adoption of new technology versus continuing existing technology, and mechanization or automation to eliminate or reduce labor expenses. A company that does not engage in such cost improvement programs is doomed to find itself with earnings substantially below the "fair return." "Regulatory lag" greatly intensifies the problem.

Energy Costs Present a New Regulatory Problem

Starting in mid-1973 the cost of sources of energy increased very rapidly. Heavy users of energy, the gas and electric companies, were forced to appeal for rate increases frequently in order to cover the increased fuel costs and *try* to earn the "fair return."

These frequent requests for rate changes tended to lead to "regulatory lag," due to the length of time to hear rate cases. Regulatory commissions in most of the states have set up procedures for automatic rate increases for cost of fuel increase adjustments in order to reduce or eliminate the "regulatory lag." The clauses for automatic rate adjustments provide for decreases in the rates as well as increases as the cost of fuel fluctuates. The "fair return" is not affected by these clauses. The significance of these automatic adjustment clauses can be visualized from a report made in March 1975, to the U.S. Senate, in which it was estimated that consumers' utility bills increased by $6.5 billion dollars in the United States in 1974 due to fuel cost adjustments.*

Social Costs of Shortsighted Regulatory Policies May Actually Defeat the Objectives of the Regulatory Agency

Anyone who has lived or traveled in some of the developing nations can testify to the frustrations that result from inadequate, inefficient, unreliable, or inferior utility service. If the water supply system operates only a few hours a day, or if the pressure is so low that little or no flow occurs when a tap is opened, the homeowners and the industrial firms will compensate by installing their own storage tanks and booster pumps. The actual monetary costs of providing, individually, the difference between an inadequate system and a satisfactory (but perhaps still inferior) system usually is much greater than the cost would have been to have provided an adequate system from the beginning.

Inadequate, unreliable, or inferior electrical service normally results in each industrial firm installing its own generation plant, with higher investment cost per unit of capacity and the loss of economies of scale inherent in large generation plants. Consumers of the products or services of these firms pay higher prices for the products or services as a result of the inferior central station electrical system.

Lack of transportation facilities deprives communities of industry and the resulting payment of spendable income in the community. Raw materials go unused, labor goes unemployed, and the area becomes an economic drain on the national or state budget to provide some semblance of relief for the population there.

Even where existing telephone service is relatively good for those fortunate enough to have an instrument, new customers may have to wait months or years to obtain service because the expansion of the service facilities lags behind the demand for service.

But such social costs are not exclusive characteristics of developing nations. Regulatory policies in industrialized nations can cause similar (although

*_Electric and Gas Utility Rate and Fuel Adjustment Clause Increases, 1974_ (Washington, D.C.: U.S. Government Printing Office, March 27, 1975). Committee Print prepared for the Subcommittee on Intergovernmental Relations and the Subcommittee on Reports, Accounting, and Management of the Committee on Government Operations, United States Senate, 94th Congress, 1st Session.

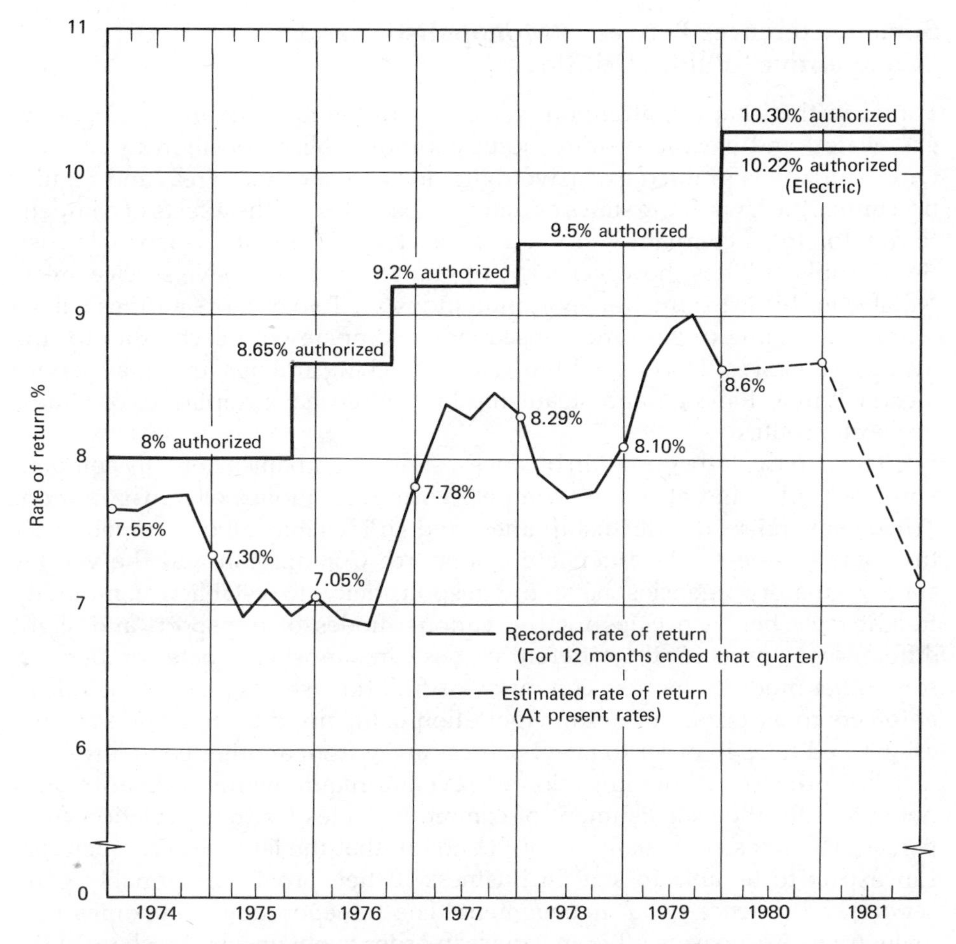

FIGURE 20–1. Rates of return found fair and reasonable for PG&E—and what was actually earned. Reproduced by permission of Pacific Gas and Electric Company.

perhaps not as extensive) adverse effects. Certain rate reductions may seem to provide short-run benefits for the existing customers but may also result in higher costs in the long run because investments to provide future growth may not be allowed in the rate base until the facilities actually are being used. Consequently, construction of service facilities may be undertaken on a piecemeal basis at a higher total cost.

Imposing an artificially low "fair return" in times of high capital costs increases the difficulty of raising new funds that are needed to take advantage of new technology. Thus, plant is kept in service beyond its economic life with resultant higher costs for maintenance and repair, and customers are deprived of more modern and more effective services. Both the customers and the utility stockholders suffer from such a policy.

Some Differences Between "Monopolistic" and "Competitive" Public Utilities

Earlier in this chapter, attention was called to the fact that most telephone, gas, water, and electric systems occupy a monopolistic position in an area. One company is granted exclusive rights to serve a certain area, and regulation under the laws of the states or nation substitutes for the effects of competition in forcing a company to render satisfactory service at a reasonable cost. Some public utilities, however, are not granted such exclusive service areas. Notable in this list is the transportation industry. Two or more airlines will be granted the same routes, and consequently compete with each other for the available business. Dozens or hundreds of trucking and bus lines may service a certain area. Railroad and steamship lines often have duplicate, or almost duplicate, routes.

Thus, in the transportation business, similar companies not only compete with each other but also must compete with other modes of transportation. Truck lines, railroads, steamship lines, and airlines may all be competing for the same business. The Interstate Commerce Commission and the various state regulatory agencies have the responsibility to establish commodity freight rates between cities for the various modes of transport, and slight differences in rates for the different modes can almost eliminate competition from other modes. Then the commissions find themselves playing the role of enforcers to assure that one transportation company does not undercut the established rates in order to take business away from another company.

The principle of minimization of revenue requirements ordinarily does not enter into the establishment of commodity rates for transportation companies. The rates are usually set high enough that the less efficient operators can expect to be able to stay in business if they provide reasonably good service and practice good management. The transportation companies normally try to out-do each other in service in order to obtain a larger share of the available business. Thus, airlines promote their food sevice, comfort of the airplanes, speed, convenient times of departure and arrival, and such as a means of gaining public acceptance. Truck lines and bus lines try to compete on the basis of door-to-door, or downtown-to-downtown service, and speed in total service time from point of origin to final destination.

Thus, in the monopolistic utilities, the engineering economy studies are usually made to find the least expensive way to provide a given service to the customer because the regulatory bodies require that approach, but the competitive utility companies perform engineering economy studies primarily for cost-cutting purposes accompanied by improved service since their rates are fixed for all the competitors. In that respect, economy studies for competing utility companies are exactly analogous to studies for any privately owned competitive business organization. The methods shown in Example 20–1 are not necessary, because the regulatory body is not really controlling the rate of return earned by the company relative to some rate base.

PROBLEMS

20–1. To provide service for an area that is rapidly being changed from farming to residential developments, a telephone company has estimated the prospective demands for a 20-year period and has drawn up several alternative plans for providing the service. Two plans appear to justify detailed economy studies. The first plan will provide underground conduits and exchange facilities immediately to take care of the entire demand for 20 years. Cables will be installed in three steps, one set now, a second set 8 years hence, and a third set 14 years hence. This plan will require an initial investment of $250,000, and each addition of cable will require an investment of $60,000. The annual O & M disbursements for the first unit will be $30,000, and each addition of cable will add $8,000 to the annual O & M disbursements.

Of the initial investment, $10,000 is for land on which to build the exchange. The land will not depreciate. $200,000 is for the exchange building and the underground conduit, both of which are assumed to have a 40-year life with zero salvage value. The remaining initial investment is for the first cable installation. Cable is assumed to have a 20-year life with a 20% salvage value.

The second plan calls for full development of the service capacity needed for 20 years at once. The total investment will be made up of $10,000 for land, $200,000 for building and conduit, and $100,000 for cable. Lives and salvage values are the same as for the stepped development program. The O & M disbursements for the full program will be $38,000 a year for the full 20 years.

Assume that the company maintains a 40/60 debt/equity ratio and that it pays 8% for the borrowed capital. The state utility commission allows the company a "fair return" of 10%. The company's applicable income tax rate is 52% and the ad valorem taxes average about 3% of the first cost of plant. Straight-line depreciation is used for all depreciable assets.

Assuming that these conditions will continue for the 20-year study period, determine the equivalent annual revenue requirements for each plan. (*Ans.* stepped = $80,882; full = $93,671)

20–2. The question often arises as to whether or not a regulated public utility company should lease equipment or buildings rather than own them. It is quite obvious that leasing is a form of borrowing and the interest rate for debt capital compared with the effective interest rate paid for the capital investment avoided by leasing is relevant to any such decision. Furthermore, the stockholders' incentive for leasing is affected by the regulatory agency's decision as to the inclusion of the value of the leased asset in the rate base.

For an example, assume that a service truck that could be purchased by a utility company for $10,000, which the company would probably keep for 4 years and then sell for about $1,000, can be leased on a 4-year lease contract on the basis of $200 a month (beginning of each month) for the first 24 months and $140 a month for the last 24 months. License, maintenance, repairs, etc., will be paid by the utility company in either case.

Compute the equivalent annual revenue required if the truck is purchased, when the debt/equity ratio is 50/50, cost of debt is 6% and "fair return" is 8%. Assume a 52% applicable income tax rate. Ad valorem taxes are included in the license fee.

Compute the equivalent annual revenue required if the truck is leased and not included in the rate base, and then compute the equivalent annual revenue required if it is included in the rate base as though financed by 100% debt and depreciated on a straight line.

Discuss the results of your computations from the customers' viewpoint and the stockholders' viewpoint. (*Ans.* purchase = $3,168; if leased = $2,149; leased and in rate base = $2,946)

20–3. A privately owned water company serving two small counties, including two small cities, has the following bond issues outstanding in early 1980:

Date of issue	*Outstanding amount*	*Maturity date*	*Interest rate payable*
July 1, 1960	$20,000,000	1980	4.4%
July 1, 1965	20,000,000	1985	4.9%
July 1, 1970	20,000,000	1995	5.8%
July 1, 1975	15,000,000	2000	8.5%

(a) What is the embedded cost of debt capital as of June 1980 before the first issue is redeemed?

(b) The company needs to borrow $20,000,000 to redeem the first issue and an additional $10,000,000 to extend and upgrade its service. The new bonds will be issued at an interest rate of 9.8%. What will be the embedded cost of debt capital after the new bonds are issued and the first issue is redeemed?

20–4. Assume that the water company in Problem 20–3 maintains a debt/equity ratio of 55/45,and that the regulatory commission considers the fair rate of return on equity capital to be 12.5%.

(a) What would be the cost of capital, debt and equity combined, that the regulatory commission would allow as the fair rate of return for rate-making purposes after the new bonds are issued in 1980 and the first bond issue has been redeemed?

(b) Suppose the regulatory commission wishes to establish the fair rate of return based on current cost of new capital. What would that rate be?

(c) Comment on how the differences in these rates would affect the market value of the water company's shares of stock and the rates it would charge the water users.

20–5. Certain parameters were given in Problem 20–1. Determine what effects the following changes in the parameters will have, individually, on the revenue requirements for the full-capacity development plan:

(a) Change the debt/equity ratio to 55/45.

(b) Change the cost of debt capital to 6% and the fair return to 9%.

(c) Comment on the effects of these changes on the rates the company would have to charge its customers.

20–6. Example 20–2 illustrated one method by which the "residual value" can be computed in the typical "coterminated plant" problem that is frequently encountered in stepped utility construction projects. Two other methods are in common use. One method is illustrated in this problem and the third is illustrated in Problem 20–7.

This method proposes that the unrecovered portion of the original investment at the end of the study period be assumed to be the "salvage value" of that installation. In Example 20–2, Plan B involves two installations, one at time zero and the second at the end of 8 years, but both are assumed to have 20-year economic lives. The study period is 20 years, which means that only 12 years of service will be rendered by the second installment. Its "salvage value" or residual value at the end of the study period can be computed as:

$$\$128{,}000(A/P,10\%,20)(P/A,10\%,8) + \$32{,}000(P/F,10\%,8) = \$95{,}138$$

If that value is taken as the prospective salvage value of the second installation at the end of the 20-year study period, a new t must be calculated for the second installation. The salvage value as a percentage of the original investment will be

$$(\$95{,}138/\$160{,}000)(100) = 59.5\%$$

If this value for the salvage value as a percentage of the first cost and a new life of 12 years is substituted in the calculations for Plan B, a new t will be found, and that will change the equivalent annual disbursement for income tax.

Using this method, compute the new equivalent annual revenue requirement for Plan B, and comment on the validity of this method compared to that used in Example 20–2.

20–7. A third method for computing the residual value in coterminated plant (see Example 20–2 and Problem 20–6) is to assume that the salvage value at the end of the study period is equal to the book value of the asset at that time. Thus, the book value of the second installation at the end of the study period of 20 years is:

$$\$160{,}000 - \$128{,}000(12/20) = \$83{,}200 \text{ or } 52\% \text{ of first cost}$$

Using that book value as the "salvage value" of the second installation after 12 years, compute a new t and the new total equivalent annual revenue required. Compare the revenue requirements as determined by the three methods illustrated in Example 20–2 and Problems 20–6 and 20–7. Comment on the logic underlying each method.

Problems 20–8 through 20–14 are based on the following problem statement:

In numerous instances companies have found themselves in the position of having to expend large amounts of money to meet certain regulations, codes, and controls. In some cases the companies have found that they had to cease operations because they could not meet the requirements and continue to operate in a profitable mode. In other cases, the companies have found it advantageous to move their operations to another city or state. The following is a hypothetical case based on an actual case. Products, locations, and monetary amounts have been changed.

The Contract Molding Company, a family-owned enterprise, has been operating for 15 years in a medium-sized city near the border between two states. That city is the primary industrial center for a large area in both states. When the company was founded, it purchased a frame building, 10 years old, on one-half acre of land for a total of $300,000. On the company's books the land was valued at $50,000 and the building at $250,000. The company invested $125,000 in equipment over a period of 3 years. The building was depreciated by the straight-line method on the basis of a remaining life of 30 years. The equipment was depreciated by the straight-line method, using a 12-year life and an estimated salvage value of $15,000.

Recently the company has been cited by both the city and OSHA for noncompliance with certain building, safety, and hazard codes. In addition, some of the chemical wastes the company has been flushing into the city's sanitary sewers have been declared toxic by HEW. The company has been given 6 months within which to comply with all of the citations. An engineering firm was employed to draw up detailed plans for renovation of the building and for treatment of the chemical wastes. These plans have been approved by the regulatory agencies, but the best bid the company has received is $190,000 for the installation of steel columns and beams, a sprinkler system, additional fire exits, and complete rewiring of the electrical system. The best bid for the chemical treatment facility is $50,000 and will involve an annual operating cost of $5,000.

Consultation with the IRS has led to the conclusion that these expenditures will increase the life of the building by at least 10 years, leading to the conclusion that the current book value of the building plus the additional investments must be depreciated over 20 years. It is assumed that the building and the treatment plant will have a zero salvage value after 20 years.

The current market value of the building "as is" is believed to be about $50,000 and the land is worth about $100,000. In 20 years the land ought to be worth about $140,000.

The company has been anticipating replacement of the equipment, which is becoming obsolete, and for that purpose has accumulated $125,000 which is currently invested in short-term investments earning about 10% before income taxes. Because the company borrowed about $200,000 initially on long-term bonds, which have recently been paid off, the company does not have the funds to meet the cost of the treatment plant and building rehabilitation. It can borrow $200,000 on a 5-year loan from the local bank at 12%, to be repaid in five equal annual year-end payments of the principal plus the interest due on the beginning-of-year balance.

The five owners of the company, who also serve as the Board of Directors, are justifiably concerned about the long-run effects of the required expenditures on the profitability of the company. One member has proposed that the company be liquidated. Another, the President and General Manager, thinks that the company should move to the adjoining state. The company sells its products to other manufacturers in eight states, and all raw materials and finished products are transported by motor freight. He believes that a relocation will not affect sales or customer relations in any way. In recent years the company's sales have been about $600,000 a year, of which about $75,000 has been taxable income. After considerable discussion the president was authorized to investigate possible locations in the other state.

The president has spent several weeks investigating possible sites for a new plant in the neighboring state and the regulations, codes, and tax structures. He is very excited about a particular site and the apparent advantages of moving the plant to that new location. His findings are summarized as follows:

The company can buy a concrete building about 50% larger than the present one on a 1.5 acre tract of land for a total of $300,000, of which $50,000 is the current value of the land. The building meets all of the local building codes and complies with OSHA regulations. A chemical treatment plant will have to be built, just as at the existing site, and at the same total cost. It will have an estimated life of 20 years, with zero salvage value, while the main building will have an economic life of 40 years with no salvage value.

Since the site is in an area dominated by agriculture, there is a large supply of unskilled and semiskilled labor available with prevailing wage scales approximately $2.00 per hour lower than those at the existing site, which average $8.00 an hour including fringe benefits.

The small city is served by several motor freight lines, so that the move should have no effect on sales or on the supply of raw materials. Service to the customers should be just as good from this location as from the exiting plant.

The company has four skilled laborers who set up, maintain, and repair the process equipment. All these persons have agreed to move to the new site if the company will pay each $5,000 moving expenses for their families and household goods and raise their wages from $15.00 to $16.50 an hour (including fringe benefits). It has 10 semi- and unskilled workers who will not be moved to the new site.

Neither state has a capital gains or loss tax, but their corporate income tax rates are different. The neighboring state charges 3% on the first $25,000 of taxable income and 5% on all over that, while the state of the existing site charges 5% on the first $50,000 and 8% on all over that. The property tax is the same in both, about 1.5% of first cost.

Since most of the equipment is rather well-worn and in some cases obsolete, the president says that it must be replaced immediately, regardless of the decision. Therefore, new equipment, costing about $140,000, will be purchased and can be delivered to either plant. The existing equipment will be sold for $15,000. The proposed new building will make it possible to arrange a more efficient layout which will help reduce operating costs and will provide space for expansion without having to make a new layout. The president estimates that the new layout, with other efficiency improvements, will reduce operating expenses about $15,000 a year. However, all of the new unskilled and semi-skilled employees will need to be trained and he estimates that training will add approximately one month's wages for them to the annual costs the first year. In addition, the president estimates that moving the office's files, dies, tools, etc., will cost about $40,000, all of which can be written off as operating expenses the first year.

The neighboring state is very anxious to have the company move to this proposed site because it will increase the state's income by the amount of the total payroll (which includes $40,000 a year for the president). Since the state estimates that 40% of the payroll will be spent on items subject to a 5% sales tax, it has offered to forgive the property tax on the plant and equipment for a period of 5 years as an added inducement to move. Finally, land values are increasing and the local real estate brokers estimate that the 1.5 acres of land will be worth $200,000 in 40 years.

A local bank near the proposed new site has agreed to loan the company as much as $250,000 at 12% to be repaid in five equal annual year-end payments of principal plus interest on the beginning-of-year debt.

All of these seemingly favorable conditions have convinced the president that a thorough economic analysis is justified.

20–8. The members of the Board of Directors assume that the total sales and existing operating expenditures will remain just about as they have been if the plant at the present site is retained, but they realize that the investments in renovation of the plant and in the chemical treatment plant are going to increase operating and maintenance expenditures. Obviously those expenditures as well as the repayment of the necessary loan with interest will affect taxable income for the company and income taxes to be paid. Prepare an after-income-tax cash flow table to show the effects on operations for the next 20 years. Make appropriate use of carry-forward provisions of the income tax laws in computing the cash flow for income taxes each year. Use the corporate income tax rates in effect at the time you work this problem. (A good reference is the current issue of the IRS Publication Number 334, *Tax Guide for Small Business*.)

20–9. Using the information contained in the cash flow table prepared in Problem 20–8, compute the prospective rate of return on the company's equity. Assume that the investment in the plant before renovation and addition of the chemical plant is the net, after-tax realizable amount if the plant were to be terminated. Also assume liquidation of the plant in 20 years.

20–10. If the decision is to renovate the existing plant, can the company meet the cash flow requirements for debt repayment and interest?

20–11. If the existing plant is to be liquidated, how much cash will the company have, after income taxes, to apply to the establishment of a plant at another site?

20–12. Considering the reevaluation of the existing plant after renovation and addition of the chemical treatment plant, how much will compliance with the regulations involved increase the equivalent uniform annual cost, before income taxes, for the company? Include the effects of the interest payments and assume an i^* before income taxes of 20%.

20–13. Prepare a complete cash flow table after income taxes for the proposed new plant in the neighboring state. All of the disbursements involved in the move will be "expensed" in the first year. Use the corporate income tax rates in effect at this time and take advantage of any carry-forward provisions of the tax rules. The chemical treatment plant must be replaced at the end of 20 years, and assume liquidation of the plant at the end of 40 years. Assume the new equipment will be depreciated on the basis of a 10-year economic life with a salvage value of $15,000. Straight-line depreciation will be used for the equipment, building, and chemical treatment plant.

20–14. The state in which the proposed new site is has offered an inducement (waiver of property taxes for 5 years) to move the plant. The state will enjoy increased sales tax revenue as a result of increase in salaries and wages in the area. Compute the estimated property tax forgone and the increase in the sales tax revenue. Find the interest rate that makes present worth of the increased revenue for 5 years equal to the present worth of the property taxes forgone. Is this a good "investment" on the part of the state? What other state revenues will be increased?

20–15. Using the cash flow table generated in Problem 20–13, and assuming an after-income-tax i^* of 14%, what is the equivalent uniform annual cash flow for the company?

20–16. The federal government and some state governments permit privately owned companies (such as power companies) to sell tax-exempt bonds to finance necessary investments in air pollution control devices required of the companies by air pollution regulatory bodies. Suppose a certain public utility company would have to offer its bonds at 10% compounded semiannually, if the interest were taxable, but can sell tax-exempt bonds at 6% compounded semiannually. Interest on debt is a deductible expense for the computation of taxable income. It is obvious that the privilege of selling tax-exempt bonds to finance required investments amounts to a subsidy by both the state and the federal governments. Assume that the incremental state corporate income tax rate is 8% and the federal rate is 46%. State income tax is a deductible item on federal income tax returns, but federal income tax is not deductible on state returns. Also assume that the bonds are to mature in 20 years. On a required investment of $500,000, how much is the present worth of the subsidy provided by both state and federal government if the company's after-tax i^* is 12%? Since this is a regulated public utility company, who is the real beneficiary of the subsidy, the company stockholders or the customers of the company? Discuss your results.

20–17. Mr. William M. Vatavuk* very kindly supplied the authors with this problem: A certain manufacturing company must install a system to control the airborne emissions (i.e., particulate matter) from the plant to a specified level in order to meet the clean air standards. Three systems have been designed to meet the requirements: (A) an electrostatic precipitator; (B) a Venturi scrubber; and (C) a fabric filter. Each of these systems involves different first costs and annual disbursements. A summary of the data for the three systems follows:

	System		
	A	*B*	*C*
First cost installed	$1,282,000	$907,000	$1,076,000
Life (years)	20	10	20
Salvage value	0	0	0
Annual disbursements			
Labor and labor extras	15,800	63,000	37,800
Maintenance	11,800	23,600	23,600
Replacement parts	0	0	11,800
Power	26,400	275,400	44,400
Water	0	18,400	0
Dust disposal	21,200	0	21,200
Water treatment	0	115,200	0
Property taxes and insurance	51,200	36,300	43,200
Total annual disbursements	$126,400	$531,900	$182,000

(a) Assume the company uses an after-tax i^* of 8%. Which system should the company select? Assume the applicable incremental income tax rate is 54%.

(b) If the after-tax i^* is 14% which system should be selected?

(c) It is recognized that all of the projected costs are subject to variation. Assume that the costs of each system could vary as much as ±30%. How would this affect your recommendation under the conditions specified in part (a)? Discuss your reasoning for your recommendation.

*Professional Engineer, Economic Analysis Branch, Office of Air Quality Planning and Standards, U.S. Environmental Protection Agency, Research Triangle Park, North Carolina.

21

CAPITAL BUDGETING

*Capital budgeting is one of the most important areas of management decision making. This is because the conception, continuing existence and growth of a successful business enterprise are entirely dependent upon the selection and implementation of sound, productive investments. The almost universal reluctance and often complete refusal of top-echelon management to delegate authority to make investment decisions amply demonstrate the widespread recognition of this importance.—R. I. Reul**

The two related aspects of capital budgeting are the sourcing of investment funds and the choice among investment proposals. We have considered certain aspects of the sourcing of investment funds in Chapters 10 and 18. However, this entire book has dealt with various aspects of the evaluation of proposed investments. In this chapter we take a further look at certain troublesome issues that arise in relating sourcing of funds to investment evaluation. We also examine certain organizational aspects of investment decisions, both at the level of top management and at the level of design.

Cost of Capital

A favorite topic for writers on business finance is the overall "cost of capital" to a business enterprise.† It is generally agreed that the objective should be to find a weighted average of the costs of all the various kinds of capital raised by the enterprise.

There is no particular difficulty in using compound interest methods of analysis to find the cost of money acquired by obligations to make fixed payments. Methods of making such analysis were discussed and illustrated in

*R. I. Reul, Sec. 4, Capital Budgeting, in *Handbook of Industrial Engineering and Management*, 2d ed., W. G. Ireson and E. L. Grant, eds. (Englewood Cliffs, N.J.: Prentice-Hall, Inc., copyright 1971), p. 166.

†For a critical view of the literature on this subject, see a paper by David Durand, "The Cost of Capital to the TK Corporation is 00.0%; or Much Ado About Very Little." This paper is included in *Decision Making Criteria for Capital Expenditures, Papers and Discussions of the Fourth Summer Symposium, Engineering Economy Division, A.S.E.E.* (Hoboken, N.J.: The Engineering Economist, Stevens Institute of Technology, 1965).

Chapters 8 and 18, with particular reference to capital raised by borrowing and long-term leases. However, in those chapters we noted that there sometimes are concealed costs associated with such raising of funds; and uncritical analysis may give a figure for cost of money that is really too low.

But there is great difficulty in finding a numerical figure for the overall cost of equity capital in business enterprise. In a corporation, such capital includes retained earnings, if any, as well as the capital raised by the sale of stock. Dozens of different formulas have been proposed by different writers to find the cost of equity capital expressed as an interest rate. When applied to specific cases, the different formulas often give quite different answers. Sometimes the answers clearly are ridiculous.

Even when a particular formula for cost of equity capital is adopted, various possible sources of difficulty remain. Figures taken from the books of account, such as book values and annual profits, are used in many of the formulas. But such accounting figures depend upon a number of arbitrary decisions made by management, including the choice of a depreciation accounting method for book purposes, the choice of a depreciation method for income tax purposes, and—in the frequent cases where different depreciation methods are used for book and tax purposes—the choice of whether or not to recognize the deferment of income taxes in the calculation of profits. Even though the accounting figures are not subject to question, the particular span of years chosen for analysis will greatly influence the computed cost of equity capital in many instances.

It was noted in Chapter 20 that some utility commissions estimate the prospective rate of return that is necessary for a public utility in order to attract equity capital. That "cost of equity capital" is then used to determine the "fair rate of return" to be allowed the utility company, which is, in effect, a weighted average cost of capital.

It is also common practice for the utility companies to maintain a fairly steady ratio of debt capital to equity capital raised through common stock and preferred stock. Such a "steady state" condition is reflected in the use of the "fair rate of return" when determining the revenue requirement for a proposed investment.

Some formulas for cost of equity capital also use market prices of a company's common stock. The fluctuation of stock prices makes the choice of the initial and terminal dates a critical matter in the application of any such formula.

For the foregoing reasons, we do not here propose any method for assigning a figure to the cost of equity capital or for finding the derived figure for the weighted average cost of capital.

Relationship of the Availability of Attractive Investment Proposals to the Sourcing of Investment Funds

The quotation from Joel Dean at the start of Chapter 18 emphasized the point that investment proposals should compete with one another on the basis of merit unrelated to the source of funds that might be deemed to apply to any particular project. Moreover, alternate sources of funds should compete on the basis of merit and the pattern of sourcing should not be determined by the specific investment proposals that are approved. Not only in the quotation from Dean but throughout this book, we have stressed the concept that separable decisions should be made separately. Generally speaking, decisions about investment proposals are separable from decisions about financing, although we noted certain types of exceptions in Chapter 18.

The issues involved in choosing a pattern for sourcing of capital funds are quite complex and are outside the scope of this book. An adequate discussion of these issues would call for a treatise on business finance—or perhaps several such treatises.

But clearly there is no point in raising new funds or in retaining funds that might legally be distributed to the owners unless such funds can be invested in a way that is expected to benefit the present owners of an enterprise. Therefore, broad policy questions regarding the amount of investment funds to be made available are related to the existence of attractive investment proposals. It is only because there are good opportunities for investment that the acquisition of new capital funds is desirable.

The cost and the availability of money that can be obtained by borrowing and long-term leasing will vary greatly from time to time and from one enterprise to another. Also, the possibility of new equity financing and the terms on which new equity funds can be secured will change from time to time and differ from enterprise to enterprise.

We have noted certain difficulties in finding a firm figure for the weighted average cost of capital to a business enterprise. If there were one agreed-on and unimpeachable figure for such a cost of capital in any particular case, if this cost did not change, and if there were no obstacles to the raising of new funds at any time, one might argue that the minimum attractive rate of return should be equal to the cost of capital. Actually, the matter is not so simple at this even under idealized assumptions. Further comments on this point are made following our discussion of the review of investment proposals at the capital budgeting level.

The Concept of an Investment Portfolio

In the mid-1960s a certain educational endowment fund was invested as follows:

Class of investments	Estimated market value as percentage of total
Cash	0.8
Bonds	
Maturing within 5 years	9.3
Maturing in from 6 to 10 years	13.7
Maturing in from 11 to 20 years	8.1
Maturing after 20 years	8.6

Common stocks	50.4
Real estate and related investments	9.1
	100.0

The common stock investments were widely diversified among industries and among companies. The bond investments were similarly diversified and also diversified among years to maturity.

The managers of this fund chose their investments to strike a balance among objectives of safety of principal, present or near-term income, and more distant future income. The various kinds of diversities in investments were related to balancing these objectives.

In a somewhat similar way, it is reasonable for decision makers at the capital budgeting level in competitive industry to view their choices among a group of competing proposals as an attempt to secure a balanced investment portfolio. The use of various other measures of performance that supplement the primary criterion of prospective rate of return can be helpful in reaching this objective.

Further Comment on the Minimum Attractive Rate of Return in Competitive Industry

In our discussion in Chapter 10 of representative values of the minimum attractive rate of return (page 177), reference was made to conversations between the authors and investment analysts who worked at the capital budgeting level. When an analyst was asked the question "What minimum attractive rate of return is used in your company?" a typical answer was "It all depends." Further questioning disclosed that the point that the analysts seemed to have in mind was that it was fairly common for some accepted projects to have lower estimated rates of return than some rejected projects.

They felt that there were various good reasons for not operating with a fixed cutoff point. Often certain important irreducibles favored some of the projects with the lower rates of return. The concept of a balanced investment portfolio might lead to choosing a mixture of higher return projects that were believed to involve relatively high risks with lower return projects that were believed to involve relatively low risks. Various supplementary or secondary decision criteria might be more favorable to some of the lower return projects; often such criteria were based on the concept of sensitivity.

In spite of there not having been a definite cutoff value of estimated rate of return that separated the accepted projects from the rejected ones, it appeared that there always had been some minimum figure for prospective rate of return below which no projects had been accepted. When we quoted representative values of the after-tax i^* in competitive industry (on page 177), we had reference to this minimum figure.

Although decision making among major investment proposals at the capital budgeting level need not be based on any cutoff value for prospective rate of return, it is practically necessary to choose a value of i^* if economy studies are to be made to compare investment alternatives in cases where decisions are to take place below the capital budgeting level in an organization. For the sake of brevity in the following discussion, we shall refer to such decisions as being made at the *design level*. To use some of the special language that has developed in connection with government projects, these are decisions at the level of "project formulation" as contrasted with "project justification."

At the *design level* the alternatives under consideration are mutually exclusive; that is, only one of the alternatives may be selected. The most attractive of the alternatives then becomes one of the proposals submitted to the capital budgeting level for consideration. As has been mentioned several times in previous chapters, the usual situation is that the decision makers have only a limited amount of capital available to invest and may not be able to finance all of the attractive proposals. The selection of those to be financed is the purpose of capital budgeting. That process will be explained later in the chapter with an example.

Implementing Policies for Economic Analysis at the Design Level

There are likely to be a great many decisions between alternatives at the design level for each investment proposal submitted in the capital budget. Every project submitted doubtless has had its subalternatives; many of the subalternatives have had their own subalternatives; and so on. The majority of the examples and problems in this book relate to the economic decisions on the level of engineering design rather than on the level of capital budgeting.

Usually it is a physical impossibility for most of the economic decisions on the design level to be reviewed at the level of capital budgeting. Moreover, in a large organization, hundreds of persons are likely to be involved in design level decisions regarding physical plant. What is needed is a feedback to these persons of the criteria for decision making established at the capital budgeting level. The quotation from Robert F. Barrell in Chapter 10 (page 175) stressed the importance of consistency of criteria throughout an organization. It is the authors' observation that Mr. Barrell is correct in his view that a sound and uniform basis for decision making on the capital budgeting level is much more common than on the design level.

If engineers are left to their own inclinations on design alternatives regarding physical plant, some may greatly overdesign in the sense of making large uneconomic increases in first costs and others may greatly underdesign in the sense of keeping first costs uneconomically low. To offset these extreme tendencies, some written material is needed stating the rules to govern economic design in the particular organization and giving some typical examples of economic comparisons of alternate designs. In some cases this material may take the form of a company manual on engineering economy. Such a manual may well be supplemented by short courses on company time for engineers and other persons concerned with decisions among investment alternatives.

Generally speaking, it is impossible to compare all subalternatives and sub-subalternatives within an organization in the way that was first illustrated in Table 10–1. For this reason the case for making economy studies by the rate of return method is not nearly as strong on the design level as on the capital budgeting level. As a practical matter, to secure rough consistency in the criteria for decision making at two levels, some figure for i^* should be established by management to be used in economy studies made at the design level. Conceivably, this figure may be different for different divisions of an organization, or for different design problems, but it should be definite for any particular case.

Once the minimum attractive rate of return is stipulated, there may be an advantage in using the annual cost method (occasionally supplemented by the present worth method) to make economy studies at the design level. This advantage is that the annual cost and present worth methods are somewhat easier to apply because of the absence of the trial-and-error calculations needed in computing rates of return. Designers may offer less resistance to the making of economy studies based on annual cost than they would offer to studies based on rate of return.

Designers may also be resistant to requirements that they make studies involving specific consideration of the income tax consequences of design decisions. Moreover, they are not always competent to judge the effect on income taxes of different alternatives. For these reasons, and because of greater simplicity, it may be desirable that many economy studies at the design level be based on a minimum attractive rate of return *before* income taxes. In general, someone who understands the income tax aspects of the subject should make a sufficient review of such studies to be sure that they are made in circumstances where the before-tax and after-tax analysis will lead to the same design decisions.

The foregoing commments regarding differences in methods for economy studies at the two levels are not applicable to regulated public utilities that make economy studies based on minimizing revenue requirements. Generally speaking, the same "fair return" should be used as the interest rate in economy studies at all levels; income taxes can be introduced in a simple manner as a percentage of first cost of plant by the type of formula explained in Chapter 20.

Comment on Weakness of the Payout Period as the Primary Criterion for Investment Decisions

In small enterprises it is common to use some variant of the payout (or payback) period as the primary criterion to compare the merits of proposed investments, particularly when the comparisons are made at the level of capital budgeting. Some large enterprises also base decisions on comparisons of payout periods.

Except for the special case where funds are so limited that *no* outlay can be made unless the money can be recovered in an extremely short time (as in Example 10–2), the payout period is *never* an appropriate way to compare a group of proposed investments. The objection is that the payout period fails to give weight to the difference in consequences of different investment proposals after the date of the payout. Thus, a proposal for an investment in jigs and fixtures might pay out in 2 years but have no useful results after the end of the 2 years. Such a proposal would provide recovery of capital with no return whatsoever. Clearly it would not be superior to a proposal for a new production machine having a longer payout period, say 4 years, but favorable enough consequences for many years thereafter to give it an overall rate of return of 20%. Some numerical examples involving payout calculations and illustrating this deficiency are given at the end of this chapter (Problems 21–1 through 21–4.)

Some analysts, attempting to correct the foregoing bias of crude payout in favor of short-lived alternatives, modify the payout calculation by computing so-called payout after depreciation, or sometimes after both depreciation and interest. Such modified payout figures are meaningless as they involve a double counting of the first cost of plant. There are, in fact, many variants of the payout method in use in industry, none of them providing a sound basis for comparing investment proposals.

We have already noted that there may be merit in using crude payout after income taxes as a supplementary criterion related to the cash budget and to judgments on the sensitivity of proposals to the estimates of the duration of their favorable consequences.

Use of Unreasonably Short Payout Periods

Sometimes the payout technique is combined with the stipulation that no proposal will be accepted unless it has an extremely short payout period, such as one or two years. Such a stipulation, if rigidly adhered to, tends to block the approval of projects that would earn excellent returns. As applied to replacement decisions, this type of stipulation tends to perpetuate the use of obsolete and uneconomical plant. George Terborgh's forceful writings on this subject were mentioned in Chapter 17.

Personal Bias as a Factor in Some Decisions on Plant Investment

Where decisions among investment alternatives are made on intuitive grounds, or where the pros and cons of the various alternatives are stated only in words rather than in monetary figures, personal bias often plays a large part in the decision making. Many psychological and personal interest factors, often unconscious, are likely to weigh against a choice for greatest long-run economy.

Particularly insidious among such factors is the desire for personal importance on the part of someone who has the responsibility for an investment decision. Such a desire may lead an individual to make the decision that increases the size of the activities under his or her own control regardless of considerations of economy.

Managers frequently favor short-lived projects over longer-lived ones even though the longer-lived project may have the prospect of providing a much higher rate or return. A manager's salary and bonuses usually depend on the annual performance of his or her division as measured by the accounting "profit" reported. The project with a long economic life may not show any real gain in profits for several years, while the short-lived project may show immediate profit improvements. Thus, the management reward system can discourage the manager from selecting a more attractive project because it will take too long for its effects to show up in the financial report. This effect is frequently cited as a reason that larger organizations are not as innovative as small, young firms. In larger organizations, managers usually try to avoid risk; a small, young firm is usually headed by an entrepreneur who will *not* be evaluated in a short period of time (a year or less) based on the profit picture. The entrepreneur can take the risk of not showing a profit for several years.

Managers with a purely financial outlook sometimes think in terms of immediate least cost. Occasionally engineers may go to the opposite extreme. They may think in terms of the most permanent structure, the latest and most mechanically efficient machine; they want the thing that is the best mechanical job or the most monumental engineering achievement. To be sure, this choice may result in long-run economy in some instances. But there are also cases where the economical thing to do is to choose the temporary installation, or the installation with a low mechanical efficiency and a correspondingly low first cost.

The managerial requirement that decisions among investment alternatives be based on formal economy studies may prevent many bad decisions that would otherwise be based on personal bias. This desirable influence of economy studies may be greatest where they are subject to the possibility of post-audit.

EXAMPLE 21–1

Relationship of Design Level Decision to Capital Budgeting Level Decision

Periodically the K&L Company, which has several divisions, requests the division directors to prepare proposals to be considered for investment in the next capital budgeting period. A special form is provided for the presentation of each proposal, and the division directors are instructed to use a before-tax i^* of 18%. Each director solicits suggestions for capital investment projects from the engineers and department managers. These suggestions may be for anything from investments in equipment to reduce operating costs to proposals for new products. In many cases there are two or more different ways by which the proposed goal could be accomplished, each different way involving different amounts of capital, annual operating disbursements, and, perhaps, sales income. The most attractive of those ways must be determined and presented to the budget committee. Their choices are "design level" decisions, or "project formulations."

One of the project divisions wished to initiate the manufacture of a special component used in its products. The component had previously been purchased from outside suppliers. The production engineers determined that there were three alternative methods of manufacture involving different degrees of automation and consequently different initial investments and operating disbursements. Since this is a rapidly changing industry the study period was set at 8 years based on the estimated time before the component would be replaced due to technological obsolesence.

The data generated in the analysis of the three production methods versus buying the components are shown in Table 21–1. Note that the volume of components varies over the years and that the fully automatic method, Method No. 3, will require approximately one additional year to obtain the equipment and have it installed and de-bugged, resulting in the necessity to purchase the components during the first year.

Computation of the equivalent uniform annual cost of providing the required components, using a before-tax i^* of 18%, indicates that Method No. 2 should be selected. Consequently the division director included Proposal K2 on the list of capital proposals submitted to the budget committee.

Applying the criterion that each increment of optional investment should earn the specified i^*, the engineers could have computed a table of differences and the prospective rate of return on each increment. Table 21–2 shows the results of such a computation.

As should have been anticipated, the $50,000 investment in Method K1 will earn more than 18% over the "Buy" alternative. Similarly, the additional

TABLE 21–1
Project proposal K, test equipment division

Year	*Buy*	*Method No. 1*	*Method No. 2*	*Method No. 3*
0		−$50,000	−$80,000	−$50,000
1	−$30,000	−15,000	−8,000	−80,000
2	−32,000	−15,500	−8,200	−5,000
3	−34,000	−16,000	−8,400	−5,100
4	−34,000	−16,000	−8,200	−5,100
5	−34,000	−16,000	−8,000	−5,000
6	−32,000	−15,000	−7,000	−4,000
7	−25,000	−11,000	−5,000	−3,000
8	−18,000	−7,000	−4,000	−2,500
8		+10,000	+20,000	+30,000
EUAC	−$30,898	−$26,251	−$25,852	−$30,512

TABLE 21–2
Computation of rate of return on each increment of investment

Year	*(K1 − Buy)*	*(K2 − K1)*	*(K3 − K2)*	*(K2 − Buy)*
0	−$50,000	−$30,000	+30,000	−$80,000
1	+15,000	+7,000	−72,000	+22,000
2	+16,500	+7,300	+3,200	+23,800
3	+18,000	+7,600	+3,300	+25,600
4	+18,000	+7,800	+3,100	+25,800
5	+18,000	+8,000	+3,000	+26,000
6	+17,000	+8,000	+3,000	+25,000
7	+14,000	+6,000	+2,000	+20,000
8	+11,000	+3,000	+1,500	+16,000
8	+10,000	+10,000	+10,000	+20,000
	+$87,5000	+$34,700	−$12,900	+$124,200
i =	29.2%	19.6%	negative	25.8%

$30,000 investment in K2 over K1 earns more than 18%, but the additional investments required for K3 over K2 will not be recovered through additional savings in disbursements.

The board of directors of the K&L Company, having decided that the total capital budget for the next period will be $250,000, received the proposals shown in Table 21–3. The proposals received were converted to after-tax cash flows, analysed, and ranked according to their after-tax prospective rates of return. Note that the after-tax prospective rate of return for proposal K2 dropped to 12.5%.

Obviously the budget committee could not fund all of the proposals without borrowing money, and the company's policies prohibited borrowing for capital investments. The committee rejected proposal D4 due to its very

TABLE 21–3
Investment proposals submitted to the budget committee

Project identification	*Project life (years)*	*Investment required*	*Rate of return (%)*	*Cumulative investment*
D4	20	$25,000	18.2	$ 25,000
B3	10	50,000	17.3	75,000
C1	12	25,000	16.1	100,000
H2	6	40,000	15.8	140,000
P3	8	40,000	15.2	180,000
Q2	6	25,000	14.7	205,000
J2	5	60,000	13.0	265,000
N3	8	50,000	12.7	315,000
K2	8	80,000	12.5	395,000
L3	10	20,000	11.5	415,000
M1	15	20,000	10.9	435,000
		$435,000		

long project life and the dynamic nature of the industry. It accepted the other proposals in the order given through J2, requiring a total investment of $240,000. The remaining $10,000 was to be held in short-term money market certificates until the next budgeting period. By its action, the committee effectively said, "The after-tax minimum attractive rate of return at this time is 13.0%."

Some Comments on Example 21–1

The provision of a before tax i^* relieved the engineers and managers from having to obtain information regarding applicable tax rates, applicability of investment tax credits, and methods of depreciation. It also helped to assure that reasonably attractive investments would be proposed so that the budget committee would have a good array of investment opportunities from which to choose with relevant information on the estimated lives of the projects and other data. That enabled the budget committee to apply any number of criteria in its decision-making process.

The provision of a standard form assured that each proposal would be presented in the same way to assist the analysts in preparing the final presentations for the budget committee. It did not and cannot assure that each person preparing a proposal will use the same degree of conservatism in making the estimates of the future consequences of the investments. The budget committee would, in all probability, interview each proposer and question the methods by which the estimates were made and the reasons underlying them. Then the estimates might be altered according to the judgment of the budget committee members. Company policies, such as the policy not to borrow money for capital investments, can be reviewed relative to the opportunities for investment, and if the opportunities are sufficiently attractive, the committee may propose changes to the board of directors.

Example of Forms to Assist in Proposal Presentation

Most organizations design special forms for use in the presentation and analysis of capital investment proposals. Many examples can be found in journals and textbooks on business finance and capital budgeting. The example presented here is an adaptation of forms developed for the Chemical Division of FMC Corporation by R. I. Reul.* Chapter 8 referred to a number of different names used for the rate of return method of analysis by different writers. Mr. Reul's *profitability index* is determined by taking the ratio of the present worth of the investments at time zero to the present worth of the receipts or benefits. Ratios are computed for different predetermined rates of return and are plotted on a graph of interest rates versus the ratios. By joining

*R. I. Reul, "Profitability Index for Investments," *Harvard Business Review*, Vol. 35, No. 4 (July–August, 1957), pp. 116–132.

the points with a smooth curve the interest rate of return, or profitability index, is found by reading the interest at which the ratio of present worths is 1.0. Figure 21–1 shows an adaptation of the analysis form and the graph developed by Mr. Reul.

Figure 21–1 has been completed as it would have been by the engineers in the product division for submission to the budget committee. The present

K&L Company

Profitability Index Form

Proposal K2 Description MAKE VS. BUY COMPONENT #147356

Investment required

T	0%	10%		25%		40%	
−2		1.21		1.56		1.96	
−1		1.10		1.25		1.40	
0	−$80,000	1.0	$80,000	1.0	$80,000	1.0	$80,000
+1		.909		.800		.714	
Total (A)	$80,000		$80,000		$80,000		$80,000
Receipts or cost reductions							
1	$22,000	.909	$19,998	.800	$17,600	.714	$15,708
2	23,800	.826	19,658	.640	15,232	.510	12,138
3	25,600	.751	19,226	.512	13,107	.364	9,318
4	25,800	.683	17,621	.410	10,578	.260	6,708
5	26,000	.621	16,146	.328	8,528	.186	4,836
6	25,000	.565	14,125	.262	6,550	.133	3,325
7	20,000	.513	10,260	.210	4,200	.095	1,900
8	36,000	.467	16,812	.168	6,048	.068	2,448
9		.424		.134		.048	
10		.386		.107		.035	
Total (B)	$204,200		$133,846		$81,843		$56,381
Ratio (A/B)	0.39		0.60		0.98		1.42

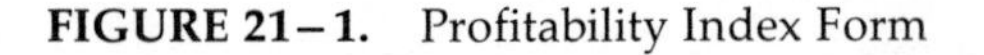

FIGURE 21–1. Profitability Index Form

worths of the anticipated cash flow stream have been computed at 0%, 10%, 25%, and 40%, and the ratio of the investment at time zero, A/B, was plotted on the graph at the bottom of the figure. The smooth curve joining the four points intersects the ratio of 1.0 at slightly more than 25%. That estimated rate of return agrees very closely with the 25.8% computed for K2 minus Buy in Table 21–2.

In both Figure 21–1 and the computations in Table 21–2, the end-of-year convention has been used. Examination of the single payment present worth factors in the interest factor tables (Appendix D) for 10%, 25%, and 40% will reveal that the factors were rounded to three significant figures. In the original article by R. I. Reul, continuous compounding factors for the present worth of specific years were used, based on *nominal* rather than *effective* interest rates. Appendix A explains the use of continuous compounding in considerable detail, and the fact that in some instances it may be desirable to use continuous compounding rather than the end-of-year convention. All that needs to be done to the Profitability Index Form is to substitute the appropriate factors for the different rates and years from Table D–30 (continuous compounding at stated effective annual interest rates) for the factors shown in Figure 21–1.

Estimation of Future Consequences

Regardless of how careful one tries to be in estimating the future cash flows that will result from an investment, it is almost inevitable that the actual cash flows will differ from the projected cash flows. Only in the case of a contractual commitment, such as the payment of interest at a specified rate and the repayment of the face value of the loan, can a person expect to be able to estimate the future cash flow with a high degree of certainty, and even then there is the possibility of default by the borrower. In most cases it is considered fortunate if the actual cash flows are within plus or minus 5% of the estimates. Thus, it does not seem necessary, or desirable, to carry computations to more than three significant figures.

The most common practice in making engineering economy studies is to use the current prices at the time the study is made in estimating future cash flows. Everyone recognizes that inflation and deflation can and usually do occur over a long period of time, but if the alternatives being compared do not differ too greatly in nature (labor, materials, supplies, etc.), it can reasonably be assumed that all of the alternatives will escalate or de-escalate at about the same rate. Under those conditions, the differences in the computed results will maintain about the same relative merit as if estimates of inflation and deflation had been made and applied to the estimates based on current prices.

If, on the other hand, there are substantial differences in the amounts and kinds of labor, materials, and so on, involved in the different alternatives, it may be worthwhile to attempt to apply inflation and deflation factors. This also applies if different elements of the project are known to inflate or deflate at substantially different rates.

Probably the best advice that can be given to anyone responsible for engineering economy studies is to perform sensitivity analyses of the various elements of the problem. This is particularly important relative to estimates of future cash flows. It is always wise to see how much the estimates could change (in both directions) without changing the indicated decision. Such a sensitivity analysis will provide the analyst with a much better insight into the degree of uncertainty that is tolerable in the decision process.

Post-Audit Decisions on Plant Investment

Two useful purposes may be served by an organized program for an audit of past investment decisions:

1. The expectation that audits will be undertaken may cause more careful estimates to be made for all economy studies. That is, estimators will take more care if they anticipate that their estimates will be checked against performance. This influence may apply to many different kinds of estimates—first costs, savings in annual operating costs, sales, prices at which products can be sold, and so on.
2. The information gained in auditing past decisions may be fed back to estimators and analysts and persons reviewing capital budgets in a way that makes it possible to prepare better estimates in the future and to do a better job of reviewing future investment proposals.

Nevertheless, a systematic program for auditing past economy studies should not be undertaken by management without a clear idea regarding the limitations of post-audits. An economy study deals with prospective differences among alternative courses of action. In principle an audit should aim to analyze the difference between what happened with the decision as actually made and what would have happened if some other different decision had been made. But it is rarely possible to be *sure* what would have happened with the other decision. Therefore, a past *decision* is not really subject to audit. Moreover, an uncritical use of accounting allocations—among activities or among different periods of time—can be as misleading in an audit as in any other economy study, especially under conditions of severe inflation or deflation.

Generally speaking, where audit procedures are used, they should be applied to various elements of a group of economy studies rather than to each study as a whole. For example, it is useful to know that particular individuals consistently over- or underestimate on certain items.

Only in a limited number of cases is it possible to make a valid audit of a past estimate of a rate of return. This may be done only for projects that are clearly separable from the other activities of an enterprise and where it is possible to carry the analysis to the date of termination of consequences of an investment. Of course, short of the terminal date, it frequently may be evident that a project is turning out approximately as estimated, or much better, or much worse.

Post-audit procedures are most readily applied to projects for which estimates have been submitted at the level of capital budgeting. In some cases they might be applied on a sampling basis to certain aspects of economy studies made on the design level.

Summary

This chapter has given a brief introduction to the topic of the steps that may be taken by management to ensure that sound decisions are made regarding proposed plant investments. The basic criteria for decision making should be consistent throughout an organization. Nevertheless, it is helpful to make a distinction between decision making at the level of capital budgeting and at the level of engineering design. It often is advantageous to use different methods of implementing decision-making criteria at different levels in an organization.

PROBLEMS

21–1. Two proposals, each requiring an immediate disbursement of $20,000, are to be compared by two methods: (a) crude payout before income taxes; and (b) rate of return before income taxes. Proposal F has a net positive cash flow of $6,000 each year for 4 years. Proposal G has a net positive cash flow of $4,000 each year for 10 years. The salvage values are zero in both cases. Comment on the results obtained from application of the two different criteria. What rate of return results from the difference in cash flows if the longer-lived proposal, Proposal G, is selected? (*Ans*. F: 3.33 years and 7.71%; G: 5 years and 15.10%; 25.7%)

21–2. Two proposals, each requiring an immediate disbursement of $16,000, are to be compared by two methods: (a) crude payout before income taxes; and (b) rate of return before income taxes. Both have a 5-year life. Proposal L has a net positive cash flow of $5,000 a year for 5 years and a zero salvage value. Proposal M has a net positive cash flow of $3,500 a year for 5 years and a $16,000 salvage value. Discuss the results of your analysis. (*Ans*. L: 3.2 years and 17.0%; M: 4.6 years and 21.9%)

21–3. The proposals described in Problem 21–2 are to be analyzed on an after-tax basis. Proposal L involves a relayout in a warehouse facility the total cost of which will be deducted from taxable income at the end of the first year. Proposal M involves the purchase of land for parking. This land will be sold at the end of 5 years at an estimated price, net of all selling costs, equal to its purchase price. Compare the two proposals by crude payout after income taxes and by rate of return after taxes. The company's applicable tax rate is 50%. How does this after-tax analysis compare with the before-tax analysis in Problem 21–2? (*Ans*. L: 3.2 years and 12.6%; M: 5 years and 10.9%)

21–4. Two proposals, each requiring an immediate disbursement of $12,000, are to be compared by two methods: (a) crude payout before income taxes; and (b) rate of return before income taxes. Both proposals have a 6-year life. Proposal R will have net positive cash flows of $5,500 in year 1 and $4,544 in year 2, decreasing by $956 each year to $720 in year 6. Proposal S will have net positive cash flows of $1,530 in year 1 and $2,580 in year 2, increasing by $1,050 each year to $6,780 in year 6. Assuming a choice must be made between the proposals, which one do you recommend? Explain your conclusions. (*Ans*. R: 2+ years and 20%; S: 4-years and 20%; select R)

21–5. A contractor's supply company is formulating its capital budget for the next year. It is a small corporation which derives most of its capital from owners' equity. While it has borrowed from banks for construction projects and maintains a line-of-credit to provide operating capital during busy periods of the year, the owners do not borrow beyond their ability to comfortably meet fixed obligations during periods of recession. In their opinion, the company is currently at that limit.

Five projects, A, B, C, D, and E, have been recommended for the coming year. Each project has two or more mutually exclusive alternative ways in which it may be implemented; i.e., Project A may be implemented by Alternative A1, A2, or A3. The projects are not interrelated. Thus, any one alternative in A, B, C, etc., may be undertaken without affecting the choice or outcome from any other project alternative. Each project has a 10-year life and no salvage value. The company uses a minimum attractive rate of return before income taxes of 20%. The initial disbursement, estimated annual savings, and rate of return for each project alternative is as follows:

Project/ alternative	*Initial disbursement*	*Annual savings*	*Before-tax rate of return*
A1	$20,000	$4,000	15.1%
A2	28,000	8,100	26.1
A3	34,000	9,300	24.2
B1	$12,000	$3,420	25.6%
B2	16,000	4,380	24.3
C1	$16,000	$2,640	10.3%
C2	21,000	14,620	17.7
D1	$12,000	$2,400	15.1%
D2	16,000	4,160	22.6
D3	19,000	4,760	21.5
D4	23,000	5,980	22.6
E1	$22,000	$7,260	30.7%
E2	30,000	9,300	28.5

(a) Make the necessary calculations to support a proposed capital budget for the company. In doing so, rank the project alternatives in order of priority based on rate of return. (*Ans*. Projects A2, B2, D4, and E2)

(b) The financial officer of the company advises that funds available for capital investment this next year will be limited to $85,000. Formulate a proposed capital budget under this circumstance. What minimum attractive rate of return before income taxes is implied by this budget limit? (*Ans*. A2, B1, D4, and E1; i^* implied = 22.6%)

21–6. The budget limitation in Problem 21–5 (b) effectively restricts certain expansion plans. On investigation, one partner in the company discovers that Project A2 requires the purchase of certain equipment costing $12,000, which could be leased. This would free $12,000 in capital investment funds which could be invested in otherwise qualified expansion projects. If leased, the company would be responsible for all expenses of maintenance, taxes, insurance, etc., just as if it owned the equipment. The least costly lease arrangement that could be made is a noncancellable 10-year lease requiring an initial payment, at time zero, of $4,000, payments of $3,000 each at the start of each of the next 4 years, and $2,000 annual payments at the start of each of the last 5 years. Although the lease expenditures will be charged to operating funds rather than to capital investment funds, only $8,000 ($12,000 – $4,000) cash will actually be freed for use elsewhere at time zero. Analyze this proposal before income taxes, and make a recommendation as to whether the lease arrangement should be made, and, if so, which project increments should be added to the budget.

21–7. Three proposals, each for the immediate disbursement of $20,000, are to be compared in crude payout before income taxes and in rate of return before income taxes. Proposal A has a net positive cash flow of $7,000 a year for 4 years. Proposal B has a net positive cash flow of $4,000 a year for 5 years. Proposal C has a net positive cash flow of $2,600 a year for 20 years. Comment on your results.

21–8. For each proposal of Problem 21–1, subtract annual straight-line depreciation from annual cash flow before income taxes. Divide the resulting figure into the investment to obtain a so-called payoff period after depreciation. Comment on your results as compared with the results of Problem 21–1.

21–9. A new product proposal in a manufacturing firm will require a $400,000 investment in land, which must be made 2 years before the date of the start of plant operation. Investment in depreciable plant of $1,200,000 must be made 1 year before the start-up date; a further investment of $1,050,000 in such plant must be made at the date of start-up. A $300,000 investment in working capital is required on the date of the start of operation. The excess of receipts over all disbursements except those for income taxes is estimated as $150,000 for the first year of operation; $300,000 the second; $450,000 the third; and $600,000 each year from the fourth through the 13th. In the 14th year this figure is estimated as $400,000, and in the 15th, $200,000. Annual disbursements for income taxes are estimated as 50% of the excess of the foregoing figures over $150,000; for example, income taxes will be $225,000 in the fourth year.

Assume that the depreciable plant will have a zero salvage value at the end of 15 years and that the investments in land and working capital will be fully recovered.

Prepare a Profitability Index Form similar to Figure 21–1 and plot the graph to determine the approximate rate of return that will be earned on the project. Verify the accuracy of the approximation by calculations using the trial-and-error method and the interest tables.

21–10. Refer to Problem 21–9. Assume that the land will require an investment of only $200,000, 2 years before the date the plant will start operating, and that the land will have a realizable value of $400,000 after the 15th year of operation. Assume that the capital gains tax rate is 28%. With no other changes in the original data prepare the Profitability Index Form as instructed in Problem 21–9.

21–11. The board of directors of the Smith Corporation has allocated $250,000 for capital investments in the next budget period. It requested proposals from the various departments with the stipulation that they should have the prospect of returning at least 18% before income taxes. Any funds not so invested will be put into money market certificates earning 12% before income taxes. If the proposals are particularly attractive, the company can borrow investment capital at a before-tax cost of 14% to be repaid in 10 years with equal annual year-end payments of principal and interest.

The following proposals were received. Note that each subproposal in a set beginning with the same letter is mutually exclusive and that only one of that set may be selected. One proposal of each set can be selected if there is sufficient capital available.

Project	*Investment required*	*Uniform annual return*	*Overall rate of return*
A1	$50,000	$12,000	20.2%
A2	100,000	25,000	21.4
A3	125,000	27,000	17.2
B1	$60,000	$14,800	21.0%
B2	80,000	16,500	15.9
C1	$40,000	$8,000	15.1%
C2	50,000	12,000	20.2
C3	75,000	16,000	16.8
C4	85,000	20,000	19.6
D1	$30,000	$7,000	19.4%
D2	38,000	8,000	16.5
D3	47,000	11,000	19.4
E1	$43,000	$10,000	19.3%
E2	60,000	14,500	20.4
E3	70,000	15,700	18.2

Each project has a 10-year life and zero salvage value.

(a) Examine each set of proposals and determine which need to be analyzed by the rate of return on the incremental investment method. Show those incremental rates of return and list the acceptable projects in decreasing order based on their prospective rates of return.

(b) Which projects should the board of directors select, if it does not borrow any money, in order to maximize the prospective rate of return on total invested capital?

(c) Which projects should the board select if it is willing to borrow any amount of money as long as the before-tax cost of borrowed money is at least 4% less than the prospective before-tax rate of return on the investments?

21–12. Refer to Figure 21–1, page 568. Change the factors from the end-of-year convention to continuous compounding using the factors in Table D–30. Prepare a new Profitability Index Form for the K2 project using the continuous compounding factors and estimate the new prospective rate of return for the project.

21–13. A company's policies are to require projects to have a crude payout of 4 years or less and to earn 12% after income tax. Two proposals to provide the same service are being considered. Proposal A will require an investment of $18,000, have an estimated life of 6 years, and provide an after-tax cash flow of $5,000 the first year and $4,700 the second, decreasing by $300 each year thereafter. It will have a zero salvage value. Proposal B will require the same investment, but will have an estimated life of 8 years with a $4,000 salvage value. It will provide an after-tax cash flow of $3,500 each year throughout its life.

Which proposal would you recommend, assuming that one must be accepted? Comment on the effects of the two criteria and the effects of a salvage value on the crude payout method of analysis.

21–14. Complete the requirements of Problem 21–9 using continuous compounding (Table D–30) instead of the end-of-year convention.

21–15. (This is suggested as a possible term report problem for a course in engineering economy.)

An economy study requires a clear definition either of some proposal for action that is to be considered or of alternative proposals that are to be compared. It then requires estimates to translate such proposals as far as possible into terms of future money receipts and disbursements, accompanied by estimates of other relevant matters (so-called irreducibles) not so translated. Financial calculations are then necessary to provide a basis for judgment. Finally there must be a decision or recommendation as to action that gives weight both to the financial calculations and to the irreducibles. In preliminary studies to determine whether or not any proposal is really promising, the result of a study may be a recommendation for or against the expenditure of time and money for a more detailed study. This problem is intended as a brief exercise in the carrying out of all of these steps.

Selection of Subject. Choose some proposal or set of alternatives for consideration. Two general types of study are possible:

1. A rough preliminary study of some proposed project or set of alternatives. The result of such a study should generally be a recommendation

that the project is or is not worthy of more detailed investigation, or a recommendation that certain alternatives be selected for detailed study. Example: Proposal for a group of small furnished apartments as housing for married students, research assistants, instructors, etc., at your college or university.

2. A complete study of some fairly simple problem of engineering alternatives.
   Example: Selection of an electric motor for a given service.

Because the time available for this report is necessarily limited, you are likely to be better satisfied with your results if you do not undertake too complicated a problem. Wherever practicable, it is a good plan to select a problem dealing with some actual situation of which you have personal knowledge. If you wish to make an economy study in some field in which you do not have access to data on an actual situation, it is necessary to set up a hypothetical case for study.

Collection of Data. Investigation of an actual case may require you to do some legwork collecting information. Any problem may involve some library work. The best sources of reference on technical papers and articles are the *Engineering Index* and the *Industrial Arts Index*. Books on cost estimating give general guidance on methods of making preliminary and detailed cost estimates. Some up-to-date price information may be found in the current issues of a number of technical and trade journals.

Report. The report should start with a clear statement of your problem. It should give your analysis of the problem and should state your definite recommendation for action. The sources of all of your estimates should be given.

APPENDIXES

A

CONTINUOUS COMPOUNDING OF INTEREST AND THE UNIFORM-FLOW CONVENTION

Throughout this book, we have used the end-of-year convention, making compound interest conversions as if cash flow occurring throughout a year were concentrated at the year end. An alternate convention was mentioned in Chapter 6—namely, that certain cash flows taking place during a year occur uniformly throughout the year. To use this alternate convention in trial-and-error calculations of unknown rates of return, it is necessary to assume continuous compounding of interest. Tables D–30 and D–31, based on continuous compounding, may be used for calculations employing this uniform-flow convention.

A Present Worth Formula That Assumes Continuous Compounding

To explain Tables D–30 and D–31, it is necessary to develop the formula for present worth of $1 flowing uniformly throughout one year.

The concept of the present worth at zero date of money flowing uniformly throughout a year may be introduced by a numerical example assuming money flowing at different specified intervals during a year. Certain aspects of the subject are illustrated to better advantage if a high interest rate is used and if the rate is specified as a nominal (rather than an effective) rate per annum. The following example assumes a nominal rate of 30% per annum and assumes a total cash flow of $1 during the year.

If the entire $1 flows at year end, the present worth at zero date is $1(*P*/*A*,30%,1) = $1(0.769) = $0.769. If there are payments of $½ at the ends of half-year intervals and if interest is compounded semiannually, the present worth at zero date is ($½)(*P*/*A*,15%,2) = ($0.50)(1.626) = $0.813. If there are payments of $1/6 at the ends of 2-month intervals and interest is compounded every 2 months, the present worth is ($1/6)(*P*/*A*,5%,6) = ($0.1667)(5.076) = $0.846. If there are monthly payments of $1/12 and interest is compounded monthly, the present worth is ($1/12)(*P*/*A*,2.5%,12) = ($0.0833)(10.258) = $0.855. It may be shown that as the frequency of payments and compounding periods increases indefinitely, the present worth approaches a limit of

\$0.864. The formula for this present worth is $\frac{e^r - 1}{re^r}$ where r is the nominal interest rate per annum.

Derivations in Chapter 4 showed that with continuous compounding, the single payment compound amount factor is e^{rn} and the single payment present worth factor is e^{-rn}. The formula for the present worth of \$1 flowing uniformly throughout a year may be derived along lines similar to the derivations in Chapter 4, as follows:

With regular end-of-period payments A and an interest rate of i per period:

$$P = A \frac{(1 + i)^n - 1}{i(1 + i)^n}$$

If \$1 is divided into m end-of-period payments a year, each payment A is $\frac{\$1}{m}$. If the nominal interest rate per annum is r, the interest rate i per compounding period is $\frac{r}{m}$. Therefore:

$$P = \frac{\$1}{m} \frac{\left(1 + \frac{r}{m}\right)^m - 1}{\frac{r}{m}\left(1 + \frac{r}{m}\right)^m}$$

Now designate $\frac{m}{r}$ by the symbol k as we did in the derivation in Chapter 4.

$$P = \$1 \frac{\left[\left(1 + \frac{1}{k}\right)^k\right]^r - 1}{r\left[\left(1 + \frac{1}{k}\right)^k\right]^r}$$

As the number of compounding periods per year, m, increases without limit, so also must k. It follows that the bracketed quantities in the numerator and denominator approach the limit e. Therefore the limiting value of P is

$$P = \$1 \frac{e^r - 1}{re^r}$$

Relationship Between Nominal and Effective Rates Per Annum Assuming Continuous Compounding

It was pointed out in Chapter 4 that the more frequent the number of compoundings during the year, the greater the difference between the values of nominal and effective rate per annum. This difference is greatest in continuous compounding, where an infinite number of compoundings is assumed. In the language of the mathematics of finance, the nominal rate r used in continuous compounding is referred to as the *force of interest*.

For high interest rates the effective rate per annum is considerably higher than the force of interest. For example, assume $1 at zero date accumulating interest for one year at a force of interest of 30%. At the end of the year the compound amount will be $\$1(e^{0.30}) = \$1.3498$. The effective rate is 34.98%, nearly 5% more than the nominal rate.

Table A-1 shows the values of force of interest to yield various integral values of effective interest rates from 1% to 50%. These values of r were used in computing Tables D–30 and D–31.

Explanation of Tables D–30 and D–31

Table D–30 (in Appendix D) converts $1 flowing uniformly throughout stated one-year periods to the corresponding present worth at zero date, with continuous compounding at effective rates from 1% to 50%.

For example, the first figure in the 30% column is 0.8796, the present worth at the start of a year of $1 flowing uniformly throughout the year. This was computed by the formula we have just derived using an r of 0.26236426.

TABLE A–1
Force of interest, r, to be used in interest formulas involving continuous compounding in order to yield various effective interest rates per annum

Effective rate per annum	*Force of interest (i.e., nominal rate compounded continuously to yield the stated effective rate)*	*Effective rate per annum*	*Force of interest (i.e., nominal rate compounded continuously to yield the stated effective rate)*
1%	0.995033%	15%	13.976194%
2%	1.980263%	20%	18.232156%
3%	2.955880%	25%	22.314355%
4%	3.922071%	30%	26.236426%
5%	4.879016%	35%	30.010459%
6%	5,826891%	40%	33.647224%
7%	6.765865%	45%	37.156356%
8%	7.696104%	50%	40.546511%
10%	9.531018%		
12%	11.332869%		

(It will be recalled that when we used an r of 0.30, the present worth was 0.846.) All subsequent figures in the 30% column are the product of 0.8796 and the appropriate single payment present worth factor. For instance, the figure of 0.4004 for the period 3 to 4 is 0.8796 $(P/F,30\%,3) = 0.8796(0.4552)$.

Table D–31 gives the present worth at zero date of $1 per year flowing uniformly through various stated periods, all starting with zero date and terminating at various dates from 1 to 100. The figures in Table D–31 may be obtained by adding the appropriate figures in the corresponding column of Table D–30. For instance, the figure 2.477 for 30% and the period 0 to 4 is the sum of figures 0.8796, 0.6766, 0.5205, and 0.4004 from Table D–30.

The Annual Interest Period Convention and Continuous Compounding

Throughout this book we have used the symbol i to represent effective interest per interest period. In most examples and problems dealing with practical situations the interest period has been one year because most businesses operate on an annual cycle for accounting, taxation, and budgetary purposes. At one time, banking institutions followed the same practice with respect to interest payments to depositors, although in recent years many other plans have been adopted. Nevertheless, the tradition of the annualizing convention persists in business practice and has been institutionalized, at least in part, in the United States through the so-called Truth in Lending Law.*

Thus, it is reasonable, when making interest conversions for studies in economy, to base analysis on an effective annual interest rate. Tables D–30 and D–31 serve this purpose for the two cases described. Table A–2 gives the mathematical formulas for some of the more important factors useful for compound interest calculations at effective interest rates involving continuous cash flows. The following definitions are used for function designation of these factors.†

$\overline{P}$ or $\overline{F}$ — An amount of money (or equivalent value) flowing continuously and uniformly during a given period.

$\overline{A}$ — An amount of money (or equivalent value) flowing continuously and uniformly during every period continuing for a specific number of periods.

By convention, $\overline{P}$ is reserved for the early investment period during the study, usually the first period which begins at time zero and concludes at time one. $\overline{F}$ may occur at any time and flows during time period t beginning at time

*This law does not require that interest be stated in terms of the true effective annual interest rate. Interest may be stated either in total dollars paid or as a nominal annual interest rate as the term is defined herein.

†Definitions, formulas, and notation have been adapted from "Manual of Standard Notation for Engineering Economy Parameters and Interest Factors," *The Engineering Economist*, Vol. 14, No. 2, Winter, 1969, published under the auspices of the Engineering Economy Division, American Society for Engineering Education. In 1972, this standard was incorporated into ANSI Std. 294.5, American National Standards Institute, New York.

TABLE A–2
Names, functional symbols, and formulas for compounding continuous uniform cash flows

Name	*Symbol*	*Formula**
Continuous compounding present worth factor (single period, continuous payment)	$(P/\overline{F},i\%,n)$	$\dfrac{i}{(1+i)^n \ln(1+i)}$
Continuous compounding compound amount factor (single period, continuous payment)	$(F/\overline{P},i\%,n)$	$\dfrac{i(1+i)^{n-1}}{\ln(1+i)}$
Continuous compounding sinking fund factor (continuous, uniform payments)	$(\overline{A}/F,i\%,n)$	$\dfrac{\ln(1+i)}{(1+i)^n-1}$
Continuous compounding capital recovery factor (continuous, uniform payments)	$(\overline{A}/P,i\%,n)$	$\dfrac{(1+i)^n \ln(1+i)}{(1+i)^n-1}$
Continuous compounding compound amount factor (continuous, uniform payments)	$(F/\overline{A},i\%,n)$	$\dfrac{(1+i)^n-1}{\ln(1+i)}$
Continuous compounding present worth factor (continuous, uniform payments)	$(P/\overline{A},i\%,n)$	$\dfrac{(1+i)^n-1}{(1+i)^n \ln(1+i)}$

*The symbol $ln(x)$ represents the natural, or Napierian, logarithm of the number x.

t – 1 and concluding at time t. The factors for the $P/\overline{F}$ formula are those of Table D–30; those for the $P/\overline{A}$ formula are in Table D–31. These formulas may be used for hand or computer calculations involving continuous cash flows.

Application of Uniform-Flow Convention in Computing Rate of Return

Consider a cost-reduction proposal to purchase certain machinery for $50,000. This proposal is to be analyzed before income taxes. It is estimated that this machinery will reduce disbursements in connection with certain operations by $10,000 a year for the next 4 years. The reduction in disbursements is estimated as $8,000 for the fifth year, $6,000 for the sixth, $4,000 for the seventh, and $2,000 for the eighth. The estimated salvage value at the end of 8 years is $25,000. Table A–3 shows calculations to determine rate of return before income taxes with the uniform-flow convention applied to each year's reduction in disbursements.

The present worth factors for the 4-year period 0 to 4 are taken from Table D–31. The present worth factors for the one-year periods 4 to 5, 5 to 6, etc., are taken from Table D–30. The present worth factors applied to the salvage value at date 8 are taken from the regular 12% and 15% tables, D–17 and D–20.

Interpolation between the present worth of +$3,070 at 12% and −$1,930 at 15% indicates a before-tax rate of return of approximately 13.8%. If the same cash flow series had been analyzed using an end-of-year convention, the computed rate of return would have been 12.4%. In general, rates of return computed with the uniform-flow convention tend to be slightly higher than with the end-of-year convention.

TABLE A–3
Present worth calculations applying uniform-flow convention

		Assuming 12%		Assuming 15%	
Time Period	*Cash flow*	*Present worth factor*	*Present worth*	*Present worth factor*	*Present worth*
0	−$50,000	1.000	−$50,000	1.000	−$50,000
0 to 4	+10,000 per year	3.216	+32,160	3.064	+30,640
4 to 5	+8,000	0.6008	+4,810	0.5336	+4,270
5 to 6	+6,000	0.5365	+3,220	0.4640	+2,780
6 to 7	+4,000	0.4790	+1,920	0.4035	+1,610
7 to 8	+2,000	0.4277	+860	0.3508	+700
8	+25,000	0.4039	+10,100	0.3269	+8,070
Totals	+$35,000		+$3,070		−$1,930

Application of the Uniform-Flow Convention to Proposed Investments

Table A-3 assumes that all the $50,000 investment was made at zero date on the time scale. As it was stipulated that the machinery was to be purchased, this assumption presumably was reasonably close to the facts.

However, if proposed fixed assets are to be constructed or otherwise acquired over a period of time, it often is reasonable to apply a uniform-flow convention to the investment as well as to the cash flow subsequent to the investment. For example, assume that the machinery in Table A-3 will be built over the one-year period prior to zero date on our time scale. The compound amount of $1 flowing uniformly throughout a year is, of course, the reciprocal of the present worth of $1 so flowing. Table A-3 might therefore be modified to show a cash flow of −$50,000 for the period −1 to 0 rather than for date 0. By dividing $50,000 by the appropriate 0-to-1 present worth factors from Table D–30, the equivalent figure at date 0 may be computed to be −$52,890 at 12% and −$53,570 at 15%. Interpolation now indicates a rate of return of 12.1%.

Comparison of Continuous Compounding Tables Based on Nominal and Effective Rates of Interest

It was pointed out in Chapter 21 that R. I. Ruel had used present worth factors based upon continuous compounding in his original profitability index paper (see page 569). Furthermore, he used the stated interest rate as the *force of interest* with the result that his effective annual rate was greater than his stated interest rate. Consequently his present worth factors were smaller than the corresponding factors in Table D–30. Figure 21–1 can be changed from the use of the year-end convention to continuous compounding, or uniform flow throughout the period, by substituting factors from Table D–30 for the single payment factors shown.

The present worth factors in Tables D–30 and D–31 are given for various effective rates. Certain differences between uniform-flow analysis using nominal and effective rates and between these analyses and the end-of-year analyses are brought out in Table A–4.

It will be observed that the figures in column A of Table A–4 are always less than the corresponding figures in column B and that the proportionate difference increases as the period becomes more distant from zero date. The differences here are due entirely to difference in interest rate. Column A, based on a nominal rate of 25%, used an effective rate of approximately 28.4%. The lower effective rate used in column B of course gives higher present worths, particularly at more distant dates.

The ratio of each figure in column B to the corresponding figure in column C is constant (0.896 to 0.800). Both columns are computed at the same interest rate, an effective 25%. The present worths in column B are higher than those in C because the cash flow is sooner—uniformly during a year rather than at year end.

The most striking comparison is between the corresponding figures in columns A and C. The present worth factors in A are greater in the first few lines of the table, approximately equal at the fifth year, and less thereafter. As time goes on, the assumption in A that cash flow occurs sooner is more than offset by the use of a higher effective interest rate.

Reasons for Using Effective Interest Rates Rather Than Nominal Rates When the Uniform-Flow Convention Is Adopted

Wherever the decision is made to use the uniform-flow convention in economy studies, there are two advantages in using continuous compounding tables such as D–30 and D–31 rather than similar tables based on nominal rates.

TABLE A–4
Comparison of certain present worth factors

Period	*Present worth factors with nominal 25% interest* A	*Present worth factors with effective 25% interest (from Table D–30)* B	*End of Year*	*Present worth factors with effective 25% interest (from Table D–24)* C
0 to 1	0.885	0.896	1	0.800
1 to 2	0.689	0.717	2	0.640
2 to 3	0.537	0.574	3	0.512
3 to 4	0.418	0.459	4	0.410
4 to 5	0.326	0.367	5	0.328
9 to 10	0.093	0.120	10	0.107
14 to 15	0.027	0.039	15	0.035
19 to 20	0.008	0.013	20	0.012
24 to 25	0.002	0.004	25	0.004

One advantage is that the reporting of prospective rates of return in terms of effective rates gives a more realistic picture of the productivity of an investment than the reporting of a nominal rate that assumes continuous compounding. For instance, it will be recalled that with continuous compounding a nominal 30% yields an effective 35% per annum. For a proposed business investment with this prospective yield, management will have a better basis for a decision making if the yield is reported as 35% rather than as a nominal 30% compounded continuously.

(This desirability of identifying effective rates applies to personal loans as well as to business investments. Both lenders and borrowers should be aware of effective rates. The officials of a bank that establishes a credit card plan calling for the payment of interest at 1½% per month are doubtless aware that the effective rate is 19.6% per annum; borrowers under this plan should also know that they are paying an effective 19.6%.)

Another advantage of using such tables as D–30 and D–31 is that these tables can be used in combination with conventional interest tables such as are found in standard handbooks and textbooks on the mathematics of finance (or in Appendix D of this book). That is, the conventional tables can be used for all conversions that do not assume uniform flow of funds. If an analysis is to be based on continuous compounding at nominal rates it is necessary to substitute for the conventional tables a complete set of continuous compounding tables applicable to all types of conversions. All factors in such specialized tables will differ from the corresponding factors in conventional tables.

The authors' observation has been that a certain amount of confusion may arise in organizations where special tables are prepared based on continuous compounding at nominal rates. Some person in the organization will use the special tables and others will use conventional tables. Difficulties may arise in checking the analysis of investment proposals and in discussion or other communications among persons using the different types of tables.

Choosing Between the End-of-Year Convention and the Uniform-Flow Convention

In most compound interest conversions in economy studies, the year is adopted as a unit of time that is not to be subdivided. Therefore *some* convention is necessary regarding receipts and disbursements that occur each year. For many items in economy studies (e.g., receipts from the sale of a product or service, routine operating disbursements), the uniform-flow convention is somewhat closer to the facts than the end-of-year convention. But there may be other items where the end-of-year convention may be a better approximation to the facts (e.g., property taxes, semiannual interest). Moreover, in many economy studies there are many receipts and disbursements that clearly apply to a point in time rather than to a period of time.

All things considered, the uniform-flow convention probably comes somewhat closer to describing the way most flow occurs than does the end-of-year convention. Nevertheless, it is desirable to recognize that either is merely a *convention* to facilitate compound interest conversions; neither is a completely accurate reflection of the way in which cash flow is expected to take place.

The authors have observed a number of instances in which the uniform-flow convention is used. But it is their impression that there are a great many organizations using the end-of-year convention for every one that uses uniform-flow.

The greater use of the end-of-year convention doubtless is based largely on grounds of its greater convenience. Standard interest tables and formulas may be used with this convention. In the numerous cases where annual cost comparisons are made, this convention has the advantage of being better adapted to such comparisons. Where compound interest methods require explanation in presenting the results of economy studies, it is easier to explain periodic compounding of interest than to explain continuous compounding.

Compound interest conversions in economy studies constitute one of the steps in providing a rational basis for decisions among alternatives. In most cases the *decisions* will be the same regardless of the convention used. That is, both conventions will normally array a series of investment proposals in the same order.

However, there are certain cases where the two conventions might lead to different recommendations for a decision between alternative investments and where the recommendation based on the uniform-flow convention will be sounder than the one based on the end-of-year convention. These cases occur particularly where an investment leading to positive cash flow concentrated in the near future is being compared with one leading to positive cash flow spread over a considerably longer period. Several examples of such cases are illustrated in the problems at the end of this appendix.

PROBLEMS

A–1. A savings and loan association offers to pay interest on deposits "from the day of deposit until the day of withdrawal." The association pays 5.75% interest compounded daily which, it claims, amounts to a true interest rate of 5.92% annually.

(a) If the association pays nominal interest at the rate of 5.75% annually, what is the effective annual interest rate under continuous compounding using the uniform flow convention? What is the mathematical formula for this computation?

(b) Calculate the effective annual interest rate assuming 365 compounding periods per year. How do the answers to (a) and (b) compare? (*Ans.* (a) 5.9185%; (b) 5.918%)

A–2. A department store offers to finance appliances at an interest rate of only 1½% per month on the unpaid balance. This, it is claimed, amounts to an Annual Percentage Rate (APR) of 18% per year. What is the effective annual interest rate? What would be the effective annual interest rate under the uniform flow, continuous compounding convention at a nominal interest rate of 18% per year? Comment on the differences between the store's claim and the results of your analysis? Which method more nearly approximates the "truth"? (*Ans.* 19.562%; 19.722%)

A–3. An analysis for economy is to be made on a certain piece of equipment having a first cost of $6,000, and economic life of 6 years, and no salvage value.

Disbursements in association with the ownership and operation are expected to be $520 each month. Annual ad valorem property taxes of $240 will be paid at the end of each year. Since this is a service business open 52 weeks a year, weekly receipts are expected to average $185. Assume the company uses an interest rate of 25% before income taxes to analyze such proposals.

(a) Use the end-of-year convention to find the net present worth of this proposal.

(b) Use the uniform-flow convention with respect to receipts and disbursements other than for property taxes to find the net present worth of this proposal.

(c) Comment on the percentage difference in the results obtained by application of the two different conventions. (*Ans.* (a) $3,266.14; (b) $4,469.42)

A–4. In order to compare the effect of different compounding periods on effective annual interest rate, consider the following situation. An investor may deposit $1,000 for exactly 1 year under the following conditions at a nominal interest rate of 12% per annum.

(a) Interest paid into the account annually.
(b) Interest paid into the account semi-annually.
(c) Interest paid into the account quarterly.
(d) Interest paid into the account monthly.
(e) Interest paid into the account weekly.
(f) Interest paid into the account daily (365 days per year).
(g) Interest paid into the account continuously from the moment of deposit to the moment of withdrawal.

Assuming the deposit and withdrawal 1 year hence are precisely timed, find the amount of the withdrawal in each case. Comment on the relative impact that adoption of these various conventions might have on an analysis for economy.

A–5. Two proposals are to be compared using (a) the uniform-flow convention and (b) the end-of-year convention. Compute the prospective before-tax rates of return for each. Proposal G involves an investment of $20,000 at zero date with zero salvage value after 3 years. The expected excess of receipts over disbursements is $9,200 each year. Proposal H involves an investment of

$20,000 with an expected life of 25 years and no salvage value. Its expected excess of receipts over disbursements is $3,700 a year.

A–6. Proposal M involves an investment of $75,000 in a project with an estimated life of 6 years and a zero salvage value. The expected excess of receipts over disbursements is $17,200 a year. Proposal N involves an investment of $75,000 in a project with an estimated life of 10 years and an estimated salvage value of $15,000. The expected excess of receipts over disbursements is $11,700 a year. Compare the prospective before-tax rates of return for these proposals using (a) the uniform-flow convention and (b) the end-of-year convention.

A–7. Compare the two proposals in Problem A–6 assuming that Proposal M involves an instantaneous payment of $75,000 when a machine is delivered at time zero, but Proposal N involves building a machine over a period of 1 year ending at time zero. Make your comparison of the prospective before-tax rate of return for each using (a) the uniform-flow convention and (b) the end-of-year convention. Comment on the differences of your results in Problems A–6 and A–7.

A–8. A bank offers an interest rate of 5.25% compounded daily from date of deposit to date of withdrawal for passbook accounts. What is the effective annual interest rate? What would the effective annual interest rate be if the interest were compounded continuously? Can you think of any reasons why the bank would advertise daily compounding rather than continuous compounding?

A–9. A state highway department is considering a proposal that a certain stretch of two-lane highway be converted into a four-lane, divided freeway. The state requires that all such projects have the prospect of earning a return for the highway users of at least 7% and that the estimated lives be 40 years. Because traffic must be maintained during the construction period, this proposed freeway will require 2 years for construction. The engineers have estimated that the project will require an investment of $20,000,000 the first year and $10,000,000 the second year.

Similarly the benefits to highway users have been estimated to be at least $1,700,000 the first year after the freeway is opened to full use and, because of increasing population and increasing costs of driving, the benefits will increase by $100,000 each year throughout the 40 years. Salvage value is assumed to be zero at the end of 40 years.

Highway economists have always used the end-of-year convention in making engineering economy studies but one of the new employees has suggested that the study should be based on the uniform-flow convention. Compute the prospective rate of return based on (a) end-of-year convention and (b) the uniform-flow convention (prepare a form such as Figure 21–1 for this purpose, and plot the results to obtain the estimated prospective rate of return). Comment on your results. Which method of analysis seems more appropriate to you?

A–10. Refer to Problem 20–17. The problem was originally solved using the end-of-year convention. Now assume that all of the annual disbursements except property taxes and insurance are made more or less uniformly throughout the year and that the initial investments (first cost) were uniformly distributed over the year preceeding the start of operation of the air quality control systems. Assuming that the $i*$ of 8% used in the problem was the effective annual rate, analyze, without computations, the effect that the use of the uniform-flow convention versus the end-of-year convention would have on the equivalent uniform annual cost of the alternatives. Then do the necessary calculations to confirm or deny your prior conclusions regarding the effects.

A–11. A regulated public utility company has been granted a fair rate of return of 10% by its regulatory commission. A new project proposal for cost reduction results in an annual revenue requirement of $1,000,000 based on the end-of-year convention. Actually the project involved the expenditure of $5,000,000 for new equipment and its installation spread over a 2-year period before the project would be in operation, half in each year. If the uniform-flow convention had been used, what would have been the original value used as the "rate base" at time zero? If there had been no change in the rate base, what would the annual revenue requirement have been if the uniform-flow convention had been used?

B

THE ANALYSIS OF PROSPECTIVE CASH FLOW SERIES THAT HAVE TWO OR MORE REVERSALS OF SIGN

Our first discussion of compound interest in Chapter 3 dealt with loan transactions. Although we examined a number of different possible cash flow patterns that occur in various types of loans, all the patterns had one characteristic in common. From the lender's viewpoint, there was an initial negative cash flow and all subsequent cash flows were positive. And, of course, from the borrower's viewpoint, an initial positive cash flow was followed by one or more negative cash flows.

Chapter 8 dealt with calculation of approximate values of unknown interest rates. Like the initial examples in Chapter 3, all of the cash flows in the examples and problems in Chapter 8 involved only one reversal of sign of cash flow. For borrowings, positive cash flows were followed by negative cash flows. For investments, negative cash flows were followed by positive cash flows. Other things being equal, low interest rates on prospective borrowings were more attractive than high ones. And, other things being equal, high rates of return on prospective investments were more attractive than low ones.

In Chapter 18, Examples 18–1 and 18–2 discussed at some length the calculations necessary to find the after-tax cost of borrowed money and of financing through leasing. In each of these cases, the difference in after-tax cash flow involved a sequence beginning with a single positive cash flow, representing the amount borrowed or provided by the lessor, followed by a sequence of negative cash flows representing repayment with interest. (Example 18–2 includes a positive cash flow in the final year resulting from recovery of the initial security deposit.) This is the usual form of a loan from the viewpoint of the borrower. Borrowing-financing cash flow patterns, however, need not all follow the pattern just described. A series of annual borrowings, positive cash flows from the viewpoint of the borrower, may be followed by either a sequence of repayments, negative cash flows, or a single repayment. An example of this type of loan is that provided by many banks under the college student loan program.

This appendix deals with analysis and evaluation of proposals that involve two or more reversals of the sign of prospective cash flow. Such proposals, in effect, involve a "borrowing" during one period and an "investment" during another period even though this mixed aspect may not always be recognized or described by these words. A practical difficulty in analysis arises because criteria of attractiveness for proposed borrowings differ from criteria of attractiveness for proposed investments.

The following discussion points out that in evaluation of such proposals, it often is desirable to use an auxiliary interest rate. For a proposal that is primarily an investment project, the auxiliary rate should be applied during the borrowing period. For a proposal that is primarily a borrowing project, the auxiliary rate should be applied during the investment period. Before examining the use of such auxiliary rates, it is helpful to observe the troubles that may arise when one makes a conventional trial-and-error solution to find the unknown interest rate in cases that have two reversals of the sign of cash flow.

EXAMPLE B–1

Computing "Solving Rate of Return" from a Proposal Involving a Delayed Investment

Facts of the Case

An oil company is offered a lease of a group of oil wells on which the primary reserves are close to exhaustion. A major condition of the contract is that the oil company agrees to undertake injection of water into the underground reservoir in order to make possible a secondary recovery at such time as the primary reserves are exhausted. The lessor will receive a standard royalty from all oil produced from the property whether from primary or secondary reserves. No immediate payment by the oil company is required.

The production department of the oil company estimates the year-by-year after-tax net cash flow from this project as shown in Table B–1. The net negative cash flow in year 5 is caused by the investment in the water flooding project. The positive cash flows in years 1 through 4 reflect the declining net receipts from primary recovery; the positive cash flows in years 6 through 11 reflect the declining net receipts from secondary recovery.*

Conventional Trial-and-Error Calculations to Find an Unknown Interest Rate

Table B–1 shows the present worths of the cash flows computed at various interest rates. The sum of the present worths starts with a positive value (+$590,000) at 0% interest. It declines to zero at an interest rate of about 28%,

*This example was suggested by an example of a similar water flooding project described in J. G. McLean's article "How to Evaluate New Capital Investments," *Harvard Business Review*, Vol. 36, No. 6 (November-December, 1958), pp. 67–68. However, the project evaluation analysis suggested in this appendix differs from that given in the McLean article.

TABLE B–1
Present worth of estimated cash flow from agreement to undertake proposed water flooding project, assuming various interest rates
(All figures in thousands of dollars)

Year	Net cash flow (and PW at 0%)	Present worth at 5%	at 10%	at 15%	at 20%	at 25%	at 30%	at 35%	at 40%	at 50%
1	+$120	+$114	+$109	+$104	+$100	+$96	+$92	+$89	+$86	+$80
2	+90	+82	+74	+68	+62	+58	+53	+49	+46	+40
3	+60	+52	+45	+39	+35	+31	+27	+24	+22	+18
4	+30	+25	+20	+17	+14	+12	+11	+9	+8	+6
5	−1,810	−1,418	−1,124	−900	−727	−593	−487	−404	−336	−238
6	+600	+448	+339	+259	+201	+157	+124	+99	+80	+53
7	+500	+355	+257	+188	+140	+105	+80	+61	+47	+29
8	+400	+271	+187	+131	+93	+67	+49	+36	+27	+16
9	+300	+193	+127	+85	+58	+40	+28	+20	+15	+8
10	+200	+123	+77	+49	+32	+21	+15	+10	+7	+3
11	+100	+58	+35	+21	+13	+9	+6	+4	+2	+1
ΣPW	+$590	+$303	+$146	+$61	+$21	+$3	−$2	−$3	+$4	+$16

is negative at 30% and 35%, rises to zero at about 37% and continues to increase at all rates above 37%. The curve of total present worth as a function of i is shown in Fig. B–1.

EXAMPLE B–2

Computing "Solving Interest Rates" For a Proposal That, In Effect, Involves a Delayed Borrowing

Facts of the Case

A company is offered a contract permitting use of a piece of land for a limited period. A major condition of the contract is that at the end of the 21st year, the property must be returned to the owner accompanied by a specified payment toward the cost of a new building. A smaller outlay must be made at once for modification of the building now on the land. During the intervening years (1 to 20, both inclusive), the company will be in a position to receive a positive net cash flow from rental of the property. The company's real estate department makes estimates of year-by-year after-tax cash flows that will result from the acceptance of this proposal.

These estimates are shown in Table B–2. The reader will note that the sum of the cash flows is negative. Moreover, a negative cash flow of $3,000,000 takes place in the 21st year, the terminal year of the contract. It is evident that even though this proposal might not be described in legal language as a borrowing, it needs to be analyzed as a borrowing or financing project. This is true in spite of the required initial investment of $700,000.

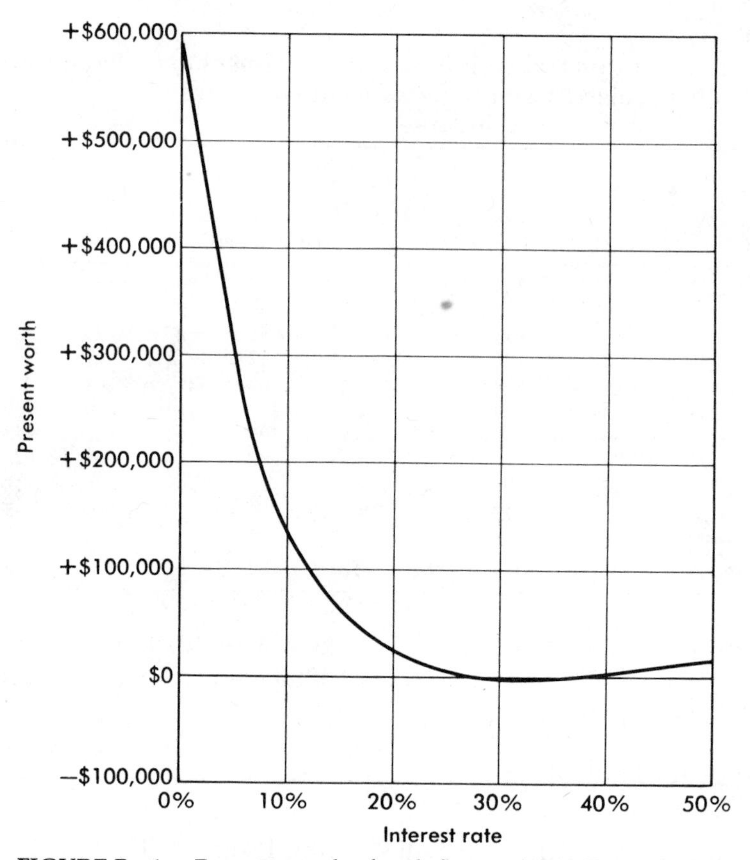

FIGURE B–1. Present worth of cash flow series in Example B–1

TABLE B–2
Present worth of estimated cash flow from proposed agreement regarding use of certain land, assuming various interest rates
(All figures in thousands of dollars)

Year	*Net cash flow (and PW at 0%)*	Present worth							
		at 2%	*at 3%*	*at 7%*	*at 10%*	*at 15%*	*at 20%*	*at 25%*	*at 30%*
0	−$700	−$700	−$700	−$700	−$700	−$700	−$700	−$700	−$700
1 to 10	+$200 per year	+1,797	+1,706	+1,405	+1,229	+1,004	+838	+714	+618
11 to 20	+$100 per year	+737	+635	+357	+237	+124	+68	+38	+22
21	−$3,000	−1,979	−1,612	−724	−405	−159	−65	−27	−12
ΣPW	−$700	−$145	+$29	+$338	+$361	+$269	+$141	+$25	−$72

Conventional Trial-and-Error Calculations to Find an Unknown Interest Rate
Table B–2 shows the present worths of the cash flows computed at various interest rates. The sum of the present worths starts with a negative value (i.e., −$700,000) at 0% interest. It increases to zero at an interest rate of about 2.8%, reaches a maximum at about 10%, after which it declines continuously, passing through zero at about 26.3%. The curve of total present worth is shown in Fig. B–2.

Some General Points Illustrated by Examples B–1 and B–2

The curves in Figs. B–1 and B–2 are typical of curves of sums of present worths of cash flow in cases where there are exactly two reversals of sign of cash flow. With ΣPW plotted as the ordinate and *i* as the abscissa, the curve will be concave upwards, as in Fig. B–1, when the signs of cash flows are in the sequence +, −, +. The curve will be convex upwards, as in Fig. B–2, when the signs of cash flows are in the sequence −, +, −.

Such curves will intersect the *i* axis in either two places or none. For example, consider a slight change in the data of Example B–1; add a positive

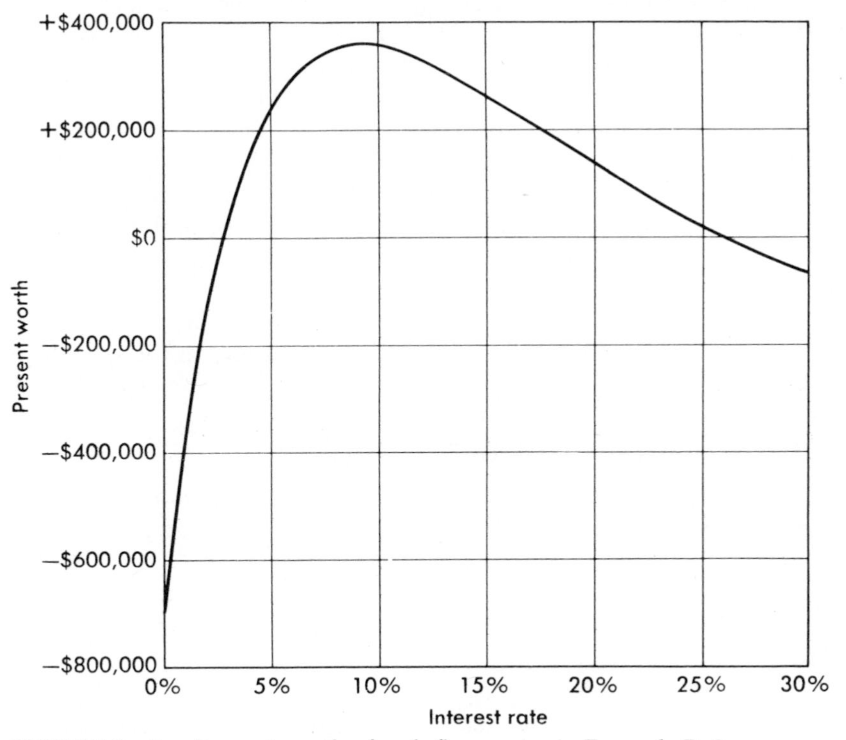

FIGURE B–2. Present worth of cash flow series in Example B–2

cash flow of \$5,000 at year 0. With such a change, all values of ΣPW in Table B–1 would be increased enough so that the curve would never intersect the *i* axis; there would be no "solving rate of return" at all.

Cash flow sequences of this type may present particular difficulties when computer routines are used for their solution. Most computer routines employ some form of search method which seeks a single answer; that is, a value of *i* for which ΣPW equals 0. Depending on the sophistication of the routine, the computer may arrive at just one of the solving values of *i*, *any* one. Unless the analyst is aware of this possibility, he or she is left believing that *the* answer has been found. And even if the analyst does recognize the problem, he or she likely does not know *which* one of the two or more values of *i* has been found.

In both examples, the two places where the curve intersects the *i* axis are at positive values of *i*. However, with different figures for cash flow, one or both of the intersections could be at negative values of *i*. For example, change the data of Example B–2 so that the required outlay at year 21 is \$2,000,000 rather than \$3,000,000. With this change, ΣPW will be +\$300,000 at $i = 0$, and there will be only one positive value of *i* (about 26.7%) where ΣPW = 0. Nevertheless, the general form of the curve will be similar to the one shown in Fig. B–2. If calculations of ΣPW are made for negative values of *i*, the curve will intersect the *i* axis at about −1.3%. (Generally speaking, negative values of solving interest rates do not have any useful meaning as a guide to decision making.)

It seems evident from the foregoing that there are difficulties in interpreting conventional compound interest analyses when prospective cash flow has two reversals of sign. Although in Examples B–1 and B–2 these difficulties arose in our attempt to find an unknown interest rate by trial-and-error calculations, we shall see that difficulties also arise in interpretation of any net present worth calculations in these and similar cases. The key to an evaluation of such proposals lies in the use of an auxiliary interest rate. We shall also see that an important aspect of the matter is the sensitivity of the conclusions of an evaluation to moderate changes in the auxiliary interest rate selected.

Analysis of Example B–1 Using an Auxiliary Interest Rate

The proposed contract in Example B–1 has two periods, a borrowing-financing period in years 1 to 5 and an investment period in years 5 to 11. However, the amount of money involved in years 5 to 11 is so much more than in the earlier years that it is clear that this proposal should be evaluated as an *investment* project. It is helpful to look at the prospective cash flows as follows:

Year	*Cash flow Series A (borrowing-financing)*	*Cash flow Series B (investment)*
1	+\$120,000	
2	+90,000	
3	+60,000	

4	+30,000	
5	$-x$	$-\$1,810,000 + x$
6		+600,000
7		+500,000
8		+400,000
9		+300,000
10		+200,000
11		+100,000

Each of these series is a conventional one with only a single reversal of the sign of cash flow. If we assign an interest rate to Series A, we can determine x; we shall then be able to solve for the unknown rate of return in Series B. The auxiliary rate assigned to the borrowing-financing period should be a rate the enterprise would deem it reasonable to pay for the funds made available during that period and should be selected with consideration given to the need for financing and to the costs of debt capital from various possible alternate sources. Auxiliary rates of 0%, 5%, and 10% give the following results (in thousands of dollars):

Auxiliary rate	x	*−$1,810 +* x	*Rate of return on investment*
0%	$300	−$1,510	14.0%
5%	348	−1,462	15.5%
10%	401	−1,409	17.4%

The foregoing figures for rate of return can be interpreted somewhat as follows:

1. Even though it is believed that there is no advantage in the earlier timing of the $300,000 of cash receipts in years 1 through 4, the water-flooding investment should be viewed as earning a 14.0% return.
2. If 0% and 10% are believed to be limiting values for the appropriate auxiliary rate, corresponding limiting values for the prospective rate of return on the water-flooding investment are 14.0% and 17.4%.
3. The figure for rate of return is somewhat less sensitive to the chosen auxiliary rate in the lower range of values of the auxiliary rate. (An increase in the auxiliary rate from 0% to 5% increases the rate of return by 1.5%, whereas an equal increase from 5% to 10% increases the rate of return by 1.9%.)

Table B–1 gave 28% and 37% as the two figures for the solving rate of return for the water flooding proposal. It should be evident that neither of these figures is useful in judging the merits of this proposal as an investment project. Unless an auxiliary rate of 28% is applied to our cash flow Series A, Series B will not show a yield of 28%. And unless an auxiliary rate of 37% is applied to Series A, Series B will not yield 37%!

These solving rates of return imply that the company would be willing to borrow capital at either 28% or 37% per annum after taxes, which is very unlikely.

Analysis of Example B–2 Using an Auxiliary Interest Rate

In Example B–2 there were positive cash flows of $200,000 a year for the first 10 years and $100,000 a year for the second 10 years. If the only offsetting negative cash flow had been the $3,000,000 at year 21, this would clearly have been an attractive financing scheme in which money was available for an extended period at a cost of 0%.

However, Example B–2 also had a negative cash flow of $700,000 at zero date. Clearly, the greater the negative cash flow at zero date, the less attractive the proposal is as a financing scheme. An analyst can evaluate this combined proposal by viewing it primarily as a borrowing-financing scheme and assuming that the positive cash flows of the first few years are applied to the recovery of a $700,000 investment at some stipulated rate of return. It is helpful to look at the prospective cash flows as follows.

Year	*Cash flow Series A (investment)*	*Cash flow Series B (borrowing-financing)*
0	−$700,000	
1 to $(y - 1)$	+200,000 per year	
y	+ x	+$200,000 − x
$(y + 1)$ to 10		+200,000 per year
11 to 20		+100,000 per year
21		−3,000,000

Each of these series is a conventional one with only a single reversal of the sign of cash flow. If we assign an interest rate to Series A, we can determine x and y; we shall then be in a position to solve for the unknown cost of money in Series B. The auxiliary rate assigned to the investment period should be the estimated rate of return that is believed to be forgone by devoting this $700,000 to a financing project rather than to a normal investment-type project. Auxiliary rates of 0%, 5%, 10%, 15%, and 20% give the following results (in thousands of dollars):

Auxiliary rate	y	x	*$200 − x*	*Cost of money during borrowing-financing period*
0%	4	$100.0	$100.0	2.6%
5%	4	188.8	11.2	3.0%
10%	5	106.3	93.7	3.7%
15%	6	68.4	131.6	4.8%
20%	7	125.0	75.0	6.9%

Because all of our cash flow figures were after income taxes, the costs of debt capital are after-tax costs. The reader may recall from Chapter 18 that after-tax costs of money obtained in certain other ways (borrowing or leasing) tend to be considerably less than before-tax costs. It is evident that the attractiveness of this particular project will depend on the auxiliary rate that is deemed to be appropriate.

The computed costs of capital are relatively insensitive to the chosen auxiliary interest rate at low values of the auxiliary rate. An increase of the auxiliary rate from 0% to 5% increased the cost of money by only 0.4%. In contrast, the answer is much more sensitive at high values of the auxiliary rate. A 5% increase in the auxiliary rate from 15% to 20% caused a 2.1% increase in the computed cost of money.

Table B–2 gave 2.8% and 26.3% as the two interest rates that made the present worth of the entire cash flow series equal to zero. Neither of these figures is appropriate for judging the merits of this project as a borrowing-financing proposal except under quite unrealistic restrictions. The low cost of money implied by the 2.8% answer implies that the auxiliary rate also is 2.8%; in other words, no alternate investment can be found for $700,000 that will yield more than 2.8%. And the high cost of money implied by the 26.3% answer implies that the auxiliary rate also is 26.3%; in other words, immediate commitment of $700,000 to this project requires the forgoing of a conventional investment with an after-tax yield of 26.3%.

Special Aspects of Certain Proposals Such as Example B–2

We evaluated the proposal in Example B–2 as a borrowing-financing proposal. The final negative cash flow of $3,000,000 made such an evaluation appropriate in spite of the negative cash flow of $700,000 at zero date.

However, a difficulty may arise in analysis for capital budgeting whenever a proposal that has a −, +, − cash flow sequence calls for evaluation as a borrowing-financing project rather than as an investment project. A *low* value of the cost of money is attractive for a financing project, whereas a *high* rate of return is attractive for an investment project. Nevertheless, the acceptance of a proposal that requires a substantial cash flow at the start may well cause the loss of a good rate of return by taking funds that otherwise could be used for a normal investment project.

In a borrowing-financing proposal that has the cash flow sequence −, +, −, it is not practicable to find a valid rate of return on the first negative cash flows that can be compared with prospective rates of return on normal investment proposals. The best that can be done is to recognize that the opportunity costs associated with the first negative cash flows are relevant in selecting the auxiliary interest rate to be used in the analysis.

Of course, not all proposals that have two reversals of the sign of cash flow are borrowing-financing proposals when the cash flow sequence is −,+,−. The classification of a project as an investment project or a borrowing-financing project should depend on the *magnitudes* of the *negative cash flows at the end of the cash flow series* in relation to the magnitudes of the preceding cash flows.

Some Mathematical Aspects of Analysis of Cash Flow Series That Have Two or More Reversals of Sign

If $\frac{1}{1+i}$ is designated as x, the usual analysis to find an unknown rate of return on a proposed investment C_0 may be thought of as a trial-and-error solution of the present worth equation:

$$-C_0 + C_1x + C_2x^2 + \ldots + C_{n-1}x^{n-1} + C_nx^n = 0 \tag{A}$$

where the values of C are the estimated cash flows for the investor at dates 0 to n and all signs after $+ C_1x$ also are plus. Similarly, the usual analysis to find an unknown interest rate associated with the borrowing of C_0 may be viewed as a trial-and-error solution of the present worth equation:

$$+C_0 - C_1x - C_2x^2 - \ldots - C_{n-1}x^{n-1} - C_nx^n = 0 \tag{B}$$

where the values of C are the stipulated cash flows for the borrower at dates 0 to n and all signs after $-C_1x$ also are minus.

Anyone familiar with the principles of algebra will recognize that such an equation has n different roots. However, some of these roots may be imaginary numbers (that is, they involve $\sqrt{-1}$) and others may be negative.

Descartes' rule of signs tells us that equations such as (A) and (B) which have only one sign change will have one positive real root. Equations with exactly two sign changes, such as those that might have been written for Examples B–1 and B–2, will have either two positive real roots or none. Equations with exactly three sign changes will have either three or one positive real root. Equations with exactly four sign changes will have either four, two, or no positive real roots. And so on.

The reader should note that Descartes' rule applies to positive real roots of an unknown that we have chosen here to represent by x. But our x is an expression for $\frac{1}{1+i}$. A positive value of x may correspond either to a positive or negative value of i. However, a negative value of x will always give a negative value of i. It follows that the number of positive values of i that will make $\Sigma PW = 0$ may be *less* than the number of positive real roots given by Descartes' rule but may never be *more*.

Equations in which the sequence of cash flows starting with zero date begins with one or more positive cash flows followed by one or more negative cash flows, and concludes with one or more positive cash flows (the +, −, + sequence) will give a present worth curve that is concave upward, such as the one in Figure B–1. Equations with the −, +, − sequence will give a curve that is convex upward, such as the one in Figure B–2. In our discussion of Examples B–1 and B–2, we noted that moderate changes in the data of the examples could shift the curves up or down so that there would be *no* value of i for which the sum of the present worths would be zero. It also was possible to change the data to shift the curves to the left so that one or both of the solving values of i would be negative (even though the x in our equation would continue to be positive).

Cash Flow Series with More Than Two Reversals of Sign

The greater the number of reversals of sign of cash flow, the greater the possible numbers of values of i for which the sum of the present worth of cash flow is equal to zero. We have seen that in the case of exactly two sign reversals, it may be true that *neither* of the solving interest rates may be a useful guide to any decision making. A similar statement applies to *all* the solving rates obtained when there are three or more reversals of sign.

The difficulties of analysis using auxiliary interest rates may be increased by an increase in the number of reversals of sign in the cash flow series. It may be harder to make the initial decision identifying a proposal as primarily an investment proposal or primarily a borrowing-financing proposal. Moreover, with only two reversals of sign, it is necessary to apply the auxiliary rate during only one time sequence. Extra sign reversals may increase the number of time sequences that require auxiliary rates and may raise the question of whether or not the same auxiliary rate is appropriate for all the different time sequences. Some of these matters are illustrated in our discussion that follows Example B–3; others are brought out in some of the problems at the end of this appendix.

Numerous papers that provide complex mathematical tests to determine whether or not a unique solution exists for equations involving two or more changes of signs have been published in a number of journals. The authors believe that these papers are of little value to analysts or decision makers, but the reader is referred to several such papers in *The Engineering Economist*.

EXAMPLE B–3

Computing "Solving Rates of Return" For a Proposal That Involves Three Reversals of Sign in the Sequence of Estimated Cash Flows

Facts of the Case

In an extractive industry, it is proposed to make an initial outlay of $180,000 to speed up recovery of certain ore. It is estimated that one consequence of this outlay will be an increase in after-tax receipts of $100,000 for each of the next 5 years. However, the additional ore recovered in earlier years will not be available for later years. It is estimated that after-tax receipts will be decreased by $100,000 for each year from 6 to 10, both inclusive. Another estimated consequence of the acceptance of this proposal will be an increase in after-tax receipts from the sale of this property of $200,000 at the end of an estimated 20-year period of ownership.

The prospective influence on cash flow of accepting this proposal is therefore:

Year	Cash flow
0	−$180,000
1 to 5	+ 100,000 per year
6 to 10	− 100,000 per year
20	+ 200,000

Conventional Trial-and-Error Calculations to Find an Unknown Interest Rate
Figure B–3 shows the sum of the present worths of the foregoing cash flows computed at various interest rates from 0% to 50%. This curve passes through zero at approximately 1.9%, 14%, and 29%.

It will be observed that the present worth declines from +$20,000 at 0% to a low point of about −$11,200 at 6%, then rises to a high point of about +$4,200 at 20%, and then declines continuously thereafter. Such a curve with two changes of direction is characteristic of present worth curves for cash flow series that have exactly three reversals of the sign of cash flow.

Moderate changes in the estimated cash flow at zero date will make great changes in the points of intersection with the i axis even though they make no

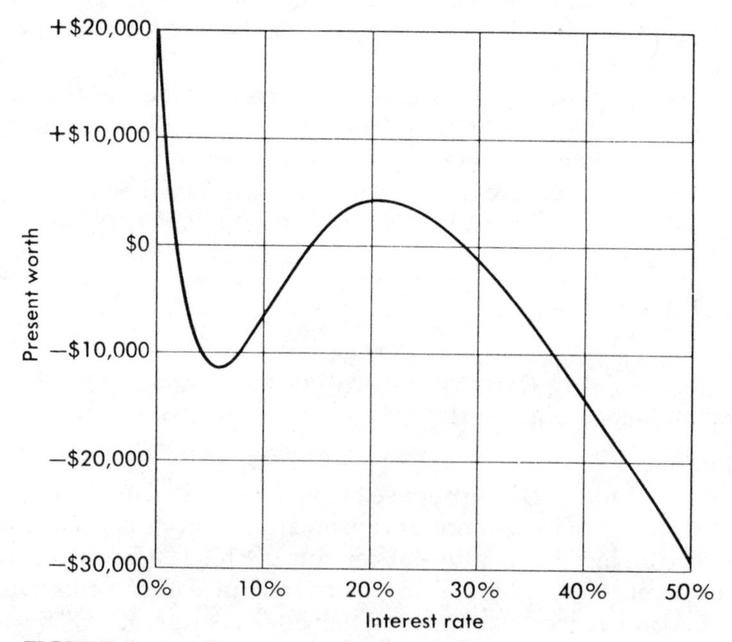

FIGURE B–3. Present worth of cash flow series in Example B–3

change in the shape of the curve. For instance, a change from −$180,000 to −$185,000 will lower the curve so that it has only one intersection with the i axis (at about 1.2%). A further change to −$205,000 will lower the curve so that there will be *no* positive value of i that makes the sum of the present worths equal to zero. (The curve would interesect the i axis at a value of i of about −0.3%.)

It seems apparent that in this case our conventional calculations are not adequate to give us satisfactory guidance on the acceptance or rejection of a proposal.

Analysis of Example B–3 Using an Auxiliary Interest Rate

The first question to be answered in our analysis is whether we want to solve for a cost of money in a proposed borrowing-financing transaction or to solve for a rate of return on a proposed investment. In Example B–3, we should aim to find a cost of money. The purpose of the project is to speed up certain receipts and the dominant cash flows are those in years 1 to 10:

Years	Cash flow
1 to 5	+$100,000 per year, a total of $500,000
6 to 10	−$100,000 per year, a total of $500,000

If the smaller cash flows at dates 0 and 20 had not been part of this project, it would have been evident that this project would provide financing at a 0% cost. Such financing would obviously have been desirable.

When we apply an auxiliary interest rate to the −$180,000 at date 0, it is as if we viewed the first $180,000 as an investment to be recovered at the stated auxiliary rate by part of the first positive cash flows during years 1 to 5. When we apply an auxiliary rate to the final +$200,000, it is as if we viewed this positive cash flow as recovery with interest of an investment made by part of the last negative cash flows of the period 6 to 10. Table B–3 illustrates the application of the foregoing concepts with auxiliary rates of 0% and 10%.

An inspection of the derived borrowing-financing series obtained with the auxiliary rate of 0% shows that the after-tax cost of this capital is negative. However, an auxiliary rate of 0% clearly is not a reasonable one to assume under any normal circumstances.

A trial-and-error solution for the borrowing-financing series that was obtained with the more reasonable auxiliary rate of 10% indicates an after-tax cost of capital of approximately 11%. This would seem much too high to be attractive in light of the 10% auxiliary rate.

We have used the same auxiliary rate for the initial period and the terminal period in both analyses. However, it is evident from inspection of Table B–3 that the conclusions of this analysis are much more sensitive to the choice of the auxiliary rate for the final period, which involves a span of at least 10 years, than to the auxiliary rate chosen for the initial period, which involves a much shorter period of time.

TABLE B–3
Application of auxiliary interest rates of 0% and 10% to analysis of the cash flow series of Example B–3 as a borrowing-financing proposal

		Auxiliary rate at 0%		*Auxiliary rate at 10%*	
Year	*Cash flow series*	*Assumed investments*	*Borrowing-financing series for analysis*	*Assumed investments*	*Borrowing-financing series for analysis*
0	−$180,000	−$180,000		−$180,000	
1	+100,000	+100,000		+100,000	
2	+100,000	+80,000	+$20,000	+100,000	
3	+100,000		+100,000	+8,580	+$91,420
4	+100,000		+100,000		+100,000
5	+100,000		+100,000		+100,000
6	−100,000		−100,000		−100,000
7	−100,000		−100,000		−100,000
8	−100,000		−100,000		−100,000
9	−100,000	−100,000			−100,000
10	−100,000	−100,000		−77,100	−22,900
20	+200,000	+200,000		+200,000	

The Cult of "Net Present Value"

Some writers on economics, finance, and operations research have taken the position that present worth (or *net present value*, as they usually call it) is the only valid method for evaluation of investment proposals. One of several arguments advanced against evaluation by prospective rate of return is that under certain circumstances two or more answers can be obtained for a "solving rate of return."

Our position in this book has been that when the methods are properly applied, sound conclusions can be obtained with a variety of methods, including equivalent uniform annual cash flow, present worth, rate of return, and analyses comparing benefits with costs either with or without the calculation of **B/C** ratios. More specifically, in the ordinary type of investment proposals in which one or more negative cash flows are followed by a sequence of positive cash flows, all of the methods will give the same conclusion on the question of whether or not a stipulated minimum attractive rate of return i^* will be obtained.

When we are confronted with the less common type of proposal in which the prospective cash flow series has two or more reversals of sign, the contention is unsound that the present worth method is superior to the conventional rate of return method. *Neither* method can be counted on to give a sound evaluation. In cases such as Examples B–1, B–2, and B–3, if an analyst should elect to use the rate of return method, there is a fair chance that he or she will observe that there are multiple solutions and therefore see that some source of difficulty exists. In contrast, if the analyst makes a present worth analysis using only one value of i^*, the existence of the difficulty doubtless will be concealed.

For instance, suppose the water flooding proposal of Example B–1 is analyzed by the present worth method using an i^* of 25%. Because the prospective cash flow series has a positive net present value at 25%, such analysis will indicate that this is an acceptable proposal. But our discussion of the application of an auxiliary interest rate to this example brought out the point that the conclusion that this investment will yield 25% is not a valid one unless it would have been appropriate for the enterprise to contract to receive the positive cash flows during the initial borrowing-financing period at an after-tax cost of 25%. Presumably such an after-tax cost of money would be deemed excessive in nearly all circumstances. Therefore, the analysis based on net present value would lead to an incorrect evaluation.

A special difficulty arises with certain prospective cash flow series that have the sequence $-$, $+$, $-$, as in Example B–2, or $-$, $+$, $-$, $+$, as in Example B–3, whenever the timing and magnitudes of the cash flows are such that the proposals ought to be evaluated as borrowing-financing rather than as investments. Because the initial cash flow is negative, there may be a tendency to evaluate such proposals as investments. Such evaluations using net present value with a stipulated value of i^* can be extremely misleading as guides to decision making. This point is illustrated in several of the problems at the end of this appendix.

Presumably, in any given case there is some after-tax cost of money above which an enterprise will be unwilling to raise money from sources that involve fixed obligations to make future payments (in other words, from other than equity sources). Let us designate this maximum acceptable after-tax cost of nonequity money as i_o. Usually i_o will be considerably less than i^*, the minimum rate of return that is deemed sufficient to justify a proposed investment.

When a single value of i is used in computing the present worth of a cash flow series that involves two or more reversals of sign, the implication is that $i_o = i^*$. This is the implication either in trial-and-error present worth calculations to find the value of i that makes the sum of the present worths equal to zero or in project evaluations by the method of net present value using a stipulated or assigned figure for i^*.

Applying Conventional Methods of Analysis to Slightly Unconventional Cash Flow Series

Nevertheless, there fortunately are many cases where it is good enough for practical purposes to use conventional methods of analysis—such as equivalent uniform annual cash flow, present worth, rate of return, excess of benefits over costs—even though a cash flow series has two or more reversals of sign. Conventional methods are satisfactory for analysis of any such proposals that clearly should be evaluated as investment projects provided the conclusions of the analysis are sufficiently insensitive to the choice of an auxiliary rate during any borrowing-financing period.

Sometimes this lack of sensitivity may be apparent merely from an inspection of the cash flow series. For instance, consider a proposal that starts with a negative cash flow followed by a series of positive cash flows that are in total considerably more than the starting negative flow. At the end there is a relatively small negative cash flow, possibly because of a negative salvage value. The small size of the final cash flow and its distance in time from zero date may make it evident that an analyst should treat this −, +, − cash flow series in the same manner that he or she treats any conventional investment proposals that have only the −, + sequence. Several problems at the end of this appendix deal with the analysis of −, +, − proposals with and without the use of auxiliary interest rates.

In some regulated public utility companies in the United States, i^* may be small enough and i_0 may be close enough to i^* for it to be reasonable to make present worth comparisons at the stipulated i^* even though there may be several reversals of sign in a cash flow series that describes the differences between two alternatives. We made such a comparison in Example 7–1 (page 97) where the column of differences in cash flow between Plans G and F showed four reversals of sign. Problem B–11 deals with a further analysis of this example.

Accounting and Income Tax Aspects of Some Proposals Where the Cash Flow Series Has Two or More Reversals of Sign

The accounting aspects of projects of the type discussed in this appendix sometimes are fairly complex and controversial. In some instances, proposals that might be deemed attractive on the basis of the type of analysis explained in this appendix may be vetoed because of the accounting treatment that will be required; there may be too great an adverse short-run influence on reported profits.

Proposals of the type discussed in this appendix are more common in the extractive industries than in other types of business enterprise. In the United States, where the mineral industry has the option of using percentage depletion, the income tax aspects of such proposals may be even more complex than the accounting aspects.

In this appendix we have deliberately avoided any discussion of these complexities by assuming after-tax cash flows in all our examples and problems.

Frequency of Occurrence of Cash Flow Series of the Types Discussed in This Appendix

It cannot be emphasized too strongly that cases such as those illustrated in Examples B–1, B–2, and B–3 are the exception rather than the rule. They occur chiefly in the mineral industries and the petroleum industry; even there they arise only in rather specialized circumstances.

PROBLEMS

B–1. Alter the data of Example B–1 in two ways. Assume that an immediate payment of $180,000 is required to secure the lease on the oil wells. Assume that the purchase of the lease does not require any agreement to undertake the water flooding project to secure the secondary recovery; the oil company can make its own decision about whether or not to undertake the water flooding at such time as the primary reserves are exhausted. Make no change in the estimates of the after-tax positive cash flows from primary recovery or from secondary recovery (if undertaken) or in the estimate of the cost of water flooding at date 5. Discuss the question of how to analyze this altered proposal, making any calculations that you consider to be relevant. (*Ans.* 31.5%; 6%)

B–2. Alter the data of Example B–1 by assuming that an immediate payment of $180,000 is required to secure the lease. Assume that the oil company must agree to undertake the water flooding at date 5, just as stipulated in Example B–1. Discuss the question of how this analysis should differ from the one appropriate for Problem B–1. Make any calculations that you believe are called for to guide the oil company's decision in this case. (*Ans.* If auxiliary rate = 0%, i = 8.8%; if auxiliary rate = 15%, i = 8.3%)

B–3. In the discussion of the application of Descartes' rule of signs on page 600, these statements appear:

"A positive value of x may correspond either to a positive or negative value of i. However, a negative value of x will always give a negative value of i."

Show that the foregoing two statements are correct. $\left(Ans.\ i = \dfrac{1 - x}{x}.\right)$

B–4. Our discussion of the analysis of the proposal in Example B–1 using several auxiliary interest rates gave rates of return of 14.0%, 15.5%, and 17.4% corresponding respectively to the auxiliary interest rates of 0%, 5%, and 10%. Make the necessary compound interest calculations to check these stated rates of return.

B–5. Our discussion of the analysis of the proposal in Example B–2 using several auxiliary interest rates gave costs of the money made available by the proposal as 2.6%, 3.7%, and 6.9% corresponding respectively to the auxiliary rates of 0%, 10%, and 20%. Make the necessary compound interest calculations to check these stated costs of money.

B–6. Figure B–3 showed the sum of the present worths of the cash flows associated with the proposal in Example B–3. Make the necessary calculations to check present worth totals for values of i of 0%, 2%, 6%, 10%, 15%, 20%, and 30%.

B–7. Cathy Jones bought the agency for a cosmetic product for $500. She had to recruit sales persons and paid them an hourly salary to sell the product. For 4 years her after-tax net cash flow was $500, but in the fifth year her net after-tax cash flow was −$1,600. Find the solving rates for her by plotting the net present worth of her cash flows at 0%, 5%, 25%, 50%, and 75%.

B–8. Since Cathy's cash flow is obviously a borrowing-financing situation, find the cost of borrowed money for her using auxiliary rates of 0%, 5%, 10%, and 15% for the investment portion of the cash flow.

B–9. The estimated after-tax cash flow associated with increased recovery of an ore is as follows:

Year	**Cash flow**
0	−$16,000
1 and 2	+ 50,000
3	−100,000

Plot the sum of the present worths of the cash flow for values of i of 0%, 10%, 15%, 50%, 100%, 200%, 250%, and 300%. (You will need to compute your own present worth factors above 50%.) At what values of i is the sum of the present worths equal to zero?

B–10. Evaluate the project in Problem B–9 as a borrowing-financing proposal, making use of auxiliary interest rates of 0%, 10%, and 15%.

B–11. Examine the cash flow series (G − F) in Table 7–1 (page 97). This cash flow series has four reversals of sign. Example 7–1 gives the sum of the present worths of this series at 0% and 7% as +$22,500 and +$2,610, respectively. Compute ΣPW at 10%. At what approximate value of i will ΣPW = 0?

Why is it evident that ΣPW will continue to decline for values of i greater than 10%?

In spite of the four reversals of sign, the ΣPW curve seems to have the same form that one would expect from a conventional investment proposal that has only one reversal of sign. Nevertheless, Descartes' rule tells us that there must be at least one more value of i that will make ΣPW equal to zero. Where will this value be? Explain your answer.

B–12. The net cash flow series related to a particular business venture is as follows:

Year	*0*	*1*	2	3	4	5
Net cash flow:	+$400	−$1,000	0	+$300	+$300	+$300

(a) This cash flow series has two solving interest rates. One, 126%, is beyond the scope of the tables in this book. Find the second solving interest rate and give an interpretation of the meaning of these two values of i. Sketch the curve of net present worth as a function of i in the range 0 to 126%.

(b) The first receipt of $400 clearly constitutes borrowing against a $1,000 disbursement at time 1. Assuming that this entrepreneur can borrow reasonable sums of money at 14% interest, what is the rate of return on the subsequent investment series?

(c) Determine the sensitivity of the rate of return on the investment series found in part (b) by using auxiliary interest rates of 0% and 20%.

B–13. This problem is based on an actual occurrence involving cash flows two orders of magnitude greater than those in this problem for a number of gas wells owned by several operating companies.

A company drilled a well in a field known to have natural gas. When the well reached a depth of about 6,000 feet, it penetrated a pocket of gas of rather limited quantity. The well was continued to about 14,000 feet where it struck a very large pocket of gas. The company decided to extract all the gas from the second level before extracting any from the first. After several years of extracting natural gas from the second level, a government agency tried to force the company to withdraw gas simultaneously from the first level. The company objected on the basis of the extra cost that would be incurred and the hazard of not recovering all the available gas from the second level. It also stated that it planned to start extracting from the first level in 3 years, at an initial investment cost of only $50,000. If it started extracting that gas now, the investment would be $70,000. The value of the recovered gas over a 3-year period would be the same in either case.

The government agency prepared the following comparative cash flow tables:

Year	*Company plan*	*Government plan*	*Difference*
0		−$70,000	−$70,000
1		+ 60,000	+ 60,000
2		+ 50,000	+ 50,000
3	−$50,000	+ 40,000	+ 90,000
4	+ 60,000		− 60,000
5	+ 50,000		− 50,000
6	+ 40,000		− 40,000

The government agency's economist solved for an interest rate that would make the present worth of the cash flows shown in the difference column equal to zero. The solution by chance turned out to be about 42.5%. The economist stated, "42.5% is enough return on the investment for any company. We will make you start extraction immediately."

(a) Plot the sum of the PW for the difference column at 10%, 15%, 20%, 40%, and 50%. Is there another solving rate? If so, what is it?

(b) At the time of this event the company's bonds were selling at about a 10% return to the bond purchaser. Use 10% as the auxiliary rate to analyze this cash flow stream as a borrowing-financing proposal.

B–14. A $50,000 structure having an estimated life of 20 years is to be compared by the present worth method with an $80,000 structure having an estimated life of 40 years. The $50,000 structure will have to be replaced at year 20 at a cost of $70,000. None of the structures will have a terminal salvage value. A consequence of the higher renewal cost will be an extra $500 of annual disbursements for income taxes for the longer-lived structure during years 21 to 40, both inclusive.

(a) At what interest rate will the present worths of disbursements for 40 years be equal for the two structures?

(b) The analysis in part (a) gives no recognition to the fact that the estimated cash flow series contains two reversals in sign. Is there a second solving rate of return that is relevant in this comparison?

(c) Assume a reasonable borrowing rate for the apparent repayment sequence of $500 per year for years 21 to 40, both inclusive. What conclusion does this lead to with respect to the remaining investment situation?

B–15. Consider the following proposal. By agreeing to pay $200 at the end of each year for 5 years a person can receive $100 at the beginning of the first year and $1,000 at the end of the fifth year. One of the two solving rates that will make the present worth of the cash flow stream zero is 189.7%. The other rate is less than 10%. Find that rate.

This obviously is an investment proposal and the person considers i^* to be 15%. What is the cost of the money obtained in the borrowing-financing period? This person can borrow money from the bank at 12%. Should the proposal be accepted?

C

THE REINVESTMENT FALLACY IN PROJECT EVALUATION

In Appendix B we used an auxiliary interest rate in addition to the unknown or solving rate. However, the two interest rates were not used during the same time period. The proposals being analyzed all combined investment during one or more periods with borrowing or financing during one or more other periods. For proposals that were primarily investment projects, the auxiliary interest rate was applied during the borrowing or financing period or periods. For proposals that were primarily borrowing or financing, the auxiliary interest rate was applied during the investment period or periods.

In this appendix, we emphasize that it is quite a different matter to use two or more interest rates for equivalence conversions during the same time period. Conclusions regarding the attractiveness of proposed investments can be badly distorted by applying two or more interest rates simultaneously. It also is possible to distort the analysis of proposed borrowing or financing schemes by using two or more rates at the same time. (An actual case of such distortion of the analysis of a financing scheme was described in Problem 18–15, page 480.)

Some Reasons Why Two or More Interest Rates Are Used for Compound Interest Conversions During the Same Time Period

Sometimes an analyst uses two or more interest rates because this method of analysis is required by company policy. Or an analyst may mistakenly believe that this technique will give useful conclusions. In either case, one aspect of the computational procedure will be the assumption of reinvestment at some stipulated interest rate. Various weaknesses in the reinvestment assumption are brought out in Example C–1 and in several of the problems at the end of this appendix.

Sometimes an analyst unintentionally uses two interest rates during the same time period as a result of substituting accounting charges for cash flows. In such cases, the auxiliary interest rate used unintentionally generally is 0%. Some possible distortions caused by making equivalence conversions at 0% are illustrated in Example C–2.

The authors have also observed cases where it seemed to them that sophisticated analysts were intentionally distorting the conclusions of an evaluation by making compound interest conversions that used two or more interest rates during the same time period. For some reason, it was desired to make a particular investment proposal appear to be better or worse than it would have seemed to be if it had been evaluated on its own merits. The analysis seemed to have been manipulated to support predetermined conclusions.

Variations on the Hoskold Method

In Chapter 8 (pp. 135–136) we discussed a purported rate of return method developed many years ago by Henry Hoskold for the valuation of mining properties.* Our example involved the following cash flows:

$$P = \$1{,}500{,}000$$
$$S = 0$$
$$A = \$391{,}000$$
$$n = 8 \text{ years}$$

The rate of return of 20% was found by solving the equation:

$$\$391{,}000(P/A,i\%,n) = \$1{,}500{,}000.$$

The Hoskold method involved the creation of a separate, and usually fictitious, conservatively invested sinking fund (at 4% interest) to provide for recovery of the investment and then solving for a rate of return on the remaining cash flow. This led to a solving rate of return of 15.2%. Thus, a portion of the positive cash flow from the project is assumed to be *reinvested* at a conservative interest rate with the remainder constituting the return on the investment.

If a salvage value is introduced into the problem, a general formulation of this method is:

$$i_s = \frac{A - (P - S)(A/F,i_r\%,n)}{P}$$

where: i_r = assumed interest rate on the reinvested sinking fund amount
i_s = solving rate of return on the remaining cash flow

(The reader will remember that an identical inappropriate formulation involving two interest rates was discussed in Chapter 11, p. 214.)

More recently, the Hoskold method has been reintroduced into some of the literature of capital investment analysis under the name Explicit Reinvestment Rate of Return method (which could have the interesting acronym

*Henry Hoskold, *Engineer's Valuing Assistant*, 2d ed. (Longman's Green & Company, New York, 1905; 1st ed., 1877).

ERROR method). Proponents of its use say that it has "the advantage of computational ease when there is a single beginning investment and there are constant receipts and disbursements each year." They recommend the use of the company's minimum attractive rate of return as the reinvestment rate, i_r.

If we assume an i^* of 18% in the previous example, the solving rate of return becomes:

$$i_s = \frac{\$391,000 - \$1,500,000(A/F,18\%,8)}{\$1,500,000} = 19.5\%$$

While this 19.5% rate is closer to the true rate of return on the investment, 20%, it continues to understate it because of the assumption that a portion of the annual cash flow is to be reinvested at a specified interest rate that is less than the investment's rate of return.

When i^*, as defined in this book, is used as the reinvestment rate in a Hoskold-type method, this method will lead to the same choices between alternatives as the equivalent uniform annual cash flow or present worth methods. However, the solving rates of return found should not be compared with solving rates of return found by the usual method recommended throughout this book. Solving rates of return using a Hoskold-type method will be lower than the true rate of return for projects that return more than i^* when i^* is used as the reinvestment rate, and will be higher for projects with a true rate of return less than i^*.

In those many cases in industry in which an i^* has not been derived, there is a temptation to use a conservative value for i_r such as the long-run average cost of borrowed money. As illustrated by the previous examples, this practice leads to even greater distortion in a solving rate of return. Problems C–7 through C–9 illustrate this point as well as several others discussed in connection with the use of two interest rates to make compound interest conversions over the same period.

EXAMPLE C–1

Misuse of an Auxiliary Interest Rate in Computing Prospective Rates of Return

Misleading Analysis of Two Investment Proposals

The previous discussion of Hoskold's method and that in Chapter 8 explained how a proposed investment sometimes is judged by calculating a rate of return that combines the consequences of two investments, the proposed one under review and an investment in an imaginary sinking fund. It was pointed out that, in general, the return from an unrelated investment—either real or imaginary—is irrelevant in judging the merits of a proposed investment under review. The basic error of the Hoskold-type analysis appears in many settings.

For instance, consider an investment proposal that has the following estimates of after-tax cash flow:*

Year	Cash flow for investment	Cash flow from excess of operating receipts over disbursements
0	−$450,000	
1		+$270,000
2		+ 220,000
3		+ 170,000
4		+ 120,000
5		+ 70,000
Totals (PW at 0%)	−$450,000	+$850,000

A conventional analysis along the lines used in this book indicates that the prospective rate of return is 33.8%. That is, if the funds indicated in the investment column were loaned and if the borrower made the payments shown in the right-hand column, the loan would be paid back with interest at 33.8%.

However, a proposed method of computing rate of return uses as an auxiliary interest rate "the average rate of return which the company is making on its investment." Assume this rate is 10%. The investment outlays are then discounted to zero date at this rate and the cash flow from the project is compounded to the final date of the study period at the same rate. At date zero the present worth of the investment obviously is $450,000. At date 5, the compound amount at 10% interest of the positive cash flows may be computed to be $1,095,900. A single payment compound amount factor for 5 years is then computed as $1,095,900 ÷ $450,000 = 2.435. Interpolation between the 18% and 20% factors indicates a return of 19.5%.

The foregoing proposed method of analysis underestimates the rate of return from the proposed investment by more than 14%. The error in principle here is the same one discussed in connection with the Hoskold formula; the computed 19.4% is, in effect, the result of two separable investments, the proposed $450,000 investment yielding 33.8% and an investment of varying amount elsewhere in the business enterprise assumed to yield 10%.

The fallacy in this type of analysis may be even more evident if we apply the method to the following estimates for another investment proposal:

Year	Cash flow for investment	Cash flow from excess of operating receipts over disbursements
0	−$200,000	
1	− 250,000	+$130,000

*Example C–1 was suggested by a method of analysis proposed in an article by R. H. Baldwin, "How to Assess Investment Proposals," *Harvard Business Review*, Vol. 37, No. 3 (May-June, 1959), pp. 98–104.

2		+ 110,000
3		+ 90,000
4		+ 70,000
5		+ 50,000
Totals (PW at 0%)	−$450,000	+$450,000

A simple inspection of the figures shows that the prospective rate of return is 0%; the $450,000 investment will be recovered with nothing left over. But if we use the 10% rate that the company is expected to make on other investments, the present worth of the investments at date zero is $427,300 and the compound amount of the positive cash flows at the end of 5 years is $572,600. If $(F/P,i\%,5)$ is computed as $572,600 ÷ $427,300 = 1.340, interpolation in our interest tables indicates a prospective rate of return of 6.0%. In effect, the investment proposal yielding 0% has been combined with the 10% assumed to be earned elsewhere in the enterprise to give the misleading conclusion that the proposal will yield 6%.

A Variation on the Baldwin Method

One variation that has been proposed in some more recent literature is called the External Rate of Return (ERR) method. It differs from that proposed by Baldwin in that all net positive cash flows are compounded to year n at the reinvestment rate, i_r. Net negative cash flows in any year are equated to this future worth using single payment compound amount factors and an unknown interest rate, i_s. The resulting equation is solved for i_s. Where there is a single investment at time zero followed only by net positive cash flows, the formulation is identical to that of Baldwin. That is, analysis of the first cash flow series in Example C–1 using this ERR method will yield the same i_s as the Baldwin method, 19.5% when the 10% reinvestment rate is assumed. Applying this method to the second cash flow series in Example C–1, the future worth at time 5 of the net positive cash flows is $382,310. A value of i_s of 3.9% is obtained by equating $\$200,000(F/P,i_s\%,5) + \$120,000(F/P,i_s\%,4) = \$382,310$ and interpolating in our interest tables between 3½% and 4%.

The reader should note that this slight variation in the treatment of positive and negative cash flows at the end of the first year resulted in a substantial difference in the solving rate of return. In both cases an investment proposal yielding 0% has been combined with the 10% assumed to be earned elsewhere in the enterprise.

Some proponents of the EER method recommend its use to circumvent the possibility of there being multiple solving rates of return when there are two or more reversals in sign in a cash flow series. As explained in Appendix B, this difficulty requires that the analyst: (1) separate the cash flow series into its investment and borrowing-financing components; and (2) determine whether the total series is primarily an investment project or a borrowing-financing scheme. We recommend against the use of any method that employs two interest rates during the same time periods.

EXAMPLE C–2

Distortion Introduced by Equivalence Conversion at 0% Interest Prior to a Present Worth Calculation

Misuse of Accounting Figures in Economy Studies

Chapter 16 explained that although allocations of expenditures among activities often are necessary for accounting purposes, such allocations occasionally are misused in analyses made to guide decisions among alternatives. It also is true that allocations among time periods that are needed in the accounts sometimes are misused in compound interest calculations made to judge whether a proposed investment will yield a stipulated rate of return.

As an example, consider a proposed investment of $100,000 in a property expected to have a life of 20 years, zero salvage value, and annual disbursements of $5,000 throughout its life. It is desired to compute the present worth of all costs, using an i^* of 15%. The correct total is figured as follows:

$$\begin{aligned} PW &= \$100{,}000 + \$5{,}000(P/A,15\%,20) \\ &= \$100{,}000 + \$5{,}000(6.259) = \$131{,}295 \end{aligned}$$

However, assume that straight-line depreciation of $5,000 a year is to be used in the accounts. The annual costs shown in the accounts will be $100,000/20 + $5,000 = $10,000. The present worth of this annual figure might conceivably be computed as

$$\begin{aligned} PW &= \$10{,}000(P/A,15\%,20) \\ &= \$10{,}000(6.259) = \$62{,}590 \end{aligned}$$

It should be obvious at a glance that this $62,590 figure that purports to be the present worth of the costs is incorrect and misleading; it makes the present worth of investment and operating costs appear to be less than the investment itself. The difficulty, of course, is that the $100,000 outlay at zero date was, in effect, converted to $5,000 a year from years 1 through 20, using an interest rate of 0%. When a 15% interest rate was used in converting the $5,000 a year back to zero date, the original $100,000 at zero date was, in a roundabout way, converted to $31,295 at zero date.

In this particular case the numerical value of the present worth made it evident that some type of error had been made. But the authors have seen conversions of this type made in analyses for industry and government where the existence of an error in principle could not have been discovered merely by looking at the results of the analyst's calculations.

Sensitivity in Relation to Matters Discussed in this Appendix

When a computed rate of return applies to the combined result of two investments, one in a project being evaluated and the other in an actual or hypothetical reinvestment fund, the computed rate will be between the rate of return on the project itself and the assumed rate of return on the reinvestment fund. In the special case where these two rates are close together, the computed rate will be almost the same as if no assumption of reinvestment had been made.

But when the two rates are far apart, their respective influences on the combined rate will be greatly affected by the cash flow pattern of the project being evaluated in relation to the assumed terminal date of the project. In certain types of cash flow patterns, the combined rate of return may depend largely on the assumed rate of return on the reinvestment fund and may be relatively insensitive to the productivity of the project that the analyst is supposed to be evaluating. The foregoing point is illustrated in Problems C–3, C–4, and C–5 at the end of this appendix.

Comparing Investment Proposals When Future Rates of Return Are Expected to Differ Substantially from Present Rates

Chapter 10 introduced the problem of choosing among investment proposals in circumstances where available funds are limited. Table 10–1 (page 167) described an assumed situation in which eight proposals requiring a total investment of $207,000 were under consideration and in which the available funds were limited to $90,000. The prospective after-tax rates of return had been computed for each proposal and were arrayed in order of descending magnitude. It was evident that the available funds would be exhausted by proposals U, Y, Z, and S, which had prospective rates of return of 15% or more. Our discussion in Chapter 10 indicated that if the maximum possible return should be desired from the available funds, it would be necessary to reject proposals X, T, V, and W, all of which had prospective rates of return of 12% or less.

The cutoff point in Table 10–1 was 15%; this was the minimum attractive rate of return because the selection of any project yielding less than 15% would force the elimination of some project having a prospective yield of 15% or more. However, our discussion in Chapter 10 gave no consideration to one aspect of capital rationing that sometimes is important. When we selected certain projects and rejected others, we gave no consideration to the duration of the favorable consequences of each project except as this duration influenced the computed rate of return. (Although our array in Table 10–1 did not show the expected pattern of future cash flow from each project, such supporting information would necessarily be available in any actual case as the source of the calculation of the respective rates of return.)

Assume that it is expected that the typical rate of return available from future investment proposals will be considerably higher or considerably lower than the rate obtainable from present proposals. To maximize the long-run rate of return on available investment funds, it is necessary to consider the duration of the expected consequences of each present proposal. In general, the prospect of lower future rates of return favors the longer-lived of the projects under present consideration; the prospect of higher future rates of return favors the shorter-lived projects (or sometimes the approval of no present projects at all).

For example, suppose that Project S in Table 10–1, yielding 15%, has a life of 2 years whereas Project X, yielding 12%, had a life of 12 years. Assume that it is believed that rates of return from projects that will be proposed in the next few years will not exceed 8%. The combination of S with its 15% and a subsequent investment yielding 8% may result in a lower overall rate of return over the next 12 years than the 12% obtainable from X.

Or, in contrast, suppose it is expected that if sufficient funds are available in a year or two, a large project can be undertaken that will yield a 25% rate of return. The acceptance of too many long-lived proposals now may make it impracticable to finance this desirable future project. A present capital rationing analysis may therefore favor projects with short capital recovery periods even though they have relatively low rates of return in order to make funds available for this attractive future project.

Occasionally circumstances arise where such forecasts of future rates of return are deemed appropriate in making decisions about present proposals for investment. If the forecasts are specific enough, it is possible to make the necessary calculations to find an overall rate of return from a proposed present investment and its successors. Such calculations would, in effect, make use of two or more interest rates. However, in most cases the best thing to do with forecasts about upward or downward trends in rates of return is to consider these forecasts on a qualitative basis, using them as irreducibles to influence capital rationing decisions in borderline cases.

PROBLEMS

C–1. An investment of $75,700 proposed to be made at zero date is expected to cause a net positive after-tax cash flow of $20,000 in the first year of the 20-year life of the asset. This positive cash flow is expected to decrease by $1,000 a year to $1,000 in the 20th year. There is no estimated terminal salvage value.

(a) At what interest rate will the positive cash flow repay the investment? (*Ans.* 20%)

(b) Compute the prospective rate of return by the proposed method illustrated in Example C–1, assuming reinvestment at 3% interest of all positive cash flows until the end of the 20-year period. (*Ans*. 7.3%)

(c) Explain the difference in the answers in (a) and (b).

C–2. An investment of $170,750 proposed to be made at zero date is expected to cause the same series of net positive after-tax cash flows described in Problem C–1. The estimated life of this asset is also 20 years with zero terminal salvage value.

(a) At what interest rate will the positive cash flow repay this investment? (*Ans*. 3%)

(b) Compute the prospective rate of return by the proposed method illustrated in Example C–1, assuming reinvestment throughout the 20-year period of all positive cash flows at the 12% average rate of return that is being earned by this business enterprise. (*Ans*. 9.2%)

(c) Explain the difference in the answers in (a) and (b).

C–3. Project A involves negative cash flow of $100,000 at zero date and positive cash flow of $65,500 at dates 1 and 2. Its consequences terminate at the end of 2 years. Project B involves the same cash flows at dates 0, 1, and 2 and also involves positive cash flows of $10,000 on dates 3, 4, 5, 6, 7, 8, 9, and 10. Its consequences terminate at the end of 10 years.

(a) For each project, find the interest rate that makes the present worth of all cash flows equal to zero. (*Ans*. A, 20%; B, 34%)

(b) In the WXY Company, rates of return on investment proposals are computed by the method illustrated in Example C–1, assuming that all positive cash flows will be reinvested at 4% interest until the terminal date of the project. With this method of evaluation, what are the computed rates of return on these two projects? (*Ans*. A, 15.6%; B, 10.6%)

(c) Project B is clearly superior to A because it involves the same cash flows at dates 0, 1, and 2, and in addition has positive cash flows of $10,000 a year from years 3 to 10. Explain what features of the evaluation method used by the WXY Company are responsible for the conclusion that B has a considerably lower rate of return than A.

C–4. Project E involves a negative cash flow of $100,000 at zero date followed by positive cash flows of $20,000 each year from years 1 to 5 and positive cash flows of $1,000 each year from 6 to 20. The consequences of this investment will terminate at the end of 20 years.

(a) What is the interest rate that makes the present worth of all cash flows equal to zero? (*Ans*. 3.5%)

(b) In the YWX Company, rates of return on proposed investments are computed by the method illustrated in Example C–1. It is assumed that all positive cash flows will be reinvested at 15% (the average rate expected to be earned by *all* investments in the company) until the terminal date of the project. A proposal will not be accepted unless its prospective rate of return calculated in this manner is at least 10%. With this method of evaluation, what is the computed rate of return on Project E? (*Ans*. 12.8%)

C–5. Project H involves a negative cash flow of $100,000 at zero date followed by a positive cash flow of $170,000 at date 5. The consequences of this investment will terminate at the end of 5 years.

(a) What is the interest rate that makes the present worth of all cash flows equal to zero? (*Ans*. 11.2%)

(b) This project is to be evaluated in the YWX Company by the method described in Problem C–4(b). With this method of evaluation, what is the computed rate of return on Project H? (*Ans*. 11.2%)

(c) The answers obtained in Problems C–4(b) and C–5(b) make it appear that Project E is preferable to Project H. Do you agree? If not, why not?

C–6. The President of the XYW Company wishes to analyze investment proposals by a method that assumes reinvestment along the lines illustrated in Example C–1. However, he recognizes a point brought out by the three preceding problems, namely, that any such method will not give a satisfactory comparison of competing projects that have substantially different terminal dates. He therefore stipulates that *all* investment proposals shall be analyzed assuming a terminal date 20 years from zero date even though the estimated cash flows directly resulting from the investment do not extend for the full 20 years.

If this rule had been followed by the WXY Company in Problem C–3(b), Project B would have shown a higher computed rate of return than Project A. And if it had been followed by the YWX Company in Problems C–4(b) and C–5(b), Project H would have shown a higher computed rate of return than Project E.

But compare Projects Q and R with the terminal date for all analysis as date 20 assuming reinvestment of all positive cash flows at 4%. Project Q involves a negative cash flow of $100,000 at zero date and a positive cash flow of $150,000 at date 1. Project R involves a negative cash flow of $100,000 at zero date and a positive cash flow of $350,000 at date 20.

Or compare Projects S and T with the terminal date for all analysis as date 20 assuming reinvestment of all positive cash flows at 15%. Project S involves a negative cash flow of $100,000 at zero date and a positive cash flow of $50,000 at date 1. Project T involves a negative cash flow of $100,000 at zero date and a positive cash flow of $680,000 at date 20.

Do you believe that the president's assumption of a common terminal date for all projects that are competing for limited funds meets all objections that can be raised to an analysis based on reinvestment of all positive cash flows at some stipulated rate? Explain your answer.

C–7. A proposed investment requires an initial disbursement of $10,000, has a 5-year economic life, and a $2,000 terminal salvage value. Annual receipts are estimated to be $5,000 per year, and annual disbursements, including disbursements for income taxes related to this proposal, are estimated to be $2,200.

(a) Find the rate of return on this proposed investment.

(b) The company requires the use of the Hoskold or ERROR method. Its prescribed rate for the sinking fund depreciation, or reinvestment, allocation is 10%, its average return on equity capital over the past 5 years. What apparent "rate of return" results from this method of evaluation?

(c) A financial officer of the company argues that the 10% sinking fund violates conservative business practices and requests that the proposal be reevaluated on the basis of a 6% sinking fund interest rate since the company may safely deposit funds in a bank at 6% interest. What "rate of return" results from the method described in part (b) assuming a 6% sinking fund reinvestment rate?

(d) Use the solving rate of return found in part (a) as the sinking fund reinvestment rate for the evaluation method of part (b) and verify that the "rate of return" resulting from the analysis is the same as this rate. It frequently is argued that this mathematical fact proves that the rate of return method implies reinvestment of recovered capital (depreciation accrued) at the solving rate of return. Discuss this statement in relation to the application of the rate of return method as presented in this book.

C–8. Assume the circumstances as described in Problem C–7 except that annual receipts are estimated to be $3,800 per year for 5 years rather than $5,000 per year.

(a) Find the rate of return on this proposed investment.

(b) Apply the analysis method described in Problem C–7(b), including the 10% reinvestment rate on the sinking fund allowance, to find the "rate of return" on this proposal.

(c) Recognizing that depreciation, in whatever form, is not a cash flow, it is recommended that the procedure described in Example C–1 be used to find the "rate of return" employing an auxiliary interest rate of 10%. Find the "rate of return" on this proposal employing the method of Example C–1.

(d) Explain the differences in your results.

C–9. Problems C–7 and C–8 investigated certain methods of analysis to demonstrate errors in reasoning. In this problem, we demonstrate how these errors can lead to incorrect decisions when data appear reasonable. The data of Problem C–7 are modified as follows:

Initial disbursement	$10,000
Annual receipts	4,200
Annual disbursements (inc. income taxes)	2,200
Estimated salvage value	0
Economic life	10 years

The company's minimum attractive rate of return after income taxes is 15%.

(a) The Hoskold or ERROR method is used for equipment analyses using a conservative sinking fund reinvestment rate of 6%. Find the "rate of return" on this proposal. Does it meet the company's standard of attractiveness?

(b) Determine the rate of return on this proposal following the procedures described throughout this book.

(c) Determine the "rate of return" by the Hoskold or ERROR method using the company's i^* of 15% as the sinking fund reinvestment rate.

(d) Find the present worth of the cash flows using the company's i^* of 15%.

(e) Comment on the difference in results, and explain why the solutions found in (b) and (d) are more appropriate.

C–10. Solve Problem C–9 using either the Baldwin or ERR method. Use 6% as the reinvestment rate in part (a) and the company's i^* of 15% in parts (c) and (d).

D

COMPOUND INTEREST TABLES

Formulas for Calculating Compound Interest Factors

Single Payment—Compound Amount Factor
$(F/P, i\%, n)$ $$(1+i)^n$$

Single Payment—Present Worth Factor
$(P/F, i\%, n)$ $$\frac{1}{(1+i)^n}$$

Sinking Fund Factor
$(A/F, i\%, n)$ $$\frac{i}{(1+i)^n - 1}$$

Capital Recovery Factor
$(A/P, i\%, n)$ $$\frac{i(1+i)^n}{(1+i)^n - 1}$$

Uniform Series—Compound Amount Factor
$(F/A, i\%, n)$ $$\frac{(1+i)^n - 1}{i}$$

Uniform Series—Present Worth Factor
$(P/A, i\%, n)$ $$\frac{(1+i)^n - 1}{i(1+i)^n}$$

Uniform Gradient—Conversion Factor
$(A/G, i\%, n)$ $$\frac{1}{i} - \frac{n}{i}\left[\frac{i}{(1+i)^n - 1}\right]$$

Uniform Gradient—Present Worth Factor
$(P/G, i\%, n)$ $$\frac{1}{i}\left[\frac{(1+i)^n - 1}{i(1+i)^n}\right] - \frac{n}{i(1+i)^n}$$

TABLE D–1
1% compound interest factors

	Single payment		*Uniform series*				*Uniform gradient*		
n	*Compound amount factor* F/P	*Present worth factor* P/F	*Sinking fund factor* A/F	*Capital recovery factor* A/P	*Compound amount factor* F/A	*Present worth factor* P/A	*Gradient conversion factor* A/G	*Present worth factor* P/G	*n*
1	1.0100	0.9901	1.000 00	1.010 00	1.000	0.990	0.000	0.000	**1**
2	1.0201	0.9803	0.497 51	0.507 51	2.010	1.970	0.498	0.980	**2**
3	1.0303	0.9706	0.330 02	0.340 02	3.030	2.941	0.993	2.921	**3**
4	1.0406	0.9610	0.246 28	0.256 28	4.060	3.902	1.488	5.804	**4**
5	1.0510	0.9515	0.196 04	0.206 04	5.101	4.853	1.980	9.610	**5**
6	1.0615	0.9420	0.162 55	0.172 55	6.152	5.795	2.471	14.321	**6**
7	1.0721	0.9327	0.138 63	0.148 63	7.214	6.728	2.960	19.917	**7**
8	1.0829	0.9235	0.120 69	0.130 69	8.286	7.652	3.448	26.381	**8**
9	1.0937	0.9143	0.106 74	0.116 74	9.369	8.566	3.934	33.696	**9**
10	1.1046	0.9053	0.095 58	0.105 58	10.462	9.471	4.418	41.843	**10**
11	1.1157	0.8963	0.086 45	0.096 45	11.567	10.368	4.901	50.807	**11**
12	1.1268	0.8874	0.078 85	0.088 85	12.683	11.255	5.381	60.569	**12**
13	1.1381	0.8787	0.072 41	0.082 41	13.809	12.134	5.861	71.113	**13**
14	1.1495	0.8700	0.066 90	0.076 90	14.947	13.004	6.338	82.422	**14**
15	1.1610	0.8613	0.062 12	0.072 12	16.097	13.865	6.814	94.481	**15**
16	1.1726	0.8528	0.057 94	0.067 94	17.258	14.718	7.289	107.273	**16**
17	1.1843	0.8444	0.054 26	0.064 26	18.430	15.562	7.761	120.783	**17**
18	1.1961	0.8360	0.050 98	0.060 98	19.615	16.398	8.232	134.996	**18**
19	1.2081	0.8277	0.048 05	0.058 05	20.811	17.226	8.702	149.895	**19**
20	1.2202	0.8195	0.045 42	0.055 42	22.019	18.046	9.169	165.466	**20**
21	1.2324	0.8114	0.043 03	0.053 03	23.239	18.857	9.635	181.695	**21**
22	1.2447	0.8034	0.040 86	0.050 86	24.472	19.660	10.100	198.566	**22**
23	1.2572	0.7954	0.038 89	0.048 89	25.716	20.456	10.563	216.066	**23**
24	1.2697	0.7876	0.037 07	0.047 07	26.973	21.243	11.024	234.180	**24**
25	1.2824	0.7798	0.035 41	0.045 41	28.243	22.023	11.483	252.894	**25**
26	1.2953	0.7720	0.033 87	0.043 87	29.526	22.795	11.941	272.196	**26**
27	1.3082	0.7644	0.032 45	0.042 45	30.821	23.560	12.397	292.070	**27**
28	1.3213	0.7568	0.031 12	0.041 12	32.129	24.316	12.852	312.505	**28**
29	1.3345	0.7493	0.029 90	0.039 90	33.450	25.066	13.304	333.486	**29**
30	1.3478	0.7419	0.028 75	0.038 75	34.785	25.808	13.756	355.002	**30**
31	1.3613	0.7346	0.027 68	0.037 68	36.133	26.542	14.205	377.039	**31**
32	1.3749	0.7273	0.026 67	0.036 67	37.494	27.270	14.653	399.586	**32**
33	1.3887	0.7201	0.025 73	0.035 73	38.869	27.990	15.099	422.629	**33**
34	1.4026	0.7130	0.024 84	0.034 84	40.258	28.703	15.544	446.157	**34**
35	1.4166	0.7059	0.024 00	0.034 00	41.660	29.409	15.987	470.158	**35**
40	1.4889	0.6717	0.020 46	0.030 46	48.886	32.835	18.178	596.856	**40**
45	1.5648	0.6391	0.017 71	0.027 71	56.481	36.095	20.327	733.704	**45**
50	1.6446	0.6080	0.015 51	0.025 51	64.463	39.196	22.436	879.418	**50**
55	1.7285	0.5785	0.013 73	0.023 73	72.852	42.147	24.505	1032.815	**55**
60	1.8167	0.5504	0.012 24	0.022 24	81.670	44.955	26.533	1192.806	**60**
65	1.9094	0.5237	0.011 00	0.021 00	90.937	47.627	28.522	1358.390	**65**
70	2.0068	0.4983	0.009 93	0.019 93	100.676	50.169	30.470	1528.647	**70**
75	2.1091	0.4741	0.009 02	0.019 02	110.913	52.587	32.379	1702.734	**75**
80	2.2167	0.4511	0.008 22	0.018 22	121.672	54.888	34.249	1879.877	**80**
85	2.3298	0.4292	0.007 52	0.017 52	132.979	57.078	36.080	2059.370	**85**
90	2.4486	0.4084	0.006 90	0.016 90	144.863	59.161	37.872	2240.567	**90**
95	2.5735	0.3886	0.006 36	0.016 36	157.354	61.143	39.626	2422.881	**95**
100	2.7048	0.3697	0.005 87	0.015 87	170.481	63.029	41.343	2605.776	**100**

TABLE D–2
1½% compound interest factors

	Single payment		Uniform series				Uniform gradient		
n	Compound amount factor F/P	Present worth factor P/F	Sinking fund factor A/F	Capital recovery factor A/P	Compound amount factor F/A	Present worth factor P/A	Gradient conversion factor A/G	Present worth factor P/G	n
1	1.0150	0.9852	1.000 00	1.015 00	1.000	0.985	0.000	0.000	**1**
2	1.0302	0.9707	0.496 28	0.511 28	2.015	1.956	0.496	0.971	**2**
3	1.0457	0.9563	0.328 38	0.343 38	3.045	2.912	0.990	2.883	**3**
4	1.0614	0.9422	0.244 44	0.259 44	4.091	3.854	1.481	5.710	**4**
5	1.0773	0.9283	0.194 09	0.209 09	5.152	4.783	1.970	9.423	**5**
6	1.0934	0.9145	0.160 53	0.175 53	6.230	5.697	2.457	13.996	**6**
7	1.1098	0.9010	0.136 56	0.151 56	7.323	6.598	2.940	19.402	**7**
8	1.1265	0.8877	0.118 58	0.133 58	8.433	7.486	3.422	25.616	**8**
9	1.1434	0.8746	0.104 61	0.119 61	9.559	8.361	3.901	32.612	**9**
10	1.1605	0.8618	0.093 43	0.108 43	10.703	9.222	4.377	40.367	**10**
11	1.1779	0.8489	0.084 29	0.099 29	11.863	10.071	4.851	48.857	**11**
12	1.1956	0.8364	0.076 68	0.091 68	13.041	10.908	5.323	58.057	**12**
13	1.2136	0.8240	0.070 24	0.085 24	14.237	11.732	5.792	67.945	**13**
14	1.2318	0.8118	0.064 72	0.079 72	15.450	12.543	6.258	78.499	**14**
15	1.2502	0.7999	0.059 94	0.074 94	16.682	13.343	6.722	89.697	**15**
16	1.2690	0.7880	0.055 77	0.070 77	17.932	14.131	7.184	101.518	**16**
17	1.2880	0.7764	0.052 08	0.067 08	19.201	14.908	7.643	113.940	**17**
18	1.3073	0.7649	0.048 81	0.063 81	20.489	15.673	8.100	126.943	**18**
19	1.3270	0.7536	0.045 88	0.060 88	21.797	16.426	8.554	140.508	**19**
20	1.3469	0.7425	0.043 25	0.058 25	23.124	17.169	9.006	154.615	**20**
21	1.3671	0.7315	0.040 87	0.055 87	24.471	17.900	9.455	169.245	**21**
22	1.3876	0.7207	0.038 70	0.053 70	25.838	18.621	9.902	184.380	**22**
23	1.4084	0.7100	0.036 73	0.051 73	27.225	19.331	10.346	200.001	**23**
24	1.4300	0.6995	0.034 92	0.049 92	28.634	20.030	10.788	216.090	**24**
25	1.4509	0.6892	0.033 26	0.048 26	30.063	20.720	11.228	232.631	**25**
26	1.4727	0.6790	0.031 73	0.046 73	31.514	21.399	11.665	249.607	**26**
27	1.4948	0.6690	0.030 32	0.045 32	32.987	22.068	12.099	267.000	**27**
28	1.5172	0.6591	0.029 00	0.044 00	34.481	22.727	12.531	284.796	**28**
29	1.5400	0.6494	0.027 78	0.042 78	35.999	23.376	12.961	302.978	**29**
30	1.5631	0.6398	0.026 64	0.041 64	37.539	24.016	13.388	321.531	**30**
31	1.5865	0.6303	0.025 57	0.040 57	39.102	24.646	13.813	340.440	**31**
32	1.6103	0.6210	0.024 58	0.039 58	40.688	25.267	14.236	359.691	**32**
33	1.6345	0.6118	0.023 64	0.038 64	42.299	25.879	14.656	379.269	**33**
34	1.6590	0.6028	0.022 76	0.037 76	43.933	26.482	15.073	399.161	**34**
35	1.6839	0.5939	0.021 93	0.036 93	45.592	27.076	15.488	419.352	**35**
40	1.8140	0.5513	0.018 43	0.033 43	54.268	29.916	17.528	524.357	**40**
45	1.9542	0.5117	0.015 72	0.030 72	63.614	32.552	19.507	635.011	**45**
50	2.1052	0.4750	0.013 57	0.028 57	73.683	35.000	21.428	749.964	**50**
55	2.2679	0.4409	0.011 83	0.026 83	84.530	37.271	23.289	868.028	**55**
60	2.4432	0.4093	0.010 39	0.025 39	96.215	39.380	25.093	988.167	**60**
65	2.6320	0.3799	0.009 19	0.024 19	108.803	41.338	26.839	1109.475	**65**
70	2.8355	0.3527	0.008 17	0.023 17	122.364	43.155	28.529	1231.166	**70**
75	3.0546	0.3274	0.007 30	0.022 30	136.973	44.842	30.163	1352.560	**75**
80	3.2907	0.3039	0.006 55	0.021 55	152.711	46.407	31.742	1473.074	**80**
85	3.5450	0.2821	0.005 89	0.020 89	169.665	47.861	33.268	1592.210	**85**
90	3.8189	0.2619	0.005 32	0.020 32	187.930	49.210	34.740	1709.544	**90**
95	4.1141	0.2431	0.004 82	0.019 82	207.606	50.462	36.160	1824.722	**95**
100	4.4320	0.2256	0.004 37	0.019 37	228.803	51.625	37.530	1937.451	**100**

TABLE D–3
2% compound interest factors

	Single payment		*Uniform series*				*Uniform gradient*		
n	*Compound amount factor F/P*	*Present worth factor P/F*	*Sinking fund factor A/F*	*Capital recovery factor A/P*	*Compound amount factor F/A*	*Present worth factor P/A*	*Gradient conversion factor A/G*	*Present worth factor P/G*	*n*
1	1.0200	0.9804	1.000 00	1.020 00	1.000	0.980	0.000	0.000	**1**
2	1.0404	0.9612	0.495 05	0.515 05	2.020	1.942	0.495	0.961	**2**
3	1.0612	0.9423	0.326 75	0.346 75	3.060	2.884	0.987	2.846	**3**
4	1.0824	0.9238	0.242 62	0.262 62	4.122	3.808	1.475	5.617	**4**
5	1.1041	0.9057	0.192 16	0.212 16	5.204	4.713	1.960	9.240	**5**
6	1.1262	0.8880	0.158 53	0.178 53	6.308	5.601	2.442	13.680	**6**
7	1.1487	0.8706	0.134 51	0.154 51	7.434	6.472	2.921	18.903	**7**
8	1.1717	0.8535	0.116 51	0.136 51	8.583	7.325	3.396	24.878	**8**
9	1.1951	0.8368	0.102 52	0.122 52	9.755	8.162	3.868	31.572	**9**
10	1.2190	0.8203	0.091 33	0.111 33	10.950	8.983	4.337	38.955	**10**
11	1.2434	0.8043	0.082 18	0.102 18	12.169	9.787	4.802	46.998	**11**
12	1.2682	0.7885	0.074 56	0.094 56	13.412	10.575	5.264	55.671	**12**
13	1.2936	0.7730	0.068 12	0.088 12	14.680	11.348	5.273	64.948	**13**
14	1.3195	0.7579	0.062 60	0.082 60	15.974	12.106	6.179	74.800	**14**
15	1.3459	0.7430	0.057 83	0.077 83	17.293	12.849	6.631	85.202	**15**
16	1.3728	0.7284	0.053 65	0.073 65	18.639	13.578	7.080	96.129	**16**
17	1.4002	0.7142	0.049 97	0.069 97	20.012	14.292	7.526	107.555	**17**
18	1.4282	0.7002	0.046 70	0.066 70	21.412	14.992	7.968	119.458	**18**
19	1.4568	0.6864	0.043 78	0.063 78	22.841	15.678	8.407	131.814	**19**
20	1.4859	0.6730	0.041 16	0.061 16	24.297	16.351	8.843	144.600	**20**
21	1.5157	0.6598	0.038 78	0.058 78	25.783	17.011	9.276	157.796	**21**
22	1.5460	0.6468	0.036 63	0.056 63	27.299	17.658	9.705	171.379	**22**
23	1.5769	0.6342	0.034 67	0.054 67	28.845	18.292	10.132	185.331	**23**
24	1.6084	0.6217	0.032 87	0.052 87	30.422	18.914	10.555	199.630	**24**
25	1.6406	0.6095	0.031 22	0.051 22	32.030	19.523	10.974	214.259	**25**
26	1.6734	0.5976	0.029 70	0.049 70	33.671	20.121	11.391	229.199	**26**
27	1.7069	0.5859	0.028 29	0.048 29	35.344	20.707	11.804	244.431	**27**
28	1.7410	0.5744	0.026 99	0.046 99	37.051	21.281	12.214	259.939	**28**
29	1.7758	0.5631	0.025 78	0.045 78	38.792	21.844	12.621	275.706	**29**
30	1.8114	0.5521	0.024 65	0.044 65	40.568	22.396	13.025	291.716	**30**
31	1.8476	0.5412	0.023 60	0.043 60	42.379	22.938	13.426	307.954	**31**
32	1.8845	0.5306	0.022 61	0.042 61	44.227	23.468	13.823	324.403	**32**
33	1.9222	0.5202	0.021 69	0.041 69	46.112	23.989	14.217	341.051	**33**
34	1.9607	0.5100	0.020 82	0.040 82	48.034	24.499	14.608	357.882	**34**
35	1.9999	0.5000	0.020 00	0.040 00	49.994	24.999	14.996	374.883	**35**
40	2.2080	0.4529	0.016 56	0.036 56	60.402	27.355	16.889	461.993	**40**
45	2.4379	0.4102	0.013 91	0.033 91	71.893	29.490	18.703	551.565	**45**
50	2.6916	0.3715	0.011 82	0.031 82	84.579	31.424	20.442	642.361	**50**
55	2.9717	0.3365	0.010 14	0.030 14	98.587	33.175	22.106	733.353	**55**
60	3.2810	0.3048	0.008 77	0.028 77	114.052	34.761	23.696	823.698	**60**
65	3.6225	0.2761	0.007 63	0.027 63	131.126	36.197	25.215	912.709	**65**
70	3.9996	0.2500	0.006 67	0.026 67	149.978	37.499	26.663	999.834	**70**
75	4.4158	0.2265	0.005 86	0.025 86	170.792	38.677	28.043	1084.639	**75**
80	4.8754	0.2051	0.005 16	0.025 16	193.772	39.745	29.357	1166.787	**80**
85	5.3829	0.1858	0.004 56	0.024 56	219.144	40.711	30.606	1246.024	**85**
90	5.9431	0.1683	0.004 05	0.024 05	247.157	41.587	31.793	1322.170	**90**
95	6.5617	0.1524	0.003 60	0.023 60	278.085	42.380	32.919	1395.103	**95**
100	7.2446	0.1380	0.003 20	0.023 20	312.232	43.098	33.986	1464.753	**100**

TABLE D–4
2½% compound interest factors

	Single payment		Uniform series				Uniform gradient		
n	Compound amount factor F/P	Present worth factor P/F	Sinking fund factor A/F	Capital recovery factor A/P	Compound amount factor F/A	Present worth factor P/A	Gradient conversion factor A/G	Present worth factor P/G	n
1	1.0250	0.9756	1.000 00	1.025 00	1.000	0.976	0.000	0.000	**1**
2	1.0506	0.9518	0.493 83	0.518 83	2.025	1.927	0.494	0.952	**2**
3	1.0769	0.9286	0.325 14	0.350 14	3.076	2.856	0.984	2.809	**3**
4	1.1038	0.9060	0.240 82	0.265 82	4.153	3.762	1.469	5.527	**4**
5	1.1314	0.8839	0.190 25	0.215 25	5.256	4.646	1.951	9.062	**5**
6	1.1597	0.8623	0.156 55	0.181 55	6.388	5.508	2.428	13.374	**6**
7	1.1887	0.8413	0.132 50	0.157 50	7.547	6.349	2.901	18.421	**7**
8	1.2184	0.8207	0.114 47	0.139 47	8.736	7.170	3.370	24.167	**8**
9	1.2489	0.8007	0.100 46	0.125 46	9.955	7.971	3.836	30.572	**9**
10	1.2801	0.7812	0.089 26	0.114 26	11.203	8.752	4.296	37.603	**10**
11	1.3121	0.7621	0.080 11	0.105 11	12.483	9.514	4.753	45.225	**11**
12	1.3449	0.7436	0.072 49	0.097 49	13.796	10.258	5.206	53.404	**12**
13	1.3785	0.7254	0.066 05	0.091 05	15.140	10.983	5.655	62.109	**13**
14	1.4130	0.7077	0.060 54	0.085 54	16.519	11.691	6.100	71.309	**14**
15	1.4483	0.6905	0.055 77	0.080 77	17.932	12.381	6.540	80.976	**15**
16	1.4845	0.6736	0.051 60	0.076 60	19.380	13.055	6.977	91.080	**16**
17	1.5216	0.6572	0.047 93	0.072 93	20.865	13.712	7.409	101.595	**17**
18	1.5597	0.6412	0.044 67	0.069 67	22.386	14.353	7.838	112.495	**18**
19	1.5987	0.6255	0.041 76	0.066 76	23.946	14.979	8.262	123.755	**19**
20	1.6386	0.6103	0.039 15	0.064 15	25.545	15.589	8.682	135.350	**20**
21	1.6796	0.5954	0.036 79	0.061 79	27.183	16.185	9.099	147.257	**21**
22	1.7216	0.5809	0.034 65	0.059 65	28.863	16.765	9.511	159.456	**22**
23	1.7646	0.5667	0.032 70	0.057 70	30.584	17.332	9.919	171.923	**23**
24	1.8087	0.5529	0.030 91	0.055 91	32.349	17.885	10.324	184.639	**24**
25	1.8539	0.5394	0.029 28	0.054 28	34.158	18.424	10.724	197.584	**25**
26	1.9003	0.5262	0.027 77	0.052 77	36.012	18.951	11.121	210.740	**26**
27	1.9478	0.5134	0.026 38	0.051 38	37.912	19.464	11.513	224.089	**27**
28	1.9965	0.5009	0.025 09	0.050 09	39.860	19.965	11.902	237.612	**28**
29	2.0464	0.4887	0.023 89	0.048 89	41.856	20.454	12.286	251.295	**29**
30	2.0976	0.4767	0.022 78	0.047 78	43.903	20.930	12.667	265.120	**30**
31	2.1500	0.4651	0.021 74	0.046 74	46.000	21.395	13.044	279.074	**31**
32	2.2038	0.4538	0.020 77	0.045 77	48.150	21.849	13.417	293.141	**32**
33	2.2589	0.4427	0.019 86	0.044 86	50.354	22.292	13.786	307.307	**33**
34	2.3153	0.4319	0.019 01	0.044 01	52.613	22.724	14.151	321.560	**34**
35	2.3732	0.4214	0.018 21	0.043 21	54.928	23.145	14.512	335.887	**35**
40	2.6851	0.3724	0.014 84	0.039 84	67.403	25.103	16.262	408.222	**40**
45	3.0379	0.3292	0.012 27	0.037 27	81.516	26.833	17.918	480.807	**45**
50	3.4371	0.2909	0.010 26	0.035 26	97.484	28.362	19.484	552.608	**50**
55	3.8888	0.2572	0.008 65	0.033 65	115.551	29.714	20.961	622.828	**55**
60	4.3998	0.2273	0.007 35	0.032 35	135.992	30.909	22.352	690.866	**60**
65	4.9780	0.2009	0.006 28	0.031 28	159.118	31.965	23.660	756.281	**65**
70	5.6321	0.1776	0.005 40	0.030 40	185.284	32.898	24.888	818.764	**70**
75	6.3722	0.1569	0.004 65	0.029 65	214.888	33.723	26.039	878.115	**75**
80	7.2100	0.1387	0.004 03	0.029 03	248.383	34.452	27.117	934.218	**80**
85	8.1570	0.1226	0.003 49	0.028 49	286.279	35.096	28.123	987.027	**85**
90	9.2289	0.1084	0.003 04	0.028 04	329.154	35.666	29.063	1036.550	**90**
95	10.4416	0.0958	0.002 65	0.027 65	377.664	36.169	29.938	1082.838	**95**
100	11.8137	0.0846	0.002 31	0.027 31	432.549	36.614	30.752	1125.975	**100**

TABLE D–5
3% compound interest factors

	Single payment		Uniform series				Uniform gradient		
n	*Compound amount factor F/P*	*Present worth factor P/F*	*Sinking fund factor A/F*	*Capital recovery factor A/P*	*Compound amount factor F/A*	*Present worth factor P/A*	*Gradient conversion factor A/G*	*Present worth factor P/G*	*n*
1	1.0300	0.9709	1.000 00	1.030 00	1.000	0.971	0.000	0.000	**1**
2	1.0609	0.9426	0.492 61	0.522 61	2.030	1.913	0.493	0.943	**2**
3	1.0927	0.9151	0.323 53	0.353 53	3.091	2.829	0.980	2.773	**3**
4	1.1255	0.8885	0.239 03	0.269 03	4.184	3.717	1.463	5.438	**4**
5	1.1593	0.8626	0.188 35	0.218 35	5.309	4.580	1.941	8.889	**5**
6	1.1941	0.8375	0.154 60	0.184 60	6.468	5.417	2.414	13.076	**6**
7	1.2299	0.8131	0.130 51	0.160 51	7.662	6.230	2.882	17.955	**7**
8	1.2668	0.7894	0.112 46	0.142 46	8.892	7.020	3.345	23.481	**8**
9	1.3048	0.7664	0.098 43	0.128 43	10.159	7.786	3.803	29.612	**9**
10	1.3439	0.7441	0.087 23	0.117 23	11.464	8.530	4.256	36.309	**10**
11	1.3842	0.7224	0.078 08	0.108 08	12.808	9.253	4.705	43.533	**11**
12	1.4258	0.7014	0.070 46	0.100 46	14.192	9.954	5.148	51.248	**12**
13	1.4685	0.6810	0.064 03	0.094 03	15.618	10.635	5.587	59.420	**13**
14	1.5126	0.6611	0.058 53	0.088 53	17.086	11.296	6.021	68.014	**14**
15	1.5580	0.6419	0.053 77	0.083 77	18.599	11.938	6.450	77.000	**15**
16	1.6047	0.6232	0.049 61	0.079 61	20.157	12.561	6.874	86.348	**16**
17	1.6528	0.6050	0.045 95	0.075 95	21.762	13.166	7.294	96.028	**17**
18	1.7024	0.5874	0.042 71	0.072 71	23.414	13.754	7.708	106.014	**18**
19	1.7535	0.5703	0.039 81	0.069 81	25.117	14.324	8.118	116.279	**19**
20	1.8061	0.5537	0.037 22	0.067 22	26.870	14.877	8.523	126.799	**20**
21	1.8603	0.5375	0.034 87	0.064 87	28.676	15.415	8.923	137.550	**21**
22	1.9161	0.5219	0.032 75	0.062 75	30.537	15.937	9.319	148.509	**22**
23	1.9736	0.5067	0.030 81	0.060 81	32.453	16.444	9.709	159.657	**23**
24	2.0328	0.4919	0.029 05	0.059 05	34.426	16.936	10.095	170.971	**24**
25	2.0938	0.4776	0.027 43	0.057 43	36.459	17.413	10.477	182.434	**25**
26	2.1566	0.4637	0.025 94	0.055 94	38.553	17.877	10.853	194.026	**26**
27	2.2213	0.4502	0.024 56	0.054 56	40.710	18.327	11.226	205.731	**27**
28	2.2879	0.4371	0.023 29	0.053 29	42.931	18.764	11.593	217.532	**28**
29	2.3566	0.4243	0.022 11	0.052 11	45.219	19.188	11.956	229.414	**29**
30	2.4273	0.4120	0.021 02	0.051 02	47.575	19.600	12.314	241.361	**30**
31	2.5001	0.4000	0.020 00	0.050 00	50.003	20.000	12.668	253.361	**31**
32	2.5751	0.3883	0.019 05	0.049 05	52.503	20.389	13.017	265.399	**32**
33	2.6523	0.3770	0.018 16	0.048 16	55.078	20.766	13.362	277.464	**33**
34	2.7319	0.3660	0.017 32	0.047 32	57.730	21.132	13.702	289.544	**34**
35	2.8139	0.3554	0.016 54	0.046 54	60.462	21.487	14.037	301.627	**35**
40	3.2620	0.3066	0.013 26	0.043 26	75.401	23.115	15.650	361.750	**40**
45	3.7816	0.2644	0.010 79	0.040 79	92.720	24.519	17.156	420.632	**45**
50	4.3839	0.2281	0.008 87	0.038 87	112.797	25.730	18.558	477.480	**50**
55	5.0821	0.1968	0.007 35	0.037 35	136.072	26.774	19.860	531.741	**55**
60	5.8916	0.1697	0.006 13	0.036 13	163.053	27.676	21.067	583.053	**60**
65	6.8300	0.1464	0.005 15	0.035 15	194.333	28.453	22.184	631.201	**65**
70	7.9178	0.1263	0.004 34	0.034 34	230.594	29.123	23.215	676.087	**70**
75	9.1789	0.1089	0.003 67	0.033 67	272.631	29.702	24.163	717.698	**75**
80	10.6409	0.0940	0.003 11	0.033 11	321.363	30.201	25.035	756.087	**80**
85	12.3357	0.0811	0.002 65	0.032 65	377.857	30.631	25.835	791.353	**85**
90	14.3005	0.0699	0.002 26	0.032 26	443.349	31.002	26.567	823.630	**90**
95	16.5782	0.0603	0.001 93	0.031 93	519.272	31.323	27.235	853.074	**95**
100	19.2186	0.0520	0.001 65	0.031 65	607.288	31.599	27.844	879.854	**100**

TABLE D–6
3½% compound interest factors

	Single payment		Uniform series				Uniform gradient		
n	*Compound amount factor F/P*	*Present worth factor P/F*	*Sinking fund factor A/F*	*Capital recovery factor A/P*	*Compound amount factor F/A*	*Present worth factor P/A*	*Gradient conversion factor A/G*	*Present worth factor P/G*	*n*
1	1.0350	0.9662	1.000 00	1.035 00	1.000	0.966	0.000	0.000	**1**
2	1.0712	0.9335	0.491 40	0.526 40	2.035	1.900	0.491	0.934	**2**
3	1.1087	0.9019	0.321 93	0.356 93	3.106	2.802	0.977	2.737	**3**
4	1.1475	0.8714	0.237 25	0.272 25	4.215	3.673	1.457	5.352	**4**
5	1.1877	0.8420	0.186 48	0.221 48	5.362	4.515	1.931	8.720	**5**
6	1.2293	0.8135	0.152 67	0.187 67	6.550	5.329	2.400	12.787	**6**
7	1.2723	0.7860	0.128 54	0.163 54	7.779	6.115	2.863	17.503	**7**
8	1.3168	0.7594	0.110 48	0.145 48	9.052	6.874	3.320	22.819	**8**
9	1.3629	0.7337	0.096 45	0.131 45	10.368	7.608	3.771	28.689	**9**
10	1.4106	0.7089	0.085 24	0.120 24	11.731	8.317	4.217	35.069	**10**
11	1.4600	0.6849	0.076 09	0.111 09	13.142	9.002	4.657	41.919	**11**
12	1.5111	0.6618	0.068 48	0.103 48	14.602	9.663	5.091	49.198	**12**
13	1.5640	0.6394	0.062 06	0.097 06	16.113	10.303	5.520	56.871	**13**
14	1.6187	0.6178	0.056 57	0.091 57	17.677	10.921	5.943	64.902	**14**
15	1.6753	0.5969	0.051 83	0.086 83	19.296	11.517	6.361	73.259	**15**
16	1.7340	0.5767	0.047 68	0.082 68	20.971	12.094	6.773	81.909	**16**
17	1.7947	0.5572	0.044 04	0.079 04	22.705	12.651	7.179	90.824	**17**
18	1.8575	0.5384	0.040 82	0.075 85	24.500	13.190	7.580	99.977	**18**
19	1.9225	0.5202	0.037 94	0.072 94	26.357	13.710	7.975	109.339	**19**
20	1.9898	0.5026	0.035 36	0.070 36	28.280	14.212	8.365	118.888	**20**
21	2.0594	0.4856	0.033 04	0.068 04	30.269	14.698	8.749	128.600	**21**
22	2.1315	0.4692	0.030 93	0.065 93	32.329	15.167	9.128	138.452	**22**
23	2.2061	0.4533	0.029 02	0.064 02	34.460	15.620	9.502	148.424	**23**
24	2.2833	0.4380	0.027 27	0.062 27	36.667	16.058	9.870	158.497	**24**
25	2.3632	0.4231	0.025 67	0.060 67	38.950	16.482	10.233	168.653	**25**
26	2.4460	0.4088	0.024 21	0.059 21	41.313	16.890	10.590	178.874	**26**
27	2.5316	0.3950	0.022 85	0.057 85	43.759	17.285	10.942	189.144	**27**
28	2.6202	0.3817	0.021 60	0.056 60	46.291	17.667	11.289	199.448	**28**
29	2.7119	0.3687	0.020 45	0.055 45	48.911	18.036	11.631	209.773	**29**
30	2.8068	0.3563	0.019 37	0.054 37	51.623	18.392	11.967	220.106	**30**
31	2.9050	0.3442	0.018 37	0.053 37	54.429	18.736	12.299	230.432	**31**
32	3.0067	0.3326	0.017 44	0.052 44	57.335	19.069	12.625	240.743	**32**
33	3.1119	0.3213	0.016 57	0.051 57	60.341	19.390	12.946	251.026	**33**
34	3.2209	0.3105	0.015 76	0.050 76	63.453	19.701	13.262	261.271	**34**
35	3.3336	0.3000	0.015 00	0.050 00	66.674	20.001	13.573	271.471	**35**
40	3.9593	0.2526	0.011 83	0.046 83	84.550	21.355	15.055	321.491	**40**
45	4.7024	0.2127	0.009 45	0.044 45	105.782	22.495	16.417	369.308	**45**
50	5.5849	0.1791	0.007 63	0.042 63	130.998	23.456	17.666	414.370	**50**
55	6.6331	0.1508	0.006 21	0.041 21	160.947	24.264	18.808	456.353	**55**
60	7.8781	0.1269	0.005 09	0.040 09	196.517	24.945	19.848	495.105	**60**
65	9.3567	0.1069	0.004 19	0.039 19	238.763	25.518	20.793	530.599	**65**
70	11.1128	0.0900	0.003 46	0.038 46	288.938	26.000	21.650	562.896	**70**
75	13.1986	0.0758	0.002 87	0.037 87	348.530	26.407	22.423	592.121	**75**
80	15.6757	0.0638	0.002 38	0.037 38	419.307	26.749	23.120	618.439	**80**
85	18.6179	0.0537	0.001 99	0.036 99	503.367	27.037	23.747	642.037	**85**
90	22.1122	0.0452	0.001 66	0.036 66	603.205	27.279	24.308	663.119	**90**
95	26.2623	0.0381	0.001 39	0.036 39	721.781	27.484	24.811	681.890	**95**
100	31.1914	0.0321	0.001 16	0.036 16	862.612	27.655	25.259	698.555	**100**

TABLE D–7
4% compound interest factors

	Single payment		*Uniform series*				*Uniform gradient*		
n	*Compound amount factor* *F/P*	*Present worth factor* *P/F*	*Sinking fund factor* *A/F*	*Capital recovery factor* *A/P*	*Compound amount factor* *F/A*	*Present worth factor* *P/A*	*Gradient conversion factor* *A/G*	*Present worth factor* *P/G*	*n*
1	1.0400	0.9615	1.000 00	1.040 00	1.000	0.962	0.000	0.000	**1**
2	1.0816	0.9246	0.490 20	0.530 20	2.040	1.886	0.490	0.925	**2**
3	1.1249	0.8890	0.320 35	0.360 35	3.122	2.775	0.974	2.703	**3**
4	1.1699	0.8548	0.235 49	0.275 49	4.246	3.630	1.451	5.267	**4**
5	1.2167	0.8219	0.184 63	0.224 63	5.416	4.452	1.922	8.555	**5**
6	1.2653	0.7903	0.150 76	0.190 76	6.633	5.242	2.386	12.506	**6**
7	1.3159	0.7599	0.126 61	0.166 61	7.898	6.002	2.843	17.066	**7**
8	1.3686	0.7307	0.108 53	0.148 53	9.214	6.733	3.294	22.181	**8**
9	1.4233	0.7026	0.094 49	0.134 49	10.583	7.435	3.739	27.801	**9**
10	1.4802	0.6756	0.083 29	0.123 29	12.006	8.111	4.177	33.881	**10**
11	1.5395	0.6496	0.074 15	0.114 15	13.486	8.760	4.609	40.377	**11**
12	1.6010	0.6246	0.066 55	0.106 55	15.026	9.385	5.034	47.248	**12**
13	1.6651	0.6006	0.060 14	0.100 14	16.627	9.986	5.453	54.455	**13**
14	1.7317	0.5775	0.054 67	0.094 67	18.292	10.563	5.866	61.962	**14**
15	1.8009	0.5553	0.049 94	0.089 94	20.024	11.118	6.272	69.735	**15**
16	1.8730	0.5339	0.045 82	0.085 82	21.825	11.652	6.672	77.744	**16**
17	1.9479	0.5134	0.042 20	0.082 20	23.698	12.166	7.066	85.958	**17**
18	2.0258	0.4936	0.038 99	0.078 99	25.645	12.659	7.453	94.350	**18**
19	2.1068	0.4746	0.036 14	0.076 14	27.671	13.134	7.834	102.893	**19**
20	2.1911	0.4564	0.033 58	0.073 58	29.778	13.590	8.209	111.565	**20**
21	2.2788	0.4388	0.031 28	0.071 28	31.969	14.029	8.578	120.341	**21**
22	2.3699	0.4220	0.029 20	0.069 20	34.248	14.451	8.941	129.202	**22**
23	2.4647	0.4057	0.027 31	0.067 31	36.618	14.857	9.297	138.128	**23**
24	2.5633	0.3901	0.025 59	0.065 59	39.083	15.247	9.648	147.101	**24**
25	2.6658	0.3751	0.024 01	0.064 01	41.646	15.622	9.993	156.104	**25**
26	2.7725	0.3607	0.022 57	0.062 57	44.312	15.983	10.331	165.121	**26**
27	2.8834	0.3468	0.021 24	0.061 24	47.084	16.330	10.664	174.138	**27**
28	2.9987	0.3335	0.020 01	0.060 01	49.968	16.663	10.991	183.142	**28**
29	3.1187	0.3207	0.018 88	0.058 88	52.966	16.984	11.312	192.121	**29**
30	3.2434	0.3083	0.017 83	0.057 83	56.085	17.292	11.627	201.062	**30**
31	3.3731	0.2965	0.016 86	0.056 86	59.328	17.588	11.937	209.956	**31**
32	3.5081	0.2851	0.015 95	0.055 95	62.701	17.874	12.241	218.792	**32**
33	3.6484	0.2741	0.015 10	0.055 10	66.210	18.148	12.540	227.563	**33**
34	3.7943	0.2636	0.014 31	0.054 31	69.858	18.411	12.832	236.261	**34**
35	3.9461	0.2534	0.013 58	0.053 58	73.652	18.665	13.120	244.877	**35**
40	4.8010	0.2083	0.010 52	0.050 52	95.026	19.793	14.477	286.530	**40**
45	5.8412	0.1712	0.008 26	0.048 26	121.029	20.720	15.705	325.403	**45**
50	7.1067	0.1407	0.006 55	0.046 55	152.667	21.482	16.812	361.164	**50**
55	8.6464	0.1157	0.005 23	0.045 23	191.159	22.109	17.807	393.689	**55**
60	10.5196	0.0951	0.004 20	0.044 20	237.991	22.623	18.697	422.997	**60**
65	12.7987	0.0781	0.003 39	0.043 39	294.968	23.047	19.491	449.201	**65**
70	15.5716	0.0642	0.002 75	0.042 75	364.290	23.395	20.196	472.479	**70**
75	18.9453	0.0528	0.002 23	0.042 23	448.631	23.680	20.821	493.041	**75**
80	23.0500	0.0434	0.001 81	0.041 81	551.245	23.915	21.372	511.116	**80**
85	28.0436	0.0357	0.001 48	0.041 48	676.090	24.109	21.857	526.938	**85**
90	34.1193	0.0293	0.001 21	0.041 21	827.983	24.267	22.283	540.737	**90**
95	41.5114	0.0241	0.000 99	0.040 99	1 012.785	24.398	22.655	552.731	**95**
100	50.5049	0.0198	0.000 81	0.040 81	1 237.624	24.505	22.980	563.125	**100**

TABLE D–8
4½% compound interest factors

	Single payment		Uniform series				Uniform gradient		
n	*Compound amount factor* F/P	*Present worth factor* P/F	*Sinking fund factor* A/F	*Capital recovery factor* A/P	*Compound amount factor* F/A	*Present worth factor* P/A	*Gradient conversion factor* A/G	*Present worth factor* P/G	*n*
1	1.0450	0.9569	1.000 00	1.045 00	1.000	0.957	0.000	0.000	**1**
2	1.0920	0.9157	0.489 00	0.534 00	2.045	1.873	0.489	0.916	**2**
3	1.1412	0.8763	0.318 77	0.363 77	3.137	2.749	0.971	2.668	**3**
4	1.1925	0.8386	0.233 74	0.278 74	4.278	3.588	1.445	5.184	**4**
5	1.2462	0.8025	0.182 79	0.227 79	5.471	4.390	1.912	8.394	**5**
6	1.3023	0.7679	0.148 88	0.193 88	6.717	5.158	2.372	12.233	**6**
7	1.3609	0.7348	0.124 70	0.169 70	8.019	5.893	2.824	16.642	**7**
8	1.4221	0.7032	0.106 61	0.151 61	9.380	6.596	3.269	21.565	**8**
9	1.4861	0.6729	0.092 57	0.137 57	10.802	7.269	3.707	26.948	**9**
10	1.5530	0.6439	0.081 38	0.126 38	12.288	7.913	4.138	32.743	**10**
11	1.6229	0.6162	0.072 25	0.117 25	13.841	8.529	4.562	38.905	**11**
12	1.6959	0.5897	0.064 67	0.109 67	15.464	9.119	4.978	45.391	**12**
13	1.7722	0.5643	0.058 28	0.103 28	17.160	9.683	5.387	52.163	**13**
14	1.8519	0.5400	0.052 82	0.097 82	18.932	10.223	5.789	59.182	**14**
15	1.9353	0.5167	0.048 11	0.093 11	20.784	10.740	6.184	66.416	**15**
16	2.0224	0.4945	0.044 02	0.089 02	22.719	11.234	6.572	73.833	**16**
17	2.1134	0.4732	0.040 42	0.085 42	24.742	11.707	6.953	81.404	**17**
18	2.2085	0.4528	0.037 24	0.082 24	26.855	12.160	7.327	89.102	**18**
19	2.3079	0.4333	0.034 41	0.079 41	29.064	12.593	7.695	96.901	**19**
20	2.4117	0.4146	0.031 88	0.076 88	31.371	13.008	8.055	104.780	**20**
21	2.5202	0.3968	0.029 60	0.074 60	33.783	13.405	8.409	112.715	**21**
22	2.6337	0.3797	0.027 55	0.072 55	36.303	13.784	8.755	120.689	**22**
23	2.7522	0.3634	0.025 68	0.070 68	38.937	14.148	9.096	128.683	**23**
24	2.8760	0.3477	0.023 99	0.068 99	41.689	14.495	9.429	136.680	**24**
25	3.0054	0.3327	0.022 44	0.067 44	44.565	14.828	9.756	144.665	**25**
26	3.1407	0.3184	0.021 02	0.066 02	47.571	15.147	10.077	152.625	**26**
27	3.2820	0.3047	0.019 72	0.064 72	50.711	15.451	10.391	160.547	**27**
28	3.4397	0.2916	0.018 52	0.063 52	53.993	15.743	10.698	168.420	**28**
29	3.5840	0.2790	0.017 41	0.062 41	57.423	16.022	10.999	176.232	**29**
30	3.7453	0.2670	0.016 39	0.061 39	61.007	16.289	11.295	183.975	**30**
31	3.9139	0.2555	0.015 44	0.060 44	64.752	16.544	11.583	191.640	**31**
32	4.0900	0.2445	0.014 56	0.059 56	68.666	16.789	11.866	199.220	**32**
33	4.2740	0.2340	0.013 74	0.058 74	72.756	17.023	12.143	206.707	**33**
34	4.4664	0.2239	0.012 98	0.057 98	77.030	17.247	12.414	214.096	**34**
35	4.6673	0.2143	0.012 27	0.057 27	81.497	17.461	12.679	221.380	**35**
40	5.8164	0.1719	0.009 34	0.054 34	107.030	18.402	13.917	256.099	**40**
45	7.2482	0.1380	0.007 20	0.052 20	138.850	19.156	15.020	287.732	**45**
50	9.0326	0.1107	0.005 60	0.050 60	178.503	19.762	15.998	316.145	**50**
55	11.2563	0.0888	0.004 39	0.049 39	227.918	20.248	16.860	341.375	**55**
60	14.0274	0.0713	0.003 45	0.048 45	289.498	20.638	17.617	363.571	**60**
65	17.4807	0.0572	0.002 73	0.047 73	366.238	20.951	18.278	382.947	**65**
70	21.7841	0.0459	0.002 17	0.047 17	461.870	21.202	18.854	399.750	**70**
75	27.1470	0.0368	0.001 72	0.046 72	581.044	21.404	19.354	414.242	**75**
80	33.8301	0.0296	0.001 37	0.046 37	729.558	21.565	19.785	426.680	**80**
85	42.1585	0.0237	0.001 09	0.046 09	914.632	21.695	20.157	437.309	**85**
90	52.5371	0.0190	0.000 87	0.045 87	1 145.269	21.799	20.476	446.359	**90**
95	65.4708	0.0153	0.000 70	0.045 70	1 432.684	21.883	20.749	454.039	**95**
100	81.5885	0.0123	0.000 56	0.045 56	1 790.856	21.950	20.981	460.538	**100**

TABLE D–9
5% compound interest factors

	Single payment		Uniform series				Uniform gradient		
n	Compound amount factor F/P	Present worth factor P/F	Sinking fund factor A/F	Capital recovery factor A/P	Compound amount factor F/A	Present worth factor P/A	Gradient conversion factor A/G	Present worth factor P/G	n
1	1.0500	0.9524	1.000 00	1.050 00	1.000	0.952	0.000	0.000	1
2	1.1025	0.9070	0.487 80	0.537 80	2.050	1.859	0.488	0.907	2
3	1.1576	0.8638	0.317 21	0.367 21	3.153	2.723	0.967	2.635	3
4	1.2155	0.8227	0.232 01	0.282 01	4.310	3.546	1.439	5.103	4
5	1.2763	0.7835	0.180 97	0.230 97	5.526	4.329	1.903	8.237	5
6	1.3401	0.7462	0.147 02	0.197 02	6.802	5.076	2.358	11.968	6
7	1.4071	0.7107	0.122 82	0.172 82	8.142	5.786	2.805	16.232	7
8	1.4775	0.6768	0.104 72	0.154 72	9.549	6.463	3.245	20.970	8
9	1.5513	0.6446	0.090 69	0.140 69	11.027	7.108	3.676	26.127	9
10	1.6289	0.6139	0.079 50	0.129 50	12.578	7.722	4.099	31.652	10
11	1.7103	0.5847	0.070 39	0.120 39	14.207	8.306	4.514	37.499	11
12	1.7959	0.5568	0.062 83	0.112 83	15.917	8.863	4.922	43.624	12
13	1.8856	0.5303	0.056 46	0.106 46	17.713	9.394	5.322	49.988	13
14	1.9800	0.5051	0.051 02	0.101 02	19.599	9.899	5.713	56.554	14
15	2.0789	0.4810	0.046 34	0.096 34	21.579	10.380	6.097	63.288	15
16	2.1829	0.4581	0.042 27	0.092 27	23.657	10.838	6.474	70.160	16
17	2.2920	0.4363	0.038 70	0.088 70	25.840	11.274	6.842	77.140	17
18	2.4066	0.4155	0.035 55	0.085 55	28.132	11.690	7.203	84.204	18
19	2.5270	0.3957	0.032 75	0.082 75	30.539	12.085	7.557	91.328	19
20	2.6533	0.3769	0.030 24	0.080 24	33.066	12.462	7.903	98.488	20
21	2.7860	0.3589	0.028 00	0.078 00	35.719	12.821	8.242	105.667	21
22	2.9253	0.3418	0.025 97	0.075 97	38.505	13.163	8.573	112.846	22
23	3.0715	0.3256	0.024 14	0.074 14	41.430	13.489	8.897	120.009	23
24	3.2251	0.3101	0.022 47	0.072 47	44.502	13.799	9.214	127.140	24
25	3.3864	0.2953	0.020 95	0.070 95	47.727	14.094	9.524	134.228	25
26	3.5557	0.2812	0.019 56	0.069 56	51.113	14.375	9.827	141.259	26
27	3.7335	0.2678	0.018 29	0.068 29	54.669	14.643	10.122	148.223	27
28	3.9201	0.2551	0.017 12	0.067 12	58.403	14.898	10.411	155.110	28
29	4.1161	0.2429	0.016 05	0.066 05	62.323	15.141	10.694	161.913	29
30	4.3219	0.2314	0.015 05	0.065 05	66.439	15.372	10.969	168.623	30
31	4.5380	0.2204	0.014 13	0.064 13	70.761	15.593	11.238	175.233	31
32	4.7649	0.2099	0.013 28	0.063 28	75.299	15.803	11.501	181.739	32
33	5.0032	0.1999	0.012 49	0.062 49	80.064	16.003	11.757	188.135	33
34	5.2533	0.1904	0.011 76	0.061 76	85.067	16.193	12.006	194.417	34
35	5.5160	0.1813	0.011 07	0.061 07	90.320	16.374	12.250	200.581	35
40	7.0400	0.1420	0.008 28	0.058 28	120.800	17.159	13.377	229.545	40
45	8.9850	0.1113	0.006 26	0.056 26	159.700	17.774	14.364	255.315	45
50	11.4674	0.0872	0.004 78	0.054 78	209.348	18.256	15.223	277.915	50
55	14.6356	0.0683	0.003 67	0.053 67	272.713	18.633	15.966	297.510	55
60	18.6792	0.0535	0.002 83	0.052 83	353.584	18.929	16.606	314.343	60
65	23.8399	0.0419	0.002 19	0.052 19	456.798	19.161	17.154	328.691	65
70	30.4264	0.0329	0.001 70	0.051 70	588.529	19.343	17.621	340.841	70
75	38.8327	0.0258	0.001 32	0.051 32	756.654	19.485	18.018	351.072	75
80	49.5614	0.0202	0.001 03	0.051 03	971.229	19.596	18.353	359.646	80
85	63.2544	0.0158	0.000 80	0.050 80	1 245.087	19.684	18.635	366.801	85
90	80.7304	0.0124	0.000 63	0.050 63	1 594.607	19.752	18.871	372.749	90
95	103.0357	0.0097	0.000 49	0.050 49	2 040.694	19.806	19.069	377.677	95
100	131.5013	0.0076	0.000 38	0.050 38	2 610.025	19.848	19.234	381.749	100

TABLE D–10
5½% compound interest factors

	Single payment		*Uniform series*				*Uniform gradient*		
n	*Compound amount factor F/P*	*Present worth factor P/F*	*Sinking fund factor A/F*	*Capital recovery factor A/P*	*Compound amount factor F/A*	*Present worth factor P/A*	*Gradient conversion factor A/G*	*Present worth factor P/G*	*n*
1	1.0550	0.9479	1.000 00	1.055 00	1.000	0.948	0.000	0.000	**1**
2	1.1130	0.8985	0.486 62	0.541 62	2.055	1.846	0.487	0.898	**2**
3	1.1742	0.8516	0.315 65	0.370 65	3.168	2.698	0.964	2.602	**3**
4	1.2388	0.8072	0.230 29	0.285 29	4.342	3.505	1.433	5.023	**4**
5	1.3070	0.7651	0.179 18	0.234 18	5.581	4.270	1.893	8.084	**5**
6	1.3788	0.7252	0.145 18	0.200 18	6.888	4.996	2.344	11.710	**6**
7	1.4547	0.6874	0.120 96	0.175 96	8.267	5.683	2.786	15.835	**7**
8	1.5347	0.6516	0.102 86	0.157 86	9.722	6.335	3.220	20.396	**8**
9	1.6191	0.6176	0.088 84	0.143 84	11.256	6.952	3.644	25.337	**9**
10	1.7081	0.5854	0.077 67	0.132 67	12.875	7.538	4.060	30.606	**10**
11	1.8021	0.5549	0.068 57	0.123 57	14.583	8.093	4.468	36.155	**11**
12	1.9012	0.5260	0.061 03	0.116 03	16.386	8.619	4.866	41.941	**12**
13	2.0058	0.4986	0.054 68	0.109 68	18.287	9.117	5.256	47.923	**13**
14	2.1161	0.4726	0.049 28	0.104 28	20.293	9.590	5.638	54.067	**14**
15	2.2325	0.4479	0.044 63	0.099 63	22.409	10.038	6.011	60.338	**15**
16	2.3553	0.4246	0.040 58	0.095 58	24.641	10.462	6.376	66.707	**16**
17	2.4848	0.4024	0.037 04	0.092 04	26.996	10.865	6.732	73.146	**17**
18	2.6215	0.3815	0.033 92	0.088 92	29.481	11.246	7.081	79.631	**18**
19	2.7656	0.3616	0.031 15	0.086 15	32.103	11.608	7.421	86.139	**19**
20	2.9178	0.3427	0.028 68	0.083 68	34.868	11.950	7.753	92.651	**20**
21	3.0782	0.3249	0.026 46	0.081 46	37.786	12.275	8.077	99.148	**21**
22	3.2475	0.3079	0.024 47	0.079 47	40.864	12.583	8.393	105.615	**22**
23	3.4262	0.2919	0.022 67	0.077 67	44.112	12.875	8.702	112.036	**23**
24	3.6146	0.2767	0.021 04	0.076 04	47.538	13.152	9.003	118.399	**24**
25	3.8134	0.2622	0.019 55	0.074 55	51.153	13.414	9.296	124.693	**25**
26	4.0231	0.2486	0.018 19	0.073 19	54.966	13.662	9.581	130.907	**26**
27	4.2444	0.2356	0.016 95	0.071 95	58.989	13.898	9.860	137.032	**27**
28	4.4778	0.2233	0.015 81	0.070 81	63.234	14.121	10.131	143.062	**28**
29	4.7241	0.2117	0.014 77	0.069 77	67.711	14.333	10.395	148.989	**29**
30	4.9840	0.2006	0.013 81	0.068 81	72.435	14.534	10.652	154.808	**30**
31	5.2581	0.1902	0.012 92	0.067 92	77.419	14.724	10.902	160.513	**31**
32	5.5473	0.1803	0.012 10	0.067 10	82.677	14.904	11.145	166.102	**32**
33	5.8524	0.1709	0.011 33	0.066 33	88.225	15.075	11.381	171.569	**33**
34	6.1742	0.1620	0.010 63	0.065 63	94.077	15.237	11.611	176.914	**34**
35	6.5138	0.1535	0.009 97	0.064 97	100.251	15.391	11.834	182.134	**35**
40	8.5133	0.1175	0.007 32	0.062 32	136.606	16.046	12.858	206.320	**40**
45	11.1266	0.0899	0.005 43	0.060 43	184.119	16.548	13.738	227.334	**45**
50	14.5420	0.0688	0.004 06	0.059 06	246.217	16.932	14.490	245.331	**50**
55	19.0058	0.0526	0.003 05	0.058 05	327.377	17.225	15.127	260.569	**55**
60	24.8398	0.0403	0.002 31	0.057 31	433.450	17.450	15.665	273.352	**60**
65	32.4646	0.0308	0.001 75	0.056 75	572.083	17.622	16.116	283.992	**65**
70	42.4299	0.0236	0.001 33	0.056 33	753.271	17.753	16.492	292.791	**70**
75	55.4542	0.0180	0.001 01	0.056 01	990.076	17.854	16.805	300.027	**75**
80	72.4764	0.0138	0.000 77	0.055 77	1 299.571	17.931	17.063	305.948	**80**
85	94.7238	0.0106	0.000 59	0.055 59	1 704.069	17.990	17.275	310.773	**85**
90	123.8002	0.0081	0.000 45	0.055 45	2 232.731	18.035	17.449	314.690	**90**
95	161.8019	0.0062	0.000 34	0.055 34	2 923.671	18.069	17.591	317.860	**95**
100	211.4686	0.0047	0.000 26	0.055 26	3 826.702	18.096	17.707	320.417	**100**

TABLE D–11
6% compound interest factors

	Single payment		*Uniform series*				*Uniform gradient*		
n	*Compound amount factor* *F/P*	*Present worth factor* *P/F*	*Sinking fund factor* *A/F*	*Capital recovery factor* *A/P*	*Compound amount factor* *F/A*	*Present worth factor* *P/A*	*Gradient conversion factor* *A/G*	*Present worth factor* *P/G*	*n*
1	1.0600	0.9434	1.000 00	1.060 00	1.000	0.943	0.000	0.000	**1**
2	1.1236	0.8900	0.485 44	0.545 44	2.060	1.833	0.485	0.890	**2**
3	1.1910	0.8396	0.314 11	0.374 11	3.184	2.673	0.961	2.569	**3**
4	1.2625	0.7921	0.228 59	0.288 59	4.375	3.465	1.427	4.946	**4**
5	1.3382	0.7473	0.177 40	0.237 40	5.637	4.212	1.884	7.935	**5**
6	1.4185	0.7050	0.143 36	0.203 36	6.975	4.917	2.330	11.459	**6**
7	1.5036	0.6651	0.119 14	0.179 14	8.394	5.582	2.768	15.450	**7**
8	1.5938	0.6274	0.101 04	0.161 04	9.897	6.210	3.195	19.842	**8**
9	1.6895	0.5919	0.087 02	0.147 02	11.491	6.802	3.613	24.577	**9**
10	1.7908	0.5584	0.075 87	0.135 87	13.181	7.360	4.022	29.602	**10**
11	1.8983	0.5268	0.066 79	0.126 79	14.972	7.887	4.421	34.870	**11**
12	2.0122	0.4970	0.059 28	0.119 28	16.870	8.384	4.811	40.337	**12**
13	2.1329	0.4688	0.052 96	0.112 96	18.882	8.853	5.192	45.963	**13**
14	2.2609	0.4423	0.047 58	0.107 58	21.015	9.295	5.564	51.713	**14**
15	2.3966	0.4173	0.042 96	0.102 96	23.276	9.712	5.926	57.555	**15**
16	2.5404	0.3936	0.038 95	0.098 95	25.673	10.106	6.279	63.459	**16**
17	2.6928	0.3714	0.035 44	0.095 44	28.213	10.477	6.624	69.401	**17**
18	2.8543	0.3503	0.032 36	0.092 36	30.906	10.828	6.960	75.357	**18**
19	3.0256	0.3305	0.029 62	0.089 62	33.760	11.158	7.287	81.306	**19**
20	3.2071	0.3118	0.027 18	0.087 18	36.786	11.470	7.605	87.230	**20**
21	3.3996	0.2942	0.025 00	0.085 00	39.993	11.764	7.915	93.114	**21**
22	3.6035	0.2775	0.023 05	0.083 05	43.392	12.042	8.217	98.941	**22**
23	3.8197	0.2618	0.021 28	0.081 28	46.996	12.303	8.510	104.701	**23**
24	4.0489	0.2470	0.019 68	0.079 68	50.816	12.550	8.795	110.381	**24**
25	4.2919	0.2330	0.018 23	0.078 23	54.865	12.783	9.072	115.973	**25**
26	4.5494	0.2198	0.016 90	0.076 90	59.156	13.003	9.341	121.468	**26**
27	4.8223	0.2074	0.015 70	0.075 70	63.706	13.211	9.603	126.860	**27**
28	5.1117	0.1956	0.014 59	0.074 59	68.528	13.406	9.857	132.142	**28**
29	5.4184	0.1846	0.013 58	0.073 58	73.640	13.591	10.103	137.310	**29**
30	5.7435	0.1741	0.012 65	0.072 65	79.058	13.765	10.342	142.359	**30**
31	6.0881	0.1643	0.011 79	0.071 79	84.802	13.929	10.574	147.286	**31**
32	6.4534	0.1550	0.011 00	0.071 00	90.890	14.084	10.799	152.090	**32**
33	6.8406	0.1462	0.010 27	0.070 27	97.343	14.230	11.017	156.768	**33**
34	7.2510	0.1379	0.009 60	0.069 60	104.184	14.368	11.228	161.319	**34**
35	7.6861	0.1301	0.008 97	0.068 97	111.435	14.498	11.432	165.743	**35**
40	10.2857	0.0972	0.006 46	0.066 46	154.762	15.046	12.359	185.957	**40**
45	13.7646	0.0727	0.004 70	0.064 70	212.744	15.456	13.141	203.110	**45**
50	18.4202	0.0543	0.003 44	0.063 44	290.336	15.762	13.796	217.457	**50**
55	24.6503	0.0406	0.002 54	0.062 54	394.172	15.991	14.341	229.322	**55**
60	32.9877	0.0303	0.001 88	0.061 88	533.128	16.161	14.791	239.043	**60**
65	44.1450	0.0227	0.001 39	0.061 39	719.083	16.289	15.160	246.945	**65**
70	59.0759	0.0169	0.001 03	0.061 03	967.932	16.385	15.461	253.327	**70**
75	79.0569	0.0126	0.000 77	0.060 77	1 300.949	16.456	15.706	258.453	**75**
80	105.7960	0.0095	0.000 57	0.060 57	1 746.600	16.509	15.903	262.549	**80**
85	141.5789	0.0071	0.000 43	0.060 43	2 342.982	16.549	16.062	265.810	**85**
90	189.4645	0.0053	0.000 32	0.060 32	3 141.075	16.579	16.189	268.395	**90**
95	253.5463	0.0039	0.000 24	0.060 24	4 209.104	16.601	16.290	270.437	**95**
100	339.3021	0.0029	0.000 18	0.060 18	5 638.368	16.618	16.371	272.047	**100**

TABLE D–12
7% compound interest factors

	Single payment		Uniform series				Uniform gradient		
n	*Compound amount factor F/P*	*Present worth factor P/F*	*Sinking fund factor A/F*	*Capital recovery factor A/P*	*Compound amount factor F/A*	*Present worth factor P/A*	*Gradient conversion factor A/G*	*Present worth factor P/G*	*n*
1	1.0700	0.9346	1.000 00	1.070 00	1.000	0.935	0.000	0.000	**1**
2	1.1449	0.8734	0.483 09	0.553 09	2.070	1.808	0.483	0.873	**2**
3	1.2250	0.8163	0.311 05	0.381 05	3.215	2.624	0.955	2.506	**3**
4	1.3108	0.7629	0.225 23	0.295 23	4.440	3.387	1.416	4.795	**4**
5	1.4026	0.7130	0.173 89	0.243 89	5.751	4.100	1.865	7.647	**5**
6	1.5007	0.6663	0.139 80	0.209 80	7.153	4.767	2.303	10.978	**6**
7	1.6058	0.6227	0.115 55	0.185 55	8.654	5.389	2.730	14.715	**7**
8	1.7182	0.5820	0.097 47	0.167 47	10.260	5.971	3.147	18.789	**8**
9	1.8385	0.5439	0.083 49	0.153 49	11.978	6.515	3.552	23.140	**9**
10	1.9672	0.5083	0.072 38	0.142 38	13.816	7.024	3.946	27.716	**10**
11	2.1049	0.4751	0.063 36	0.133 36	15.784	7.499	4.330	32.466	**11**
12	2.2522	0.4440	0.055 90	0.125 90	17.888	7.943	4.703	37.351	**12**
13	2.4098	0.4150	0.049 65	0.119 65	20.141	8.358	5.065	42.330	**13**
14	2.5785	0.3878	0.044 34	0.114 34	22.550	8.745	5.417	47.372	**14**
15	2.7590	0.3624	0.039 79	0.109 79	25.129	9.108	5.758	52.446	**15**
16	2.9522	0.3387	0.035 86	0.105 86	27.888	9.447	6.090	57.527	**16**
17	3.1588	0.3166	0.032 43	0.102 43	30.840	9.763	6.411	62.592	**17**
18	3.3799	0.2959	0.029 41	0.099 41	33.999	10.059	6.722	67.622	**18**
19	3.6165	0.2765	0.026 75	0.096 75	37.379	10.336	7.024	72.599	**19**
20	3.8697	0.2584	0.024 39	0.094 39	40.995	10.594	7.316	77.509	**20**
21	4.1406	0.2415	0.022 29	0.092 29	44.865	10.836	7.599	82.339	**21**
22	4.4304	0.2257	0.020 41	0.090 41	49.006	11.061	7.872	87.079	**22**
23	4.7405	0.2109	0.018 71	0.088 71	53.436	11.272	8.137	91.720	**23**
24	5.0724	0.1971	0.017 19	0.087 19	58.177	11.469	8.392	96.255	**24**
25	5.4274	0.1842	0.015 81	0.085 81	63.249	11.654	8.639	100.676	**25**
26	5.8074	0.1722	0.014 56	0.084 56	68.676	11.826	8.877	104.981	**26**
27	6.2139	0.1609	0.013 43	0.083 43	74.484	11.987	9.107	109.166	**27**
28	6.6488	0.1504	0.012 39	0.082 39	80.698	12.137	9.329	113.226	**28**
29	7.1143	0.1406	0.011 45	0.081 45	87.347	12.278	9.543	117.162	**29**
30	7.6123	0.1314	0.010 59	0.080 59	94.461	12.409	9.749	120.972	**30**
31	8.1451	0.1228	0.009 80	0.079 80	102.073	12.532	9.947	124.655	**31**
32	8.7153	0.1147	0.009 07	0.079 07	110.218	12.647	10.138	128.212	**32**
33	9.3253	0.1072	0.008 41	0.078 41	118.933	12.754	10.322	131.643	**33**
34	9.9781	0.1002	0.007 80	0.077 80	128.259	12.854	10.499	134.951	**34**
35	10.6766	0.0937	0.007 23	0.077 23	138.237	12.948	10.669	138.135	**35**
40	14.9745	0.0668	0.005 01	0.075 01	199.635	13.332	11.423	152.293	**40**
45	21.0025	0.0476	0.003 50	0.073 50	285.749	13.606	12.036	163.756	**45**
50	29.4570	0.0339	0.002 46	0.072 46	406.529	13.801	12.529	172.905	**50**
55	41.3150	0.0242	0.001 74	0.071 74	575.929	13.940	12.921	180.124	**55**
60	57.9464	0.0173	0.001 23	0.071 23	813.520	14.039	13.232	185.768	**60**
65	81.2729	0.0123	0.000 87	0.070 87	1 146.755	14.110	13.476	190.145	**65**
70	113.9894	0.0088	0.000 62	0.070 62	1 614.134	14.160	13.666	193.519	**70**
75	159.8760	0.0063	0.000 44	0.070 44	2 269.657	14.196	13.814	196.104	**75**
80	224.2344	0.0045	0.000 31	0.070 31	3 189.063	14.222	13.927	198.075	**80**
85	314.5003	0.0032	0.000 22	0.070 22	4 478.576	14.240	14.015	199.572	**85**
90	441.1030	0.0023	0.000 16	0.070 16	6 287.185	14.253	14.081	200.704	**90**
95	618.6697	0.0016	0.000 11	0.070 11	8 823.854	14.263	14.132	201.558	**95**
100	867.7163	0.0012	0.000 08	0.070 08	12 381.662	14.269	14.170	202.200	**100**

TABLE D–13
8% compound interest factors

	Single payment		*Uniform series*				*Uniform gradient*		
n	*Compound amount factor F/P*	*Present worth factor P/F*	*Sinking fund factor A/F*	*Capital recovery factor A/P*	*Compound amount factor F/A*	*Present worth factor P/A*	*Gradient conversion factor A/G*	*Present worth factor P/G*	*n*
1	1.0800	0.9259	1.000 00	1.080 00	1.000	0.926	0.000	0.000	**1**
2	1.1664	0.8573	0.480 77	0.560 77	2.080	1.783	0.481	0.857	**2**
3	1.2597	0.7938	0.308 03	0.388 03	3.246	2.577	0.949	2.445	**3**
4	1.3605	0.7350	0.221 92	0.301 92	4.506	3.312	1.404	4.650	**4**
5	1.4693	0.6806	0.170 46	0.250 46	5.867	3.993	1.846	7.372	**5**
6	1.5869	0.6302	0.136 32	0.216 32	7.336	4.623	2.276	10.523	**6**
7	1.7138	0.5835	0.112 07	0.192 07	8.923	5.206	2.694	14.024	**7**
8	1.8509	0.5403	0.094 01	0.174 01	10.637	5.747	3.099	17.806	**8**
9	1.9990	0.5002	0.080 08	0.160 08	12.488	6.247	3.491	21.808	**9**
10	2.1589	0.4632	0.069 03	0.149 03	14.487	6.710	3.871	25.977	**10**
11	2.3316	0.4289	0.060 08	0.140 08	16.645	7.139	4.240	30.266	**11**
12	2.5182	0.3971	0.052 70	0.132 70	18.977	7.536	4.596	34.634	**12**
13	2.7196	0.3677	0.046 52	0.126 52	21.495	7.904	4.940	39.046	**13**
14	2.9372	0.3405	0.041 30	0.121 30	24.215	8.244	5.273	43.472	**14**
15	3.1722	0.3152	0.036 83	0.116 83	27.152	8.559	5.594	47.886	**15**
16	3.4259	0.2919	0.032 98	0.112 98	30.324	8.851	5.905	52.264	**16**
17	3.7000	0.2703	0.029 63	0.109 63	33.750	9.122	6.204	56.588	**17**
18	3.9960	0.2502	0.026 70	0.106 70	37.450	9.372	6.492	60.843	**18**
19	4.3157	0.2317	0.024 13	0.104 13	41.446	9.604	6.770	65.013	**19**
20	4.6610	0.2145	0.021 85	0.101 85	45.762	9.818	7.037	69.090	**20**
21	5.0338	0.1987	0.019 83	0.099 83	50.423	10.017	7.294	73.063	**21**
22	5.4365	0.1839	0.018 03	0.098 03	55.457	10.201	7.541	76.926	**22**
23	5.8715	0.1703	0.016 42	0.096 42	60.893	10.371	7.779	80.673	**23**
24	6.3412	0.1577	0.014 98	0.094 98	66.765	10.529	8.007	84.300	**24**
25	6.8485	0.1460	0.013 68	0.093 68	73.106	10.675	8.225	87.804	**25**
26	7.3964	0.1352	0.012 51	0.092 51	79.954	10.810	8.435	91.184	**26**
27	7.9881	0.1252	0.011 45	0.091 45	87.351	10.935	8.636	94.439	**27**
28	8.6271	0.1159	0.010 49	0.090 49	95.339	11.051	8.829	97.569	**28**
29	9.3173	0.1073	0.009 62	0.089 62	103.966	11.158	9.013	100.574	**29**
30	10.0627	0.0994	0.008 83	0.088 83	113.283	11.258	9.190	103.456	**30**
31	10.8677	0.0920	0.008 11	0.088 11	123.346	11.350	9.358	106.216	**31**
32	11.7371	0.0852	0.007 45	0.087 45	134.214	11.435	9.520	108.857	**32**
33	12.6760	0.0789	0.006 85	0.086 85	145.951	11.514	9.674	111.382	**33**
34	13.6901	0.0730	0.006 30	0.086 30	158.627	11.587	9.821	113.792	**34**
35	14.7853	0.0676	0.005 80	0.085 80	172.317	11.655	9.961	116.092	**35**
40	21.7245	0.0460	0.003 86	0.083 86	259.057	11.925	10.570	126.042	**40**
45	31.9204	0.0313	0.002 59	0.082 59	386.506	12.108	11.045	133.733	**45**
50	46.9016	0.0213	0.001 74	0.081 74	573.770	12.233	11.411	139.593	**50**
55	68.9139	0.0145	0.001 18	0.081 18	848.923	12.319	11.690	144.006	**55**
60	101.2571	0.0099	0.000 80	0.080 80	1 253.213	12.377	11.902	147.300	**60**
65	148.7798	0.0067	0.000 54	0.080 54	1 847.248	12.416	12.060	149.739	**65**
70	218.6064	0.0046	0.000 37	0.080 37	2 720.080	12.443	12.178	151.533	**70**
75	321.2045	0.0031	0.000 25	0.080 25	4 002.557	12.461	12.266	152.845	**75**
80	471.9548	0.0021	0.000 17	0.080 17	5 886.935	12.474	12.330	153.800	**80**
85	693.4565	0.0014	0.000 12	0.080 12	8 655.706	12.482	12.377	154.492	**85**
90	1 018.9151	0.0010	0.000 08	0.080 08	12 723.939	12.488	12.412	154.993	**90**
95	1 497.1205	0.0007	0.000 05	0.080 05	18 701.507	12.492	12.437	155.352	**95**
100	2 199.7613	0.0005	0.000 04	0.080 04	27 484.516	12.494	12.455	155.611	**100**

TABLE D–14
9% compound interest factors

	Single payment		Uniform series				Uniform gradient		
n	Compound amount factor F/P	Present worth factor P/F	Sinking fund factor A/F	Capital recovery factor A/P	Compound amount factor F/A	Present worth factor P/A	Gradient conversion factor A/G	Present worth factor P/G	n
1	1.0900	0.9174	1.000 00	1.090 00	1.000	0.917	0.000	0.000	**1**
2	1.1881	0.8417	0.478 47	0.568 47	2.090	1.759	0.478	0.842	**2**
3	1.2950	0.7722	0.305 05	0.395 05	3.278	2.531	0.943	2.386	**3**
4	1.4116	0.7084	0.218 67	0.308 67	4.573	3.240	1.393	4.511	**4**
5	1.5386	0.6499	0.167 09	0.257 09	5.985	3.890	1.828	7.111	**5**
6	1.6771	0.5963	0.132 92	0.222 92	7.523	4.486	2.250	10.092	**6**
7	1.8280	0.5470	0.108 69	0.198 69	9.200	5.033	2.657	13.375	**7**
8	1.9926	0.5019	0.090 67	0.180 67	11.028	5.535	3.051	16.888	**8**
9	2.1719	0.4604	0.076 80	0.166 80	13.021	5.995	3.431	20.571	**9**
10	2.3674	0.4224	0.065 82	0.155 82	15.193	6.418	3.798	24.373	**10**
11	2.5804	0.3875	0.056 95	0.146 95	17.560	6.805	4.151	28.248	**11**
12	2.8127	0.3555	0.049 65	0.139 65	20.141	7.161	4.491	32.159	**12**
13	3.0658	0.3262	0.043 57	0.133 57	22.953	7.487	4.818	36.073	**13**
14	3.3417	0.2992	0.038 43	0.128 43	26.019	7.786	5.133	39.963	**14**
15	3.6425	0.2745	0.034 06	0.124 06	29.361	8.061	5.435	43.807	**15**
16	3.9703	0.2519	0.030 30	0.120 30	33.003	8.313	5.724	47.585	**16**
17	4.3276	0.2311	0.027 05	0.117 05	36.974	8.544	6.002	51.282	**17**
18	4.7171	0.2120	0.024 21	0.114 21	41.301	8.756	6.269	54.886	**18**
19	5.1417	0.1945	0.021 73	0.111 73	46.018	8.950	6.524	58.387	**19**
20	5.6044	0.1784	0.019 55	0.109 55	51.160	9.129	6.767	61.777	**20**
21	6.1088	0.1637	0.017 62	0.107 62	56.765	9.292	7.001	65.051	**21**
22	6.6586	0.1502	0.015 90	0.105 90	62.873	9.442	7.223	68.205	**22**
23	7.2579	0.1378	0.014 38	0.104 38	69.532	9.580	7.436	71.236	**23**
24	7.9111	0.1264	0.013 02	0.103 02	76.790	9.707	7.638	74.143	**24**
25	8.6231	0.1160	0.011 81	0.101 81	84.701	9.823	7.832	76.926	**25**
26	9.3992	0.1064	0.010 72	0.100 72	93.324	9.929	8.016	79.586	**26**
27	10.2451	0.0976	0.009 73	0.099 73	102.723	10.027	8.191	82.124	**27**
28	11.1671	0.0895	0.008 85	0.098 85	112.968	10.116	8.357	84.542	**28**
29	12.1722	0.0822	0.008 06	0.098 06	124.135	10.198	8.515	86.842	**29**
30	13.2677	0.0753	0.007 34	0.097 34	136.308	10.274	8.666	89.028	**30**
31	14.4618	0.0691	0.006 69	0.096 69	149.575	10.343	8.808	91.102	**31**
32	15.7633	0.0634	0.006 10	0.096 10	164.037	10.406	8.944	93.069	**32**
33	17.1820	0.0582	0.005 56	0.095 56	179.800	10.464	9.072	94.931	**33**
34	18.7284	0.0534	0.005 08	0.095 08	196.982	10.518	9.193	96.693	**34**
35	20.4140	0.0490	0.004 64	0.094 64	215.711	10.567	9.308	98.359	**35**
40	31.4094	0.0318	0.002 96	0.092 96	337.882	10.757	9.796	105.376	**40**
45	48.3273	0.0207	0.001 90	0.091 90	525.859	10.881	10.160	110.556	**45**
50	74.3575	0.0134	0.001 23	0.091 23	815.084	10.962	10.430	114.325	**50**
55	114.4083	0.0087	0.000 79	0.090 79	1 260.092	11.014	10.626	117.036	**55**
60	176.0313	0.0057	0.000 51	0.090 51	1 944.792	11.048	10.768	118.968	**60**
65	270.8460	0.0037	0.000 33	0.090 33	2 998.288	11.070	10.870	120.334	**65**
70	416.7301	0.0024	0.000 22	0.090 22	4 619.223	11.084	10.943	121.294	**70**
75	641.1909	0.0016	0.000 14	0.090 14	7 113.232	11.094	10.994	121.965	**75**
80	986.5517	0.0010	0.000 09	0.090 09	10 950.574	11.100	11.030	122.431	**80**
85	1 517.9320	0.0007	0.000 06	0.090 06	16 854.800	11.104	11.055	122.753	**85**
90	2 235.5266	0.0004	0.000 04	0.090 04	25 939.184	11.106	11.073	122.976	**90**
95	3 593.4971	0.0003	0.000 03	0.090 03	39 916.635	11.108	11.085	123.129	**95**
100	5 529.0408	0.0002	0.000 02	0.090 02	61 422.675	11.109	11.093	123.234	**100**

TABLE D–15
10% compound interest factors

	Single payment		*Uniform series*				*Uniform gradient*		
n	*Compound amount factor F/P*	*Present worth factor P/F*	*Sinking fund factor A/F*	*Capital recovery factor A/P*	*Compound amount factor F/A*	*Present worth factor P/A*	*Gradient conversion factor A/G*	*Present worth factor P/G*	*n*
1	1.1000	0.9091	1.000 00	1.100 00	1.000	0.909	0.000	0.000	**1**
2	1.2100	0.8264	0.476 19	0.576 19	2.100	1.736	0.476	0.826	**2**
3	1.3310	0.7513	0.302 11	0.402 11	3.310	2.487	0.937	2.329	**3**
4	1.4641	0.6830	0.215 47	0.315 47	4.641	3.170	1.381	4.378	**4**
5	1.6105	0.6209	0.163 80	0.263 80	6.105	3.791	1.810	6.862	**5**
6	1.7716	0.5645	0.129 61	0.229 61	7.716	4.355	2.224	9.684	**6**
7	1.9487	0.5132	0.105 41	0.205 41	9.487	4.868	2.622	12.763	**7**
8	2.1436	0.4665	0.087 44	0.187 44	11.436	5.335	3.004	16.029	**8**
9	2.3579	0.4241	0.073 64	0.173 64	13.579	5.759	3.372	19.421	**9**
10	2.5937	0.3855	0.062 75	0.162 75	15.937	6.144	3.725	22.891	**10**
11	2.8531	0.3505	0.053 96	0.153 96	18.531	6.495	4.064	26.396	**11**
12	3.1384	0.3186	0.046 76	0.146 76	21.384	6.814	4.388	29.901	**12**
13	3.4523	0.2897	0.040 78	0.140 78	24.523	7.103	4.699	33.377	**13**
14	3.7975	0.2633	0.035 75	0.135 75	27.975	7.367	4.996	36.800	**14**
15	4.1772	0.2394	0.031 47	0.131 47	31.772	7.606	5.279	40.152	**15**
16	4.5950	0.2176	0.027 82	0.127 82	35.950	7.824	5.549	43.416	**16**
17	5.0545	0.1978	0.024 66	0.124 66	40.545	8.022	5.807	46.582	**17**
18	5.5599	0.1799	0.021 93	0.121 93	45.599	8.201	6.053	49.640	**18**
19	6.1159	0.1635	0.019 55	0.119 55	51.159	8.365	6.286	52.583	**19**
20	6.7275	0.1486	0.017 46	0.117 46	57.275	8.514	6.508	55.407	**20**
21	7.4002	0.1351	0.015 62	0.115 62	64.002	8.649	6.719	58.110	**21**
22	8.1403	0.1228	0.014 01	0.114 01	71.403	8.772	6.919	60.689	**22**
23	8.9543	0.1117	0.012 57	0.112 57	79.543	8.883	7.108	63.146	**23**
24	9.8497	0.1015	0.011 30	0.111 30	88.497	8.985	7.288	65.481	**24**
25	10.8347	0.0923	0.010 17	0.110 17	98.347	9.077	7.458	67.696	**25**
26	11.9182	0.0839	0.009 16	0.109 16	109.182	9.161	7.619	69.794	**26**
27	13.1100	0.0763	0.008 26	0.108 26	121.100	9.237	7.770	71.777	**27**
28	14.4210	0.0693	0.007 45	0.107 45	134.210	9.307	7.914	73.650	**28**
29	15.8631	0.0630	0.006 73	0.106 73	148.631	9.370	8.049	75.415	**29**
30	17.4494	0.0573	0.006 08	0.106 08	164.494	9.427	8.176	77.077	**30**
31	19.1943	0.0521	0.005 50	0.105 50	181.943	9.479	8.296	78.640	**31**
32	21.1138	0.0474	0.004 97	0.104 97	201.138	9.526	8.409	80.108	**32**
33	23.2252	0.0431	0.004 50	0.104 50	222.252	9.569	8.515	81.486	**33**
34	25.5477	0.0391	0.004 07	0.104 07	245.477	9.609	8.615	82.777	**34**
35	28.1024	0.0356	0.003 69	0.103 69	271.024	9.644	8.709	83.987	**35**
40	45.2593	0.0221	0.002 26	0.102 26	442.593	9.779	9.096	88.953	**40**
45	72.8905	0.0137	0.001 39	0.101 39	718.905	9.863	9.374	92.454	**45**
50	117.3909	0.0085	0.000 86	0.100 86	1 163.909	9.915	9.570	94.889	**50**
55	189.0591	0.0053	0.000 53	0.100 53	1 880.591	9.947	9.708	96.562	**55**
60	304.4816	0.0033	0.000 33	0.100 33	3 034.816	9.967	9.802	97.701	**60**
65	490.3707	0.0020	0.000 20	0.100 20	4 893.707	9.980	9.867	98.471	**65**
70	789.7470	0.0013	0.000 13	0.100 13	7 887.470	9.987	9.911	98.987	**70**
75	1 271.8952	0.0008	0.000 08	0.100 08	12 708.954	9.992	9.941	99.332	**75**
80	2 048.4002	0.0005	0.000 05	0.100 05	20 474.002	9.995	9.961	99.561	**80**
85	3 298.9690	0.0003	0.000 03	0.100 03	32 979.690	9.997	9.974	99.712	**85**
90	5 313.0226	0.0002	0.000 02	0.100 02	53 120.226	9.998	9.983	99.812	**90**
95	8 556.6760	0.0001	0.000 01	0.100 01	85 556.760	9.999	9.989	99.877	**95**
100	13 780.6123	0.0001	0.000 01	0.100 01	137 796.123	9.999	9.993	99.920	**100**

TABLE D–16
11% compound interest factors

n	Single payment: Compound amount factor F/P	Present worth factor P/F	Uniform series: Sinking fund factor A/F	Capital recovery factor A/P	Compound amount factor F/A	Present worth factor P/A	Uniform gradient: Gradient conversion factor A/G	Present worth factor P/G	n
1	1.1100	0.9009	1.000 00	1.110 00	1.000	0.901	0.000	0.000	1
2	1.2321	0.8116	0.473 93	0.583 93	2.110	1.713	0.474	0.812	2
3	1.3676	0.7312	0.299 21	0.409 21	3.342	2.444	0.931	2.274	3
4	1.5181	0.6587	0.212 33	0.322 33	4.710	3.102	1.370	4.250	4
5	1.6851	0.5935	0.160 57	0.270 57	6.228	3.696	1.792	6.624	5
6	1.8704	0.5346	0.126 38	0.236 38	7.913	4.231	2.198	9.297	6
7	2.0762	0.4817	0.102 22	0.212 22	9.783	4.712	2.586	12.187	7
8	2.3045	0.4339	0.084 32	0.194 32	11.859	5.146	2.958	15.225	8
9	2.5581	0.3909	0.070 60	0.180 60	14.164	5.537	3.314	18.352	9
10	2.8394	0.3522	0.059 80	0.169 80	16.722	5.889	3.654	21.522	10
11	3.1518	0.3173	0.051 12	0.161 12	19.561	6.207	3.979	24.695	11
12	3.4984	0.2858	0.044 03	0.154 03	22.713	6.492	4.288	27.839	12
13	3.8833	0.2575	0.038 15	0.148 15	26.212	6.750	4.582	30.929	13
14	4.3104	0.2320	0.033 23	0.143 23	30.095	6.982	4.862	33.945	14
15	4.7846	0.2090	0.029 07	0.139 07	34.405	7.191	5.127	36.871	15
16	5.3109	0.1883	0.025 52	0.135 52	39.190	7.379	5.379	39.695	16
17	5.8951	0.1696	0.022 47	0.132 47	44.501	7.549	5.618	42.409	17
18	6.5436	0.1528	0.019 84	0.129 84	50.396	7.702	5.844	45.007	18
19	7.2633	0.1377	0.017 56	0.127 56	56.939	7.839	6.057	47.486	19
20	8.0623	0.1240	0.015 58	0.125 58	64.203	7.963	6.259	49.842	20
21	8.9492	0.1117	0.013 84	0.123 84	72.265	8.075	6.449	52.077	21
22	9.9336	0.1007	0.012 31	0.122 31	81.214	8.176	6.628	54.191	22
23	11.0263	0.0907	0.010 97	0.120 97	91.148	8.266	6.797	56.186	23
24	12.2392	0.0817	0.009 79	0.119 79	102.174	8.348	6.956	58.066	24
25	13.5855	0.0736	0.008 74	0.118 74	114.413	8.422	7.104	59.832	25
26	15.0799	0.0663	0.007 81	0.117 81	127.999	8.488	7.244	61.490	26
27	16.7386	0.0597	0.006 99	0.116 99	143.079	8.548	7.375	63.043	27
28	18.5799	0.0538	0.006 26	0.116 26	159.817	8.602	7.498	64.497	28
29	20.6237	0.0485	0.005 61	0.115 61	178.397	8.650	7.613	65.854	29
30	22.8923	0.0437	0.005 02	0.115 02	199.021	8.694	7.721	67.121	30
31	25.4104	0.0394	0.004 51	0.114 51	221.913	8.733	7.821	68.302	31
32	28.2056	0.0355	0.004 04	0.114 04	247.324	8.769	7.915	69.401	32
33	31.3082	0.0319	0.003 63	0.113 63	275.529	8.801	8.002	70.423	33
34	34.7521	0.0288	0.003 26	0.113 26	306.837	8.829	8.084	71.372	34
35	38.5749	0.0259	0.002 93	0.112 93	341.590	8.855	8.159	72.254	35
40	65.0009	0.0154	0.001 72	0.111 72	581.826	8.951	8.466	75.779	40
45	109.5302	0.0091	0.001 01	0.111 01	986.639	9.008	8.676	78.155	45
50	184.5648	0.0054	0.000 60	0.110 60	1 688.771	9.042	8.819	79.734	50
∞				0.110 00		9.091	9.091	82.645	∞

TABLE D–17
12% compound interest factors

	Single payment		Uniform series				Uniform gradient		
n	*Compound amount factor F/P*	*Present worth factor P/F*	*Sinking fund factor A/F*	*Capital recovery factor A/P*	*Compound amount factor F/A*	*Present worth factor P/A*	*Gradient conversion factor A/G*	*Present worth factor P/G*	*n*
1	1.1200	0.8929	1.000 00	1.120 00	1.000	0.893	0.000	0.000	**1**
2	1.2544	0.7972	0.471 70	0.591 70	2.120	1.690	0.472	0.797	**2**
3	1.4049	0.7118	0.296 35	0.416 35	3.374	2.402	0.925	2.221	**3**
4	1.5735	0.6355	0.209 23	0.329 23	4.779	3.037	1.359	4.127	**4**
5	1.7623	0.5674	0.157 41	0.277 41	6.353	3.605	1.775	6.397	**5**
6	1.9738	0.5066	0.123 23	0.243 23	8.115	4.111	2.172	8.930	**6**
7	2.2107	0.4523	0.099 12	0.219 12	10.089	4.564	2.551	11.644	**7**
8	2.4760	0.4039	0.081 30	0.201 30	12.300	4.968	2.913	14.471	**8**
9	2.7731	0.3606	0.067 68	0.187 68	14.776	5.328	3.257	17.356	**9**
10	3.1058	0.3220	0.056 98	0.176 98	17.549	5.650	3.585	20.254	**10**
11	3.4785	0.2875	0.048 42	0.168 42	20.655	5.938	3.895	23.129	**11**
12	3.8960	0.2567	0.041 44	0.161 44	24.133	6.194	4.190	25.952	**12**
13	4.3635	0.2292	0.035 68	0.155 68	28.029	6.424	4.468	28.702	**13**
14	4.8871	0.2046	0.030 87	0.150 87	32.393	6.628	4.732	31.362	**14**
15	5.4736	0.1827	0.026 82	0.146 82	37.280	6.811	4.980	33.920	**15**
16	6.1304	0.1631	0.023 39	0.143 39	42.753	6.974	5.215	36.367	**16**
17	6.8660	0.1456	0.020 46	0.140 46	48.884	7.120	5.435	38.697	**17**
18	7.6900	0.1300	0.017 94	0.137 94	55.750	7.250	5.643	40.908	**18**
19	8.6128	0.1161	0.015 76	0.135 76	63.440	7.366	5.838	42.998	**19**
20	9.6463	0.1037	0.013 88	0.133 88	72.052	7.469	6.020	44.968	**20**
21	10.8038	0.0926	0.012 24	0.132 24	81.699	7.562	6.191	46.819	**21**
22	12.1003	0.0826	0.010 81	0.130 81	92.503	7.645	6.351	48.554	**22**
23	13.5523	0.0738	0.009 56	0.129 56	104.603	7.718	6.501	50.178	**23**
24	15.1786	0.0659	0.008 46	0.128 46	118.155	7.784	6.641	51.693	**24**
25	17.0001	0.0588	0.007 50	0.127 50	133.334	7.843	6.771	53.105	**25**
26	19.0401	0.0525	0.006 65	0.126 65	150.334	7.896	6.892	54.418	**26**
27	21.3249	0.0469	0.005 90	0.125 90	169.374	7.943	7.005	55.637	**27**
28	23.8839	0.0419	0.005 24	0.125 24	190.699	7.984	7.110	56.767	**28**
29	26.7499	0.0374	0.004 66	0.124 66	214.583	8.022	7.207	57.814	**29**
30	29.9599	0.0334	0.004 14	0.124 14	241.333	8.055	7.297	58.782	**30**
31	33.5551	0.0298	0.003 69	0.123 69	271.292	8.085	7.381	59.676	**31**
32	37.5817	0.0266	0.003 28	0.123 28	304.847	8.112	7.459	60.501	**32**
33	42.0915	0.0238	0.002 92	0.122 92	342.429	8.135	7.530	61.261	**33**
34	47.1425	0.0212	0.002 60	0.122 60	384.520	8.157	7.596	61.961	**34**
35	52.7996	0.0189	0.002 32	0.122 32	431.663	8.176	7.658	62.605	**35**
40	93.0510	0.0107	0.001 30	0.121 30	767.091	8.244	7.899	65.116	**40**
45	163.9876	0.0061	0.000 74	0.120 74	1 358.230	8.283	8.057	66.734	**45**
50	289.0022	0.0035	0.000 42	0.120 42	2 400.018	8.305	8.160	67.762	**50**
∞				0.120 00		8.333	8.333	69.444	∞

TABLE D–18
13% compound interest factors

	Single payment		*Uniform series*				*Uniform gradient*		
n	*Compound amount factor F/P*	*Present worth factor P/F*	*Sinking fund factor A/F*	*Capital recovery factor A/P*	*Compound amount factor F/A*	*Present worth factor P/A*	*Gradient conversion factor A/G*	*Present worth factor P/G*	*n*
1	1.1300	0.8850	1.000 00	1.130 00	1.000	0.885	0.000	0.000	**1**
2	1.2769	0.7831	0.469 48	0.599 48	2.130	1.668	0.469	0.783	**2**
3	1.4429	0.6931	0.293 52	0.423 52	3.407	2.361	0.919	2.169	**3**
4	1.6305	0.6133	0.206 19	0.336 19	4.850	2.974	1.348	4.009	**4**
5	1.8424	0.5428	0.154 31	0.284 31	6.480	3.517	1.757	6.180	**5**
6	2.0820	0.4803	0.120 15	0.250 15	8.323	3.998	2.147	8.582	**6**
7	2.3526	0.4251	0.096 11	0.226 11	10.405	4.423	2.517	11.132	**7**
8	2.6584	0.3762	0.078 39	0.208 39	12.757	4.799	2.869	13.765	**8**
9	3.0040	0.3329	0.064 87	0.194 87	15.416	5.132	3.201	16.428	**9**
10	3.3946	0.2946	0.054 29	0.184 29	18.420	5.426	3.516	19.080	**10**
11	3.8359	0.2607	0.045 84	0.175 84	21.814	5.687	3.813	21.687	**11**
12	4.3345	0.2307	0.038 99	0.168 99	25.650	5.918	4.094	24.224	**12**
13	4.8980	0.2042	0.033 35	0.163 35	29.985	6.122	4.357	26.674	**13**
14	5.5348	0.1807	0.028 67	0.158 67	34.883	6.302	4.605	29.023	**14**
15	6.2543	0.1599	0.024 74	0.154 74	40.417	6.462	4.837	31.262	**15**
16	7.0673	0.1415	0.021 43	0.151 43	46.672	6.604	5.055	33.384	**16**
17	7.9861	0.1252	0.018 61	0.148 61	53.739	6.729	5.259	35.388	**17**
18	9.0243	0.1108	0.016 20	0.146 20	61.725	6.840	5.449	37.271	**18**
19	10.1974	0.0981	0.014 13	0144 13	70.749	6.938	5.627	39.037	**19**
20	11.5231	0.0868	0.012 35	0.142 35	80.947	7.025	5.792	40.685	**20**
21	13.0211	0.0768	0.010 81	0.140 81	92.470	7.102	5.945	42.221	**21**
22	14.7138	0.0680	0.009 48	0.139 48	105.491	7.170	6.088	43.649	**22**
23	16.6266	0.0601	0.008 32	0.138 32	120.205	7.230	6.220	44.972	**23**
24	18.7881	0.0532	0.007 31	0.137 31	136.831	7.283	6.343	46.196	**24**
25	21.2305	0.0471	0.006 43	0.136 43	155.620	7.330	6.457	47.326	**25**
26	23.9905	0.0417	0.005 65	0.135 65	176.850	7.372	6.561	48.369	**26**
27	27.1093	0.0369	0.004 98	0.134 98	200.841	7.409	6.658	49.328	**27**
28	30.6335	0.0329	0.004 39	0.134 39	227.950	7.441	6.747	50.209	**28**
29	34.6158	0.0289	0.003 87	0.133 87	258.583	7.470	6.830	51.018	**29**
30	39.1159	0.0256	0.003 41	0.133 41	293.199	7.496	6.905	51.759	**30**
31	44.2010	0.0226	0.003 01	0.133 01	332.315	7.518	6.975	52.438	**31**
32	49.9471	0.0200	0.002 66	0.132 66	376.516	7.538	7.039	53.059	**32**
33	56.4402	0.0177	0.002 34	0.132 34	426.463	7.556	7.097	53.626	**33**
34	63.7774	0.0157	0.002 07	0.132 07	482.903	7.572	7.151	54.143	**34**
35	72.0685	0.0139	0.001 83	0.131 83	546.681	7.586	7.200	54.615	**35**
40	132.7816	0.0075	0.000 99	0.130 99	1 013.704	7.634	7.389	56.409	**40**
45	244.6414	0.0041	0.000 53	0.130 53	1 874.165	7.661	7.508	57.515	**45**
50	450.7359	0.0022	0.000 29	0.130 29	3 459.507	7.675	7.581	58.187	**50**
∞				0.130 00		7.692	7.692	59.172	∞

TABLE D–19
14% compound interest factors

	Single payment		*Uniform series*				*Uniform gradient*		
n	*Compound amount factor F/P*	*Present worth factor P/F*	*Sinking fund factor A/F*	*Capital recovery factor A/P*	*Compound amount factor F/A*	*Present worth factor P/A*	*Gradient conversion factor A/G*	*Present worth factor P/G*	*n*
1	1.1400	0.8772	1.000 00	1.140 00	1.000	0.877	0.000	0.000	**1**
2	1.2996	0.7695	0.467 29	0.607 29	2.140	1.647	0.467	0.769	**2**
3	1.4815	0.6750	0.290 73	0.430 73	3.440	2.322	0.913	2.119	**3**
4	1.6890	0.5921	0.203 20	0.343 20	4.921	2.914	1.337	3.896	**4**
5	1.9254	0.5194	0.151 28	0.291 28	6.610	3.433	1.740	5.973	**5**
6	2.1950	0.4556	0.117 16	0.257 16	8.536	3.889	2.122	8.251	**6**
7	2.5023	0.3996	0.093 19	0.233 19	10.730	4.288	2.483	10.649	**7**
8	2.8526	0.3506	0.075 57	0.215 57	13.233	4.639	2.825	13.103	**8**
9	3.2519	0.3075	0.062 17	0.202 17	16.085	4.946	3.146	15.563	**9**
10	3.7072	0.2697	0.051 71	0.191 71	19.337	5.216	3.449	17.991	**10**
11	4.2262	0.2366	0.043 39	0.183 39	23.045	5.453	3.733	20.357	**11**
12	4.8179	0.2076	0.036 67	0.176 67	27.271	5.660	4.000	22.640	**12**
13	5.4924	0.1821	0.031 16	0.171 16	32.089	5.842	4.249	24.825	**13**
14	6.2613	0.1597	0.026 61	0.166 61	37.581	6.002	4.482	26.901	**14**
15	7.1379	0.1401	0.022 81	0.162 81	43.842	6.142	4.699	28.862	**15**
16	8.1372	0.1229	0.019 62	0.159 62	50.980	6.265	4.901	30.706	**16**
17	9.2765	0.1078	0.016 92	0.156 92	59.118	6.373	5.089	32.430	**17**
18	10.5752	0.0946	0.014 62	0.154 62	68.394	6.467	5.263	34.038	**18**
19	12.0557	0.0829	0.012 66	0.152 66	78.969	6.550	5.424	35.531	**19**
20	13.7435	0.0728	0.010 99	0.150 99	91.025	6.623	5.573	36.914	**20**
21	15.6676	0.0638	0.009 54	0.149 54	104.768	6.687	5.711	38.190	**21**
22	17.8610	0.0560	0.008 30	0.148 30	120.436	6.743	5.838	39.366	**22**
23	20.3616	0.0491	0.007 23	0.147 23	138.297	2.792	5.955	40.446	**23**
24	23.2122	0.0431	0.006 30	0.146 30	158.659	6.835	6.062	41.437	**24**
25	26.4619	0.0378	0.005 50	0.145 50	181.871	6.873	6.161	42.344	**25**
26	30.1666	0.0331	0.004 80	0.144 80	208.333	6.906	6.251	43.173	**26**
27	34.3899	0.0291	0.004 19	0.144 19	238.499	6.935	6.334	43.929	**27**
28	39.2045	0.0255	0.003 66	0.143 66	272.889	6.961	6.410	44.618	**28**
29	44.6931	0.0224	0.003 20	0.143 20	312.094	6.983	6.479	45.244	**29**
30	50.9502	0.0196	0.002 80	0.142 80	356.787	7.003	6.542	45.813	**30**
31	58.0832	0.0172	0.002 45	0.142 45	407.737	7.020	6.600	46.330	**31**
32	66.2148	0.0151	0.002 15	0.142 15	465.820	7.035	6.652	46.798	**32**
33	75.4849	0.0132	0.001 88	0.141 88	532.035	7.048	6.700	47.222	**33**
34	86.0528	0.0116	0.001 65	0.141 65	607.520	7.060	6.743	47.605	**34**
35	98.1002	0.0102	0.001 44	0.141 44	693.573	7.070	6.782	47.952	**35**
40	188.8835	0.0053	0.000 75	0.140 75	1 342.025	7.105	6.930	49.238	**40**
45	363.6791	0.0027	0.000 39	0.140 39	2 590.565	7.123	7.019	49.996	**45**
50	700.2330	0.0014	0.000 20	0.140 20	4 994.521	7.133	7.071	50.438	**50**
∞				0.140 00		7.143	7.143	51.020	∞

TABLE D–20
15% compound interest factors

	Single payment		*Uniform series*				*Uniform gradient*		
n	*Compound amount factor F/P*	*Present worth factor P/F*	*Sinking fund factor A/F*	*Capital recovery factor A/P*	*Compound amount factor F/A*	*Present worth factor P/A*	*Gradient conversion factor A/G*	*Present worth factor P/G*	*n*
1	1.1500	0.8696	1.000 00	1.150 00	1.000	0.870	0.000	0.000	**1**
2	1.3225	0.7561	0.465 12	0.615 12	2.150	1.626	0.465	0.756	**2**
3	1.5209	0.6575	0.287 98	0.437 98	3.472	2.283	0.907	2.071	**3**
4	1.7490	0.5718	0.200 26	0.350 27	4.993	2.855	1.326	3.786	**4**
5	2.0114	0.4972	0.148 32	0.298 32	6.742	3.352	1.723	5.775	**5**
6	2.3131	0.4323	0.114 24	0.264 24	8.754	3.784	2.097	7.937	**6**
7	2.6600	0.3759	0.090 36	0.240 36	11.067	4.160	2.450	10.192	**7**
8	3.0590	0.3269	0.072 85	0.222 85	13.727	4.487	2.781	12.481	**8**
9	3.5179	0.2843	0.059 57	0.209 57	16.786	4.772	3.092	14.755	**9**
10	4.0456	0.2472	0.049 25	0.199 25	20.304	5.019	3.383	16.979	**10**
11	4.6527	0.2149	0.041 07	0.191 07	24.349	2.234	3.655	19.129	**11**
12	5.3503	0.1869	0.034 48	0.184 48	29.002	5.421	3.908	21.185	**12**
13	6.1528	0.1625	0.029 11	0.179 11	34.352	5.583	4.144	23.135	**13**
14	7.0757	0.1413	0.024 69	0.174 69	40.505	5.724	4.362	24.972	**14**
15	8.1371	0.1229	0.021 02	0.171 02	47.580	5.847	4.565	26.693	**15**
16	9.3576	0.1069	0.017 95	0.167 95	55.717	5.954	4.752	28.296	**16**
17	10.7613	0.0929	0.015 37	0.165 37	65.075	6.047	4.925	29.783	**17**
18	12.3755	0.0808	0.013 19	0.163 19	75.836	6.128	5.084	31.156	**18**
19	14.2318	0.0703	0.011 34	0.161 34	88.212	6.198	5.231	32.421	**19**
20	16.3665	0.0611	0.009 76	0.159 76	102.444	6.259	5.365	33.582	**20**
21	18.8215	0.0531	0.008 42	0.158 42	118.810	6.312	5.488	34.645	**21**
22	21.6447	0.0462	0.007 27	0.157 27	137.632	6.359	5.601	35.615	**22**
23	24.8915	0.0402	0.006 28	0.156 28	159.276	6.399	5.704	36.499	**23**
24	28.6252	0.0349	0.005 43	0.155 43	184.168	6.434	5.798	37.302	**24**
25	32.9190	0.0304	0.004 70	0.154 70	212.793	6.464	5.883	38.031	**25**
26	37.8568	0.0264	0.004 07	0.154 07	245.712	6.491	5.961	38.692	**26**
27	43.5353	0.0230	0.003 53	0.153 53	283.569	6.514	6.032	39.289	**27**
28	50.0656	0.0200	0.003 06	0.153 06	327.104	6.534	6.096	39.828	**28**
29	57.5755	0.0174	0.002 65	0.152 65	377.170	6.551	6.154	40.315	**29**
30	66.2118	0.0151	0.002 30	0.152 30	434.745	6.566	6.207	40.753	**30**
31	76.1435	0.0131	0.002 00	0.152 00	500.957	6.579	6.254	41.147	**31**
32	87.5651	0.0114	0.001 73	0.151 73	577.100	6.591	6.297	41.501	**32**
33	100.6998	0.0099	0.001 50	0.151 50	664.666	6.600	6.336	41.818	**33**
34	115.8048	0.0086	0.001 31	0.151 31	765.365	6.609	6.371	42.103	**34**
35	133.1755	0.0075	0.001 13	0.151 13	881.170	6.617	6.402	42.359	**35**
40	267.8635	0.0037	0.000 56	0.150 56	1 779.090	6.642	6.517	43.283	**40**
45	538.7693	0.0019	0.000 28	0.150 28	3 585.128	6.654	6.583	43.805	**45**
50	1 083.6574	0.0009	0.000 14	0.150 14	7 217.716	6.661	6.620	44.096	**50**
∞				0.150 00		6.667	6.667	44.444	∞

TABLE D–21
16% compound interest factors

	Single payment		Uniform series				Uniform gradient		
n	*Compound amount factor F/P*	*Present worth factor P/F*	*Sinking fund factor A/F*	*Capital recovery factor A/P*	*Compound amount factor F/A*	*Present worth factor P/A*	*Gradient conversion factor A/G*	*Present worth factor P/G*	*n*
1	1.1600	0.8621	1.000 00	1.160 00	1.000	0.862	0.000	0.000	**1**
2	1.3456	0.7432	0.462 96	0.622 96	2.160	1.605	0.463	0.743	**2**
3	1.5609	0.6407	0.285 26	0.445 26	3.506	2.246	0.901	2.024	**3**
4	1.8106	0.5523	0.197 38	0.357 38	5.066	2.798	1.316	3.681	**4**
5	2.1003	0.4761	0.145 41	0.305 41	6.877	3.274	1.706	5.586	**5**
6	2.4364	0.4104	0.111 39	0.271 39	8.977	3.685	2.073	7.638	**6**
7	2.8262	0.3538	0.087 61	0.247 61	11.414	4.039	2.417	9.761	**7**
8	3.2784	0.3050	0.070 22	0.230 22	14.240	4.344	2.739	11.896	**8**
9	3.8030	0.2630	0.057 08	0.217 08	17.519	4.607	3.039	14.000	**9**
10	4.4114	0.2267	0.046 90	0.206 90	21.321	4.833	3.319	16.040	**10**
11	5.1173	0.1954	0.038 86	0.198 86	25.733	5.234	3.578	17.994	**11**
12	5.9360	0.1685	0.032 41	0.192 41	30.850	5.197	3.819	19.847	**12**
13	6.8858	0.1452	0.027 18	0.187 18	36.786	5.342	4.041	21.590	**13**
14	7.9875	0.1252	0.022 90	0.182 90	43.672	5.468	4.246	23.217	**14**
15	9.2655	0.1079	0.019 36	0.179 36	51.660	5.575	4.435	24.728	**15**
16	10.7480	0.0930	0.016 41	0.176 41	60.925	5.668	4.609	26.124	**16**
17	12.4677	0.0802	0.013 95	0.173 95	71.673	5.749	4.768	27.407	**17**
18	14.4625	0.0691	0.011 88	0.171 88	84.141	5.818	4.913	28.583	**18**
19	16.7765	0.0596	0.010 14	0.170 14	98.603	5.877	5.046	29.656	**19**
20	19.4608	0.0514	0.008 67	0.168 67	115.380	5.929	5.167	30.632	**20**
21	22.5745	0.0443	0.007 42	0.167 42	134.841	5.973	5.277	31.518	**21**
22	26.1864	0.0382	0.006 35	0.166 35	157.415	6.011	5.377	32.320	**22**
23	30.3762	0.0329	0.005 45	0.165 45	183.601	6.044	5.467	33.044	**23**
24	35.2364	0.0284	0.004 67	0.164 67	213.978	6.073	5.549	33.697	**24**
25	40.8742	0.0245	0.004 01	0.164 01	249.214	6.097	5.623	34.284	**25**
26	47.4141	0.0211	0.003 45	0.163 45	290.088	6.118	5.690	34.811	**26**
27	55.0004	0.0182	0.002 96	0.162 96	337.502	6.136	5.750	35.284	**27**
28	63.8004	0.0157	0.002 55	0.162 55	392.503	6.152	5.804	35.707	**28**
29	74.0085	0.0135	0.002 19	0.162 19	456.303	6.166	5.853	36.086	**29**
30	85.8499	0.0116	0.001 89	0.161 89	530.312	6.177	5.896	36.423	**30**
31	99.5859	0.0100	0.001 62	0.161 62	616.162	6.187	5.936	36.725	**31**
32	115.5196	0.0087	0.001 40	0.161 40	715.747	6.196	5.971	36.993	**32**
33	134.0027	0.0075	0.001 20	0.161 20	831.267	6.203	6.002	37.232	**33**
34	155.4432	0.0064	0.001 04	0.161 04	965.270	6.210	6.030	37.444	**34**
35	180.3141	0.0055	0.000 89	0.160 89	1 120.713	6.215	6.055	37.633	**35**
40	378.7212	0.0026	0.000 42	0.160 42	2 360.757	6.233	6.144	38.299	**40**
45	795.4438	0.0013	0.000 20	0.160 20	4 965.274	6.242	6.193	38.660	**45**
50	1 670.7038	0.0006	0.000 10	0.160 10	10 435.649	6.246	6.220	38.852	**50**
∞				0.160 00		6.250	6.250	39.063	∞

TABLE D–22
18% compound interest factors

	Single payment		*Uniform series*				*Uniform gradient*		
n	*Compound amount factor* *F/P*	*Present worth factor* *P/F*	*Sinking fund factor* *A/F*	*Capital recovery factor* *A/P*	*Compound amount factor* *F/A*	*Present worth factor* *P/A*	*Gradient conversion factor* *A/G*	*Present worth factor* *P/G*	*n*
1	1.1800	0.8475	1.000 00	1.180 00	1.000	0.847	0.000	0.000	**1**
2	1.3924	0.7182	0.458 72	0.638 72	2.180	1.566	0.459	0.718	**2**
3	1.6430	0.6086	0.279 92	0.459 92	3.572	2.174	0.890	1.935	**3**
4	1.9388	0.5158	0.191 74	0.371 74	5.215	2.690	1.295	3.483	**4**
5	2.2878	0.4371	0.139 78	0.319 78	7.154	3.127	1.673	5.231	**5**
6	2.6996	0.3704	0.105 91	0.285 91	9.442	3.498	2.025	7.083	**6**
7	3.1855	0.3139	0.082 36	0.262 36	12.142	3.812	2.353	8.967	**7**
8	3.7589	0.2660	0.065 24	0.245 24	15.327	4.078	2.656	10.829	**8**
9	4.4355	0.2255	0.052 39	0.232 39	19.086	4.303	2.936	12.633	**9**
10	5.2338	0.1911	0.042 51	0.222 51	23.521	4.494	3.194	14.352	**10**
11	6.1759	0.1619	0.034 78	0.214 78	28.755	4.656	3.430	15.972	**11**
12	7.2876	0.1372	0.028 63	0.208 63	34.931	4.793	3.647	17.481	**12**
13	8.5994	0.1163	0.023 69	0.203 69	42.219	4.910	3.845	18.877	**13**
14	10.1472	0.0985	0.019 68	0.199 68	50.818	5.008	4.025	20.158	**14**
15	11.9737	0.0835	0.016 40	0.196 40	60.965	5.092	4.189	21.327	**15**
16	14.1290	0.0708	0.013 71	0.193 71	72.939	5.162	4.337	22.389	**16**
17	16.6722	0.0600	0.011 49	0.191 49	87.068	5.222	4.471	23.348	**17**
18	19.6733	0.0508	0.009 64	0.189 64	103.740	5.273	4.592	24.212	**18**
19	23.2144	0.0431	0.008 10	0.188 10	123.414	5.316	4.700	24.988	**19**
20	27.3930	0.0365	0.006 82	0.186 82	146.628	5.353	4.798	25.681	**20**
21	32.3238	0.0309	0.005 75	0.185 75	174.021	5.384	4.885	26.300	**21**
22	38.1421	0.0262	0.004 85	0.184 85	206.345	5.410	4.963	26.851	**22**
23	45.0076	0.0222	0.004 09	0.184 09	244.487	5.432	5.033	27.339	**23**
24	53.1090	0.0188	0.003 45	0.183 45	289.494	5.451	5.095	27.772	**24**
25	62.6686	0.0160	0.002 92	0.182 92	342.603	5.467	5.150	28.155	**25**
26	73.9490	0.0135	0.002 47	0.182 47	405.272	5.480	5.199	28.494	**26**
27	87.2598	0.0115	0.002 09	0.182 09	479.221	5.492	5.243	28.791	**27**
28	102.9665	0.0097	0.001 77	0.181 77	566.481	5.502	5.281	29.054	**28**
29	121.5005	0.0082	0.001 49	0.181 49	669.447	5.510	5.315	29.284	**29**
30	143.3706	0.0070	0.001 26	0.181 26	790.948	5.517	5.345	29.486	**30**
31	169.1774	0.0059	0.001 07	0.181 07	934.319	5.523	5.371	29.664	**31**
32	199.6293	0.0050	0.000 91	0.180 91	1 103.496	5.528	5.394	29.819	**32**
33	235.5625	0.0042	0.000 77	0.180 77	1 303.125	5.532	5.415	29.955	**33**
34	277.9638	0.0036	0.000 65	0.180 65	1 538.688	5.536	5.433	30.074	**34**
35	327.9973	0.0030	0.000 55	0.180 55	1 816.652	5.539	5.449	30.177	**35**
40	750.3783	0.0013	0.000 24	0.180 24	4 163.213	5.548	5.502	30.527	**40**
45	1 716.6839	0.0006	0.000 10	0.180 10	9 531.577	5.552	5.529	30.701	**45**
50	3 927.3569	0.0003	0.000 05	0.180 05	21 813.094	5.554	5.543	30.786	**50**
∞				0.180 00		5.556	5.556	30.864	∞

TABLE D–23
20% compound interest factors

	Single payment		*Uniform series*				*Uniform gradient*		
n	*Compound amount factor F/P*	*Present worth factor P/F*	*Sinking fund factor A/F*	*Capital recovery factor A/P*	*Compound amount factor F/A*	*Present worth factor P/A*	*Gradient conversion factor A/G*	*Present worth factor P/G*	*n*
1	1.2000	0.8333	1.000 00	1.200 00	1.000	0.833	0.000	0.000	**1**
2	1.4400	0.6944	0.454 55	0.654 55	2.200	1.528	0.455	0.694	**2**
3	1.7280	0.5787	0.274 73	0.474 73	3.640	2.106	0.879	1.852	**3**
4	2.0736	0.4823	0.186 29	0.386 29	5.368	2.589	1.274	3.299	**4**
5	2.4883	0.4019	0.134 38	0.334 38	7.442	2.991	1.641	4.906	**5**
6	2.9860	0.3349	0.100 71	0.300 71	9.930	3.326	1.979	6.581	**6**
7	3.5832	0.2791	0.077 42	0.277 42	12.916	3.605	2.290	8.255	**7**
8	4.2998	0.2326	0.060 61	0.260 61	16.499	3.837	2.576	9.883	**8**
9	5.1598	0.1938	0.048 08	0.248 08	20.799	4.031	2.836	11.434	**9**
10	6.1917	0.1615	0.038 52	0.238 52	25.959	4.192	3.074	12.887	**10**
11	7.4301	0.1346	0.031 10	0.231 10	32.150	4.327	3.289	14.233	**11**
12	8.9161	0.1122	0.025 26	0.225 26	39.581	4.439	3.484	15.467	**12**
13	10.6993	0.0935	0.020 62	0.220 62	48.497	4.533	3.660	16.588	**13**
14	12.8392	0.0779	0.016 89	0.216 89	59.196	4.611	3.817	17.601	**14**
15	15.4070	0.0649	0.013 88	0.213 88	72.035	4.675	3.959	18.509	**15**
16	18.4884	0.0541	0.011 44	0.211 44	87.442	4.730	4.085	19.321	**16**
17	22.1861	0.0451	0.009 44	0.209 44	105.931	4.775	4.198	20.042	**17**
18	26.6233	0.0376	0.007 81	0.207 81	128.117	4.812	4.298	20.680	**18**
19	31.9480	0.0313	0.006 46	0.206 46	154.740	4.844	4.386	21.244	**19**
20	38.3376	0.0261	0.005 36	0.205 36	186.688	4.870	4.464	21.739	**20**
21	46.0051	0.0217	0.004 44	0.204 44	225.026	4.891	4.533	22.174	**21**
22	55.2061	0.0181	0.003 69	0.203 69	271.031	4.909	4.594	22.555	**22**
23	66.2474	0.0151	0.003 07	0.203 07	326.237	4.925	4.647	22.887	**23**
24	79.4968	0.0126	0.002 55	0.202 55	392.484	4.937	4.694	23.176	**24**
25	95.3962	0.0105	0.002 12	0.202 12	471.981	4.948	4.735	23.428	**25**
26	114.4755	0.0087	0.001 76	0.201 76	567.377	4.956	4.771	23.646	**26**
27	137.3706	0.0073	0.001 47	0.201 47	681.853	4.964	4.802	23.835	**27**
28	164.8447	0.0061	0.001 22	0.201 22	819.223	4.970	4.829	23.999	**28**
29	197.8136	0.0051	0.001 02	0.201 02	984.068	4.975	4.853	24.141	**29**
30	237.3763	0.0042	0.000 85	0.200 85	1 181.882	4.979	4.873	24.263	**30**
31	284.8516	0.0035	0.000 70	0.200 70	1 419.258	4.982	4.891	24.368	**31**
32	341.8219	0.0029	0.000 59	0.200 59	1 704.109	4.985	4.906	24.459	**32**
33	410.1863	0.0024	0.000 49	0.200 49	2 045.931	4.988	4.919	24.537	**33**
34	492.2235	0.0020	0.000 41	0.200 41	2 456.118	4.990	4.931	24.604	**34**
35	590.6682	0.0017	0.000 34	0.200 34	2 948.341	4.992	4.941	24.661	**35**
40	1 469.7716	0.0007	0.000 14	0.200 14	7 343.858	4.997	4.973	24.847	**40**
45	3 657.2620	0.0003	0.000 05	0.200 05	18 281.310	4.999	4.988	24.932	**45**
50	9 100.4382	0.0001	0.000 02	0.200 02	45 497.191	4.999	4.995	24.970	**50**
∞				0.200 00		5.000	5.000	25.000	∞

TABLE D–24
25% compound interest factors

	Single payment		*Uniform series*				*Uniform gradient*		
n	*Compound amount factor* *F/P*	*Present worth factor* *P/F*	*Sinking fund factor* *A/F*	*Capital recovery factor* *A/P*	*Compound amount factor* *F/A*	*Present worth factor* *P/A*	*Gradient conversion factor* *A/G*	*Present worth factor* *P/G*	*n*
1	1.2500	0.8000	1.000 00	1.250 00	1.000	0.800	0.000	0.000	**1**
2	1.5625	0.6400	0.444 44	0.694 44	2.250	1.440	0.444	0.640	**2**
3	1.9531	0.5120	0.262 30	0.512 30	3.813	1.952	0.852	1.664	**3**
4	2.4414	0.4096	0.173 44	0.423 44	5.766	2.362	1.225	2.893	**4**
5	3.0518	0.3277	0.121 85	0.371 85	8.207	2.689	1.563	4.204	**5**
6	3.8147	0.2621	0.088 82	0.338 82	11.259	2.951	1.868	5.514	**6**
7	4.7684	0.2097	0.066 34	0.316 34	15.073	3.161	2.142	6.773	**7**
8	5.9605	0.1678	0.050 40	0.300 40	19.842	3.329	2.387	7.947	**8**
9	7.4506	0.1342	0.038 76	0.288 76	25.802	3.463	2.605	9.021	**9**
10	9.3132	0.1074	0.030 07	0.280 07	33.253	3.571	2.797	9.987	**10**
11	11.6415	0.0859	0.023 49	0.273 49	42.566	3.656	2.966	10.846	**11**
12	14.5519	0.0687	0.018 45	0.268 45	54.208	3.725	3.115	11.602	**12**
13	18.1899	0.0550	0.014 54	0.264 54	68.760	3.780	3.244	12.262	**13**
14	22.7374	0.0440	0.011 50	0.261 50	86.949	3.824	3.356	12.833	**14**
15	28.4217	0.0352	0.009 12	0.259 12	109.687	3.859	3.453	13.326	**15**
16	35.5271	0.0281	0.007 24	0.257 24	138.109	3.887	3.537	13.748	**16**
17	44.4089	0.0225	0.005 76	0.255 76	173.636	3.910	3.608	14.108	**17**
18	55.5112	0.0180	0.004 59	0.254 59	218.045	3.928	3.670	14.415	**18**
19	69.3889	0.0144	0.003 66	0.253 66	273.556	3.942	3.722	14.674	**19**
20	86.7362	0.0115	0.002 92	0.252 92	342.945	3.954	3.767	14.893	**20**
21	108.4202	0.0092	0.002 33	0.252 33	429.681	3.963	3.805	15.078	**21**
22	135.5253	0.0074	0.001 86	0.251 86	538.101	3.970	3.836	15.233	**22**
23	169.4066	0.0059	0.001 48	0.251 48	673.626	3.976	3.863	15.362	**23**
24	211.7582	0.0047	0.001 19	0.251 19	843.033	3.981	3.886	15.471	**24**
25	264.6978	0.0038	0.000 95	0.250 95	1 054.791	3.985	3.905	15.562	**25**
26	330.8722	0.0030	0.000 76	0.250 76	1 319.489	3.988	3.921	15.637	**26**
27	413.5903	0.0024	0.000 61	0.250 61	1 650.361	3.990	3.935	15.700	**27**
28	516.9879	0.0019	0.000 48	0.250 48	2 063.952	3.992	3.946	15.752	**28**
29	646.2349	0.0015	0.000 39	0.250 39	2 580.939	3.994	3.955	15.796	**29**
30	807.7936	0.0012	0.000 31	0.250 31	3 227.174	3.995	3.963	15.832	**30**
31	1 009.7420	0.0010	0.000 25	0.250 25	4 034.968	3.996	3.969	15.861	**31**
32	1 262.1774	0.0008	0.000 20	0.250 20	5 044.710	3.997	3.975	15.886	**32**
33	1 577.7218	0.0006	0.000 16	0.250 16	6 306.887	3.997	3.979	15.906	**33**
34	1 972.1523	0.0005	0.000 13	0.250 13	7 884.609	3.998	3.983	15.923	**34**
35	2 465.1903	0.0004	0.000 10	0.250 10	9 856.761	3.998	3.986	15.937	**35**
40	7 523.1638	0.0001	0.000 03	0.250 03	30 088.655	3.999	3.995	15.977	**40**
45	22 958.8740	0.0001	0.000 01	0.250 01	91 831.496	4.000	3.998	15.991	**45**
50	70 064.9232	0.0000	0.000 00	0.250 00	280 255.693	4.000	3.999	15.997	**50**
∞				0.250 00		4.000	4.000	16.000	∞

TABLE D–25
30% compound interest factors

	Single payment		Uniform series				Uniform gradient		
n	*Compound amount factor F/P*	*Present worth factor P/F*	*Sinking fund factor A/F*	*Capital recovery factor A/P*	*Compound amount factor F/A*	*Present worth factor P/A*	*Gradient conversion factor A/G*	*Present worth factor P/G*	*n*
1	1.3000	0.7692	1.000 00	1.300 00	1.000	0.769	0.000	0.000	**1**
2	1.6900	0.5917	0.434 78	0.734 78	2.300	1.361	0.435	0.592	**2**
3	2.1970	0.4552	0.250 63	0.550 63	3.990	1.816	0.827	1.502	**3**
4	2.8561	0.3501	0.161 63	0.461 63	6.187	2.166	1.178	2.552	**4**
5	3.7129	0.2693	0.110 58	0.410 58	9.043	2.436	1.490	3.630	**5**
6	4.8268	0.2072	0.078 39	0.378 39	12.756	2.643	1.765	4.666	**6**
7	6.2749	0.1594	0.056 87	0.356 87	17.583	2.802	2.006	5.622	**7**
8	8.1573	0.1226	0.041 92	0.341 92	23.858	2.925	2.216	6.480	**8**
9	10.6045	0.0943	0.031 24	0.331 24	32.015	3.019	2.396	7.234	**9**
10	13.7858	0.0725	0.023 46	0.323 46	42.619	3.092	2.551	7.887	**10**
11	17.9216	0.0558	0.017 73	0.317 73	56.405	3.147	2.683	8.445	**11**
12	23.2981	0.0429	0.013 45	0.313 45	74.327	3.190	2.795	8.917	**12**
13	30.2875	0.0330	0.010 24	0.310 24	97.625	3.223	2.889	9.314	**13**
14	39.3738	0.0254	0.007 82	0.307 82	127.913	3.249	2.969	9.644	**14**
15	51.1859	0.0195	0.005 98	0.305 98	167.286	3.268	3.034	9.917	**15**
16	66.5417	0.0150	0.004 58	0.304 58	218.472	3.283	3.089	10.143	**16**
17	86.5042	0.0116	0.003 51	0.303 51	285.014	3.295	3.135	10.328	**17**
18	112.4554	0.0089	0.002 69	0.302 69	371.518	3.304	3.172	10.479	**18**
19	146.1920	0.0068	0.002 07	0.302 07	483.973	3.311	3.202	10.602	**19**
20	190.0496	0.0053	0.001 59	0.301 59	630.165	3.316	3.228	10.702	**20**
21	247.0645	0.0040	0.001 22	0.301 22	820.215	3.320	3.248	10.783	**21**
22	321.1839	0.0031	0.000 94	0.300 94	1 067.280	3.323	3.265	10.848	**22**
23	417.5391	0.0024	0.000 72	0.300 72	1 388.464	3.325	3.278	10.901	**23**
24	542.8008	0.0018	0.000 55	0.300 55	1 806.003	3.327	3.289	10.943	**24**
25	705.6410	0.0014	0.000 43	0.300 43	2 348.803	3.329	3.298	10.977	**25**
26	917.3333	0.0011	0.000 33	0.300 33	3 054.444	3.330	3.305	11.005	**26**
27	1 192.5333	0.0008	0.000 25	0.300 25	3 971.778	3.331	3.311	11.026	**27**
28	1 550.2933	0.0006	0.000 19	0.300 19	5 164.311	3.331	3.315	11.044	**28**
29	2 015.3813	0.0005	0.000 15	0.300 15	6 714.604	3.332	3.319	11.058	**29**
30	2 619.9956	0.0004	0.000 11	0.300 11	8 729.985	3.332	3.332	11.069	**30**
31	3 405.9943	0.0003	0.000 09	0.300 09	11 349.981	3.332	3.324	11.078	**31**
32	4 427.7926	0.0002	0.000 07	0.300 07	14 755.975	3.333	3.326	11.085	**32**
33	5 756.1304	0.0002	0.000 05	0.300 05	19 183.768	3.333	3.328	11.090	**33**
34	7 482.9696	0.0001	0.000 04	0.300 04	24 939.899	3.333	3.329	11.094	**34**
35	9 727.8604	0.0001	0.000 03	0.300 03	32 422.868	3.333	3.330	11.098	**35**
∞				0.300 00		3.333	3.333	11.111	∞

TABLE D–26
35% compound interest factors

	Single payment		Uniform series				Uniform gradient		
n	Compound amount factor F/P	Present worth factor P/F	Sinking fund factor A/F	Capital recovery factor A/P	Compound amount factor F/A	Present worth factor P/A	Gradient conversion factor A/G	Present worth factor P/G	*n*
1	1.3500	0.7407	1.000 00	1.350 00	1.000	0.741	0.000	0.000	**1**
2	1.8225	0.5487	0.425 53	0.775 53	2.350	1.289	0.426	0.549	**2**
3	2.4604	0.4064	0.239 66	0.589 66	4.172	1.696	0.803	1.362	**3**
4	3.3215	0.3011	0.150 76	0.500 76	6.633	1.997	1.134	2.265	**4**
5	4.4840	0.2230	0.100 46	0.450 46	9.954	2.220	1.422	3.157	**5**
6	6.0534	0.1652	0.069 26	0.419 26	14.438	2.385	1.670	3.983	**6**
7	8.1722	0.1224	0.048 80	0.398 80	20.492	2.507	1.881	4.717	**7**
8	11.0324	0.0906	0.034 89	0.384 89	28.664	2.598	2.060	5.352	**8**
9	14.8937	0.0671	0.025 19	0.375 19	39.696	2.665	2.209	5.889	**9**
10	20.1066	0.0497	0.018 32	0.368 32	54.590	2.715	2.334	6.336	**10**
11	27.1439	0.0368	0.013 39	0.363 39	74.697	2.752	2.436	6.705	**11**
12	36.6442	0.0273	0.009 82	0.359 82	101.841	2.779	2.520	7.005	**12**
13	49.4697	0.0202	0.007 22	0.357 22	138.485	2.799	2.589	7.247	**13**
14	66.7841	0.0150	0.005 32	0.355 32	187.954	2.814	2.644	7.442	**14**
15	90.1585	0.0111	0.003 93	0.353 93	254.738	2.825	2.689	7.597	**15**
16	121.7139	0.0082	0.002 90	0.352 90	344.897	2.834	2.725	7.721	**16**
17	164.3138	0.0061	0.002 14	0.352 14	466.611	2.840	2.753	7.818	**17**
18	221.8236	0.0045	0.001 59	0.351 58	630.925	2.844	2.776	7.895	**18**
19	299.4619	0.0033	0.001 17	0.351 17	852.748	2.848	2.793	7.955	**19**
20	404.2736	0.0025	0.000 87	0.350 87	1 152.210	2.850	2.808	8.002	**20**
21	545.7693	0.0018	0.000 64	0.350 64	1 556.484	2.852	2.819	8.038	**21**
22	736.7886	0.0014	0.000 48	0.350 48	2 102.253	2.853	2.827	8.067	**22**
23	994.6646	0.0010	0.000 35	0.350 35	2 839.042	2.854	2.834	8.089	**23**
24	1 342 7973	0.0007	0.000 26	0.350 26	3 833.706	2.855	2.839	8.106	**24**
25	1 812.7763	0.0006	0.000 19	0.350 19	5 176.504	2.856	2.843	8.119	**25**
26	2 447.2480	0.0004	0.000 14	0.350 14	6 989.280	2.856	2.847	8.130	**26**
27	3 303.7848	0.0003	0.000 11	0.350 11	9 436.528	2.856	2.849	8.137	**27**
28	4 460.1095	0.0002	0.000 08	0.350 08	12 740.313	2.857	2.851	8.143	**28**
29	6 021.1478	0.0002	0.000 06	0.350 06	17 200.422	2.857	2.852	8.148	**29**
30	8 128.5495	0.0001	0.000 04	0.350 04	23 221.570	2.857	2.853	8.152	**30**
31	10 973.5418	0.0001	0.000 03	0.350 03	31 350.120	2.857	2.854	8.154	**31**
32	14 814.2815	0.0001	0.000 02	0.350 02	42 323.661	2.857	2.855	8.157	**32**
33	19 999.2800	0.0001	0.000 02	0.350 02	57 137.943	2.857	2.855	8.158	**33**
34	26 999.0280	0.0000	0.000 01	0.350 01	77 137.223	2.857	2.856	8.159	**34**
35	36 448.6878		0.000 01	0.350 01	104 136.251	2.857	2.856	8.160	**35**
∞				0.350 00		2.857	2.857	8.163	∞

TABLE D–27
40% compound interest factors

	Single payment		*Uniform series*				*Uniform gradient*		
n	*Compound amount factor F/P*	*Present worth factor P/F*	*Sinking fund factor A/F*	*Capital recovery factor A/P*	*Compound amount factor F/A*	*Present worth factor P/A*	*Gradient conversion factor A/G*	*Present worth factor P/G*	*n*
1	1.4000	0.7143	1.000 00	1.400 00	1.000	0.714	0.000	0.000	**1**
2	1.9600	0.5102	0.416 67	0.816 67	2.400	1.224	0.417	0.510	**2**
3	2.7440	0.3644	0.229 36	0.629 36	4.360	1.589	0.780	1.239	**3**
4	3.8416	0.2603	0.140 77	0.540 77	7.104	1.849	1.092	2.020	**4**
5	5.3782	0.1859	0.091 36	0.491 36	10.946	2.035	1.358	2.764	**5**
6	7.5295	0.1328	0.061 26	0.461 26	16.324	2.168	1.581	3.428	**6**
7	10.5414	0.0949	0.041 92	0.441 92	23.853	2.263	1.766	3.997	**7**
8	14.7579	0.0678	0.029 07	0.429 07	34.395	2.331	1.919	4.471	**8**
9	20.6610	0.0484	0.020 34	0.420 34	49.153	2.379	2.042	4.858	**9**
10	28.9255	0.0346	0.014 32	0.414 32	69.814	2.414	2.142	5.170	**10**
11	40.4957	0.0247	0.010 13	0.410 13	98.739	2.438	2.221	5.417	**11**
12	56.6939	0.0176	0.007 18	0.407 18	139.235	2.456	2.285	5.611	**12**
13	79.3715	0.0126	0.005 10	0.405 10	195.929	2.469	2.334	5.762	**13**
14	111.1201	0.0090	0.003 63	0.403 63	275.300	2.478	2.373	5.879	**14**
15	155.5681	0.0064	0.002 59	0.402 59	386.420	2.484	2.403	5.969	**15**
16	217.7953	0.0046	0.001 85	0.401 85	541.988	2.489	2.426	6.038	**16**
17	304.9135	0.0033	0.001 32	0.401 32	759.784	2.492	2.444	6.090	**17**
18	426.8789	0.0023	0.000 94	0.400 94	1 064.697	2.494	2.458	6.130	**18**
19	597.6304	0.0017	0.000 67	0.400 67	1 491.576	2.496	2.468	6.160	**19**
20	836.6826	0.0012	0.000 48	0.400 48	2 089.206	2.497	2.476	6.183	**20**
21	1 171.3554	0.0009	0.000 34	0.400 34	2 925.889	2.498	2.482	6.200	**21**
22	1 639.8976	0.0006	0.000 24	0.400 24	4 097.245	2.498	2.487	6.213	**22**
23	2 295.8569	0.0004	0.000 17	0.400 17	5 737.142	2.499	2.490	6.222	**23**
24	3 214.1997	0.0003	0.000 12	0.400 12	8 032.999	2.499	2.493	6.229	**24**
25	4 499.8796	0.0002	0.000 09	0.400 09	11 247.199	2.499	2.494	6.235	**25**
26	6 299.8314	0.0002	0.000 06	0.400 06	15 747.079	2.500	2.496	6.239	**26**
27	8 819.7640	0.0001	0.000 05	0.400 05	22 046.910	2.500	2.497	6.242	**27**
28	12 347.6696	0.0001	0.000 03	0.400 03	30 866.674	2.500	2.498	6.244	**28**
29	17 286.7374	0.0001	0.000 02	0.400 02	43 214.343	2.500	2.498	6.245	**29**
30	24 201.4324	0.0000	0.000 01	0.400 02	60 501.081	2.500	2.499	6.247	**30**
31	33 882.0053	...	0.000 01	0.400 01	84 702.513	2.500	2.499	6.248	**31**
32	47 434.8074	...	0.000 01	0.400 01	118 584.519	2.500	2.499	6.248	**32**
33	66 408.7304	...	0.000 01	0.400 01	166 019.326	2.500	2.500	6.249	**33**
34	92 972.2225	...	0.000 00	0.400 00	232 428.056	2.500	2.500	6.249	**34**
35	130 161.1116	...	...	0.400 00	325 400.279	2.500	2.500	6.249	**35**
∞				0.400 00		2.500	2.500	6.250	∞

TABLE D–28
45% compound interest factors

	Single payment		Uniform series				Uniform gradient		
n	*Compound amount factor F/P*	*Present worth factor P/F*	*Sinking fund factor A/F*	*Capital recovery factor A/P*	*Compound amount factor F/A*	*Present worth factor P/A*	*Gradient conversion factor A/G*	*Present worth factor P/G*	*n*
1	1.4500	0.6897	1.000 00	1.450 00	1.000	0.690	0.000	0.000	**1**
2	2.1025	0.4756	0.408 16	0.858 16	2.450	1.165	0.408	0.476	**2**
3	3.0486	0.3280	0.219 66	0.669 66	4.552	1.493	0.758	1.132	**3**
4	4.4205	0.2262	0.131 56	0.581 56	7.601	1.720	1.053	1.810	**4**
5	6.4097	0.1560	0.083 18	0.533 18	12.022	1.876	1.298	2.434	**5**
6	9.2941	0.1076	0.054 26	0.504 26	18.431	1.983	1.499	2.972	**6**
7	13.4765	0.0742	0.036 07	0.486 07	27.725	2.057	1.661	3.418	**7**
8	19.5409	0.0512	0.024 27	0.474 27	41.202	2.109	1.791	3.776	**8**
9	28.3343	0.0353	0.016 46	0.466 46	60.743	2.144	1.893	4.058	**9**
10	41.0847	0.0243	0.011 23	0.461 23	89.077	2.168	1.973	4.277	**10**
11	59.5728	0.0168	0.007 68	0.457 68	130.162	2.185	2.034	4.445	**11**
12	86.3806	0.0116	0.005 27	0.455 27	189.735	2.196	2.082	4.572	**12**
13	125.2518	0.0080	0.003 62	0.453 62	276.115	2.204	2.118	4.668	**13**
14	181.6151	0.0055	0.002 49	0.452 49	401.367	2.210	2.145	4.740	**14**
15	263.3419	0.0038	0.001 72	0.451 72	582.982	2.214	2.165	4.793	**15**
16	381.8458	0.0026	0.001 18	0.451 18	846.324	2.216	2.180	4.832	**16**
17	553.6764	0.0018	0.000 81	0.450 81	1 228.170	2.218	2.191	4.861	**17**
18	802.8308	0.0012	0.000 56	0.450 56	1 781.846	2.219	2.200	4.882	**18**
19	1 164.1047	0.0009	0.000 39	0.450 39	2 584.677	2.220	2.206	4.898	**19**
20	1 687.9518	0.0006	0.000 27	0.450 27	3 748.782	2.221	2.210	4.909	**20**
21	2 447.5301	0.0004	0.000 18	0.450 18	5 436.734	2.221	2.214	4.917	**21**
22	3 548.9187	0.0003	0.000 13	0.450 13	7 884.264	2.222	2.216	4.923	**22**
23	5 145.9321	0.0002	0.000 09	0.450 09	11 433.182	2.222	2.218	4.927	**23**
24	7 461.6015	0.0001	0.000 06	0.450 06	16 579.115	2.222	2.219	4.930	**24**
25	10 819.3222	0.0001	0.000 04	0.450 04	24 040.716	2.222	2.220	4.933	**25**
26	15 688.0173	0.0001	0.000 03	0.450 03	34 860.038	2.222	2.221	4.934	**26**
27	22 747.6250	0.0000	0.000 02	0.450 02	50 548.056	2.222	2.221	4.935	**27**
28	32 984.0563	...	0.000 01	0.450 01	73 295.681	2.222	2.221	4.936	**28**
29	47 826.8816	...	0.000 01	0.450 01	106 279.737	2.222	2.222	4.937	**29**
30	69 348.9783	...	0.000 01	0.450 01	154 106.618	2.222	2.222	4.937	**30**
∞				0.450 00		2.222	2.222	4.938	∞

TABLE D–29
50% compound interest factors

	Single payment		Uniform series				Uniform gradient		
n	*Compound amount factor F/P*	*Present worth factor P/F*	*Sinking fund factor A/F*	*Capital recovery factor A/P*	*Compound amount factor F/A*	*Present worth factor P/A*	*Gradient conversion factor A/G*	*Present worth factor P/G*	*n*
1	1.5000	0.6667	1.000 00	1.500 00	1.000	0.667	0.000	0.000	**1**
2	2.2500	0.4444	0.400 00	0.900 00	2.500	1.111	0.400	0.444	**2**
3	3.3750	0.2963	0.210 53	0.710 53	4.750	1.407	0.737	1.037	**3**
4	5.0625	0.1975	0.123 08	0.623 08	8.125	1.605	1.015	1.630	**4**
5	7.5938	0.1317	0.075 83	0.575 83	13.188	1.737	1.242	2.156	**5**
6	11.3906	0.0878	0.048 12	0.548 12	20.781	1.824	1.423	2.595	**6**
7	17.0859	0.0585	0.031 08	0.531 08	32.172	1.883	1.565	2.947	**7**
8	25.6289	0.0390	0.020 30	0.520 30	49.258	1.922	1.675	3.220	**8**
9	38.4434	0.0260	0.013 35	0.513 35	74.887	1.948	1.760	3.428	**9**
10	57.6650	0.0173	0.008 82	0.508 82	113.330	1.965	1.824	3.584	**10**
11	86.4976	0.0116	0.005 85	0.505 85	170.995	1.977	1.871	3.699	**11**
12	129.7463	0.0077	0.003 88	0.503 88	257.493	1.985	1.907	3.784	**12**
13	194.6195	0.0051	0.002 58	0.502 58	387.239	1.990	1.933	3.846	**13**
14	291.9293	0.0034	0.001 72	0.501 72	581.859	1.993	1.952	3.890	**14**
15	437.8939	0.0023	0.001 14	0.501 14	873.788	1.995	1.966	3.922	**15**
16	656.8408	0.0015	0.000 76	0.500 76	1 311.682	1.997	1.976	3.945	**16**
17	985.2613	0.0010	0.000 51	0.500 51	1 968.523	1.998	1.983	3.961	**17**
18	1 477.8919	0.0007	0.000 34	0.500 34	2 953.784	1.999	1.988	3.973	**18**
19	2 216.8378	0.0005	0.000 23	0.500 23	4 431.676	1.999	1.991	3.981	**19**
20	3 325.2567	0.0003	0.000 15	0.500 15	6 648.513	1.999	1.994	3.987	**20**
21	4 987.8851	0.0002	0.000 10	0.500 10	9 973.770	2.000	1.996	3.991	**21**
22	7 481.8276	0.0001	0.000 07	0.500 07	14 961.655	2.000	1.997	3.994	**22**
23	11 222.7415	0.0001	0.000 04	0.500 04	22 443.483	2.000	1.998	3.996	**23**
24	16 834.1122	0.0001	0.000 03	0.500 03	33 666.224	2.000	1.999	3.997	**24**
25	25 251.1683	0.0000	0.000 02	0.500 02	50 500.337	2.000	1.999	3.998	**25**
∞				0.500 00		2.000	2.000	4.000	∞

TABLE D–30
Present worth at zero date of $1 flowing uniformly throughout one-year periods ($P/\bar{F}$)
This table assumes continuous compounding of interest at various stated effective rates per annum

Period	*1%*	*2%*	*3%*	*4%*	*5%*	*6%*	*7%*	*8%*	*10%*
0 to **1**	0.9950	0.9902	0.9854	0.9806	0.9760	0.9714	0.9669	0.9625	0.9538
1 to **2**	0.9852	0.9707	0.9567	0.9429	0.9295	0.9164	0.9037	0.8912	0.8671
2 to **3**	0.9754	0.9517	0.9288	0.9067	0.8853	0.8646	0.8445	0.8252	0.7883
3 to **4**	0.9658	0.9331	0.9017	0.8718	0.8431	0.8156	0.7893	0.7641	0.7166
4 to **5**	0.9562	0.9148	0.8755	0.8383	0.8030	0.7695	0.7377	0.7075	0.6515
5 to **6**	0.9467	0.8968	0.8500	0.8060	0.7647	0.7259	0.6894	0.6551	0.5922
6 to **7**	0.9374	0.8792	0.8252	0.7750	0.7283	0.6848	0.6443	0.6065	0.5384
7 to **8**	0.9281	0.8620	0.8012	0.7452	0.6936	0.6461	0.6021	0.5616	0.4895
8 to **9**	0.9189	0.8451	0.7779	0.7165	0.6606	0.6095	0.5628	0.5200	0.4450
9 to **10**	0.9098	0.8285	0.7552	0.6890	0.6291	0.5750	0.5259	0.4815	0.4045
10 to **11**	0.9008	0.8123	0.7332	0.6625	0.5992	0.5424	0.4915	0.4458	0.3677
11 to **12**	0.8919	0.7964	0.7118	0.6370	0.5706	0.5117	0.4594	0.4128	0.3343
12 to **13**	0.8830	0.7807	0.6911	0.6125	0.5435	0.4828	0.4293	0.3822	0.3039
13 to **14**	0.8743	0.7654	0.6710	0.5889	0.5176	0.4554	0.4012	0.3539	0.2763
14 to **15**	0.8656	0.7504	0.6514	0.5663	0.4929	0.4297	0.3750	0.3277	0.2512
15 to **16**	0.8571	0.7357	0.6325	0.5445	0.4695	0.4053	0.3505	0.3034	0.2283
16 to **17**	0.8486	0.7213	0.6140	0.5236	0.4471	0.3824	0.3275	0.2809	0.2076
17 to **18**	0.8402	0.7071	0.5962	0.5034	0.4258	0.3608	0.3061	0.2601	0.1887
18 to **19**	0.8319	0.6933	0.5788	0.4841	0.4055	0.3403	0.2861	0.2409	0.1716
19 to **20**	0.8236	0.6797	0.5619	0.4655	0.3862	0.3211	0.2674	0.2230	0.1560
20 to **21**	0.8155	0.6664	0.5456	0.4476	0.3678	0.3029	0.2499	0.2065	0.1418
21 to **22**	0.8074	0.6533	0.5297	0.4303	0.3503	0.2857	0.2335	0.1912	0.1289
22 to **23**	0.7994	0.6405	0.5143	0.4138	0.3336	0.2696	0.2182	0.1770	0.1172
23 to **24**	0.7915	0.6279	0.4993	0.3979	0.3178	0.2543	0.2040	0.1639	0.1065
24 to **25**	0.7837	0.6156	0.4847	0.3826	0.3026	0.2399	0.1906	0.1518	0.0968

TABLE D–30 ***Continued***
Present worth at zero date of $1 flowing uniformly throughout one-year periods (P/$\bar{F}$)
This table assumes continuous compounding of interest at various stated effective rates per annum

Period	*1%*	*2%*	*3%*	*4%*	*5%*	*6%*	*7%*	*8%*	*10%*
25 to **26**	0.7759	0.6035	0.4706	0.3679	0.2882	0.2263	0.1782	0.1405	0.0880
26 to **27**	0.7682	0.5917	0.4569	0.3537	0.2745	0.2135	0.1665	0.1301	0.0800
27 to **28**	0.7606	0.5801	0.4436	0.3401	0.2614	0.2014	0.1556	0.1205	0.0728
28 to **29**	0.7531	0.5687	0.4307	0.3270	0.2490	0.1900	0.1454	0.1116	0.0661
29 to **30**	0.7456	0.5576	0.4181	0.3144	0.2371	0.1793	0.1359	0.1033	0.0601
30 to **31**	0.7382	0.5466	0.4060	0.3024	0.2258	0.1691	0.1270	0.0956	0.0547
31 to **32**	0.7309	0.5359	0.3941	0.2907	0.2151	0.1596	0.1187	0.0886	0.0497
32 to **33**	0.7237	0.5254	0.3827	0.2795	0.2048	0.1505	0.1109	0.0820	0.0452
33 to **34**	0.7165	0.5151	0.3715	0.2688	0.1951	0.1420	0.1037	0.0759	0.0411
34 to **35**	0.7094	0.5050	0.3607	0.2585	0.1858	0.1340	0.0969	0.0703	0.0373
35 to **36**	0.7024	0.4951	0.3502	0.2485	0.1769	0.1264	0.0906	0.0651	0.0339
36 to **37**	0.6955	0.4854	0.3400	0.2390	0.1685	0.1192	0.0846	0.0603	0.0309
37 to **38**	0.6886	0.4759	0.3301	0.2298	0.1605	0.1125	0.0791	0.0558	0.0281
38 to **39**	0.6818	0.4666	0.3205	0.2209	0.1528	0.1061	0.0739	0.0517	0.0255
39 to **40**	0.6750	0.4574	0.3111	0.2124	0.1456	0.1001	0.0691	0.0478	0.0232
40 to **41**	0.6683	0.4484	0.3021	0.2043	0.1386	0.0944	0.0646	0.0443	0.0211
41 to **42**	0.6617	0.4396	0.2933	0.1964	0.1320	0.0891	0.0603	0.0410	0.0192
42 to **43**	0.6552	0.4310	0.2847	0.1888	0.1257	0.0841	0.0564	0.0380	0.0174
43 to **44**	0.6487	0.4226	0.2764	0.1816	0.1198	0.0793	0.0527	0.0352	0.0158
44 to **45**	0.6422	0.4143	0.2684	0.1746	0.1141	0.0748	0.0493	0.0326	0.0144
45 to **46**	0.6359	0.4062	0.2606	0.1679	0.1086	0.0706	0.0460	0.0302	0.0131
46 to **47**	0.6296	0.3982	0.2530	0.1614	0.1035	0.0666	0.0430	0.0279	0.0119
47 to **48**	0.6234	0.3904	0.2456	0.1552	0.0985	0.0628	0.0402	0.0259	0.0108
48 to **49**	0.6172	0.3827	0.2385	0.1492	0.0938	0.0593	0.0376	0.0239	0.0098
49 to **50**	0.6111	0.3752	0.2315	0.1435	0.0894	0.0559	0.0351	0.0222	0.0089

TABLE D–30 *Continued*
Present worth at zero date of $1 flowing uniformly throughout one-year periods (P/F̄)
This table assumes continuous compounding of interest at various stated effective rates per annum

Period	*12%*	*15%*	*20%*	*25%*	*30%*	*35%*	*40%*	*45%*	*50%*
0 to **1**	0.9454	0.9333	0.9141	0.8963	0.8796	0.8639	0.8491	0.8352	0.8221
1 to **2**	0.8441	0.8115	0.7618	0.7170	0.6766	0.6399	0.6065	0.5760	0.5481
2 to **3**	0.7537	0.7057	0.6348	0.5736	0.5205	0.4740	0.4332	0.3973	0.3654
3 to **4**	0.6729	0.6136	0.5290	0.4589	0.4004	0.3511	0.3095	0.2740	0.2436
4 to **5**	0.6008	0.5336	0.4408	0.3671	0.3080	0.2601	0.2210	0.1889	0.1624
5 to **6**	0.5365	0.4640	0.3674	0.2937	0.2369	0.1927	0.1579	0.1303	0.1083
6 to **7**	0.4790	0.4035	0.3061	0.2350	0.1822	0.1427	0.1128	0.0899	0.0722
7 to **8**	0.4277	0.3508	0.2551	0.1880	0.1402	0.1057	0.0806	0.0620	0.0481
8 to **9**	0.3818	0.3051	0.2126	0.1504	0.1078	0.0783	0.0575	0.0427	0.0321
9 to **10**	0.3409	0.2653	0.1772	0.1203	0.0829	0.0580	0.0411	0.0295	0.0214
10 to **11**	0.3044	0.2307	0.1476	0.0962	0.0638	0.0430	0.0294	0.0203	0.0143
11 to **12**	0.2718	0.2006	0.1230	0.0770	0.0491	0.0318	0.0210	0.0140	0.0095
12 to **13**	0.2427	0.1744	0.1025	0.0616	0.0378	0.0236	0.0150	0.0097	0.0063
13 to **14**	0.2167	0.1517	0.0854	0.0493	0.0290	0.0175	0.0107	0.0067	0.0042
14 to **15**	0.1935	0.1319	0.0712	0.0394	0.0223	0.0129	0.0076	0.0046	0.0028
15 to **16**	0.1727	0.1147	0.0593	0.0315	0.0172	0.0096	0.0055	0.0032	0.0019
16 to **17**	0.1542	0.0997	0.0494	0.0252	0.0132	0.0071	0.0039	0.0022	0.0013
17 to **18**	0.1377	0.0867	0.0412	0.0202	0.0102	0.0053	0.0028	0.0015	0.0008
18 to **19**	0.1229	0.0754	0.0343	0.0161	0.0078	0.0039	0.0020	0.0010	0.0006
19 to **20**	0.1098	0.0656	0.0286	0.0129	0.0060	0.0029	0.0014	0.0007	0.0004
20 to **21**	0.0980	0.0570	0.0238	0.0103	0.0046	0.0021	0.0010	0.0005	0.0002
21 to **22**	0.0875	0.0496	0.0199	0.0083	0.0036	0.0016	0.0007	0.0003	0.0002
22 to **23**	0.0781	0.0431	0.0166	0.0066	0.0027	0.0012	0.0005	0.0002	0.0001
23 to **24**	0.0698	0.0375	0.0138	0.0053	0.0021	0.0009	0.0004	0.0002	0.0001
24 to **25**	0.0623	0.0326	0.0115	0.0042	0.0016	0.0006	0.0003	0.0001	

TABLE D–30 *Continued*
Present worth at zero date of $1 flowing uniformly throughout one-year periods (P/F̄)
This table assumes continuous compounding of interest at various stated effective rates per annum

Period	*12%*	*15%*	*20%*	*25%*	*30%*	*35%*	*40%*	*45%*	*50%*
25 to **26**	0.0556	0.0284	0.0096	0.0034	0.0012	0.0005	0.0002	0.0001	
26 to **27**	0.0497	0.0247	0.0080	0.0027	0.0010	0.0004	0.0001	0.0001	
27 to **28**	0.0443	0.0214	0.0067	0.0022	0.0007	0.0003	0.0001		
28 to **29**	0.0396	0.0186	0.0055	0.0017	0.0006	0.0002	0.0001		
29 to **30**	0.0353	0.0162	0.0046	0.0014	0.0004	0.0001			
30 to **31**	0.0316	0.0141	0.0039	0.0011	0.0003	0.0001			
31 to **32**	0.0282	0.0123	0.0032	0.0009	0.0003	0.0001			
32 to **33**	0.0252	0.0107	0.0027	0.0007	0.0002	0.0001			
33 to **34**	0.0225	0.0093	0.0022	0.0006	0.0002				
34 to **35**	0.0201	0.0081	0.0019	0.0005	0.0001				
35 to **36**	0.0179	0.0070	0.0015	0.0004	0.0001				
36 to **37**	0.0160	0.0061	0.0013	0.0003	0.0001				
37 to **38**	0.0143	0.0053	0.0011	0.0002	0.0001				
38 to **39**	0.0127	0.0046	0.0009	0.0002					
39 to **40**	0.0114	0.0040	0.0007	0.0001					
40 to **41**	0.0102	0.0035	0.0006	0.0001					
41 to **42**	0.0091	0.0030	0.0005	0.0001					
42 to **43**	0.0081	0.0026	0.0004	0.0001					
43 to **44**	0.0072	0.0023	0.0004	0.0001					
44 to **45**	0.0065	0.0020	0.0003						
45 to **46**	0.0058	0.0017	0.0002						
46 to **47**	0.0051	0.0015	0.0002						
47 to **48**	0.0046	0.0013	0.0002						
48 to **49**	0.0041	0.0011	0.0001						
49 to **50**	0.0037	0.0010	0.0001						

TABLE D–31
Present worth at zero date of $1 flowing uniformly throughout stated periods starting at zero date (P/Ā)
This table assumes continuous compounding of interest at various stated effective rates per annum

Period	*1%*	*2%*	*3%*	*4%*	*5%*	*6%*	*7%*	*8%*	*10%*
0 to **1**	0.995	0.990	0.985	0.981	0.976	0.971	0.967	0.962	0.954
0 to **2**	1.980	1.961	1.942	1.924	1.906	1.888	1.871	1.854	1.821
0 to **3**	2.956	2.913	2.871	2.830	2.791	2.752	2.715	2.679	2.609
0 to **4**	3.921	3.846	3.773	3.702	3.634	3.568	3.504	3.443	3.326
0 to **5**	4.878	4.760	4.648	4.540	4.437	4.338	4.242	4.150	3.977
0 to **6**	5.824	5.657	5.498	5.346	5.202	5.063	4.931	4.805	4.570
0 to **7**	6.762	6.536	6.323	6.121	5.930	5.748	5.576	5.412	5.108
0 to **8**	7.690	7.398	7.124	6.867	6.623	6.394	6.178	5.974	5.597
0 to **9**	8.609	8.244	7.902	7.583	7.284	7.004	6.741	6.494	6.042
0 to **10**	9.519	9.072	8.658	8.272	7.913	7.579	7.267	6.975	6.447
0 to **11**	10.419	9.884	9.391	8.935	8.512	8.121	7.758	7.421	6.815
0 to **12**	11.311	10.681	10.103	9.572	9.083	8.633	8.218	7.834	7.149
0 to **13**	12.194	11.461	10.794	10.184	9.627	9.116	8.647	8.216	7.453
0 to **14**	13.069	12.227	11.465	10.773	10.144	9.571	9.048	8.570	7.729
0 to **15**	13.934	12.977	12.116	11.339	10.637	10.001	9.423	8.897	7.980
0 to **16**	14.791	13.713	12.749	11.884	11.107	10.406	9.774	9.201	8.209
0 to **17**	15.640	14.434	13.363	12.407	11.554	10.789	10.101	9.482	8.416
0 to **18**	16.480	15.141	13.959	12.911	11.979	11.149	10.407	9.742	8.605
0 to **19**	17.312	15.835	14.538	13.395	12.385	11.490	10.693	9.983	8.777
0 to **20**	18.136	16.514	15.100	13.860	12.771	11.811	10.961	10.206	8.932
0 to **21**	18.951	17.181	15.645	14.308	13.139	12.114	11.210	10.412	9.074
0 to **22**	19.759	17.834	16.175	14.738	13.489	12.399	11.444	10.604	9.203
0 to **23**	20.558	18.475	16.689	15.152	13.823	12.669	11.662	10.781	9.320
0 to **24**	21.349	19.102	17.188	15.550	14.141	12.923	11.866	10.945	9.427
0 to **25**	22.133	19.718	17.673	15.932	14.443	13.163	12.057	11.096	9.524

TABLE D–31 ***Continued***
Present worth at zero date of $1 flowing uniformly throughout stated periods starting at zero date (P/$\bar{A}$)
This table assumes continuous compounding of interest at various stated effective rates per annum

Period	*1%*	*2%*	*3%*	*4%*	*5%*	*6%*	*7%*	*8%*	*10%*
0 to **26**	22.909	20.322	18.144	16.300	14.732	13.389	12.235	11.237	9.612
0 to **27**	23.677	20.913	18.601	16.654	15.006	13.603	12.402	11.367	9.692
0 to **28**	24.438	21.493	19.044	16.994	15.268	13.804	12.557	11.487	9.765
0 to **29**	25.191	22.062	19.475	17.321	15.517	13.994	12.703	11.599	9.831
0 to **30**	25.937	22.620	19.893	17.636	15.754	14.174	12.838	11.702	9.891
0 to **31**	26.675	23.166	20.299	17.938	15.979	14.343	12.965	11.798	9.945
0 to **32**	27.406	23.702	20.693	18.229	16.195	14.502	13.084	11.887	9.995
0 to **33**	28.129	24.228	21.076	18.508	16.399	14.653	13.195	11.969	10.040
0 to **34**	28.846	24.743	21.447	18.777	16.594	14.795	13.299	12.044	10.081
0 to **35**	29.555	25.248	21.808	19.035	16.780	14.929	13.396	12.115	10.119
0 to **40**	32.999	27.628	23.460	20.186	17.585	15.493	13.793	12.395	10.260
0 to **45**	36.275	29.784	24.885	21.132	18.215	15.915	14.076	12.587	10.348
0 to **50**	39.392	31.737	26.114	21.909	18.709	16.230	14.278	12.717	10.403
0 to **55**	42.358	33.505	27.174	22.548	19.096	16.466	14.422	12.805	10.437
0 to **60**	45.179	35.107	28.089	23.073	19.399	16.642	14.525	12.865	10.458
0 to **65**	47.864	36.558	28.878	23.505	19.636	16.773	14.598	12.906	10.471
0 to **70**	50.419	37.872	29.558	23.859	19.822	16.871	14.650	12.934	10.479
0 to **75**	52.850	39.063	30.145	24.151	19.968	16.945	14.688	12.953	10.484
0 to **80**	55.162	40.141	30.652	24.391	20.082	17.000	14.714	12.966	10.487
0 to **85**	57.363	41.117	31.088	24.588	20.172	17.041	14.733	12.975	10.489
0 to **90**	59.456	42.001	31.465	24.749	20.242	17.071	14.747	12.981	10.490
0 to **95**	64.448	42.802	31.790	24.883	20.297	17.094	14.756	12.985	10.491
0 to **100**	63.344	43.528	32.071	24.992	20.340	17.111	14.763	12.988	10.491

TABLE D–31 *Continued*
Present worth at zero date of $1 flowing uniformly throughout stated periods starting at zero date $(P/\bar{A})$
This table assumes continuous compounding of interest at various stated effective rates per annum

Period	*12%*	*15%*	*20%*	*25%*	*30%*	*35%*	*40%*	*45%*	*50%*
0 to **1**	0.945	0.933	0.914	0.896	0.880	0.864	0.849	0.835	0.822
0 to **2**	1.790	1.745	1.676	1.613	1.556	1.504	1.456	1.411	1.370
0 to **3**	2.543	2.450	2.311	2.187	2.077	1.978	1.889	1.809	1.736
0 to **4**	3.216	3.064	2.840	2.646	2.477	2.329	2.198	2.083	1.979
0 to **5**	3.817	3.598	3.281	3.013	2.785	2.589	2.419	2.271	2.142
0 to **6**	4.353	4.062	3.648	3.307	3.022	2.782	2.577	2.402	2.250
0 to **7**	4.832	4.465	3.954	3.542	3.204	2.924	2.690	2.492	2.322
0 to **8**	5.260	4.816	4.209	3.730	3.344	3.030	2.771	2.554	2.370
0 to **9**	5.642	5.121	4.422	3.880	3.452	3.108	2.828	2.596	2.402
0 to **10**	5.983	5.386	4.599	4.000	3.535	3.166	2.869	2.626	4.424
0 to **11**	6.287	5.617	4.747	4.096	3.599	3.209	2.899	2.646	2.438
0 to **12**	6.559	5.818	4.870	4.173	3.648	3.241	2.920	2.660	2.447
0 to **13**	6.802	5.992	4.972	4.235	3.686	3.265	2.935	2.670	2.454
0 to **14**	7.018	6.144	5.058	4.284	3.715	3.282	2.945	2.677	2.458
0 to **15**	7.212	6.276	5.129	4.324	3.737	3.295	2.953	2.681	2.461
0 to **16**	7.385	6.390	5.188	4.355	3.754	3.305	2.958	2.684	2.463
0 to **17**	7.539	6.490	5.238	4.381	3.767	3.312	2.962	2.686	2.464
0 to **18**	7.676	6.577	5.279	4.401	3.778	3.317	2.965	2.688	2.465
0 to **19**	7.799	6.652	5.313	4.417	3.785	3.321	2.967	2.689	2.465
0 to **20**	7.909	6.718	5.342	4.430	3.791	3.324	2.968	2.690	2.466
0 to **21**	8.007	6.775	5.366	4.440	3.796	3.326	2.969	2.690	2.466
0 to **22**	8.095	6.824	5.385	4.448	3.800	3.328	2.970	2.691	2.466
0 to **23**	8.173	6.868	5.402	4.455	3.802	3.329	2.971	2.691	2.466
0 to **24**	8.243	6.905	5.416	4.460	3.804	3.330	2.971	2.691	2.466
0 to **25**	8.305	6.938	5.427	4.465	3.806	3.330	2.971	2.691	2.466

TABLE D–31 *Continued*
Present worth at zero date of $1 flowing uniformly throughout stated periods starting at zero date (P/$\bar{A}$)
This table assumes continuous compounding of interest at various stated effective rates per annum

Period	*12%*	*15%*	*20%*	*25%*	*30%*	*35%*	*40%*	*45%*	*50%*
0 to **26**	8.360	6.966	5.437	4.468	3.807	3.331	2.972	2.691	2.466
0 to **27**	8.410	6.991	5.445	4.471	3.808	3.331	2.972	2.691	2.466
0 to **28**	8.454	7.012	5.452	4.473	3.809	3.331	2.972	2.691	2.466
0 to **29**	8.494	7.031	5.457	4.474	3.810	3.332	2.972	2.691	2.466
0 to **30**	8.529	7.047	5.462	4.476	3.810	3.332	2.972	2.691	2.466
0 to **31**	8.561	7.061	5.466	4.477	3.810	3.332	2.972	2.691	2.466
0 to **32**	8.589	7.073	5.469	4.478	3.811	3.332	2.972	2.691	2.466
0 to **33**	8.614	7.084	5.471	4.479	3.811	3.332	2.972	2.691	2.466
0 to **34**	8.637	7.093	5.474	4.479	3.811	3.332	2.972	2.691	2.466
0 to **35**	8.657	7.101	5.476	4.480	3.811	3.332	2.972	2.691	2.466
0 to **40**	8.729	7.128	5.481	4.481	3.811	3.332	2.972	2.691	2.466
0 to **45**	8.770	7.142	5.483	4.481	3.811	3.332	2.972	2.691	2.466
0 to **50**	8.793	7.148	5.484	4.481	3.811	3.332	2.972	2.691	2.466
0 to **55**	8.807	7.152	5.485	4.481	3.811	3.332	2.972	2.691	2.466
0 to **60**	8.814	7.153	5.485	4.481	3.811	3.332	2.972	2.691	2.466
0 to **65**	8.818	7.154	5.485	4.481	3.811	3.332	2.972	2.691	2.466
0 to **70**	8.821	7.155	5.485	4.481	3.811	3.332	2.972	2.691	2.466
0 to **75**	8.822	7.155	4.485	4.481	3.811	3.332	2.972	2.691	2.466
0 to **80**	8.823	7.155	5.485	4.481	3.811	3.332	2.972	2.691	2.466
0 to **85**	8.823	7.155	5.485	4.481	3.811	3.332	2.972	2.691	2.466
0 to **90**	8.824	7.155	5.485	4.481	3.811	3.332	2.972	2.691	2.466
0 to **95**	8.824	7.155	5.485	4.481	3.811	3.332	2.972	2.691	2.466
0 to **100**	8.824	7.155	5.485	4.481	3.811	3.332	2.972	2.691	2.466

E

SELECTED REFERENCES

AASHO. *Road User Benefit Analysis for Highway Improvements*. Washington, D.C.: American Association of State Highway Officials, 1960. Among highway engineers, this was called the "Red Book."

AASHTO. *Manual on User Benefit Analysis of Highway and Bus-Transit Improvements*. Washington, D.C.: American Association of State Highway and Transportation Officials, 1978.

ARBELAEZ, E. V., and RODRIGUEZ, H. G. *Naciones de Ingenieria Economica*. Medellin, Colombia; Universidad Nacional de Colombia, 1968.

AT&T Co., ENGINEERING DEPARTMENT. *Engineering Economy*, 2d ed. New York: American Telephone and Telegraph Co., 1963.

BARISH, N. N. *Economic Analysis for Engineering and Managerial Decision Making*. New York: McGraw-Hill Book Co., Inc., 1978.

BAUMOL, W. J. *Economic Theory and Operations Analysis*, 4th ed. Englewood Cliffs, N.J.: Prentice-Hall, Inc., 1977.

BIERMAN, H., and SMIDT, S. *The Capital Budgeting Decision*, 4th ed. New York: The Macmillan Co., 1975.

BONBRIGHT, J. C. *Principles of Public Utility Rates*. New York: Columbia University Press, 1961.

———. *Valuation of Property*. New York: McGraw-Hill Book Co., Inc., 1937. Reprinted in two vols., Charlottesville, Va.: Michie Co., 1965.

BUSSEY, LYNN E. *The Economic Analysis of Industrial Projects*. Englewood Cliffs, N.J.: Prentice-Hall, Inc., 1978.

CANADA, J. R., and WHITE, J. A. *Capital Investment Decision Analysis for Management and Engineering*. Englewood Cliffs, N.J.: Prentice-Hall, Inc., 1980.

CLARK, J. M. *Studies in the Economics of Overhead Costs*. Chicago: University of Chicago Press, 1923.

COUGHLAN, J. D., and STRAND, W. K. *Depreciation: Accounting, Taxes, and Business Decisions*. New York: The Ronald Press Co., 1969.

DEAN, JOEL. *Capital Budgeting*. New York: Columbia University Press, 1951.

———. *Managerial Economics*. Englewood Cliffs, N.J.: Prentice-Hall, Inc., 1951.

———. *Sec. 2, Managerial Economics* in *Handbook of Industrial Engineering and Management*, 2d ed., (W. G. Ireson and E. L. Grant, eds.). Englewood Cliffs, N.J.: Prentice-Hall, Inc., 1971.

DE FARO, CLOVIS. *Engenharia Economica Elementos*. Rio de Janeiro, Brazil: APEC Editora, S.A., 1969.

———. *Matematica Financeira*, 2nd ed. Rio de Janerio, Brazil: APEC Editora, S.A., 1970.

DE FARO, CLOVIS, and CARREIRO FILHO, J. P. *Tabelas Complementacao Ao Livro Matematica Financeiro*. Rio de Janeiro, Brazil: APEC Editora, S.A. (no date given.)

DE GARMO, E. P., CANADA, J. R. and SULLIVAN, W. G. *Engineering Economy*, 6th ed. New York: The Macmillan Co., 1979.

DIVISION OF WATER RESOURCES DEVELOPMENT, ECONOMIC COMMISSION FOR ASIA AND THE FAR EAST. *Manual of Standards and Criteria for Planning Water Resource Projects*, Water Resources Series No. 26, United Nations Publication Sales Number: 64. II, F. 12. New York: United Nations, 1964.

ENGINEERING ECONOMIST, THE. A quarterly journal jointly published by the Engineering Economy Divisions of the American Society for Engineering Education and the American Institute of Industrial Engineers. Published at AIIE, Norcross, Ga., 30092; first issue was in 1955.

ENGINEERING NEWS-RECORD. *Construction Costs*. Published annually; secure the most recent issue.

ENGLISH, J. M., ed. *Cost Effectiveness: Economic Evaluation of Engineered Systems*. New York: John Wiley & Sons, Inc., 1968.

———, ed. *Workshop on Economics in Engineering Systems*. Los Angeles: Department of Engineering Reports Group, University of California, Los Angeles, Cal., 1968.

FABRYCKY, W. J., and THUESEN, G. J. *Economic Decision Analysis*, 2d ed. Englewood Cliffs, N.J.: Prentice-Hall, Inc., 1980.

FISH, J. C. L. *Engineering Economics*, 2d ed. New York: McGraw-Hill Book Co., Inc., 1923. The first edition of this book, published in 1915, was the first general textbook on what later came to be called engineering economy.

FLEISCHER, G. A. *Capital Allocation Theory*. New York: Appleton-Century-Crofts, Inc., 1969.

FLEISCHER, G. A., ed. *Risk and Uncertainty: Non-Deterministic Decision Making in Engineering Economy*, Publication No. 2 in the monograph series of the Engineering Economy Division. Norcross, Ga.: American Institute of Industrial Engineers, Inc., 1975.

GRANT, E. L., and BELL, L. F. *Basic Accounting and Cost Accounting*, 2d ed. New York: McGraw-Hill Book Co., Inc., 1964.

GRANT, E. L., and LEAVENWORTH, R. S. *Statistical Quality Control*, 5th ed. New York: McGraw-Hill Book Co., Inc., 1980.

GRANT, E. L., and NORTON, P. T., JR. *Depreciation*, rev. printing. New York: The Ronald Press Co., 1955.

HAPPEL, J. and JORDAN, D. *Chemical Process Economics*, 2d ed. New York: Marcel Dekker, Inc., 1975.

Harvard Business Review. *Capital Investment, Part I*, a volume of reprints of 15 articles, 1954–64. Cambridge, Mass.: Harvard Business Review.

———. *Capital Investment, Part II*, a volume of reprints of 14 articles, 1965–68. Cambridge, Mass.: Harvard Business Review.

Higgins, Robert C. *Financial Management: Theory and Applications*. Chicago: Scientific Research Associates, 1977.

Hillier, F. S., and Lieberman, G. J. *Introduction to Operations Research*, 2d ed. San Francisco: Holden-Day, Inc., 1974.

Hirshleifer, J., DeHaven, J. C., and Milliman, J. W. *Water Supply: Economics, Technology, and Policy*. Chicago: University of Chicago Press, 1960.

James, L. D., ed. *Man and Water: The Social Sciences in Management of Water Resources*. Lexington, Ky.: University Press of Kentucky, 1974.

———, and Lee, R. R. *Economics of Water Resources Planning*. New York: McGraw-Hill Book Co., Inc., 1970.

Jeynes, P. H. *Profitability and Economic Choice*. Ames, Iowa: Iowa State University Press, 1968.

Jones, D. N., and Dovell, Susan. *Electric and Gas Utility Rate and Fuel Adjustment Clause Increases, 1974*. Washington, D.C.: U.S. Government Printing Office. Committee Print, 94th Congress, 1st Session, March 27, 1975.

Kendall, M. G., ed. *Cost–Benefit Analysis*. New York: American Elsevier Publishing Co., 1971.

Krutilla, J. V., and Eckstein, Otto. *Multiple Purpose Development Studies in Applied Economic Analysis*. Baltimore: Johns Hopkins Press, 1958.

Kurtz, Max. *Engineering Economics for Professional Engineers' Examinations*, 2d ed. New York: McGraw-Hill Book Co., Inc., 1975.

Lesourne, J. *Cost–Benefit Analysis and Economic Theory*. New York: American Elsevier Publishing Co. Translated from French by Mrs. A. Silvey, 1975.

Lesser, Arthur, Jr., ed. Summer Symposium Papers of the Engineering Economy Division, American Society for Engineering Education: *Planning and Justifying Capital Expenditures*, 1959; *Applications of Economic Evaluation in Industry*, 1962; *Decision-Making Criteria for Capital Expenditures*, 1965; *Economic Analysis of Complex Programs*, 1968; *Decision and Risk Analysis*, 1971; *Evaluations in an Uncertain Future: Technology and Impact Assessment*, 1974.

Linsley, R. K., and Franzini, J. B. *Water-Resources Engineering*, 3d ed. New York: McGraw-Hill Book Co., Inc., 1979.

Lopez Leautaud, J. L. *Elementos de Evaluacion Economica*. Monterrey, Mexico: Impresos y Tesis (no date given).

Luce, R. D., and Raiffa, H. *Games and Decisions*. New York: John Wiley & Sons, Inc., 1957.

Lüder, Klaus. *Investitionskontrolle*. Wiesbaden, Germany: Betriebswirtschftlicher Verlag Dr. Th. Gabler, 1969.

Machinery and Allied Products Institute. *MAPI Replacement Manual*. Washington, D.C.: Machinery and Allied Products Institute, 1950.

———. *Leasing of Industrial Equipment, A MAPI Symposium*. Washington, D.C.: Machinery and Allied Products Institute and the Council for Technological Advancement, 1965.

MAO, JAMES. *Quantitative Analysis of Financial Decisions*. New York: The Macmillan Co., 1969.

MARSTON, A., WINFREY, R., and HEMPSTEAD, J. C. *Engineering Valuation and Depreciation*. 2d ed. Ames, Iowa: Iowa State University Press, 1953.

MASSÉ, PIERRE. *Optimal Investment Decisions*. Englewood Cliffs, N.J.: Prentice-Hall, Inc., 1962.

MAYNARD, H. B., ed. *Industrial Engineering Handbook*, 3d ed. New York: McGraw-Hill Book Co., Inc., 1971.

MCKEAN, R. N. *Efficiency in Government Through Systems Analysis*. New York: John Wiley & Sons, Inc., 1958.

MORRIS, W. T. *The Analysis of Managerial Decisions*. Homewood, Ill.: Richard D. Irwin, Inc., 1964.

———. *Decision Analysis*. Columbus, Ohio: Grid Publishing, Inc., 1977.

NEWNAN, DONALD G. *Engineering Economic Analysis*, rev. ed. San Jose, Cal.: Engineering Press, 1980.

———. *Economic Analysis for the Professional Engineering Examination*. San Jose, Cal.: Engineering Press, 1978.

NORTON, P. T., JR. *Economic Lot Sizes in Manufacturing*. Virginia Polytechnic Institute Bulletin No. 31, Blacksburg, Va., 1934.

———. *Sec. 3, Engineering Economy* in *Handbook of Industrial Engineering and Management*, 2d ed. (W. G. Ireson and E. L. Grant, eds.). Englewood Cliffs, N.J.: Prentice-Hall, Inc., 1971.

———. *The Selection and Replacement of Manufacturing Equipment*. Virginia Polytechnic Institute Bulletin No. 32, Blacksburg, Va., 1934.

NOVICK, DAVID, ed. *Program Budgeting: Program Analysis and the Federal Budget*, 2d ed. Cambridge, Mass.: Harvard University Press, 1967.

OAKFORD, R. V. *Capital Budgeting: A Quantitative Evaluation of Investment Alternatives*. New York: John Wiley & Sons, Inc., 1970.

ODIER, LIONEL. *The Economic Benefits of Road Construction and Improvements*, trans. Noel Lindsay from *Les Intérêts Economiques des Travaux Routiers*. Paris: Publications ESTOUP, 1963.

OGLESBY, C. H. and HICKS, R. G. *Highway Engineering*, 4th ed. New York: John Wiley & Sons, Inc., 1982.

PETERS, M. S., and TIMMERHAUS, K. D. *Plant Design and Economics for Chemical Enginners*, 2d ed. New York: McGraw-Hill Book Co., Inc., 1968.

QUADE, E. S., and BOUCHER, W. I., eds. *Systems Analysis and Policy Planning: Applications in Defense*. New York: American Elsevier Publishing Co., 1968.

RADFORD, K. J. *Managerial Decision Making*. Reston, Va.: Reston Publishing Co., Inc., 1975.

RAIFFA, HOWARD. *Decision Analysis: Introductory Lectures on Choices Under Uncertainty*. Reading, Mass.: Addison-Wesley, 1968.

———, and SCHLAIFER, R. *Applied Statistical Decision Theory*. Boston: Harvard Business School, 1961.

REUL, R. I. *Sec. 4, Capital Budgeting* in *Handbook of Industrial Engineering and Management*, 2d ed. (W. G. Ireson and E. L. Grant, eds.). Englewood Cliffs, N.J.: Prentice-Hall, Inc., 1971.

RICHMOND, S. B. *Operations Research for Management Decisions*. New York: John Wiley & Sons, Inc., 1968.

RIGGS, J. L. *Economic Decision Models for Engineers and Managers*. New York: McGraw-Hill Book Co., Inc., 1968.

———. *Engineering Economics*. New York: McGraw-Hill Book Co., Inc., 1977.

ROSE, L. M. *Engineering Investment Decisions: Planning Under Uncertainty*. Amsterdam: Elsevier Scientific Publishing Co., 1976.

SCHLAIFER, ROBERT. *Probability and Statistics for Business Decisions*. New York: McGraw-Hill Book Co., Inc., 1959.

SCHNEIDER, ERICH. *Wirtschaftlichkeits-rechnung*. Tübingen, Germany: J. C. B. Mohr (Paul Siebeck), 1957.

SCHWEYER, H. E. *Analytic Models for Managerial and Engineering Economics*. New York: D. Van Nostrand Reinhold Publishing Co., 1964.

SHARPE, W. F. *Introduction to Managerial Economics*. New York: Columbia University Press, 1973.

SMITH, G. W. *Engineering Economy: The Analysis of Capital Expenditures*, 2d ed. Ames, Iowa: The Iowa State University Press, 1973.

SMITH, S. C., and CASTLE, E. *Economics and Public Policy in Water Resource Development*. Ames, Iowa: Iowa State University Press, 1964.

SOLOMON, EZRA. *The Theory of Financial Management*. New York: Columbia University Press, 1963.

———, ed. *The Management of Corporate Capital*. New York: The Macmillan Co., 1959.

SOLOMONS, DAVID, ed. *Studies in Cost Analysis*, 2nd ed. London: Sweet & Maxwell Limited, 1968.

STEINER, H. M. *Public and Private Investments—Socioeconomic Analysis*. New York: John Wiley & Sons, Inc., 1980.

SUBCOMMITTEE ON EVALUATION STANDARDS OF INTER-AGENCY COMMITTEE ON WATER RESOURCES. *Proposed Practices for Economic Analysis of River Basin Projects*. Washington, D.C.: Government Printing Office, 1958. In many writings on the economics of public works, this useful pamphlet is referred to as the "Green Book."

SWALM, R. O. *Capital Expenditure Analysis–A Bibliography*. Syracuse, N.Y.: R. O. Swalm, Syracuse University, 1967. This comprehensive bibliography also was published in *The Engineering Economist*, Vol. 13, No. 2 (Winter, 1968), pp. 105–129.

TARQUIN, A. J., and BLANK, L. T. *Engineering Economy: A Behavioral Approach*. New York: McGraw-Hill Book Co., Inc., 1976.

TAYLOR, G. A. *Managerial and Engineering Economy*, 3d ed. New York: D. Van Nostrand Reinhold Co., 1980.

TERBORGH, GEORGE. *Business Investment Management*. Washington, D.C.: Machinery and Allied Products Institute, 1967.

———. *Business Investment Policy*. Washington, D.C.: Machinery and Allied Products Institute, 1958.

———. *Dynamic Equipment Policy*. Washington, D.C.: Machinery and Allied Products Institute, 1949.

———. *Realistic Depreciation Policy*. Washington, D.C.: Machinery and Allied Products Institute, 1954.

THUESEN, H. G., FABRYCKY, W. J., and THUESEN, G. J. *Engineering Economy*, 5th ed. Englewood Cliffs, N.J.: Prentice-Hall, Inc., 1977.

TILANUS, C. B., ed. *Quantitative Methods in Budgeting*. Leiden, Netherlands: Martinus Nijhoff Social Sciences Division, 1976.

UNITED STATES GOVERNMENT. *The Analysis and Evaluation of Public Expenditures: The PPB System*. Washington, D.C.: U.S. Government Printing Office, 1969. Vol. 1, Parts I, II, and III; Vol. 2, Part IV; Vol. 3, Parts V and VI.

———. *Benefit–Cost Analyses of Federal Programs, A Compendium of Papers*. Washington, D.C.: U.S. Government Printing Office, 1973. Joint Committee Print, 92nd Congress, 2nd Session.

———. *The Economics of Federal Subsidy Programs*. Hearings before the Subcommittee on Priorities and Economy in Government. Washington, D.C.: U.S. Government Printing Office, January 13, 14, and 17, 1972. 92nd Congress, 1st Session.

———. *Improving National Productivity*. Hearings before the Subcommittee on Priorities and Economy in Government. Washington, D.C.: U.S. Government Printing Office, 1972. 92nd Congress, 2nd Session.

———. *Program Budgeting: Program Analysis and the Federal Budget*. Washington, D.C.: U.S. Government Printing Office, 1964. A report prepared and copyrighted by the RAND Corp.

WAGNER, HARVEY M. *Principles of Management Science: With Applications to Executive Decisions*, 2d ed. Englewood Cliffs, N.J.: Prentice-Hall, Inc., 1975.

———. *Principles of Operations Research: With Applications to Managerial Decisions*, 2d ed. Englewood Cliffs, N.J.: Prentice-Hall, Inc., 1975.

WEINGARTNER, H. M. *Mathematical Programming and the Analysis of Capital Budgeting Problems*. Englewood Cliffs, N.J.: Prentice-Hall, Inc., 1963.

WELLINGTON, A. M. *The Economic Theory of Railway Location*, 2d ed. New York: John Wiley & Sons, Inc., 1887.

White, J. A., Agee, M. H., and Case, K. E. *Principles of Engineering Economic Analysis*. New York: John Wiley & Sons, Inc., 1977.

Williams, B. R., and Scott, W. P. *Investment Proposals and Decisions*. London: George Allen and Unwin, Ltd., 1965.

Winfrey, Robley. *Economic Analysis for Highways*. Scranton, Pa.: International Textbook Co., 1969.

———. *Statistical Analysis of Industrial Property Retirements, Bulletin 125*. Ames, Iowa: Iowa State University Engineering Experiment Station, 1935.

Zanobetti, Dino. *Economia dell'Ingegneria*. Bologna, Italy: Casa Editrice Prof. Riccardo Pàtron, 1966.

F

NOTES ON CERTAIN CHANGES IN THE INCOME TAX LAWS OF THE UNITED STATES

In this book the authors have tried to present a general approach to the relationship between income taxation and economy studies. Our objective has been to explain methods of analysis that can be useful to decision makers wherever income taxes are levied. Nevertheless, it has been helpful to use a particular set of tax laws to illustrate a number of points. In Chapter 12, and in several of the subsequent chapters, we have made reference to various aspects of the federal tax laws and regulations that applied to 1980 taxable income in the United States.

The Economic Recovery Tax Act of 1981 (passed in August of that year) made major changes in the federal tax laws of the United States. (In the remainder of this appendix we call this the *1981 tax act*.) This appendix describes certain tax law changes that will influence many of the engineering economy studies made in the United States after the passage of this legislation. Our comments deal chiefly with tax rates for individuals and corporations, with depreciation deductions from taxable income, and with investment tax credits.

Changes in Tax Rates

Individual income taxes were reduced by 5% on October 1, 1981, by 10% more on July 1, 1982, and by a further 10% on July 1, 1983. Starting in 1985 the spread of the tax brackets, the personal exemptions, and the standard deductions (so-called zero bracket amounts) were all scheduled to be indexed to reflect changes in price levels. (On pages 329, 330, and 331 we discussed the desirability of indexing personal income taxes during periods of inflation.) In the opinion of the authors this stipulation that individual income taxes are to be indexed will turn out to be the most important of the many changes made by the 1981 tax act.

After individual income taxes are indexed, taxpayers no longer will be moved into higher and higher tax brackets solely because of inflation. Such movement into higher brackets during 1981–1984 will occur only to the extent that the tax reductions during these years do not fully compensate for the inflation that takes place. The main consequence of all of this will be to promote equity in the levying of individual income taxes. An incidental effect will be to simplify economy studies for individuals by making it unnecessary to forecast rising applicable tax rates caused by so-called bracket creep.

The highest bracket in the various tax rate schedules for individual income taxes was reduced from 70% to 50% for 1982 and thereafter. (One 1980 schedule applicable to certain single taxpayers was shown in Table 12–8, page 244.) This change really applied particularly to investment income because, as pointed out on page 243, income defined as "personal service income" already was subject to a maximum tax rate of 50%. The change had the desirable effect of reducing the incentive for investment in so-called tax shelters by persons in the highest tax brackets. Many of the tax shelter investments that had been caused by the 70% incremental tax rates were less useful from a social viewpoint than the more conventional investments in the productive assets of industry that they had been displacing.

The change in the tax rate schedule for corporations was much less than the changes in the various schedules for individuals. The 1980 schedule (which we show as Table 12–7 on page 242) remained in effect for 1981. For 1982 the 17% rate for the first bracket of taxable income from zero to $25,000 was reduced to 16% and the second bracket from $25,000 to $50,000 was reduced from 20% to 19%. For 1983 there were further reductions in these brackets to 15% and 18%, respectively. No changes were made in the rates for brackets over $50,000. For 1983 the tax on taxable income over $100,000 was $25,750 plus 46% of the excess of income over $100,000.

Because it is the incremental rates that are relevant in economy studies, the 1981 tax law really made no change in the applicable tax rates that were appropriate wherever income was more than $50,000. For corporations with taxable income below $100,000, the modest reductions in the two lower brackets of the tax rate schedule were not enough to compensate for the bracket creep likely to be caused by inflation.

Depreciation

From the passage of the first modern income tax law in 1913, the tax laws and regulations through the year 1980 gave at least lip service to the concept that allowable "useful lives" of assets should depend on the length of service of the assets for the particular taxpayer. All this was changed by the Accelerated Cost Recovery System (ACRS) in the 1981 tax act, which divided all depreciable assets into only four classes. Under ACRS, all eligible depreciable assets had statutory recovery periods of 3, 5, 10, or 15 years for tax purposes. The percentage of cost that could be deducted from taxable income for each year was stipulated for each class. The depreciation changes made in the 1981 tax act were made retroactive to January 1, 1981 for all assets placed in service then or later.

Most eligible personal property was in the 5-year class. This included most machinery and equipment, furniture and fixtures, and transportation equipment. A few short-lived items, such as cars and light-duty trucks, were in the 3-year class, which also included research and experimentation equipment. The 10-year class was mostly public utility property that, under the Asset Depreciation Range system that had been in effect before 1981, had been assigned "midpoint" lives of 18.5 to 25 years. There were two 15-year classes subject to slightly different rules. One consisted of certain public utility property that previously had been assigned lives of more than 25 years. The other class was depreciable real property such as buildings and structures.

The depreciation rates assigned to the 5-year class placed in service during years 1981 through 1984 were 15%, 22%, 21%, 21%, and 21%, respectively, for years 1, 2, 3, 4, and 5. For assets placed in service in 1985 the rates were 18%, 33%, 25%, 16%, and 8%, respectively. For assets placed in service in 1986 and thereafter, the rates were 20%, 32%, 24%, 16%, and 8%, respectively. Annual percentages were specified in a similar manner for the 3-year, 10-year, and 15-year classes. For the years 1981–1984 the percentages were approximately those that would be obtained from 150% declining-balance depreciation accounting for the initial years followed by a switch to straight-line in the final years. For 1985 they approximated 175% declining-balance for the initial years followed by years-digits for the final years. For 1986 and thereafter they approximated 200% declining-balance followed by years-digits. In all cases they assumed the use of a "half-year convention" during the year of acquisition. (See page 222 for a reference to this averaging convention.)

Under certain restrictions, taxpayers could elect to use straight-line depreciation over the same or certain stipulated longer lives. For example, for 5-year property a taxpayer could elect to use straight-line depreciation over 5 years, 12 years, or 25 years. Presumably, taxpayers would choose such slower rates of write-off only in cases where higher incremental tax rates were expected over the coming years.

For all depreciable assets, depreciation deductions were to be computed as if all *prospective* salvage values were zero. If, later, there should turn out to be an *actual* net salvage value when an asset was disposed of, and if this asset's actual salvage should exceed its remaining book value for income tax purposes, there would, of course, be a "gain on disposal" that would be taxable.

The 1981 tax act permitted a taxpayer to make a 100% write-off of $5,000 of selected assets (not including buildings) in the year of acquisition. (The favorable consequences of such a write-off were illustrated in Example 12–6, page 236.) Assets written off in this way were ineligible for any investment tax credit. The $5,000 figure applied to years 1982 and 1983. The allowable total increased to $7,500 for years 1984 and 1985, and to $10,000 for 1986 and thereafter. This provision was intended to be helpful to small business. (But see Problems F–15 through F–20 for some limitations on its usefulness small corporations that are in the lower income brackets.)

We pointed out on pages 324 through 329 that when inflation occurs, and when depreciation deductions from taxable income are based on the costs of assets at lower price levels, there is overtaxation in the sense that the true tax rate is greater than the stipulated rate. In one particular case cited on page 328, a 6% inflation rate caused an actual average tax rate of 56% even though the stated rate was 49% throughout the period. The 1981 tax act made no change in this adverse feature of the conventional rules for determining taxable income during periods of inflation.

Investment Tax Credit

Under the 1981 tax act, eligible assets in the 5, 10, and 15 year classes were entitled to a 10% investment tax credit. This involved no change for the assets in the 10 and 15 year classes. Although assets with 5-year estimated lives previously had been allowed only a 6⅔% credit, most of the assets in the new 5-year class previously had been written off in at least 7 years and therefore had been eligible for the earlier 10% credit. Eligible assets in the new 3-year class now were allowed a 6% investment tax credit.

Subject to a number of restrictions, the 1981 tax act made investment tax credits available for rehabilitation expenditures on older nonresidential buildings. For buildings at least 30 years old that were to be used for industrial or commercial purposes, the credit was 15% of the cost of rehabilitation. If such buildings were at least 40 years old, the credit was 20% of the rehabilitation costs. There also was a 25% credit available for rehabilitation of certain buildings classified as "historic sites."

Example F–1 applies provisions of the 1981 tax act to the facts of Example 12–2 (page 232)

EXAMPLE F–1

Prospective Rate of Return After Income Taxes for Property in the 5-Year Class

Facts of the Case
The proposed assets in Example 12–2 are assumed to be eligible for the depreciation percentages permissible for 5-year assets during the years 1981–1984. A 10% investment tax credit is allowable.

Analysis
Table F–1 shows the calculation of after-tax cash flow. Conventional trial-and-error calculations indicate that the prospective after-tax rate of return is approximately 16.2%. Similar calculations will show that the depreciation allowances permissible in 1985 will give a prospective after-tax rate of return of approximately 16.9%; with the allowances permissible starting in 1986, the rate is approximately 17.0%.

TABLE F–1
Estimation of cash flow after income taxes, Example F–1

Year	*Cash flow before income taxes*	*Write-off of initial outlay for tax purposes*	*Influence on taxable income*	*Influence of income taxes on cash flow* −0.49C	*Cash flow after income taxes* (A + D)
	A	B	C	D	E
0	−$110,000			+$11,000*	−$99,000
1	+26,300	−$16,500	+$9,800	−4,802	+21,498
2	+26,300	−24,200	+2,100	−1,029	+25,271
3	+26,300	−23,100	+3,200	−1,568	+24,732
4	+26,300	−23,100	+3,200	−1,568	+24,732
5	+26,300	−23,100	+3,200	−1,568	+24,732
6	+26,300	0	+26,300	−12,887	+13,413
7	+26,300	0	+26,300	−12,887	+13,413
8	+26,300	0	+26,300	−12,887	+13,413
9	+26,300	0	+26,300	−12,887	+13,413
10	+26,300	0	+26,300	−12,887	+13,413
Totals	+$153,000	−$110,000	+$153,000	−$63,970	+$89,030

*Investment tax credit of 10% of investment.

Some Comments on ACRS in Relation to Economy Studies

We pointed out in Chapter 12 that technological progress may be delayed when many proposed capital investments are made unattractive by the combination of high income tax rates and required slow rates of write-off for tax purposes. This undesirable condition developed in the United States after the end of World War II. The requirement of slow write-off had taken place during the depression years of the 1930s when tax rates were relatively low. The war raised tax rates to a high level in the 1940s.

Although there have been a few ups and downs, tax rates in brackets that were relevant to business investment have continued for some 40 years at essentially the high wartime levels; tax rates never went back to the relatively low levels of the 1930s. In fact, the relevant corporate tax rate actually increased over the years; in 1949 it was 38% for all taxable income over $50,000, and in 1982 it was 46% for all taxable income above $100,000. (It should be noted that because of inflation the $100,000 in 1982 had approximately the same purchasing power as $25,000 in 1949.)

Therefore, the tax obstacles to capital formation were not being reduced by lowering the relevant incremental tax rates. Another possible way to reduce them was to shorten permissible write-off periods. The first step in this direction took place in 1954 with the authorization of the 200% declining-balance and the sum-of-years-digits depreciation methods as optional alternatives to straight-line. (See pages 200–202.) A second step was the ADR system authorized in 1971. (See page 260.) ADR was liberalized a bit over its years. ACRS may be viewed as a third step; as already mentioned, this was established with a scheduled liberalization in 1985 and another in 1986.

The investment tax credit, first established in 1962 and liberalized in 1975, was a different tax policy also having the effect of reducing the tax obstacles to capital formation.

Several points relative to ACRS are brought out in the problems at the end of this appendix. Answers are given to all of the numerical problems to permit readers to check their own solutions. All the problems that involve depreciation are similar to Example F–1 in assuming 5-year property; we have pointed out that most machinery and equipment falls in this class.

The problems illustrate the differences in prospective rates of return caused by the depreciation allowances permissible in different years. The reader will note that the 1985 allowances give appreciably better results than the ones available for 1981–1984. On the other hand, the allowances permissible in 1986 and thereafter give a relatively small improvement over 1985.

In Example F–1 and in all the problems that call for after-tax evaluation of proposed investments, the write-off period for tax purposes is shorter than the study period. Presumably this will become the common condition in after-tax economy studies in the United States. A possible simplification of economy studies for 5-year property during 1981–1984 is to compute taxes as if multiple-straight-line depreciation were to be used at a 20% rate for the first five years followed by 0% for all subsequent years. In Problem F–1(d), this simplification gives 16.4% as the rate of return; this is not far from the 16.2% obtained in Problem F–2 where the stipulated rates are used. The comparable figures where the two methods are used in Problems F–9(d) and F–10 are 14.2% and 14.0%.

Problems F–3 and F–11 bring out the point that the uniform-flow convention gives substantially the same results as the end-of-year-convention if the reasonable assumption is made that uniform flow is assumed for the investment and investment tax credit as well as for the other cash flow throughout the study period.

In levying their respective income taxes, many of the states of the United States follow the federal rules for determining taxable income whenever possible. (For constitutional reasons there necessarily are a few differences.) In these states a change in federal rules, such as the depreciation rate change made under ACRS, will automatically cause the same change in the state rules. Conformity to federal rules reduces a state's expenses in making tax audits and collections. It also reduces taxpayers' expenses in keeping separate sets of books for state and federal income tax purposes.

Nevertheless, a number of states have their own rules, which often differ from one another as well as from the federal rules. Persons who are responsible for preparing tax returns must follow each taxing agency's own rules no matter how these rules differ among agencies. In contrast, analysts who make after-tax economy studies for taxpayers have no legal necessity to make their economy studies in any particular way. Where rules differ, an analyst can decide whether it is good enough for practical purposes to simplify an economy study by assuming that a given state's rules for computing taxable income are the same as the federal rules. This is a question of sensitivity. Problems F–13 and F–14 illustrate certain aspects of this topic with reference to ACRS.

The "Averaging Convention" Built into the 1981 Tax Act

One year is the standard accounting period used by business enterprises. The one-year period is used not only for business accounting but also for computing taxable income.

It is common for new assets to be placed in service at irregular intervals throughout the accounting and tax year. For each new asset the question arises as to what proportion of a full year's depreciation charge should be made during the year it is put in service. If this charge should be based on the ratio of the time remaining in the year to the length of the year, each new asset might have a different depreciation percentage. A fair amount of expensive clerical labor is avoided by the use of some *averaging convention*. The most popular convention has been to charge all newly placed assets with half a year's depreciation regardless of the time of the year when they are placed in service.

Before 1981 the Internal Revenue Service accepted averaging conventions when it seemed that their use would not "distort" taxable income, provided such conventions were used consistently. On its page 40, the 1980 Tax Guide for Small Business (which we quoted in another connection on our page 469) discussed this topic in part as follows.

> If you use an averaging convention, you must consistently follow it for those accounts for which you adopt it. You may not use the averaging convention to figure depreciation for property that is large and unusual that would substantially distort your depreciation deduction for the year. You must depreciate these assets separately from the multiple asset accounts and use their own useful lives. You would start to depreciate this type of asset when you place it in service and stop when you retire it from service.

In contrast, the 1981 tax act stipulated depreciation percentages for the year of placement of *all* assets in the 3, 5, and 10 year classes. In effect, these percentages incorporated a liberal averaging convention for all assets in each class. The percentages were to be used for the year of placement regardless of the date of placement within the tax year. It appears that no question of the distortion of taxable income will arise under the 1981 tax act for assets in these classes.

Although we mentioned averaging conventions briefly on page 222, their use has not been discussed or illustrated elsewhere in the main body of this book. Because they do not influence before-tax cash flow, they are not relevant in economy studies made before income taxes. The explanation of after-tax economy studies is simplified by assuming that the tax consequences of an investment in depreciable assets start from the moment the assets are placed in service. Until the passage of the 1981 tax act, this assumption was fairly close to the facts in the United States.

Certain Other Aspects of the 1981 Tax Act

In effect, the act makes certain stipulations about the accounting of United States corporations. They are required to compute their "earnings and profits" using straight-line depreciation with "recovery periods" as follows:

3-year property	5 years
5-year property	12 years
10-year property	25 years
15-year property	35 years

The half-year convention is stipulated for all property except 15-year real property. Any prospective salvage value is to be disregarded.

In general, the examples and problems in this book have illustrated economy studies for business enterprises that are currently paying income taxes and expect to continue to pay such taxes into the foreseeable future. If new investments in assets are entitled to depreciation deductions and investment tax credits, the investments have a so-called tax benefit in the sense that the deductions and credits reduce taxes that otherwise would have to be paid. Corporations that are not currently earning taxable income have no such immediate tax benefits from proposed investments; they must wait until some uncertain future time when they do have taxable income. At that time they will be able to use the deductions and tax credits only to the extent that the law permits losses and tax credits to be carried forward for tax purposes.

In drawing up the 1981 tax act, Congress recognized that it was in the public interest for corporations currently losing money to have an incentive to make technological improvements. The laws governing the tax aspects of leasing were liberalized so that, under a rather complex set of restrictions, it was possible for deductions and credits to be transferred to a profitable corporation that financed a lease. In effect, corporations losing money were allowed to peddle their entitlements to deductions and credits. (Of course, once transferred, the depreciation deductions and investment tax credits would not be available to a lessee if its operations later became profitable.)

The 1981 act also increased from 7 to 15 years the maximum period for carryover (that is, carry forward) of the business losses of one year to offset the taxable profit of some subsequent year.

PROBLEMS

Use the end-of-year convention unless the uniform-flow convention is specified.

F-1. A proposed $180,000 investment in machinery at date zero will lead to estimated net positive cash flow before income taxes of $46,800 a year for years 1 through 8. The expected applicable tax rate is 46% for the entire period. The estimated net salvage value of the machinery is zero at the end of 8 years. There will be a 10% investment tax credit at year zero. Compute (a) the prospective rate of return before income taxes, (b) the prospective after-tax rate of return with straight-line depreciation over 8 years, (c) the after-tax rate with years-digits depreciation over 8 years, and (d) the after-tax rate with multiple-straight-line depreciation at 20% for the first 5 years and 0% for the final 3 years. (*Ans.* 19.9%; 14.6%; 16.1%; 16.4%)

F-2. Consider the proposed investment description in Problem F-1. Depreciation deductions for income tax purposes will be 15%, 22%, 21%, 21%, and 21%, respectively, for years 1, 2, 3, 4, and 5. What is the prospective after-tax rate of return? (*Ans.* 16.2%)

F-3. Solve Problem F-2 using the uniform-flow convention rather than the end-of-year convention. Apply the uniform-flow convention not only to the cash flow in years 1 through 8 but also to the investment and the investment tax credit in year zero. (See the discussion of this procedure on page 584.) This assumes that the investment occurs throughout year zero and that the expected tax credit influences the quarterly payments of estimated income taxes for the year. (*Ans.* 16.4%)

F-4. Solve Problem F-2 changing the depreciation deductions for income tax purposes to 18%, 33%, 25%, 16%, and 8%, respectively, for years 1, 2, 3, 4, and 5. (*Ans.* 17.0%)

F-5. Solve Problem F-2 changing the depreciation deductions for income tax purposes to 20%, 32%, 24%, 16%, and 8%, respectively, for years 1, 2, 3, 4, and 5. (*Ans.* 17.1%)

F-6. Example F-1 stated that the after-tax rate of return would be approximately 16.9% with the depreciation percentages permissible in 1985, and approximately 17.0% with the ones permissible in 1986 and thereafter. Show calculations to check these figures.

F-7. A proposed $165,000 investment in machinery at date zero will lead to estimated net positive cash flow before income taxes of $36,850 a year for years 1 through 10. The expected applicable tax rate is 46% for the entire period. The estimated net salvage value of the machinery is zero at the end of 10 years. There will be a 10% investment tax credit at year zero. Compute (a) the prospective rate of return before income taxes, (b) the prospective after-tax rate of return with straight-line depreciation over 10 years, (c) the after-tax rate with years-digits depreciation over 10 years, and (d) the after-tax rate with multiple-straight-line depreciation at 20% for the first 5 years and 0% for the final 5 years. (*Ans.* 18.1%; 13.1%; 14.6%; 15.5%)

F-8. Consider the proposed investment described in Problem F-7. Depreciation deductions for income tax purposes will be 15%, 22%, 21%, 21%, and 21%, respectively, for years 1, 2, 3, 4, and 5. What is the prospective after-tax rate of return? (*Ans.* 15.3%)

F-9. A proposed $180,000 investment in machinery at date zero will lead to estimated net positive cash flow before income taxes of $32,000 a year for years 1 through 15. The expected applicable tax rate is 46% for the entire period. The estimated net salvage value of the machinery is zero at the end of 15 years. There will be a 10% investment tax credit at year zero. Compute (a) the prospective rate of return before income taxes, (b) the prospective after-tax rate of return with straight-line depreciation over 15 years, (c) the after-tax rate with years-digits depreciation over 15 years, and (d) the after-tax rate with multiple-straight-line depreciation at 20% for the first 5 years and 0% for the final 10 years. (*Ans.* 15.8%; 11.2%; 12.4%; 14.2%)

F-10. Consider the proposed investment described in Problem F-9. Depreciation deductions for income tax purposes will be 15%, 22%, 21%, 21%, and 21%, respectively, for years 1, 2, 3, 4, and 5. What is the prospective after-tax rate of return? (*Ans.* 14.0%)

F-11. Solve Problem F-10 using the uniform-flow convention rather than the end-of-year convention. Apply the uniform-flow convention not only to the cash flow in years 1 through 15 but also to the investment and investment tax credit in year zero. (*Ans.* 14.1%)

F-12. Solve Problem F-10 changing the depreciation deductions for income tax purposes to 20%, 32%, 24%, 16%, and 8%, respectively, for years 1, 2, 3, 4, and 5. (*Ans.* 14.6%)

F-13. In Problems F-9 and F-10 the applicable tax rate was stated as 46%. In 1982 in the United States such a 46% rate would have been appropriate for a corporation for which annual income subject to federal taxes was expected to exceed $100,000 each year provided the corporation had all its operations in a state that did not have a corporate income tax.

Alter Problem F-10 so that there will be both an applicable federal tax rate of 46% and an applicable state tax rate of 10%. Assume that state taxes are deductible on the federal return but federal taxes are not deductible on the state return. Therefore the combined applicable tax rate is 51.4%. (See the explanation of the combination of such rates on pages 246 and 247.) The state will accept the same depreciation figures each year that are used on the federal return. The state does not allow any investment tax credit. Compute the prospective after-tax rate of return. (*Ans.* 13.4%)

F-14. Change Problem F-13 by assuming that the depreciation figures specified for the federal return are not acceptable to the state, which insists on straight-line depreciation over a 15-year life. Compute the prospective after-tax rate of return. (*Ans.* 13.0%)

F-15. One limitation on the attractiveness of the provision permitting a 100% write-off of certain selected assets during their year of acquisition is that any assets so written off are not eligible for the investment tax credit. Losing the tax credit is particularly bad for taxpayers in the lower tax brackets because the credit is a reduction of the *tax* rather than merely of the *taxable income*.

Consider a proposed $5,000 investment that could be written off in five years using the rates illustrated in Example F-1. The investment will cause an estimated positive cash flow before income taxes of $1,100 a year for 10 years. The applicable tax rate is 15%. (According to the tax act of 1981 this will be the rate for the first $25,000 of corporate income for 1983 and thereafter.) What is the prospective after-tax rate of return (a) if this investment is selected for 100% immediate write-off with no investment tax credit, and (b) if it is written off in five years and a 10% tax credit is taken? (*Ans.* 17.7%; 19.1%)

F-16. Assume that a taxpayer with an applicable tax rate of 15% makes a $5,000 investment in machinery. The taxpayer can choose between (a) writing off the $5,000 in year zero with no investment tax credit, and (b) taking a 10% tax credit in year zero and writing off 15%, 22%, 21%, 21%, and 21%, respectively, over the first five years. Show that the differences in cash flow that will be caused by selecting (a) rather than (b) will be 0, +$250; 1, −$112.5; 2, −$165; 3, −$157.5; 4, −$157.5; 5, −$157.5. Note that these differences (which existed in Problem F-15) are independent of the merits of the investment. Express as an interest rate the after-tax cost of the $250 made available at zero date if (a) should be selected rather than (b). (*Ans.* 49%)

F-17. The high cost of choosing the immediate write-off in Problem F-16 depended on the relatively low applicable tax rate of 15%. Make a similar analysis to find the interest rate that expresses the after-tax cost of the money obtained by choosing an immediate write-off when the expected applicable tax rate is 30%. (*Ans.* 14.6%)

F-18. Solve the preceding problem using an applicable tax rate of 46%. (*Ans.* 8.4%)

F-19. A business taxpayer expects an increase in taxable income within a year or two. This will cause a higher incremental tax rate. Should this prospect of a higher applicable tax rate in the near future influence the decision about electing 100% write-off for eligible assets? If so, how? Explain your answer.

F-20. Problem 12-52 (pages 274 and 275) described a provision for "additional first year depreciation" that was included in the federal income tax laws of the United States for many years through 1980. This provision was eliminated by the tax act of 1981 in favor of a limited permission for 100% write-off in the year of acquisition. Problem 12-52 dealt with a decision in 1980 by a sole proprietor in the construction industry. In 1982 he would no longer be allowed 20% additional first year depreciation on $20,000 of purchased equipment. However, he would be permitted an immediate write-off of $5,000 of certain purchased assets; this 100% write-off would make the chosen assets ineligible for a 10% investment tax credit.

From the point of view of this sole proprietor, does it seem to you that the 1982 rule is preferable to the 1980 one? Use a numerical illustration to support your answer. Assume that his applicable tax rate will be 40% in 1982 and thereafter.

F-21. With reference to the changes in corporate taxes made by the tax act of 1981, the text states that "for corporations with taxable incomes below $100,000, the modest reduction in the two lower brackets of the tax rate schedule was not enough to compensate for the bracket creep likely to be caused by inflation." Make up a numerical example to illustrate this point.

F-22. Problem 14-33 (page 347) dealt with United States federal income taxes for two individuals, Sam Smith and Jennifer Jones. In year zero, when Sam and Jennifer had adjusted gross incomes of $16,000 and $30,000, respectively, they were subject to the tax rate schedule shown in Table 12-8 (page 244). Now assume that year zero is, in fact, 1980. Assume also that the inflation rate for 1981 is 10% and that adjusted gross incomes for Sam and Jennifer increased by 10% to $17,600 and $33,000, respectively. Individual income taxes were reduced by 5% effective October 1, 1981. To implement this stated tax reduction it was necessary to calculate what a taxpayer would have owed under the appropriate 1980 tax rate schedule and then to subtract 1¼% (that is, one-fourth of 5%) of the tax so calculated. By what percentages did federal income taxes for Sam and Jennifer increase from 1980 to 1981? (*Ans.* Sam, 16.9%; Jennifer 16.1%)

INDEX